MODERN EXPERIMENTAL ORGANIC CHEMISTRY

THIRD EDITION

Royston M. Roberts
John C. Gilbert
University of Texas, Austin

Lynn B. Rodewald
Alan S. Wingrove
Towson State University

HOLT, RINEHART AND WINSTON
New York   Chicago   San Francisco   Atlanta   Dallas
Montreal   Toronto   London   Sydney

The first two editions were entitled *An Introduction to Modern Experimental Organic Chemistry.*

Acquiring Editor: *Thom Gorman*

Managing Editor: *Jean Shindler*

Production Manager: *Robert Ballinger*

Design Supervisor: *Renee Davis*

Text Design: *Barbara Bert*

Cover Design: *Fred Charles*

Library of Congress Cataloging in Publication Data

Main entry under title:
Modern experimental organic chemistry.

  Second ed. published in 1974 under title:
An introduction to modern experimental organic
chemistry.
  Bibliography: p.
  Includes index.
  1. Chemistry, Organic—Laboratory manuals.
I. Roberts, Royston M.   II. Title: An introduction
to modern experimental organic chemistry.
QD261.I5 1979   547'.0028   78–10752
ISBN 0-03-044391-1

Printed in the United States of America
9 0 1 2 039 9 8 7 6 5 4 3 2 1

PREFACE

The third edition of this book continues our original goal of providing students with an essentially self-contained laboratory *textbook* that requires little or no supplementation for complete understanding of the experiments. Thus laboratory procedures are preceded by a thorough discussion of the theoretical as well as the practical aspects of the experiment. Moreover, the incorporation of modern spectroscopy into an elementary organic textbook—an innovative feature of the first edition—has been expanded.

This revision may be briefly characterized as follows: (1) all experiments that have been proved satisfactory and valuable by thousands of students using the first two editions have been retained, (2) any experiments found to be less satisfactory or useful have either been removed or improved, (3) new textual material and experiments have been added, (4) all material and experiments have been updated, and (5) *safety in the laboratory* has been strongly emphasized.

Certain sections have been deleted or modified in the current edition. The chapter on alkynes, which involved a procedure requiring mercuric sulfate, has been eliminated because of concern over the toxicity of mercury. The preparation of triphenylmethyl radical has also been removed because of the difficulty with reproducibility by inexperienced students. The section on studies of molecular models has been deleted because a number of persons expressed the view that such studies are not appropriate laboratory work (although this section was enthusiastically suggested by some users of the first edition!). The experiments on alkylations of benzene have been replaced with an alkylation of *p*-xylene because of the growing concern over the toxicity and possible carcinogenicity of benzene. In fact *the use of benzene has been eliminated throughout the book,* except in one instance where it is utilized as a cosolvent in the thin-layer chromatographic analysis of derivatized amino acids. Some of the classical tests for carbohydrates have been removed from Chapter 22. The isolation of tropylium iodide (Chapter 18) has been made more dependable by a simple modification of the experimental procedure.

New material has been inserted throughout the third edition. There is a new section on *uv absorption spectroscopy* in Chapter 4, and uv spectra and exercises now appear in appropriate places throughout the book. A *malonic ester synthesis* has been added in Chapter 14; the enolate anion of dimethyl malonate is alkylated with 1,3-dibromopropane to form dimethyl cyclobutane-1,1-dicarboxylate, which can be converted to the corresponding diacid and thence to cyclobutane carboxylic acid.

Chapter 11 illustrates the interesting new technique of *phase transfer catalysis* in the preparation and reaction of dichlorocarbene. The increasing importance of photochemical techniques is exemplified by the *photochemical reduction-dimerization* of 4-methylbenzophenone to produce 4,4'-dimethylbenzopinacol; this has been added in Chapter 20. The latter compound is used in a *pinacol-pinacolone rearrangement* experiment, allowing a determination of migratory aptitudes of aryl groups.

A new chapter on *polymers* (Chapter 21) includes a new procedure for the addition polymerization of styrene that permits the use of a safer catalyst than benzoyl peroxide, the one specified by most other laboratory manuals and by previous editions of this book. Condensation polymerization is illustrated by the "*Nylon Rope Trick,*" a spectacular example of polymerization at the interface of two liquid layers. The preparation of a *rigid polyurethane foam* is also described in Chapter 21, illustrating an important new industrial application of polymer chemistry. In response to a number of suggestions from those who have used the previous editions, the chapter on qualitative organic analysis has been expanded to include a *general scheme for the separation of mixtures* of unknown compounds. Moreover, the nitrous acid test for amines has been replaced with modified Simons and Ramini tests to avoid the production of potentially carcinogenic nitrosamines. A new appendix describes *laboratory techniques for heating and stirring.*

The entire text has been updated, specifically the chapters on spectroscopy, organometallic chemistry, and the use of organic chemistry literature.

A significant development in recent years has been the growing awareness of the toxic and carcinogenic nature of many organic compounds. In preparing this revised edition, we felt a strong sense of responsibility to inform students of the potential danger inherent in handling certain substances. In the earlier editions we emphasized the usual precautions to be taken in the handling of flammable, corrosive, and noxious organic chemicals. Now, in view of recent warnings about the toxic and carcinogenic properties of compounds formerly thought to be rather innocuous (such as benzene, acetonitrile, nitrosamines, and carbon tetrachloride), we felt more stringent precautions should be taken in laboratory courses to protect students from contact with these substances.

We have considered every chemical used as substrate, reagent, or solvent in the experiments, and we have avoided use of the potentially dangerous ones as far as is practical. For example, other solvents have been substituted for benzene except in one experiment. In every experiment involving a possibly toxic substance, we have called for its handling in a protected system or in a fume hood. To dramatize safety precautions of all kinds, we have added a Do It Safely section at the beginning of most Experimental Procedures. We believe that in this edition we have provided a text for the safest possible course that will still offer a satisfactory introduction to experimental organic chemistry.

We would like to call attention to the *Instructor's Manual,* which is available without charge from the publisher. This manual contains lists of recommended desk equipment, shelf reagents and solvents, lists of required chemicals and equipment—by chapter and alphabetically, notes on the Experimental Procedures—including time requirements, answers to selected exercises, and suggestions for optional experiments taken from recent issues of the *Journal of Chemical Education.*

We are pleased to acknowledge helpful comments from many colleagues who have used the previous editions of this book, and we earnestly solicit comments on this edition. Special thanks are due to Aubrey Skinner and Robert G. Landolt for their assistance in updating the literature chapter and to Kevin Cann, Sophon Roengsumran, John Ginascol, and Jim Walter for development and testing of some of the new procedures.

We are particularly grateful to the following individuals who reviewed the manuscript and offered suggestions for its improvement: Kenneth K. Andersen, University of New Hampshire; Victor Badding, Manhattan College; Richard Bozak, California State University at Hayward; Donald Brundage, University of Toledo; Michael DeGregorio, College of San Mateo; Sal Gimelli, Fairleigh Dickinson University; Lars H. Hellberg, San Diego State University; Eugene N. Losey, Elmhurst College; M. G. Newton, University of Georgia; David Allan Owen, University of Oklahoma; Wesley Pearson, St. Olaf College; J. Christopher Phillips, University of Detroit; Howard C. Price, Marshall University; Bonnie Sandel, Franklin and Marshall College; Gene Schaumberg, California State College, Sonoma; Suzanne Slayden, George Mason University; Peter A. Smith, University of Michigan; Charles J. Thoman, S. J., University of Scranton; James G. Traynham, Louisiana State University; Thomas G. Waddell, University of Tennessee; Karl Wiegers, University of Illinois.

Austin, Texas
Baltimore, Maryland
September 1978

R. M. R.
J. C. G.
L. B. R.
A. S. W.

CONTENTS

Preface   iii

Introduction   1

1 PHYSICAL CONSTANTS OF ORGANIC COMPOUNDS 7

1.1 Melting Point of a Pure Substance   8
1.2 Effect of Impurities on Melting Points; Mixtures   9
1.3 Other Kinds of Melting Behavior of Mixtures   11
1.4 Micro Melting-Point Methods   12
1.5 Boiling Points   18
1.6 Index of Refraction   21
1.7 Density   22

2 SEPARATION AND PURIFICATION OF ORGANIC COMPOUNDS / Distillation, Recrystallization, and Sublimation 24

2.1 Simple Distillation   24
2.2 Fractional Distillation   27
2.3 Fractional Distillation Columns   29
2.4 Fractional Distillation of Nonideal Solutions   32
2.5 Distillation under Reduced Pressure   40
2.6 Steam Distillation   44
2.7 Recrystallization   49
2.8 Sublimation   60

3 SEPARATION AND PURIFICATION OF ORGANIC COMPOUNDS / Phase Distribution: Extraction and Chromatography 64

3.1 Extraction   64
3.2 Technique of Simple Extraction   70
3.3 Chromatography   74
3.4 Gas Chromatography   75
3.5 Column Chromatography   83

3.6   High-Pressure Liquid Chromatography   88

3.7   Thin-Layer Chromatography   90

3.8   Dry-Column Chromatography   94

3.9   Paper Chromatography   94

4   SPECTROSCOPIC METHODS OF IDENTIFICATION AND STRUCTURE PROOF   97

4.1   Infrared Spectroscopy   99

4.2   Nuclear Magnetic Resonance Spectroscopy   107

4.3   Ultraviolet and Visible Spectroscopy   117

5   ALKANES / Free Radical Halogenation   126

5.1   Chlorination by Means of Sulfuryl Chloride   126

5.2   Bromination: Relative Ease of Substitution of Hydrogen in Different Environments   135

6   ALKENES / Preparations and Reactions   140

6.1   Dehydrohalogenation of Alkyl Halides   141

6.2   Dehydration of Alcohols   151

6.3   Addition Reactions of Alkenes   164

7   ELECTROPHILIC AROMATIC SUBSTITUTION   174

7.1   Introduction to Electrophilic Aromatic Substitution   174

7.2   Friedel-Crafts Alkylation of *p*-Xylene with 1-Bromopropane   176

7.3   Nitration of Bromobenzene   184

7.4   Relative Rates of Electrophilic Aromatic Substitution   190

8   DIENES / The Diels-Alder Reaction   199

9   KINETIC AND EQUILIBRIUM CONTROL OF A REACTION   210

10   NUCLEOPHILIC SUBSTITUTION   220

10.1   Nucleophilic Substitution at Saturated Carbon   220

10.2   Chemical Kinetics: Evidence for Nucleophilic Substitution Mechanisms   229

11 CARBENES AND ARYNES / Highly Reactive Intermediates ... 237

11.1   Carbenes ... 238
11.2   Arynes ... 251

12 ORGANOMETALLIC CHEMISTRY ... 257

12.1   Introduction ... 257
12.2   The Grignard Reagent: Its Preparation and Reactions ... 258
12.3   The Organolithium Reagent ... 270
12.4   Organocopper Reagents ... 271

13 OXIDATION REACTIONS OF ALCOHOLS AND CARBONYL COMPOUNDS ... 277

13.1   Preparation of Aldehydes and Ketones by Oxidation of Alcohols ... 277
13.2   Classification of Alcohols and Carbonyl Compounds by Means of Chromic Acid ... 285
13.3   Base-catalyzed Oxidation-Reduction of Aldehydes: The Cannizzaro Reaction ... 287

14 SOME TYPICAL REACTIONS OF CARBONYL COMPOUNDS ... 290

14.1   Reactions Involving Nucleophilic Addition to the Carbonyl Group ... 290
14.2   Reactions of Carbonyl Compounds Involving the α- and β-Carbon Atoms ... 308
14.3   Identification of an Alcohol or Carbonyl Compound by Means of Three Reagents ... 327

15 CARBOXYLIC ACIDS AND THEIR DERIVATIVES ... 328

15.1   Carboxylic Acids ... 329
15.2   Carboxylic Acid Esters ... 330
15.3   Carboxylic Acid Halides (Acyl Halides) and Anhydrides ... 332
15.4   Carboxylic Acid Amides and Nitriles ... 334

16 HETEROCYCLIC SYNTHESIS ... 344

16.1   2-Aminothiazole ... 345
16.2   4-Aminoquinolines ... 349

17 MULTISTEP ORGANIC SYNTHESES — 359

17.1 Introduction — 359
17.2 The Synthesis of Sulfathiazole — 361
17.3 The Synthesis of 1-Bromo-3-chloro-5-iodobenzene — 377
17.4 Additional Multistep Synthetic Sequences — 389

18 NONBENZENOID AROMATIC COMPOUNDS — 391

18.1 Aromaticity — 391
18.2 Ferrocene — 395
18.3 Tropylium Iodide — 401

19 ISOMERISM AND OPTICAL ACTIVITY / Resolution of Racemic α-Phenylethylamine — 405

19.1 Polarimetry — 409
19.2 Resolution of Racemic α-Phenylethylamine — 411
19.3 Optional Projects — 414

20 PHOTOCHEMICAL DIMERIZATION OF 4-METHYLBENZOPHENONE; PINACOL-PINACOLONE REARRANGEMENT — 417

20.1 Preparation and Photochemical Reaction of 4-Methylbenzophenone — 417
20.2 Pinacol-Pinacolone Rearrangement of 4,4'-Dimethyl-benzopinacol — 420

21 POLYMERS — 431

21.1 Addition Polymers — 431
21.2 Condensation Polymers — 441
21.3 Polyurethanes — 446

22 CARBOHYDRATES — 451

22.1 Introduction — 451
22.2 Mutarotation of Glucose — 453
22.3 The Hydrolysis of Sucrose — 458
22.4 Isolation of α,α-Trehalose — 460

23 AMINO ACIDS AND PEPTIDES 463

23.1 Introduction 463
23.2 Analysis of Amino Acids 465
23.3 Determination of Primary Structure 474

24 NATURAL PRODUCTS / Classification, Isolation, and Characterization 482

24.1 Introduction 482
24.2 Citral from Lemon Grass Oil 485
24.3 Piperine from Black Pepper 489

25 IDENTIFICATION OF ORGANIC COMPOUNDS 495

25.1 Separation of Mixtures of Organic Compounds 495
25.2 Classical Qualitative Organic Analysis Procedure for Identification of a Pure Compound 500
25.3 Modern Spectroscopic Methods of Analysis 533
25.4 Tables of Derivatives 538

26 THE LITERATURE OF ORGANIC CHEMISTRY 550

APPENDIX 1 DRYING AGENTS / Desiccants 563
APPENDIX 2 THE LABORATORY NOTEBOOK 568
APPENDIX 3 COMPILATION OF PMR ABSORPTIONS 574
APPENDIX 4 TABLE OF CHARACTERISTIC IR FREQUENCIES 576
APPENDIX 5 HEATING AND STIRRING TECHNIQUES IN THE LABORATORY 582

Index 591

MODERN
EXPERIMENTAL
ORGANIC
CHEMISTRY

INTRODUCTION

The laboratory part of an introductory course in organic chemistry is complementary to the lecture part; this is where you learn firsthand that the compounds and reactions described in lectures are not merely abstract notations. In addition, many of the theoretical concepts discussed in the lectures are amenable to experimentation at even an introductory level, and you will find that collecting and interpreting your own data will add reality to the theoretical framework of the subject.

Probably the most important purpose of the laboratory, however, is the introduction it provides to the experimental techniques of the practicing organic chemist. Learning to handle organic chemicals and manipulate apparatus is obviously a necessary part of a chemist's education. Less obvious, but equally important, is the learning of the scientific approach to laboratory work, which results in what is called good experimental technique. Some suggestions to aid in developing good technique are given in the following paragraphs.

Never begin any experiment unless you understand the overall purpose of the experiment and the reasons for each operation involved. This requires *studying* (not just reading) an experiment *before coming to the laboratory.* For this reason laboratory experiments are always assigned several days in advance, except under unusual circumstances. You will find that not only will your performance in laboratory be better if you are well-prepared but also you will benefit much more from the experiments, in knowledge as well as in grades.

Neatness is an important part of good technique. Carelessness in handling chemicals may not only lead to poor results but is often unsafe. Similarly, careless assembly of apparatus is not only aesthetically displeasing but is also dangerous.

The procedures given in this book are meant to be followed closely. There is usually a reason why each operation is to be carried out exactly as described, although the reason may not at first be obvious to the beginning student. In the earlier chapters of the book the operations required in the experimental procedures are described rather specifically, but in later chapters, after experimental techniques have become more familiar to you, the directions are less specific. In this way we hope to avoid your feeling that you are following "cookbook" directions. As you gain more practice in the techniques of the organic laboratory, you may even formulate alternative procedures for performing a reaction, purifying a product, and so on. However, for the sake of safety it will be wise for you to check any original plans with your instructor. In more advanced courses you may expect to use your chemical initiative more freely.

In some of the experimental procedures, stars($\star$) have been inserted to indicate points at which the experiment may be interrupted without affecting the results unfavorably. This has been done to help instructors and students plan to make the best use of laboratory time. For example, some experiments may be started toward the end of a laboratory period, carried through to a starred point ($\star$), and then discontinued until the next laboratory period, the reaction mixture being safely stored in the desk in the interim. In cases where it is already obvious from the text that the experiment may or should be interrupted, stars are not inserted.

One of the most valuable characteristics that you can develop as an experimentalist is being *observant* during all stages of an experiment. Was there a fleeting color change when a drop of reagent was added to the solution? Was a precipitate formed? Is the reaction exothermic? Such observations may seem insignificant at the time but may later prove to be vital to the correct interpretation of an experimental result. All such observations should be *recorded* in a notebook.

A permanent record of all laboratory work should be kept in a *bound notebook,* approximately 8×10 in., rather than in a loose-leaf or spiral notebook. Pages of a bound book are less likely to tear out accidentally and be lost. Always write in ink. It will be easier to keep the notes neat and legible if the pages are lined horizontally. (Although notebooks with vertical as well as horizontal lines are sometimes recommended for laboratory notes, particularly in physical chemistry, they are less desirable for organic chemistry laboratory notes. In the special cases where graphs are necessary, they may be pasted in.) Use the first page as a title page, and reserve two additional pages to be filled in as a table of contents. The pages should be numbered throughout the notebook.

Your instructor will probably have specific directions regarding the format of your notebook and of any reports you may be asked to submit. In the case of the notebook these may include a preliminary "write-up," in which you give the title of the experiment, reference(s), complete equations for the main and side reactions, and calculations of the theoretical yield in synthetic experiments. You may be allowed some flexibility in style, but proper recording of experimental results does not allow great literary license. Appendix 2 contains suggested formats for your notebook for both preparative and investigative types of experiment.

The overriding requirements of good experimental description are accuracy and completeness of observation and recording. The results should usually be summarized and conclusions should be drawn from each experiment. If the results are obviously not those expected, an explanation should be given.

SAFETY IN THE LABORATORY

Public and professional concerns over the potential health-related dangers of chemicals have grown in recent years. The chemical laboratory is indeed a potentially dangerous place; it contains flammable liquids, fragile glassware, and corrosive and poisonous chemicals. Nevertheless, if proper precautions are taken and safe procedure is followed, it is no more dangerous than an ordinary kitchen or bathroom. To an experienced laboratory worker, "proper precaution and safe procedure" is synonymous with knowledge and awareness of the properties of the substances handled and of the limitations that should be placed upon the use of various types of equipment, as well as an element of common sense. However, an inexperienced student cannot be expected to possess extensive knowledge of what constitutes safe laboratory operation. This must be learned and developed in the same way as are the theoretical aspects of organic chemistry in the lecture part of this course.

We have therefore included a series of sections entitled "Do It Safely," which are featured throughout the book. These sections accompany most of the experiments, and they are somewhat more detailed for those experiments likely to be performed in the first semester of organic laboratory. In many instances the dangerous aspects of specific chemicals are discussed. The purpose of these sections is to inform you of some precautions always taken by an experienced worker in the laboratory. We hope you read and accept this feature of the book with the same attitude with which it is written, that is, with a *positive* attitude toward developing safe technique and the knowledge that will lead to what might be called "common sense in the laboratory." Some important *general* safety considerations are discussed below. You should read them carefully now and review them prior to each laboratory period until they become "second nature" to you.

DO IT SAFELY

1. When you first receive laboratory equipment, examine the glassware closely for small cracks, chips, or other imperfections that might weaken it. It is particularly important to check round-bottomed flasks and condensers (see Equipment Commonly Used in the Organic Laboratory, pictured inside the back cover) carefully. Cracks in a round-bottomed flask may cause it to break during use, spilling quantities of potentially dangerous and flammable chemicals. Cracks at the ring seals of condensers where the inner tube and the water jacket are joined may allow water to drain into a flask containing reagents, which may react violently with water. Replace any such imperfect glassware immediately. Take the time to properly store your equipment in your locker or drawer. Carelessly stored glassware may become cracked as a result of opening and closing drawers. *Develop the habit of examining your glassware before each use.*

2. Nearly all organic substances are **flammable.** It is best to avoid the use of flames whenever possible. There may be times, however, when the use of laboratory burners is necessary, as in the distillation of particularly high-boiling liquids. Refer to Appendix 5.

When flames are used, you should take care to observe the following precautions:

a. A burner should be used to heat a flammable liquid only when it is contained in a flask protected by a condenser (see Figures 2.2 and 7.1). *Never* heat a flammable liquid in an *open* container with a burner; if possible, use a steam bath or an electrical heating device instead. In Table 2.1 the flammability of many commonly used solvents is indicated; if a solvent is not listed there, however, you should not assume that a substance is nonflammable. Consult your instructor.

    b. Avoid pouring flammable liquids from one container to another within several feet of a flame. Do not pour waste flammables into the center trough of your work bench; they may be carried near a flame farther down the bench.

    3. Most of the chemicals that you will encounter in the laboratory are *at least* mildly toxic, and many may be corrosive or caustic. The reasons for the following guidelines should be apparent.

    a. Never *taste* anything in the laboratory unless your instructor specifically tells you to do so.

    b. Always wear safety glasses or goggles when in the laboratory to avoid possibly irreversible eye damage from splashing chemicals in case of an accident. Remember that even when you are not actually working with chemicals, another student may have an accident in which you may become involved. *Contact lenses are particularly dangerous in a laboratory environment and should not be worn.*

    c. Avoid the inhalation of vapors of the solvents and other substances that you use or encounter. Use a fume hood insofar as possible when pouring and handling volatile substances and for carrying out reactions in which noxious gases are released. If this is not feasible, affix appropriate gas traps (described among the Experimental Procedures) to your apparatus.

    d. Use appropriate equipment and glassware to handle and transfer chemicals from one container to another. Generally avoid allowing chemicals to come into contact with your skin. It is prudent to keep a pair of ordinary kitchen rubber gloves in your drawer and to use them when you are measuring and transferring particularly harmful liquids and solutions. Most chemicals can be removed from the surface of the skin by washing thoroughly with soap and warm water. This should be done promptly following any contamination. Avoid using organic solvents such as acetone or ethanol to remove chemicals from your skin; such solvents may actually hasten the absorption of the chemical through the skin.

    4. Familiarize yourself with the layout of your laboratory. Note particularly the location of fire extinguishers, fire blankets, safety showers, and eye-wash fountains. Your instructor will explain the operation and purpose for each of these safety devices. Read the section entitled "First Aid in Case of Accident," which is provided inside the front cover of the book.

GLASSWARE: PRECAUTIONS AND CLEANING

    The cardinal rule in handling laboratory glassware is *never apply undue pressure or strain to any piece of glassware.*

    This rule applies to insertion of thermometers or glass tubes into rubber stoppers, rubber tubes, or corks. "If you have to force it, don't do it!" Make the hole larger or use a smaller piece of glass. Lubricate the rubber or cork with a little water or glycerol. Always grasp the glass part very close to the rubber or cork part when pushing the glass into it.

    These directions will become less necessary as more and more laboratories are making the transition to all ground-glass-jointed apparatus. However, the cardinal rule stated above applies to this glassware as well; take care that strain does not develop because of carelessly positioned components. Strained glassware will break at times when heated, or simply upon standing.

If ground-glass-jointed glassware is available, it is important that the joints be properly lubricated so that they do not "freeze" and become difficult, if not impossible, to separate. Proper lubrication can be accomplished by putting a thin layer of grease on opposite sides of the male joint, mating the two joints, and then rotating them together to cover the surfaces of the joints with a thin coating of grease. The quantity of lubricant used is important since too much grease may cause ultimate contamination of the reaction whereas too little will permit the joints to freeze.

Glassware is most easily cleaned *immediately* after use. Most chemical residues can be removed by washing the glassware with detergent and water or with common organic solvents such as toluene or acetone. (*Caution:* Do *not* use acetone to clean apparatus that contains residual amounts of bromine since a powerful lachrymator, bromoacetone, may thus be formed.)

More stubborn residues may require the use of powerful chemical cleaning solutions such as chromic acid (concentrated sulfuric acid and chromic anhydride) or potassium hydroxide in ethanol. Gentle warming of the organic solvent or cleaning solution on a steam bath will generally hasten the cleaning process. Before you resort to the use of any cleaning agents other than detergent and water, consult your instructor for permission and directions concerning their safe handling.

Brown stains of manganese dioxide on glassware may sometimes be encountered; these can generally be removed by rinsing the apparatus with a 30% aqueous solution of sodium bisulfite.

It is good practice to wipe off any lubricant from ground-glass-jointed glassware with a towel or tissue wetted with a solvent such as acetone or hexane before washing with detergent and water. Otherwise the lubricant will stick to the brush used to scrub the glassware and be carried by the brush onto all surfaces it touches.

DISPOSAL OF CHEMICALS

Your laboratory should be supplied with containers for the proper disposal of waste chemicals. Liquid and solid organic wastes should be placed in appropriately labeled containers. In this way environmental pollution is minimized, and the buildup of solids and flammable liquids in drain traps and pipes is avoided. Aqueous solutions may be discarded in the sink using running water. Do not discard aqueous waste into the organic liquid waste container because the acidic, basic, or other types of solutes possibly present in such solutions may initiate chemical reactions among the organic wastes in the container.

Solid inorganic wastes such as desiccants (see Appendix 1) should be discarded in an appropriate container. This procedure will provide for a cleaner and safer laboratory and will protect the janatorial staff from inadvertently being exposed to powdery and perhaps dangerous chemicals.

Many laboratories have a special place to discard broken glassware. Ask your instructor if this is the case in your laboratory.

REFERENCES

1. N. I. Sax, *Dangerous Properties of Industrial Materials,* Reinhold Publishing Corporation, New York, 1957.
2. *Merck Index of Chemicals and Drugs,* 9th ed., Merck and Company, Rahway, N. J., 1976.
3. *Handbook of Laboratory Safety,* 2d ed., The Chemical Rubber Company, Cleveland, Ohio, 1971.
4. *Manual of Hazardous Chemical Reactions,* 4th ed., National Fire Protection Association, Boston, 1971.
5. *Suspected Carcinogens,* 2d ed., National Institute for Occupational Safety and Health, 1976.

1

PHYSICAL CONSTANTS OF ORGANIC COMPOUNDS

The physical constants of organic compounds are numerical values associated with the measurable properties of these substances. So long as accurate measurements are made under specified conditions such as temperature and pressure, these properties are invariant and useful in the identification and characterization of the substances encountered in the laboratory. Among the more commonly measured physical properties are melting point (mp), boiling point (bp), index of refraction (n), and density (d). This chapter includes discussion and experiments involving measurement of these constants. Beginning with Chapter 4, we see how more recently developed spectroscopic techniques are extremely useful in the identification and structural determination of organic compounds.

Although it is certainly true that more than one compound might exhibit the same constant for one or two of the common physical properties, it is extremely unlikely that two compounds will display the same constants for several measurable properties. A list of physical constants is therefore regarded as a highly useful characterization of a substance. Several extensive compilations of the physical constants of organic compounds are available.[1] One of the most convenient is the Chemical Rubber Company's *Handbook of Chemistry and Physics*, in which a very large number of inorganic and organic compounds and their known physical constants and properties are tabulated.

It must be emphasized that physical constants are useful only in the identification of *previously known* compounds, because it is not possible to predict accurately such properties as melting point and boiling point of *newly* synthesized or isolated compounds. Moreover, the observed melting point or boiling point of a substance may provide information about the purity of the sample under consideration. Other properties such as color, odor, and crystal form are also useful.

[1] See Chapter 27.

1.1 Melting Point of a Pure Substance

The melting point of a substance is defined as the temperature at which the liquid and solid phases exist in equilibrium with one another without change in temperature. If heat is added to a mixture of the solid and liquid phases of a pure substance at the melting point, ideally no rise in temperature will occur until all the solid has been converted to liquid (melted); if heat is removed from such a mixture, the temperature will not drop until all the liquid has been converted to solid (frozen). Thus the melting point and freezing point of a pure substance are identical. The relationship between phase composition, total supplied heat, and temperature for a pure compound is shown in Figure 1.1. It should be understood that *heat* is being supplied to the compound at a constant rate, and thus that the elapsed time of heating is a cumulative measure of the supplied heat. At the lower temperature (below the melting point) the compound exists in the solid phase and the addition of heat causes the temperature of the solid to rise. As the melting point is reached, the first small amount of liquid appears; equilibrium is established between the solid and liquid phases. As heat continues to be added, the temperature does not change, the additional heat causing conversion of solid to liquid, with both phases remaining in equilibrium, however. As the last of the solid melts, the heat subsequently supplied causes the temperature to rise linearly at a rate which depends on the rate of heat supply.

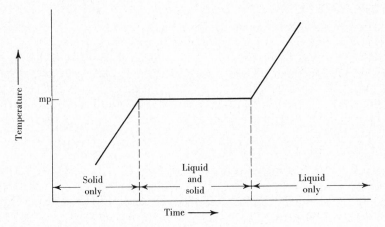

FIGURE 1.1 Phase changes with time and temperature.

We may describe the interconversion of the liquid and solid phases in terms of their respective vapor pressures, which are directly related to the rates at which the molecules pass from one phase to the other. For simplicity we may consider only transfer of molecules between the solid and liquid phases. The term vapor pressure may then seem to be a misnomer, but the same "escaping tendency" that produces the measurable equilibrium vapor pressure is responsible for the direct transfer of molecules between the solid and liquid phases. Figure 1.2 represents these transfers

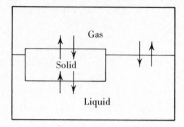

FIGURE 1.2 Phase equilibria.

between the three phases by means of reversible arrows. At the melting point, the vapor pressures of the solid and liquid phases are equal, and no *net* transfer of molecules from one phase to the other occurs unless there is a change in the heat content of the system.

1.2 Effect of Impurities on Melting Points; Mixtures

Consider a mixture of solid and liquid A at its melting temperature. If a small amount of a second pure material, B, is dissolved in the liquid A, solid A will begin to melt. This is because the addition of B lowers the vapor pressure of liquid A (Raoult's law), and the vapor pressure of solid A remains the same; thus the rate at which molecules of solid A pass into the liquid phase is greater than the rate of the reverse process. If the temperature is kept constant by supplying the heat required for this melting process, all of solid A will melt; if no heat is added, the temperature will drop because heat energy is consumed by the melting solid. Since the vapor pressure of a solid decreases more rapidly than the vapor pressure of its solution as the temperature drops, they will become equal at a lower temperature, and equilibrium will be established again at that temperature. This lower temperature then represents the freezing and melting point of the mixture of A and B, that is, "impure A." The more of the impurity B that is added, the lower will be the melting point of the mixture.

Binary mixtures of most organic substances exhibit this kind of behavior; that is, the addition of a second substance progressively lowers the melting point of the first, as may be represented by the melting point–composition diagram of Figure 1.3. In this diagram, a represents the melting point of pure substance A and b the melting point of pure substance B. Point F represents the temperature (f) at which crystals of pure A are in equilibrium with a molten solution of B (20%) in A (80%). If a mixture of this composition is prepared and melted, and the temperature is then lowered to· f, pure crystals of A will form in the presence of the melt. As additional heat is removed, A will continue to crystallize, reducing the percentage of A in the melt. The equilibrium temperature will correspondingly decrease because of the reduced vapor pressure of A in the molten solution. When the equilibrium temperature reaches e, at which point the melt has the composition indicated at E, then and *only then* does B, the impurity, also begin to crystallize. A and B will crystallize in the constant ratio 60% A to 40% B, and the melt composition will no longer change. During this stage

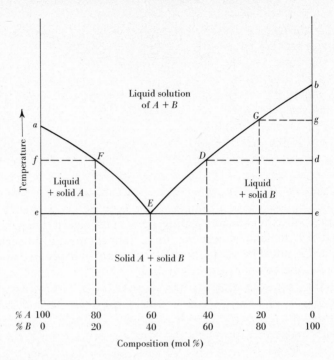

FIGURE 1.3 Melting point–composition diagram.

the melt of the specified composition is in equilibrium with both solid *A* and solid *B*. The temperature remains constant until the entire sample has crystallized, and then again begins to fall as the solid cools. Point *E* is called the *eutectic point,* and it defines the composition at which *A* and *B* can cocrystallize (eutectic mixture) and the temperature (*e*) at which the eutectic mixture freezes (and melts).

Now consider the reverse process, the melting of a solid mixture of 80% *A* and 20% *B*, which is of more interest to the organic chemist. Heat is applied and the temperature rises. When it reaches *e*, *A* and *B* will melt together at a constant ratio (eutectic composition), and the temperature will remain constant (remember the eutectic melt is in equilibrium with both solid *A* and solid *B*). Eventually *B* will be entirely melted (as it is the minor component), leaving only solid *A* in equilibrium with a melt of eutectic composition. As more heat is applied, the remaining *A* will continue to melt, raising the percentage of *A* in the molten solution above the eutectic composition. Since the vapor pressure of *A* in the solution is thus increased, the temperature at which solid *A* is in equilibrium with the molten solution will rise, and the relationship between the equilibrium temperature and the composition of the molten solution will be represented by the curve *EF.* When the temperature reaches *f,* the last of *A* melts. Thus this impure sample *A* exhibits a depressed melting "point" occurring over a relatively broad range (*e–f*). Notice that we have been considering the case of substance *A* with impurity *B*. If the composition of the solid had been to the right of point *E* in Figure 1.3, we would conversely speak of substance *B* with impurity A, and the rising temperature during the melting process would follow curve ED or EG, with melting range of e–d or e–g.

It may be noted that the actual melting *range* of a mixture containing 20% impurity may be broader than that of a mixture containing 40% impurity; for example, in Figure 1.3, compare the range e–g with e–d. In practice, however, particularly in the capillary tube method given in the procedure, it is very difficult to observe the initial melting point. If the amount of impurity is small, the amount of liquid produced in the early stages (near the eutectic temperature) is very small. However, the temperatures at which the last crystals disappear (d, g) can be determined quite accurately, so that the mixture containing the smaller amount of impurity will have both a higher final melting point and a narrower (*observed*) range.

It should be noted that a sample whose composition is exactly that of the eutectic mixture will exhibit a *sharp* melting point at the eutectic temperature. Thus a eutectic mixture is sometimes mistaken for a pure compound. In such a case the sample may be identified by adding a small amount of either component (assuming one knows what the components are) and observing that the melting point rises.

The melting point depression produced by the introduction of an impurity into a pure compound may be used to advantage in identification of that compound. This important procedure is known as a *mixture–melting point*[2] determination. It is perhaps best presented by an example. Assume that a compound melts at 133°, and you suspect it is either urea or *trans*-cinnamic acid (both melt at 133°). If the compound is mixed intimately with urea and the melting point of this mixture is found to be lower than either the pure compound or pure urea, then urea is acting as an impurity, and the compound cannot be urea. If the mixture–melting point is identical to that of the pure compound and of pure urea, then the compound is tentatively identified as urea. It should be noted that one can exclude possible compounds by this method with far more certainty than one can select them, but the procedure is nevertheless occasionally useful in making identifications when samples of likely possibilities are on hand.

1.3 Other Kinds of Melting Behavior of Mixtures

Although the melting temperature behavior (eutectic formation) just described is typical of most binary mixtures, there are many exceptions to this general pattern. A second component does not always lower the melting temperature of an organic compound. Figure 1.4a–c illustrates some unusual melting point–composition diagrams. In Figure 1.4a the eutectic composition contains so little of component B that even a fairly small amount of B as impurity may actually raise the observed melting point of A. Figure 1.4b illustrates the formation of a compound (C) from A and B, and Figure 1.4c is representative of one type of solid solution formation. For further information consult either reference 1 at the end of this chapter or a physical chemistry textbook.

---

[2] Frequently referred to (ungrammatically) as a "mixed–melting point" determination.

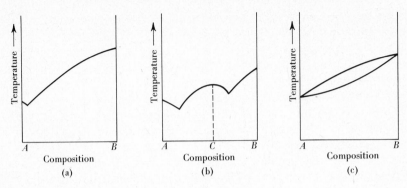

FIGURE 1.4   Unusual melting point–composition diagrams.

1.4   Micro Melting-Point Methods

The determination of an exact melting point of an organic compound requires enough material so that equilibrium can be established between the liquid and solid phases and the temperature at the equilibrium state measured, usually by means of repeated cooling and heating cycles which show a plateau in a temperature-time plot. The amount of material required for such procedures is often more than the chemist has available, so micro methods have been developed that are not so exact but which are convenient and require negligible amounts of sample.[3] The most commonly used method is the capillary tube melting-point procedure, described in detail later. The *microscopic hot stage*[4] and simpler variations of it are also popular; although the equipment is somewhat more expensive, still smaller quantities of crystals may be satisfactorily utilized. In all micro methods the melting point is actually determined as a melting *range,* encompassed by the temperature at which the sample is first observed to begin to melt and the temperature at which the melting process is complete. When properly carried out with a *pure* crystalline compound, a range of no more than 0.5–1.0° should be observed. However, few commercially obtained compounds exhibit such a narrow melting range.

A wide variety of types of heating devices have been used in connection with capillary-tube, melting-point determinations, ranging from a simple beaker containing a high-boiling liquid heated by a burner and stirred manually to elaborate electrically heated and mechanically stirred devices. The Thomas-Hoover melting-point apparatus pictured in Figure 1.5 is an example of the latter variety and illustrates its general features. Inside the casing is a container of either mineral oil or (better) silicone oil into which is immersed an electrical resistance heater. By varying the voltage across the heater by means of the large knob on the front of the apparatus, the oil may be heated at a controlled rate. A motor-driven stirrer is controlled by the knob at the bottom of the instrument. Some models are fitted with a movable magnifying lens system in order to view the thermometer better while simultaneously

[3]It must be admitted that there is less emphasis on extremely precise melting-point determinations since the development of spectroscopic methods for characterization of organic compounds (see Chapter 4).

[4]The Kofler hot-stage technique is described on page 155 of reference 1 at the end of this chapter.

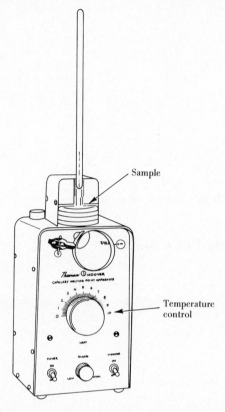

Sample

Temperature
control

FIGURE 1.5   Thomas-Hoover melting-point apparatus. (Courtesy of Arthur H. Thomas
Company)

viewing the sample in the capillary tube. The Mel-Temp apparatus shown in Figure
1.6 utilizes a heated metal block rather than a liquid for heat transfer to the capillary
tube. A thermometer inserted into a hole bored into the block gives the temperature
of the block and capillary tube. Again, heating is accomplished by voltage control of
a heating element within the block. Further description and discussion of these types
of devices are available in reference 1 at the end of the chapter.

The proper use of a simple type of melting-point device, the Thiele tube (Figure
1.7), is described in the following procedure; naturally, the melting-point determinations may just as well be performed with other devices such as those just discussed if
they are available. Note that the Thiele tube is shaped so that heat applied to the
sidearm by a burner is distributed to all parts of the vessel by convection currents so
that additional stirring is not required. Temperature control by adjustment of the
burner may seem difficult at first but can be mastered with practice.

Although in most instances the melting range of a substance may simply and
accurately be determined by careful observation so long as the heating rate is slow

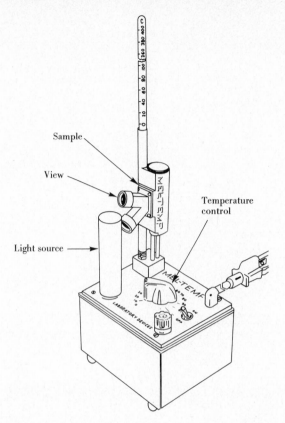

FIGURE 1.6   Mel-Temp melting-point apparatus. (Courtesy of Laboratory Devices)

(see part A of Experimental Procedure), unusual melting characteristics are occasionally observed. Most likely to be encountered are *softening* and *decomposition*. Many organic compounds undergo a change in crystal structure just before melting, usually as a consequence of the release of solvent of crystallization. The sample takes on a softer, perhaps "wet," appearance, which may also be accompanied by shrinkage of the sample in the tube. Observance of these types of changes in the sample should *not* be interpreted as the beginning of the melting process; wait for the first tiny drop of liquid to appear.

Some compounds decompose on melting. When this happens, discoloration of the sample is usually evident. The decomposition products constitute impurities in the sample, and the melting point is actually lowered as a result of such decomposition. When this behavior is observed, the melting point should be reported in such a way as to denote its occurrence, for example, 183° d or 183° (dec).

Inasmuch as the accuracy in any type of temperature measurement depends ultimately on the thermometer, you may be advised by your instructor to calibrate your thermometer. In some temperature ranges a given thermometer may provide an accurate reading; in others the reading may be a degree or two off, either high or low.

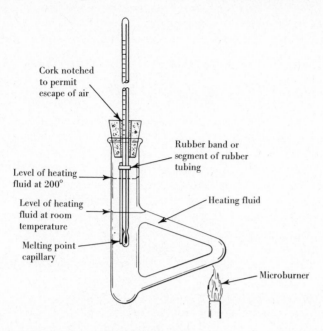

Cork notched
to permit
escape of air

Rubber band or
segment of rubber
tubing

Level of heating
fluid at 200°

Heating fluid

Level of heating
fluid at room
temperature

Melting point
capillary

Microburner

FIGURE 1.7 Thiele melting-point apparatus.

Accordingly, directions for calibration of a thermometer are included in part B of Experimental Procedure.

EXPERIMENTAL PROCEDURE

DO IT SAFELY

1. If you are using a burner in this experiment, take care that no flammable solvents are nearby. If your hair is long, you should consider tying it back in order to keep it safely away from the flame. Be careful to keep the rubber tubing leading to the burner safely away from the flame. Turn off the burner when it is not actually in use.

2. Mineral and silicone oils may not be safely heated if they have become contaminated with even a few drops of water. Heating these oils above 100° (the boiling point of water) may produce spattering of hot oil as a result of steam formation. Fire can also result as the oil comes in contact with open flames. Examine the base of your Thiele tube for evidence of water droplets. If any are observed, either change the oil or exchange tubes, giving the contaminated tube to your instructor.

3. Do not heat mineral oil (a mixture of high-boiling hydrocarbons) above about 200° because there is the possibility of spontaneous ignition, particularly when a burner is being used. Silicone oils may be heated to about 300°.

4. In this and all other experiments, be careful that chemicals do not come in contact with your skin. If you spill any chemicals, clean them up immediately with a brush or a paper towel.

A. Determination of Capillary-Tube Melting Points

General Procedure. Place the sample in a closed-end capillary tube supplied by your instructor. The easiest way to fill the capillary is to place a bit of the *powdered* sample on a small watch glass or the bottom of an inverted beaker and to tap the open end of the capillary into the powder a few times. The solid may be made to "filter" down to the closed end of the tube by inverting it and lightly scratching the tube with a file, or by briskly tapping the closed end of the tube onto a solid surface (as one would a pencil). The solid should be tightly packed in the tube to optimize heat transfer through the sample; this can best be accomplished by dropping the capillary down a larger piece of glass tubing about one meter in length and held vertically to a solid surface. The size of the sample should be such that the capillary is filled to a depth of no more than 2–3 mm after the compacting process.

Attach the capillary to a thermometer by means of a small rubber band (conveniently obtained as a slice of ordinary rubber tubing). The sample itself should be directly adjacent to the bulb of the thermometer. The rubber band should be positioned such that even at 200° it will remain above the level of the heating fluid (see Figure 1.5). This accomplished, place the thermometer into the heating vessel, and support it by means of a bored cork cut away on one side so as to make visible the thermometer markings in that vicinity. This cut also serves the purpose of making the apparatus an open system. *Never heat a closed system.* By applying heat from a small Bunsen burner (microburner) raise the temperature of the heating fluid slowly (about 2° per minute). Note the temperature at which melting is first observed and the temperature at which the last of the solid melts, and record these as the melting range of the solid. *Because one can obviously spend a great deal of time waiting for the melting point of a high-melting solid by slow heating, it is usually convenient to prepare two samples of the solid under consideration and determine the approximate melting point of the first by heating rapidly. Then let the heating fluid cool to 10–15° below this approximate point and insert the second tube and reheat slowly to obtain an accurate melting point.*

The observed melting point depends on a number of factors: sample size, state of subdivision of the sample, and heating rate, as well as purity and identity of the sample. The first three cause the observed melting point to differ from the actual melting point because of the time lag in heat transfer from heating fluid to sample and conduction within the sample. Furthermore, if heating is too fast, the thermometer reading will lag behind the actual temperature of the heating fluid.

Mixture–Melting Points. In the preparation of a sample for mixture–melting point determination, it is important that the two components be thoroughly and intimately mixed. This is best accomplished by grinding them together by means of a small mortar and pestle. If these are not available, a small watch glass and test tube may be substituted. Be careful, however, not to apply too much pressure to the test tube, because it may break since it is more fragile than a pestle.

Experiments. **1.** Select one or two compounds from a list of supplied compounds of *known* melting point. Determine the melting points (ranges) for each of these substances, using the capillary melting-point procedure. Repeat as necessary until you are able confidently and accurately to complete these measurements.

2. Using one of the compounds whose melting range was determined in part A, introduce about 5–10% of a second substance as an impurity. Thoroughly mix the two components of the mixture and determine the melting range of the sample in order to verify the anticipated effects of impurities on the melting range of a substance.

3. Accurately determine and report the melting range of an unknown sample supplied by your instructor. This determination may be considered as a practical laboratory quiz pertaining to your mastery of this technique.

At the completion of this experiment and at your instructor's direction, either return any unconsumed samples or dispose of them in an appropriately labeled container for solid organic wastes. Do not discard these solids loosely in a waste can or wash them down the drain. Successful control of chemical pollution begins at "home."

B. Calibration of Thermometer

Calibration of a thermometer involves the measurement of temperature at a series of known points within the range of the thermometer, and the comparison of these actual readings with expected temperatures. The difference between the actual and the expected readings provides a correction which must be applied to the actual thermometer reading in order to correct for thermometer error. Most simply, these corrections are plotted in your notebook as deviations from zero (where the thermometer is accurate) versus the temperature over the range of the thermometer. Thus at a glance you can tell, for example, that at about 130° your thermometer gives readings that are 2° too low, or that at 190° the readings are about 1.5° too high. These corrections are valid only for the thermometer used in the calibration; if the thermometer is broken, a new one must be calibrated. These corrections should then be applied to all temperature measurements taken during the course.

To calibrate a thermometer, *carefully* determine the melting points of a series of standard substances. A list of suitable standards is provided in Table 1.1. Your

TABLE 1.1 STANDARDS FOR THERMOMETER CALIBRATION

Compound	Melting Point (°C)
Ice water	0
3-Phenylpropanoic acid	48.6
Acetamide	82.3
Acetanilide	114
Benzamide	133
Salicylic acid	159
4-Chloroacetanilide	179
3,5-Dinitrobenzoic acid	205

instructor may suggest other or additional standards to be used. The temperatures given in Table 1.1 correspond to the upper limit of the melting range for pure samples of these standards.

1.5 Boiling Points

The molecules of a liquid are in constant motion. When at the surface, some of these molecules are able to escape into the space above the liquid. Consider a liquid contained in a closed *evacuated* vessel. The number of molecules in the gas phase can increase until the rate at which molecules reenter the liquid becomes equal to the rate of their escape; the rate of reentry is proportional to the number of molecules in the gas phase. At this point no further *net* change is observed in the system, and it is said to be in a state of dynamic equilibrium.

The molecules in the gas phase are in rapid motion and are continually colliding with the walls of the vessel, resulting in the exertion of pressure against the walls. The magnitude of this pressure at a given temperature is called the *equilibrium vapor pressure* of the particular liquid substance at that temperature. This vapor pressure is temperature dependent, as shown in Figure 1.8. This dependence is easily understood in terms of the escaping tendency of molecules from the liquid. As the temperature increases, the average kinetic energy of the molecules increases, facilitating their escape into the gas phase. The rate of reentry also increases, and equilibrium is soon established at the higher temperature. However, the number of molecules in the gas phase is now larger than at the lower temperature, so the vapor pressure is greater.

Now consider a liquid sample at a given temperature, placed in an open container so that molecules of the vapor phase above the liquid may escape its confines. The vapor above such a sample is composed of molecules of air as well as of the sample. The *total* (external) pressure above the liquid is, according to Dalton's

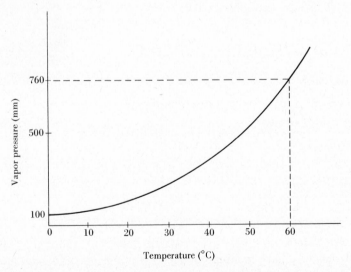

FIGURE 1.8 Dependence of vapor pressure on temperature.

law of partial pressures, equal to the sum of the partial pressures of the sample and of the air (see equation 1). The partial pressure of the sample is equal to its *equilibrium*

$$P_{\text{total}} = P_{\text{sample}} + P_{\text{air}} \tag{1}$$

vapor pressure at the given temperature. If the temperature is raised (thus increasing the equilibrium vapor pressure of the sample), the number of molecules from the sample in the space immediately above the liquid will increase, in effect displacing some of the air. At the higher temperature the partial pressure of the sample will be a larger percentage of the total pressure. The same trend will continue as the temperature is further increased, until the equilibrium vapor pressure becomes equal to the external pressure, at which point all the air will have been effectively displaced from the vessel. Further evaportion will have the effect of displacing gaseous mole-cules of the sample. Consideration of these facts leads to the conclusion that the equilibrium vapor pressure of the sample has an upper limit dictated by the external pressure. At this point the rate of evaporation increases dramatically (as evidenced by the formation of bubbles in the liquid), and the process commonly known as boiling occurs. *The temperature at which the vapor pressure of a liquid just equals the external pressure* is defined as the *boiling point* of the liquid. As the observed boiling point is obviously directly dependent on the external pressure, in reporting boiling points it is necessary to state the external pressure, for example, "bp 152° (752 mm)." The *standard boiling point* is typically measured at atmospheric pressure, 760 mm of Hg. The standard boiling point of the substance in Figure 1.8 is shown by the dotted lines to be 60°.

Boiling points are useful for identification of liquids and some low-melting solids. Higher melting solids usually boil at temperatures too high to be measured conveniently. Many compounds, notably the higher boiling ones, decompose at their standard boiling points.

Impurities, both volatile and nonvolatile, affect boiling points. These effects are discussed in Chapter 2.

EXPERIMENTAL PROCEDURE

DO IT SAFELY

1. Pay particular attention to the precautions for use of burners given in the Do It Safely section on page 15.

2. Spilled liquids should be carefully absorbed into a paper towel which is then discarded as directed by your instructor. Do not allow organic liquids to come into contact with your skin. If this happens, wash the affected area thoroughly with warm soap and water.

Micro Boiling Points. A simple micro boiling-point apparatus may be constructed from two 1-mm capillary melting-point tubes and 4-mm soft glass tubing in the following way. Seal the ends of two capillary tubes in a flame and join the tubes at

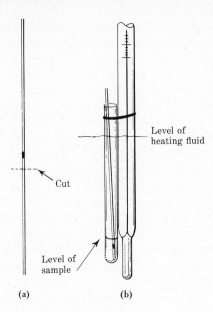

FIGURE 1.9 Micro boiling-point apparatus.

the seal. Make a clean cut about 3–4 mm from the joint, as shown in Figure 1.9a. Seal a piece of 4-mm soft glass tubing at one end and cut it to a length about 1 cm shorter than the prepared capillary ebullition tube.

Attach the 4-mm tube to a thermometer with a rubber ring, with the rubber ring near the top of the tube and the bottom of the tube even with the mercury bulb of the thermometer. Place about two drops of the liquid for which a boiling point is to be determined in the bottom of the tube by means of a capillary pipet. Introduce the capillary ebullition tube as shown in Figure 1.9b. If the liquid level of the sample is below the joint seal of the capillary, add enough more sample to bring it above the seal.

Immerse the thermometer and attached tubes in a heating bath (Thiele tube or other melting-point apparatus), *taking care that the rubber ring is above the liquid level.* Raise the temperature rather quickly until a rapid and continuous stream of bubbles comes out of the capillary ebullition tube. (A decided change from the slow evolution of bubbles caused by thermal expansion of the trapped air will be seen when the boiling temperature of the liquid is reached.) Discontinue heating at this point. As the bath is allowed to cool down slowly, the rate of bubbling will decrease. At the moment the bubbling ceases entirely and the liquid begins to rise into the capillary, note the temperature of the thermometer. This is the boiling point of the liquid sample.

Remove the capillary ebullition tube and expel the liquid from the small end by gentle shaking. Replace it in the sample tube and repeat the heating and cooling procedure. With a little practice and care the observed boiling points may be reproduced to within 1 or 2°.

Your instructor will provide the liquids for which you should determine the boiling points.

1.6   Index of Refraction

The velocity of light propagation is not constant in all media. One direct and observable consequence of this variation is the bending of a beam of light as it passes through the interface of two different media. The light bends because its "speed" changes. (The effect is commonly observed; most people are aware that an object seen under water—a fish, for example—is not where it seems to be because of the refraction at the interface between the water and the air.) The angle of refraction (angle of bending) is dependent on the density of the media, the types of molecules present, the temperature of the media, and the wavelength of the light. If one medium such as air is chosen as a standard and all angles of refraction are measured at the same temperature with the same wavelength of light, then the measured angle is a property of the second medium at the interface and may be used as an identifying characteristic of the substance composing that medium.

The value actually reported as a physical constant of a liquid is not the angle of refraction, but the *index* of refraction (n), which is defined by the equation

$$n = \sin i / \sin p \tag{2}$$

where i is the angle which incident light (in air) makes with a perpendicular to the interface and p is the angle which the refracted light (in the liquid) makes with the perpendicular (see Figure 1.10). Since monochromatic light gives more precise values than white light, most indices of refraction are reported for the D line of sodium ($\lambda = 589.3$ nm). The use of this frequency is indicated by a subscript, and the temperature by a superscript, so that a typical notation is $n_D^{20} = 1.3330$. In the Abbé refractometer, the instrument most frequently used by organic chemists, white light is used as a source, but compensation by a prism system gives indices for the sodium D line.

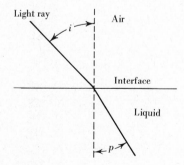

FIGURE 1.10 Refraction of light.

The standard temperature for measurement of refractive index is 20°. If the temperature of measurement is not 20°, a temperature correction should be applied. Although the magnitude of this correction may vary slightly according to the class of compound under consideration, an approximate correction of 0.00045 per degree may conveniently be used for most organic liquids. Inasmuch as the index of refraction *decreases* with *increasing* temperature, the correction must be *added* to the observed index if the actual temperature of measurement is above 20°, and *subtracted* from the observed index if the temperature of measurement is below 20°.

Impurities affect the index of refraction, and comparison of the observed index with the known index for the pure compound can be used as an indication of purity as well as for identification.

1.7 Density

The density of a substance is used less often for identification than are the physical constants which have already been discussed. Density is effectively a measure of the concentration of matter and is generally measured and reported in units of grams per milliliter at 20°. The density of a liquid is rather easily measured by making use of a small container, called a *pycnometer,* whose volume is precisely known. The container is weighed accurately and filled with the liquid whose density is to be determined. After bringing the pycnometer and its contents to 20° or some other desired temperature and readjusting the volume of the contents, if necessary, a second weighing will then yield the accurate weight of the substance whose volume is known, and the density may be calculated by dividing the weight in grams by the volume in milliliters.

Densities of solids are more difficult to determine. One way in which this measurement is often obtained is known as the *flotation method.* The density of the solid is determined as equal to the density of a liquid (measured as above) *in which* a single crystal of the solid will remain suspended, neither rising to the top nor settling to the bottom (Figure 1.11). Such a liquid may be obtained by first finding a suitable

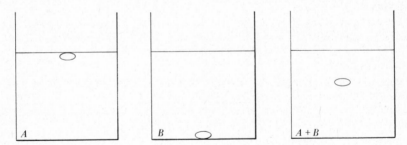

FIGURE 1.11 Density of solids by the flotation method.

liquid, *A,* in which the crystal rises to the top (liquid density is greater than the crystal density). A second liquid, *B,* is found in which the crystal sinks to the bottom. A mixture of *A* and *B* is then prepared in which the crystal remains suspended. The final measurement is made by determining the density of this liquid mixture, which is equal to the density of the solid. Of course, only liquids in which the solid is insoluble are suitable.

EXERCISES

1. Describe errors in procedure which may cause an observed capillary melting point of a pure compound (a) to be *lower* than the correct mp, (b) to be *higher* than the correct mp, (c) to be *broad* in range (over several degrees).
2. Why is filter paper usually a poor material on which to powder a solid sample before introducing it into a capillary mp tube?
3. Criticize the following statements: (a) An impurity always lowers the mp of an organic compound. (b) A sharp mp for a crystalline organic substance always indicates a pure single compound. (c) If the addition of a sample of compound *A* to compound *X* does not lower the mp of *X*, *X* must be identical with *A*. (d) If the addition of a sample of compound *A* lowers the mp of compound *X*, *X* and *A* cannot be identical.
4. Give an explanation for the observed marked increase in rate of bubble evolution as the boiling point is reached in a micro bp determination.
5. Why is the micro bp taken as the temperature at the time the sample liquid begins to rise into the capillary tube? Justify this procedure in terms of the definition of boiling point.
6. Using an Abbé refractometer, the index of refraction of 2-butanol at 23° is observed to be 1.3965. Apply the temperature correction discussed in Section 1.6 to determine the standard index of refraction at 20°. Check your answer against the index reported in the *Handbook of Chemistry and Physics* (Chemical Rubber Company).

REFERENCES

MELTING POINT
1. E. L. Skau and J. C. Arthur, Jr., in *Technique of Chemistry,* A. Weissberger and B. W. Rossiter, editors, Wiley-Interscience, New York, 1971, Vol 1, Part 5, Chapter 3, pp. 105 f.

INDEX OF REFRACTION
2. N. Bauer, K. Fajans, and S. Z. Lewin, in *Technique of Organic Chemistry,* A. Weissberger, editor, 3d ed., Interscience Publishers, New York, 1960, Vol. 1, Part II, Chapter 18.

DENSITY
3. N. Bauer and S. Z. Lewin, in *Technique of Chemistry,* A. Weissberger and B. W. Rossiter, editors, Wiley-Interscience, New York, 1972, Vol. 1, Part 4, Chapter 2.

2

SEPARATION AND PURIFICATION OF ORGANIC COMPOUNDS
Distillation, Recrystallization, and Sublimation

It is rare good fortune when the practicing organic chemist obtains a desired organic compound in pure form, whether that substance is the desired product of a chemical reaction or is isolated from some natural source. For example, a chemical transformation designed to produce a particular target compound almost invariably yields a reaction mixture that contains a number of contaminants. These include products of side reactions that proceed concurrently with the main reaction, varying amounts of unchanged starting materials, inorganic materials, and solvents. The chemist devotes a great deal of effort to the separation of the desired product from such impurities. Even chemicals purchased commercially, especially organic liquids, are often not more than 95% pure, owing to the expense to the supplier of the final purification process or to decomposition prior to use. The purpose of this chapter and the following one is to present an introduction to the theory and practice of the most important methods used by the modern chemist to separate and purify organic compounds.

2.1 Simple Distillation

The most commonly used method of purifying *liquids* is distillation, a process that consists of vaporizing the liquid by heating and condensing the vapor in a separate vessel to yield a *distillate*. If the impurities present in the initial liquid are nonvolatile, they will be left behind in the distillation residue, and a *simple distillation* will effect the purification. If the impurities are volatile, a *fractional distillation* will be required.

When a nonvolatile impurity such as sugar is added to a pure liquid, it has the effect of reducing the vapor pressure of the liquid. This is because the presence of the nonvolatile component effectively lowers the concentration of the volatile constituent; that is, no longer are all the molecules at the surface those of the volatile substance, and thus its ability to vaporize is lowered. The effect of a nonvolatile constituent on the vapor pressure of a mixture is shown in Figure 2.1. In this figure curve 1 corresponds to the vapor pressure-temperature dependence for a pure liquid. The curve intersects the 760-mm line at 60°. Curve 2 is for the same liquid which now, however, contains a nonvolatile impurity. Note that the vapor pressure at any temperature is reduced by a constant amount by the presence of the impurity (Raoult's law). The temperature at which this curve intersects the 760-mm line is higher because of the lower vapor pressure, and the temperature of the boiling solution is higher, 65°.

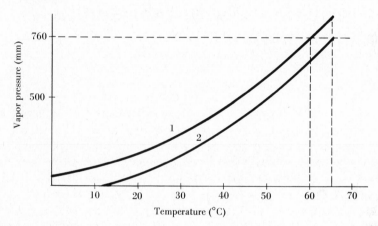

FIGURE 2.1 Dependence of vapor pressure on temperature.

A simple distillation may be carried out in an apparatus such as that illustrated in Figure 2.2. The thermometer is placed in the position shown for the purpose of determining the boiling temperature of the distillate (the material being collected in the receiver). In the case of a pure liquid, the temperature shown on the thermometer, the *head temperature,* will be identical to the temperature of the boiling liquid in the distilling vessel, the *pot temperature,* if the liquid is not superheated. The head temperature, which thus corresponds to the boiling point of the liquid, will remain constant throughout the process of distillation.

When a nonvolatile impurity is present in a liquid being distilled, the head temperature will be the same as for the pure liquid, since the material condensing on the thermometer bulb is uncontaminated by the impurity. However, the pot temperature will be elevated owing to the decreased vapor pressure of the solution. The pot temperature will also rise during the course of the distillation because the concentration of impurity will increase steadily as the volatile component is distilled away,

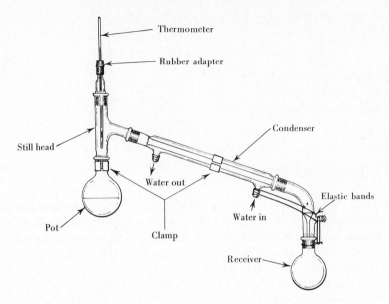

FIGURE 2.2 Typical apparatus for simple distillation, atmospheric pressure or vacuum.

further lowering the vapor pressure of the liquid. However, the head temperature will remain constant, as in the case of a pure liquid.

The quantitative relation between vapor pressure and composition of homogeneous liquid mixtures (solutions) is known as Raoult's law and may be expressed as in equation 1. P_R represents the partial pressure of component R and, at a given

$$P_R = P_R^0 N_R \tag{1}$$

temperature, is equal to the vapor pressure of pure R at that temperature (P_R^0) times the mole fraction of R in the mixture (N_R). The mole fraction of R is defined as the fraction of *all* molecules present that are molecules of R; it is obtained by dividing the number of moles of R in a mixture by the sum of the number of moles of all components (equation 2).

$$N_R = \frac{n_R}{n_R + n_S + n_T + \cdots} \tag{2}$$

It should be noted that the partial vapor pressure of R above an ideal[1] solution containing R depends only on its mole fraction and is in no way dependent on the vapor pressures of the other components. If all components other than R are nonvolatile, the total vapor pressure of the mixture will be equal to the partial pressure of R, since the vapor pressure of nonvolatile compounds may be taken as zero. Thus the distillate from such a mixture will always be pure R. If, however, two or more of the components are volatile, then the total vapor pressure will equal the sum of the

[1]The definition of an ideal solution is given in the following paragraph.

partial vapor pressures of each of these volatile components (Dalton's law, equation 3, where R, S, and T refer to the volatile components only). The process of distilling

$$P_{\text{total}} = P_R + P_S + P_T + \cdots \tag{3}$$

such a liquid mixture is significantly different from that of simple distillation, because here the distillate may contain each of the volatile components. Separation in this case will require fractional distillation.

2.2 Fractional Distillation

For simplicity we shall consider here only binary, ideal solutions, which contain two volatile components, R and S.[2] Ideal solutions are defined as those in which the interactions between *like* molecules are the same as those between *unlike* molecules. Only ideal solutions obey Raoult's law strictly, but many organic solutions approximate the behavior of ideal solutions. Some important examples exhibiting significant deviation from ideality are discussed in Section 2.4.

Because vapor pressure is a measure of the ease with which molecules escape the surface of a liquid, the number of molecules of component R in a given volume of the vapor above a liquid mixture of R and S is proportional to the partial vapor pressure of component R. The same is true for component S, as may be seen from equation 4, where N'_R/N'_S is the ratio of the mole fractions of R and S in the *vapor* phase.

$$\frac{N'_R}{N'_S} = \frac{P_R}{P_S} = \frac{P_R^0 N_R}{P_S^0 N_S} \tag{4}$$

The mole fraction of each component may be calculated from the equations $N'_R = P_R/(P_R + P_S)$ and $N'_S = P_S/(P_R + P_S)$. The partial vapor pressures, P_R and P_S, are determined by the composition of the liquid solution (Raoult's law). Since the solution will boil when the sum of the partial vapor pressures of R and S is equal to the external pressure, the boiling temperature of the solution is determined by its composition.

The relationship between temperature and the composition of the liquid and vapor phases of ideal binary solutions may be best illustrated by a diagram such as that of Figure 2.3 for mixtures of benzene (bp 80°) and toluene (bp 111°). The lower curve gives the boiling points of all mixtures of these two compounds, and the upper curve, calculated using Raoult's law, gives the composition of the vapor phase in equilibrium with the boiling liquid phase at the same temperature. For example, a liquid mixture with the composition 58 mol % benzene, 42 mol % toluene (*A*, Figure 2.3) boils at 90°, and the vapor phase in equilibrium with it has the composition 78 mol % benzene, 22 mol % toluene (*B*, Figure 2.3). Note that the *vapor phase is much richer in the more volatile component than the boiling liquid with which it is in*

---

[2]Solutions containing more than two volatile components are often encountered, but their behavior on distillation may be understood by extension of the principles given for binary systems.

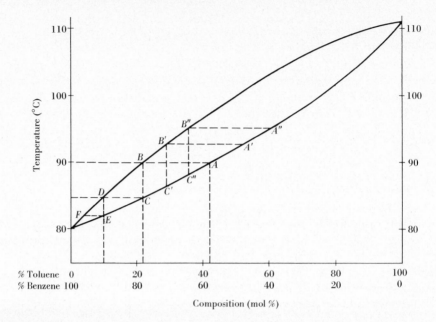

FIGURE 2.3 Boiling point–composition diagram for the system benzene-toluene.

equilibrium, at any given temperature. For example, if component R is more volatile than component S, so that

$$\frac{P_R^0}{P_S^0} > 1$$

then, from equation 4,

$$\frac{N_R'}{N_S'} > \frac{N_R}{N_S}$$

This is the basis of fractional distillation, as is discussed in the following paragraphs.

If a mixture of composition A (Figure 2.3) is distilled, the first few drops of distillate (produced by *condensation* of some of the vapor phase) will have composition B, much richer in benzene than the original mixture from which it was distilled. Conversely, the liquid remaining in the distilling flask will be depleted in benzene and enriched in toluene, since more benzene than toluene was removed, and the boiling point will rise (from A to A', for example). If distillation is continued, the boiling point of the mixture will continue to rise (from A' to A'', and so on) until eventually it approaches or reaches the boiling point of toluene. Meanwhile the composition of the distillate will change from B to B' to B'', and so on, until at the end it will be nearly pure toluene.

Now let us return our attention to the first few drops of distillate which had the composition B. If these are collected separately and then redistilled, the boiling point will be the temperature at C, 85°. If only the first small fraction of distillate from C were collected, it would have the composition D, 90 mol % benzene, 10 mol % toluene. Repetition of this process could theoretically (but impractically) give a *very small amount* of pure benzene. Similarly, by collecting the *last* small fraction of each distillation and redistilling in the same stepwise manner, one could obtain a very small amount of pure toluene. By collecting larger amounts of initial and final distillation fractions, practical amounts of materials can be separated, but a proportionately larger number of individual simple distillations is required. Such a procedure is extremely tedious and time-consuming. Fortunately the repeated distillations can be accomplished almost automatically in a single step by means of a *fractional distillation column,* the theory and use of which is described in the following section.

2.3   Fractional Distillation Columns

There are a large number of types of fractional distillation columns, but all can be described in terms of a few fundamental characteristics. The column provides a vertical path through which the vapor must pass from distilling flask to condenser (Figure 2.4); this path is significantly longer than in a simple distilling apparatus. As vapor from the distilling flask passes up the column, some of it condenses. *If the lower part of the column is maintained at a higher temperature than the upper part of the column,* as the condensate drains down the column it will be partially revaporized. The uncondensed vapor, together with that produced by revaporization of condensate, rises higher in the column and goes through a series of condensations and

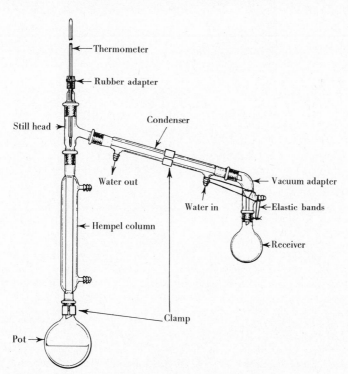

FIGURE 2.4 Apparatus for fractional distillation, atmospheric pressure or vacuum.

vaporizations. These amount to repeated distillations[3] and, as described with reference to Figure 2.3, the vapor phase produced in each step becomes richer in the more volatile component. The condensates, which drain down the column, are at each level richer in the less volatile component than the vapor with which they are in contact. Under ideal conditions, equilibrium becomes established throughout the column between the liquid and vapor phases, with the vapor phase at the top consisting almost entirely of the more volatile component and the liquid phase at the bottom being very rich in the less volatile component. The most important requirements for producing this state are (1) intimate and extensive contact between the liquid and vapor phases in the column, (2) maintenance of the proper temperature gradient along the column, (3) sufficient length of the column, and (4) sufficient difference in the boiling points of the components of the liquid mixture.

If the first two requirements are well met, compounds having small differences in boiling points can be separated satisfactorily with a long column, since the length of the column required and the difference in boiling points of the components are reciprocally related. The most common way of providing the necessary contact between liquid and vapor phases is to fill the column with some inert material providing a large surface area, for example, glass, ceramic, or metal pieces in a variety of shapes (helices, "saddles," woven mesh, and so forth.) Figure 2.5a shows a Hempel

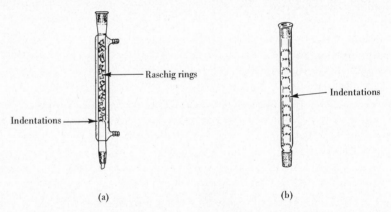

(a)   (b)

FIGURE 2.5 Fractional distillation columns. (a) Hempel column filled with Raschig rings. (b) Vigreux column.

---

[3]Each step along the route *A-B-C-D-E-F* of Figure 2.3 represents a single ideal distillation. Because one type of fractional distillation column, a "bubble-plate" column, was designed to bring about one such step for each "plate" it contains, the efficiency of any fractional distillation column is often described in terms of its equivalency to such a column in "theoretical plates." Another index of design efficiency of a fractional distillation column is the HETP (height equivalent to a theoretical plate), which is the vertical length of a column that is necessary to obtain a separation efficiency of one theoretical plate; that is, a column 60 cm long with an efficiency of 30 plates has an HETP value of 2 cm. Such a column would usually be better for research purposes than a 60-plate column which is 300 cm long (HETP = 5 cm) because of the smaller liquid capacity and "holdup" of the shorter column. "Holdup" refers to the condensate that remains in a column during and after a distillation. When small amounts of material are to be distilled, a column must be chosen which has an efficiency (HETP) adequate for the desired separation and a moderate holdup.

column packed with Raschig rings (approximately 6 mm lengths of glass tubing). Such a column will have from 2 to 4 theoretical plates (see footnote 3) depending on whether or not the distillation is carried out sufficiently slowly to maintain equilibrium conditions in the column. An alternative type of fractional distillation column, which is useful for small-scale distillations where low holdup (see footnote 3) is desirable, is the Vigreux column shown in Figure 2.5b. With the Vigreux column, indentations in the side of the column serve to increase the surface area of contact between liquid and vapor. Although a Vigreux column about the same length as a standard Hempel column will have only about 1 to 2 theoretical plates, and consequently will be less efficient, it has the advantage of a holdup of less than 1 mL compared to the holdup (and consequent loss of distillate) of 2–3 mL for the Raschig ring-filled Hempel column.

A particularly effective type of fractional distillation apparatus is the *spinning band column.* In this type of column, a helical band of metal or Teflon rotating at high speed within the column serves to force condensed liquid *down* the column while at the same time leaving a very thin film of liquid on the inside surface of the column. This provides for *rapid* and therefore very effective equilibration between liquid and vapor. A well-insulated spinning band column of 60 cm in length may have a holdup of only about 0.2 mL and yet have a rating of 125 theoretical plates, allowing it to separate components of only 2° difference in boiling points.

Maintenance of the proper temperature gradient in the column is a particularly important requirement for an effective fractional distillation. Ideally the temperature at the bottom of the column is approximately equal to the boiling temperature of the solution in the pot, and it decreases continually in the column until it reaches the boiling point of the more volatile component at the head. The significance of the temperature gradient may be readily visualized by reference to Figure 2.3, where with each succeeding step the boiling temperature of the condensate decreases, for example, A (90°) to C (85°) to E (82°). The necessary temperature gradient from pot to still head will, in most distillations, be established automatically by the condensing vapors if the rate of distillation is properly adjusted. Frequently this gradient can be maintained only by insulating the column with asbestos, glass wool, or, most effectively, with a silver-coated vacuum jacket. The insulation helps to reduce heat losses from the column to the atmosphere. For longer columns, particularly when poorer insulation is used, it is often necessary to supply additional heat by means of electrical resistance wire wrapped about the column. Even when insulated, if the rate of heating of the pot is too low, an insufficient amount of vapor may be produced to heat the column. In such a case, little or no condensate will reach the head and the rate of heating must be increased, but should be kept below that which will cause flooding of the column.

A factor intimately related to the temperature gradient in the column is the rate of heating of the pot and the rate at which vapor is removed at the still head. If the heating is vigorous and the vapor is removed too rapidly, the whole column will heat up almost uniformly and there will be no fractionation (separation of components). On the other hand, if the pot is heated too vigorously and if vapor is removed too slowly at the top, the column will "flood" with returning condensate. Proper operation of a fractional distillation column requires judicious control of heating and "reflux ratio"—the ratio of the amount of vapor condensed and returned down the column to the amount taken off as distillate at the still head in the same time period. In general the higher the reflux ratio, the more efficient the fractionation.

The most convenient way to measure and control the reflux ratio is to use a still head equipped for total reflux and variable "takeoff." For example, with a still head of this type (see Figure 2.6), the number of drops of condensate falling from the tip of the condenser (P) in a given length of time gives a measure of the total reflux rate, and the number of drops allowed to pass through the stopcock (S) and fall into the receiver at R in the same time period is a measure of the takeoff rate. If one drop of distillate is taken off for every ten drops of total reflux, the reflux ratio is 9 to 1. For very precise fractionation, reflux ratios of 100 to 1 may sometimes be used with an efficient column.

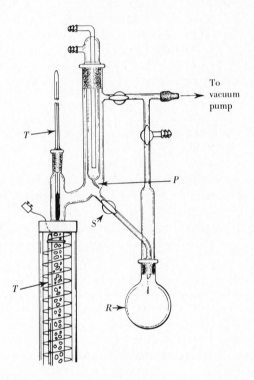

FIGURE 2.6 Total reflux-partial takeoff still head.

2.4 Fractional Distillation of Nonideal Solutions

Although most homogeneous liquid mixtures behave as ideal solutions, there are many known examples in which the behavior is nonideal; in these solutions the unlike molecules are not indifferent to one another's presence. The resultant deviations from Raoult's law occur in either of two directions: Some solutions display greater vapor pressures than expected and are said to exhibit *positive deviation;*

others display lower vapor pressures than expected and are said to exhibit *negative deviation.*

In the case of positive deviation the forces of attraction between the different molecules of the two components are weaker than those between the identical molecules of each component, with the result that, in a certain composition range, the combined vapor pressure of the two components is greater than the vapor pressure of the pure, more volatile component. Hence mixtures in this composition range (between X and Y in Figure 2.7) have boiling temperatures *lower* than the boiling temperature of either pure component. The *minimum-boiling* mixture in this range (the composition of Z, Figure 2.7) must be considered as if it were a third component. It has a constant boiling point because the vapor in equilibrium with the liquid has the same composition as the liquid itself. Such a mixture is called a *minimum-boiling azeotropic mixture,* or *azeotrope.*[4] Fractional distillation of such mixtures will not yield both of the components in pure form but only the azeotrope and the component present in excess of the azeotropic composition. It is for this reason that pure ethanol cannot be obtained by fractional distillation of aqueous solutions containing less than 95.57% ethanol (the azeotropic composition), even though the boiling point of this azeotrope is only 0.15° lower than the boiling point of pure ethanol.[5]

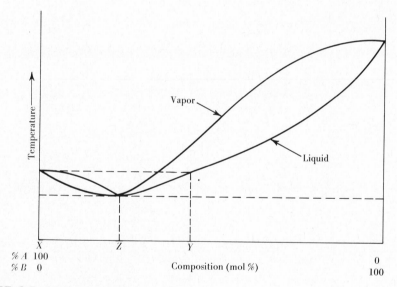

FIGURE 2.7 Minimum-boiling azeotrope.

[4] This discussion is limited to binary mixtures. However, ternary mixtures are known that exhibit the same property, for example, water (7.4%), ethanol (18.5%) and benzene (74.1%).

[5] Since optimum fractional distillations of aqueous solutions of ethanol containing less than 95.57% yield this azeotropic mixture, "95% ethyl alcohol" is the common form of commercial solvent ethanol. Pure or "absolute" ethanol can be obtained by chemical removal of the water or by a distillation procedure involving the use of a ternary mixture of ethanol, water, and benzene (see footnote 4). This procedure is described in the references at the end of the chapter.

In the case of negative deviation from Raoult's law, the forces of attraction between the different molecules of the two components are stronger than those between the identical molecules of each component, with the result that, in a certain composition range, the combined vapor pressure of the two components is less than the vapor pressure of the pure, less volatile component (see Figure 2.8). Hence

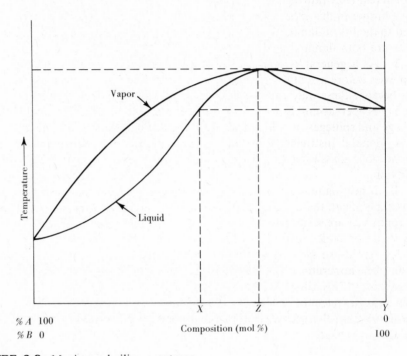

FIGURE 2.8 Maximum-boiling azeotrope.

mixtures in this composition range (between X and Y in Figure 2.8) boil at even *higher* temperatures than the boiling temperature of the pure, higher boiling component. There is one particular composition corresponding to a *maximum-boiling azeotrope* (*Z*, Figure 2.8). Fractional distillation of mixtures of any composition other than that of the azeotrope will result in elimination from the mixture of whichever of the two components is in excess of the azeotropic composition (*Z*), so the composition of the pot residue will approach that of *Z*. As an illustration, the maximum-boiling azeotrope of formic acid (bp 100.7°) and water boils at 107.3°; any mixture having other than the azeotropic composition (77.5% formic acid, 22.5% water) will yield this mixture as residue on fractional distillation.

Data on other azeotropes of both types may be found in the references at the end of the chapter; extensive tables are given in the handbooks listed there.

EXPERIMENTAL PROCEDURE

DO IT SAFELY

1.   Although we advise against the use of burners in the organic laboratory except in relatively rare instances, if the conditions in your laboratory necessitate your using a burner in this experiment, refer to part 1 of the Do It Safely section on page 15 and to the list of flammable solvents given on page 51. Ethyl acetate and *n*-butyl acetate are both flammable.

2.   As a general rule, *always* examine your glassware for cracks and other weaknesses before assembly. Look particularly for "star cracks" in round-bottomed flasks because these may cause a flask to break while it is being heated.

3.   Proper assembly of glassware is important in order to avoid possible breakage and spillages, as well as leakage of the distillate vapors into the room. Read part A, General Instructions, on the assembly of glassware. Have your instructor examine your glassware after it is assembled and before you proceed with the distillation.

4.   If heating mantles or oil baths (see Appendix 5) are to be used for heating in this experiment, there are possible attendant dangers. Be sure the water hoses to your condenser are securely in place in order to avoid having them "pop" loose and spray water onto electrical connections, into the heating mantle, or into *hot* oil.

5.   Avoid excessive inhalation of organic vapors at any time.

6.   The apparatus being used is open to the atmosphere at the receiving end of the condenser. This allows for pressure equalization. At no time in the laboratory should a *closed* system be heated. If pressure equalization is not allowed for, material expansion within the system will result in elevated pressures and may cause the apparatus to *explode*.

This experiment is designed to demonstrate in its several parts both the procedures of carrying out simple and fractional distillations and the relative efficiencies of different types of apparatus in separating mixtures of volatile components. Detailed directions are given for the purification of ethyl acetate by simple distillation and the fractional separation of mixtures of ethyl acetate and *n*-butyl acetate. Take care to note the differences in procedure for the simple distillation of a liquid having only a single volatile component and for the fractional distillation of a mixture of two or more volatile components. In the former case, less complex apparatus may be used; compare, for example, Figures 2.2 and 2.4. Furthermore, when fractional separation is not required, a much higher rate of distillation may be employed. The distillation of ethyl acetate may be quite conveniently and satisfactorily carried out at the rapid rate of 2 to 4 drops of distillate per second, whereas the fractional distillation of ethyl acetate and *n*-butyl acetate is effected at the more moderate rate of only 1 drop every 1 to 2 sec. This slower rate is necessary to maintain equilibrium conditions and the proper temperature gradient in the distillation column. In simple distillation the entire distillate is collected in a single receiving flask, whereas in fractional distillation, receiving flasks are changed as the composition of the distillate changes, as indicated by the head temperature.

A.   General Instructions: Assembly of Glassware

The following general discussion pertains to the assembly of glassware for experiments throughout this text. Proper assembly of apparatus is an important part of good laboratory technique, particularly with regard to laboratory safety. For example, a loosely jointed distillation assembly may "leak" flammable vapors that are likely to be ignited by a nearby Bunsen burner; or a too rigidly clamped assembly may be stressed to the point of breaking during a laboratory operation, thus causing, in addition to the physical danger of broken glass, the spillage of flammable or caustic chemicals. It is possible here to provide only general guidelines to proper procedure; it is advisable in the first few experiments to ask the instructor to check your glassware for proper and safe assembly. The experience gained in these first experiments should provide you with the "feel" for correct usage of your equipment. Beyond the guidelines given here, common sense is the watchword. Think about what you are to do before you do it.

In all experiments where the typical glassware assembly to be used is depicted in a figure, it should be noted that the "clamping points" have been marked. The general tendency among beginning students is to overclamp their apparatus; this frequently, although not necessarily, results in an unsafe situation, because the more clamps that are used, the greater the likelihood of stressing and possibly breaking a glass joint between two clamps. Furthermore, mispositioning of a clamp may not provide sufficient stability for the apparatus.

When applying a clamp, it is important that the jaws of the clamp be aligned parallel to the piece of glass being clamped; this will enable the clamp to be tightened without twisting the glass and either breaking it or pulling another joint loose. *Do not tighten the clamp until you are sure that the piece of glassware is correctly positioned and the clamp is properly aligned.*

It is generally most efficient to put together a complex assembly from the bottom up. For example, consider assembly of the fractional distillation apparatus shown in Figure 2.4. Clamp the distillation flask (pot) and then add the Hempel column and still head. Adjust the clamp so that the column is vertical. Now place the condenser in position and clamp it at the indicated position. Be sure that this clamp is properly aligned and vertically positioned *before* tightening it. Several minor adjustments may be needed. It is not necessary that this clamp be *rigidly* tight, only that it prevent the joint between the condenser and still head from slipping under normal usage. The vacuum adapter may be added and held in place by means of a strong rubber band, as shown in Figure 2.4. Likewise, by twisting a short piece of stiff wire about the neck of the receiving flask, the flask may be fastened to the vacuum adapter by a rubber band. Rubber has a tendency to deteriorate in the presence of laboratory vapors; be sure that the rubber bands are uncracked and retain sufficient elasticity and strength.

This use of rubber bands is acceptable in many instances; however, you should be careful not to depend on them to support relatively heavy weight. For example, additional support should be provided for flasks of 250-mL capacity or larger, as well as 100-mL flasks that may be expected to become more than half-full during the course of a distillation. In case of doubt it is better, of course, to be overly cautious. Instead of using another clamp, use an iron ring holding a piece of wire gauze to support the flask from underneath.

B. Experiments

1. Simple distillation. In a 100-mL round-bottomed flask place 40 mL of ethyl acetate and two or three boiling chips to ensure smooth boiling. Arrange the apparatus for simple distillation according to Figure 2.2. Use a receiving flask of sufficient size to collect the anticipated volume of distillate. The position of the thermometer bulb is particularly important; the *top* of the bulb should be on a level with the *bottom* of the sidearm of the still head. Have your instructor check your assembly.

Using the method of heating suggested by your instructor (see also Appendix 5), begin heating the distillation flask. As soon as the liquid begins to boil *and the condensing vapors have reached the thermometer bulb,* regulate the heat supply so that distillation continues steadily at a rate of *2 to 4 drops per second.*

As soon as the distillation rate is adjusted and the thermometer has reached a steady temperature, note and record the head temperature. Continue the distillation, periodically recording the head temperature, until only 2 or 3 mL of ethyl acetate remain in the distillation flask, and then discontinue heating. Note and record the distillation range of ethyl acetate that you have observed. Inform your instructor when you have completed the experiment and unless you have other instructions, return the ethyl acetate, distilled and undistilled, to the bottle marked "Recovered Ethyl Acetate." (It is not only economically advantageous to recover chemicals that may be reused, but also well worth the effort to avoid, insofar as is possible, the environmental abuse which results from discarding chemicals down the drain.)

2. Fractional distillation. Although this experiment provides detailed directions for the separation by fractional distillation of mixtures of ethyl acetate and *n*-butyl acetate, similar results may be obtained with other mixtures such as carbon tetrachloride-toluene and methanol-water. Alternatively or additionally, your instructor may wish to give you a multicomponent mixture for distillation which is (to you) of unknown composition. Regardless of your knowledge of the boiling points of the various components of the mixture you are to separate, successful completion of this experiment depends on your understanding that the distillate should be collected into *separate* receiving flasks, as the thermometer readings indicate different temperature *plateaus.* Thus a two-component mixture will require *three* receiving flasks, two for the main fractions and one for an intermediate fraction as the temperature rises from the first plateau to the second, and a three-component mixture will require *five* receiving flasks, three for the main fractions and two for the intermediate fractions between the three plateaus. If you know the composition of the mixture assigned to you, ascertain the necessary boiling points from a suitable handbook.

Your instructor will inform you if you are to follow procedure I or procedure II, given later. Procedure II is advantageous if you are to repeat this experiment with different equipment, since it provides the data in graphic form. If you are to follow procedure II, before coming to the laboratory prepare in your notebook a table for recording approximately 25 successive measurements of temperature and total accumulated volume of distillate.

In a 100-mL round-bottomed flask place 25 mL of ethyl acetate, 25 mL of *n*-butyl acetate, and two or three boiling chips to ensure smooth boiling. Assemble the apparatus for fractional distillation according to Figure 2.4, using a packed Hempel or similar column. The packing may consist of glass beads, Raschig rings, or glass helices (listed in order of increasing efficiency). Alternatively, a very satisfactory substitute for glass helices may be made from a copper (or, better but more expensive, stainless steel) cleaning sponge. The sponge is pulled apart and stuffed into the glass column rather compactly with the aid of a pencil or a spatula. *In doing this or in removing any packing from a Hempel column, be careful not to break off the glass support indentations at the base of the column.* (*Note:* The rapid corrosion of copper by organic halides makes this type of packing material unsuitable for a mixture containing carbon tetrachloride or other similar halogen-containing components.) Do not pack the column so tightly that vapors cannot pass through it.

In this assembly the position of the thermometer bulb is particularly important; the *top* of the thermometer bulb should be on a level with the *bottom* of the sidearm of the distillation head. Have your instructor check your assembly to be sure that all connections are tight.

Prepare three containers of 50-mL capacity as receiving flasks and label them *A, B,* and *C.* If you are following procedure I, these may be bottles or Erlenmeyer flasks; if procedure II, use three 50-mL graduated cylinders. (No matter which type of container is used, the tip of the vacuum adapter should extend inside the neck of the container in order to minimize evaporation of distillate.)

Using the method of heating suggested by your instructor (see also Appendix 5), begin heating the distillation flask. As soon as the mixture begins to boil and the vapors have reached the thermometer, regulate the heat so that distillation continues steadily at a rate of *only* about *1 drop of distillate every 1 to 2 sec.* Continue with either procedure I or procedure II.

Procedure I. Collect the first distillate in receiver *A.* When the head temperature reaches 81°, change to receiver *B,* and at 123° change to receiver *C.* (Recall that these temperature cuts are only for a mixture of ethyl acetate and *n*-butyl acetate.) Continue the distillation until about 1 or 2 mL of liquid remain in the pot and then discontinue heating. With a graduated cylinder measure the volumes of the distillation fractions in receivers *A, B,* and *C* and record them in your notebook. Allow the liquid in the column to drain into the pot (distillation flask) and measure and record the volume of this residue. Either submit or save 1-mL samples of each fraction *A, B,* and *C* for gas chromatographic analysis if you have been asked to do so.[6]

Procedure II. Collect the first distillate in graduated cylinder *A.* As soon as the distillation rate is adjusted to a rate of approximately 1 drop every 1 to 2 sec, note and record the head temperature and the total accumulated volume of distillate in the

---

[6]Columns containing as the stationary phase either silicone gum rubber or SF-96 give good separation of ethyl acetate and *n*-butyl acetate. (See Section 3.4 for discussion of gas chromatography.)

receiving cylinder. Continue the distillation, recording the temperature and total volume for each additional increment of approximately 2 mL of distillate. When the temperature reaches 81°, change to cylinder *B;* when it reaches 123°, change to cylinder *C*. (Recall that these temperature cuts are for mixtures of ethyl acetate and *n*-butyl acetate only.) Although receivers have been changed, continue to record the temperature and *total* accumulated volume of distillate for 2-mL increments until about 1 or 2 mL of liquid remain in the pot (distillation flask) and then discontinue heating. Record the volume of each of the three distillation cuts taken. Allow the liquid in the column to drain into the pot and measure and record the volume of this residue. Submit or save 1-mL samples of each fraction *A, B,* or *C* for gas chromatographic analysis if you have been asked to do so.[6] Plot on graph paper the head temperature versus the total accumulated volume of distillate.

3. Fractional Distillation Using Alternate Equipment. If you have been asked to compare the results of fractional distillation using different equipment, combine with the pot residue the three fractions obtained in the first experiment, after setting aside 1-mL samples for analysis. Use this mixture for the second distillation. Repeat exactly the procedure provided for fractional distillation, using instead either the simple distillation apparatus, Figure 2.2, or the fractional distillation apparatus, Figure 2.4, *without the packing.* If you are following procedure II, it is most convenient to plot on the same piece of graph paper the data from two or more distillations so that the curves obtained for each distillation involving different equipment will overlay one another. Compare the results for each type of distillation equipment, drawing any conclusions suggested by the data concerning the relative efficiencies of the various columns.

When you have finished the experiment, return all samples of ethyl acetate and *n*-butyl acetate, distilled or undistilled, to the bottles provided for them, unless you have been instructed otherwise.

EXERCISES

1. Explain why a packed fractional distillation column is more efficient at separating two closely boiling liquids than an unpacked column.
2. If heat is supplied to the distillation flask too rapidly, the ability to separate two liquids by fractional distillation may be drastically reduced. In terms of the general theory of distillation presented in the discussion, explain why this is so.
3. In the distillation of certain liquids such as benzene and toluene, the first few milliliters of distillate are frequently "cloudy." To what might you attribute such an observation? (*Hint:* See Section 2.4.)
4. Explain why the column of a fractional distillation apparatus should be aligned as close to a vertical configuration as possible.
5. Explain the role of the boiling stones that are normally added to a liquid which is to be heated to boiling.
6. The bulb of the thermometer placed at the head of a distillation apparatus should be adjacent to the exit to the condenser. Explain the effect on the temperature

reading of placement of the thermometer bulb (a) below the exit to the condenser and (b) above the exit.

7. (a) A mixture of 80 mol % n-propylcyclohexane and 20 mol % n-propylbenzene is distilled through a *simple distillation apparatus* (assume that no fractionation occurs during the distillation). The boiling temperature is found to be 157.3° as the first *small* amount of distillate is collected. The standard vapor pressures of n-propylcyclohexane and n-propylbenzene are known to be 769 mm and 725 mm, respectively, at 157.3°. Calculate the percentage of each of the two components in the first few drops of distillate.

   (b) A mixture of 80 mol % benzene and 20 mol % toluene is distilled under exactly the same conditions as in part a. Using Figure 2.3, determine the distillation temperature and the percentage composition of the first few drops of distillate.

   (c) The standard boiling points of n-propylcyclohexane and n-propylbenzene are 156.9° and 159°, respectively. Compare the distillation results in parts a and b. Which of the two mixtures would require the more efficient fractional distillation column for separation of the components? Why?

8. The still head of the fractional distillation assembly shown in Figure 2.6 is designed to allow receiving flasks to be changed during a vacuum distillation without disrupting the distillation by the readmission of air into the column. Determine the sequential manipulation of stopcocks necessary to accomplish this task. (*Hint:* See last paragraph of Section 2.5.)

9. Examine the boiling point–composition diagram for mixtures of toluene and benzene given in Figure 2.3.

   (a) Assume you are given a mixture of these two liquids of composition 80 mol % toluene and 20 mol % benzene and that it is necessary to effect a fractional distillation which will afford *at least some* benzene of greater than 99% purity. What would be the *minimum* number of theoretical plates required in the fractional distillation column chosen to accomplish this separation?

   (b) Assume that you are given a 20-cm Vigreux column having an HETP (see footnote 3) of 10 cm in order to distil a mixture of 58 mol % benzene and 42 mol % toluene. What would be the composition of the first small amount of distillate which you obtained?

2.5 Distillation under Reduced Pressure

The boiling point of a liquid is known to be that temperature at which the total vapor pressure is equal to the external pressure. It is most convenient to distil liquids under conditions such that the external pressure is the atmospheric pressure. In many instances, however, boiling temperatures at this pressure are higher than desirable, because the compound being distilled may decompose, oxidize, or undergo molecular rearrangement at temperatures below its normal boiling point. Sometimes impurities present may catalyze such reactions at higher temperatures. These problems may often be alleviated by carrying out the distillation at pressures less than atmospheric, because under these conditions the boiling temperature is, of course, lower; the technique is commonly called *vacuum* distillation.

Reduced pressures may be obtained by connecting an aspirator ("water pump") or a mechanical oil pump to the distillation apparatus. An aspirator will commonly reduce the pressure to about 25 mm and an oil pump to below 1 mm. The exact pressure obtained in the lower range by an oil pump is highly dependent on the condition of the pump and its oil and on the tightness of the connections of the distillation apparatus. The vacuum produced by a water aspirator is limited by the vapor pressure, and hence by the temperature of the water issuing from the water lines; in cold climates, pressures as low as 8–10 mm may be obtained.

Two useful *approximations* of the effect of lowered pressure on boiling points are the following:

1. Reduction from atmospheric pressure to 25 mm lowers the boiling point of a high-boiling compound (250–300°) *by* about 100–125°.
2. Below 25 mm, each time the pressure is halved, the boiling point is lowered about 10°. More accurate estimates of boiling points at various pressures may be made by the use of charts and nomographs found in references at the end of this chapter. Calculations may also be made using integrated forms of the Clausius-Clapeyron equation, as described in physical chemistry textbooks.

The typical apparatus for vacuum distillation is pictured in Figure 2.9. The distilling flask, which should not be more than half-full, is heated by means of an oil bath or some other suitable means. The flask should be immersed to a depth above the level of the liquid in the flask as an aid in preventing bumping problems. The flask should *never* be heated directly with a flame because this causes localized hot spots and intensifies the problem of bumping. The oil bath may be heated with a burner or by means of electrical resistance coils immersed in the oil and connected to a variable transformer for temperature control. The vacuum adapter is connected to a safety flask which serves not only as a trap to prevent backup of water into the

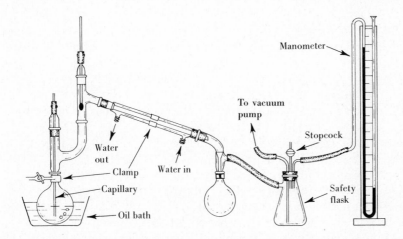

FIGURE 2.9 Typical apparatus for vacuum distillation.

apparatus from the water pump in case of loss of water pressure, but also as a vacuum manifold, providing vacuum connections to the apparatus and manometer and bearing a vacuum release valve (stopcock). The manometer provides for measurement of the pressure at which the distillation is being effected. The reliability of such measurements of pressure is dependent on the rate of distillation, as discussed later. These measurements are an important and integral part of a reported boiling point at reduced pressure, for example, benzaldehyde: bp 180°(760 mm), 87°(35 mm), and should be taken with care.

Two significant problems are encountered in vacuum distillation which only rarely complicate atmospheric distillation. These are both created by the fact that the volume of vapor formed by the volatilization of a given amount of liquid is pressure dependent. Thus, for example, the volume of vapor formed from vaporization of a drop of liquid will be approximately 20 times as great at 38 mm as at 760 mm. Serious bumping problems may occur in the distilling flask during distillation under reduced pressure as *large* bubbles explosively escape from the liquid, giving rise to vigorous, and sometimes violent, splashing and splattering. The insertion of the Claisen connecting tube between the distillation flask and still head (Figure 2.9) is a partial solution to this problem. There is no direct "line-of-flight" path for liquid to splash from the flask into the side arm leading to the condenser. Boiling stones are generally ineffective in preventing bumping under reduced pressure. The usual method for promoting regular and even ebullition is to provide a thin, flexible capillary tube extending into the boiling flask. This tube allows a fine stream of air bubbles to be introduced at the bottom of the flask, where they serve as nuclei for the regular production of vapor bubbles. Because the volume of air introduced is small compared to the evacuating capacity of a water or oil pump, this "leak" has no significant effect on the pressure of the system. The capillary is drawn from a piece of 6-mm soft glass tubing and should be fine enough to allow only a slow stream of fine bubbles when air is blown through it into a test tube containing acetone. Alternatively, if the equipment is available, a magnetic stirring bar (a cylindrical bar magnet covered with Teflon or glass) in the flask may be employed with a magnetic stirrer. The stirrer consists of a larger, motor-driven bar magnet; magnetic coupling of the magnets allows the motor indirectly to spin the stirring bar in the flask. Other stratagems against bumping, although frequently not successful, consist of filling the volume of the distilling flask above the liquid with glass wool (finely spun glass), or insertion into the flask of relatively long (so that they will stand vertically) dry pine applicator sticks, such as those used for medicinal cotton swabs.

The second problem with lower vapor densities at reduced pressure is that the measured pressure is greatly affected by the rate of distillation. Since a drop of condensate forms from a much larger volume of vapor at low pressure than at atmospheric pressure, the *velocity* of vapor molecules entering the condenser, even at moderate distillation rates, is tremendously increased by a reduction of pressure. The back pressure occasioned by high velocities of vapor provides higher pressure in the distillation column than that read on the manometer, which is beyond the condenser and receiver and not affected by the precondensed vapor. The difference between actual and apparent pressure is directly dependent on the rates of distillation and application of heat to the boiling flask. Therefore important requirements for proper conduct of a vacuum distillation are to maintain a *slow* but steady distillation rate and to avoid superheating of the vapor by maintaining the oil bath at a temperature no more than 15–25° higher than the head temperature. This problem is accentuated at still lower pressures.

The following paragraphs provide the general procedure for carrying out a vacuum distillation using equipment such as that shown in Figure 2.9:

1.   *Caution:* Never use glassware with cracks or thin-walled vessels, especially those with flat bottoms, such as Erlenmeyer flasks, in a system to be evacuated. Even with systems of only moderate size under water-pump evacuation, pressures of many hundreds of pounds may be exerted on the exterior surfaces of the assembly. Weak points may yield to implosion; the in-rushing air will shatter the glassware in a manner little different from an explosion. A very real additional danger is that of burns from the hot oil of the bath. *Examine the glassware carefully and always wear safety glasses.*

2.   Lubricate and seal all glass joints during assembly as an aid against air leaks. Check the rubber fittings holding the thermometer and capillary in place to ensure that they are tight. The neoprene fittings normally used with the thermometer adapters may be replaced, if necessary, with short pieces of heavy-walled vacuum tubing. Rubber stoppers are not advisable, since rubber in direct contact with the hot vapors during distillation may cause contamination. A three-holed rubber stopper in the safety flask should fit snugly and provide tight connections to the pieces of glass tubing. *Heavy-walled tubing must be used for all vacuum connections.* Before placing liquid in the pot, test the completely assembled apparatus to make certain that the system is tight.

3.   Place the liquid to be distilled in the flask. Make sure the capillary tip extends nearly to the bottom of the flask and turn on the water pump. The release valve on the safety flask should be *open.* Do not heat the flask until the system is fully evacuated. *Slowly* close the release valve, being ready to reopen it if necessary. If the liquid contains small quantities of low-boiling solvents—it very likely will if the liquid has been recently obtained from a solution by evaporation of solvent—foaming and bumping will almost certainly occur in the flask. If this happens, reopen the valve until the foaming abates. This may have to be done several times until the solvent has all been evaporated. As soon as the surface of the liquid is relatively quiet and the system is fully evacuated (check the manometer for constancy of pressure), begin the heating of the oil bath. Maintain a bath temperature as low as possible to provide a *slow* rate of distillation.

4.   If, as in fractional distillation, it is necessary to use multiple receivers to collect fractions of different boiling ranges, the distillation will need to be interrupted to change flasks. Lower the oil bath with caution and allow the distillation flask to cool a little. *Slowly* open the vacuum release valve to readmit air to the system. Change receivers, close the release valve, and when evacuation is complete, raise the oil bath and continue. This operation frequently results in a change of pressure in the fully evacuated system. It is good practice periodically to monitor and record in your notebook the head temperature and the pressure, particularly just before and after changing flasks.

5.   At the end of the distillation lower the oil bath, allow the pot to cool somewhat, slowly release the vacuum, and turn off the water pump.

The most inconvenient aspect of this procedure is the disruption of the distillation in order to change receiving flasks. Improved apparatus, although seldom available in the introductory organic laboratory, is designed to eliminate this problem. For example, the "cow" receiver shown in Figure 2.10 bears four receiving flasks (one not in view). These flasks are successively employed by rotating them into the receiving position. A second example is shown as part of the fractional distillation apparatus in Figure 2.6. The "fraction cutter" is included as an integral part of the still head and cold-finger condenser. Although this type of receiver is a little more complex to use, it is more advantageous because it does not limit the number of receiving flasks that may be used without disrupting the distillation. The proper sequential manipulation of stopcocks allows (1) isolation of the receiving end of the system while maintaining evacuation of the column, (2) release of vacuum in the receiver to change flasks, (3) reevacuation of the receiver while the column is isolated from the pump, and (4) reconnection of the receiver and column for continuation of the distillation.

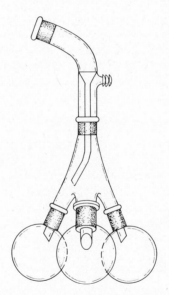

FIGURE 2.10   Multiple-flasked receiver for vacuum distillation.

2.6   Steam Distillation

The separation and purification of *volatile* organic compounds that are immiscible with water or nearly so is often accomplished by steam distillation, a technique which involves the codistillation of a mixture of water and organic substances. The virtues and limitations of this technique can best be illustrated by consideration of the principles which underlie steam distillation.

The partial pressure, P_i, at a given temperature of each component, i, of a mixture of immiscible, volatile substances is equal to the vapor pressure, P_i^0, of the pure compound at the same temperature (equation 5) and does not depend on the

$$P_i = P_i^0 \tag{5}$$

mole fraction of the compound in the mixture; that is, each component of the mixture vaporizes independently of the others. This behavior is in sharp contrast to that exhibited by solutions of miscible liquids, for which the partial pressure of each constituent depends on its mole fraction in the solution (Raoult's law, equation 1). Now, the total pressure, P_T, of a solution (mixture) of gases, according to Dalton's law (equation 3) is equal to the sum of the partial pressures of the constituent gases so that the total vapor pressure of a mixture of immiscible, volatile compounds is given by equation 6.

$$P_T = P_a^0 + P_b^0 + \cdots P_i^0 \tag{6}$$

Note from this expression that the total vapor pressure of the mixture at any temperature is always greater than the vapor pressure of even the most volatile component at that temperature, owing to the contribution of the vapor pressures of the other constituents of the mixture. The boiling temperature of a mixture of immiscible compounds must then be *lower* than that of the lowest boiling component.

Demonstration of the principles just outlined is available from discussion of the steam distillation of a mixture of water (bp 100°) and bromobenzene (bp 156°), substances that are insoluble in one another. The vapor pressure versus temperature plot for a mixture of these substances, Figure 2.11, along with the corresponding plots

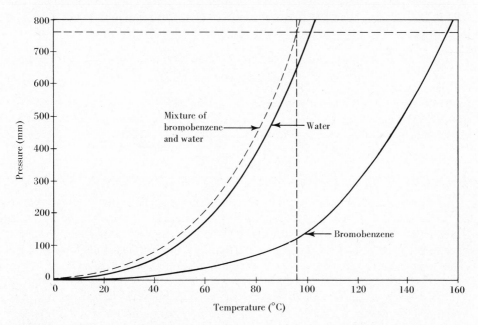

FIGURE 2.11 Vapor pressure versus temperature plots for bromobenzene, water, and a mixture of bromobenzene and water.

for the pure liquids, shows that the mixture should boil at about 95°, the temperature at which the total vapor pressure equals atmospheric pressure. As would be predicted from theory, this temperature is below the boiling point of water, the lowest boiling component in this example. The ability to distil a compound at the relatively low temperature of 100° or less by means of a steam distillation is often of great use, particularly in the purification of substances that are heat-sensitive and which would decompose at higher temperatures. It is useful also in the separation of compounds from reaction mixtures which contain large amounts of nonvolatile residues such as the notorious "tars" so often formed during the course of an organic reaction.

The composition of the condensate from a steam distillation depends upon the molecular weights of the compounds being distilled and upon their respective vapor pressures at the temperature at which the mixture distils. Consider a mixture of the two immiscible components, A and B. If the vapors of A and B behave as ideal gases, the ideal gas law can be applied, and the following two expressions obtained:

$$P_A^0 V_A = (g_A/M_A)(RT) \quad \text{and} \quad P_B^0 V_B = (g_B/M_B)(RT) \tag{7}$$

where P^0 is the vapor pressure of the pure liquid, V is the volume in which the gas is contained, g is the weight in grams of the component in the gas phase, M is its molecular weight, R is the universal gas constant, and T is the absolute temperature (°K). Dividing the first equation by the second, one obtains

$$\frac{P_A^0 V_A}{P_B^0 V_B} = \frac{g_A M_B (RT)}{g_B M_A (RT)} \tag{8}$$

Because the RT factors in the numerator and the denominator are identical and because the volume in which the gases are contained is the same for both ($V_A = V_B$), the expression just given becomes

$$\frac{\text{grams of } A}{\text{grams of } B} = \frac{(P_A^0)(\text{molecular weight of } A)}{(P_B^0)(\text{molecular weight of } B)} \tag{9}$$

For a mixture of bromobenzene and water, which have vapor pressures of 120 and 640 mm, respectively, at 95° (see Figure 2.11), the composition of the distillate would be calculated from equation 9 as follows:

$$\frac{g_{\text{bromobenzene}}}{g_{\text{water}}} = \frac{(120)(157)}{(640)(18)} = \frac{1.64}{1}$$

Consequently, on the basis of weight, more bromobenzene than water is contained in the steam distillate, even though the vapor pressure of the bromobenzene is much lower at the temperature of the distillation. Because organic compounds generally have molecular weights much higher than that of water, it is possible to steam distil compounds having vapor pressures of only about 5 mm at 100° with a fair efficiency on a weight-to-weight basis. Even solids can often be purified by steam distillation.

A steam distillation is generally accomplished in one of two ways. The first, and usually most efficient, method involves placing the organic compounds to be distilled in a round-bottomed flask fitted with a Claisen head equipped with a still head connected to a water-cooled condenser (Figure 2.12). The Claisen head helps to

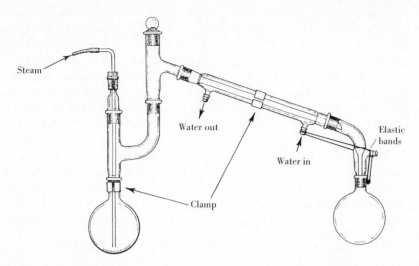

FIGURE 2.12 Apparatus for steam distillation. The long tube is replaced by a stopper if steam is generated by direct heating.

prevent splattering of the mixture into the condenser during distillation. Steam can be produced externally in a generator such as that shown in Figure 2.13 and then introduced into the bottom of the distillation vessel via a tube (Figure 2.12), or it can be obtained from a laboratory steam line. If the latter source is used, a trap is usually placed between the line and the distillation flask to allow removal of any water present in the steam (Figure 2.14). When an external source of steam is used, water may condense in the distillation flask, filling it to undesirable levels. This problem can generally be circumvented by gently heating the flask with a Bunsen burner.[7]

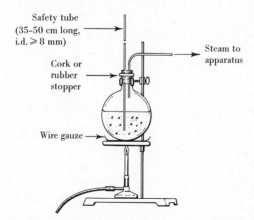

FIGURE 2.13 A steam generator. The round bottom flask is initially half-filled with water, and boiling chips are added before heating. The safety tube serves to relieve internal pressure if steam is generated at too rapid a rate.

[7] Alternatively, heating mantles may be used, if they are available. See Appendix 5.

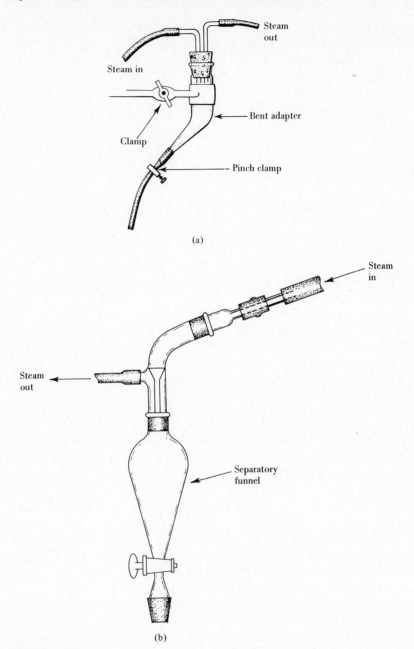

Steam out

Steam in

Bent adapter

Clamp

Pinch clamp

(a)

Steam in

Steam out

Separatory funnel

(b)

FIGURE 2.14 Water traps.

If only a small quantity of steam is required to distil the mixture completely, a second method of steam distillation may be employed. This method utilizes the direct addition of water into the distillation flask along with the organic compounds to be distilled. The flask is then heated directly with a Bunsen burner (see footnote 7) to effect the steam distillation. This method is generally not applicable for distillations in which large amounts of steam are needed; one would have to replenish the supply of water in the flask or use an inconveniently large flask.

In summary, steam distillation provides a method for separation of volatile liquid and solid organic compounds which are insoluble in water, or nearly so, from nonvolatile compounds under comparatively mild conditions. The technique obviously is not applicable to substances which decompose on prolonged contact with steam or hot water, react with water, or have a vapor pressure of 5 mm or less at 100°.

2.7 Recrystallization

One of the most valuable and useful techniques to be mastered by the organic chemist is that of *recrystallization*. Many organic compounds are solids and the procedure of recrystallization is the technique of choice in most instances in effecting their purification. Even in those instances in which an organic solid has been "purified" by sublimation (Section 2.8) or chromatography (Chapter 3), a careful worker will frequently perform a final recrystallization of the material to achieve added confidence in its purity.

Essentially the process is one in which the crystal structure is completely disrupted, either by fusion (melting) or by dissolution of the solid, and then the crystals are allowed to regrow so that the impurities are left either in the melt or in the solution. The fusion method is seldom used, because the crystals are usually formed in the presence of a rather viscous oil (containing the impurities) from which they are difficult to separate.

In solution recrystallization, advantage is taken of the fact that nearly all solids are *more soluble in a hot solvent than in a cold solvent.* The upper limit of temperature is dictated by the boiling point of the solvent. If the crystals are dissolved in a quantity of hot solvent which is insufficient to dissolve them when *cold,* and if that hot solution is then allowed to cool, it should be anticipated that crystals will precipitate from the cooling solution to the extent of the difference in solubility between the temperature extremes. The high-temperature extreme is limited, of course, to the boiling point of the solvent being used; the low temperature is commonly determined by convenience, an ice-water bath often being used, although the freezing point of the solvent can be the determining factor. If the impurities present in the original crystals have dissolved and *remain dissolved* after the solution is cooled, filtration of the crystals which have formed on cooling should then provide purified material. Or if the impurities remain undissolved in the hot solution and are filtered from it *before* it is allowed to cool, the crystals which subsequently form on cooling should be more pure than the original crystals. The technique of recrystallization is not always so simple, but this scenario should supply some understanding of the general principles of the procedure.

Application of the technique of solution recrystallization involves several steps: (1) selection of an appropriate solvent, (2) dissolution of the solid to be purified in the solvent near or at its boiling point, (3) filtration of the hot solution to remove insoluble impurities, (4) crystallization from the solution as it cools, (5) filtration of the purified crystals from the cooled supernatant solution (the "mother liquors"), (6) washing the crystals to remove the adhering solution, and (7) drying the crystals.

Each individual step in the overall procedure is now discussed separately and in some detail.

Selecting a Solvent. A solvent must satisfy certain criteria in order to be used as a recrystallization solvent. (1) Its temperature coefficients for the solute and impurities should be favorable; that is, the compound being purified should ideally be quite soluble in the hot solvent but somewhat insoluble in the cold (this will minimize losses), and the impurities should remain at least moderately soluble in the cold solvent. Another possibility here is that the impurities be insoluble in the hot solution, from which they may be filtered. (2) The boiling point of the solvent should be low enough so that it can be easily removed from the crystals in the final drying step. (3) It is generally preferable that the boiling point of the solvent be lower than the melting point of the solute. (4) The solvent should not react chemically with the compound being purified.

If the compound has been previously studied, the chemical literature will generally give information concerning a suitable solvent. If the compound has not been studied, it will be necessary to resort to trial-and-error methods using *small* amounts of material. Some general solubility principles should be kept in mind if this needs to be done. Normally, polar compounds are insoluble in nonpolar solvents and soluble in polar solvents. Conversely, nonpolar compounds are more soluble in nonpolar solvents. These solubility relationships are frequently summarized with the phrase "like dissolves like." Therefore a highly polar compound is unlikely to be soluble in a hot nonpolar solvent but may well be too soluble in a cold, very polar solvent, so that a solvent of intermediate polarity will be optimum.

Some appreciation for the wide range of polarity of the common solvents may be gained by consulting the dielectric constants for these solvents listed in Table 2.1. Those solvents with dielectric constants in the 2–3 range should be regarded as nonpolar, and those with constants above about 10 as polar. The remaining solvents in the 3–10 range are of intermediate polarity.

Occasionally, mixed solvents are found to work well. In this case the solubility of a compound in one solvent is reduced by the addition of a second solvent in which it is much less soluble. Some frequently used mixed solvent pairs are 95% ethanol-water, toluene-petroleum ether,[8] acetic acid-water, diethyl ether-alcohol, and diethyl ether-petroleum ether. Such solvent mixtures as toluene-ethanol are infrequently used, and then only with absolute ethanol, since the presence of water causes solvent separation, particularly on cooling; toluene and aqueous ethanol are not fully miscible.

Solution. The following operations are best carried out within a ventilation hood to avoid inhalation of solvent vapors. All organic solvents are either flammable or toxic to some extent, or both. The solid to be purified is placed in an appropriately sized Erlenmeyer flask, along with a few milliliters of the desired solvent. It is good laboratory technique to save a few crystals of the impure solid; they may be needed

---

[8] Petroleum ether is a mixture of aliphatic hydrocarbons obtained from petroleum refining. Such a mixture may vary in composition and boiling range, depending on the distillation "cut" taken. In order to define the liquid being used, the boiling range is usually given, for example, petroleum ether (bp 60–80°). The name *ligroin* is occasionally used in place of petroleum ether.

TABLE 2.1 SOLVENTS FOR RECRYSTALLIZATION[a,b]

Solvent	Boiling Point	Freezing Point[c]	Water Soluble	Dielectric Constant	Flammable	Specific Gravity[d]
Water*[e]	100°	0°	—	78.54	No	1.000
95% Ethanol*	78°		Yes	24.6	Yes	
Methanol	65°		Yes	32.63	Yes	
Petroleum ether*	Variable		No	1.9	Yes	About 0.7
Cyclohexane	81°	6°	No	2.02	Yes	0.779
Toluene	111°		No	2.38	Yes	0.867
Diethyl ether	35°		Slightly	4.34	Yes	0.714
Tetrahydrofuran	65°		Yes	7.58	Yes	
1,4-Dioxane	101°	11°	Yes	2.21	Yes	
Dichloromethane	41°		No	9.08	No	1.335
Chloroform*	61°		No	4.81	No	1.492
Carbon tetrachloride	77°		No	2.23	No	1.594
Ethyl acetate*	77°		Yes	6.02	Yes	
Acetone	56°		Yes	20.7	Yes	
Acetic acid	118°	17°	Yes	6.15	Yes	

[a] Benzene has been purposefully omitted from this list, owing to its toxicity. Cyclohexane can often be successfully substituted for it.

[b] As a general rule, avoid use of chlorocarbon solvents such as dichloromethane, chloroform, and carbon tetrachloride, *if another equally good solvent can be found.* If not, take care to avoid excessive inhalation of their vapors.

[c] Freezing points not listed are below 0°.

[d] Only the specific gravities of water-insoluble solvents are included.

[e] The solvents marked with asterisks are those which should normally be employed first in a trial-and-error search for the best recrystallization solvent.

later to induce crystallization if problems are encountered at that step. With *constant* stirring to prevent bumping of the boiling mixture (boiling chips are not very effective in the presence of quantities of undissolved solids), the mixture is then heated to boiling, preferably using a steam bath. More solvent is added in *small* portions to the boiling mixture until just enough boiling solvent is present to dissolve the solid. It is generally desirable at this point to add from 2 to 5% additional solvent to prevent premature crystallization during the hot filtration, if this step appears necessary. A large excess of solvent must be avoided in order to maximize the recovery of purified crystals. Solids remain soluble to some extent even in cool solution, and the recovery will be reduced by an amount which depends on both this solubility and the quantity of solvent present. If near the end of the dissolution it is apparent that additional solvent is not dissolving any more of the solid, *particularly when only a relatively small quantity of solid remains,* enough solvent has probably been added. The remaining solid is likely to consist of insoluble impurities and may be removed in the hot filtration step. Many beginning laboratory students obtain poor results in their recrystallizations because, when trying to dissolve the last traces of solid, quantities of solvent are added which are far in excess of that needed. During the dissolution of the impure solid, time should be allowed between each small addition of fresh solvent, for some solids dissolve only slowly.

When mixed solvents are used, they must, of course, be miscible. The crystals being purified should be soluble in one of the solvents, but insoluble or only slightly soluble in the other. The crystals are first dissolved in that pure, boiling solvent in which they are soluble. The second solvent is then added to the boiling solution until the solution turns *cloudy*. The second solvent decreases the dissolving ability of the solvent medium; when the solubility limit is reached, the solute begins to come out of solution, resulting in a cloudy appearance. If the second solvent has a lower boiling point than the first, the solution is cooled to below the boiling point of the second solvent before it is added. Finally, more of the first solvent is added *dropwise* until the solution just becomes clear again. Occasionally it is advantageous to carry out the hot filtration step before adding the second solvent to prevent crystallization during filtration.

If colored impurities are present, these may often be removed by adding a small amount of decolorizing carbon to the hot (*not boiling*) solution. Seldom is more decolorizing carbon needed than that which can be held on the tip of a small spatula. The impurities, especially colored ones, adsorb on the surface of the carbon particles and are removed during filtration. If too much carbon is used, some of the substance being purified will be adsorbed and subsequently lost in this step. After adding the carbon, the solution should be heated to boiling for a few minutes, while being continuously stirred or swirled to prevent bumping of the boiling mixture.

Hot Filtration. To remove insoluble impurities (including dust and decolorizing carbon, if used), the hot solution is filtered by gravity filtration. If no insoluble impurities are present and the solution is clear, this step may usually be omitted. Suction filtration is not desirable, because evaporation of the hot solvent under reduced pressure will both cool and concentrate the solution, resulting in premature crystallization. A short-stemmed or stemless glass funnel and a fluted filter paper should be used for filtration into a second Erlenmeyer flask (Figure 2.15). The use of *fluted* filter paper allows for more rapid filtration. The top of the paper should not extend above the top of the funnel.

Be sure to pour the hot liquid on the upper portion of the filter paper in order to maximize the efficiency of the filtration. In this way the solution will come in contact with a larger area of the filter paper, and thus will be filtered more rapidly. Be careful, however, not to allow any solution to pass between the edge of the paper and the funnel.

One of several possible ways of folding a fluted filter is shown in Figure 2.16. Fold the paper in half, and then into quarters. Fold edge 2 into 3 to form edge 4, and then 1 into 3 to form 5 (a). Now fold 2 into 5 to form 6, and 1 into 4 to form 7 (b). Continue by folding 2 into 4 to form 8, and 1 into 5 to form 9 (c); the paper now appears as in (d). Note that all folds have been in the same direction. Do not crease the folds tightly at the center because this might weaken the paper and cause it to tear during filtration. Now make new folds *in the opposite direction* between 1 and 9, 9 and 5, 5 and 7, and so on, giving the paper a fanlike appearance (e). Open the paper (f)

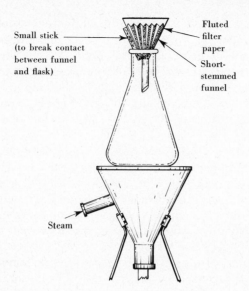

Small stick
(to break contact
between funnel
and flask)

Fluted
filter
paper

Short-
stemmed
funnel

Steam

FIGURE 2.15 Apparatus for effecting hot filtration.

and fold each of the sections 1 and 2 in half with reverse folds to form (g). The paper is now ready to use.

Crystallization from the solution occasionally occurs in the filter paper or on the surface of the funnel. This is most conveniently avoided by adding 2 or 3 mL of the recrystallization solvent to the receiving flask and heating to boiling (Figure 2.15). The condensing vapors will heat the funnel, preventing crystallization. For low-boiling solvents, a steam bath may be used; for solvents boiling higher than about

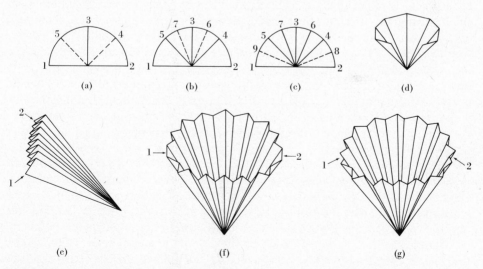

FIGURE 2.16 Folding of a fluted filter.

90°, an electrically heated oil bath is preferable. A Bunsen burner can be used *only* if the solvent is nonflammable.

Decolorizing carbon, being very finely divided, is not always completely removed by this filtration. If it is not, a small amount of "filter-aid" (a finely divided silica that is easily filterable) may be added to the filtrate. The solution is heated again just to boiling, and refiltered using a fresh fluted filter paper. The carbon is trapped in the filter-aid and thus removed.

Crystallization. The hot filtrate is allowed to cool slowly by standing at room temperature. Crystallization should then result. Rapid cooling by immersion in water, for example, is undesirable because the crystals formed will tend to be quite small. Their large surface area may then facilitate *adsorption* of impurities from the solution. Generally the solution should not be agitated while cooling, since this will also lead to formation of small crystals. However, formation of very large crystals (larger than approximately 2 mm) may cause *occlusion* (trapping) of solution *within* the crystals. Such crystals are difficult to dry and will, when dried, have deposits of impurities in them. If large crystals seem to be forming, agitation may be used to lower the average crystal size. Judgment of proper crystal size will become easier with experience.

If crystallization does not occur after cooling, the supersaturated solution can usually be made to yield crystals by *seeding*. A tiny crystal of the original solid is added to the cooled solution. Crystals will usually form quite rapidly. Another method that may be used to induce crystal formation is to scratch the inside surface of the flask at or *just above the surface of the solution* with a glass rod.

Occasionally the solute will separate from the solution as an "oil" rather than as crystals. Because this type of precipitation is not as selective as crystallization, these oils generally contain significant amounts of impurities, and their formation is undesirable. There are two somewhat different types of oiling problems. (1) Oils may persist on full cooling with no evidence of crystallization. In these cases the remedy which most frequently works is to scratch the oil with a glass stirring rod against the side of the flask in the presence of the mother liquor; this frequently will induce crystallization. Failing this, a few *small* seed crystals of the original impure solid may be added to the oil and the mixture allowed to stand for a time, perhaps until the next laboratory period. If this does not work it may be necessary to separate the oil from the mother liquor and crystallize it from a different solvent. (2) With certain compounds (acetanilide from water, for example), cooling of the hot concentrated solution results in the separation of an oil; at a lower temperature this oil freezes into a compact mass, while, at this same temperature or a little lower, relatively pure crystals separate from the residual solution. This oil is not pure liquid solute, but a liquid solution of solute, solvent, and perhaps other impurities, whose freezing point lies below the temperature at which it separated from solution, even though the crystalline solute may have a melting point lying above the boiling temperature of the solution. When this occurs, the problem may usually be remedied by reheating the mixture to boiling, adding a few milliliters of additional solvent, and again allowing the solution to cool. This will either solve the problem or at least reduce the amount of oil formed. In the latter case the procedure may be repeated.

After filtering the crystals (see discussion of Filtration), a second "crop" of crystals can usually be obtained by further cooling with ice water or by concentrating the solution by boiling away some of the solvent and cooling. The crystals obtained as a second (or even third) crop may not be as pure as the first. Melting points of the crystals obtained in various crops can be used to determine their purity.

Filtration.   The cool mixture of crystals and solution is now filtered by vacuum filtration, using a Büchner funnel and a vacuum filter flask attached to an aspirator or house vacuum line through a trap, as shown in Figure 2.17. The trap prevents water from the aspirator from backing up into the filter flask in case of loss of water pressure. The flask used for the trap should be a *heavy-walled* Erlenmeyer flask or bottle, or a second vacuum filter flask. If a filter flask is used, employ a two-holed stopper and attach to the side arm the tube leading to the aspirator. Before filtration, the filter paper, which should be of a size to lay *flat* on the funnel plate, should be wetted with the solvent in order to "seal" it to the funnel.

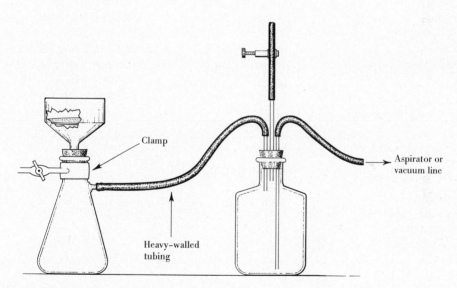

FIGURE 2.17   Apparatus for vacuum filtration.

A stirring rod or spatula may be used as an aid in transferring the crystals to the funnel. The last quantity of crystals may be transferred by washing them out of the flask with some of the filtrate (mother liquor). As soon as the mother liquor has passed through the filter, release the suction by opening the screw clamp or stopcock on the trap. Wash the crystals to remove adhering mother liquor, containing impurities, by adding to the funnel *cold* fresh solvent just sufficient to cover the crystals. Close the trap to reapply suction and to remove the wash solvent from the crystals. Press the crystals as dry as possible on the filter plate under suction with a cork or a spatula.

Drying the Crystals.   Most of the solvent may be evaporated by allowing the aspirator to pull air through the mass of crystals on the funnel for a few minutes. By means of a spatula the crystals are then transferred to a clean watch glass. Complete drying is accomplished by allowing them to air-dry for a few hours. If necessary, the drying process may be accelerated by placing the watch glass in an oven (*Caution:* The temperature of the oven should be at least 20° below the melting point of the crystals), or by placing the crystals in a vacuum desiccator. A workable facsimile of a vacuum desiccator may be assembled by replacing the Büchner funnel (Figure 2.17) with a solid rubber stopper. The crystals may be laid on the bottom of the flask, or, better, if the quantity of crystals is not too great, they may be placed in a test tube around the mouth of which is wrapped a piece of filter paper held in place with a rubber band. The test tube is placed in the desiccator. If the crystals are left on the bottom of the flask, take caution when releasing the pressure at the trap that the in-rushing air does not "blow" the crystals about the flask.

Heat drying may not be used for crystals that sublime readily at atmospheric pressure (see Section 2.8).

EXPERIMENTAL PROCEDURE

DO IT SAFELY

1.   Protect yourself and your neighbors by not using a burner in this experiment, because most of you will be using flammable solvents.

2.   If you are using a hot plate, do not turn it too high. A more moderate setting will prevent too rapid heating and the consequent bumping and splashing of hot solvents and materials from the flask during heating. Also, hot plates should not be used for heating volatile and flammable solvents (see Appendix 5); for these types of solvents, use of a steam bath is preferable.

3.   Avoid excessive inhalation of solvent vapors. If a ventilation hood is not available for your use, clamp an *inverted* funnel in place just above the Erlenmeyer flasks in which you will be heating solvents, and attach this funnel by means of rubber tubing to a water aspirator or to a house vacuum line. This will serve to lower the concentration of vapors in your work area.

4.   During the portions of this experiment when you are pouring or transferring solutions, either wear rubber gloves or be particularly careful to avoid getting these solutions on your skin. Organic compounds are much more rapidly absorbed through the skin when they are in solution, particularly in water-soluble solvents such as ethanol, acetone, and others. It is for this reason also that you should never rinse organic materials off your skin with solvents such as acetone; instead, wash your hands thoroughly with hot water and soap.

5.   Do not add decolorizing carbon to a *boiling* solution because it may cause the mixture to boil out of the container.

A. Selection of Solvent for Recrystallization

Heat a beaker of water to a gentle boil on a hot plate and use the hot water to serve as a heating medium; replenish the water as necessary. Alternatively, a steam bath may be used (see Appendix 5).

Place about 20 mg (a small spatula-tip full) of finely crushed resorcinol in a small test tube and add about 0.5 mL of water to the tube. Stir with a glass rod and determine whether resorcinol is soluble in water at room temperature. Record your observations in your notebook, using the following definitions: *soluble*—20 mg of solute will dissolve in 0.5 mL of solvent; *slightly soluble*—some but not all of the 20 mg of solute will dissolve in 0.5 mL of solvent; *insoluble*—none of the solute appears to dissolve. If the resorcinol is not completely soluble at room temperature, place the test tube in the hot-water bath, and with stirring or swirling of the tube observe whether resorcinol is soluble in hot water. Record your observations on the basis of the definitions.

Repeat the solubility test for resorcinol, using ethanol and then petroleum ether (bp 60–80°), recording your observations in your notebook. If additional practice is desired, determine in like fashion the solubility properties of naphthalene, benzoic acid, and acetanilide in water, in ethanol, and in petroleum ether.

If any of these solutes is soluble in the hot solvent but only slightly soluble or insoluble in the cold solvent, allow the hot solution to cool slowly to room temperature and compare the quantity, size, color, and form of the resulting crystals with the original solid material. Note which solvent you would consider best suited for recrystallization of each of the solutes.

If you are given an unknown for recrystallization, the preceding procedure should be followed in order to ascertain the most appropriate solvent for recrystallization of the unknown substance. Table 2.1 provides a listing of the solvents most generally useful for recrystallization of organic crystalline solids, with some descriptive information concerning each. It is usually not necessary to test *all* the solvents, but you should consider trying at least those solvents in the table which are denoted with an asterisk.

B. Recrystallization of Impure Solids

Carefully read the introductory paragraphs of Section 2.7 to become familiar with the techniques for carrying out each of the steps in the recrystallization procedure. This information should allow you to avoid or surmount most of the problems that may be encountered in effecting the purification of impure solids by recrystallization. Only abbreviated directions are provided for the experiments that follow.

Obtain one or more samples of solids for recrystallization. Among the impure solids that might be assigned are (1) benzoic acid, (2) acetanilide, (3) naphthalene, or (4) an unknown compound. If you are assigned an unknown, determine from part A the appropriate solvent to use, and then proceed with part B, following the general instructions given on pages 50–56.

Benzoic Acid. Place 1 g of impure benzoic acid in a clean 50-mL Erlenmeyer flask. Measure 25 mL of water in a graduated cylinder and add a 10-mL portion of it to the benzoic acid. Heat to gentle boiling with a microburner or hot plate. Add, as necessary, water in 1-mL portions until no more solid appears to dissolve in the boiling solution. Record the total volume of water used. No more than 20 mL should be required.

Since pure benzoic acid is colorless, a colored solution should be treated with decolorizing carbon. (*Caution:* Do not add decolorizing carbon to a *boiling* solution!) Cool the solution slightly, add approximately 0.1 g of carbon, and reheat to boiling for a few minutes. To aid in the filtration of the finely divided carbon, allow the solution to cool slightly, add a small amount of filter-aid, and reheat.

Perform a hot filtration according to the directions provided in the paragraph on that subject with reference to Figure 2.14. Rinse the empty flask with 1 or 2 mL of *hot* water and filter this wash solution into the main solution. If the filtered solution remains colored, repeat the treatment with decolorizing carbon. Cover the flask with a watch glass or inverted beaker,[*] and allow the filtrate to stand undisturbed until it has cooled to room temperature and no more crystals form. To complete the crystallization place the flask in ice water for at least 15 min.

Collect the white crystals by suction filtration and wash the filter cake with two small portions of *cold* water. With a clean spatula or cork press the crystals as dry as possible on the funnel. Spread the crystals onto a piece of filter paper or, better, a watch glass and allow them to air-dry completely. Determine the weight and melting point of the purified product. Calculate the percent recovery.

Acetanilide.   Place 5 g of impure acetanilide in a 250-mL Erlenmeyer flask. Measure 100 mL of water into a graduated cylinder and add a 50-mL portion to the crude acetanilide. Boil the mixture gently with the aid of a burner or hot plate.

Note the formation of an oil layer consisting of a solution of water in acetanilide which forms a separate phase. This second liquid phase forms at temperatures only above 83° in mixtures whose compositions lie between 5.2% and 87% acetanilide. Acetanilide, however, is sufficiently soluble in water at temperatures slightly above 100°,[9] the boiling point of the solution, to form solutions of greater than 5.2% acetanilide. Thus, a boiling solution prepared with the *minimum* quantity of water to effect solution will yield an oil on cooling to 83°, and crystals below 83° (see the paragraph on Crystallization: discussion of oiling).

Continue adding water in small portions (3–5 mL) to the boiling solution until the oil has completely dissolved. Note that any solid present at this point must consist of insoluble impurities. (The interested student may wish to allow this solution to cool to about 50° to verify the phase properties of acetanilide-water discussed. If this is done, reheat the solution to boiling following these observations and continue.) Once the acetanilide has just dissolved, add an additional 5 mL of water to prevent formation of oil during the crystallization step. If oil does form at that time, reheat the solution and add a little more water. Record the total volume of water used.

Allow the solution to cool below boiling and add about 0.1 g of decolorizing carbon; with stirring, gently boil the solution for a few minutes. Cool the solution, add a small amount of filter-aid, stirring thoroughly. Reheat to boiling, and perform a hot

[9]The solubility of acetanilide in water is 5.5 g per 100 mL at 100° and 0.53 g per 100 mL at 0°.

filtration according to the directions given in the paragraph on this subject with reference to Figure 2.15. Cover the flask with a watch glass or, better, an inverted beaker,★ and allow the filtrate to stand undisturbed while cooling to room temperature; when crystallization apparently is complete, cool the flask in ice water for at least 15 min to complete the crystallization.

Collect the crystals by vacuum filtration and wash the filter cake with two small portions of *cold* water. With a spatula or cork press the crystals as dry as possible on the funnel. Spread the crystals onto a piece of filter paper or, better, a watch glass and allow them to air-dry completely. Determine the weight and melting point of the purified acetanilide. Calculate the percent recovery.

Naphthalene. Naphthalene may be conveniently recrystallized from either methanol, ethanol, or 2-propanol. Because these solvents are all either toxic or flammable, or both, proper precautions should be taken. Operations through the hot filtration step should be carried out in a ventilation hood. If this is not available, an inverted funnel connected by tubing to an aspirator or vacuum line and positioned *no more than an inch or two* above the mouth of the flask in which the solvent is being heated will afford reasonable protection.

Place 5 g of impure naphthalene in a 250-mL Erlenmeyer flask and dissolve it in the minimum required amount of boiling alcohol. (*Caution:* Use a steam bath for heating; do not use a burner.) Add 2 or 3 mL of additional solvent. A colored solution should be treated with decolorizing carbon. Perform a hot filtration, and cover the flask containing the hot filtrate with either a watch glass or an inverted beaker,★ and allow the filtrate to stand undisturbed while cooling to room temperature. Collect the crystals by vacuum filtration, wash with two small portions of *cold* solvent, and press dry. Transfer the crystals to a piece of filter paper or a watch glass and allow them to air-dry. Determine the weight and melting point of the purified naphthalene. Calculate the percent recovery.

EXERCISES

1. List each of the steps in the systematic procedure for recrystallization. Indicate briefly the functional purpose of each of these steps in accomplishing the purification of the originally impure solid.
2. The goal of the recrystallization procedure is to obtain *purified* material with a *maximized recovery.* For each of the items listed, explain why this goal would be adversely affected.
    (a) In the solution step, an unnecessarily large volume of solvent is used.
    (b) The crystals obtained by vacuum filtration are not washed with fresh cold solvent before drying. This step is omitted.
    (c) The crystals referred to in (b) are washed with fresh *hot* solvent.
    (d) A large excess of decolorizing carbon is used.
    (e) Crystals are obtained by breaking up the solidified mass of an oil which originally separated from the hot solution.
    (f) Crystallization is accelerated by immediately placing the flask of hot solution in ice water.

3. A second crop of crystals may be obtained by concentrating the vacuum filtrate and cooling. Why is this crop of crystals probably less pure than the first crop?
4. Explain why the rate of dissolution of a crystalline substance may depend on the *size* of its crystals.
5. The solubility of benzoic acid at 0° is 0.02 g per 100 mL of water and that of acetanilide is 0.53 g per 100 mL of water. If you performed either of these recrystallizations, calculate, with reference to the total volume of water used in preparing the hot solution, the amount of material in your experiment which was unrecoverable by virtue of its solubility at 0°.
6. Assuming that either solvent is otherwise acceptable in a given instance, what advantages does ethanol have over 1-octanol as a crystallization solvent? hexane over pentane? water over methanol?
7. Look up the solubility of benzoic acid in hot water. According to the published solubility, what is the minimum amount of water in which 1 g of benzoic acid can be dissolved?
8. Why is it important to:
    (a) break the vacuum before turning off the water pump when employing the equipment shown in Figure 2.17 (vacuum filtration)?
    (b) avoid the inhalation of vapors of organic solvents?
    (c) know the position and procedure of operation of the nearest fire extinguisher when employing diethyl ether as a crystallization solvent?
    (d) use a *fluted* filter paper for hot filtration?

2.8 Sublimation

An alternative to crystallization for the purification of some solids is the process of *sublimation*. This method uses to advantage the differing vapor pressures of solids in a way analogous to simple distillation. The impure sample is vaporized directly from the solid state by heating it at a temperature below the melting point and the vapor is then condensed (crystallized) directly to the solid state on a cold surface; both processes occur *without* the intermediacy of the liquid state. Figure 2.18 shows a

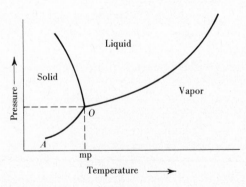

FIGURE 2.18 Single-component phase diagram.

typical phase diagram, relating the solid, liquid, and vapor states of a substance with pressure and temperature. Note that with conditions of temperature and pressure below those depicted at point O, the liquid state cannot exist (that is, is thermodynamically unstable). The vapor pressure of the solid at any temperature below the melting point is given by the curve OA. It is the equilibrium between solid and vapor, represented by this curve, which is of importance for sublimation.

Two types of sublimation apparatus commonly used are shown in Figure 2.19.

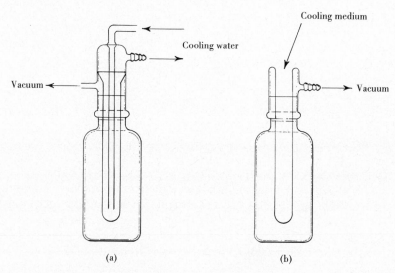

FIGURE 2.19 Sublimation apparatus.

They have two characteristics in common: (1) a chamber that may be evacuated by attachment to a vacuum pump (the impure sample is placed at the bottom of this chamber) and (2) a fingerlike projection in the center of this chamber that may be cooled to provide a surface on which the sublimed crystals may form. The "cold-finger" in (a) is cooled by circulating water, and that in (b) by a medium such as Dry Ice in acetone, or ice water. Figure 2.20 shows a simple type of sublimation apparatus that may be assembled inexpensively from two test tubes (one with a side arm), rubber stoppers, and glass tubing.

For effective purification using this technique, two criteria must be met: (1) the solid must have a relatively high vapor pressure, and (2) the impurities must exhibit vapor pressures substantially different from the primary material (preferably lower). Should the first criterion *not* be met, the time necessary to pass any significant quantity of material through the vaporization-crystallization sequence would usually be prohibitively long. In this case, either recrystallization (Section 2.7) or chromatographic techniques (Chapter 3) must be used.

The "mechanics" of purification by sublimation may be simply described as follows. The impure solid is heated to a temperature higher than that of the surface

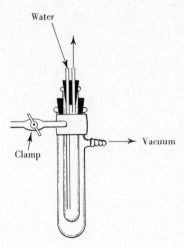

FIGURE 2.20 Test tube sublimator.

(cold-finger) on which the pure material is to be condensed but lower than the melting point. Since vapor pressure and thermodynamic instability of the solid vary directly with the temperature (see curve *OA,* Figure 2.18), material will be transferred via the vapor phase to the colder surface. Crystals developing on the cold surface will tend to be very pure, because the molecules of the impurities will usually not be incorporated into the growing crystal structure and therefore will not condense on the cold surface. Should the vapor pressure of an impurity be similar to that of the material being purified, sublimation will not be an effective method of purification, since crystals of both substances will tend to form on the cold surface (criterion 2).

Very few organic compounds exhibit vapor pressures adequate for sublimation at atmospheric pressure. Generally, reduced pressure (vacuum) is needed to increase the rate of evaporation of the solid. This process is analogous to the use of vacuum distillation for high-boiling liquids.

Sublimation is generally restricted to relatively nonpolar substances having fairly symmetrical structures. In these cases, crystal forces are lower and vapor pressures tend to be higher. The ease with which a molecule may escape from the solid to the vapor phase is determined by the strength of the intermolecular attractive forces between the molecules of the solid. The most important attractive force is of an electrostatic nature. Symmetrical structures will have relatively symmetrical distributions of electron density, and as such will have smaller dipole moments than less symmetrical structures whose electron distribution will be more polarized. A smaller dipole moment implies a higher vapor pressure. Van der Waals forces are also important, but are generally less important than electrostatic attractions. In general, van der Waals forces increase in magnitude with increasing molecular weight, and thus large molecules, even if symmetrical, are less adaptable to purification by sublimation.

REFERENCES

DISTILLATION

1. *Technique of Organic Chemistry,* A. Weissberger, editor, Interscience Publishers, New York, 1965, Vol. IV.
2. *Azeotropic Data,* American Chemical Society, Washington, D. C. 1952.
3. T. Earl Jordan, *Vapor Pressure of Organic Compounds,* Interscience Publishers, New York, 1954.

RECRYSTALLIZATION

4. R. Stuart Tipson, in *Technique of Organic Chemistry,* A Weissberger, editor, 2d ed., Interscience Publishers, New York, 1956, Vol. III, Part I, Chapter 3.

SUBLIMATION

5. R. Stuart Tipson, in *Technique of Organic Chemistry,* A Weissberger, editor, Interscience Publishers, New York, 1965, Vol. IV, Chapter 8.

3

SEPARATION AND PURIFICATION OF ORGANIC COMPOUNDS
Phase Distribution: Extraction and Chromatography

Two widely applicable and relatively simple methods of separating a desired compound from its impurities, or of isolating each of the individual components of a mixture, are extraction and chromatography, both of which are based on the principle of *phase distribution*. A substance may establish an equilibrium distribution between two *insoluble* phases with which it is in contact, in a ratio dependent upon its relative stability in each of those phases. The techniques being discussed involve the *selective* removal of one or more of the components of a gaseous, liquid, or solid mixture by contact of the mixture with a second phase.

The process on which such a distribution depends may be one of two varieties: (1) *partitioning,* based on differing relative solubilities of the components in immiscible solvents (selective dissolution), and (2) *adsorption,* based on the selective adherence of the components of a liquid or gaseous mixture to the surface of a solid phase. The various techniques of chromatography involve both of these processes, whereas the techniques of extraction involve only the former.

3.1 Extraction

A most widely employed method of separating organic compounds from mixtures in which they are found or produced is that of liquid-liquid extraction. In fact, virtually every organic reaction requires extraction at some stage in the purification of its products.

In its simplest form, extraction involves the distribution of a solute between two immiscible solvents. The distribution is expressed quantitatively in terms of the *distribution* (or *partition*) *coefficient, K* (equation 1). This expression indicates that a solute, *A,* in contact with a mixture of two immiscible liquids, *S* and *S'*, will be distributed (partitioned) between the liquids so that at equilibrium the ratio of concentrations of *A* in each phase will be constant, at constant temperature.

$$K = \frac{\text{concentration of } A \text{ in } S}{\text{concentration of } A \text{ in } S'} \tag{1}$$

Ideally, the distribution coefficient of *A* is equal to the ratio of the individual solubilities of *A* in pure *S* and in pure *S'*. In practice, however, this correspondence is generally only approximate, since no two liquids are completely immiscible. The extent to which they dissolve in each other alters their solvent characteristics and thus slightly affects the value of *K.*

It is evident that for *A* to dissolve completely in one or the other of two immiscible liquids, the value of *K* must be infinity or zero.[1] Neither of these limiting values is actually attained. However, so long as *K* is larger than 1.0, and the volume of solvent *S* is equal to or larger than the volume of solvent *S'* (see equation 3), the solute will be found in *greater amounts* in solvent *S.* The amount of solute that will remain in the other solvent, *S'*, will depend on the value of *K.*

Equation 1 may be rewritten as shown in equations 2 and 3, given below:

$$K = \frac{\text{grams of } A \text{ in } S/\text{mL of } S}{\text{grams of } A \text{ in } S'/\text{mL of } S'} \tag{2}$$

$$K = \frac{\text{grams of } A \text{ in } S}{\text{grams of } A \text{ in } S'} \times \frac{\text{mL of } S'}{\text{mL of } S} \tag{3}$$

Note that when the volumes of *S* and *S'* are equal,[2] the ratio of the grams of *A* in *S* and in *S'* will equal the value of *K.* If the volume of *S* is doubled and the volume of *S'* kept the same, the ratio of the grams of *A* in *S* to the grams of *A* in *S'* will be *increased* by a factor of two. This must necessarily follow because *K* is a *constant.* Therefore, if *A* is to be recovered by extraction into solvent *S,* the amount of *A* recovered will be increased by using larger quantities of solvent *S.*

A further consequence of the distribution law (equation 1) is of practical importance in performing an extraction. If a given total volume of solvent *S* is to be used to separate a solute from its solution in *S'*, it can be shown to be more efficient to effect several successive extractions with portions of that volume than one extraction with the full volume of solvent. Thus *more* butyric acid will be removed from water solution by two successive extractions with 50-mL portions of diethyl ether than will be removed in a single extraction with 100 mL of diethyl ether. Three successive extractions with 33-mL portions would be still more efficient (see Exercises). There is, however, a point beyond which the further effort of additional extractions no longer yields a commensurate return. The larger the distribution coefficient, the fewer the

---

[1] For simplicity we shall define *S* as that solvent in which the solute is more soluble. Therefore in the remaining discussion the value of *K* will by definition always be greater than 1.0.

[2] To be strictly correct, the volumes of *solution* should be used in this expression. However, when the solutions are reasonably dilute, volumes of *solvent* may be used without appreciable error.

number of repetitive extractions that are necessary to separate the solute effectively. This is an important consideration because it is desirable to keep the total volume of extracting solvent to a minimum, not only for reasons of expense but also because of the time involved in eventual removal of the solvent by distillation.

Consider now a solution of two compounds in solvent S'. It should be evident that for effective separation of these two compounds by extraction with solvent S, the distribution coefficient of one should be significantly greater than 1.0, whereas the distribution coefficient of the other should be significantly smaller than 1.0. If these conditions are met, one compound will be mainly distributed in solvent S and the other in solvent S', at equilibrium. Physical separation of the two liquid layers would then result in at least a partial separation of the two compounds.

When the coefficients are of similar magnitude, separation by the extraction technique may be quite ineffective, since the relative concentrations of the compounds in each of the two liquid phases may be little changed from those of the original mixture. In this case other methods of separation should be used, such as adsorption chromatography (Section 3.5), fractional distillation (Section 2.2), or fractional crystallization. The specialized techniques of *fractional extraction* and *countercurrent distribution* can also be used to separate compounds with similar but not identical distribution coefficients. For the interested student details of these techniques are supplied in the references at the end of the chapter.

When choosing an extracting solvent for the isolation of a component from a solution, some general principles should be kept in mind. (1) The solvent must of course be immiscible with the solvent of the solution. (2) The solvent chosen must have the most favorable distribution coefficient for the component which is to be separated and must have unfavorable coefficients for the impurities or other components. (Since distribution coefficients for most compounds in various solvents are not available, extraction solvents used in experiments in this book will be suggested to you. However, as you gain experience and understanding about the solubility principles of organic compounds, you should be able to suggest appropriate solvents.) (3) A solvent must be chosen that does not react chemically with the components of the mixture, just as in recrystallization. (4) The solvent should be readily separable from the solute, following the extraction. Usually the solvent is removed by distillation, so relatively volatile solvents are advantageous.

Acid and Base Extractions.   The solubility characteristics of organic acids and bases in water are quite dramatically affected by the adjustment of pH. For example, except for the lower molecular weight members, most organic acids, such as carboxylic acids and phenols, are either insoluble or only slightly soluble in water. However, these same acids are found to be soluble in dilute aqueous sodium hydroxide solution in which the pH is decidedly *above* 7. The reason for this behavior is illustrated in equation 4. The organic acid undergoes *deprotonation* in an acid-base reaction with

$$\underset{\substack{\text{(Insoluble in} \\ \text{H}_2\text{O)}}}{R-\overset{\displaystyle O}{\overset{\|}{C}}-O-H} + NaOH \longrightarrow \underset{\text{(Soluble in H}_2\text{O)}}{R-\overset{\displaystyle O}{\overset{\|}{C}}-O^{\ominus}Na^{\oplus}} + H_2O \qquad (4)$$

the sodium hydroxide to produce its corresponding conjugate base, which with the sodium ion constitutes a *salt*. Because of its ionic character, the salt of the organic acid is soluble in water even though the acid itself is not. If this alkaline solution is then neutralized or made slightly acidic by the subsequent addition of a mineral acid such as hydrochloric acid, the conjugate base will be reprotonated to form the original organic acid which, because of its insolubility in water, will precipitate from solution (equation 5). The organic acid may then be recovered from this heterogeneous mixture, usually by filtration.

$$
\underset{\substack{\text{(Soluble in } H_2O)}}{R-\overset{\displaystyle O}{\overset{\|}{C}}-O^{\ominus}Na^{\oplus}} + HCl \longrightarrow \underset{\substack{\text{(Insoluble in} \\ H_2O)}}{R-\overset{\displaystyle O}{\overset{\|}{C}}-O-H} + Na^{\oplus}Cl^{\ominus} \tag{5}
$$

Similarly, an organic base that is insoluble in water (pH 7) will generally be found to be quite soluble in dilute hydrochloric acid solution in which the pH is decidedly *below* 7. In this case the increase in solubility of the organic base in the aqueous medium relies on the *protonation* of that base by the HCl to produce the ionic and therefore water-soluble conjugate acid, a substituted ammonium ion, which, with the chloride ion, constitutes a salt (equation 6). If this acidic solution is then neutralized or made slightly basic by the addition of aqueous sodium hydroxide, the conjugate acid will be *deprotonated* to produce the original organic base, which is insoluble in water and precipitates from the solution, leading to its recovery (equation 7).

$$
\underset{\substack{\text{(Insoluble in} \\ H_2O)}}{R-\overset{\displaystyle ..}{N}H_2} + HCl \longrightarrow \underset{\substack{\text{(Soluble in} \\ H_2O)}}{R-\underset{\oplus}{\overset{\displaystyle H}{\overset{|}{N}}}H_2 \ Cl^{\ominus}} \tag{6}
$$

$$
\underset{\substack{\text{(Soluble in} \\ H_2O)}}{R-\underset{\oplus}{\overset{\displaystyle H}{\overset{|}{N}}}H_2 \ Cl^{\ominus}} + NaOH \longrightarrow \underset{\substack{\text{(Insoluble in} \\ H_2O)}}{R-\overset{\displaystyle ..}{N}H_2} + H_2O + Na^{\oplus}Cl^{\ominus} \tag{7}
$$

Thus an organic acid may be selectively removed from a mixture containing other neutral or basic materials by dissolving that mixture in an organic solvent such as diethyl ether or dichloromethane and then extracting the solution with dilute sodium hydroxide solution, which will *selectively* remove the organic acid into the aqueous phase through formation of its conjugate base. Or an organic base may be selectively removed from a similar mixture by extraction with dilute hydrochloric acid solution.

Continuous Extraction.   Another type of experimental problem often encountered is that involving separation of one component that is only slightly soluble in the extracting solvent from a mixture whose other components are essentially insoluble. Large quantities of solvent would have to be used in order to effect the separation in only one or two extractions, and the handling of such quantities may be extremely unwieldy. Alternatively, it would be tedious to do a very large number of extractions with smaller quantities of solvent. The method of *continuous extraction,* in which a relatively small volume of extracting solvent is used, is a possible solution to this problem. By means of specialized apparatus, the solution of extracting solvent and solute is continuously separated into a boiling flask from the mixture being extracted. The solution is subjected to continuous distillation and the condensed distillate returned as fresh extracting solvent to the extraction vessel and reused. In the process the extracted material builds up in an increasingly concentrated solution in the boiling flask. This is because more dilute solution is continuously draining into the flask, while, at the same time, the solvent is being distilled away.

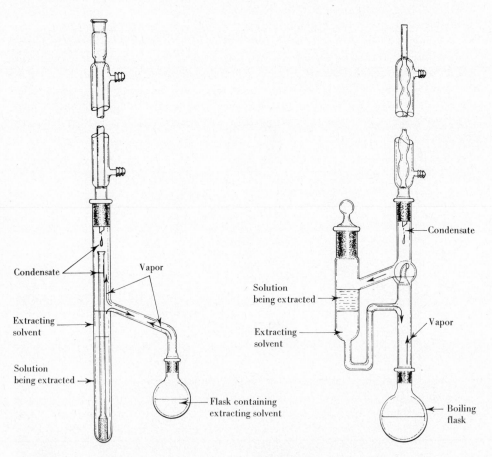

FIGURE 3.1   Light-solvent extractor.   **FIGURE 3.2**   Heavy-solvent extractor.

For continuous liquid-liquid extraction the solvent (as condensate from the condenser) is made to pass either up or down, depending on relative densities, through the solution containing the desired compound. If the solvent is less dense than the solution being extracted, an apparatus such as that shown in Figure 3.1 may be used. If it is more dense, apparatus such as that shown in Figure 3.2 may be used.

For separation of the components of a solid mixture by continuous solid-liquid extraction, a Soxhlet extraction apparatus (Figure 3.3) is convenient. The solid is placed in a *porous* thimble in the chamber, as shown, and the extracting solvent in the boiling flask below. The solvent is heated to reflux, and the distillate, as it drops from the condenser, collects in the chamber. By coming in contact with the solid in the thimble, the liquid effects the extraction. After the chamber fills to the level of the upper reach of the siphon arm, the solution empties from this chamber into the boiling flask by a siphoning action. This process may be continued automatically and without attendance for as long as is necessary for effective removal of the desired component, which will then be contained with the solvent in the boiling flask.

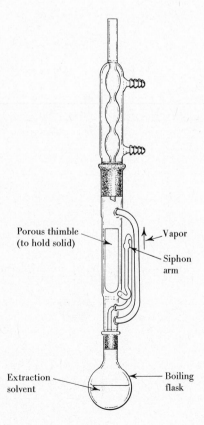

Porous thimble (to hold solid)

Vapor

Siphon arm

Extraction solvent

Boiling flask

FIGURE 3.3 Soxhlet extractor.

3.2   Technique of Simple Extraction

The performance of extraction as a technique is one that is encountered widely in the organic laboratory. It is commonly encountered in the isolation and purification of the products of nearly all organic reactions. Because of the relative frequency in which you will need to use this procedure, the use of a separatory funnel is discussed in some detail.

Separatory Funnels.   This type of glassware is available in several different shapes, from almost spherical to elongated pear shape (Figure 3.4). The more elongated the funnel, the longer the time required for the two liquid phases to separate once shaken together. When the liquids have similar densities, the flatter (more spherical) separatory funnels are preferable, because otherwise, a needless amount of time may be wasted in waiting for the phases to separate. The funnels have a stopcock at the bottom through which the contents may be drained.

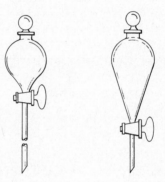

FIGURE 3.4   Separatory funnels.

The Extraction.   To perform an extraction, the solution is placed in a separatory funnel (with the stopcock closed!) and to this is added a quantity of the extracting solvent. The funnel should not be too full (three-fourths of the height of the funnel is a reasonable maximum). The upper opening of the funnel is stoppered either with a ground-glass stopper, for which most separatory funnels are fitted, or with a rubber stopper. The funnel is held during the shaking process in a rather specific manner, which will, when mastered, be found to be quite efficient. If right-handed, set the stopper at the top of the funnel against the base of the index finger of the left hand and grasp the funnel with the first two fingers and the thumb. The funnel should be turned so that the first two fingers of the right hand can be curled around the handle of the stopcock. In such a manner the stopper and stopcock can be held tightly in place during the shaking process (see Figure 3.5). Left-handed people would, of course, find it easier to use the opposite hand for each position.

The funnel and its contents are shaken vigorously, so that the two immiscible liquids are mixed as intimately as possible. The purpose of the shaking process is to

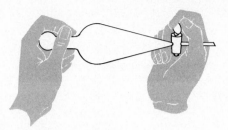

FIGURE 3.5 Efficient method of holding separatory funnel.

increase greatly the surface area of contact between the solvents so that the equilibrium distribution of solute between the solvents will be attained in a relatively short time. Every few seconds it is important to "vent" the funnel by inverting it (stopcock up) and carefully opening the stopcock to release any pressure that may have developed inside. This is particularly important when solvents of low-boiling points are used, or if an acid solution is extracted with sodium bicarbonate solution (CO_2 is released). If this is not done, the stopper may well be blown free, and the contents of the funnel lost. If the funnel is held as described, the stopcock may be opened by twisting the fingers curled around it, without readjusting the grip on the funnel. At the end of the shaking period (if the shaking is vigorous, 1 or 2 min are usually sufficient), the funnel is vented a final time, supported on a ring[3] and the layers allowed to separate. The two liquid layers are then separated by carefully drawing the lower layer into a flask through the stopcock.

Layer Identification. As a general rule the layers will separate so that the solvent of greater density will be on the bottom. Thus knowledge of the densities of solvents to be employed is useful in identification of the layers. (The densities of several water-insoluble solvents may be found in Table 2.1.) This rule, however, is not foolproof, because the nature and concentration of solute may be such as to invert the relative densities of the two solvents. A common mistake is to confuse the identity of the two layers in the funnel. Often one layer is discarded, and it is later found that the wrong layer was saved. *It is suggested, as a point of caution, that both layers always be saved until there is no doubt about the identity of each.* In almost all instances one of the layers is aqueous and the other an organic solution. A very good method of ascertaining which layer is which is to withdraw a few drops of the upper layer with a dropper and add these drops to about 0.5 mL of water in a test tube. If the upper layer is water, these drops will be miscible with the water in the test tube. If the upper layer is the organic layer, the droplets will remain undissolved and will be visually evident in the test tube.

Emulsions. Occasionally the two immiscible liquids will not separate cleanly into two layers (after shaking), but instead will form an *emulsion,* which is a result of a

[3]When using an iron ring to support a funnel, it is best to cover the ring with a length of rubber tubing to prevent breakage. This may be accomplished by slicing the tubing on one side and then slipping the tubing over the ring. Copper wire may be used to fix the tubing permanently in place.

colloidal mixing of the two layers. When one encounters an emulsion during an experiment, it frequently can be a very frustrating experience, because there are no infallible procedures to employ in "breaking" emulsions. When an emulsion is left unattended, the layers may sometimes separate after an extended time; however, it is usually more expedient to attempt one or more of the following remedies. (1) Add a few milliliters of brine (*saturated* aqueous sodium chloride solution) to the funnel and reshake the contents. The increase in the ionic strength of the water layer will sometimes force a break in the emulsion. This procedure might be repeated, but if it does not work the second time, try the following. (2) Filter the heterogeneous mixture and return the filtrate to the separatory funnel. Sometimes emulsions are caused by small amounts of gummy organic materials which are present. Removal of these materials onto the filter paper may allow the layers to separate. (3) Add a *small* quantity of a water-soluble detergent to the mixture and reshake the mixture. This approach is not so desirable as the first two, particularly if the desired materials are in the water layer, because it adds an impurity (the detergent), which must be separated later. (4) Finally, if difficult and unresponsive emulsions are produced, it may be necessary to choose an entirely different extraction solvent to avoid this problem.

Small quantities of insoluble material often collect at the phase interface, making it difficult to observe the true boundary between layers. Typically this difficulty is solved by removing any solids along with the undesired liquid layer; however, inevitably a small amount of the desired layer is lost by this procedure. An alternate procedure, filtration of the mixture before separating layers, is appropriate in extreme cases.

EXPERIMENTAL PROCEDURE

DO IT SAFELY

1. You should develop a continuing respect for the chemicals and reagents you use during the performance of experiments in an organic laboratory. Learn to avoid getting these materials on your hands. We recommend the use of rubber gloves for the following experiment. Depending on the type of rubber glove you are wearing, wet glassware may feel slippery. Because of the advantage and protection rubber gloves offer, take extra care in handling your equipment. If you should accidentally allow any chemicals to come in contact with your skin during the experiment, wash the affected areas with soap and warm water.

2. In order to avoid the development of gas pressure be sure to vent the separatory funnel frequently during any extractions or washings that you do. If sufficient pressure builds up during shaking, the stopper may be blown free, allowing the contents of the funnel to be splashed onto either you or your neighbors.

3. Wear your safety glasses.

This experiment will demonstrate the ability to separate various types of organic compounds by control of the pH of the extraction medium.

Prepare or obtain a mixture containing 2 g of each of the following compounds: benzoic acid, *p*-bromoaniline, and naphthalene. Dissolve this mixture into about 100 mL of dichloromethane. If *small* quantities of solid do not dissolve, filter the solution into a separatory funnel. Extract this solution two times, using 30-mL portions of 6 *M* hydrochloric acid each time. Combine the aqueous layers from each extraction into the *same* labeled flask. Return the organic layer, which has now been extracted twice with hydrochloric acid solution, to the separatory funnel and extract it twice more, using 30-mL portions of 3 *M* sodium hydroxide solution each time. Combine these aqueous layers into a second labeled flask.

Dry the organic layer by adding 4–5 g of *anhydrous* sodium sulfate to the liquid contained in a third labeled flask.[4] Let this mixture stand for about 0.5 hr, swirling it occasionally.

While the organic layer is drying over sodium sulfate, separately neutralize each of the aqueous extracts; cool each of the mixtures with an ice-water bath before neutralization. Use 6 *M* sodium hydroxide solution to neutralize the acid washes, and 6 *M* hydrochloric acid to neutralize the base washes. Follow the neutralizations with the aid of litmus paper.

Upon neutralization, precipitates should be observed in each flask. Isolate each of the precipitates separately by means of vacuum filtration. Each of the resulting solids should be washed *within the Büchner funnel* with *cold* distilled water. Preweigh (tare) three small labeled vials and place in two of them the solid materials thus far obtained. Either allow these to air-dry until the next laboratory period[5] or place them in an oven held at 90° for 1.5 hr.

Separate the organic layer from the sodium sulfate by gravity filtration and remove the solvent by simple distillation, using a steam bath for heating. Even if no crystals are evident in the boiling flask, discontinue heating when only a small amount of material remains and allow the flask to cool. Attach it to a vacuum source such as an aspirator pump for a few minutes in order to remove the last traces of solvent, transfer the solid residue to the third tared vial, and allow it to air-dry (do not use oven).

After the solids have dried, reweigh each of the vials to obtain the weights of the separated materials. Determine the melting points of each of the products obtained. For final purification, your instructor may wish you to recrystallize each of the isolated substances. The melting points of the pure original components of the mixture are: benzoic acid, mp 121–122°; *p*-bromoaniline, mp 65–66°; and naphthalene, mp 80–81°.

EXERCISES

1. If, when extracting an aqueous solution with an organic solvent, you are uncertain of which layer in the separatory funnel is the organic layer, how could you quickly settle the issue?

[4] For a discussion of drying agents, see Appendix 1.

[5] If these materials are allowed to air-dry in the laboratory desk until the next period, the vials should be covered loosely with a piece of paper to prevent contamination of the contents with dust. Do not cap the vials.

2. In the extraction experiment why was it suggested that *cold* water be used to rinse the solids in the Büchner funnel?

3. From the results of the extraction experiment, what can you conclude about the properties of the compounds used? Write chemical equations showing any changes that occurred during the extraction procedure.

4. Which layer (upper or lower) will each of the following solvents usually form if used to extract an aqueous solution: diethyl ether? chloroform? acetone? hexane?

5. Given 500 mL of an aqueous solution containing 8 g of compound *A*, from which it is desired to separate *A*, how many grams of *A* could be removed in a single extraction with 150 mL of diethyl ether? (Assume the distribution coefficient, diethyl ether:water, to equal 3.0.) How many *total* grams can be removed with three successive extractions of 50 mL each? [*Note:* In solving problems of this type, one should recognize that equation 3 applies to the situation pertaining *after* equilibrium is reached; for example, a practical form of the equation is

$$K = \frac{x}{a-x} \times \frac{\text{mL of } S'}{\text{mL of } S}$$

where a = grams of *A* originally present in water (S') and x = grams of *A* present in diethyl ether (S) after extraction.]

3.3   Chromatography

As mentioned in the introductory section, chromatography is based on the general principles of phase distribution. Reduced to its fundamentals the method involves the *selective removal* of the components of one phase from that phase as it is *flowing* past (or through) a second *stationary* phase. The removal of a component by the stationary phase is an equilibrium process, and the molecules of that component reenter the moving phase. Separation of two or more components in the moving phase will result when the equilibrium constants for the distribution of these components between the two phases differ. Expressed simply, the more tenaciously one component is held by the stationary phase, the higher the percentage of molecules of that component that are held *immobile*. A second component, less strongly held, will have a higher percentage of molecules in the *mobile* phase than will the first component. Therefore, on the average, the molecules of the component that is held less strongly will move over the stationary phase (in the direction of flow) at a higher rate than the other, resulting in a migration of the components into separate regions (bands) of the stationary phase (see, for example, Figure 3.11).

The separation between the bands is linearly related to the distance traveled on the column. In general the longer the distance, the greater the separation. For example, if one band moves 10 cm along the pathway and a second band moves 5 cm during the same time interval, the separation between the centers of the bands will be 5 cm. At a later time, if the first band has moved 100 cm, the second will have moved only 50 cm, and the separation will have increased to 50 cm. It should be remembered that the separation of a mixture by phase distribution requires that the components of the mixture have different distribution coefficients; if these coefficients are similar, only partial separation of the components into individual bands will occur, unless, of course, the path length is increased to give the components "time" to migrate apart.

There are four important types of chromatography based on the principles just discussed. These are gas chromatography (more specifically, gas-liquid partition chromatography, glpc), column chromatography, thin-layer chromatography (tlc), and paper chromatography.

3.4 Gas Chromatography

In gas chromatography, the mixture to be separated is *vaporized* and carried along a column by a flowing *inert* gas such as nitrogen or helium (the carrier gas). The gaseous mixture is the *mobile* phase. The column is packed with a solid, finely divided substance, on the surface of which is coated a liquid of relatively low volatility. This liquid serves as the *stationary* phase. Because of selective phase distribution of the components of the mixture between the mobile and stationary phases, these components may move through the column at different rates, and thus be separated. The physical process involved in the separation of the components of a mixture in the glpc column is the *partitioning* of the components between the gas and liquid phases.

A large variety of gas chromatographs of different types are commercially available. However, the basic features of these instruments are quite similar and are represented by the schematic shown in Figure 3.6. Parts 1–5 are needed to supply dry carrier gas at a controlled flow rate. The column (7) is connected to the gas supply and is contained within an oven (8); the temperature within the oven is controlled by a thermostat and heating elements. The sample to be separated is introduced into the

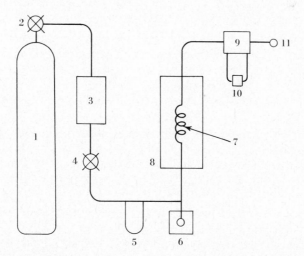

FIGURE 3.6 Schematic diagram of apparatus for gas chromatography. (1) Carrier gas supply. (2) Pressure-reducing valve. (3) Dessicant. (4) Fine-control valve. (5) Flowmeter. (6) Heated injection port. (7) Column. (8) Oven. (9) Detector. (10) Electronic recorder. (11) Exit port.

flow system at the injection port (6), which is individually heated to facilitate vaporization of the sample. The vaporized sample is then swept into the column by the carrier gas. As the sample passes through the column, its components separate into individual "bands" in the carrier gas, which then pass through the detector (9). The detector produces an electronic signal whose voltage is proportional to the amount of material different from the carrier gas itself present in the gas stream. The recorder (10) plots this voltage as a function of time to give the gas chromatogram (see Figure 3.8, for example). The vapors then pass from the detector into either the atmosphere or a collecting system at the exit port (11).

The elapsed time necessary for a given component to pass from the injection port to the detector is known as its *retention time.* Four important experimental factors affect the retention time of a compound: (1) the *nature* of the stationary liquid phase, (2) the *length* of the column, (3) the *temperature* at which the column is maintained, and (4) the *rate of flow* of the inert gas. *The retention time is independent of the presence (or absence) of other components in the mixture.* For any given column and set of conditions (temperature and flow rate), the retention time is a property of the compound in question and may be used to identify it.[6]

The choice of a stationary liquid phase best suited for a glpc experiment depends for the most part on the types of compounds to be separated. Although a very large number of liquid phases are available, only a few (Table 3.1) are widely used. The best choice will be that liquid giving rise to the largest differences in the partition coefficients of the components to be separated. Obviously, then, the solubility principles of organic compounds need to be considered.

The solubility of a gas in a liquid depends to some extent on its vapor pressure. In general, lower boiling, more highly volatile components will move through the column faster and exit sooner than those of lower volatility, because, to a rough approximation, the higher the vapor pressure of a gas, the lower its solubility in a liquid. Factors other than volatility (and thus, indirectly, molecular weight) are important, however, in determining the rate at which a compound will move through the column. The solubility of a substance in a liquid is also influenced by polar interactions between the molecules of solute and solvent, for example, hydrogen bonding and other electrostatic interactions.

An age-old rule of chemistry is "like dissolves like." This would suggest that if it is desired to separate nonpolar types of compounds, a nonpolar liquid substrate should be used. On the other hand, if the components to be separated are of the polar variety, a nonpolar substrate would not be a very good choice, because none of the components would be very soluble and all would tend to pass unhindered through the column with little if any separation. Thus for polar samples it is best to use polar liquid phases.

The most common means of supporting the liquid phase within the column is as a thin film coating on an inert solid support. The support should be of small, evenly meshed granules, so that a large surface area is available. This provides a corre-

[6]For the accurate determinations necessary in many research laboratories, *retention volumes* are used rather than retention times. These measurements are more difficult to perform, but are more reproducible than retention times. They are not a measure of the time necessary for a component to pass through the apparatus, but instead a measure of the *volume* of inert gas needed to carry the component through.

TABLE 3.1 GLPC STATIONARY PHASES

Liquid Phase	Type	Property	Maximum Temperature Limit, °C	Used for Separating
Squalane	Hydrocarbon grease	Nonpolar	100	Hydrocarbons, general application
Apiezon-L	Hydrocarbon grease	Nonpolar	300	Hydrocarbons, general application
Carbowax 20M	Hydrocarbon wax	Polar	250	Alcohols, C_6–C_{18} aldehydes, sulfur compounds
DC-550	Silicone oil	Intermediate polarity	275	C_1–C_5 aldehydes, C_6–up hydrocarbons, halogen compounds
QF-1	Silicone (fluoro)	Intermediate polarity	250	Polyalcohols, alkaloids, halogen compounds, pesticides, steroids
SE-30	Silicone gum rubber	Nonpolar	375	C_5–C_{10} hydrocarbons, pesticides, steroids
Diethyleneglycol succinate (DEGS)	Polyester	Polar	190	Esters, fatty acids
Butanediol succinate	Polyester	Intermediate polarity	225	Esters, fatty acids

spondingly large film area in contact with the vapor phase, which is necessary for efficient separation. Some common types of solid supports are given in Table 3.2. The liquid is coated on the solid support by dissolving the liquid in a suitable low-boiling solvent and mixing this solution with the solid. The low-boiling solvent is then evaporated, leaving the solid granules evenly coated, and the column is filled with these coated granules.

TABLE 3.2 SOLID SUPPORTS

Chromosorb P	Pink diatomaceous earth (surface area: 4–6 m^2/g)
Chromosorb W	White diatomaceous earth (surface area: 1–3.5 m^2/g)
Crushed Firebrick	
Chromosorb T	40/60 mesh Teflon 6

An alternative method of supporting the liquid phase is used in capillary columns of the Golay type. These are very long (300 m is not unusual) and of very small diameter (0.1–0.2 mm) columns. In these columns the liquid is coated directly on the inner walls of the tubing. These types of columns are highly efficient (and relatively expensive).

In general, column efficiency increases with increasing path length and decreases with increasing diameter. As suggested in the introductory discussion, the separation between bands increases with increasing path length. Therefore, with a longer column, the likelihood of separating two components will be increased (their retention time difference will be larger). The diameter of the column will affect the band *width*. A smaller diameter column will give rise to narrower bands and greater efficiency. It is evident that with a small band separation (measured from the band centers), wide bands are more likely to *overlap* (Figure 3.7a) than narrow bands (Figure 3.7b). Therefore a smaller diameter column will effect a more efficient separation, that is, the *resolution* will be greater.

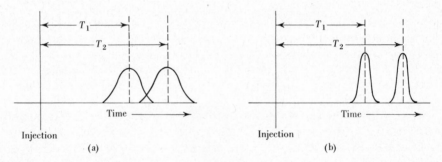

FIGURE 3.7   Effect of band width on resolution.

The temperature at which the column is maintained is another important factor in a glpc experiment. Retention times may be shortened by using higher temperatures, because the solubility of gases in liquids decreases with increasing temperature. The partition coefficients are thus affected, and the bands move through the column at a faster rate. This is sometimes desirable because of the convenience of a faster experiment, but it will also usually mean that the band separation will decrease, resulting in lower resolution.

For each liquid phase there is a maximum temperature limit. This point is determined by the stability and volatility of the liquid being used. At higher temperatures the liquid phase may volatilize to some extent and be carried off the column by the carrier gas, which is of course undesirable. With some liquids there is a minimum temperature, governed by the melting point of the substance or its viscosity. If the liquid should be partially solidified or quite viscous, it will be very inefficient at dissolving the components of the gaseous mixture.

Just as with higher temperatures, higher flow rates of the carrier gas will also cause retention times to decrease. In spite of the decreased resolution obtained at higher temperatures and flow rates, these conditions are sometimes necessary for substances otherwise having very long retention times.

There are two basic types of experiments to which gas chromatography may be applied: (1) qualitative and/or quantitative analysis of the sample and (2) preparative experiments for the purpose of separating and purifying the components of the sample (preparative glpc).

Under a given set of conditions (column, temperature, flow rate, and so forth), the retention time of a given compound is a property of that compound and may be used to identify it. Often, identification may be made by injecting and obtaining the retention time, under the same conditions, of material *known* to be the compound in question. If the retention time of the known compound matches that of a peak of the chromatogram, this is *necessary* evidence that the two compounds may be identical; it is not *sufficient* evidence, however, because more than one compound may exhibit the same retention time. Since one usually has some indication of what compounds are present in the mixture, the identification is reasonably certain. Furthermore, if it is assumed that the voltage output of the detector (and therefore the pen response of the recorder) is proportional to the mole fraction of the material being detected in the vapor, then the relative *areas* under two peaks of the chromatogram will equal the percentage ratio of these two components in the mixture. Although the relative detector response may not necessarily correspond exactly to the relative mole fractions of two components in a mixture, the deviation is usually fairly small. Thus peak-area measurements will provide a reasonably accurate *quantitative* analysis of a mixture.

An example of the analytical use of glpc is shown in Figure 3.8. These sets of peaks represent gas chromatographic analyses of the distillation fractions of a mixture of ethyl acetate and *n*-butyl acetate similar to those obtained in the distillation experiment of Chapter 2. The notations *A*, *B*, and *C* refer to the three fractions taken in that experiment. Comparison of retention times with those of pure ethyl acetate and pure *n*-butyl acetate indicates that the first peak (lower retention time) in each case is ethyl acetate and the other is *n*-butyl acetate. Comparison of the relative areas for the peaks of fractions *A* and *C* using the packed column with the results for fractions *A* and *C* for the unpacked column demonstrates the greater efficiency of the packed fractional distillation column. Clearly the introduction of packing in the Hempel column has increased the number of theoretical plates in the column.

There are a number of methods of obtaining the peak areas necessary for quantitative evaluation of the chromatogram. (1) The detector output may be *electronically integrated* to provide a digital readout of the area automatically. However, this instrumentation is relatively expensive and is not likely to be encountered outside of sophisticated research laboratories. (2) The areas may be measured by means of a *planimeter* (a tracing device that provides a number which is proportional to the area of a region whose boundary has been traced). (3) When peaks are *symmetrical* (such as those shown in Figure 3.8), their areas may be geometrically approximated by considering them as *triangles*. Thus the area of a symmetrical peak is approximately equal to its height times its width at half-height (this is illustrated in Figure 3.9). (4) Since chart paper is of reasonably uniform thickness and density, peaks may be carefully cut out with scissors and their areas assumed

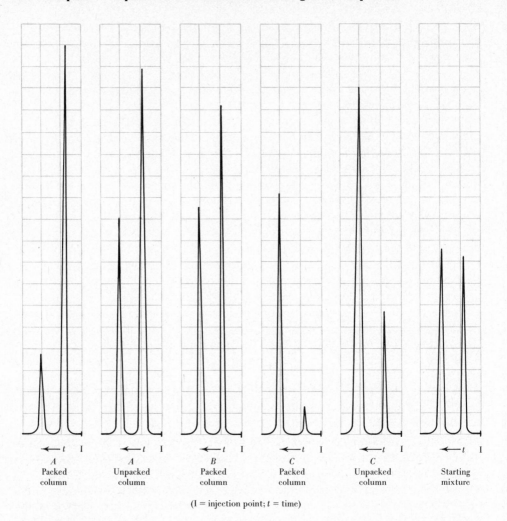

(I = injection point; t = time)

FIGURE 3.8 Gas chromatographic analysis of the distillation fractions from the distillation experiment of Chapter 2.

proportional to their weight, as measured on an analytical balance. Since one usually wishes to save the original chromatogram as a permanent record, it is best to cut the peaks from a photocopy of the chromatogram.

Once the peak areas have been obtained, the percentage of each component in the mixture may be calculated as the area of the peak corresponding to that component, expressed as a percentage of the sum of the areas of all peaks in the chromatogram. Sample calculations of this type are shown in Figure 3.9.

When particularly accurate quantitative analyses are required, one must remain aware that a given detector is not necessarily equally sensitive or responsive to each of the components of a mixture. Thus even though two components may be present in equal amounts in a sample mixture, their respective peaks in the chromatogram may have slightly different areas. As an example, the chromatogram of a 50:50 (wt:wt)

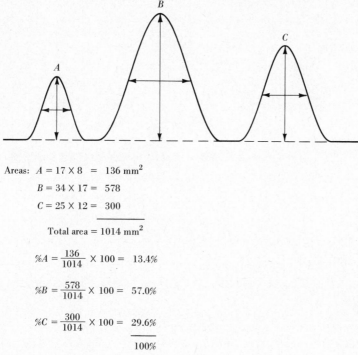

Areas: $A = 17 \times 8 = 136 \text{ mm}^2$

$B = 34 \times 17 = 578$

$C = 25 \times 12 = 300$

Total area $= 1014 \text{ mm}^2$

$\%A = \dfrac{136}{1014} \times 100 = 13.4\%$

$\%B = \dfrac{578}{1014} \times 100 = 57.0\%$

$\%C = \dfrac{300}{1014} \times 100 = 29.6\%$

100%

FIGURE 3.9 Determination of percentage composition of a mixture by gas chromatography.

mixture of toluene and benzene does not show equally sized peaks, but instead shows a peak for toluene which is about 2% smaller than the peak for benzene. (This value is for a thermal conductivity detector; the difference is 1% when a flame ionization detector is used.) To obtain an accurate analysis, the peak area for toluene must be multiplied by 1.02 before calculating the percentages of the components of the mixture.

The relative response factors of many common organic compounds with different detectors are given in various reference sources, such as reference 8 at the end of this chapter. The factors for a few of the materials that students may encounter in this course are provided in Table 3.3. These factors are determined by preparing a standard solution containing known *weights* of the substances whose factors are desired and a known weight of a standard substance (benzene for those compounds given in Table 3.3). The density of this solution is determined so that one can calculate the exact weight of each component, including the standard, which will be injected in, for example, a 1-microliter (μL) sample. The areas of the various peaks are then measured, and each is divided by the exact weight of that component in the injected sample. The inverse of each of the resulting numbers is then normalized to the standard by dividing each by the similarly attained number for the standard.

In order to analyze quantitatively a mixture of substances whose weight factors are known, the peak area for each component of the mixture is *multiplied* by the weight factor for that particular compound, and the resulting *corrected* areas are

TABLE 3.3   WEIGHT (W$_f$) AND MOLAR (M$_f$) CORRECTION FACTORS FOR SOME REPRESENTATIVE SUBSTANCES[a]

Substance	THERMAL CONDUCTIVITY		FLAME IONIZATION	
	W$_f$	M$_f$	W$_f$	M$_f$
Benzene	1.00	1.00	1.00	1.00
Toluene	1.02	0.86	1.01	0.86
Ethylbenzene	1.05	0.77	1.02	0.75
Isopropylbenzene	1.09	0.71	1.03	0.67
Ethyl acetate	1.01	0.895	1.69	1.50
n-Butyl acetate	1.10	0.74	1.48	0.995
Heptane	0.90	0.70	1.10	0.86
o-Xylene	1.08	0.79	1.02	0.75
m-Xylene	1.04	0.765	1.02	0.75
p-Xylene	1.04	0.765	1.02	0.75
Ethanol	0.82	1.39	1.77	3.00
Water	0.71	3.08	—	—

[a] Taken from reference 8 at the end of this chapter.

utilized in calculating the percentage composition of the mixture according to the procedure outlined in Figure 3.9. Note that the use of these factors provides the composition of a *weight percentage* basis. To obtain mole factors, M$_f$, which would provide the composition on a mole percentage basis, the weight factors are divided by the molecular weights of each component of the standard solution and the resulting numbers normalized to the standard.

A sample calculation utilizing these correction factors is provided here in the analysis of a mixture of ethanol, heptane, benzene, and ethyl acetate on a gas chromatograph equipped with a thermal conductivity detector. The last column shows the percentage composition calculated directly from the measured peak areas, without correction for detector response. The dramatic differences in the calculated composition, with and without this correction, as noted in the last two columns, should underscore the necessity of this correction when accurate quantitative results are desired.

Compound	Area (A) (mm^2)	M$_f$	A × M$_f$	mol % $\left(\dfrac{A \times M_f}{194} \times 100\right)$	Uncorrected % $\left(\dfrac{A}{207.1} \times 100\right)$
Ethanol	44.0	1.39	61.16	31.5%	21.2%
Heptane	78.0	0.70	54.6	28.1%	37.7%
Benzene	23.2	1.00	23.2	11.9%	11.2%
Ethyl acetate	61.9	0.895	55.4	28.5%	29.9%
	Total = 207.1		Total = 194	100%	100%

In *preparative glpc,* the purified components of the mixture are obtained by collecting (and condensing) the vapors as they come from the exit port. The recorder indicates when the material is passing through the detector; by making allowance for the time required for the material to pass from the detector to the exit port, one can estimate when to collect the condensed vapor. As different peaks are observed, different collection vessels may be used.

Many instruments are designed specifically for analytical determinations. Relatively small columns are used, and thus only very small samples may be injected (approximately $0.1–5 \, \mu L$). With such sample sizes, it is evident that a very large number of repetitive injections would be needed before any sufficient quantity of the components could be collected in a preparative experiment. Instruments designed specifically for preparative experiments are also available; these utilize much larger columns (some of the larger ones are 30 cm in diameter), in order to handle larger samples. Larger quantities of purified materials can thus be obtained with fewer repetitions of the injection-separation-collection cycle.

EXERCISES

1. Refer to the glpc traces given in Figure 3.8. These are analyses of the various fractions taken in the fractional distillation of the mixture of ethyl acetate and *n*-butyl acetate. Utilizing the glpc correction factors (thermal conductivity detector) provided in Table 3.3, determine both the weight percent and the mole percent compositions of the fractions *A, B,* and *C* obtained using a packed distillation column.
2. Benzene (20 g, 0.25 mol) is subjected to Friedel-Crafts alkylation with 1-chloropropane (19.6 g, 0.25 mol) and $AlCl_3$. The product (24.0 g) is subjected to analysis on a gas chromatograph equipped with a thermal conductivity detector. The chromatogram shows two product peaks identified as *n*-propylbenzene (area = $60 \, mm^2$; $W_f = 1.06$)   and   isopropylbenzene   (area = $108 \, mm^2$; $W_f = 1.09$). Calculate the percent yield of each of the two isomeric products obtained in this reaction. [Since each of the products has the same molecular weight (120), the use of weight factors gives both weight and mole percent composition.]

3.5   Column Chromatography

Column chromatography involves distribution of substances between liquid and solid phases and would therefore be classified as a type of solid-liquid adsorption chromatography. The stationary phase is a solid, which separates the components of a liquid passing through it by selective adsorption on its surface. The types of interactions that cause adsorption are the same as those which cause attractions between any molecules, that is, electrostatic attraction, complexation, hydrogen bonding, van der Waals forces, and so forth.

A column such as that shown in Figure 3.10 is used to separate a mixture by column chromatography. The column is packed with an "active" solid (the stationary phase) such as alumina or silica gel, and a small liquid sample[7] is applied at the top. The sample will become adsorbed initially at the top of the column. An eluting

---

[7]If the mixture is a solid, it must be dissolved in a minimum quantity of solvent and then applied.

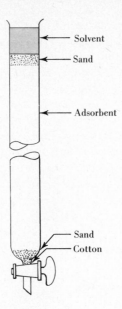

FIGURE 3.10 Chromatography column.

solvent is then allowed to flow through the column; this mobile liquid phase will carry with it the components of the mixture. Owing to the selective adsorption power of the solid phase, however, the components may move down the column at different rates. A more weakly adsorbed compound will be eluted more rapidly than a more strongly adsorbed compound, because the former will have a higher percentage of molecules in the mobile phase. The process is seen to be quite analogous to that operating in glpc. The progressive separation of the components is depicted in Figure 3.11.

The separated components may be recovered in two ways. (1) The solid packing may be extruded and the portion of the solid containing the desired band cut out and extracted with an appropriate solvent. (2) Solvent can be passed through the column until the bands are eluted from the bottom of the column and collected in different containers, since they will be eluted at different times. The second method is the more commonly used; the first is rather difficult because the solid must be extruded from the column in one piece.

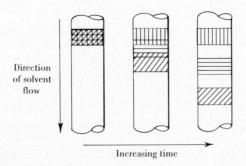

FIGURE 3.11 Development of the chromatogram.

With colored materials the bands may be directly observed as they pass down the column. The word chromatography was originally coined to describe this technique from such observations (Gr. *chromatos,* a color). With colorless materials the changes cannot be observed directly. Many materials fluoresce when irradiated with ultraviolet light, however, and this provides a method of observing the bands in these cases. Usually the progress of a column chromatographic experiment is followed by collecting a series of fractions of eluent of constant volume, for example, 25 mL. The solvent is then evaporated from each to see if any solute is present. If the volume of each fraction is kept relatively small—for example, less than 10% of the volume of the column—the different bands will usually be obtained in different flasks, although each component may be distributed among several flasks. Another convenient method of following the separation is to analyze the eluent at intervals by thin-layer chromatography (Section 3.7).

The composition of the material eluted from the column may be ascertained by determination of its physical constants (Chapter 1) or its spectra (Chapter 4) and comparison of these with those of known materials.

A few of the solid adsorbents commonly used include alumina, silica gel, Florisil, charcoal, magnesia, calcium carbonate, starch, and sugar. The organic chemist usually finds alumina, silica gel, and Florisil (activated magnesium silicate) of the greatest utility.

Alumina (Al_2O_3) is a highly active, strongly adsorbing, polar compound coming in three forms: neutral, and base and acid washed. Basic and acidic alumina offer good separating power for acids and bases, respectively. For compounds sensitive to chemical reaction under acidic or basic conditions, neutral alumina should be used. Being highly polar itself, alumina adsorbs polar compounds quite tenaciously, so that they may be difficult to elute from the column. The activity (adsorptivity) of alumina may be reduced by additions of small amounts of water; the weight percentage of water present determines the activity grade of the alumina. Silica gel and Florisil are also polar, but less so than alumina.

For the greatest effectiveness, the solid adsorbent should be of uniform particle size and of high *specific area,* a property which promotes more rapid equilibrium of the solute between the two phases. This is important for producing narrow bands. Good grades of alumina and silica gel have very high specific areas, of the order of several hundred m^2/g.

The strength of adsorption depends on the adsorbate as well as on the adsorbent. It has been found that the strength of adsorption for compounds having the following types of polar functional groups increases on any given adsorbent in the order noted:

$$\text{Cl}-, \text{Br}-, \text{I}- \; < \; \text{C}{=}\text{C} \; < \; -\text{OCH}_3 \; < \; -\text{CO}_2\text{R} \; < \; \text{C}{=}\text{O} \; < \; -\text{CHO}$$

$$< \; -\text{SH} \; < \; -\text{NH}_2 \; < \; -\text{OH} \; < \; -\text{CO}_2\text{H}$$

The nature of the liquid phases (solvents) to be used is an important consideration when designing a chromatographic experiment. The solvent may also be adsorbed on the solid, thereby competing with the solute for the adsorptive sites on the surface. If the solvent is more polar and more strongly adsorbed than the components of the mixture, these components will remain almost entirely in the mobile liquid phase and little separation will occur during the experiment. For effective separation, then, the eluting solvent must be significantly less polar than the components of the mixture. Furthermore, the components must be soluble in the solvent; if they are not, they will remain permanently adsorbed on the stationary phase of the column. The eluting powers of various solvents, that is, their ability to move a given substance down a column, are generally found to occur in the order shown.

Hexane
Carbon tetrachloride
Toluene   Increasing
Benzene   eluting
Dichloromethane   power
Chloroform
Diethyl ether
Ethyl acetate
Acetone
1-Propanol
Ethanol
Methanol
Water

In a *simple elution* experiment the sample[8] is placed on the column and a *single* solvent is used throughout the separation. The optimum solvent choice will be that which produces the greatest band separation. Since the best solvent will most likely be found only by trial and error, it is sometimes convenient to use the techniques of thin-layer chromatography (Section 3.7) in the selection of a solvent for column chromatography. A series of thin-layer chromatographic experiments using various solvents may be performed in a relatively short time. The best solvent, or mixture of solvents, found in this way will usually be appropriate for the column chromatography.

A procedure known as *stepwise* (or *fractional*) *elution* is most commonly used. In this method a *series* of increasingly more polar solvents is used to develop the chromatogram. Starting with a nonpolar solvent (usually hexane), one band may move down and off the column while the others remain very near the top. Ideally, then, the solvent would be changed to one of slightly greater polarity in the hope that one more band will be eluted while the others remain behind. If too large a jump in polarity is attempted, all the remaining bands may come off at once. Therefore small systematic increases in solvent polarity should be effected at each step. This is best accomplished not by changing solvents entirely but by using *mixed* solvents. For

---

[8]A sample size of approximately 1 g of sample per 25 g of adsorbent is satisfactory.

example, an appropriate quantity[9] of hexane could be passed through the column, followed by a quantity of solvent having the composition 95% hexane—5% chloroform. Solvent mixtures containing 10, 15, 20, 40, and 80% chloroform could then be used in succession, followed by pure chloroform. A similar series utilizing chloroform with diethyl ether or acetone could then be employed. This technique has proved to be quite efficient with regard both to column resolution and to time invested and is very commonly employed in research laboratories.

The method of packing the column is very important because a poorly packed column will have minimal resolution. The packing should be homogeneous and should not have entrapped air or vapor bubbles. The best way to achieve an evenly packed column is described in the experiment below. Column sizes are variable, but the same considerations hold as for glpc: the larger the diameter, the longer the column must be.

EXPERIMENTAL PROCEDURE

A. Preparation of Column

Clamp in a vertical position a 50-mL buret with its stopcock closed but ungreased. Fill the buret to approximately the 40-mL mark with 30–60° petroleum ether, and insert a small plug of glass wool into the bottom of the buret by means of a long piece of glass tubing. Introduce into the buret enough clean sand to form a 1-cm layer on top of the glass wool plug. As the sand settles through the liquid, any entrapped air bubbles will be allowed to escape. *Slowly* add 15 g of alumina to the buret while continually tapping the column with a "tapper" made from a pencil and a one-holed rubber stopper. The agitation of the column while the alumina is being sifted in assures an evenly packed column. With a little additional petroleum ether, wash down the inner walls of the buret to loosen any adhered alumina. In order to protect the packed alumina, introduce a 1-cm layer of sand at the top of the column. Open the stopcock and allow the solvent to drain just to the top of the top layer of sand.[10] The column, as shown in Figure 3.10, is now ready to receive the sample.

B. Separation and Purification of *anti*-Azobenzene

$$N=N$$
$$C_6H_5 \qquad C_6H_5$$
syn-Azobenzene

$$C_6H_5$$
$$N=N$$
$$C_6H_5$$
anti-Azobenzene

[9] A rule of thumb to use in determining the amount of solvent to pass through before changing to one of higher polarity is to use a volume equal to approximately three times the packed volume of the column. There are exceptions, but this is a reasonable suggestion in the absence of more specific knowledge.

[10] At no time during the experiment should the solvent be allowed to drain below this level. Air bubbles and channels would be introduced into the alumina which would give rise to ragged bands during the development of the column.

Obtain 1–2 mL of a half-saturated solution of commercial azobenzene in petroleum ether. Also obtain a small amount of the solid commercial azobenzene and accurately determine its melting range. To a column prepared as described in part A using 15 g of alumina, apply 1.0 mL of the azobenzene solution. The solution may be transferred to the column with a pipet, allowing the solution to run evenly down the inside surface of the buret. Open the stopcock and allow the liquid to drain to the top of the sand. In a like manner introduce 1–2 mL of fresh petroleum ether to wash down the inner surface of the buret. Again allow the liquid to drain to the top of the sand. Fill the buret with fresh petroleum ether. Open the stopcock; as the solvent runs through the column, a broad orange band of the *anti*-isomer of azobenzene should be observed to move slowly down the column. A narrow yellow band of the *syn*-isomer should remain very near the top of the column. Allow petroleum ether to run through the column until the orange band is eluted and collected at the bottom of the column in a small Erlenmeyer flask. Remove most of the solvent by simple distillation and then transfer the final 1–2 mL of the concentrated eluent to a small round-bottomed flask. Attach the flask to an aspirator or house vacuum line in order to remove the last small amount of solvent under evacuation. When the crystals of the purified *anti*-azobenzene are completely dry, accurately determine their melting range.

If desired, after the *anti*-isomer has eluted from the column, the buret may be refilled with methanol and the *syn*-isomer eluted from the column. Typically, however, an insufficient amount of the *syn*-isomer is obtained for adequate characterization.

EXERCISES

1. Why is it preferable to use an ungreased rather than a greased stopcock on a buret used for column chromatography?
2. Why should care be exercised in the preparation of the column to prevent air bubbles from being trapped in the adsorbent?
3. Why does *syn*-azobenzene move down the column faster when methanol is used as an eluent than when petroleum ether is used?

3.6   High-Pressure Liquid Chromatography

In the classical applications of liquid chromatography, *low flow rates* of the mobile liquid phase over the stationary solid phase are necessary for reasonable efficiency in the chromatographic separation. This requirement is directly related to the *slow rates of diffusion* which are characteristic in liquid phases. If flow rates are too high, the desired chromatographic equilibration of the components of the sample between the two phases is neither attained nor maintained, and resolution of the components is lost. Thus the conditions required for efficient separations generally result in rather time-consuming experiments.

One approach to the solution of this problem is to decrease the *distance* through which molecules must diffuse. This may be accomplished by using much smaller particles for column packings. Not only does this dramatically increase the total surface area of the stationary phase, but the particles pack more tightly, resulting in a significant reduction of the interstitial volumes of liquid between the particles. Under these conditions equilibration between phases is established in a much shorter time, allowing much higher flow rates. Concomitantly, however, the much more tightly packed solid leads to a restriction of flow so that inlet pressures at the head of the column must be greatly increased in order to attain enhanced flow rates. In modern high-pressure liquid chromatographs, pulse-free liquid pumping systems are used to provide inlet pressures up to 10,000 psi with small (2–3 mm diameter) columns. The liquid issues from the column at atmospheric pressure, which allows for the employment of a variety of detectors and for the convenient collection of eluent fractions.

Packing materials of uniformly sized particles as small as 10 microns are available for use in high-pressure liquid chromatographic columns. With smaller particle size the solid packs so tightly in the column that it becomes nearly impermeable to liquid flow. These types of packing materials, such as alumina or silica gel, are discussed in Section 3.5. A very popular type of packing material now in vogue consists of glass beads 30–50 μ in diameter coated with a layer (1–2 μ) of porous material. These coated beads are referred to as *pellicular beads*. Pellicular beads either may be used directly as a solid stationary phase, or the porous material may in turn be coated with a very thin layer of liquid stationary phase. Depending on the nature of this liquid phase and on the solvent characteristics of the mobile liquid phase, the liquid coating may be rinsed away during use of the column. This problem may be circumvented by chemically bonding the liquid phase to the surface of the glass support. For example, the surface of porous silica beads may be esterified by reaction with various alcohols. Unless these types of bonds are chemically reactive to components of the eluting liquid, for example, hydrolysis by water, this experimental stratagem works very well.

Because of the small columns and higher flow rates used in high-pressure liquid chromatography, the chromatographic system must be designed to incorporate detectors and collection apparatus of small dead volume. ("Dead volume" refers to those volumes within the flow system and subsequent to the column in which the components of two closely separated bands may remix, either through diffusion or because of the turbulence of flow.) This requirement might be most appreciated by recognizing that from a highly efficient liquid chromatograph column the band of one component of the sample may elute within a total volume of the order of 50 μL! Thus the diameter and length of tubing leading from the column to the detector should be kept as small as conveniently possible, and the cavity of the detector cell should be 5 μL or less (10% of the band volume).

Detection systems are designed with flow-through detection cells in which some property of the solution is monitored continuously as the eluent flows through the cell. This results ultimately in a variable voltage signal which is fed to a recorder providing a chromatogram similar to a gas chromatogram (see Section 3.4). Components are identified by their retention times, as in gas chromatography. Examples of properties of the sample frequently monitored in detection systems are ultraviolet-visible absorptivity, fluorescence, and refractive index.

The improvement in operating time achieved by the new high-pressure systems is significant. Separations that would require many hours in gravity columns may be made in a matter of minutes in the high-pressure columns. There are advantages over other forms of chromatography as well. Efficiency is much higher than in thin-layer chromatography (Section 3.7). Whereas gas chromatography is not practical for high molecular weight, relatively nonvolatile compounds (molecular weight greater than 200), high-pressure column chromatography can be applied to high molecular weight compounds with the same order of efficiency as that of gas chromatography for low molecular weight compounds. Another valuable aspect is the fact that high-pressure liquid chromatography is normally carried out at room temperature, so that there is no danger of decompositions or molecular rearrangements which may be induced in thermally unstable compounds by gas chromatography.

High-pressure liquid chromatography is a relatively new development in the analysis and separation of organic compounds—a development that is becoming extremely valuable, particularly in the field of natural products. It has been credited with making significant contributions to the synthesis of Vitamin B_{12}, a complex, high molecular weight compound, and to the isolation of the sex attractant of the American cockroach.

3.7 Thin-Layer Chromatography

Thin-layer chromatography involves the same principles as column chromatography; it is a form of solid-liquid adsorption chromatography. In this case, however, the solid adsorbent is spread as a thin layer (approximately 250μ) on a plate of glass or rigid plastic. A drop of the solution to be separated is placed near one edge of the plate and the plate is placed in a container (developing chamber) with enough of the eluting solvent to come to a level just below the "spot." The solvent migrates *up* the plate, carrying with it the components of the mixture at different rates. The result may then be a series of spots on the plate, falling on a line perpendicular to the solvent level in the container (see, for example, Figure 3.12b).

This chromatographic technique is very easy and rapid to perform. It lends itself well to the routine analysis of mixture composition, and may also be used to advantage in determining the best eluting solvent for subsequent column chromatography.

The same solid adsorbents used for column chromatography may be employed for tlc; silica gel and alumina are the most widely used. The adsorbent is usually mixed with a small amount of "binder"—for example, plaster of paris, calcium sulfate, or starch—to ensure proper adherence of the adsorbent to the plate. The plates may be prepared before use, or commercially available prelayered plastic sheets may be used.

The relative eluting abilities of the various solvents that may be used are the same as those given in Section 3.5. It should be remembered that the eluting power required of the solvent is directly related to the strength of adsorption of the components of the mixture on the adsorbent.

A distinct advantage of tlc is the very small quantity of sample required. A

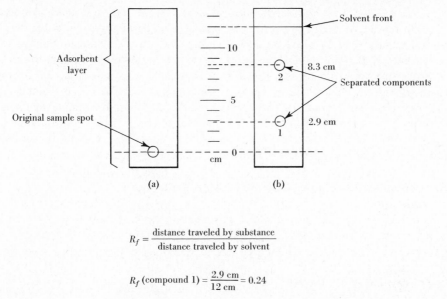

$$R_f = \frac{\text{distance traveled by substance}}{\text{distance traveled by solvent}}$$

$$R_f \text{ (compound 1)} = \frac{2.9 \text{ cm}}{12 \text{ cm}} = 0.24$$

$$R_f \text{ (compound 2)} = \frac{8.3 \text{ cm}}{12 \text{ cm}} = 0.69$$

FIGURE 3.12 Thin-layer chromatogram: (a) original plate; (b) developed chromatogram.

lower limit of detection of 10^{-9} g is possible in some cases. However, sample sizes as large as 0.5 mg may also be used. With the larger samples preparative experiments may be conducted by scraping the individual spots from the plate and eluting (extracting) them with an appropriate solvent. Of course, it would be necessary to repeat this process a number of times in order to obtain even several *milli*grams of material. Such a procedure would offer a method of identification of the various spots, however, because enough could be collected to obtain spectra of the individual components (see Chapter 4).

Detection of spots on the chromatogram is easy for colored materials, and a number of procedures are available for locating spots of colorless materials. For example, irradiation of the plate with ultraviolet light will permit location of the spots of compounds that fluoresce. Alternatively, the solid adsorbent may be impregnated with an otherwise inert, fluorescent substance. Spots of materials that absorb ultraviolet light but do not fluoresce will show up as black spots against the fluorescing background when the plate is irradiated with ultraviolet light. Other detecting agents are more often used. These agents may be sprayed onto the chromatograms, causing the spots to become readily apparent. Examples of detecting agents used in this way are sulfuric acid, which causes many organic compounds to char, and potassium permanganate solution. Iodine is another popular detecting agent. In this case the plate is placed in a vessel whose atmosphere is saturated with iodine vapor. Iodine is adsorbed by many organic compounds, and their spots on the chromatogram become colored (usually brown).

Under a given set of conditions (adsorbent, solvent, layer thickness, and homogeneity) the rate of movement of a compound with respect to the rate of movement of the solvent front, R_f, is a property of that compound. The value is determined by measuring the distance traveled by a substance from a starting line (Figure 3.12). The property has the same significance as retention volume (see footnote 6) in a glpc experiment.

EXPERIMENTAL PROCEDURE

A. Separation of *syn*- and *anti*-Azobenzenes

Obtain from your instructor a 10-cm strip of silica gel chromatogram sheet (without fluorescent indicator) and about 0.5 mL of a 10% toluene solution of commercial azobenzene. Place a spot of this solution on the tlc plate about 1 cm from one edge and about 2 cm from the bottom, using a capillary tube to apply the spot. The spot should be 1 or 2 mm in diameter. Allow the spot to dry and then expose the plate to sunlight for one or two hours. Alternatively, the plate may be placed beneath a sunlamp for about 20 min. After this time apply another spot of the *original* solution on the plate at the same distance from the bottom as the first and leave about 1 cm between the spots. Again, allow the plate to dry.

A wide-mouth bottle with a tightly fitting screw-top cap may be used as a developing chamber. Alternatively, a beaker covered with a watch glass may be used. Prepare a 9:1 (by volume) mixture of hexane:chloroform and place a 1-cm layer of this mixture in the bottom of the developing chamber. Fold a piece of filter paper, as shown in Figure 3.13a, and place it into the developing chamber, as shown in Figure 3.13b. Saturate the chamber with the vapors of the solvent by shaking. This inhibits the evaporation of solvent from the plate during the development of the chromatogram. The piece of filter paper aids in the maintenance of this saturated state.

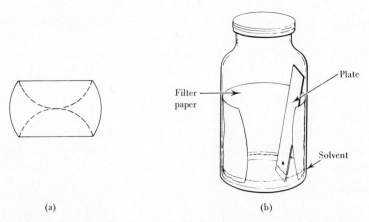

(a)   (b)

FIGURE 3.13 TLC chamber.

Place the chromatogram plate in the chamber, being careful not to splash the solvent onto the plate. The spots *must be above* the solvent level. Allow solvent to climb to within about 1.5 cm of the top of the plate and then remove the plate and allow it to air-dry. Note the number of spots arising from each of the two original spots and compare the intensities of the two spots corresponding to *syn*-azobenzene (the spots nearest the starting point).

B. Separation of Green Leaf Pigments

Place in a mortar several fresh spinach leaves[11] and a few milliliters of a 2:1 mixture of petroleum ether and ethanol and grind the leaves well with a pestle. By means of a pipet, transfer the liquid extract to a small separatory funnel and *swirl* with an equal volume of water; shaking may cause formation of an emulsion. Separate and discard the lower aqueous phase. Repeat the water washing twice, discarding the aqueous phase each time. The water washing serves to remove the ethanol as well as other water-soluble materials which have been extracted from the leaves. Transfer the petroleum ether layer to a small Erlenmeyer flask and add 2 g of *anhydrous* sodium sulfate. After a few minutes decant the solution from the sodium sulfate, and if the solution is not deeply and darkly colored, concentrate it by evaporating part of the petroleum ether, using a gentle stream of air.

Obtain from your instructor a 10-cm strip of silica gel chromatogram sheet (without fluorescent indicator). Better results will be obtained if these sheets have been dried for a few hours in an oven at 110°; this procedure removes moisture that is adsorbed on the silica gel and produces a surface that is more effective at adsorbing and separating the components of the mixture being analyzed. Place a spot of the pigment solution on the sheet about 1.5 cm from one end, using a capillary tube to apply the spot. Avoid allowing the spot to diffuse to a diameter of more than 1–2 mm during application of the sample. Allow the spot to dry, and develop the chromatogram according to the general directions given in the second paragraph of part A, but use chloroform as the developing solvent.

It is sometimes possible to observe as many as eight colored spots. In order of decreasing R_f values, these spots have been identified as the carotenes (two spots, orange), chlorophyll *a* (blue-green), chlorophyll *b* (green), and the xanthophylls (four spots, yellow).

Calculate the R_f values of any spots observed on your developed plate. Also, as an aid in maintenance of a permanent record of the plate, draw to scale a picture of the developed plate in your notebook.

EXERCISES

1. Which of the two isomers of azobenzene would you expect to be the more thermodynamically stable and why?

---

[11] Green leaves from a variety of sources might be used, such as grass and various types of trees and shrubs. The interested student may wish to complete this experiment with more than one type of leaf and to compare the results.

2. From the results of the tlc experiment with the azobenzenes, describe the role of sunlight.
3. In a tlc experiment why must the *spot* not be immersed in the solvent in the developing chamber?
4. Explain why the solvent must not be allowed to evaporate from the plate during the development.

3.8   Dry-Column Chromatography

Experience with the solid-liquid chromatographic methods described in the previous two sections permits the following generalizations. (1) Thin-layer chromatography has the advantage of being a convenient and rapid analytical technique but has the disadvantage of not being readily adaptable to preparative-scale separations. (2) Column chromatography permits large-scale separations but suffers from the disadvantage that it is commonly quite time-consuming, 4 hr or more sometimes being required to elute the column. Dry-column chromatography is a technique that has been developed to take advantage of the speed afforded by thin-layer chromatography and of the scale offered by column chromatography.

In this method the solid adsorbent is poured into a dry glass or nylon column and tamped to achieve uniform packing.[12] The mixture to be separated is adsorbed on a separate portion of adsorbent, and this is added to the top of the column. Solvent is then allowed to descend the column until it has just reached the bottom of the column. At this point, development of the column is terminated, the adsorbent is removed from the column, and the bands of separated components are isolated by extraction of the adsorbent with an appropriate solvent.

In general it is not convenient to use a series of solvents with this type of chromatography; rather, a pure solvent or a single mixture of solvents is preferable. The choice of solvent is greatly facilitated by the observation that the solvent which provides the best separation of the components of mixture on a tlc plate will also give the best separation with a column. Thus a number of pure solvents and mixtures of them can be tested rapidly on the plates to determine the solvent of choice for the dry-column separation.

An experiment utilizing dry-column chromatography is included in Section 18.2.

3.9   Paper Chromatography

Paper chromatography bears a resemblance to thin-layer chromatography (Section 3.7) but is slightly different in principle. Small spots of the mixture to be separated are placed near the bottom of a strip of paper; the end of the paper strip (but not the spot) is placed in solvent, and as the solvent ascends the paper, the components of the mixture are separated into individual spots. Paper chromatogra-

---

[12]As a general rule, about 70 g of adsorbent is required per gram of mixture to be separated.

phy involves distribution of the sample between a polar liquid phase (normally water), which is strongly adsorbed on the cellulose fibers of the paper, and an eluting solvent. It is thus an example of liquid-liquid partition chromatography.

Compounds may be identified by comparison of their R_f values (defined in Section 3.7). R_f values are far more reproducible with paper chromatography than with tlc and thus are more valuable in this application.

The great utility of paper chromatography (and tlc as well) is in the rapid analysis of reaction mixtures rather than in separation of compounds on a useful preparative scale. The minute quantities of sample required to accomplish an analysis and the ease and rapidity with which qualitative analysis of reaction mixtures can be performed make this chromatographic technique a valuable tool for the organic chemist and especially the biochemist.

The application of paper chromatography to the separation and identification of α-amino acids is described in Section 23.2.

EXPERIMENTAL PROCEDURE

A. Analysis of Inks

Cut a 3 × 12-cm strip of filter paper. Apply spots of two different ink solutions (washable black, washable blue, and permanent emerald green are provided) to the paper at a distance approximately 1 cm from the end and 5 mm from each edge. The amount of solution used should be such that spots of only 1–2 mm in diameter are formed. Allow the paper to dry. Place a pencil mark at the edge of the strip at the level of these spots. Thread the paper through a slot in a stiff piece of cardboard, and hang the strip in a vertical configuration within a wide-mouth bottle containing a 1-cm layer of water. The stiff piece of paper should cover the mouth of the bottle, making the bottle as airtight as possible (Figure 3.14). Do not let the strip of paper

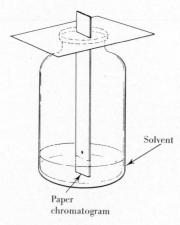

Solvent

Paper
chromatogram

FIGURE 3.14 Paper chromatography chamber.

touch the sides of the bottle; only the bottom of the strip (the end having the spots of ink) is to be in contact with the solvent. The spots should be above the water surface and not in contact with it. Leave the strip in the solvent until the solvent has climbed about 10 cm up the paper. Then remove the strip, place a pencil mark on it at the solvent front, and allow the paper to dry. The developed chromatogram should exhibit several colored spots, each of which corresponds to an organic dye which was a component of one or the other of the two inks. Calculate the R_f value for each dye.

Repeat this experiment with a *mixture* of the two inks. Calculate R_f values for each of the spots obtained. Using the R_f values obtained from the first part of the experiment, identify each of the spots on the chromatogram of the mixture of inks with one or the other of the original ink solutions.

B.   Alternative Experiment

As an alternative to the use of inks in part A, aqueous solutions of various food coloring agents may be used.

EXERCISES

1.   Why should the spot on the paper not be below the level of solvent in the developing chamber?
2.   What might be the consequence of putting a *large* spot of sample on the strip of paper?
3.   Why is it important to have the developing chamber sealed as tightly as possible?

REFERENCES

EXTRACTION

1.   L. C. Craig and D. Craig, in *Technique of Organic Chemistry,* 2d ed., A. Weissberger, editor, Interscience Publishers, New York, 1956, Vol. III, Part 1, Chapter 2.

CHROMATOGRAPHY

2.   R. Stock and C. B. F. Rice, *Chromatographic Methods,* Reinhold Publishing Corp., New York, 1963.
3.   *Chromatography,* 2d ed., E. Heftmann, editor, Reinhold Publishing Corp., New York, 1967.
4.   O. E. Schupp III, "Gas Chromatography," in *Technique of Organic Chemistry,* E. S. Perry and A. Weissberger, editors, Interscience Publishers, New York, 1968, Vol. XIII.
5.   H. G. Cassidy, in *Technique of Organic Chemistry,* A. Weissberger, editor, Interscience Publishers, New York, 1957, Vol. X, Chapter 8.
6.   J. G. Kirchner, "Thin-Layer Chromatography," in *Technique of Organic Chemistry,* E. S. Perry and A. Weissberger, editors, Interscience Publishers, New York, 1967, Vol. XII.
7.   Louis F. Fieser, "Thin-Layer Chromatography," *Chemistry,* **37,** 23 (1964).
8.   H. M. McNair and E. J. Bonelli, *Basic Gas Chromatography,* 3d ed., Varian Aerograph, Walnut Creek, Calif., 1967.

4

SPECTROSCOPIC METHODS OF IDENTIFICATION AND STRUCTURE PROOF

An important part of the science of organic chemistry is determination of the exact structures of products formed in chemical reactions. Formerly this was a very time-consuming and often impossible task for the classical organic chemist, because the primary method of determining the structures of compounds usually involved conversion by chemical reaction of the "unknown" substance to a compound having a "known" structure. The procedure is now known to have often led to erroneous conclusions concerning the identity of the unknown, because the reactions employed to convert the unknown to a known compound produced structural rearrangements which went unrecognized by the investigator. The modern organic chemist fortunately has available various spectroscopic techniques that supplement and even replace the more classical chemical methods of proof of structure and which permit much more rapid analysis of unknowns. It is noteworthy that the development of spectroscopy as a tool for the elucidation of structures of organic compounds has, more than anything else, contributed to the extremely rapid growth of organic chemistry in the past 30 years. Consequently it is vitally important for the student to learn how to analyze and to interpret the information available from infrared (ir), nuclear magnetic resonance (nmr), ultraviolet (uv) and visible spectroscopy, and mass spectrometry (ms). All but the last of these techniques are discussed in this chapter, with most emphasis being given to the first two, owing to their more ready availability and generally greater utility to the undergraduate organic chemistry course. Additional information about each technique is available in the references cited at the end of this chapter. The interested student can also profit by reading one or more of the references on mass spectrometry also cited at the end of the chapter.

All the various spectral methods to be discussed in this chapter depend upon molecular absorption of radiation to produce a specific type of molecular excitation. The type of excitation that is induced depends upon the energy of the radiation incident on the absorbing molecule. For example, ultraviolet and visible spectroscopy normally involve excitation of electrons from an orbital of the ground electronic state to an orbital of the next higher electronic state of the molecule, for example, the transitions associated with the energies E_1, E_2, and E_3 of Figure 4.1. Infrared spectroscopy, in contrast, depends upon transitions among different molecular vibrational-rotational levels with the *same* electronic state (usually the ground state), examples being the processes of energies E_4 and E_5. Finally, nmr spectroscopy involves realignment of the spins of atomic nuclei in a magnetic field as a result of absorption of energy.

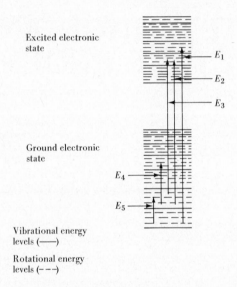

FIGURE 4.1 Rotational, vibrational, and electronic energy levels and transitions between them.

As indicated in Figure 4.1, electronic transitions are of higher energy than those involving vibrational-rotational changes, and consequently require light of greater energy in order to occur. The energy (in kcal/mol) associated with radiation of a particular frequency (v) or wavelength (λ) is readily determined by application of equations 1a and 1b, respectively, in which N is Avogadro's number, h is Planck's constant (6.6256×10^{-27} erg $\cdot$ sec), c is the velocity of light (cm/sec), and v and λ would be measured, respectively, in Hertz (1.0 Hz = 1.0 cycle/sec) and nanometers (nm = 10^{-9} m).

$$E = Nhv \tag{1a}$$

$$E = Nhc/\lambda \tag{1b}$$

In summary, the various types of spectroscopy involve some sort of molecular excitation initiated by absorption of radiation; it is fundamentally the corresponding energy required to stimulate the transition from one molecular state to another that differentiates the methods. Table 4.1 presents the wavelengths of the various spectroscopic techniques of interest, the energies associated with them, and the basic molecular phenomena that are being induced. Reference to this chart as you learn more about the various methods should aid you in understanding the relationships among them.

TABLE 4.1   RELATIONSHIP BETWEEN WAVELENGTH, ENERGY, AND MOLECULAR PHENOMENON

Wavelength (nm)	Energy (kcal/mol)	Type of Spectroscopy	Molecular Phenomenon
200	143 ⎫		Excitation of valence electrons
	⎬	Ultraviolet	from a filled to an unfilled
320	89 ⎭		orbital, for example, $\pi \rightarrow \pi^*$
400	71.5 ⎫		Same as above, except that
	⎬	Visible	absorption of light occurs in a
750	38.1 ⎭		region that is visible (colored)
			to the human eye
2000	14 ⎫		Stretching and bending
16000	2 ⎭	Infrared	of interatomic bonds
5×10^7	5.7×10^{-4} ⎫		Realignment of spins of
	⎬	NMR	atomic nuclei in an
2.1×10^{13}	1.3×10^{-9} ⎭		applied magnetic field

4.1   Infrared Spectroscopy

The spectra most generally available for aid in elucidation of the structure of an organic compound are those obtained by irradiation of the compound with light from the infrared (ir) region, 5000 to 500 cm^{-1} (reciprocal centimeters or wave numbers),[1] of the electromagnetic spectrum. The interaction of ir light with an organic molecule can be explained qualitatively by imagining that molecular bonds between atoms are analogous to springs. These molecular springs are constantly undergoing stretching and bending motions (Figure 4.2) at frequencies which depend upon the masses of the atoms involved and upon the type of chemical bond joining the atoms. Since the frequencies of the various vibrations of the molecule correspond to those of ir radiation, absorption of the radiation occurs, producing an increase in the *amplitude* of the molecular vibrational modes. No irreversible change in the molecule results, however, because the energy gained by the molecule in the form of light is soon lost in the form of heat. By plotting the percent transmittance, defined as the ratio of the intensity of the light passing through the sample, *I,* to the intensity of the light

---

[1]To convert frequency, ν, of radiation from dimensions of Hertz to cm^{-1}, ν is divided by c, the velocity of light, expressed in cm/sec. The wavelength, λ, in microns, μ, is obtained by dividing 10,000 by ν (in cm^{-1}). The position of an absorption band can thus be expressed in either microns or wave numbers; the latter is preferred and is used in this textbook.

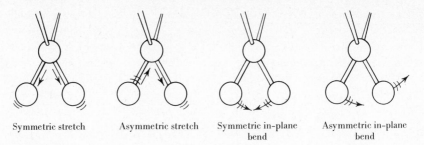

Symmetric stretch   Asymmetric stretch   Symmetric in-plane   Asymmetric in-plane
                                                                bend   bend

FIGURE 4.2   Some vibrational modes of a group of atoms.

striking the sample, I_0, multiplied by 100 (equation 2), versus the frequency of the radiation, an ir spectrum such as that shown in Figure 4.3 is obtained.

$$\%T = \frac{I}{I_0} \times 100 \tag{2}$$

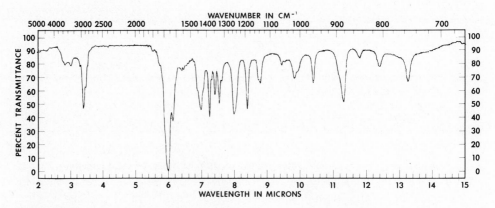

FIGURE 4.3   IR spectrum of 3-methyl-2-cyclohexenone.

As a general rule, absorptions between 5000 and 1250 cm^{-1} involve vibrational excitations of particular functional groups, such as carbonyl, cyano, and nitro, present in the molecule. Thus the stretching mode of the carbon-oxygen double bond of a ketone usually appears in the region 1680–1760 cm^{-1} (see Figure 4.3). Analysis of the portion of the ir spectrum from 5000–1250 cm^{-1} can therefore provide much information concerning the presence or absence of a number of functional groups. A tabulation of the normal frequency range of absorption for a number of groups is presented in Table 4.2. A more complete listing of infrared absorption bands is contained in Appendix 4.

The absorption maxima observed between 1250 and 500 cm^{-1} in the ir spectrum cannot usually be associated with vibrational excitation of a particular func-

TABLE 4.2   INFRARED ABSORPTION RANGES OF FUNCTIONAL GROUPS

Bond	Type of Compound	Frequency Range, cm^{-1}	Intensity
C—H	Alkane	2850–2970	Strong
		1340–1470	Strong
C—H	Alkenes 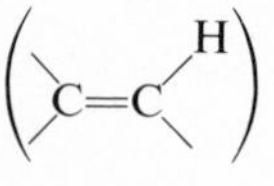 $\left(\begin{array}{c} \diagup \\ C = C \diagdown^{H} \end{array} \right)$	3010–3095	Medium
		675–995	Strong
C—H	Alkynes (—C≡C—H)	3300	Strong
C—H	Aromatic rings	3010–3100	Medium
		690–900	Strong
O—H	Monomeric alcohols, phenols	3590–3650	Variable
	Hydrogen-bonded alcohols, phenols	3200–3600	Variable, sometimes broad
	Monomeric carboxylic acids	3500–3650	Medium
	Hydrogen-bonded carboxylic acids	2500–2700	Broad
N—H	Amines, amides	3300–3500	Medium
C=C	Alkenes	1610–1680	Variable
C=C	Aromatic rings	1500–1600	Variable
C≡C	Alkynes	2100–2260	Variable
C—N	Amines, amides	1180–1360	Strong
C≡N	Nitriles	2210–2280	Strong
C—O	Alcohols, ethers, carboxylic acids, esters	1050–1300	Strong
C=O	Aldehydes, ketones, carboxylic acids, esters	1690–1760	Strong
NO_2	Nitro compounds	1500–1570	Strong
		1300–1370	Strong

tional group, but rather are the result of a complex vibrational-rotational excitation of the *entire* molecule. Accordingly the spectrum from 1250–500 cm^{-1} is characteristic and *unique* for every compound, and the apt description "fingerprint region" is often applied to this portion of the spectrum. It is considered highly unlikely that two different organic compounds would exhibit identical spectra in the fingerprint region, even though their spectra might be quite similar from 5000–1250 cm^{-1} (see Figure 4.4).

The apparently unique character of the ir spectrum of each different organic compound has led to the acceptance of the hypothesis that if two spectra are completely identical as to positions and intensities of absorption maxima—that is, if the spectra are completely superimposable—the spectra must be of the same compound. Comparison of spectra of unknown compounds with those of known compounds can therefore be one very useful technique for structure elucidation.

When the student is confronted with the ir spectrum of a compound for which identification is being attempted, he or she should first try to determine what functional groups are present in the compound by careful analysis of the region, 5000–1250 cm^{-1}. Subsequently the student may be able to assign a few of the bands

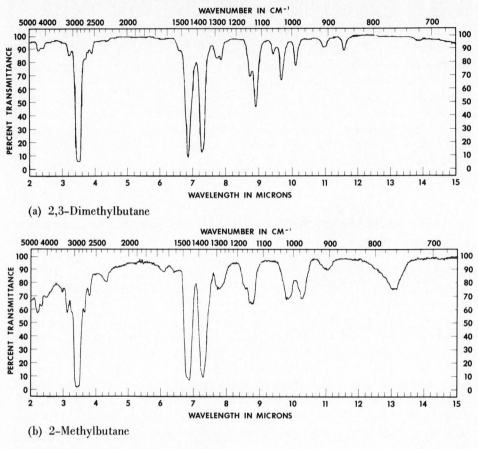

(a) 2,3-Dimethylbutane

(b) 2-Methylbutane

FIGURE 4.4   IR spectra of methylbutanes.

in the fingerprint region of the spectrum in order to define further the structure of the unknown substance. At this point it should be possible to formulate several potential structures for the unknown and to differentiate among the several possibilities by comparison of their published spectra with that of the unknown. Several catalogs of ir spectra are currently available, and some of the more useful and comprehensive are listed in the references at the end of this chapter.

A majority of the possibilities may also be eliminated by a comparison of the physical properties of the unknown, for example, mp, bp, and index of refraction, with those of the unknown, or by their nuclear magnetic resonance or ultraviolet spectra (see Sections 4.2 and 4.3, respectively). If, however, reference spectra of some of the substances considered as possibilities are not available, the information derived from the analysis of the ir spectrum of the unknown may be utilized to suggest a specific chemical reaction that would provide a solid derivative, the melting point of which is published in the chemical literature. The identity of the derivative of the unknown with that of the known would then depend on a comparison of their melting points, or, ideally, a melting point of a mixture of the derivative of the unknown and an authentic sample of the same derivative obtained from the known compound.

Experimentally, ir spectra can be obtained of samples contained in the gas, liquid (either of a pure, "neat," substance or of a solution), or solid phase. Only the general techniques necessary for the preparation for ir spectra of solid and liquid samples will be considered here, since the apparatus required for trapping gaseous samples and transferring them to an ir cell is somewhat too specialized for use in an introductory organic chemistry course. The student should see reference 2 for a thorough discussion of various ways of preparing samples for infrared analysis.

A brief discussion of the factors which determine the intensity of absorption in an ir spectrum is appropriate here. The amount of absorption, termed absorbance, A, at a given frequency is defined by equation 3, in which I and I_0 are, respectively, the

$$A = \log_{10} \frac{I_0}{I} = kcl \tag{3}$$

amounts of light of the given frequency transmitted by and impinging on the sample, k is the absorptivity of the sample at that frequency, l is the path length, in cm, of the cell in which the sample is contained, and c is the concentration, in g/L, of the solute in the solution. Although the absorptivity is a constant, its magnitude is frequency dependent, which results in the observation of maxima and minima in the spectrum. The absorptivity is also somewhat characteristic of the functional group absorbing the radiation. Thus the absorptivity of the stretching mode of a carbonyl group is generally greater than that of a carbon-carbon double bond, as can be seen by comparison of the intensities of the corresponding bands in the spectrum reproduced in Figure 4.3 (the bands at 1630 and 1690 cm^{-1}).

Sample Preparation. The absorptivity of a given molecule is not subject to experimental modification, so only two variables can be changed in order to control the intensity of the ir spectrum of a liquid sample: the path length, l, of the cell containing the sample, and the concentration, c, of the solute in a solution. For spectra of *neat liquids*, of course, only the former alternative is available, and it is found that path lengths of 0.025–0.030 mm are appropriate. As a general rule, *solutions* having 5–10 weight percent of solute and contained in a 0.1-mm cell yield suitable spectra. *Solid samples*, however, need not be dissolved in a solvent in order for an ir spectrum to be obtained. Two of the most common alternative techniques used for preparing solids for examination are the *mull* method and the *potassium bromide (KBr) pellet* method.

A mull is simply a dispersion of finely divided solid in a carrier fluid, commonly mineral oil (Nujol) or perfluorokerosene. Each of these liquids has a reasonably high viscosity, which aids in the maintenance of the dispersion, and each has relatively few absorption bands in its ir spectrum, as shown in Figure 4.5. A mull is prepared by first thoroughly grinding the solid between glass plates or with the aid of a mortar and pestle made of agate, then adding a few drops of the suspending fluid and continuing the grinding until a smooth paste is formed. This paste is placed on a transparent plate (see the discussion on windows near the end of this section), a second plate is positioned above the first, and the resulting "sandwich" is transferred to a cell holder (Figure 4.6).

Failure to grind the sample finely enough will result in production of a poor quality ir spectrum. Naturally, even good quality spectra will contain the absorptions

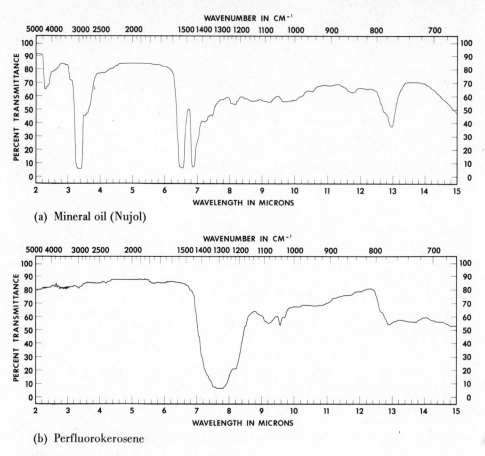

FIGURE 4.5   IR spectra of mulling fluids.

of the mulling fluid (see Figure 4.7, for example) in addition to the absorptions of the compound of which the mull was prepared. However, note that in Figure 4.5 the regions of strong absorption by mineral oil, 2800–3000 cm^{-1} and 1300–1500 cm^{-1}, correspond to locations where perfluorokerosene does *not* absorb, so that essentially all portions of the ir spectrum of a solid can be observed if two spectra, one of each type of mull, are obtained.

A potassium bromide (KBr) pellet is a transparent disk prepared by subjecting an intimate mixture of dry powdered KBr and the solid sample to high pressure (8–20 tons per square inch), generally under vacuum to minimize the presence of atmospheric moisture. The heat generated under these conditions of high pressure causes the mixture to fuse; the resulting disk can then be placed in a holder that may be attached to the spectrophotometer. The intimate mixture of KBr and the sample is prepared by thorough grinding with a mortar and pestle made of agate.

It is important to realize that KBr is extremely hygroscopic, and it is ordinarily impossible to prepare a disk which is free of water. Consequently, spectra of solids

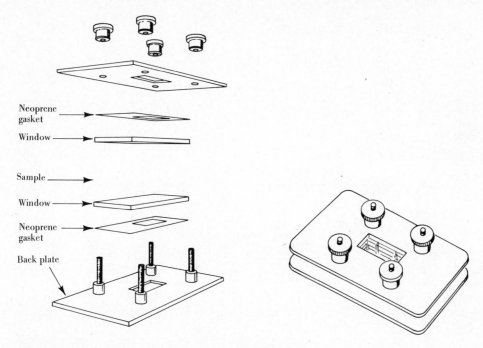

Neoprene gasket

Window

Sample

Window

Neoprene gasket

Back plate

FIGURE 4.6 Components of a demountable IR cell.

taken as KBr pellets almost invariably show absorption bands for water, and these bands may partially or completely obscure N—H and O—H stretching bands, rendering this region of the spectrum difficult to interpret. The presence of bands due to water is graphically illustrated by comparing Figures 4.8a and b.

Other factors to be considered with respect to spectra of solids contained as KBr disks is that heat-sensitive compounds may decompose and that certain compounds may react with the KBr during the fusion process.

Neither the mull nor the KBr disk technique is applicable to quantitative work, because it is not possible to prepare the mull or the disk reproducibly for the same sample. Nevertheless, these techniques are extremely valuable for qualitative purposes.

The windows of most types of ir cells are composed of a clear fused salt such as sodium chloride or potassium bromide, since these materials are transparent to ir radiation. As a result, contact of the cells with moisture must be scrupulously avoided or the windows will soon become cloudy and will transmit very little light. Thus avoid breathing directly onto the windows, touching their faces with your fingers, or using samples which contain moisture. Furthermore, the windows and cells should be cleaned with *dry* solvent, such as carbon tetrachloride or chloroform. (It is unwise to use any hydroxylic solvents such as alcohols to clean cells, since these normally contain some water.) The cells are then dried under a slow stream of dry nitrogen and stored in a desiccator.

Infrared spectroscopy provides valuable information concerning the functional

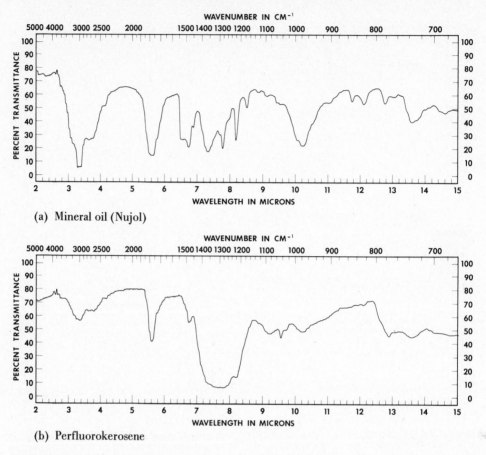

FIGURE 4.7   IR spectra of cyclobutane-1,1-dicarboxylic acid as mulls in mineral oil and perfluorokerosene.

groups present in molecules and serves as a means of identifying an unknown compound if a reference spectrum of the compound is available.[2] In those cases, and there are many, in which the ir spectrum alone does not provide sufficient information to permit determination of structure, other chemical or spectroscopic techniques must be employed.

EXERCISE

Assign as many bands as possible in the spectra presented in Figures 4.3, 4.4, 4.7, and 4.8 to the functional groups responsible for the absorptions.

[2] In addition to its use as a tool for *qualitative analysis* of organic compounds, ir spectroscopy is also utilized for *quantitative analysis* of mixtures of known substances. A discussion of this latter technique can be found in reference 2 at the end of this chapter.

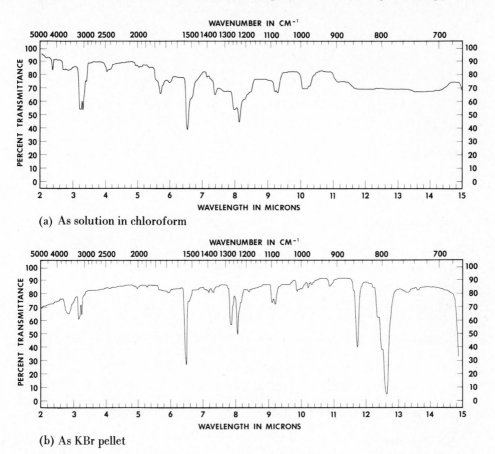

(a) As solution in chloroform

(b) As KBr pellet

FIGURE 4.8 IR spectra of triptycene in chloroform solution and as KBr pellet.

4.2 Nuclear Magnetic Resonance Spectroscopy

Although ir spectroscopy has been used profitably by organic chemists since the early 1950s, nuclear magnetic resonance (nmr) spectroscopy became generally available only in the early 1960s. This extremely important experimental technique is based on the interesting property of *nuclear spin* exhibited by various nuclei, some of which are commonly found in organic compounds. These include hydrogen (^{1}H), fluorine (^{19}F), phosphorus (^{31}P), and the minor isotope of carbon, ^{13}C. Our development of the theory and interpretation of this type of spectroscopy will focus on *proton*[3] magnetic resonance (pmr) spectra.

The hydrogen nucleus behaves as if it were a spherical body having a uniform

[3] During the presentation of this discussion it will be convenient to refer occasionally to the hydrogen nucleus simply as a hydrogen or as a proton. These latter designations are not strictly correct, since we are referring to the nucleus of a covalently bound hydrogen and not to the free atom or the charged ion, H$^+$.

distribution of charge and a mechanical spin about an axis. The spin of the nucleus produces circulation of charge which in turn generates a nuclear magnetic dipole, the direction or moment of which is *colinear* with the axis of nuclear spin (Figure 4.9a).

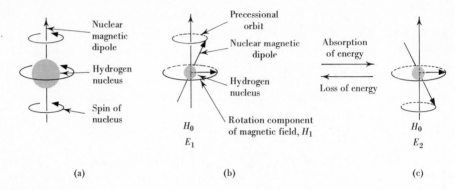

(a)   (b)   (c)

FIGURE 4.9   Spin properties of the hydrogen nucleus.

    When the spinning nucleus is placed in an external magnetic field, H_0, the magnetic dipole, and hence the axis of spin, becomes oriented either *with* the applied field, state E_1 (Figure 4.9b), or *against* it, state E_2. The energy state E_1 represents the more stable alignment. It is very important to realize that the magnetic dipole does not become aligned either precisely parallel or antiparallel to H_0, but rather traces out a circular path about the axis defined by H_0, in a process called *precession*. The phenomenon of nuclear precession can be likened to the motion observed for a spinning gyroscope or toy top that has not been oriented exactly parallel to the earth's gravitational field.

    The frequency, ω_0, of the precession is linearly related to the strength of the applied field, H_0, as is expressed in equation 4, in which the constant, γ_H, is the

$$\omega_0 = \gamma_H H_0 \tag{4}$$

magnetogyric ratio for the hydrogen nucleus. Thus ω_0 increases as H_0 increases. Now a transition between the energy states E_1 and E_2 can be accomplished if the nucleus, precessing at a frequency, ω_0, in a given field, H_0, is subjected to a second, oscillating magnetic field, H_1, which has a rotation component perpendicular to H_0 (Figure 4.9b). The frequency of rotation of the magnetic component can be varied by changing the frequency, ν, of the oscillator used to produce the oscillating magnetic field, H_1.

    When the frequency of rotation becomes equal to the precessional frequency, ω_0, *resonance* between the precessing nucleus and H_1 will be achieved. At this point the nucleus in state E_1 can absorb the electromagnetic energy provided by the oscillator and undergo a "flip" of its spin into the higher energy state, E_2. The absorption of energy can be detected electronically and recorded as a peak on a chart.

The oscillator frequency required to produce the desired nuclear magnetic transition depends upon the strength of the applied field, H_0, as is expressed in equation 5. It turns out that if H_0 is of an experimentally convenient magnitude,

$$\nu = \frac{\gamma_H H_0}{2\pi} \tag{5}$$

about 14,000 gauss, ν must be of the order of 60 megaHertz (MHz),[4] in order to produce resonance. This value is in the radio-frequency (rf) region of the electromagnetic spectrum, so an rf oscillator is required.

Inspection of equation 5 should make it apparent that the condition of resonance for a proton could be achieved either by holding H_0 constant and varying ν, or by maintaining ν at a constant value and changing H_0. The latter approach is more practical and is employed in all but the most technically advanced, research level spectrometers.

A typical pmr spectrum, that of 1-nitropropane, is presented in Figure 4.10. In the spectrum, the strength of the applied magnetic field, H_0, *increases* as one goes from left to right, the *upfield* direction. Since the frequency of the radio oscillator is held constant, the *energy* required to produce the transition between the spin states, E_1 and E_2, also *increases* from left to right.[5] The three groups of peaks, A, B, and C, all result from absorption of energy by hydrogens, but the amount of energy required to realign the nuclei in the magnetic field—that is, the energy necessary to produce

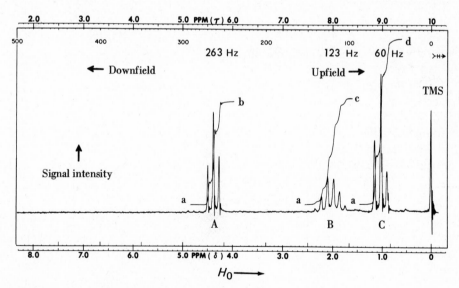

FIGURE 4.10 PMR spectrum of 1-nitropropane.

[4] Hertz (Hz) is a unit synonymous with cycles per second (cps). The term megaHertz (MHz) therefore is equivalent to 10^6 cps.

[5] Both ν and H_0 are linearly related to energy through the expression $E = h\gamma_H H_0/2\pi$.

resonance—is clearly not identical for all the protons in the molecule. This is surprising, because equation 5 indicates that all hydrogen nuclei should undergo spin flip at the same H_0 (at a given value of ν) so that only a *single* peak would be expected.

The qualitative explanation for the observation of more than one peak is as follows: The three groups, A, B, and C, represent three types of hydrogens which are rendered nonequivalent by virtue of the slightly different magnetic environments in which they exist. The magnetic environment of each proton in the molecule is dependent upon two major factors: (1) the externally applied field, H_0, which has a uniform strength over the entire molecule and therefore cannot produce nonequivalence of the different hydrogens; (2) a magnetic field which is induced by circulation of the electrons of the molecule and which, because the electron density varies over the molecule, is *not* uniform. This latter field results in the magnetic nonequivalence of the protons. Those hydrogen nuclei which experience higher fluxes of the induced magnetic field of the molecule will—because this field will act to oppose the external, applied field—be more highly shielded from the external field. In other words, more highly shielded nuclei will experience an *effective* field, H_e, which is *less* than the applied field, so that *higher* applied fields will be necessary to produce the desired nuclear spin transitions. The appearance of more than one group of peaks in the spectrum therefore reflects the difference in the flux of the magnetic field induced within the molecule by circulation of electrons.

The positions of peaks in the pmr spectrum of a compound are routinely measured relative to the position of the peak produced by the protons of a standard compound, tetramethylsilane (TMS), $(CH_3)_4Si$. In reference to the spectrum of Figure 4.10 the groups of peaks are found to be centered at 60, 123, and 263 Hz *downfield* from the TMS peak when an rf oscillator set at 60 MHz is employed. These values represent the *chemical shifts* relative to TMS of the three types of hydrogen nuclei in 1-nitropropane.

The chemical shift of a hydrogen determined in units of Hz is *dependent* on the frequency at which the oscillator is operating. It has been found useful to convert such shifts into units *independent* of the oscillator frequency. This is accomplished by dividing the chemical shift, expressed in Hz, by the frequency, in Hz, of the oscillator and then, to arrive at more convenient numbers, multiplying the result by 10^6 (equation 6). Chemical shifts obtained in this manner are values on a standard scale,

$$\delta = \frac{\text{chemical shift in Hz}}{\text{oscillator frequency in Hz}} \times 10^6 \tag{6}$$

termed the delta (δ) scale, which has the peak for TMS at 0.0δ, peaks downfield from TMS at values of $+\delta$, and peaks upfield from TMS at values of $-\delta$.[6] The oscillator frequency-independent chemical shifts, δ, of the groups of peaks observed for 1-nitropropane are as follows:

Group A: $(60\ \text{Hz})/(60 \times 10^6\ \text{Hz}) = 1.00 \times 10^{-6}$ or 1 part per million (ppm)
$\qquad (1.00 \times 10^{-6})(10^6) = 1.00\ \delta_A$
Group B: $[(123)/(60 \times 10^6)](10^6) = 2.05\ \delta_B$
Group C: $[(263)/(60 \times 10^6)](10^6) = 4.38\ \delta_C$

[6] Actually the sign convention should be exactly opposite to that given, so that peaks downfield from TMS would be at values of $-\delta$. However, common practice corresponds to the convention described above.

A second standard scale, values of which are also independent of oscillator frequency, has been derived in which the TMS peak is defined as appearing at 10 tau (τ).[7] To convert a chemical shift given on the delta scale to a value on the tau scale, one simply subtracts the shift as measured on the delta scale from 10. The groups of peaks in the spectrum of 1-nitropropane, for example, appear at $10 - 1.0 = 9.0\,\tau$, $10 - 2.05 = 7.95\,\tau$, and $10 - 4.38 = 5.62\,\tau$. Because both these standard scales are currently in use, it is important to know how they are derived and how they can be interconverted. The delta scale is the more widely used of the two, however.

The magnitude of the chemical shift of a proton attached to a given type of functional group is relatively constant so that determination of the shifts of the various protons of an unknown compound will often provide valuable information concerning the functional groups present in the compound. Tables are available which correlate chemical shifts with structural features; a partial compilation is presented in Table 4.3. Appendix 3 contains a more extensive listing of the observed chemical shifts for a large variety of specific types of hydrogens.

TABLE 4.3   TYPICAL CHEMICAL SHIFTS OF HYDROGENS ATTACHED TO VARIOUS TYPES OF FUNCTIONAL GROUPS

Type of Hydrogen	Chemical Shift, τ	ppm δ	Type of Hydrogen	Chemical Shift, τ	ppm δ
Cyclopropane	9.6–10.0	0.0–0.4	ICH	6–8	2–4
RCH_3	9.1	0.9	OCH (alcohols,		
R_2CH_2	8.7	1.3	ethers)	6–6.7	3.3–4
			O—CH	4.7	5.3
			$\quad$ |		
R_3CH	8.5	1.5	$\quad$ O		
C=CH	4.1–5.4	4.6–5.9	OCH (esters)	5.9–6.3	3.7–4.1
CCH	7–8	2–3	RO_2CCH	7.4–8	2–2.6
ArH	1.5–4	6–8.5	$\quad$ O		
ArCH	7–7.8	2.2–3	$\quad$ ‖		
C=CCH$_3$	8.3	1.7	RCCH	7.3–8	2–2.7
C≡CCH$_3$	8.2	1.8	$\quad$ O		
			$\quad$ ‖		
FCH	5–5.6	4–4.5	RCH	0–1	9–10
ClCH	6–7	3–4	ROH	4.5–9	1–5.5
Cl_2CH	4.2	5.8	ArOH	−2 to 6	4–12
BrCH	6–7.5	2.5–4	RCOOH	−2 to −0.5	10.5–12
O_2NCH	5.4–5.8	4.2–4.6	RNH_2	5–9	1–5

Note in the tabulations the relatively narrow range, 10 ppm (600 Hz), over which occur the chemical shifts of hydrogen when involved in the types of bonds commonly found in organic compounds. The apparently low sensitivity of the chemical shift of hydrogen as a function of its magnetic environment often leads to undesirable complexities in interpretation of spectra, owing to overlap of peaks, for example.

[7] The standard value of 10 is chosen because the protons of most organic compounds absorb within 600 Hz downfield of TMS when a 60-MHz oscillator is used.

The peaks within each group of absorptions in the spectrum of 1-nitropropane are the result of *spin-spin splitting*, a phenomenon which results because there can be magnetic interaction, termed *coupling*, between magnetically nonequivalent hydrogen nuclei in the molecule. As a rule, only nonequivalent nuclei on the same carbon or on adjacent carbons will interact with one another to produce spin-spin splitting. Thus if we label the three types of hydrogens as shown in Figure 4.11, we would expect the nuclei designated H_a to couple with those labeled H_b and these latter protons to interact with both the nuclei H_a *and* H_c. However, the H_c protons should couple only with the H_b protons.

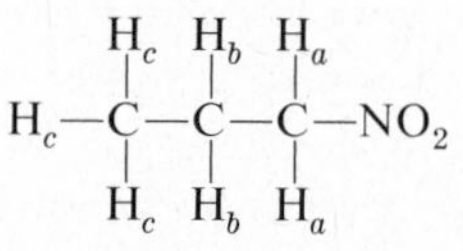

$$H_c-\overset{\displaystyle \overset{H_c}{|}}{\underset{\displaystyle \underset{H_c}{|}}{C}}-\overset{\displaystyle \overset{H_b}{|}}{\underset{\displaystyle \underset{H_b}{|}}{C}}-\overset{\displaystyle \overset{H_a}{|}}{\underset{\displaystyle \underset{H_a}{|}}{C}}-NO_2$$

FIGURE 4.11   Types of hydrogens in 1-nitropropane.

The magnitude of the spin-spin splitting between hydrogen nuclei can be measured in Hz, and the values obtained are called *coupling constants, J.* In contrast to chemical shifts, the magnitude of coupling constants is *independent* of oscillator frequency, so units of Hz (or cps, see footnote 4) are used with such values. Some typical hydrogen-hydrogen coupling constants, J_{HH}, are given in Figure 4.12.

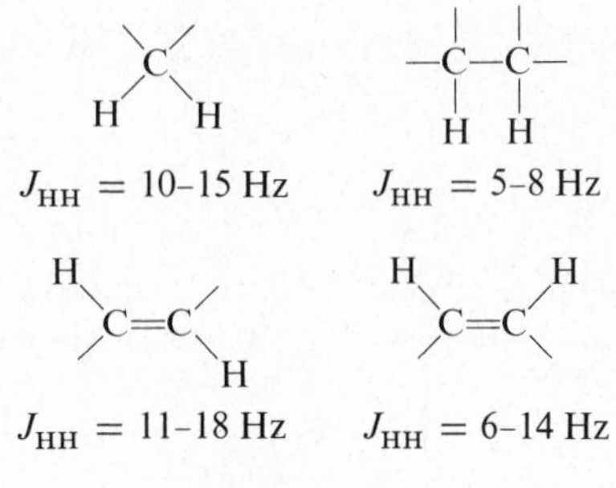

$$J_{HH} = 10\text{--}15 \text{ Hz} \qquad J_{HH} = 5\text{--}8 \text{ Hz}$$

$$J_{HH} = 11\text{--}18 \text{ Hz} \qquad J_{HH} = 6\text{--}14 \text{ Hz}$$

FIGURE 4.12   Some typical coupling constants.

The number of peaks into which the pmr absorption of a given hydrogen nucleus, H_i, will be divided because of spin-spin splitting depends in a linear manner on the number of protons coupling with H_i. Thus if N protons, all of which have the same coupling constant, J, with a hydrogen, H_i, are coupled with this nucleus, the absorption of H_i will appear as a group of $N + 1$ peaks.[8] If the coupling constants are not identical, the spectrum may be much more complicated, but a detailed discussion of such cases is not appropriate here.

[8]This statement is an oversimplification and is strictly true only in those cases in which the ratio of the chemical shift, expressed in Hz, between coupled protons and their coupling constant is greater than 5 to 10. This point is further discussed later in this section with regard to the interpretation of Figure 4.14. For detailed discussion of this point, see also references 1, 3, and 4 at the end of this chapter.

Applying the $N + 1$ rule to the different hydrogens of our reference compound, 1-nitropropane, the H_a protons should appear as a triplet, the H_b protons as a sextet *if* $J_{ab} = J_{bc}$, and the H_c protons as a triplet. These expectations are confirmed in the observed spectrum. It should be apparent that the multiplicity of the pmr absorption of a given hydrogen, H_i, can be utilized to determine the number of protons adjacent to it, a fact which is quite useful for purposes of analyzing spectra.

The relative numbers of different types of protons can be determined by integration, electronic or manual, of the absorption bands of the spectrum, since the areas under the bands are a linear function of the number of hydrogen nuclei producing the band. For example, in the spectrum of Figure 4.10 the relative areas of the three bands are obtained by measuring the distances a–b, a–c, and a–d. The ratio of these heights (if measured in millimeters) is $28:28:44$, so that the relative abundance of the corresponding protons is determined to be $2.0:2.0:3.1$. This is within experimental error of the absolute ratio of $2.0:2.0:3.0$ anticipated for 1-nitropropane.

It should now be clear that analysis of a pmr spectrum can yield a wealth of information. Determination of the *chemical shift* of each peak or group of peaks allows one to speculate about the *type of functional group* to which the hydrogen nucleus producing the absorption is attached. The *spin-spin splitting* pattern provides knowledge concerning the *number of nearest neighbors* of the hydrogen. *Integration* permits an evaluation of the *relative numbers of each type* of hydrogen present in the molecule.

When attempting to analyze the pmr spectrum of an unknown compound, it is profitable to begin by determining the relative numbers of different types of hydrogen present by measurement of the integrated peak areas. If the molecular formula of the unknown is available, the *relative* ratios can be converted to *absolute* numbers of the hydrogens present.

The second step in the analysis is an attempt to define the functional groups present in the molecule by determination of the chemical shifts of the different hydrogens. Further knowledge concerning the molecular environment of a given type of hydrogen can be derived by analysis of the spin-spin splitting patterns. At this point, it should be possible to propose for the unknown one or more complete, or at least partial, structures which are consistent with the spectrum.

If an infrared spectrum of the unknown is also available, it can be analyzed as described in Section 4.1. This analysis should provide additional information concerning the nature of functional groups present, and this information may then be used to supplement or to confirm that derived from analysis of the pmr spectrum.

The structures of many unknowns whose molecular formulas are available can often be derived immediately by combination of the information available from the ir and pmr spectra.[9] However, proper interpretation of the spectra requires considerable practice on the part of the beginning student of spectroscopy. In order to provide an opportunity for such practice in interpretation, the spectra of many of the organic compounds encountered as reagents or products in reactions described in this text

---

[9] Structural assignments based solely on interpretation of spectra must be confirmed either by comparison of the spectra of the unknown with those of the known compound, which has been synthesized in an unequivocal fashion, or by more classical means such as determination of mixture-melting points of solid derivatives (see Section 1.2).

have been reproduced at the end of each experiment. As these spectra are labeled, you will not be attempting to assign a structure to the substance producing them, but will be trying to interpret the spectra in terms of the known structure. As complete analysis as is possible should be included as part of the write-up of each experiment.

As an example of the kind of analysis that can be done, consider the spectra of 2-methyl-1-propanol, $(CH_3)_2CHCH_2OH$, given in Figure 4.13. Looking first at the ir spectrum (Figure 4.13a), three prominent bands are present in the region of 5000–1250 cm^{-1}. By referring to Table 4.2, it is possible to assign the absorptions at about 2900 and 1465 cm^{-1} to the C—H bonds of the molecule. The broad and intense band at 3250 cm^{-1} undoubtedly is due to the O—H function in the alcohol. The position

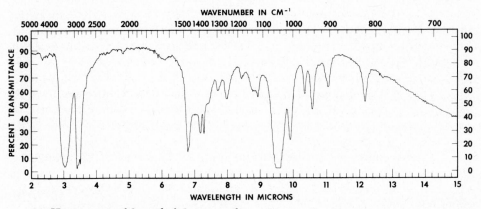

(a) IR spectrum of 2-methyl-1-propanol

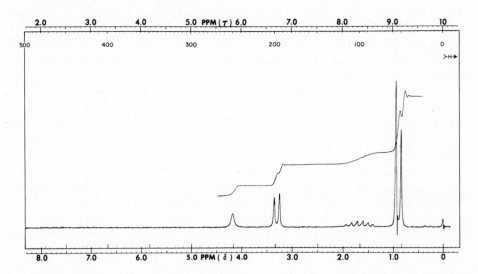

(b) PMR spectrum of 2-methyl-1-propanol

FIGURE 4.13   Spectra of 2-methyl-1-propanol.

and breadth of the band indicates that the alcohol is hydrogen-bonded. Finally the very strong absorption at 1050 cm^{-1} is characteristic of a C—O single bond. As would be expected on the basis of the structure of 2-methyl-1-propanol, no absorptions are seen for carbonyl functions, aromatic rings, or carbon-carbon multiple bonds.

Turning now to the pmr spectrum (Figure 4.13b), the first point of interest is the integration. The areas of the peaks, going *upfield,* are in the ratio (measured in millimeters) 2.8:5.6:3.0:15, which is equal to the ratio 1.0:2.0:1.1:5.4. Since the total number of hydrogens in 2-methyl-1-propanol is 10, the latter ratio is within experimental error of the absolute ratio.

It is now possible to complete the analysis of the spectrum by assigning absorptions to specific types of hydrogens, basing the assignment on chemical shift and integrated area. The doublet at about 0.85 δ has the appropriate relative area, 6, and a chemical shift consistent with six hydrogens of the two methyl groups of the alcohol. The multiplicity of methyl absorption is due to coupling with the lone methine (tertiary) hydrogen, $(CH_3)_2CH$. The multiplet from 1.4–2.0 δ can be assigned to the methine hydrogen by consideration of the area, 1, the chemical shift, which is consistent with the nucleus being tertiary, and the multiplicity. The methine hydrogen is coupled with no less than eight other hydrogens. The doublet at about 3.3 δ has the area, 2, the chemical shift and multiplicity anticipated for the methylene group of 2-methyl-1-propanol. This leaves the somewhat broad singlet at 4.2 δ to be assigned to the hydroxylic hydrogen.[10]

Spectra in which the multiplicities of the various hydrogens can successfully be predicted by the $N + 1$ rule are commonly referred to as "first order," and the pmr spectra of 1-nitropropane and 2-methyl-1-propanol fall in this category. The splitting patterns recorded in many of the spectra contained in this book, however, will be much more complex and essentially uninterpretable according to the rule. In many cases this will be the consequence of the presence in a compound of a variety of functionally similar hydrogens having very similar magnetic environments, and therefore similar chemical shifts. This situation is exemplified by the methylene hydrogens of 1-butanol (Figure 4.14). Consideration of substituent effects allows assignment of the triplet centered at 3.5 δ to the methylene hydrogens at C-1 of the alcohol. However, it clearly is not possible to make separate assignments of the various bands in the region 1.1–1.7 δ to the two pairs of skeletally distinct methylene hydrogens at C-2 and C-3 responsible for these resonances. Rather, one simply assigns the multiplet centered at about 1.4 δ as representing both of the pairs at C-2 and C-3 and does not attempt to differentiate them.

Another complicating factor in applying the $N + 1$ rule is illustrated in Figure 4.14. The high-field resonance centered at about 0.9 δ can be assigned to the methyl group with confidence on the basis of the integration of this band and the anticipated chemical shift of methyl hydrogens. Nevertheless, the methyl group is not the uncomplicated triplet that would have been expected from application of the

---

[10] It may seem strange that the hydroxylic proton appears as a singlet, albeit a broad singlet, since one would expect it to be coupled to the methylene hydrogens. Such coupling can, in fact, be observed in highly purified alcohols or in solvents such as dimethylsulfoxide, but for our purposes it can be assumed that no spin-spin splitting will be detected between such a proton and protons on the carbon atom bonded to the oxygen.

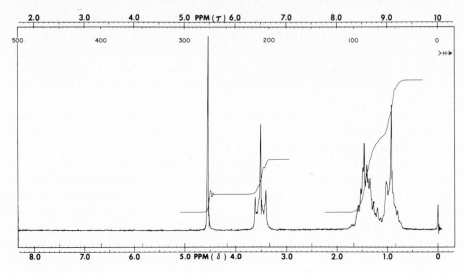

FIGURE 4.14   PMR spectrum of 1-butanol.

$N + 1$ rule. The theoretical basis for this departure from expectation is beyond the scope of this discussion, but in essence the failure of the rule to predict multiplicities arises whenever the difference in the chemical shift between coupled hydrogens is similar to, or not much larger in magnitude than, the value of the coupling constant between those hydrogens; in other words, if $\Delta v/J$ (Δv is the difference in Hz between the chemical shifts of coupled hydrogens) is less than 5 to 10, that portion of the pmr spectrum will no longer be first order and have multiplicities predictable from the $N + 1$ rule. In the case of 1-butanol the ratio for the methyl group and the methylene hydrogens with which it is coupled is about 5.

Sample Preparation.   The pmr spectra of greatest use to organic chemists are those obtained on liquids contained in special precision-ground glass tubes, 5 × 178 mm in size. Most commonly the substrate of interest, whether it is solid, liquid, or gaseous, is dissolved in an appropriate solvent, although spectra can also be obtained on neat liquids if their viscosities are not too high. With too viscous liquids, broadened rather than sharp bands are observed in the spectra, resulting in a loss of resolution between peaks. Approximately 0.3 mL of a 10–20% solution (by weight) is normally required.[11] Carbon tetrachloride is an excellent choice for the solvent, since it has no protons that would produce absorption peaks in addition to those of the solute. For samples that are insoluble in carbon tetrachloride, benzene, 7.67 δ (chemical shift), chloroform, 7.27 δ, dichloromethane, 5.35 δ, nitromethane, 4.33 δ, acetonitrile, 2.00 δ, or acetone, 2.17 δ may be suitable solvents if their proton resonances do not occur in a region of absorption of the solute. Many deuterated solvents, such as chloroform-d ($CDCl_3$), acetone-d_6 (CD_3COCD_3), and benzene-d_6 (C_6D_6), are available. Their use

---

[11] Sophisticated modern spectrometers, however, can provide usable pmr spectra on solutions containing a milligram or even less of solute.

circumvents the problem of resonance absorptions of solvent protons in the pmr spectrum. Even though deuterium (^{2}H) possesses the property of nuclear spin, the experimental conditions for achieving the resonance condition differ sufficiently from those required for hydrogen, so deuterium absorptions do not interfere with the pmr spectrum. The chemical shift of deuterium is vastly different from that of hydrogen.

Spectra of poor quality can result if a neat liquid sample or a solution contains undissolved solids such as the sample itself or contaminants (*even dust*). Moreover, trace amounts of ferromagnetic impurities picked up from contact of the sample or solvent with metals—for example, the iron of a "tin" can—cause havoc that results in observation of very broad and weak absorptions of the sample. If any solid material *at all* can be seen in the sample, it should be filtered. The simplest and most efficient method for filtering very small volumes is to insert a small plug of glass wool into a disposable pipet and to filter the sample directly into the nmr tube through the plug. After use the nmr tubes should be well cleaned and dried; they should be stored either in a closed container or with open end downward, in order to avoid dust settling inside the tube.

EXERCISES

1. The pmr spectra of the three fractions obtained in a distillation of a mixture of benzene and toluene consist of a sharp singlet and a multiplet in the region of 7.2 δ, and a singlet at 2.3 δ. The integrated intensities of the downfield and upfield absorptions are given below for each fraction. Calculate the percentage of toluene present in each fraction.

	Downfield	Upfield
Fraction 1	165	9
Fraction 2	192	30
Fraction 3	167	81

2. The ir and pmr spectra of a compound, $C_{12}H_{18}$, are given in Figure 4.15. Interpret both spectra as completely as possible and attempt to assign a structure or structures to the unknown on the basis of your interpretation.

4.3   Ultraviolet and Visible Spectroscopy

Ultraviolet (uv) spectroscopy and visible spectroscopy, as noted at the beginning of this chapter, both result from the same basic molecular phenomenon, namely excitation of an electron from a lower energy to a higher energy electronic state (see Figure 4.1). Reference to Table 4.1 indicates that the energy required for such an excitation ranges from about 38 to over 100 kcal/mol and involves light in the wavelength range of 750 to 200 nm.

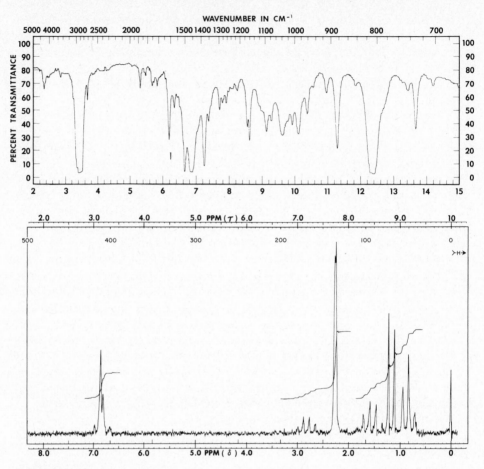

FIGURE 4.15   Spectra of $C_{12}H_{18}$.

The two types of electronic excitations occurring in this range that are of greatest interest to an organic chemist are those involving promotion into an *antibonding* molecular orbital of an electron that originally was an occupant either of a *nonbonding* molecular orbital (an *n* electron) or of a *bonding* molecular orbital (a σ or π electron). More specifically, however, the energy requirements for promotion of σ-type electrons are too high to be observed in the 750–200 nm range, so that this type of molecular phenomenon is not a feature of uv-visible spectra. Thus uv-visible spectroscopy is limited to the promotional transitions of *n*- and π-type electrons.

Furthermore, since π-type antibonding molecular orbitals, designated π*, are considerably lower in energy than the corresponding σ* orbitals, the transitions normally stimulated by light in the uv-visible range involve population of the π* state. To summarize, organic chemists generally are most interested in electronic transitions that are classified as $n \rightarrow \pi^*$ and $\pi \rightarrow \pi^*$; these are represented schematically in Figure 4.16.

A typical uv spectrum, that of 4-methyl-3-penten-2-one, a molecule that con-

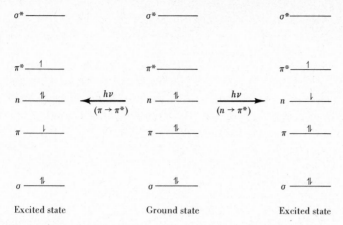

FIGURE 4.16 Electronic transitions.

tains both *n*- and π-type electrons, is shown in Figure 4.17. Note that the absorption bands are rather broad, with relatively poorly defined maxima. The diffuseness of the spectrum is a consequence of the fact that electronic transitions can occur from a variety of vibrational and rotational levels of the ground electronic state into a number of different such levels of the excited electronic state. Thus, although the transitions themselves are quantized (that is, only certain of them are theoretically permissible) and therefore should appear as sharp "lines," the fact that closely spaced vibrational-rotational levels give rise to closely spaced lines (compare E_1, E_2, and E_3 in Figure 4.1) causes coalescence of the discrete absorptions into a band *envelope* to produce the broad bands observed experimentally.

The most important characteristics of a spectrum are the location of any

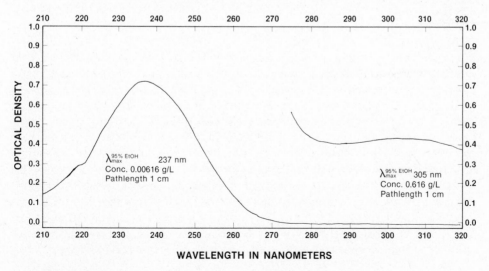

FIGURE 4.17 UV spectrum of 4-methyl-3-penten-2-one.

maxima and their corresponding intensities. Reference to Figure 4.17 shows two maxima, of differing intensities, at 238 and 314 nm. The location and intensity of a maximum are characteristic for a particular type of electronic excitation, or, for our purposes, are diagnostic for a specific type of *chromophore,* a term that refers to the functional group responsible for absorption of the incident light. In the case of the spectrum in Figure 4.17, it is the α, β-unsaturated carbonyl moiety that is the chromophore; the longer wavelength, less intense maximum at 314 nm is assigned to the $n \rightarrow \pi^*$ excitation, whereas the other maximum at 238 nm results from the $\pi \rightarrow \pi^*$ process.[12] It thus turns out to be possible to detect the presence of various functional groups if they serve as chromophores in the uv-visible region of the electromagnetic spectrum.

A compilation of a few of the chromophores commonly encountered by organic chemists, along with the wavelengths and intensities of their absorption maxima, is provided in Table 4.4. The measure of intensity, ε (or its log), is defined as the *molar absorptivity,* which essentially is a measure of the probability that a particular electronic transition will occur. Again, because the transitions are quantized, some are theoretically "allowed" and others are "forbidden"; the latter nevertheless can be experimentally observed in some cases, for reasons that need not presently concern us. These forbidden transitions, when they occur, do so with lower probability than the allowed transitions and are therefore observed as bands of lower intensity. Reference to Figure 4.17 and to Table 4.4 suggests that $\pi \rightarrow \pi^*$ excitations are allowed, whereas $n \rightarrow \pi^*$ are not, and this in fact corresponds to the prediction from theory. It should be evident, then, that knowledge of the location and intensity of maxima in the uv-visible region could be valuable to identification of functional groups and thus to structure elucidation.

The experimentally observed spectrum is subject to a number of variables, among which are solvent, concentration of the solution being examined, and the path-length of the cell through which the light must pass. The amount of light absorbed by a particular solution is quantitatively defined by the Beer-Lambert law (equation 7),

$$A = \varepsilon l c \qquad (7)$$

where A is the *absorbance,* l is the cell pathlength (in cm), and c is the concentration of the light-absorbing solute (in moles/liter).[13] It is clear from this expression that a plot of ε, or, more conveniently, log ε, versus wavelength will provide a standard spectrum in which the experimental variables of concentration and cell path will have been removed. However, the presentation of uv spectra in this book will be according to the manner in which they are observed experimentally—that is, as plots of *absorbance* versus wavelength. The concentration of the solution studied, as well as the pathlength of the cell, will be shown on the spectrum; note that in some instances a single spectrum will show traces made at more than one concentration so that both strong and weak absorbances will be discernible (see Figure 4.17).

---

[12]The $\pi \rightarrow \pi^*$ transition is normally observed only in *conjugated* carbonyl compounds. Even though the carbonyl group itself has both nonbonding and π-bonding types of electrons, when the carbonyl is not conjugated, such as in simple ketones, the $\pi \rightarrow \pi^*$ excitation generally occurs at a wavelength of less than 200 nm, and is thus out of the usual uv range.

[13]Compare this equation with equation 3.

TABLE 4.4 SOME COMMON ULTRAVIOLET-ACTIVE CHROMOPHORES AND THEIR PROPERTIES

Chromophore	Wavelength of λ_{max} (nm)	Type of Excitation	log ε_{max}
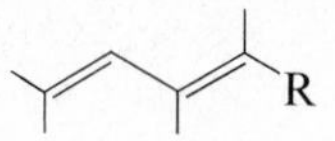 (diene with R)	217–230	$\pi \to \pi^*$	4.0–4.3
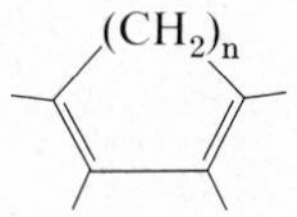 (cyclic diene, $(CH_2)_n$)	240–280	$\pi \to \pi^*$	3.5–4.0
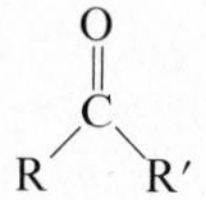 ketone O=C, R R'	270–300	$n \to \pi^*$	1.1–1.3
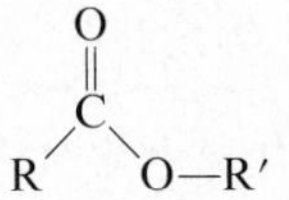 ester O=C, R O—R' (R' = H, alkyl)	200–235	$n \to \pi^*$	1.0–1.7
(enone)	215–250	$\pi \to \pi^*$	4.0–4.3
	310–330	$n \to \pi^*$	1.3–1.5
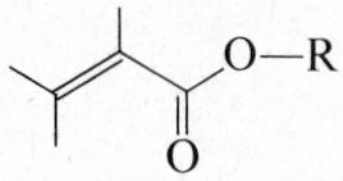 O—R, O (R = H, alkyl)	200–240	$\pi \to \pi^*$	4.0–4.1
benzene-G (G = one or more alkyl groups)	256–272	$\pi \to \pi^*$	2.3–2.5
	200–210	$\pi \to \pi^*$	3.9–4.1
R—Het— (Het = heteroatom)	265–290	$\pi \to \pi^*$	2.3–3.4
	210–230	$\pi \to \pi^*$	3.8–4.0

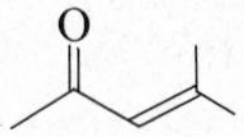

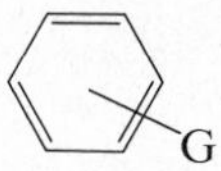

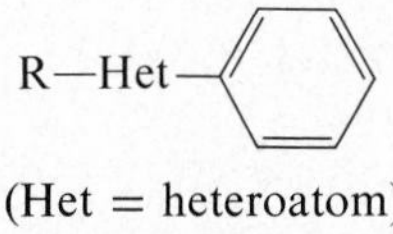

Often the entire uv-visible spectrum of a compound is not reported; rather only the wavelength and intensity of any maxima are given, along with the solvent in which the measurement was made. The crucial information contained in Figure 4.17 might thus be expressed in the following way:

$$\lambda_{max}^{95\%EtOH} \ 238 \text{ nm, log } 4.1; \ 305 \text{ nm, log } 1.8$$

The solvent should be stated, since the values of both λ_{max} and ε are solvent-dependent, for reasons that need not concern us here.[14]

[14] See reference 4 for discussions of solvent effects and reference 7 for a recent interpretation of the origins of such effects.

If the molecular weight of the substance being studied is not known, the molar absorptivity, ε, cannot be determined. This follows, since one would not be able to express the concentration c in *moles* per liter of solution. In this instance the intensity of an absorption must be evaluated in a manner different from that described by equation 7. One way for doing so is given in equation 8, in which A and l have their usual meaning, but c is measured in grams per 100 mL of solution. E, like ε, is a characteristic measure of the absorptivity of the unknown substance.

$$E_{1\,\text{cm}}^{1\%} = \frac{A}{lc} \tag{8}$$

Examination of equation 7 makes it clear that if ε, l, and A are all known for a particular sample, it is possible to determine c, the concentration of the absorbing species in solution. The value of ε can be determined experimentally by preparing solutions of the species of known concentrations and then determining A for each of these solutions in a cell of known pathlength. Given then that ε for a species and l are known and constant values, any change in A for a given sample as a function of time *must* result from a variation in c. Monitoring A for a particular solution over time will permit determination of the time dependence of the concentration of the absorbing species; in other words, the kinetics of appearance or disappearance of the species can be evaluated, and reaction rates can thus be determined.[15] This particular use of uv-visible spectroscopy is applied to quantification of the relative rates of electrophilic bromination of substituted benzenes (equation 9) in Section 7.4. The success of the measurements depends on the disappearance of the absorption due to molecular bromine, whose maximum is at 400 nm in the visible region.

$$\text{G} - \bigcirc + \text{Br}_2 \longrightarrow \text{G} - \bigcirc^{\text{Br}} + \text{HBr} \tag{9}$$

The use of uv-visible spectroscopy as an aid in elucidation of structure of unknowns depends on knowledge of the values of λ_{max} and ε that are characteristic of a chromophore. (To evaluate the latter, of course, the molecular weight must be known, and a variety of methods, preeminent of which is mass spectrometry, are available for making this measurement.) As noted previously, Table 4.4 contains a partial listing of the absorption characteristics of several chromophores, and more extensive compilations are available in references cited at the end of this chapter. Application of this information can supplement and validate conclusions about structural features drawn from other spectroscopic techniques and/or from chemical tests for functionality.

As a simple example, suppose an unknown substance gave ir absorptions at 1715 and 1610 cm^{-1}, decolorized a solution of bromine in carbon tetrachloride, and produced a 2,4-dinitrophenylhydrazone when treated with 2,4-dinitrophenylhydrazine. These data are consistent with the presence of a carbon-carbon double bond and a carbonyl function. A uv spectrum of the compound would permit determination of

---

[15] If the absorption curves of reactant and product overlap, the determination of the concentration versus time of each species may be somewhat more complex but can still be determined.

whether these two groupings were conjugated with one another (see footnote 12), as two maxima, one considerably more intense than the other, would be expected if they were; otherwise, only a weak maximum, due to the $n \rightarrow \pi^*$ excitation of the carbonyl group, would be observed in the vicinity of 300 nm. No maximum for the carbon-carbon double bond would be present unless this function were conjugated with some other functional group such as another double bond.

It should be noted that empirical generalizations have been developed that permit prediction of λ_{max} as a function of the nature of substituents and, in some instances, their location on the chromophore of interest. Although it is not appropriate to outline the rules here, suffice it to say that they permit prediction of the relative locations of λ_{max} in compounds as closely related as **1** and **2**.[16]

It is relatively easy to prepare samples for analysis by uv-visible spectroscopy. Most spectra are obtained on solutions, although cells appropriate for gaseous samples are available. Commonly the cells for uv work are constructed of quartz and have a pathlength of 1 cm, whereas those for visible spectroscopy are made of the less expensive borosilicate glass; this type of glass is opaque to light in the uv region and is therefore not suitable for uv cells. These cells require approximately 3 mL of solution to be filled.

A variety of organic solvents, as well as water, can be employed for uv-visible spectroscopy. They all share the property of not absorbing significantly at wavelengths greater than about 220 nm, as is indicated in Table 4.5. The wavelengths given in the table constitute the so-called "cutoff point" for the respective solvents. Below the given wavelength the solvent begins to absorb appreciably, so that this solvent may not be used for a solute which absorbs at wavelengths below that point. Although "technical" and "reagent" grade solvents must usually be purified before use in spectroscopy owing to the presence of light-absorbing impurities, more expensive "spectral" grades are available which can be used directly from their containers. Obviously the choice of which solvent to use for a particular solute will depend not only on the solubility properties but also on the absence of chemical reaction between solvent and solute.

The concentration of the solution to be examined should be such that the observed value of A is in the range of about 0.3–1.5; this permits the greatest accuracy in making the measurement. To accomplish this, the value of ε for any chromophores present should be estimated (see Table 4.4) so that the concentration, c, required to give an A in the range suggested can be calculated with the aid of equation 7. Generally speaking, 0.01–0.001 M solutions will give absorbances of appropriate magnitude for excitations of low intensity (log ε of about 1.0); dilutions of this

---

[16] For elaboration of these rules, see reference 4.

TABLE 4.5 SOLVENTS FOR UV-VISIBLE
SPECTROSCOPY

Solvent	Useful Spectral Range (Lower Limit)
Acetonitrile	<200 nm
Chloroform	245
Cyclohexane	205
95% Ethanol	205
Hexane	200
Methanol	205
Water	200

original solution will then permit more intense absorptions to be observed in the desired range for A.

It should be evident that careful weighing of solute, quantitative transfer of it to a volumetric flask, and accurate dilution with solvent are necessary to achieve precise measurements of A, and therefore ε. Furthermore, the accidental introduction into the solution of even a minute amount of an intensely absorbing impurity can have a dramatic effect on the observed spectrum, so that great care should be exercised in the handling and cleaning of all apparatus associated with the preparation of the solution; the cells to be used should be thoroughly rinsed with the solvent being used, both before and after use, in order to minimize the development of contamination in the sample. Quartz cells to be used for uv work should never be rinsed with acetone for cleaning and drying purposes. Trace residues of acid or base catalysts on the surface of the quartz may, through an acid- or base-catalyzed aldol condensation, lead to the formation of trace quantities of 4-methyl-3-penten-2-one (Figure 4.17), whose presence in the cell may invalidate precise measurements of ε, when such work must be done.

EXERCISES

1. For the uv spectrum of Figure 4.17, assign the chromophores responsible for each maximum and calculate the value(s) of λ_{max} and of $\log \varepsilon$.
2. Replot Figure 4.17 using values $\log \varepsilon$ rather than absorbance (A) along the ordinate.
3. Repeat exercise 1 for each of the uv spectra reproduced in the book (see Figures 9.5, 17.11, 17.13, 20.6, 21.6, and 24.3).

REFERENCES

1. J. R. Dyer, *Applications of Absorption Spectroscopy of Organic Compounds,* Prentice-Hall, Englewood Cliffs, N. J., 1965 (paperback).
2. R. T. Conley, *Infrared Spectroscopy,* Allyn and Bacon, Boston, 1966.

3. D. J. Pasto and C. R. Johnson, *Organic Structure Determination,* Prentice-Hall, Engle-wood Cliffs, N. J., 1969. Ir, nmr, and uv-visible spectroscopy and mass spectrometry.

4. R. M. Silverstein, G. C. Bassler, and T. C. Morrill, *Spectrometric Identification of Organic Compounds,* 3d ed., John Wiley & Sons, New York, 1974. Ir, nmr, and uv spectroscopy and mass spectrometry.

5. F. W. McLafferty, *Interpretation of Mass Spectrometry,* 2d ed., W. A. Benjamin, Menlo Park, Calif., 1973.

6. I. Fleming and D. H. Williams, *Spectroscopic Methods in Organic Chemistry,* McGraw-Hill Book Company, New York, 1966.

7. P. Haberfield et al., *Journal of the American Chemical Society,* **99,** 6828 (1977).

5

ALKANES
Free Radical Halogenation

5.1 Chlorination by Means of Sulfuryl Chloride

In general, saturated hydrocarbons are chemically inert even in the presence of very strong acids or bases, and the organic chemistry of these substances is therefore rather limited. Some chemical reactions have been developed, however, by which hydrocarbons can be converted to organic substances suitable for use in a variety of reactions. For example, a hydrocarbon, RH, can be transformed to a nitroalkane, RNO_2, or to a hydroperoxide, ROOH, on reaction with nitrogen tetroxide or molecular oxygen, respectively.

This experiment involves the production of alkyl chlorides, RCl, from hydrocarbons. No reaction occurs when chlorine gas and a hydrocarbon are simply mixed at room temperature, but ultraviolet irradiation or heating (200–400°) of the mixture affords an alkyl chloride and hydrogen chloride (equation 1). The photochemical or

$$RH + Cl_2 \xrightarrow[h\nu]{\text{heat or}} RCl + HCl \tag{1}$$

thermal activation of the mixture is required in order to convert some molecular chlorine into chlorine atoms. The generation of chlorine atoms, which are free radicals, is essential to the initiation of the reaction between the hydrocarbon and chlorine to give the alkyl chloride.

It is somewhat easier experimentally to generate chlorine atoms by use of an initiator, a substance that will decompose to free radicals under relatively mild conditions (equation 2). The free radicals, In·, arising from decomposition of the

$$In\!-\!In \xrightarrow[h\nu]{\text{heat or}} In\cdot + In\cdot \tag{2}$$

initiator will react with molecular chlorine to produce InCl and chlorine atoms (equation 3), and these chlorine atoms will then react with the hydrocarbon to

$$In\cdot + Cl\!-\!Cl \longrightarrow InCl + Cl\cdot \tag{3}$$

produce an alkyl chloride. In this experiment the usual procedure for chlorinating hydrocarbons is modified by the use of sulfuryl chloride, SO_2Cl_2, in place of chlorine gas, in the interest of safety and convenience in the undergraduate laboratory.

The mechanism of the free radical chlorination of hydrocarbons is fundamentally the same whether sulfuryl chloride or molecular chlorine is employed and can be divided into three different phases: *initiation, propagation,* and *termination.* In the

Initiation

$$CH_3-\underset{\underset{CH_3}{|}}{\overset{\overset{CN}{|}}{C}}-N{=}N-\underset{\underset{CH_3}{|}}{\overset{\overset{CN}{|}}{C}}-CH_3 \xrightarrow{80\text{--}100°} N_2 + 2\,CH_3-\underset{\underset{CH_3}{|}}{\overset{\overset{CN}{|}}{C}}\cdot \qquad (4)$$

$$(CH_3-\underset{\underset{CH_3}{|}}{\overset{\overset{CN}{|}}{C}}\cdot = In\cdot)$$

$$In\cdot + Cl-\underset{\underset{O}{\overset{||}{}}}{\overset{\overset{O}{||}}{S}}-Cl \longrightarrow InCl + \cdot\underset{\underset{O}{\overset{||}{}}}{\overset{\overset{O}{||}}{S}}-Cl \qquad (5)$$

$$\cdot\underset{\underset{O}{\overset{||}{}}}{\overset{\overset{O}{||}}{S}}-Cl \longrightarrow \underset{\underset{O}{\overset{||}{}}}{\overset{\overset{O}{||}}{S}} + Cl\cdot \qquad (6)$$

Propagation

$$Cl\cdot + RH \longrightarrow R\cdot + HCl \qquad (7)$$

$$R\cdot + ClSO_2Cl \longrightarrow RCl + \cdot SO_2Cl \qquad (8)$$

$$\cdot SO_2Cl \longrightarrow SO_2 + Cl\cdot \qquad (9)$$

Termination

$$Cl\cdot + Cl\cdot \longrightarrow Cl_2 \qquad (10)$$

$$R\cdot + Cl\cdot \longrightarrow RCl \qquad (11)$$

$$R\cdot + R\cdot \longrightarrow R-R \qquad (12)$$

experiment the first step of the initiation phase is the homolytic cleavage[1] of azobisisobutyronitrile[2] into nitrogen and the initiator free radicals $(CH_3)_2C(CN)\cdot$ (equation 4), which is accomplished by heating at 80–100°. The initiator radicals then attack sulfuryl chloride molecules to produce SO_2 and chlorine atoms (equations 5 and 6).

The propagation steps involve abstraction of a hydrogen atom from the hydrocarbon by a chlorine atom to produce a new free radical, $R\cdot$, which can attack the sulfuryl chloride to produce the alkyl chloride and to regenerate the radical, $\cdot SO_2Cl$, the precursor of chlorine atoms (equations 7–9). A chain reaction is thus produced.

---

[1] Homolytic cleavage may be defined as rupture of a bond in a fashion such that the bonding electrons become distributed equally between the atoms originally linked by the bond. The alternate manner of distribution in which the bonding electrons are not equally divided between the atoms is termed heterolytic cleavage.

[2] This compound is also known as 2,2′-azobis[2-methylpropionitrile].

Once initiated, the chain reaction could in principle continue until either of the principal reagents, sulfuryl chloride or the hydrocarbon, depending on which is the *limiting reagent,*[3] is exhausted. In practice, however, the termination reactions (equations 10–12, for example) intervene to interrupt the free radical-chain process so that initiation must be continued throughout the reaction.

It is worthwhile to note some characteristics of the various steps in a free radical-chain mechanism. In general an *initiation step* involves the formation of a low concentration of free radicals from molecules and results in an increase in the concentration of free radicals present in the system. A *propagation step* produces no *net* change in the concentration of radicals present, whereas a *termination step* gives a decrease in their concentration.

Free radical halogenation of hydrocarbons generally produces mixtures of several isomers and polyhalogenated products which are difficult to separate into pure components. However, chlorination of hydrocarbons is a useful industrial process in cases where pure individual alkyl chlorides are not required. For example, *n*-dodecane may be chlorinated to give a mixture of monochlorododecanes (equation 13). These alkyl chlorides may be converted by the Friedel-Crafts reaction

$$R-CH_2-CH_2-CH_2-CH_3 \xrightarrow{Cl_2} R-CH_2-CH_2-CH_2-CH_2Cl$$

$$+ R-CH_2-CH_2-\underset{\underset{Cl}{|}}{CH}-CH_3 + R-CH_2-\underset{\underset{Cl}{|}}{CH}-CH_2-CH_3$$

$$(R = n\text{-}C_8H_{17})$$

$$+ R-\underset{\underset{Cl}{|}}{CH}-CH_2-CH_2-CH_3 + \cdots \tag{13}$$

(Section 7.2) into a mixture of phenyldodecanes, which are useful as intermediates in the manufacture of biodegradable detergents.

For the preparation of pure individual alkyl halides, methods other than halogenation of saturated hydrocarbons are commonly used. The reaction of a hydrogen halide with an alkene (equation 14) or an alcohol (equation 15) provides the corresponding alkyl halide. Alcohols can also be transformed to alkyl halides by

$$\underset{R}{\overset{R}{>}}C=C\underset{R}{\overset{R}{<}} + HX \longrightarrow R-\underset{\underset{X}{|}}{\overset{\overset{R}{|}}{C}}-\underset{\underset{H}{|}}{\overset{\overset{R}{|}}{C}}-R \tag{14}$$

$$R-OH + HX \longrightarrow R-X + H_2O \tag{15}$$

reaction with a thionyl halide, SOX_2 ($X = Cl$ or Br), followed by thermal decomposition of the intermediate halosulfite ester (equation 16). These reactions of alcohols are discussed further in Section 10.1.

$$RO-H + X-\underset{\underset{O}{\|}}{S}-X \xrightarrow{-HX} RO-\underset{\underset{O}{\|}}{S}-X \xrightarrow{heat} R-X + SO_2 \tag{16}$$

---

[3] See Appendix 2 for an explanation of this term.

EXPERIMENTAL PROCEDURE

DO IT SAFELY

1.   The sulfuryl chloride used in this experiment reacts rather violently with water. Be sure that your glassware is *dry.* Take especial care to avoid getting sulfuryl chloride on your skin and do not breathe its vapors. We recommend wearing rubber gloves when transferring this substance and suggest that it be weighed out in a ventilation hood.

2.   The hydrocarbons used in this experiment are *flammable,* and the use of burners should be avoided if possible.

3.   When the reaction mixture is washed with aqueous sodium carbonate, carbon dioxide is generated in the separatory funnel. Be sure to vent the funnel frequently when shaking it, in order to relieve any gas pressure that develops.

This experiment provides for the chlorination of three compounds: cyclohexane, heptane, and 1-chlorobutane. Optimally, each student may do all three parts,

$$
\begin{array}{c}
\text{CH}_2 \\
\text{CH}_2 \quad \text{CH}_2 \\
\text{CH}_2 \quad \text{CH}_2 \\
\text{CH}_2 \\
\text{Cyclohexane}
\end{array}
\qquad
\begin{array}{c}
\text{CH}_3(\text{CH}_2)_5\text{CH}_3 \\
\text{Heptane}
\end{array}
\qquad
\begin{array}{c}
\text{CH}_3\text{CH}_2\text{CH}_2\text{CH}_2\text{Cl} \\
\text{1-Chlorobutane}
\end{array}
$$

since the number, type, and proportion of products from each of the three compounds illustrate important principles. Alternatively, the three compounds may be distributed among the class, and the students may compare their results.

A.   Cyclohexane

Fit a 100-mL round-bottomed flask with a water-cooled reflux condenser which is in turn fitted at the top with a vacuum adapter connected as shown in Figure 5.1 to serve as a trap for the SO_2 and HCl produced in the reaction. Place 33.6 g (43.3 mL, 0.400 mol) of cyclohexane,[4] 27.0 g (16.2 mL, 0.200 mol) of sulfuryl chloride, and 0.1 g of azobisisobutyronitrile in the flask, and weigh the flask and its contents. Connect the condenser and trap, turn on the aspirator so as to produce a *gentle* flow of air through the trap, and heat the mixture to a gentle reflux for 20 min, using either a heating mantle or an oil bath. Cool the reaction mixture, disconnect the flask from the condenser, and weigh it and its contents.[*] If the loss of weight is less than theoretical,[5] add another small portion of the azobisisobutyronitrile and heat the mixture to reflux for 10 min.

---

[4] If the starting material has been stored in a metal container, it may be necessary to perform a simple distillation to remove traces of metal ions, which can inhibit the reaction.

[5] The theoretical loss of weight is calculated on the basis that 1 mol each of HCl and of SO_2 is evolved for each mole of sulfuryl chloride consumed. This calculation should be made before coming to laboratory.

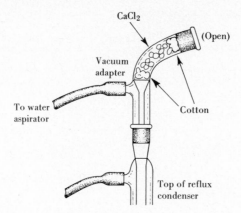

FIGURE 5.1   Details of gas trap arrangement.

After the theoretical amount of weight has been lost, cool the reaction mixture and *cautiously* pour it into 50 mL of *ice* water. Transfer the resulting two-phase solution to a separatory funnel and separate the layers.★ If separation into two layers does not occur readily, add some sodium chloride to the separatory funnel and shake. Wash the organic layer in the funnel with 0.5 *M* sodium carbonate solution until the aqueous washes are basic to litmus paper. Wash it once again with water and then dry it over about 3 g of *anhydrous* calcium chloride.[6] Filter the dried solution into a 100-mL flask fitted with a glass-packed column for fractional distillation and carefully distil in order to separate the chlorinated products from unchanged starting material. Suggested boiling ranges are as follows:

Fraction 1: Ambient–85°.

Fraction 2: 85–145°.

Weigh each of the distillation cuts and submit them for analysis by gas chromatography. Estimate the approximate yields of the chlorinated products, taking into account the amount of unchanged starting material recovered.

B.   Heptane

Follow the directions given in part A, using 40.0 g (58.5 mL, 0.400 mol) of heptane (see footnote 4) in place of cyclohexane. Suggested boiling ranges are as follows:

Fraction 1: Ambient–105°

Fraction 2: 105–160°

C.   1-Chlorobutane

Follow the directions given in part A, using 37.0 g (41.8 mL, 0.400 mol) of 1-chlorobutane (see footnote 4) in place of cyclohexane. Suggested boiling ranges are

[6]See Appendix 1 for a discussion of drying agents.

as follows:

Fraction 1: Ambient–82°

Fraction 2: 82–165°

EXERCISES

1. Suggest at least one termination process for halogenation with sulfuryl chloride reagent which has no counterpart in halogenation with molecular chlorine.
2. What is the reason for using less than the amount of sulfuryl chloride theoretically required to convert all the starting materials to monochlorinated products?
3. From which of the three starting materials should it be easiest to prepare a single pure monochlorinated product? Why?
4. What factors determine the proportion of monochlorinated isomers of heptane? of 1-chlorobutane?
5. Why is only a catalytic amount of initiator used?
6. Calculate the heat of reaction for the reaction between cyclohexane and chlorine

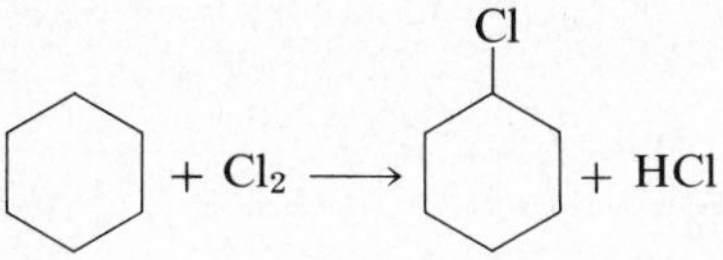

to yield chlorocyclohexane and hydrogen chloride, using the following bond energies (in kcal/mol):

C—H, 98.7; C—Cl, 81; Cl—Cl, 58.0; H—Cl, 103.2

7. Calculate the percent of each monochlorination product expected from heptane based on a relative reactivity of primary (1°): secondary (2°): tertiary (3°) hydrogens of $1.0:3.3:4.4$. Referring to Figure 5.2, calculate the observed ratio of $1°:2°$ chloroheptanes and compare the result with the anticipated theoretical ratio.
8. Chlorination of propene leads to high yields of 3-chloro-1-propene to the exclusion of products of substitution of the vinyl hydrogens. Explain why this is so.
9. Write out the expected products of free radical reaction of SO_2Cl_2 with 1-methylcyclohexane. Predict which chlorinated isomer would be formed in highest yield.
10. Referring to Figure 5.3, calculate the percentage of each dichlorobutane present in fraction 2 (do not include the area of the peak due to 1-chlorobutane when making the calculation). Explain why the observed ratio of products does not agree with that predicted by use of the relative reactivities given in Exercise 7.
11. Figure 5.4 gives the pmr spectrum of a mixture of dichlorobutanes obtained by free radical chlorination of 1-chlorobutane.
    (a) Assign the multiplet at 5.80 δ, the doublet at 1.55 δ, and the triplet at 1.07 δ

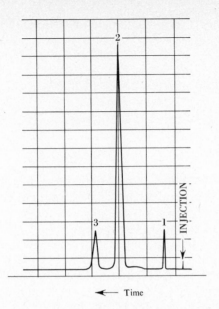

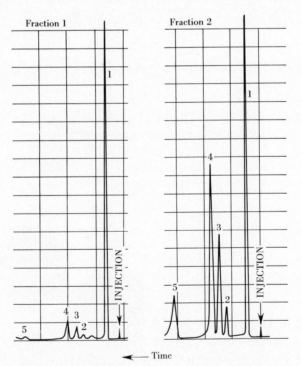

FIGURE 5.2 Gas chromatogram of fraction 2 resulting from chlorination of heptane. Peak 1: heptane; peak 2: 2-, 3-, 4-chloroheptanes; bp 46° (19.5 mm), 48.3° (21 mm), and 48.9° (21 mm), respectively; peak 3: 1-chloroheptane bp 61.4° (27 mm); 158.5–159.5° (769 mm); column and conditions: 1.5 m, 5% silicone elastomer on Chromosorb W; 80°, 40 mL/min.

FIGURE 5.3 Gas chromatograms of fractions 1 and 2 resulting from chlorination of 1-chlorobutane. Peak 1: 1-chlorobutane; peak 2: 1,1-dichlorobutane (bp 114–115°); peak 3: 1,2-dichlorobutane (bp 121–123°); peak 4: 1,3-dichlorobutane (bp 131–133°); peak 5: 1,4-dichlorobutane (bp 161–163°); column and conditions: 1.5 m, 5% silicone elastomer on Chromosorb W; 60°, 40 mL/min.

to the hydrogen nuclei of the three different isomers from which they arise. (*Hint:* Write the structures of the isomeric 1,*x*-dichlorobutanes and predict the multiplicity and approximate chemical shift of each group of hydrogens.)

(b)   Calculate the approximate percentage of each isomer present in the mixture.

REFERENCES

1. M. S. Kharasch and H. C. Brown, *Journal of the American Chemical Society,* **61,** 2142 (1939).
2. P. C. Reeves, *Journal of Chemical Education,* **48,** 636 (1971).

SPECTRA OF STARTING MATERIALS AND PRODUCTS

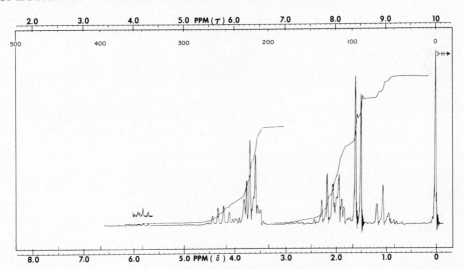

FIGURE 5.4   PMR spectrum of mixture of 1,*x*-dichlorobutanes.

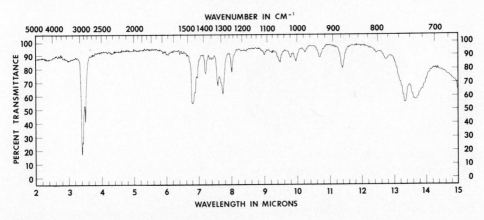

FIGURE 5.5   IR spectrum of 1-chlorobutane.

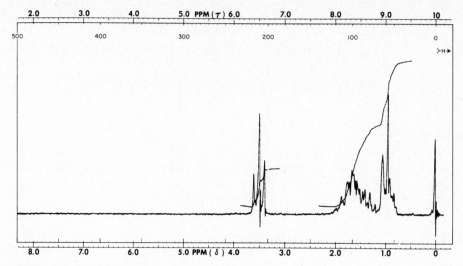

FIGURE 5.6 PMR spectrum of 1-chlorobutane.

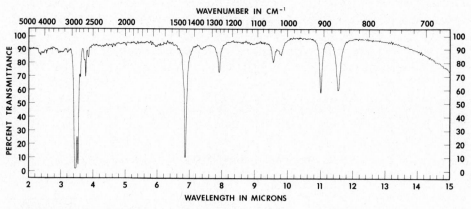

FIGURE 5.7 IR spectrum of cyclohexane.

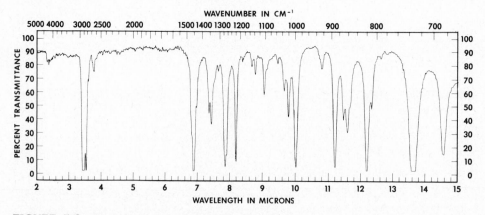

FIGURE 5.8 IR spectrum of chlorocyclohexane.

5.2 Bromination: Relative Ease of Substitution of Hydrogen in Different Environments

The preceding experiment allows a comparison of the reactivity of hydrogen atoms bonded to primary (1°) and secondary (2°) carbon atoms[7] toward substitution by chlorine atoms on the basis of the relative amounts of primary and secondary chloroheptanes produced from heptane. In the present experiment the relative rates of reaction of bromine with several different hydrocarbons containing primary, secondary, and tertiary hydrogens will be determined. The hydrogens may also be described as being aliphatic, aromatic, or benzylic.[8] We may combine these terms and speak of primary, secondary, and tertiary aliphatic or benzylic hydrogens. Only one type of aromatic hydrogen is possible. By careful consideration of your results from this experiment, you should be able to deduce an order of reactivity for seven different types of hydrogens.

Experimentally, the relative rates of bromination of the various hydrocarbons are measured by the lengths of time required for the bromine color to be discharged in the reactions, which are conducted under similar experimental conditions. Bromine is a less reactive and hence a more selective reagent than chlorine, so that the rates of bromination of the various hydrocarbons are sufficiently different to allow a qualitative order of reactivity to be determined quite easily.

The reactions are to be carried out in carbon tetrachloride solution under three different conditions: (1) at room temperature without special illumination, (2) at room temperature with strong illumination, and (3) at about 50° with strong illumination. Under all these conditions, substitution of hydrogen by bromine occurs by a free radical mechanism analogous to that of the chlorination reactions of Section 5.1. In this case the initiation step is the thermal and/or photochemical dissociation of bromine molecules into bromine atoms (radicals) as shown in equation 17. The propagation steps are those of equations 18 and 19. The termination steps are exactly analogous to those of equations 10–12 in Section 5.1.

$$Br_2 \xrightarrow[\text{or } h\nu]{\text{heat}} 2\,Br\cdot \tag{17}$$

$$Br\cdot + RH \longrightarrow HBr + R\cdot \tag{18}$$

$$R\cdot + Br_2 \longrightarrow RBr + Br\cdot \tag{19}$$

EXPERIMENTAL PROCEDURE

DO IT SAFELY
---

1. *Bromine is a hazardous chemical.* Do not breathe its vapors or allow it to come into contact with the skin because it may cause serious chemical burns. All operations involving the transfer of the pure liquid or its solutions should be carried

---

[7] In the discussion we shall follow the slightly inaccurate but convenient practice of referring to a hydrogen atom bonded to a primary carbon atom as a "primary hydrogen" and so forth.

[8] A benzylic hydrogen is one attached to a carbon atom bonded to an aromatic ring.

out in a ventilation hood, and rubber gloves should be worn. If you get bromine on your skin, wash the area quickly with soap and warm water and soak the skin in 0.6 M sodium thiosulfate solution (for up to 3 hr if the burn is particularly serious).

2.   Bromine reacts with acetone to produce a powerful lachrymator. Do not wash glassware containing residual bromine with acetone. At the end of the experiment add small quantities of cyclohexene to any test tubes having residual bromine. Discard the solutions in an organic liquid-waste container only after the characteristic color of bromine has been discharged.

3.   Avoid excessive inhalation of the vapors of any of the materials being used in this experiment. You should consider using an inverted funnel attached to a vacuum source and placed over the test tubes in order to lower the concentration of vapors in your area.

[*Note to the Instructor:* We recommend that the 1 M bromine in carbon tetrachloride solution be dispensed from communal burets (with Teflon stopcocks) placed in ventilation hoods. This will simplify the precise measurement of 1-mL portions of the solution and provide for greater safety in transferring the bromine.]

Construct a table in your notebook with the following headings: *Hydrocarbon, Types of Hydrogen* ($1°, 2°$, and so on); *Conditions* (with the subheadings, $25°$; $25°$, *hv*; $50°$, *hv*). In each of six 18×100-mm test tubes place a solution of 1 mL of hydrocarbon in 5 mL of carbon tetrachloride, using each of the following hydrocarbons: toluene, ethylbenzene, isopropylbenzene (cumene), *t*-butylbenzene, cyclohexane, and methylcyclohexane. These test tubes should be labeled to avoid confusion and mixing

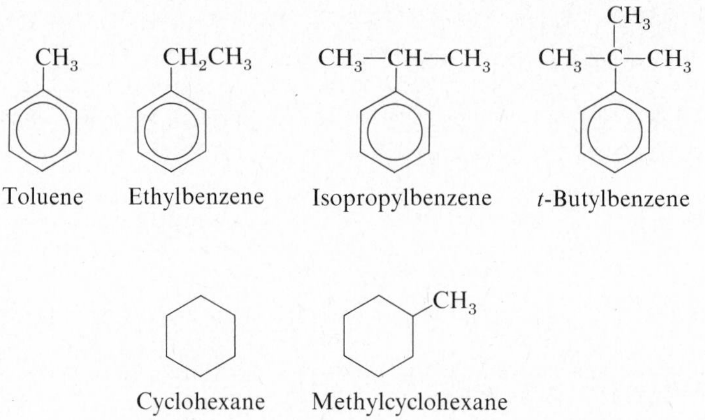

Toluene   Ethylbenzene   Isopropylbenzene   *t*-Butylbenzene

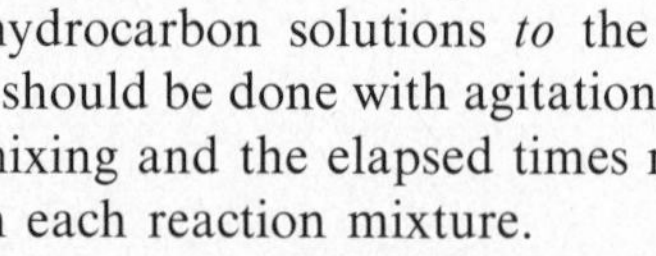

Cyclohexane   Methylcyclohexane

up of results. In each of six other test tubes, place 1 mL of 1 M bromine in carbon tetrachloride. Add the hydrocarbon solutions *to* the bromine solutions in rapid succession. This addition should be done with agitation to ensure good mixing. Note and record the time of mixing and the elapsed times required for the red bromine color to be discharged in each reaction mixture.

In some cases, because the rates of bromination are quite high, it is advisable to repeat the determination in order to be confident that the *relative* orders of decoloration are known. After several minutes have elapsed, place those mixtures in which the bromine color remains into a beaker of water which is kept at $50°$ on a steam bath or hot plate.[9] Suspend an *unfrosted* 100- or 150-watt light bulb over the test tubes at a distance of 10-13 cm. Continue to observe and to record the times at which the color is discharged in each tube. You may discontinue the experiment when only one colored solution remains because it is obvious that this hydrocarbon reacts the slowest.

On the basis of your results, do each of the exercises below.

EXERCISES

1. Arrange the six hydrocarbons in increasing order of reactivity toward bromination.
2. On the basis of the order of reactivity of the hydrocarbons, deduce the order of reactivity of the seven different types of hydrogens found in these compounds, that is, (1) primary aliphatic, (2) secondary aliphatic, (3) tertiary aliphatic, (4) primary benzylic, (5) secondary benzylic, (6) tertiary benzylic, and (7) aromatic.
3. Clearly explain how you arrived at your sequence in Exercise 2.

SPECTRA OF STARTING MATERIALS

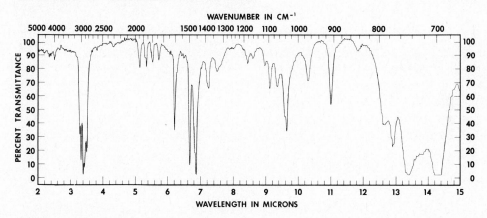

FIGURE 5.9 IR spectrum of ethylbenzene.

[9] Do not heat the water much above 50° because carbon tetrachloride has a boiling point of 77°.

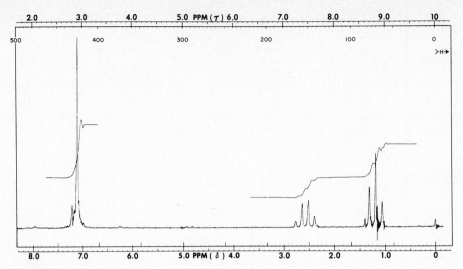

FIGURE 5.10 PMR spectrum of ethylbenzene.

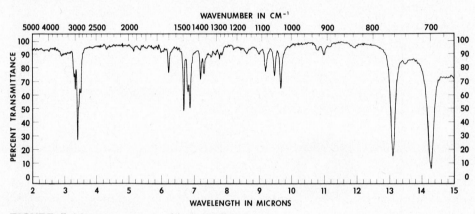

FIGURE 5.11 IR spectrum of isopropylbenzene.

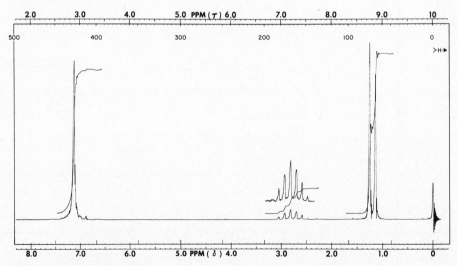

FIGURE 5.12 PMR spectrum of isopropylbenzene.

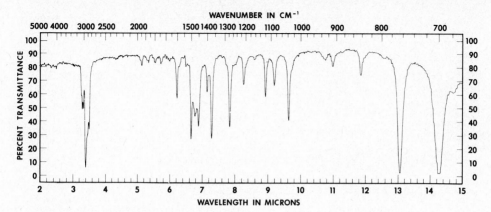

FIGURE 5.13 IR spectrum of *t*-butylbenzene.

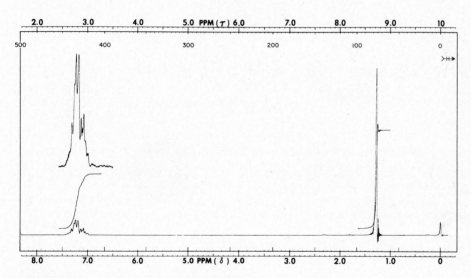

FIGURE 5.14 PMR spectrum of *t*-butylbenzene.

6

ALKENES
Preparations and Reactions

Alkenes, alcohols, alkyl halides—and indeed most other organic compounds—are subject to a much greater variety of chemical reactions than are alkanes, owing to the presence of reactive functional groups in these compounds. Alkyl halides, **1**, for example, are more reactive than alkanes because the greater *electronegativity* of the halogen atom, relative to hydrogen and carbon, serves to *polarize* the carbon-halogen bond generating positive charge on the carbon atom bearing the halogen atom. The polarization enhances the reactivity of the molecule primarily in two ways. (1) The substituted carbon atom is subject to attack by nucleophiles, which are electron-rich and frequently anionic species, leading to displacement of the halogen with the

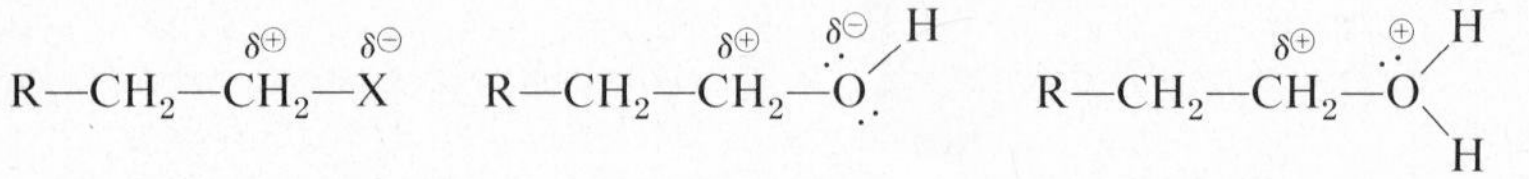

$$R\text{—}CH_2\text{—}CH_2\text{—}X \qquad R\text{—}CH_2\text{—}CH_2\text{—}\overset{\delta\ominus}{\underset{..}{O}}\!\!\diagup^{H} \qquad R\text{—}CH_2\text{—}CH_2\text{—}\overset{\oplus}{O}\diagup^{H}_{\diagdown H}$$

$$\mathbf{1} \qquad\qquad\qquad \mathbf{2} \qquad\qquad\qquad \mathbf{3}$$

formation of substitution products (this type of reaction is examined in detail in Section 10.1). (2) The *acidity* of hydrogen atoms near the carbon-halogen bond is enhanced, making the molecule susceptible to reaction with strong bases (reactions of this type will be introduced in Section 6.1).[1] Furthermore, the partial negative charge on the halogen atom of **1** produces significant electrostatic attraction for surrounding polar *solvent* molecules. These interactions weaken the carbon-halogen bond, facilitating both types of reactions mentioned. None of these features influence the reactivity of alkanes, because carbon and hydrogen are of about the same electronegativity, and hence there is little to no polarization within the molecules.

---

[1] Also, the nonbonding electron pairs on the halogen atom are available for complexation with Lewis acids, species which have *empty* valence shell orbitals. For example, interaction of $AlCl_3$, a Lewis acid, with an alkyl halide plays a central role in the Friedel-Crafts alkylation of aromatic hydrocarbons (see Section 7.2).

Alcohols, **2**, have enhanced reactivity relative to alkanes primarily because of the presence of the nonbonding electron pairs on the oxygen atom. These pairs interact with Lewis acids (see footnote 1) and other electrophiles (electron-deficient species which seek to react with electron-rich sites). The simplest such interaction is the reaction of an alcohol, **2**, with a proton-donating acid to produce an oxonium ion, **3**. The positively charged oxygen atom in **3** serves much the same role as the halogen atom in **1**: a positive charge on the adjacent carbon atom is induced, leading to substitution reactions (see Chapter 10), and the acidity of nearby hydrogens is increased, leading to dehydration reactions, which are the subject of Section 6.2.

The ways in which alkenes react as a result of the presence of highly polarizable pi-bonding electrons are discussed in Section 6.3.

6.1 Dehydrohalogenation of Alkyl Halides

As noted in the preceding section, the presence of the halogen atom in an alkyl halide enhances the acidity of nearby hydrogen atoms. The effect decreases the farther the hydrogen is from the carbon-halogen bond, so that the α-hydrogens of **1** are more acidic than the β-hydrogens. However, there are in general no *low-energy* pathways stemming from abstraction of an α-hydrogen by base that ultimately lead to stable products; therefore reaction by this pathway is not observed under most conditions. In contrast, the abstraction of a β-hydrogen leads smoothly *in one step* to the formation of a carbon-carbon pi bond with the extrusion of the halogen atom as halide, as indicated by the arrows in equation 1, and is a commonly observed transformation.

$$\text{(1)}$$

This type of reaction is usually referred to as an E2 process, where E stands for elimination and 2 refers to the molecularity of the rate-limiting step of the reaction. In this case two species, the alkyl halide and the base B:, must collide in order to pass through the transition state of the rate-limiting step, and the molecularity is therefore two. The rate of the reaction is equal to a proportionality constant k_2 (the rate constant) times the product of the concentrations of each of these species (equation 2).

$$\text{Rate} = k_2[\text{alkyl halide}][\text{B:}] \tag{2}$$

A second type of mechanism for an elimination reaction is shown in equation 3. Here the carbon-halogen bond is broken in a slow step to provide a carbocation (**4**),[2] from which the base then rapidly abstracts a β-hydrogen to provide the alkene (**5**).

---

[2] We shall use the term "carbocation" to refer to a species containing a trivalent, positively charged carbon atom, in order to conform with recent attempts to systematize the nomenclature of organic ions. This term should be accepted as being synonymous with "carbonium ion," a term that remains in use in many textbooks.

The first step of this two-step process is the rate-limiting step, and the rate of the reaction is therefore proportional *only* to the concentration of the alkyl halide (equation 4).[3] Reactions following this type of elimination mechanism (a unimolecular E1 process) are encountered in Section 6.2. It is important here only to note that

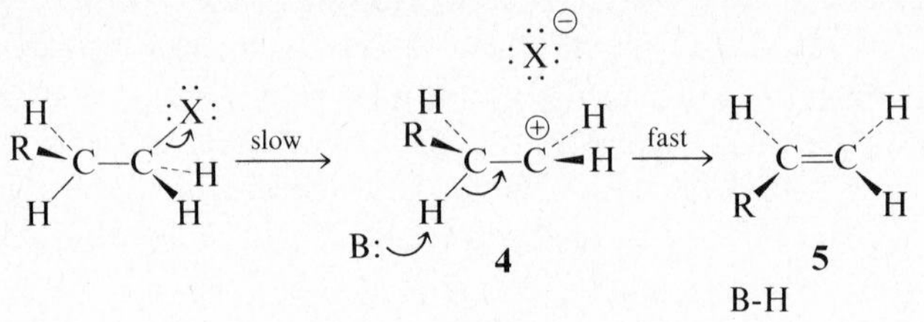

$$\text{Rate} = k_1 \text{[alkyl halide]} \tag{4}$$

if these two processes were proceeding simultaneously, the E2 pathway could be favored by using an excess of base, thereby making that pathway kinetically more competitive. This is because the rate of the E1 reaction will be unchanged by the increase in base concentration, whereas that of the E2 pathway will increase.

A variety of functionalities *other* than halogen may serve as leaving groups in elimination reactions. Among these are sulfonate esters, **6**, dimethylsulfonium, **7**, and trimethylammonium, **8**. Such groups, after departing, become species that are

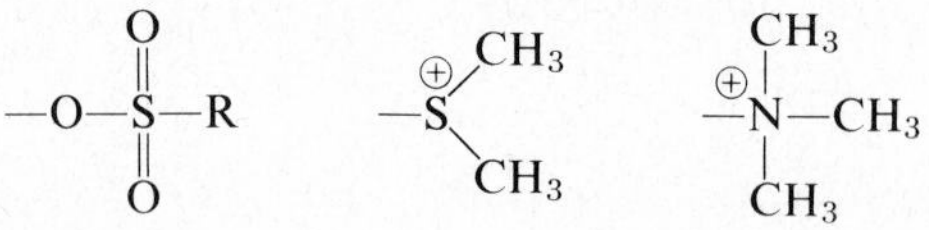

relatively *weak* bases, a property characteristic of all so-called "good" leaving groups. This is because the ability of a functional group to leave is related to its propensity for accepting the electron pair which bonds it to the molecule; the ability to accept an electron pair varies *inversely* with the base strength of the species formed.

Substitution reactions (equation 5) may sometimes compete with elimination reactions, decreasing the yield of the desired alkene. The degree to which this side

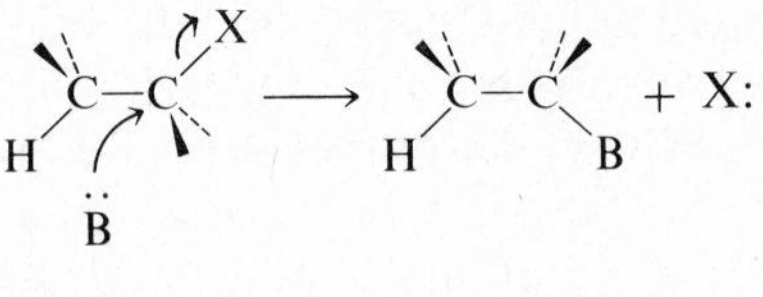

reaction is important depends on the natures of the substrate and the base employed. For steric reasons the competition is most important when the carbon bearing the leaving group is primary, becomes less so when it is secondary, and is usually unimportant when it is tertiary.

---

[3] Molecules of solvent are involved, because through the process of solvation they aid in the departure of the halogen in the rate-limiting step. However, the concentration of solvent molecules remains unchanged throughout the course of the reaction, and therefore the concentration of solvent is incorporated as an arithmetic constant into the value of k_1.

The use of strong bases enhances elimination at the expense of substitution. Consequently, to effect E2 reactions, base-solvent combinations such as alkoxide (RO^-) in alcohol, amide ion (H_2N^-) in benzene or diethyl ether, or potassium hydroxide in ethanol are commonly used. The competition between substitution and elimination is discussed in more detail in Section 10.1.

In cases where the leaving group is unsymmetrically located on the alkyl carbon skeleton, elimination may occur in two different directions. For example, 2-chloro-2-methylbutane yields both 2-methyl-2-butene (**9**) and 2-methyl-1-butene (**10**). As these

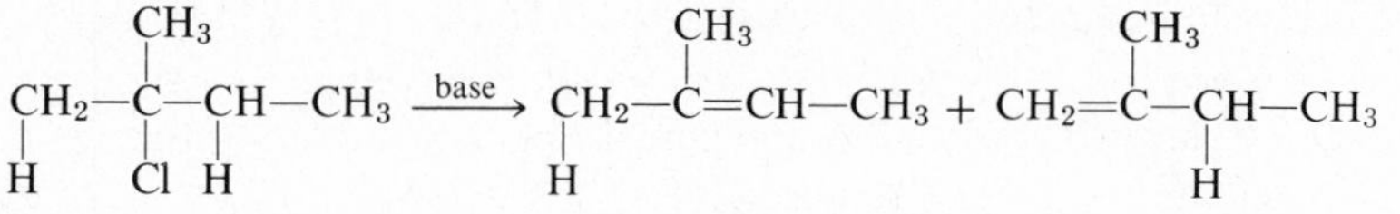

$$
\underset{\substack{|\\ H}}{\overset{\substack{CH_3\\ |}}{CH_2}}-\underset{\substack{|\\ Cl}}{\overset{\substack{CH_3\\ |}}{C}}-\underset{\substack{|\\ H}}{CH}-CH_3 \xrightarrow{\text{base}} \underset{\substack{|\\ H}}{\overset{\substack{CH_3\\ |}}{CH_2}}-C=CH-CH_3 + CH_2=\underset{\substack{|\\ H}}{\overset{\substack{CH_3\\ |}}{C}}-CH-CH_3 \qquad (6)
$$

9   10

reactions are *irreversible* under the experimental conditions, the alkenes **9** and **10** may be considered to be the products of two competing elimination reactions, and their proportions are thus subject to the relative rates of those two reactions (*kinetic control,* see Chapter 9). Their proportions are therefore determined by the relative free energies of their respective transition states. The transition states for elimination in each direction for 2-chloro-2-methylbutane are shown below (**9‡** and **10‡**).[4] It is

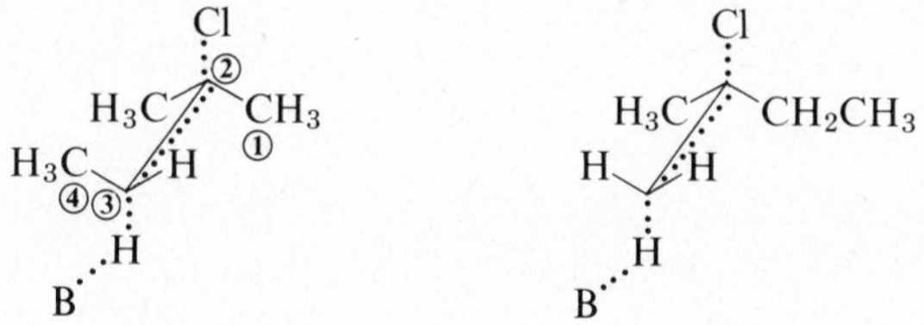

9‡   10‡

important to note that the HCCCl grouping is pictured in a *trans*-coplanar arrangement, since it is in this geometry that E2 processes occur most readily. If the dihedral angle between the β-hydrogen and the leaving group is much different from 180°, the energy of activation for elimination increases substantially.

When there are no complicating factors (primarily steric, see discussion in following paragraphs), the predominant product in an E2-type elimination is found to be the more highly substituted ethylene. Recall that increasing the number of alkyl substituents on the double bond increases the stability of alkenes (decreases free energy). This is interpreted to mean that the free energy of the partial double bond is the major contributor to the total free energy of the transition state. Thus the free energy of activation for the formation of **9** is less than that for the formation of **10**, and **9** should be the preferred product in the 2-chloro-2-methylbutane example. Prediction of the major product is not always this simple, however, because there may be other complicating factors.

---

[4]The symbol ‡ is frequently used by organic chemists to refer to a transition state.

The relative free energies of the transition states of the competing reactions may also be influenced by steric crowding. This crowding will tend to increase the free energy of the transition state and thus the energy requirements of the reaction. If steric effects are more important along the pathway to one of the products than along the pathway to the other, the proportion of products formed may be quite dramatically affected. This is particularly so when the pathway to the more highly substituted ethylene suffers from the greater steric crowding, in which case it may well be the *minor* product.

Unfavorable steric interactions in the transition state may be caused by steric interference of the substituents on the HCCL grouping (L stands for the leaving group) to the approach of the base (B:), particularly when the base and/or the substituents are bulky. Substituents attached to the carbon from which the hydrogen is removed will cause greater hindrance to the base than will substituents on the carbon bearing the leaving group. Therefore, because of the methyl group attached to carbon 3, the transition state **9** should be more adversely affected than **10** if a more bulky base is used to effect the elimination. If, then, the base were large enough, the lesser substituted ethylene could become the major product.

Similar trends toward favoring the lesser substituted ethylene would also be observed by using a larger leaving group L[5] or by changing the substrate so that the size of the group(s) crowding the base is increased.

The following series of experiments is designed to demonstrate the steric effect as a molecular parameter of the E2 reaction and the general techniques of performing base-induced elimination reactions.[6]

EXPERIMENTAL PROCEDURE[7]

DO IT SAFELY

1. The majority of materials used in this experiment are highly flammable. *Use no flames.*

2. All precautions listed in the Do It Safely section for distillation on page 35 should be followed here. Pay particular attention to those regarding the assembly and integrity of your glassware.

---

[5] Steric effects caused by increasing the size of the leaving group make it more difficult to attain the *trans*-coplanar arrangement of HCCL leading to the more substituted ethylene. In examples in which the dimethylsulfonium or trimethylammonium group functions as the leaving group, the lesser substituted ethylene is usually the major product, regardless of the nature of the base.

[6] These experiments are based in part on the work of H. C. Brown and I. Moritani, *Journal of the American Chemical Society,* **75,** 4112 (1953).

[7] It is suggested that students do the experiment with different bases that are assigned at random. The results may then be collected, averaged for each base, and these results presented to the class in order to observe the trend of the effect with base size.

3. The solutions used in this experiment are *highly caustic.* Take care not to allow them to come into contact with your skin. If this should happen, flood the area with water and then neutralize the area with a solution of *dilute* (about 1%) acetic acid. We recommend that you wear rubber gloves while preparing and transferring solutions in this experiment.

4. If it is necessary for you to handle sodium metal during the experiment, remember that sodium metal reacts violently with water with the formation, and possible *explosive* combustion, of hydrogen gas. Use only *dry* containers, forceps, and so forth. Do not handle pieces of sodium metal with your bare fingers.

5. If you are to handle solid sodium methoxide or potassium *t*-butoxide, avoid spilling these during the weighing process. Although they will be hydrolyzed rather rapidly in moist air, the resulting solution will be strongly alkaline. Clean up spillages with a water-soaked paper towel, and then wash your hands.

A.   Elimination with Alcoholic Potassium Hydroxide

Place 0.075 mol of potassium hydroxide—5.0 g, correcting for the fact that commercial potassium hydroxide contains approximately 15% by weight of water (see Exercises)—and 50 mL of *absolute* ethanol in a *dry* 100-mL round-bottomed flask. Fit a calcium chloride drying tube to the flask and warm the mixture on a steam bath until the potassium hydroxide has dissolved. Cool the flask to room temperature, using an ice-water bath, and add 0.050 mol (5.3 g, 6.2 mL) of 2-chloro-2-methylbu-tane and a few boiling chips to the flask.

Continuation. Equip the flask for fractional distillation as shown in Figure 2.4. If you are using glassware equipped with ground-glass joints, be sure to lubricate the joint connecting the Hempel column to the flask with a hydrocarbon or silicone grease. (Lubrication of ground-glass joints is good laboratory practice at all times; it is particularly important in this experiment because the strong bases being used may cause the joints to "freeze.") In order to increase the efficiency of the Hempel column as a reflux condenser, fill it with Raschig rings or other packing. Fit the vacuum adapter holding the receiving flask with a calcium chloride drying tube and immerse the receiving flask in an ice-water bath. Circulate water through the jacket of the *Hempel column* during the period of reflux for this reaction. Heat the reaction mixture at a *gentle* reflux for a period of 2 hr. This is sufficient time to allow the reaction to go to about 95% completion. By the end of this time some solid should have precipi-tated, and may cause some bumping.★

At the end of the reflux period, cool the reaction mixture with an ice-water bath, turn off the cooling water, and remove the water hoses from the Hempel column, allowing the water to drain from the jacket. Reconnect the cooling water leads to the condenser so that the apparatus is now set for fractional distillation. If any of the low-boiling products has condensed in the receiving flask during the reflux period, allow it to remain, while continuing to cool the flask in an ice-water bath. Distil the product mixture, collecting all distillate boiling below 45° (2-methyl-1-butene, bp 31°; 2-methyl-2-butene, bp 38°). Transfer the product to a preweighed sample bottle with a *tight-fitting* stopper or cap and determine the yield. Perform qualitative tests which will demonstrate the presence of alkenes in the distillate (Chapter 25, p. 507). Submit a sample of your product for glpc analysis, and after receiving the results, calculate the relative percentage of the two isomeric alkenes formed. Typical glpc tracings of the products from this elimination and from one in which potassium *t*-butoxide was used as base are shown in Figure 6.1. Figure 6.8 shows the pmr spectrum of the product mixture from a typical experiment.

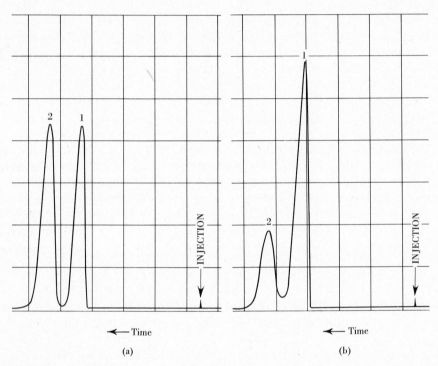

FIGURE 6.1 Typical glpc traces of the products of elimination of 2-chloro-2-methylbutane.[8] Assignments: peak 1: 2-methyl-1-butene; peak 2: 2-methyl-2-butene. (a) Elimination with KOH, showing approximately 44% 2-methyl-1-butene. (b) Elimination with $KOC(CH_3)_3$, showing approximately 76% 2-methyl-1-butene.

B.   Elimination with Sodium Methoxide

Prepare a solution of sodium methoxide in *anhydrous* methanol by either of the following methods. (1) If solid sodium methoxide is available, prepare the base solution by dissolving 0.075 mol (4.1 g) of the solid in 50 mL of *anhydrous* methanol contained in a dry 100-mL round-bottomed flask. Use a clean *dry* spatula for transferring the methoxide, taking care to perform the weighing operation as rapidly

[8]These analyses were performed at 45° on a 3-m column of 30% silicone gum rubber supported on Chromosorb P.

as possible to minimize the reaction between the methoxide and atmospheric water vapor (equation 7). The bottle from which the sodium methoxide is taken should be

$$NaOCH_3 + H_2O \longrightarrow HOCH_3 + NaOH \tag{7}$$

kept *tightly* closed. (2) The sodium methoxide solution may also be prepared by dissolving 0.075 g-atom (1.7 g) of sodium metal in 50 mL of *anhydrous* methanol. (*Caution: No flames; hydrogen is evolved*!) In order to weigh the sodium metal, place a 50-mL beaker containing 15 mL of *dry* toluene on a balance and determine its weight. Add to this weight on the balance an amount of weight equal to the weight of sodium desired. Sodium metal is normally stored under mineral oil for protection from the air (water and oxygen). Use a small knife or spatula (*dry*) to cut a small piece of sodium metal about the size of a pea. Impale this piece of sodium on the tip of the knife and "swish" it in a second beaker containing toluene to remove the mineral oil. Remove the piece of sodium, blot it with a *dry* paper towel, and transfer it to the beaker on the balance. Continue until the correct amount of sodium has been obtained. Place 50 mL of *anhydrous* methanol in a 100-mL round-bottomed flask, and with forceps remove a piece of sodium from the beaker of toluene, again blotting it to remove toluene, and add it to the methanol. Wait until the reaction subsides, then continue adding the sodium one piece at a time until it is all added. Attach a calcium chloride drying tube to the flask and allow time for the sodium to go completely into solution.

If the base solution prepared by either of these above methods is warm, cool it to room temperature in a water bath, and add 0.050 mol (5.3 g, 6.2 mL) of 2-chloro-2-methylbutane to the flask along with a few boiling chips. Complete the experiment from this point by following the directions in the paragraph headed Continuation in part A.

C. Elimination with Sodium Ethoxide

Prepare a solution of sodium ethoxide in *absolute* ethanol by dissolving 0.075 g atom (1.7 g) of sodium metal in 50 mL of *absolute* ethanol contained in a *dry* 100-mL round-bottomed flask. The procedure for handling the sodium metal is described in part B. Cool the solution to room temperature and add 0.050 mol (5.3 g, 6.2 mL) of 2-chloro-2-methylbutane to the flask. Add a few boiling chips, and complete the experiment from this point by following the directions under Continuation in part A.

D. Elimination with Potassium *t*-Butoxide

Place 0.075 mol (8.4 g) of solid potassium *t*-butoxide and 50-mL of *anhydrous* *t*-butyl alcohol in a *dry* 100-mL round-bottomed flask. Use the same precautions for handling potassium *t*-butoxide as used for solid sodium methoxide described in part B. Attach a calcium chloride drying tube and warm the flask in a water bath to dissolve the solid. It may not all dissolve, but continue even if some solid remains in the flask. Cool the flask to room temperature with an ice-water bath, and add 0.050 mol (5.3 g, 6.2 mL) of 2-chloro-2-methylbutane and a few boiling chips to the flask. Complete the experiment from this point by following the directions in the Continuation section of part A. A typical glpc tracing and a typical pmr spectrum of the products of this elimination are shown in Figures 6.1 and 6.9, respectively.

EXERCISES

1. What would be the expected results of the alkoxide-induced eliminations if the alcohol solvents employed were wet with water?
2. What is the solid material that precipitates as the eliminations proceed?
3. Why does the excess of base used in these eliminations favor the E2 elimination as opposed to the E1 elimination? (*Hint:* Consider the rate law expressions for the E1 and E2 processes.)
4. If all the elimination reactions in the experimental section had proceeded by the E1 mechanism, would the results have been different from those actually obtained? Why?
5. From the results of the experiments just performed and/or from the data in Figure 6.1, what conclusions can be drawn concerning the relative sizes of the bases employed?
6. What differences might you expect in product distributions for the eliminations of 2-chloro-2-methylbutane and 2-chloro-2,3-dimethylbutane with excess sodium methoxide in methanol solution?
7. If the leaving group in the 2-methyl-2-butyl system were larger than a methyl group, why would 2-methyl-1-butene be expected to be formed in greater amounts than if the leaving group were smaller than methyl, regardless of the base used? Use "sawhorse" structural formulas in your explanation.
8. What is the reason for packing the Hempel column with pieces of broken-glass tubing in these experiments?
9. Referring to Figures 6.5, 6.7, 6.8, and 6.9, calculate the percentage compositions of the mixtures of isomeric methylbutenes obtained from reaction of 2-chloro-2-methylbutane with potassium hydroxide and potassium *t*-butoxide.
10. Commercial potassium hydroxide contains approximately 15% of weight of water. Verify that to obtain 0.075 mol of potassium hydroxide one must weigh out 5.0 g of the commercial material.

SPECTRA OF STARTING MATERIALS AND PRODUCTS

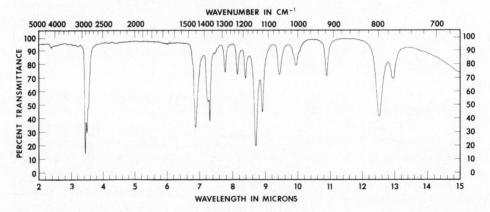

FIGURE 6.2 IR spectrum of 2-chloro-2-methylbutane.

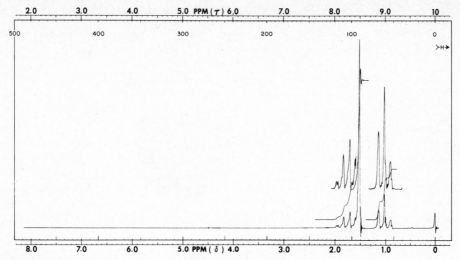

FIGURE 6.3 PMR spectrum of 2-chloro-2-methylbutane.

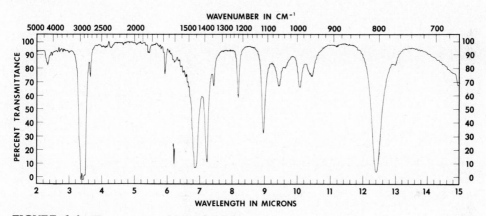

FIGURE 6.4 IR spectrum of 2-methyl-2-butene.

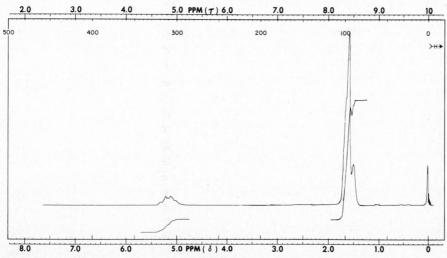

FIGURE 6.5 PMR spectrum of 2-methyl-2-butene.

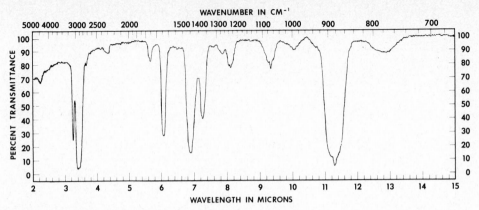

FIGURE 6.6 IR spectrum of 2-methyl-1-butene.

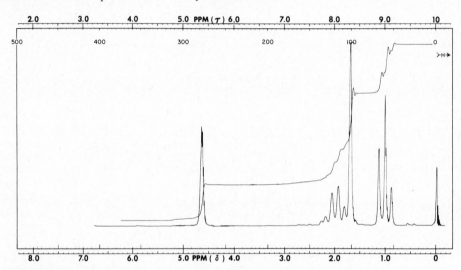

FIGURE 6.7 PMR spectrum of 2-methyl-1-butene.

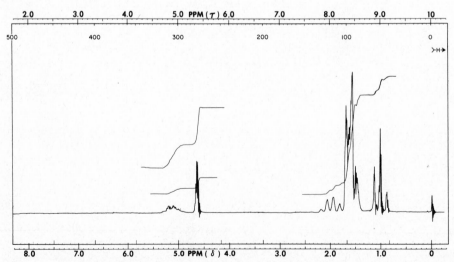

FIGURE 6.8 PMR spectrum of the product mixture from the elimination of 2-chloro-2-methylbutane with potassium hydroxide.

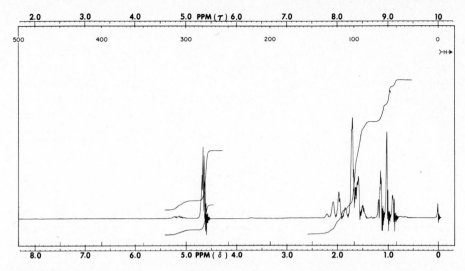

FIGURE 6.9 PMR spectrum of the product mixture from the elimination of 2-chloro-2-methylbutane with potassium *t*-butoxide.

6.2 Dehydration of Alcohols

The acid-catalyzed dehydration of an alcohol can usually be classified mechanistically as an elimination of the E1 type (see Section 6.1). This reaction (equation 8) involves the initial protonation of the hydroxyl oxygen atom to form an *oxonium ion* (**11**). The next step, kinetically a first-order (unimolecular) process, involves the

$$\text{(8)}$$

endothermic decomposition of the oxonium ion to a carbocation (**12**) (see footnote 2) and water. The carbocation then loses a proton from a carbon adjacent to the charged atom, yielding the alkene. This final loss of the proton is very probably aided by a molecule of water or alcohol in the reaction medium acting as a base, as shown in equation 8.

The second step leading to the carbocation is the rate-determining step (RDS) for the overall process and therefore controls the rate of dehydration of a given alcohol. The activated complex at the transition state for this step is shown here as **13**.

13

The free energy of the activated complex is predominantly influenced by the developing positive charge on the carbon atom as the carbon-oxygen bond breaks. Therefore the energies of activation for dehydration of various types of alcohols parallel the order of stability of the carbocations formed in each case. Since carbocation stability increases with an increasing number of alkyl or aryl substituents, tertiary carbocations are more stable than secondary, which in turn are more stable than

$$
\underset{\oplus}{CH_3-\overset{\overset{\displaystyle CH_3}{|}}{C}-CH_3} > \underset{\oplus}{CH_3-\overset{\overset{\displaystyle H}{|}}{C}-CH_3} > \underset{\oplus}{CH_3-\overset{\overset{\displaystyle H}{|}}{C}-H}
$$

$\longleftarrow$
increasing stability

primary. This directly relates to the observation that tertiary alcohols undergo dehydration more rapidly and at lower temperatures than the other types of alcohols. To reiterate, this is because the most stable of the three carbocations is formed in the rate-determining step in the dehydration of tertiary alcohols.

It should be noted that each of the steps along the reaction pathway is reversible; under the conditions of the experiment, the alkene may be rehydrated to alcohol. In order to carry the elimination to completion (100% conversion), the alkene may be distilled from the reaction mixture as it is formed. This has the effect of shifting the equilibrium to the right and maximizing the yield of alkene. *The constant removal from a reaction mixture of products as they are formed is a technique often utilized to afford high yields of products from reversible reactions.*

The same overall result could potentially arise by developing a method for continuous removal of water from the reaction mixture. However, in the experiments included at the end of this section, it is more advantageous to remove the alkenes. This is not only because the alkenes formed in each experiment are of lower boiling point than water but also because they would be produced in reduced yield if left in the presence of the sulfuric acid catalyst, which promotes the formation of polymeric products.

An E1 elimination is usually accompanied to some extent by substitution reactions. These competing reactions involve combination of the intermediate carbocation with the anion of the acid used for catalysis or some other anion. For example, with hydrochloric acid the chloride ion may react to give some chloroalkane, a product in which chlorine has replaced the hydroxyl group. The extent to which substitution products will be formed depends on the amount and nature of the acid used. However, since the formation of substitution products is frequently reversible under the reaction conditions, these side reactions will not always constitute a practical problem in affecting the yield of an elimination, assuming the alkene is distilled as it is formed.

Just as with the E2 reaction (Section 6.1), it is not at all uncommon for two or more isomeric alkenes to be formed. If the charge of the carbocation is unsymmetrically located on the carbon skeleton and hydrogens are attached to two of the adjacent carbon atoms, different products may be formed, depending on which of the hydrogens is lost. This possibility is illustrated in equation 9.

$$R-\underset{\underset{H_A}{|}}{\overset{\overset{H}{|}}{C}}-CH-\underset{\underset{H_B}{|}}{\overset{\overset{R''}{|}}{C}}-R'$$

$$\downarrow H^{\oplus}, \; -H_2O \qquad\qquad (9)$$

$$\underset{14}{\underset{R}{\overset{H}{\diagdown}}C=CH-\underset{\underset{H_B}{|}}{\overset{\overset{R''}{|}}{C}}-R'} \xleftarrow{-H_A^{\oplus}} R-\underset{\underset{H_A}{|}}{\overset{\overset{H}{|}}{C}}-\overset{\oplus}{C}H-\underset{\underset{H_B}{|}}{\overset{\overset{R''}{|}}{C}}-R' \xrightarrow{-H_B^{\oplus}} \underset{15}{R-\underset{\underset{H_A}{|}}{\overset{\overset{H}{|}}{C}}-CH=C\underset{\diagdown R'}{\overset{\diagup R''}{}}}$$

The proportion of the two alkenes (**14** and **15**) formed is dependent upon the relative activation energies of the two competing elimination processes. These activation energies are directly related to the free energies of the two transition states, which may be represented by the general formula

$$\underset{H \, \delta^{\oplus}}{\overset{}{C}} \mathrel{\ldots} \overset{}{C} \underset{\delta^{\oplus}}{}$$

In these transition states the C—H bond is partially broken and the C—C double bond is partially formed. As each of the two transition states has partial double-bond character, it is reasonable that their relative energies will parallel those of the corresponding alkenes, which of course have full double bonds. Therefore the more energetically favorable transition state will lead to the more stable alkene. Alkene stability usually increases as the number of alkyl substituents on the double bond is increased. Thus the more highly branched alkene (more highly substituted ethylene) is normally formed in higher yield when competing E1 eliminations are possible.

In many cases the number of kinds of alkenes formed as products in E1 reactions cannot be explained entirely by the type of reaction "choice" described. In general, carbocations are highly susceptible to rearrangement if through the process a more stable ion is produced. Indeed, some rearrangement may occur even when the new carbocation is less stable than the first, although the extent to which it occurs may be slight. Such rearrangements are accomplished by the migration of either an alkyl anion (R : ⁻) or a hydride ion (H : ⁻) from an adjacent carbon atom to the carbon having the formal positive charge. Loss of a proton from the new cationic intermediate may then lead to other isomeric alkenes.

Recalling that carbocationic stability increases from primary to secondary to tertiary, a migration that leads from a secondary or primary carbocation to a tertiary carbocation should be anticipated to be quite favorable. For example, the neopentyl cation (**16**), a primary ion, rearranges by methyl migration to give the *t*-pentyl cation (**17**) as the predominant species. The same "driving force" (tendency to minimize free energy) will also cause the isobutyl cation (**18**), a primary carbocation, to rearrange to the *t*-butyl cation (**19**) by hydride migration. In this case one should expect very little

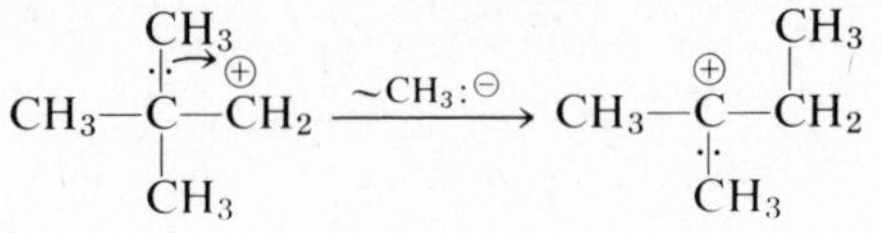

$$CH_3-\underset{\underset{CH_3}{|}}{\overset{\overset{CH_3}{|}}{C}}-\overset{\oplus}{CH_2} \xrightarrow{\sim CH_3:^{\ominus}} CH_3-\underset{\underset{CH_3}{|}}{\overset{\overset{CH_3}{|}}{\overset{\oplus}{C}}}-CH_2 \tag{10}$$

$$\underset{\textbf{16}}{} \qquad\qquad \underset{\textbf{17}}{}$$

methyl migration, because this would lead to a secondary ion, whereas hydride migration leads to the even more stable tertiary ion.

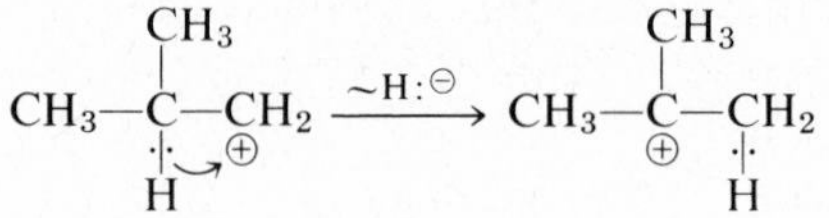

$$CH_3-\underset{\underset{H}{|}}{\overset{\overset{CH_3}{|}}{C}}-\overset{\oplus}{CH_2} \xrightarrow{\sim H:^{\ominus}} CH_3-\underset{\underset{H}{|}}{\overset{\overset{CH_3}{|}}{\overset{\oplus}{C}}}-CH_2 \tag{11}$$

$$\underset{\textbf{18}}{} \qquad\qquad \underset{\textbf{19}}{}$$

Experimental procedures for the dehydration of 4-methyl-2-pentanol and cyclohexanol are presented below. 4-Methyl-2-pentanol yields a mixture of isomeric alkenes, including 4-methyl-1-pentene **(20)**, *trans*-4-methyl-2-pentene **(21)**, *cis*-4-methyl-2-pentene **(22)**, 2-methyl-2-pentene **(23)**, and 2-methyl-1-pentene **(24)**. Products **23** and **24** are typically found to comprise approximately 45% of the product

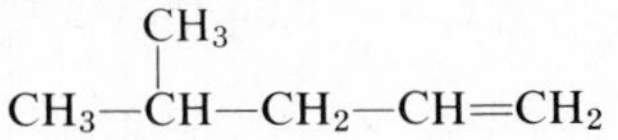

$$CH_3-\underset{\underset{}{\overset{\overset{CH_3}{|}}{C}H}}{}-CH_2-CH=CH_2 \qquad\qquad CH_3-\underset{\overset{CH_3}{|}}{CH}\overset{\displaystyle H}{\underset{\displaystyle CH_3}{\overset{}{}}}$$

$$\underset{\textbf{20}}{} \qquad\qquad\qquad \underset{\textbf{21}}{}$$

$$CH_3-\underset{\overset{CH_3}{|}}{CH}\,\,CH_3 \qquad\qquad CH_3\,\,CH_2-CH_3$$

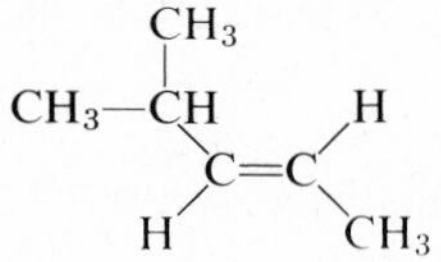

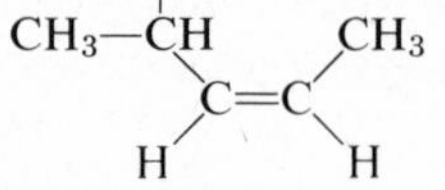

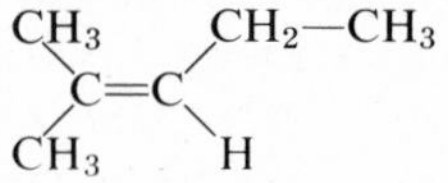

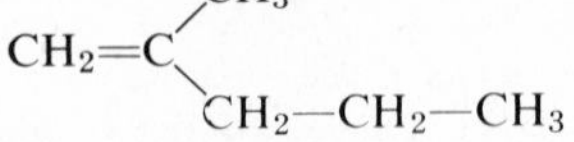

$$\underset{\textbf{22}}{} \qquad\qquad\qquad \underset{\textbf{23}}{}$$

$$CH_2=\underset{\overset{}{\underset{CH_2-CH_2-CH_3}{}}}{\overset{CH_3}{C}}$$

$$\underset{\textbf{24}}{}$$

mixture, showing that rearrangement is an important phenomenon. These products arise by loss of a proton from the carbocation formed by two successive hydride migrations (equation 12).

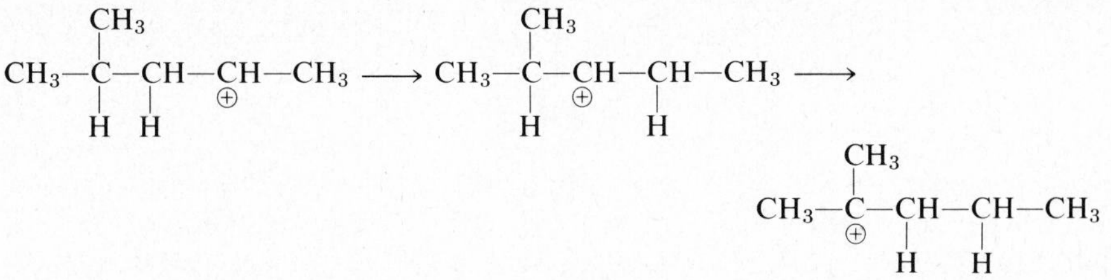

$$
\begin{array}{c}
\underset{\displaystyle \overset{\displaystyle CH_3}{|}}{CH_3-C-CH-CH-CH_3} \longrightarrow \underset{\displaystyle \overset{\displaystyle CH_3}{|}}{CH_3-C-CH-CH-CH_3} \longrightarrow
\end{array}
$$

$$
\underset{\displaystyle \overset{\displaystyle CH_3}{|}}{CH_3-C-CH-CH-CH_3} \tag{12}
$$

The rather complex reaction mixture may be analyzed by gas chromatography (see Section 3.5). By comparing retention times of authentic samples of the methylpentenes with those of the components of the mixture, the peaks in the gas chromatogram of the mixture can be assigned to the different methylpentenes. By measuring the relative areas under the peaks, the percentage of each component of the mixture may be obtained.

Cyclohexanol undergoes acid-catalyzed dehydration without rearrangement to yield a single product, cyclohexene. After purification the reaction product may be identified by comparing its ir spectrum with that of pure cyclohexene.

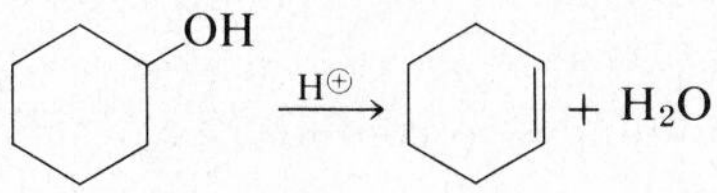

$$
\overset{OH}{\bigcirc} \xrightarrow{\;H^{\oplus}\;} \bigcirc + H_2O \tag{13}
$$

EXPERIMENTAL PROCEDURE

DO IT SAFELY

1. The materials used in this experiment are flammable, particularly the alkene products. Avoid the use of flames.

2. The precautions in the Do It Safely section on page 35 regarding distillation also apply here; read that section carefully.

3. There are several operations within the experiment which require you to pour, transfer, and weigh chemicals and reagents which you should not get on your hands. We recommend wearing rubber gloves during these operations, particularly during the work-up and washing steps with a separatory funnel.

4. If you do the tests for unsaturation on the products of these experiments, read the caution regarding *bromine as a hazardous chemical* in the Do It Safely section on page 135.

A. Dehydration of 4-Methyl-2-pentanol

Place 20 g (25 mL, 0.20 mol) of 4-methyl-2-pentanol in a 100-mL round-bottomed flask and to this add 10 mL of 9 *M* sulfuric acid.[9] Thoroughly mix the

---

[9]85% phosphoric acid is somewhat less satisfactory. The reaction is slower and the yields are lower. Use 5 mL of this acid if your instructor should desire you to do so.

contents of the flask by swirling. Add two or three boiling chips, and assemble the flask for fractional distillation according to Figure 2.4. To increase its efficiency the Hempel column should be filled with Raschig rings or other packing. The receiving flask should be immersed in an ice-water bath. Heat the reaction flask with a heating mantle or an oil bath, collecting all distillates, *but keeping the head temperature below 90°*. If the reaction mixture is not heated too strongly, the head temperature will remain at about 60–70° for most of the reaction. When only about 10 mL of liquid remains in the distillation flask, discontinue heating.

Transfer the organic distillate to a small separatory funnel, add 5–10 mL of 3 *M* sodium hydroxide solution and shake well, being cautious to vent the funnel from time to time.

Transfer the organic layer to a dry 50-mL Erlenmeyer flask and add 1–2 g of *anhydrous* calcium chloride.★ Occasionally swirl the mixture during a period of 10 min and then decant the dried organic mixture into a 100-mL distilling flask. Add boiling stones and distil the mixture through a simple distillation apparatus. Collect the fraction boiling between 53 and 69° in a preweighed receiver cooled in ice water. Weigh the product and calculate the yield obtained. If the procedure is carried out properly, a yield of 75–85% may be anticipated.★ The expected products of the reaction and their boiling points are 4-methyl-1-pentene (53.9°), *trans*-4-methyl-2-pentene (58.6°), *cis*-4-methyl-2-pentene (56.4°), 2-methyl-1-pentene (61°), and 2-methyl-2-pentene (67.3°).

Test the distillate for unsaturation, using both the bromine in carbon tetrachloride and the Baeyer tests (see Chapter 25, page 507). Submit a small sample of your product for glpc analysis. Figure 6.10 shows a typical glpc tracing of the products of this reaction. Calculate the percentage composition of your product mixture. Also submit another small sample for ir spectroscopic analysis. By comparing the absorption peaks from the ir spectrum obtained with those from spectra of each of the expected products, identify as many of the components as you are able (see Figures 6.13, 6.15, 6.17, 6.19, and 6.21).

B.   Dehydration of Cyclohexanol

Place 20.0 g (21.2 mL, 0.200 mol) of cyclohexanol and 10 mL of 9 *M* sulfuric acid (see footnote 9) in a 100-mL round-bottomed flask, and mix the contents well. Add two or three boiling chips and assemble the flask for fractional distillation according to Figure 2.4. To increase its efficiency the Hempel column should be filled with Raschig rings or other packing. The receiving flask should be immersed in an ice-water bath. Heat the reaction mixture with a heating mantle or oil bath, collecting all distillates, but keeping the head temperature below 90°. If the reaction mixture is not heated too strongly, the head temperature will remain below 80° for most of the reaction. When only about 10 mL of liquid remain in the distillation flask, discontinue heating.

Transfer the organic distillate to a small separatory funnel, and shake well with about 10 mL of 3 *M* sodium hydroxide solution, being cautious to vent the funnel from time to time. Transfer the organic layer to a dry 50-mL Erlenmeyer flask, and add 1–2 g of *anhydrous* sodium sulfate.★ Occasionally swirl the flask during a period

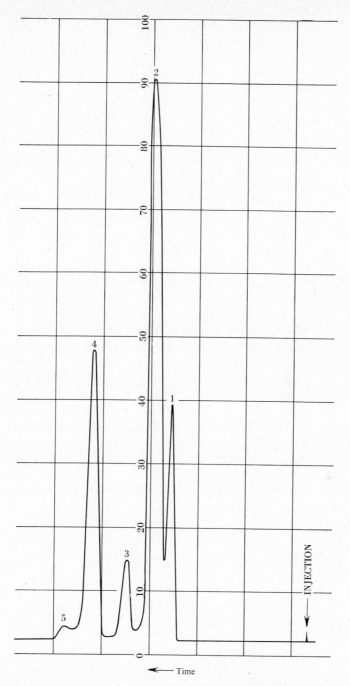

← Time

FIGURE 6.10 GLPC tracing of the product mixture from dehydration of 4-methyl-2-pentanol.[10,11]

[10] The peaks in Figure 6.10 have been assigned as follows: (1) 4-methyl-1-pentene (14.4%), (2) *cis-* and *trans*-4-methyl-2-pentene (combined 52.1%), (3) 2-methyl-1-pentene (6.5%), (4) 2-methyl-2-pentene (26.9%), and (5) unidentified.

[11] This analysis was made with a 2.4-m column packed with 5% SF-96 on 60/80 Chromosorb W at 35°.

of about 5 min, then decant the liquid into another 50-mL Erlenmeyer flask, and add a fresh 1–2 g portion of *anhydrous* sodium sulfate. Swirl occasionally during the next 5 min or so and filter the liquid into a *dry* 50-mL round-bottomed flask. (The product *must* be dry at this stage in order to obtain pure cyclohexene, since water and cyclohexene form a minimum-boiling azeotrope.) Distil the crude cyclohexene through a simple distillation apparatus, and collect the product as a fraction boiling between 80 and 85° in a preweighed ice-cooled receiver. A calcium chloride drying tube should be attached to the side arm of the vacuum adapter.★ Calculate the yield of the reaction. Perform both the bromine in carbon tetrachloride and the Baeyer tests for unsaturation on your product (see Chapter 25, p. 507). Submit a sample of your product for ir analysis. Compare your spectrum with that of authentic cyclo-hexene (Figure 6.25).

EXERCISES

1. Suggest the reason for washing the distillates in each of the preceding experiments with dilute sodium hydroxide.
2. Why is the distillation temperature kept below 90° during the preceding dehydration experiments?
3. Which of the following primary alcohols would be most likely to dehydrate by the E1 mechanism? by the E2 mechanism? Explain.

$$CH_3CH_2OH \qquad (CH_3)_3CCH_2OH$$

4. Give a detailed mechanism explaining each of the products obtained from the dehydration of 4-methyl-2-pentanol.
5. Near the end of the dehydration of 4-methyl-2-pentanol, a white solid may precipitate from the reaction mixture. What is the solid likely to be?
6. Give structures for the products of dehydration of each of the following alcohols. For each, order the products with respect to preference of formation.

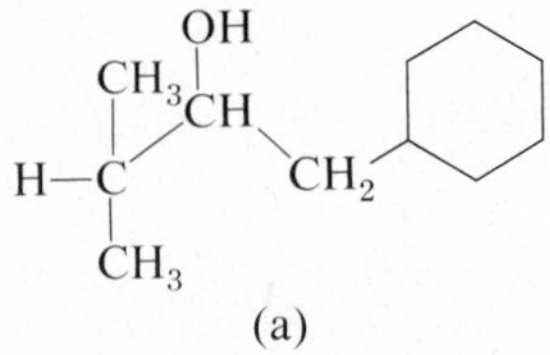

(a)

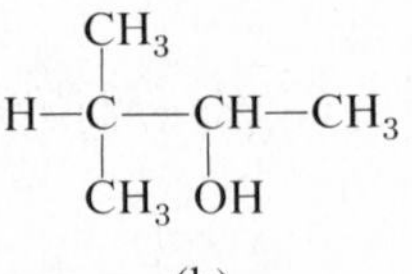

(b)

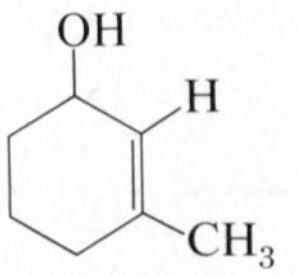

(c)

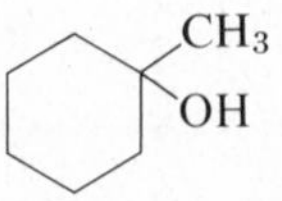

(d)

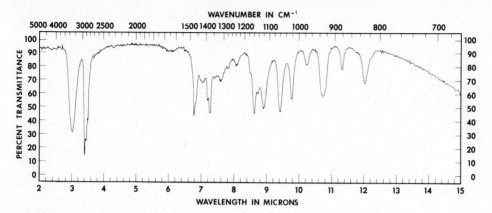

FIGURE 6.11 IR spectrum of 4-methyl-2-pentanol.

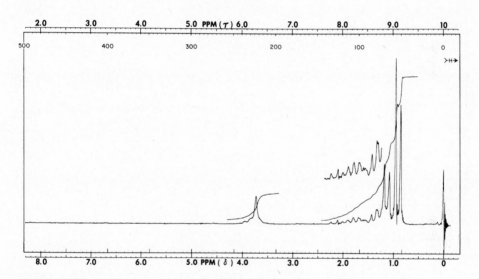

FIGURE 6.12 PMR spectrum of 4-methyl-2-pentanol.

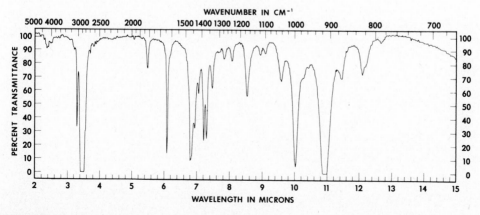

FIGURE 6.13 IR spectrum of 4-methyl-1-pentene.

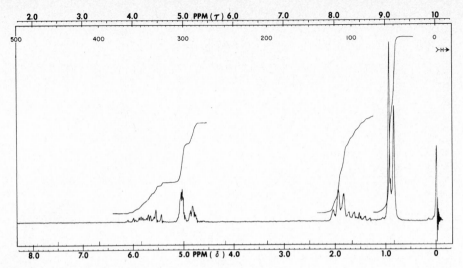

FIGURE 6.14 PMR spectrum of 4-methyl-1-pentene.

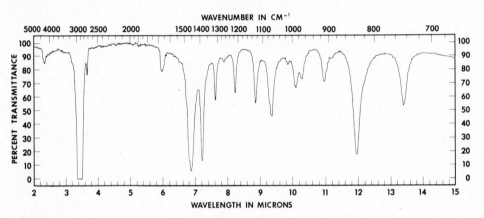

FIGURE 6.15 IR spectrum of 2-methyl-2-pentene.

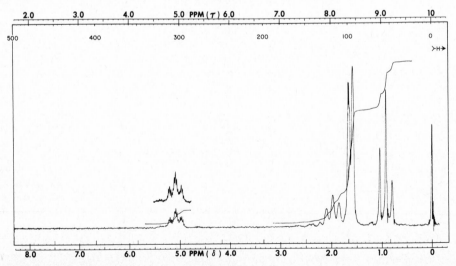

FIGURE 6.16 PMR spectrum of 2-methyl-2-pentene.

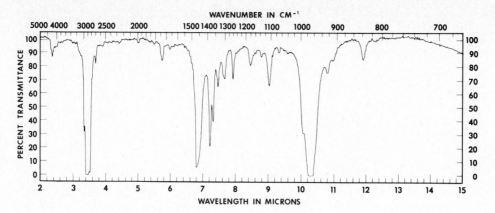

FIGURE 6.17 IR spectrum of *trans*-4-methyl-2-pentene.

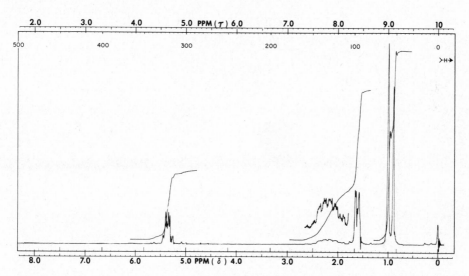

FIGURE 6.18 PMR spectrum of *trans*-4-methyl-2-pentene.

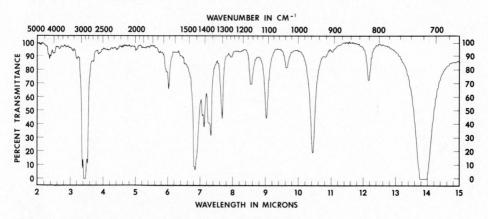

FIGURE 6.19 IR spectrum of *cis*-4-methyl-2-pentene.

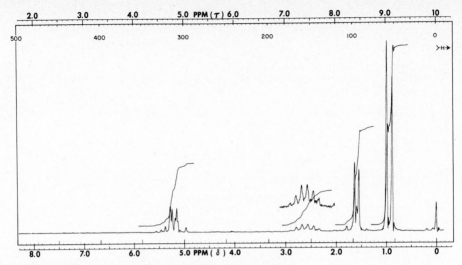

FIGURE 6.20 PMR spectrum of *cis*-4-methyl-2-pentene.

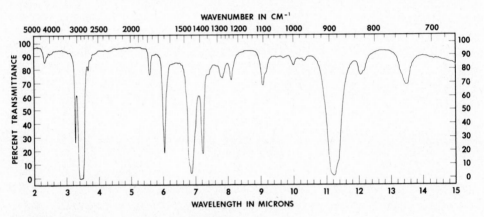

FIGURE 6.21 IR spectrum of 2-methyl-1-pentene.

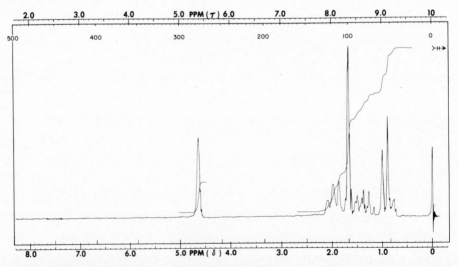

FIGURE 6.22 PMR spectrum of 2-methyl-1-pentene.

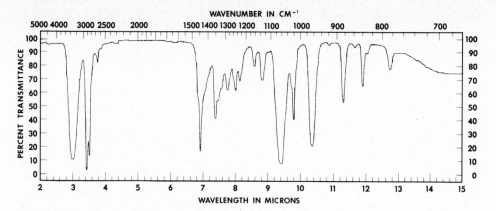

FIGURE 6.23 IR spectrum of cyclohexanol.

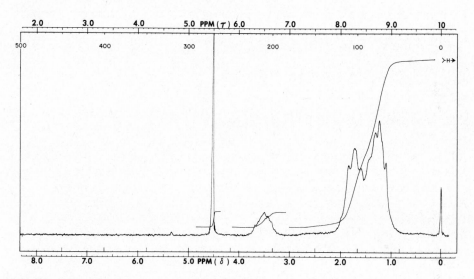

FIGURE 6.24 PMR spectrum of cyclohexanol.

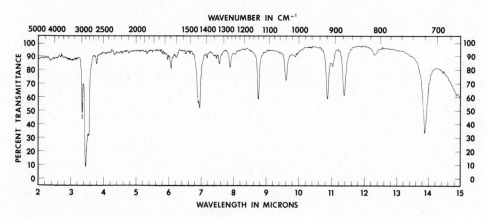

FIGURE 6.25 IR spectrum of cyclohexene.

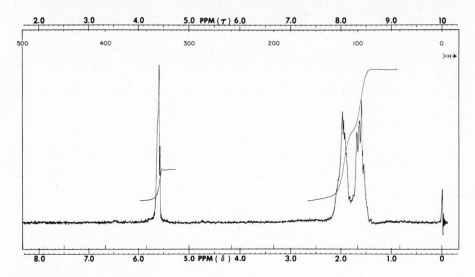

FIGURE 6.26   PMR spectrum of cyclohexene.

6.3   Addition Reactions of Alkenes

Alkenes as a class are very useful starting materials for organic syntheses, since these compounds undergo a large variety of reactions involving the carbon-carbon pi bond. In most instances the chemistry of the double bond of alkenes amounts to the *addition* of a reagent, X—Y, across the pi bond so that a saturated molecule results (equation 14).

$$R_2C{=}CR_2 + X{-}Y \rightarrow R_2\underset{X}{C}{-}\underset{Y}{C}R_2 \tag{14}$$

The rich chemistry of alkenes is for the most part dependent on one or both of two factors. (1) The pi bond is somewhat weaker (about 65 kcal/mol) than typical sigma bonds (80–100 kcal/mol), so that most of the reactions of alkenes are exothermic. (2) The pi-bonding electrons are more loosely held to the bonded carbon atoms than are sigma electrons, making them much more polarizable; the pi-electronic "cloud" is thus readily distorted through electrostatic interaction with an approaching reagent, which enhances the reactivity of the alkene toward attack. These features are not necessarily important in *all* reactions of alkenes, but they are a significant influence in many of the reactions which alkenes undergo.

The detailed mechanism through which the addition proceeds is dependent upon both the nature of X—Y and the conditions under which the reaction is accomplished. One of the more common mechanisms of addition of a reagent to an alkene involves the attack of an electrophile, E^+, an electron-deficient species, on the double bond to give an intermediate onium ion (**25**), which reacts with a nucleophilic reagent, Nu^-, to give the product **26** (equation 15). For unsymmetrical X—Y compounds, Nu is the more electronegative of the two atoms. The bromination of alkenes, a reaction used both as a qualitative test for unsaturation as described in Chapter 25 and as a quantitative measure of the amount of unsaturation, generally proceeds by this mechanism. The direction of addition of E—Nu to an unsymmetrical

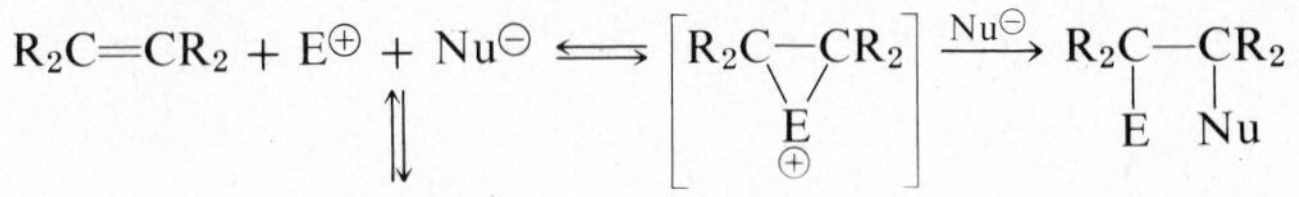

$$R_2C{=}CR_2 + E^{\oplus} + Nu^{\ominus} \Longleftrightarrow \left[\underset{\underset{\oplus}{E}}{R_2C{-}CR_2} \right] \xrightarrow{Nu^{\ominus}} \underset{E\ \ Nu}{R_2C{-}CR_2} \qquad (15)$$

$$\underset{E-Nu}{\Updownarrow} \qquad\qquad\qquad \underset{\mathbf{25}}{} \qquad\qquad \underset{\mathbf{26}}{}$$

alkene is such as to result in bond formation between the electrophile, E^+, and the carbon atom bearing the larger number of hydrogen atoms [Markownikoff's rule, for example, equation 16]. The stereochemistry of the addition is such that E and Nu are *trans* to one another, as shown by the observation that *trans*-1,2-dibromocyclohexane **(28)** is obtained from the addition of bromine to cyclohexene.

$$R_2C{=}CH_2 + HBr \xrightarrow{CCl_4} \underset{\underset{\mathbf{27}}{Br\ \ H}}{R_2C{-}CH_2} \qquad (16)$$

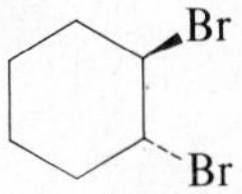

28

A related stepwise mechanism, which is observed less commonly, involves initial attack on the alkene by a nucleophile, Nu^-, to produce the carbanion **29**, which then reacts with E^+ to provide **30** (equation 17). In fact, addition by this mechanism occurs readily only if the double bond bears substituents such as cyano (—C≡N), nitro (—NO_2), or carbonyl (—C—), which are capable of stabilizing a negative charge. The
  ‖
  O

$$R_2C{=}CR_2 + Nu^{\ominus} \Longleftrightarrow \underset{\underset{\mathbf{29}}{Nu}}{R_2C{-}\overset{\ominus}{C}R_2} \xrightarrow[\text{(or E—Nu)}]{E^{\oplus}} \underset{\underset{\mathbf{30}}{Nu\ \ E}}{R_2C{-}CR_2} \qquad (17)$$

1,4-addition of aniline to benzalacetophenone (equation 18), an experiment described in Chapter 14, follows a mechanism of this type.

$$C_6H_5CH{=}CHCOC_6H_5 + C_6H_5NH_2 \longrightarrow \underset{}{\overset{NHC_6H_5}{\underset{|}{C_6H_5CH{-}CH_2COC_6H_5}}} \qquad (18)$$

A reagent, X—Y, can often be added to an alkene by a free radical, stepwise process if free radical initiators are present or if the reaction mixture is exposed to light. In such reactions the radical, $X\cdot$, adds to the double bond to yield an intermediate radical (**31**). The subsequent reaction between **31** and $X{-}Y$ provides the product (**26**) and regenerates $\mathbf{X}\cdot$ (equation 19). With unsymmetrical alkenes, X

$$R_2C{=}CR_2 + X\cdot \longrightarrow \underset{31}{R_2\overset{\displaystyle |}{\underset{\displaystyle X}{C}}{-}\dot{C}R_2} \xrightarrow{X{-}Y} \underset{26}{R_2\overset{\displaystyle |}{\underset{\displaystyle X}{C}}{-}\overset{\displaystyle |}{\underset{\displaystyle Y}{C}}R_2 + X\cdot} \tag{19}$$

becomes bonded to that carbon bearing the larger number of hydrogen atoms to yield the more stable of the two possible radicals as an intermediate (equation 20).

$$R_2C{=}CH_2 + HBr \xrightarrow[\text{(initiator)}]{\text{peroxides}} \underset{32}{R_2\overset{\displaystyle |}{\underset{\displaystyle H}{C}}{-}\overset{\displaystyle |}{\underset{\displaystyle Br}{C}}H_2} \tag{20}$$

Thus the products of polar and of free radical addition of hydrobromic acid to an alkene, **27** and **32**, respectively, are different, since bromine atom, $Br\cdot$, is the chain carrier in the latter process. One must therefore carefully control reaction conditions in order to obtain pure products when free radical and polar addition can produce different products. Such control may still be important even in those instances in which the *direction* of addition is of no importance, since the *stereoselectivity* of free radical addition is generally low, and products resulting from both *trans* and *cis* addition of $X{-}Y$ to the double bond can be obtained.

Reaction conditions which favor *electrophilic* addition are low temperatures, use of polar solvents such as water and alcohols, presence of ionic salts like sodium or potassium bromide in the bromination reaction, and absence of light and of peroxides. In contrast, *free radical* addition is favored by performing the reaction in the gas phase or in nonpolar solvents in the presence of strong light or peroxides.

A final point concerning additions to alkenes is noteworthy. We have now formulated three general mechanisms of addition of a reagent, $X{-}Y$, to an alkene, mechanisms which are distinguished by the type of species, electrophile, nucleophile, or free radical, involved in the initial step of the reaction. This first step, at least in principle, could be *reversible,* and in the case of appropriately substituted alkenes could lead to geometric isomerization (equation 21). Thus the intermediates **29** and

$$A^* + \underset{R_1}{\overset{R}{\diagdown}}C{=}C\underset{R_1}{\overset{R}{\diagup}} \rightleftharpoons \left[\underset{R_1}{\overset{R}{\diagdown}}\overset{*}{C}{-}C\overset{R}{\underset{R_1}{\diagup}}A\right] \rightleftharpoons \left[\underset{R_1}{\overset{R}{\diagdown}}\overset{*}{C}{-}C\overset{R_1}{\underset{R}{\diagup}}A\right] \rightleftharpoons \underset{R_1}{\overset{R}{\diagdown}}C{=}C\underset{R}{\overset{R_1}{\diagup}} + A^* \tag{21}$$

$(*= \odot,\oplus,\ominus)$

31, which are generated by addition of a nucleophile and of a free radical, respectively, to an alkene, are free to undergo *rotation* about the single bond *between* the carbons which were originally joined by a double bond. Regeneration of the alkene could produce a mixture of geometric isomers. Geometric isomerization might also be a consequence of reversible addition of an electrophile to an alkene if the intermediate so formed, **25**, exists in equilibrium with a carbocation (equation 22) which would be free to undergo rotation about the necessary carbon-carbon bond to produce isomerization (equation 21, * = ⊕).

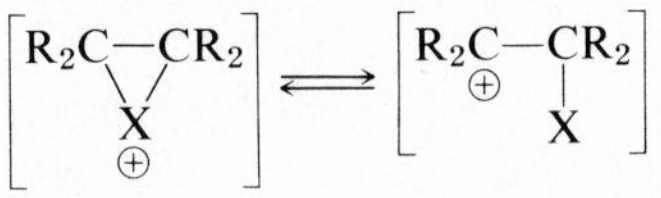

$$\left[\begin{array}{c} R_2C-CR_2 \\ \diagdown V \diagup \\ X \\ \oplus \end{array}\right] \rightleftharpoons \left[\begin{array}{c} R_2C-CR_2 \\ \oplus \quad | \\ X \end{array}\right] \qquad (22)$$

25

As a specific example, consider the possible consequences of the addition of bromide ion to dimethyl maleate **(33)**, a *cis* isomer (equation 23). This process would give the stabilized anion **34** which could undergo rotation about the carbon-carbon bond to afford **35**. Loss of bromide ion from **35** provides the geometric isomer of **33**, dimethyl fumarate **(36)** (equation 23):

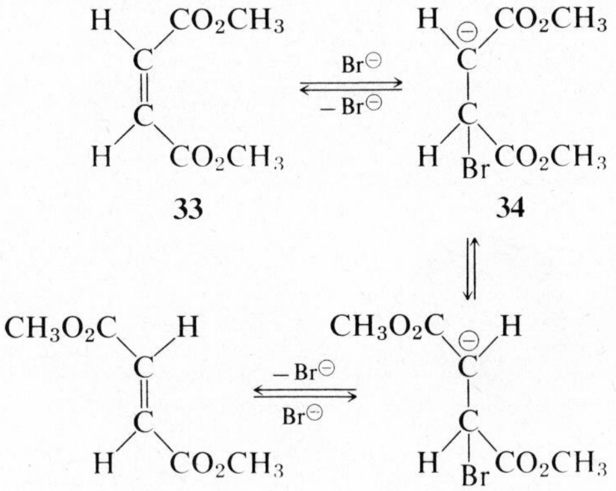

$$ (23) $$

As described in the following Experimental Procedure, the conversion of dimethyl maleate to dimethyl fumarate is accomplished by treatment of the maleate with a dilute solution of bromine in carbon tetrachloride. A potentially important reaction, the attack of Br_2 on the double bond of either the maleate or the fumarate, is relatively unimportant because the presence of two electron-withdrawing carbomethoxy groups, $-CO_2CH_3$, on the double bond decreases the electron density of the bond sufficiently to preclude facile attack of an electrophile. Therefore the mechanism of the geometric isomerization may be either nucleophilic or free radical. Observations made in the experiment should provide a basis for a decision between the two.

A fourth type of mechanism for the reaction of the reagent X—Y with an alkene is concerted addition (equation 24). As might be anticipated, the stereochemistry of the reaction is such as to provide *cis* addition of the reagent. Some reactions of this general mechanistic type are hydroxylation, hydroboration, ozonolysis, hydrogenation, and the Diels-Alder reaction. All these reactions are of considerable synthetic utility and examples of some of them are described elsewhere in this text (see Experimental Procedure and Chapter 8).

$$R_2C{=}CR_2 + X{-}Y \longrightarrow \left[\begin{array}{c} R_2C{\cdots}CR_2 \\ | \quad | \\ X{\cdots}Y \end{array}\right] \longrightarrow \begin{array}{c} R_2C{-}CR_2 \\ | \quad | \\ X \quad Y \end{array} \qquad (24)$$

In the experiment that follows, an alkene will be converted to an alkane by hydrogenation ($X = Y = H$). This reaction can be accomplished in several ways, but generally it involves use of hydrogen gas either at atmospheric pressure or greater and a noble metal such as palladium, platinum, or nickel as the catalyst. For our purposes it will be convenient to utilize sodium borohydride as the *in situ* source of hydrogen (equation 25) and to generate the catalyst by reduction of platinic chloride with hydrogen in the presence of decolorizing carbon, which serves as a solid support for the finely divided metallic platinum produced (equation 26).

$$NaBH_4 + HCl \xrightarrow{\;3\ H_2O\;} 4\ H_2 + B(OH)_3 + NaCl \qquad (25)$$

$$H_2PtCl_6 + 2\ H_2 \longrightarrow Pt° + 6\ HCl \qquad (26)$$

EXPERIMENTAL PROCEDURE

A.   Geometric Isomerization of Dimethyl Maleate

DO IT SAFELY

Bromine is a hazardous chemical. Do not breathe its vapors or allow it to come into contact with the skin because it may cause serious chemical burns. All operations involving the transfer of solutions of bromine should be carried out in a ventilation hood; it is prudent to wear rubber gloves. If you get bromine on your skin, wash the area quickly with soap and warm water and soak the skin in 0.6 M sodium thiosulfate solution (up to 3 hr if the burn is particularly serious).

Place 1.5 mL of dimethyl maleate in each of three 150-mm test tubes and with a dropper add enough of a 0.6 M solution of bromine in carbon tetrachloride to *two* of the tubes so that an orange solution results. Add an equal volume of carbon tetra-chloride to the third test tube. Stopper all the test tubes and place one of the tubes containing bromine in the dark and expose the other two tubes to strong light. If decoloration of a solution should occur, add an additional portion of the bromine solution. After 30 min cool all three solutions in ice water, observe in which test tube(s) crystals appear, and isolate the precipitate by vacuum filtration. Wash the crystals free of bromine with a little *cold* carbon tetrachloride and press them as dry as possible on the filter disk. Recrystallize the product from ethanol and determine its melting point and weight. The reported melting point for dimethyl fumarate is 101–102°.

Add a few drops of cyclohexene as needed to any solutions or containers in which the color of bromine is evident in order to dispel that color, and then discard all solutions in the bottle labeled "organic liquid waste."

B. Hydrogenation of 4-Cyclohexene-*cis*-1,2-dicarboxylic Acid

---

DO IT SAFELY

---

1. Hydrogen gas is extremely flammable and may be explosively ignited. Use no flames, and for your own safety do not allow your neighbors to do so.

2. If, in the isolation of the crude product, you proceed to remove the ether solvent by evaporation in the hood, *do not use a hot plate.* Rather, as indicated in the procedure, use a steam bath. One of the authors once had a rather serious fire in the laboratory when dense ether vapors billowed onto a hot plate and were ignited by the very hot surface of the electrical heating element. Although a little more time is involved, it is best to remove the ether by simple distillation, using a steam bath.

3. As in all experiments of this type, avoid skin contamination by the solutions you are handling. In several of the operations the use of rubber gloves is advisable. If you do get these solutions on your hands, wash them thoroughly with warm water and soap. In the case of sodium borohydride solution, rinse the area with 1% acetic acid solution; in case of acid burns, apply a paste of sodium bicarbonate to the area for a few minutes.

Prepare a reaction vessel for hydrogenation by tying a heavy-walled balloon to the sidearm of a 125-mL filter flask. Also prepare a 1 M aqueous solution of sodium borohydride by dissolving 0.4 g (0.01 mol) of sodium borohydride in 10 mL of water and adding 0.1 g of sodium hydroxide as a stabilizer. Place 10 mL of water, 1 mL of a 5% solution of chloroplatinic acid ($H_2PtCl_6 \cdot 6H_2O$), and 0.5 g of decolorizing carbon in the reaction flask and add, with swirling, 3 mL of the 1 M sodium borohydride solution. Allow the resulting slurry to stand for 5 min to permit formation of the catalyst. During this time dissolve 1 g (0.006 mol) of 4-cyclohexene-*cis*-1,2-dicar-boxylic acid (see Chapter 8) by heating it with 10 mL of water.

Pour 4 mL of *concentrated* hydrochloric acid into the catalyst-containing reaction flask, and then add the hot aqueous solution of the diacid to the flask. Seal the flask with a serum cap, and wire the cap securely in place. Draw 1.5 mL of the 1 M sodium borohydride solution into a plastic syringe, push the needle of the syringe through the serum cap, and inject the solution dropwise while swirling the flask. If the balloon on the sidearm of the flask becomes inflated, stop the addition of sodium borohydride solution until deflation occurs. When the syringe is empty, remove it from the flask, and refill it with an additional 1.5 mL of solution. The dropwise addition of this further quantity of sodium borohydride should cause the balloon to inflate and to remain inflated, indicating a positive pressure of hydrogen in the system. Remove the syringe from the serum cap and allow the flask to stand, occasionally swirling for 5 min. Heat and swirl the flask on the steam cone until the balloon deflates to a constant size and then heat for an additional 5 min. A total of no more than 20 min of heating should be required.

Release the pressure from the reaction flask by pushing through the serum cap the needle of a syringe from which the barrel has been removed, filter the hot reaction mixture by suction, and place the filter paper in a container reserved for "recovered catalyst."＊ Cool the filtrate and extract it with three 25-mL portions of *technical* ether. Combine the extracts, wash them with 10 mL of saturated sodium chloride solution, and then filter the organic solution through a cone of sodium sulfate into a tared round-bottomed flask. Remove the ether by simple distillation, discarding the distillate in the organic liquid-waste bottle. The product will remain as a solid residue. Allow the crude product to air-dry, and then determine its weight and melting point.

Transfer the bulk of the crude diacid by scraping it into a 25-mL Erlenmeyer flask. Add about 2 mL of water to the larger flask, heat to dissolve any residual diacid, and pour the hot solution into the smaller flask. Bring all the diacid into solution at the boiling point by adding no more than two additional milliliters of water. After solution has been accomplished, add 3 drops of *concentrated* hydrochloric acid to decrease the solubility of the diacid in the solution, allow the mixture to cool to room temperature, and then place the flask in an ice-water bath to effect more complete crystallization of product. Isolate the product and determine its weight and melting point. The reported melting point is 192°. Ascertain that the product is saturated by performing the qualitative tests for unsaturation described in Chapter 25, page 507. Also determine whether hydrogenation has affected the acidic nature of the molecule by testing an aqueous solution of the diacid with litmus paper.

EXERCISES

GEOMETRIC ISOMERIZATION

1. Write a reasonable mechanism for the bromine-catalyzed geometrical isomerization of dimethyl maleate to dimethyl fumarate observed in this experiment.
2. What is the function of light in the isomerization reaction?
3. What is the purpose of exposing a sample of dimethyl maleate containing no bromine to light?
4. Suggest a reason why decoloration of the solution of bromine and dimethyl maleate is slow.
5. Offer an explanation for the observation that maleic anhydride (**37**) does not isomerize to **38** under the influence of bromine and light.

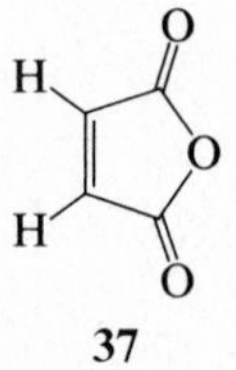

37 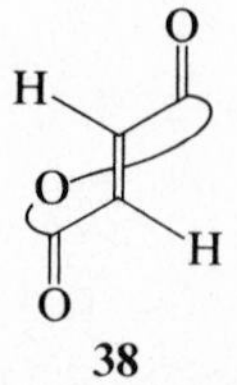   38

6. Figure 6.27 is a pmr spectrum of a mixture of dimethyl fumarate and dimethyl maleate. Calculate the percentage of each isomer in this mixture.

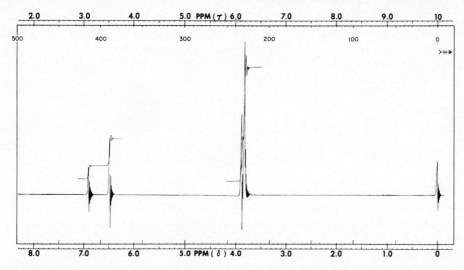

FIGURE 6.27 PMR spectrum of a mixture of dimethyl fumarate and dimethyl maleate.

HYDROGENATION

1. Draw the structure of the product of catalytic hydrogenation of (a) 1,2-dimethylcyclohexene, (b) *cis*-2,3-dideuterio-2-butene, (c) *trans*-2,3-dideuterio-2-butene.

2. Why does the addition of hydrochloric acid to an aqueous solution of a carboxylic acid decrease the solubility of the carboxylic acid in water?

SPECTRA OF STARTING MATERIALS AND PRODUCTS

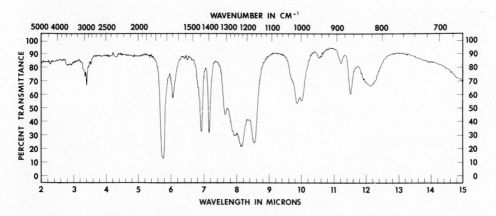

FIGURE 6.28 IR spectrum of dimethyl maleate.

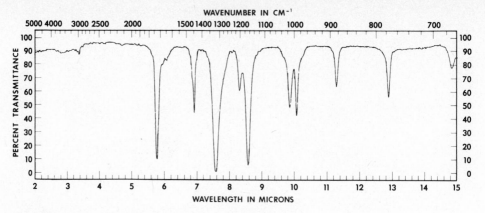

FIGURE 6.29 IR spectrum of dimethyl fumarate.

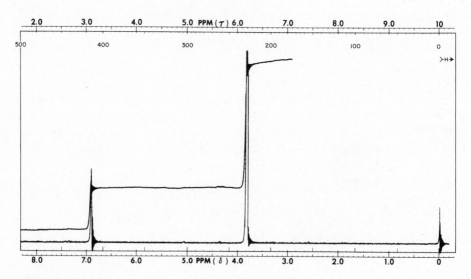

FIGURE 6.30 PMR spectrum of dimethyl fumarate.

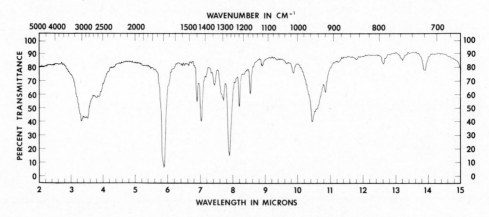

FIGURE 6.31 IR spectrum of 4-cyclohexene-*cis*-1,2-dicarboxylic acid.

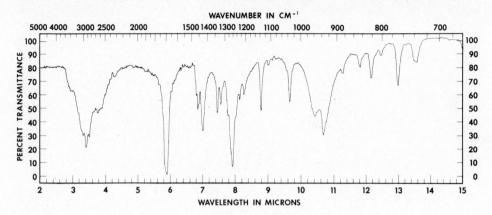

FIGURE 6.32 IR spectrum of *cis*-cyclohexane-1,2-dicarboxylic acid.

7

ELECTROPHILIC AROMATIC SUBSTITUTION

7.1 Introduction to Electrophilic Aromatic Substitution

Electrophilic aromatic substitution is a very important part of organic chemistry, because the introduction of many functional groups onto an aromatic ring is accomplished in this manner. The general form of this reaction may be represented by equation 1, in which Ar—H is an aromatic compound (an "arene") and E^+ is any

$$Ar—H + E^{\oplus} \longrightarrow Ar—E + H^{\oplus} \tag{1}$$

one of a number of different electrophiles that may replace H. Although equation 1 represents the overall net reaction correctly, it is greatly oversimplified. For example, the electrophile must usually be generated from the starting materials during the course of the reaction.

The chemical kinetics of electrophilic aromatic substitution reactions have been studied extensively. For many but not all[1] cases the rate of substitution has been found to be first order in arene and first order in electrophile:

$$\text{rate} = k_2[Ar—H][E^{\oplus}] \tag{2}$$

Because of its bimolecular nature it is often termed an S_E2 reaction (S = substitution, E = electrophilic, and 2 = bimolecular). A general mechanism for the S_E2 reaction is given below. This mechanism has been substantiated by considerable evidence, especially that coming from isotope effect studies.

---

[1] No attempt is made here to differentiate specifically between the systems which follow the second-order rate expression and those which do not; in general, however, activated arenes do not follow this rate expression.

Step 1: Formation of electrophile (an equilibrium reaction):

$$E-Nu \xrightleftharpoons{\text{catalyst}} E^{\oplus} + Nu^{\ominus} \qquad (3)$$

Step 2: Reaction of electrophile with arene (slow step):

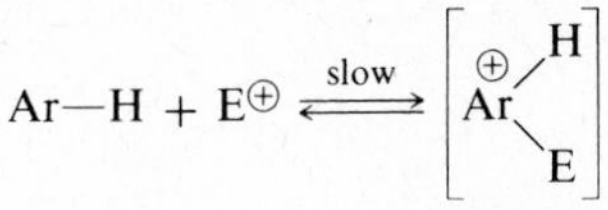

$$Ar-H + E^{\oplus} \xrightleftharpoons{\text{slow}} \left[Ar \overset{\oplus}{\underset{E}{\overset{\diagup H}{\diagdown}}} \right] \qquad (4)$$

Intermediate

Step 3: Loss of proton to give product (fast step):

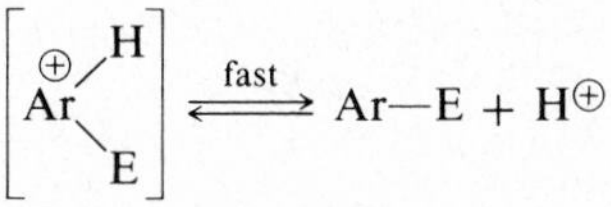

$$\left[Ar \overset{\oplus}{\underset{E}{\overset{\diagup H}{\diagdown}}} \right] \xrightleftharpoons{\text{fast}} Ar-E + H^{\oplus} \qquad (5)$$

The electrophile is most often produced by the reaction between a catalyst and a compound which contains a potential electrophile (equation 3). The second-order nature of the reaction may be seen to be related to equation 4, in which *two* molecules react to give the intermediate. If indeed the kinetics are second order, then this must be the slow, rate-limiting step in the overall reaction; the loss of a proton (equation 5) which follows must be fast relative to the reaction of equation 4.

A listing of various common electrophiles, the conditions under which they are produced, and the reactions which they undergo is given in Table 7.1, along with references to other parts of this book where experiments utilizing them are given. The specific mechanisms, along with detailed information and side reactions, will accompany each of the experiments in the text.

TABLE 7.1   EXAMPLES OF ELECTROPHILIC AROMATIC SUBSTITUTION REACTIONS

Type of Reaction	Electrophile, E^+	Electrophile Precursor	Catalyst, If Any[a]	Structure of Product	Reference to This Reaction
Friedel-Crafts alkylation	R^+	$R-X$	AlX_3	$Ar-R$	Section 7.2
Friedel-Crafts acylation	$\overset{+}{R-C}=O$	$R-\overset{\overset{\displaystyle O}{\|}}{C}-X$	AlX_3	$Ar-\overset{\overset{\displaystyle O}{\|}}{C}-R$	Section 17.1, 18.2
Chlorination	Cl^+	$Cl-Cl$	None	ArCl	Section 17.3
Bromination	Br^+	$Br-Br$	None	ArBr	Section 17.3
Iodination	I^+	$I-Cl$	None	ArI	Section 17.3
Nitration	NO_2^+	$HO-NO_2$	H_2SO_4	$ArNO_2$	Section 7.3
Sulfonation	SO_3	$Cl-SO_3-H$	None	$ArSO_3H$	Section 17.2

[a] Catalysts listed are those used in the experiments referred to in the table; in those reactions where no catalyst is indicated, other examples than those specifically referred to frequently require catalysts.

7.2 Friedel-Crafts Alkylation of *p*-Xylene with 1-Bromopropane

The Friedel-Crafts alkylation reaction is one of the classic types of electrophilic aromatic substitution and as such has been subjected to extensive mechanistic study. It is also of great industrial importance as the most versatile method for attaching alkyl side chains to aromatic rings. The two main limitations to this reaction as a synthetic tool are (1) the difficulty of preventing the introduction of more than one alkyl group onto an aromatic ring (owing to the activating effect of the first group introduced) and (2) the occurrence of rearrangements of the alkyl group. The first difficulty can often be resolved by using a large excess of the arene (note the proportion of *p*-xylene to 1-bromopropane used in this experiment). There has been considerable confusion about the second limitation, the occurrence and extent of alkyl group rearrangement; it has sometimes been exaggerated and sometimes overlooked, especially in early work. This aspect of the reaction is discussed further in the following paragraphs.

In the generally accepted mechanism of alkylation (equations 6, 7, and 8), the electrophile is a carbocation or a polarized complex, shown for simplicity as R^+. The reaction is catalyzed by a Lewis acid ($AlCl_3$) which assists in pulling the chlorine atom from the alkyl chloride to give R^+ and $AlCl_4^-$. Observe that the $AlCl_3$ does indeed serve as a catalyst because it is regenerated in the final step.

$$R—Cl + AlCl_3 \rightleftharpoons R^\oplus + AlCl_4^\ominus \qquad (6)$$

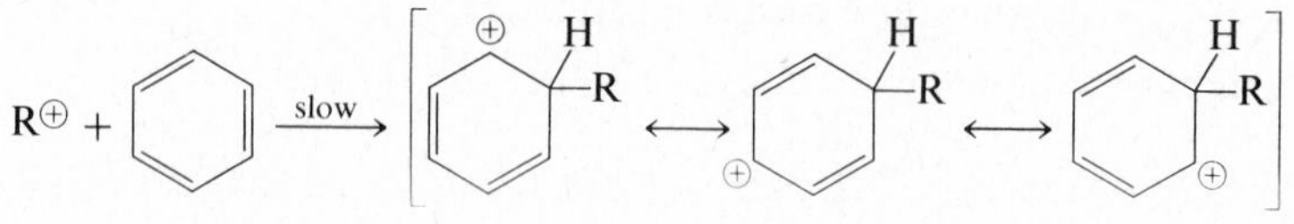

This resonance-stabilized intermediate may be represented by the single symbol

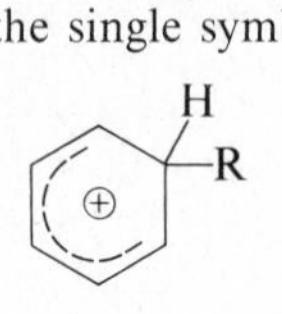

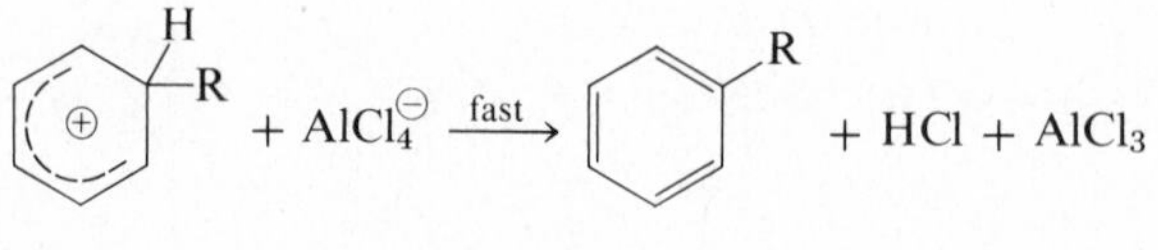

When a carbocation is involved as an intermediate, rearrangement is to be expected in some but not necessarily all cases. (For rearrangements of carbocations that accompany elimination reactions, see Section 6.2.) For example, rearrangement of an unstable carbocation to a more stable one is to be expected, but a rearrangement that involves little gain in stability (decrease in energy) or requires a higher energy intermediate is not necessarily to be expected. Thus an alkylation with a secondary butyl halide **1** (Figure 7.1), which involves a secondary carbocation **2** as an

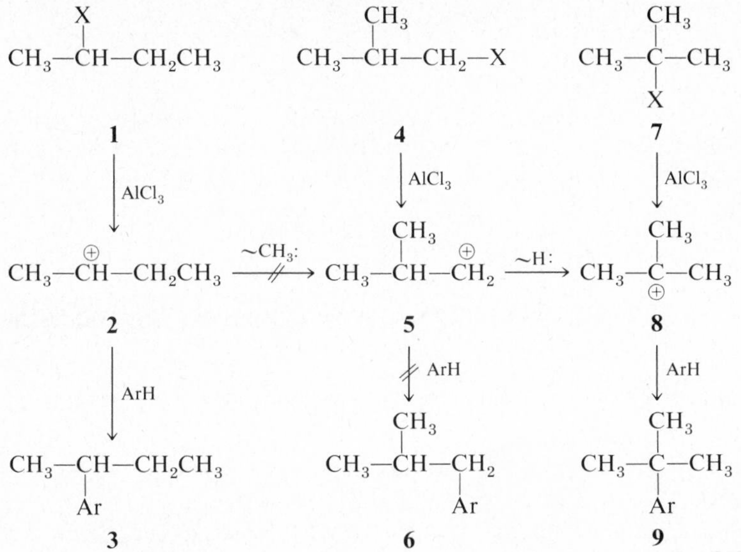

$$X$$
$$CH_3{-}CH{-}CH_2CH_3 \qquad CH_3{-}CH{-}CH_2{-}X \qquad CH_3{-}C{-}CH_3$$

with CH_3 above middle structure and CH_3 above / X below the right structure.

1   **4**   **7**

| AlCl$_3$ | AlCl$_3$ | AlCl$_3$ |

$$CH_3{-}\overset{\oplus}{C}H{-}CH_2CH_3 \xrightarrow[\;]{\;\sim CH_3:\;}\!\!\!// \;\; CH_3{-}CH{-}\overset{\oplus}{C}H_2 \xrightarrow{\;\sim H:\;} CH_3{-}\overset{\oplus}{C}{-}CH_3$$

2   **5**   **8**

| ArH | ArH (crossed) | ArH |

$$CH_3{-}CH{-}CH_2CH_3 \qquad CH_3{-}CH{-}CH_2 \qquad CH_3{-}C{-}CH_3$$
$$\underset{Ar}{} \qquad \underset{Ar}{} \qquad \underset{Ar}{}$$

3   **6**   **9**

FIGURE 7.1 Alkylations with butyl halides.

intermediate, gives only a *sec*-butylarene (**3**); no rearrangement to a *t*-butyl derivative (**9**) occurs because of the high energy of the primary isobutyl carbocation **5**, which is required as an intermediate between **2** and **8**. On the other hand, an isobutyl halide (**4**) gives only *t*-butylarene (**9**) and *no* "unrearranged" isobutylarene (**6**). The intermediate produced from **4** and AlCl$_3$, although it actually may not be a true primary carbocation (**5**) but rather a polarized complex (**5a**), is capable of rearrangement to

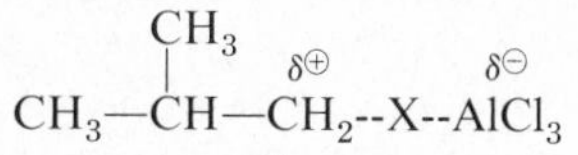

$$CH_3{-}CH{-}CH_2{-}{-}\overset{\delta\ominus}{X}{-}{-}AlCl_3$$

with CH$_3$ above and $\delta\oplus$ marking.

5a

the stable tertiary carbocation so rapidly that direct alkylation by **5** or **5a** does not compete successfully, and none of the unrearranged product is formed. As might be expected, *t*-butyl halides (**7**) give only *t*-butylarenes (**9**).

The extent of rearrangement accompanying alkylation is also dependent on factors such as the nature of the arene (when other than benzene), the temperature, the solvent (if any), and the nature and concentration of the catalyst. The present experiment relates to the first of these factors, the effect of using different arenes as substrates in alkylation by the same alkyl halide.

The reaction of 1-bromopropane with benzene and aluminum chloride gives a mixture of *n*-propylbenzene (**12**) and isopropylbenzene (**13**, cumene), with only a small variation in the proportion of the isomers at different temperatures (Figure 7.2). Apparently, the rate of rearrangement of the primary propyl carbocation **10** to the secondary propyl ion **11** is not as rapid as the rate of conversion of the primary butyl carbocation **5** to the *tertiary* butyl ion **8**, so that direct reaction of **10** with benzene can compete with some success with the rearrangement of **10** to **11**. One might expect that

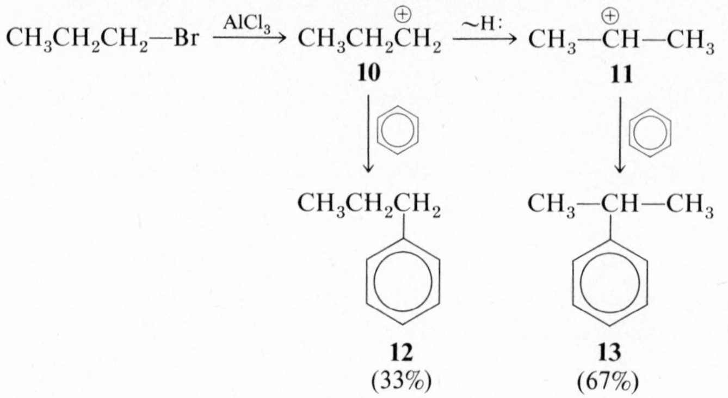

$$CH_3CH_2CH_2\!-\!Br \xrightarrow{\text{AlCl}_3} CH_3CH_2\overset{\oplus}{CH_2} \xrightarrow{\sim H:} CH_3\!-\!\overset{\oplus}{CH}\!-\!CH_3$$

10   **11**

$$CH_3CH_2CH_2$$

$$CH_3\!-\!CH\!-\!CH_3$$

12   **13**
(33%)   (67%)

FIGURE 7.2   Alkylation of benzene with 1-bromopropane.

an arene which is more reactive than benzene toward electrophilic substitution (that is, a more nucleophilic arene) would compete better than benzene for reaction with the primary carbocation **10** before it rearranges to **11**. Methyl groups are known to activate the benzene ring toward electrophilic substitution;[2] hence the reaction of 1-bromopropane with toluene, *p*-xylene, and mesitylene might be expected to give increasing proportions of *n*-propylarene:isopropylarene. This prediction is based on

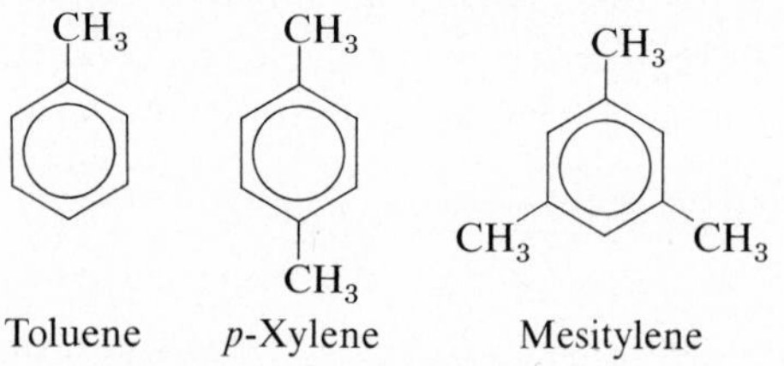

$$CH_3$$ $$CH_3$$ $$CH_3$$

$$CH_3$$ $$CH_3 \qquad CH_3$$

$$CH_3$$

Toluene   *p*-Xylene   Mesitylene

the reasonable assumption that the rate of rearrangement of intermediate **10** to **11** (k_1 in Figure 7.3) will be little different in the presence of benzene, *p*-xylene, or mesitylene, whereas the rate of reaction of **10** with *p*-xylene (k_2) or mesitylene will be significantly faster than with benzene.

In this experiment we have chosen to study the reaction of 1-bromopropane with *p*-xylene for several practical reasons. *p*-Xylene is chosen rather than toluene because it is more nucleophilic than toluene and because alkylation of *p*-xylene does not give rise to orientation isomers. Note that propylation of toluene would yield *ortho-*, *meta,* and *para*-propyltoluenes, which would complicate the determination of the ratio of *n*-propyltoluenes to isopropyltoluenes. Mesitylene is less satisfactory than *p*-xylene because alkylation between two methyl groups on the ring involves some steric hindrance, which introduces an additional factor. The choice of 1-bromo-propane rather than 1-chloropropane is solely on the basis of economics, owing to the rather unusual circumstance that this particular bromide is less expensive than the corresponding chloride.

[2] Refer to Table 7.2. The methyl group of toluene increases the rate of substitution by a factor of 3 to 600 times the rate of reaction of benzene, depending on the nature of the electrophile.

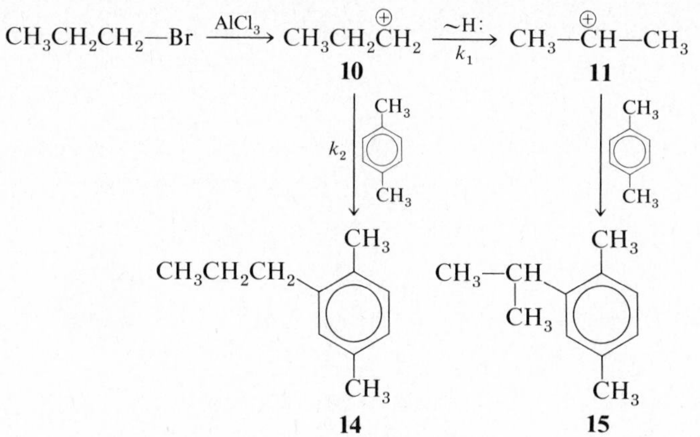

$$CH_3CH_2CH_2-Br \xrightarrow{AlCl_3} CH_3CH_2\overset{\oplus}{C}H_2 \xrightarrow[k_1]{\sim H:} CH_3-\overset{\oplus}{C}H-CH_3$$

10   11

FIGURE 7.3   Alkylation of *p*-xylene with 1-bromopropane.

Mixtures of *n*-propyl-*p*-xylene (**14**, 1,4-dimethyl-2-*n*-propylbenzene) and iso-propyl-*p*-xylene (**15**, 1,4-dimethyl-2-isopropylbenzene) can be analyzed conveniently by gas chromatography and by ir and pmr spectroscopy. It may be interesting to compare your results with those reported in the reference given at the end of this section.

EXPERIMENTAL PROCEDURE

Alkylation of *p*-Xylene with 1-Bromopropane

DO IT SAFELY

---

1.   Anhydrous aluminum chloride reacts vigorously with water, even the moisture on your hands, producing fumes of hydrogen chloride. Do not allow it to touch your skin. The reagent bottle and a scale should be placed in a hood so that you can quickly weigh out the required amount and place it in your round-bottomed flask *in the hood.*

2.   *p*-Xylene is flammable. Make sure that your apparatus is assembled correctly and have your instructor inspect it before you begin the distillation. Since the boiling temperature of *p*-xylene and the propylxylenes is higher than 100°, steam heating is not adequate and electric heating is desirable. If a gas burner is used, take care to keep the flame away from the distillate.

Equip a 250-mL round-bottomed flask with a Claisen connecting tube, a water-cooled condenser, a gas trap, and an addition funnel as shown in Figure 7.4. Place 2.7 g (0.020 mol) of anhydrous powdered aluminum chloride in the flask and

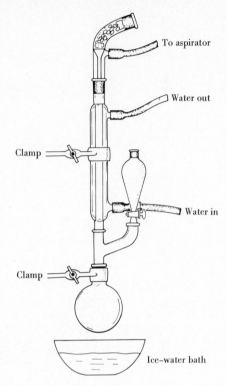

FIGURE 7.4   Apparatus for Friedel-Crafts alkylation.

immediately cover it with 50 mL of *p*-xylene. Measure 18.2 mL (24.6 g, 0.200 mol) of 1-bromopropane in a 100-mL graduated cylinder and pour it into the separatory funnel (stopcock closed). Prepare an ice-water bath so that it will be ready should it become necessary to cool the reaction mixture. Turn on the water to the reflux condenser and to the aspirator and then begin adding the 1-bromopropane dropwise to the mixture of *p*-xylene and aluminum chloride. Loosen the clamp holder of the lower clamp so that the reaction mixture can be swirled gently from time to time as the 1-bromopropane is being added.[3] If the evolution of hydrogen bromide becomes too vigorous, raise the ice-water bath so as to cool the reaction mixture, and reduce the rate at which the 1-bromopropane is being added. The addition should take about 15 min. After all the 1-bromopropane has been added, allow another 45 min for reaction, swirling the reaction mixture every few minutes. Pour the mixture into a 250-mL beaker containing 40 g of crushed ice. After stirring the mixture until all the ice has melted, pour it into a 125-mL separatory funnel and drain off the lower (aqueous) layer and discard it. Pour the organic layer into an Erlenmeyer flask containing about 3 g of *anhydrous* calcium chloride (4–6 mesh) and swirl it for about 5 min.* Filter the dried solution into a 100-mL round-bottomed flask and equip the flask for fractional distillation with either a Vigreux or an unpacked Hempel column

---

[3] If magnetic stirring is available, of course it will not be necessary to swirl the reaction mixture manually.

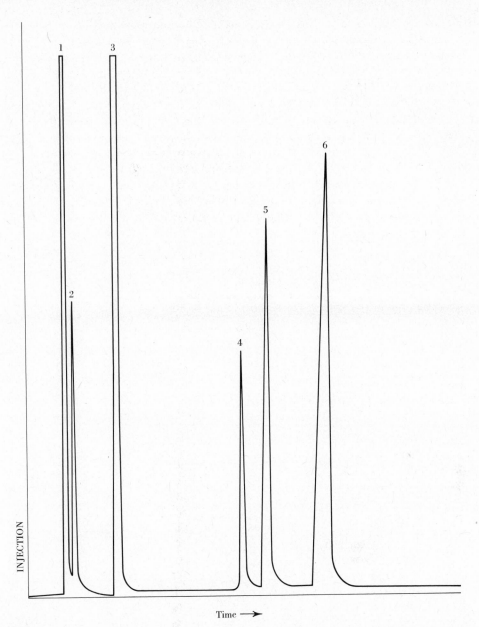

FIGURE 7.5 GLPC trace of reaction mixture from alkylation of *p*-xylene with 1-bromo-propane. The peaks are identified as (1) diethyl ether (solvent), (2) benzene and/or toluene (present as impurity in *p*-xylene), (3) *p*-xylene, (4) isopropyl-*p*-xylene, (5) *n*-propyl-*p*-xylene, and (6) 1,2,4-trichlorobenzene (added to mixture as internal standard). Column: 15-m, 2% silicone oil on DC-550 HiPak.

(Figure 2.4). Using either electrical or gas heating and after making sure that all connections are tight, distil the excess *p*-xylene and forerun into one receiver (labeled A) until the distillation temperature reaches 180°. Discontinue heating and allow any liquid in the column to drain into the flask.★ Replace the column with a simple distilling head (Figure 2.2) and resume distillation, collecting the fraction boiling between 180 and 207° in a second receiver (labeled B). If any material boiling below 180° is obtained in the second distillation, add it to the material in the first receiver. When the distillation temperature begins to rise above 207°, stop the heating, allow the residue to cool, and pour it into a receptacle provided by your instructor.

Record the weights of fractions A and B. Submit or save samples of each fraction for gas chromatographic, ir, and/or pmr spectroscopic analysis (see Figures 7.5, 7.6, and 7.7). Using the weights of fractions A and B and the analytical results, calculate the yield of propylxylenes, and estimate the proportion of the isomers.

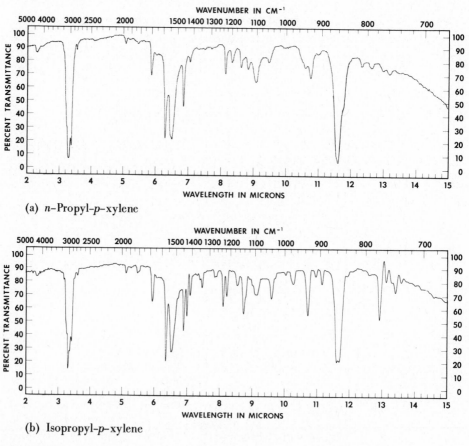

(a) *n*–Propyl–*p*–xylene

(b) Isopropyl–*p*–xylene

FIGURE 7.6 IR spectra of the propyl-*p*-xylenes.

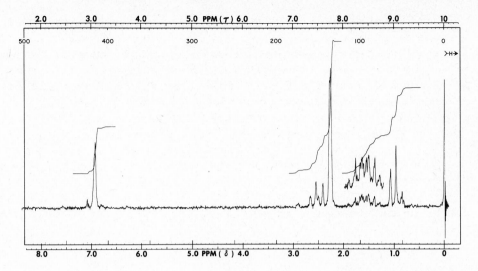

(a) *n*–Propyl–*p*–xylene

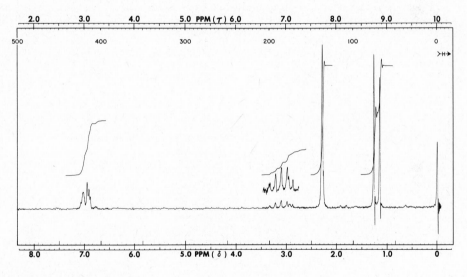

(b) Isopropyl–*p*–xylene

FIGURE 7.7   PMR spectra of the propyl-*p*-xylenes.

EXERCISES

1. What compound(s) do you think might be present in the higher-boiling residue remaining in the distillation flask after collection of the propylxylene isomers?
2. Alkylation of toluene with 1-bromopropane gives a mixture of four isomeric propyltoluenes. Write formulas for these compounds, and on the basis of the

results of your alkylation of *p*-xylene predict the relative amounts in which they would be formed.

3. Suggest a procedure for preparing pure *n*-propyl-*p*-xylene, that is, without the necessity of separating it from isopropyl-*p*-xylene.

4. What products are expected from alkylation of benzene with each of the following alkyl halides, using aluminum chloride as a catalyst?
   (a) 1-chlorobutane   (c) 2-methyl-1-chloropropane
   (b) 2-chloropentane   (d) 2-methyl-2-chloropropane

REFERENCE

R. M. Roberts and D. Shiengthong, *Journal of the American Chemical Society,* **86,** 2851 (1964).

7.3   Nitration of Bromobenzene

Reaction of an aromatic compound such as benzene with a mixture of concentrated sulfuric and nitric acids leads to the introduction of a nitro group onto the aromatic ring. This transformation is an example of an electrophilic aromatic substi-

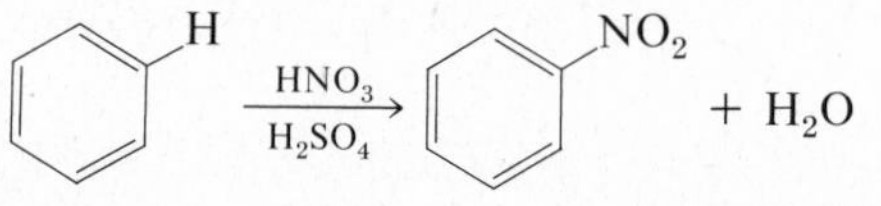

$$\text{C}_6\text{H}_5\text{H} \xrightarrow[\text{H}_2\text{SO}_4]{\text{HNO}_3} \text{C}_6\text{H}_5\text{NO}_2 + \text{H}_2\text{O} \qquad (9)$$

tution reaction (Section 7.1) in which the electrophilic species is the nitronium ion, $^+\text{NO}_2$. This cation is produced by reaction of sulfuric acid with the weaker nitric acid, as shown in equation 10. This process is reminiscent of the rate-determining step of dehydration of alcohols (Section 6.2) in which an oxonium ion fragments with loss of water to produce a carbocation. Notice that with respect to sulfuric acid, nitric acid is acting as a *base,* not an acid!

$$\text{HO}-\text{NO}_2 \xrightarrow{\text{H}_2\text{SO}_4} \overset{\text{H}}{\underset{\text{H}}{\text{O}}}\overset{\oplus}{-}\text{NO}_2 \longrightarrow \text{H}_2\text{O} + {}^{\oplus}\text{NO}_2 \qquad (10)$$

The rate-determining step in the nitration reaction is that in which the nitronium ion reacts with the aromatic ring, localizing a pair of pi electrons in the formation of a sigma bond to form the resonance stabilized intermediate **16**. In the

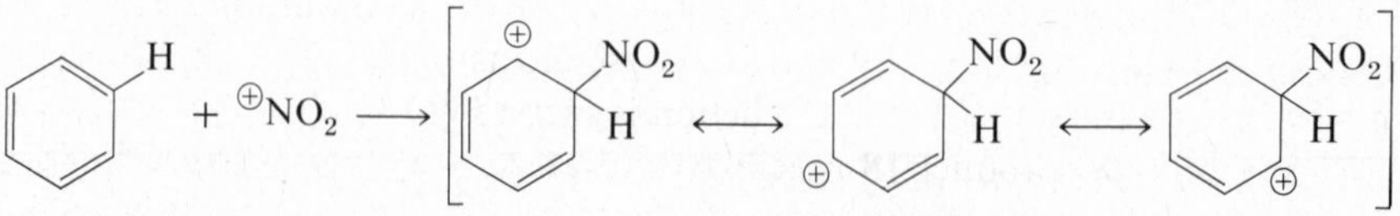

$$\text{C}_6\text{H}_5\text{H} + {}^{\oplus}\text{NO}_2 \longrightarrow \left[\; \cdots \longleftrightarrow \cdots \longleftrightarrow \cdots \; \right] \qquad (11)$$

16

final step the intermediate rapidly loses a proton, probably to a molecule of water, to regenerate the aromaticity of the ring.

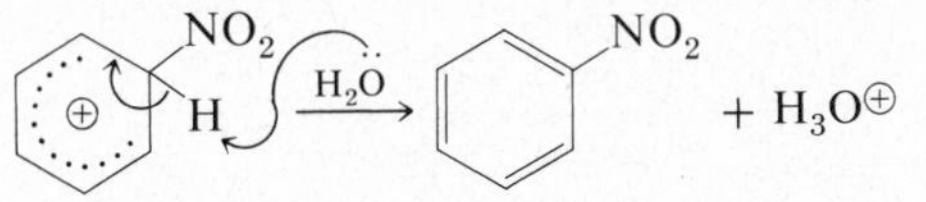

$$\text{(12)}$$

The reaction of the nitronium ion with a substituted benzene such as bromobenzene (17), the substrate used in this experiment, is similar in its mechanism to that of benzene, but more complex in its detailed analysis. Owing to the presence of the bromine atom on the ring, the sixfold symmetry of benzene is destroyed so that there are three different sites for the attack of the nitronium ion, each leading to a different product (equation 13). For electronic reasons, attack at either the *ortho-* or the

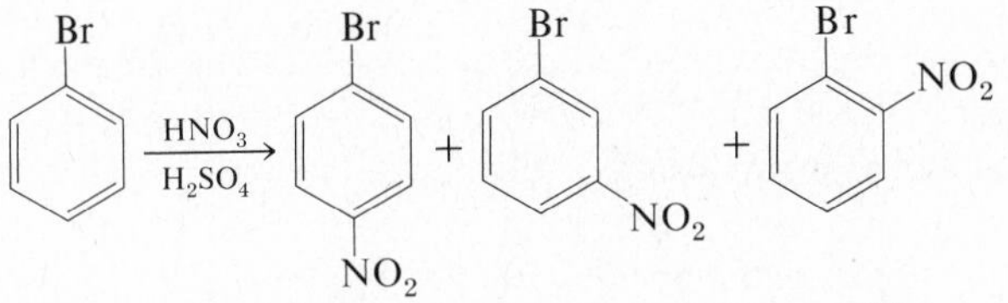

$$\text{(13)}$$

17   18   19

para-position occurs with a lower energy of activation than for attack at the *meta-* position. This is because an electron pair on the bromine atom in the intermediate produced by attack at either of these positions may be delocalized, providing *additional* resonance stabilization of the intermediate. (Note the additional resonance structure in **20**, as compared to **16**.) This additional stabilization by bromine is *not* provided in the intermediate resulting from *meta*-attack. Thus bromine is an example

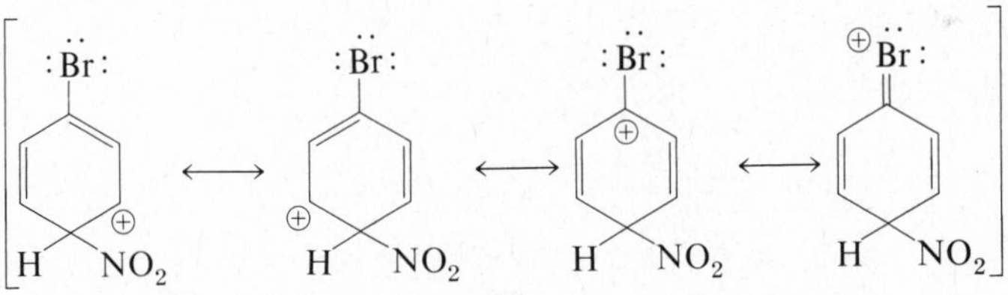

20

of an *ortho,para*-directing group. Typically, the *para*-substituted product **18** predominates in the reaction mixture, owing to the *steric* inhibition by the large bromine atom to the approach of the electrophile to the *ortho* position.

The introduction of a second nitro group to provide 2,4-dinitrobromobenzene (**21**, equation 14) is possible in principle. However, under the conditions in which the reaction is conducted in our experiment, very little dinitration is observed. Owing to the strong *deactivating* effect of the *first* nitro group introduced onto the ring, the energy of activation for the introduction of the second is raised. Thus by keeping below 60° the temperature at which the reaction is carried out, dinitration is largely avoided. Also, the major product **18** precipitates from the reaction mixture, which effectively removes it from the reaction and further reduces the extent of dinitration.

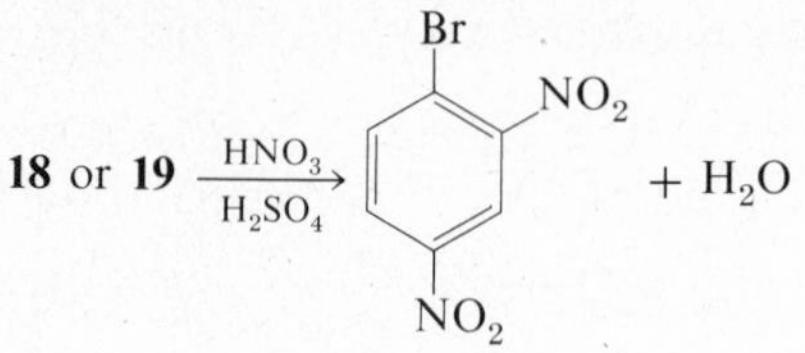

$$18 \text{ or } 19 \xrightarrow[\text{H}_2\text{SO}_4]{\text{HNO}_3} \quad \underset{\textbf{21}}{\text{[structure]}} \quad + \text{ H}_2\text{O} \qquad (14)$$

4-Nitrobromobenzene **(18)** has a more symmetrical structure than 2-nitro-bromobenzene **(19)**. Consequently it is somewhat less polar and therefore less soluble than the *ortho*-isomer. This is dramatically demonstrated in ethanol in which the *ortho*-isomer is very soluble at room temperature but the *para*-isomer is soluble only to the extent of 1.2 g per 100 mL of ethanol. Advantage may be taken of this difference in solubility in order to separate these isomers by the technique of *fractional crystallization*. This is accomplished by dissolution of the product mixture from nitration of bromobenzene into hot 95% ethanol, followed by cooling of the hot solution. The less soluble *para*-isomer selectively crystallizes from the cooling solution, from which it may then be filtered. By concentrating the mother liquors, a second crop of 4-nitrobromobenzene may be obtained. Ordinarily, in a fractional crystallization procedure, once the less soluble component of the mixture has been mostly removed from the solution, the other component (such as the *ortho*-isomer in this experiment) is induced to crystallize. In this experiment, however, owing to the very low melting point (mp 40–41°) of 2-nitrobromobenzene, it is very difficult to induce its crystallization in the presence of impurities. We have therefore elected the option of isolating 2-nitrobromobenzene by column chromatography.

EXPERIMENTAL PROCEDURE

DO IT SAFELY

1.   Concentrated sulfuric and nitric acids may each cause severe chemical burns if they are allowed to come into contact with your skin. Take proper precautions not to allow this to happen. Watch carefully for drips and runs on the outside surface of reagent bottles and graduated cylinders when you pick them up. Wash any affected area immediately with cold water, and apply 0.6 M sodium bicarbonate solution to the area.

2.   The nitrobromobenzenes produced in the experiment are irritating to sensitive skin areas. If you should have these materials on your hands and then accidentally rub or touch your face, you may notice a slight stinging and sensitivity in that area. Apply clean mineral oil to a soft paper or cloth towel, and gently swab the area.

A. Nitration of Bromobenzene

Prepare a mixture of 28.5 g (20.0 mL) of concentrated nitric acid and 37.0 g (20.0 mL) of concentrated sulfuric acid in a 250-mL round-bottomed flask, and cool it to room temperature by means of a water bath. Equip the flask with a Claisen connecting tube fitted with a water-cooled condenser and a thermometer which extends into the flask, as shown in Figure 7.8. Through the top of the condenser add 15.7 g (10.5 mL, 0.100 mol) of bromobenzene to the flask in 2–3 mL portions over a period of about 15 min. Loosen the clamp to the flask and shake the flask vigorously and frequently during the addition. Do not allow the temperature of the reaction mixture to rise above 50–55° during the addition; the temperature may be controlled by allowing more time between the addition of successive portions of bromobenzene and by cooling the reaction flask as necessary with an ice-water bath.

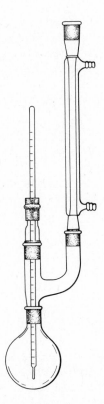

FIGURE 7.8 Apparatus for nitration of bromobenzene.

After the addition is complete and the exothermic reaction has subsided, heat the flask with a steam bath for 30 min, keeping the temperature of the reaction mixture below 60°. Cool the flask to room temperature and then pour the reaction mixture into 200 mL of cold water in a 500-mL beaker. Isolate the crude nitrobromobenzene by suction filtration. Wash the filter cake thoroughly with cold water and allow the crystals to drain under suction until nearly dry.★

Transfer the crystals to a 250-mL Erlenmeyer flask along with 80 mL of 95% ethyl alcohol. Heat this mixture to boiling in order to dissolve the crude product. Set the flask aside and allow the contents to cool slowly to room temperature. Isolate the nearly pure crystals of 4-nitrobromobenzene by suction filtration. Wash the crystals with a little *ice-cold* alcohol, allowing the washes to drain into the filter flask with the mother liquors. When the crystals have dried, obtain their weight and melting point. About 10–12 g of 4-nitrobromobenzene may be expected.★

Evaporate the mother liquors to a volume of 45–50 mL on a steam bath, preferably in the hood, and allow the solution to cool to room temperature. A second crop of 1–2 g of 4-nitrobromobenzene may be obtained in this way. It may either be combined with the first crop and the whole recrystallized, or it may be submitted separately.

Further concentrate the mother liquors from the second crop to a volume of 15–20 mL. Cooling will result in the formation of an oil which contains 2-nitrobromobenzene **(19)**. Separate the oil from the two-phase mixture by means of a pipet. The 2-nitrobromobenzene may be isolated and purified by column chromatography, as described in part B.

B.   Thin-Layer and Column Chromatography

In two small vials prepare solutions of 4-nitrobromobenzene and of the oil containing 2-nitrobromobenzene in about 0.5 mL of chloroform. Obtain a 3 × 8-cm strip of silica gel chromatogram sheet (without fluorescent indicator). Following the procedure of Section 3.7, apply spots of each of these two solutions. Allow the spots to dry and develop the chromatogram in a tlc chamber with 9:1 (v:v) hexane:chloroform as the solvent.

When the solvent has climbed to within about 2 cm of the top of the plate, remove the developed chromatogram from the chamber, quickly mark the solvent front with a pencil, and allow the plate to dry. The spots may be visualized by placing the dry plate in a chamber whose atmosphere is saturated with iodine vapor. Calculate the R_f values of the spots observed and identify them as either 2- or 4-nitrobromobenzene. (A small orange spot may be observed very near the origin for the oil; this spot is identified as 2,4-dinitrobromobenzene.)

Following the procedure given in Section 3.5, prepare a column using a 50-mL buret and 15–20 g of silica gel. (Unless a larger column is available to you, it will not be possible to submit to column chromatography the entire sample of oil obtained in part A.) Apply a 0.5-g sample of the oil containing the 2-nitrobromobenzene to the head of the column and rinse the inside of the buret with 1 mL of chloroform. Open the stopcock and allow the liquid to drain just to the top of the sand. Fill the buret with 9:1 (v:v) hexane:chloroform and elute the column until a total of 100 mL of the solvent has passed through the column. (Do not allow the level of liquid to drain below the sand at the top of the column.) Collect the eluent in 20- to 25-mL fractions in a series of numbered 50-mL Erlenmeyer flasks. After 100 mL of the 9:1 solvent has passed through the column, continue by passing 50 mL of a 4:1 (v:v) hexane:chloroform solution through the column and collecting 20- to 25-mL fractions.

Evaporate each of the fractions obtained to dryness on a steam bath in the hood. Characterize any solid residues obtained as either 2- or 4-nitrobromobenzene through either melting point determinations or tlc analysis. 2-Nitrobromobenzene has a reported melting point of 40–41°; the melting point of 4-nitrobromobenzene is 125–126°.

EXERCISES

1. Explain why 4-nitrobromobenzene **(18)** predominates in the product mixture over 2-nitrobromobenzene **(19)**.
2. Using resonance structures, show why no detectable 3-nitrobromobenzene is found in the product mixture.
3. Explain how tlc may be used to select the most appropriate solvent for use in a column chromatographic separation.
4. Explain why 4-nitrobromobenzene **(18)** has a larger R_f than the 2-isomer, **19**, even though it is the lesser soluble of the two.
5. What is the brown gas that is observed when sulfuric and nitric acids are mixed?

SPECTRA OF PRODUCTS

The ir spectrum of bromobenzene is included in Figure 12.5.

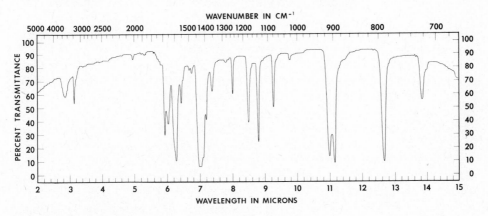

FIGURE 7.9   IR spectrum of 4-nitrobromobenzene.

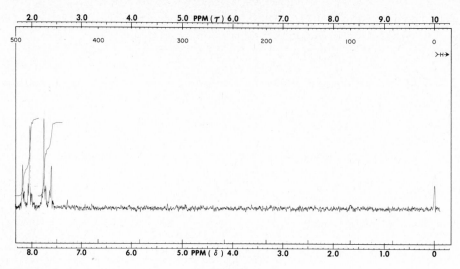

FIGURE 7.10 PMR spectrum of 4-nitrobromobenzene.

7.4   Relative Rates of Electrophilic Aromatic Substitution

Electrophilic aromatic substitution reactions are among the best understood of all organic reactions from the mechanistic standpoint. Most of the information obtained about aromatic substitution is internally consistent. The qualitative principles of the reaction are usually discussed in organic chemistry textbooks. These include the effect of a substituent already present in an aromatic compound on the reactivity of that compound toward a new, entering substituent and on the orientation of the entering substituent (*ortho, meta,* or *para*). Normally, however, the subject of reactivity is discussed in terms of activation or deactivation of the molecule with respect to benzene, and little if any information is given about the *quantitative* differences in rates and reactivities of substituted aromatic compounds. The purpose of this discussion and accompanying experimental work is to give the student some feeling for the magnitude of the actual differences in reactivity.

Table 7.2 lists some data which have been collected for some common types of electrophilic substitution reactions. Most of the reactions were carried out under identical conditions, so that a direct comparison of the numbers is reasonably valid. As mentioned in the footnote of Table 7.2 the rates are for toluene *relative* to benzene; for example, bromination is about 605 times more rapid for toluene than for benzene. Indeed it can be seen that all electrophilic substitution reactions of toluene are faster than those for benzene, which supports the idea presented in most textbooks that the methyl group *activates* an aromatic ring and makes it more reactive toward an electrophile.

Table 7.3 gives some data for the bromination of different monosubstituted benzene derivatives. The rates are again relative to benzene (R = H in C_6H_5R), so that these results indicate the relative reactivities of the various substituted benzenes.

TABLE 7.2[a] **RELATIVE RATES OF REACTIVITY OF TOLUENE TOWARD ELECTROPHILIC SUBSTITUTION**

Type of Reaction	Relative Rate[b]
Bromination	605
Chlorination	340
Acetylation	128
Sulfonation	31.0
Nitration	24.5
Methylation	2.9

[a] From L. M. Stock and H. C. Brown, *Advances in Physical Organic Chemistry*, V. Gold, editor, Academic Press, New York, 1963, Vol. I, pp. 50–52.

[b] The rates here are relative rates, where the rate of substitution on benzene has been taken as 1.0. Mathematically, the relative rate may be defined as:

$$\text{relative rate} = \frac{\text{rate of substitution on toluene}}{\text{rate of substitution on benzene}}$$

The values quoted were obtained at 25° or have been corrected to 25° by extrapolation.

TABLE 7.3[a] **RATES OF BROMINATION FOR VARIOUS AROMATIC COMPOUNDS**

Substituent R in C_6H_5R	Relative Rate[b]
—OH	6.1×10^{11}
—OCH$_3$	1.8×10^9
—NHCOCH$_3$	2.1×10^8
—OC$_6$H$_5$	1.7×10^7
—C$_6$H$_5$	1.0×10^3
—CH$_3$	6.05×10^2
—C$_2$H$_5$	4.6×10^2
—CH(CH$_3$)$_2$	2.6×10^2
—C(CH$_3$)$_3$	1.38×10^2
—H	1.0 (standard)
—X (Cl, Br)	0.5 (estimated value only)
—CO$_2$H	7.5×10^{-3}
—N$^+$(CH$_3$)$_3$	1.6×10^{-5}
—NO$_2$	1.6×10^{-5}

[a] From L. M. Stock and H. C. Brown, *Advances in Physical Organic Chemistry,* V. Gold, editor, Academic Press, New York, 1963, Vol. I, pp. 62–80.

[b] Rates are relative rates, where the rate of substitution on benzene has been taken as 1.0. The values were obtained at 25° with bromine in aqueous acetic acid or have been corrected to 25° by extrapolation.

Those exhibiting a rate greater than 1.0 have substituents already present that activate the aromatic ring toward further substitution, whereas those whose rate is less than 1.0 have substituents already present which deactivate the ring.

It can be seen from the data in Table 7.3 *how much* more activating one group is compared to another. In particular it should be pointed out that phenol (R = OH) is *about 10^{16} times more reactive* toward bromine than is nitrobenzene (R = NO$_2$). Differences in reactivity can be seen only from data of this sort, and the student is advised to examine Table 7.3 carefully.

As has been stated before, the two most important steps in electrophilic aromatic substitution reactions are (1) the attack of the electrophile on the electron-rich aromatic ring and (2) loss of a proton (see equations 4 and 5 in Section 7.1). The intimate details of the mechanism, particularly with regard to the rate expression, differ with various substrates and reaction conditions. Part of the problem lies in the role and concentration of the catalyst in these reactions, and some differences in mechanism are believed to be related to the speed with which the electrophile, E$^+$, is generated and consumed.

The measurement of reaction rates makes it necessary to review some of the factors which influence them (see Section 10.2 for a brief summary). Some of the variable factors are temperature, concentration of reactants, solvent composition, and nature and concentration of catalysts. Since it is possible to change any of these variables while keeping all other factors constant, the effect of all the variables can be determined.

In this experiment it will be possible for the student to measure rates and rate differences for an electrophilic aromatic substitution reaction. The reaction chosen for study here is the bromination by elemental bromine of a series of typical aromatic compounds. The rate expression for this reaction is given by equation 15, in which Br$_2$

$$\text{Rate} = k_2[\text{ArH}][\text{Br}_2] \tag{15}$$

has been inserted in place of E$^+$ of equation 2 in Section 7.1, since bromine is clearly the precursor to the actual electrophile. The rate expression could be more complex under certain conditions. Bromination of less reactive aromatic compounds requires the use of a Lewis acid catalyst for reaction at a reasonable rate. Water, in *trace* amounts, has been found to accelerate the rate of these electrophilic reactions. However, by avoiding the use of a catalyst and by carrying out the brominations in a mixed solvent composed of 90% acetic acid and 10% water,[4] both of these complications are excluded from the rate studies. The rate expression of equation 15 states that the rate of the reaction is proportional to the concentration of the aromatic substrate and to the concentration of bromine, and as written is a second-order rate expression. Although a student could, in principle, work with this rate expression, the calculations are much more difficult and time-consuming than are those for first-order reactions. Thus the experimental procedure is designed to provide a "pseudo-first-order" reaction to simplify the calculations. This is accomplished by maintaining the concentration of the aromatic substrate at a high level so that it does not change

---

[4]This much water in the solvent ensures a concentration which remains essentially constant during the course of the reaction and hence does not enter into the rate expression.

appreciably throughout the course of the reaction. Under these conditions the rate expression becomes

$$\text{rate} = k_1[\text{Br}_2] \tag{16}$$

where k_1 is a new rate constant which will contain all the factors of constancy for the reagents and solvents.[5]

The integrated form of equation 16 is

$$[\text{Br}_2]_t = [\text{Br}_2]_0 e^{-k_1 t} \tag{17}$$

which can be rearranged to

$$\ln \frac{[\text{Br}_2]_0}{[\text{Br}_2]_t} = k_1 t \tag{18}$$

or

$$2.303 \log \frac{[\text{Br}_2]_0}{[\text{Br}_2]_t} = k_1 t \tag{19}$$

where $[\text{Br}_2]_0$ = initial concentration of bromine, and $[\text{Br}_2]_t$ = concentration of bromine at any time t.[6] From the experimental data, one can prepare a graph of $\log [\text{Br}_2]_0/[\text{Br}_2]_t$ versus time, t. If the reaction is first order, the graph should show a reasonably straight line with a slope of $k_1/2.303$ if common logarithms are used. It is important to note that $[\text{Br}_2]_t$ is the concentration of bromine *remaining* at time t, and not the amount of bromine which has reacted.

The reaction temperature and initial concentration of reactants must be controlled. The use of a large water bath in the experiment and careful preparation of the reagents will ensure that these factors are kept constant.

One of the main reasons for choosing bromination reactions for carrying out rate studies is that it is easy to follow the rate of disappearance of the bromine color as the reagent reacts with the aromatic substrate. The virtue of choosing reaction conditions so that the rate of the reaction is proportional only to the concentration of the bromine thus becomes quite apparent.

An important limitation must be placed on the types of aromatic compounds to be studied. Because no Lewis acid catalyst is present, the study is necessarily limited to compounds that are more reactive than benzene. Benzene and compounds of similar or lower reactivity would react so slowly with bromine in the absence of a catalyst that it would not be possible to obtain good kinetic data in a single laboratory period. This limitation presents no serious problem, since it will still be possible to

---

[5] Rate constant k_1 is called a *pseudo*-first-order rate constant. Since all other factors are being held constant, the reaction rate will *appear* to depend only on the concentration of bromine, as the rate expression indicates. *Pseudo*-rate constants in general refer to reactions in which the concentration of one or more reactants which appear in the rate expression are chosen so that they do not appreciably change as the reaction proceeds. Thus they remain constant and are automatically incorporated in the pseudo-rate constant.

[6] See Section 10.2 for a more detailed discussion of these equations for first-order rate constants. The equations are very similar to those presented for the hydrolysis of tertiary halides.

determine the relative reactivities of a series of compounds and to illustrate the principles involved in these sorts of measurements.

Experiments are described which allow one to obtain qualitative differences in reactivity. Depending on the facilities available, either semiquantitative or quantitative measurements of absolute rates may also be determined.

The qualitative experiments involve preparing solutions of various substrates in 90% acetic acid containing bromine. From the lengths of time required for the bromine color to be discharged, one can deduce the relative reactivities.

The semiquantitative experiments involve preparing a series of solutions of 90% acetic acid which contain known concentrations of bromine. If one then prepares a solution containing an excess of substrate and a known concentration of bromine, a rough estimate of the rate may be made by visual comparison of the color of the reaction mixtures with those of the known solutions.

The quantitative experiments require the use of an inexpensive visible/ultraviolet spectrophotometer,[7] which allows accurate measurement of the *rate of disappearance* of bromine. This instrument is nonrecording, but the data can be read from a meter. If a compound strongly absorbs light at a given wavelength, then the ability of light to pass through a sample of that compound is diminished. The spectrophotometer measures the amount of light which has been absorbed, this value being called absorbance. Absorbance is defined as

$$\text{absorbance } (A) = \log \frac{I_0}{I} = \varepsilon c l \tag{20}$$

and is often called *optical density*.[8] In the experiment you will use the same cell for all measurements for a given aromatic substrate, so that for a given kinetic run the values of ε and l in equation 20 remain constant. The absorbing species is bromine in all instances. Therefore, with reference to equation 21, the following equivalencies may be written:

$$A_0 = \varepsilon[\text{Br}_2]_0 l$$
$$A_t = \varepsilon[\text{Br}_2]_t l$$

These may be combined with equation 19 to give equation 21. This equation is useful in that it relates a series of measurements of absorbance to the rate constant k_1. By measuring the rate constants for several substrates one may quantitatively evaluate the reactivities of these substrates toward bromination. Additional instructions for analyzing the data collected are provided within the experimental procedure.

$$2.303 \log \frac{A_0}{A_t} = k_1 t \tag{21}$$

---

[7] Several relatively inexpensive instruments are available, such as the Bausch and Lomb Spectronic 70. A discussion of uv-visible spectroscopy is provided in Section 4.3.

[8] See Section 4.3 for a discussion of equation 20 and a definition of its terms.

EXPERIMENTAL PROCEDURE

DO IT SAFELY

Bromine is a hazardous chemical which may produce severe chemical burns in contact with the skin. Even though the solutions used in the experiments are dilute, take proper precautions. Wash your hands and soak any affected area for a few minutes in a 0.6 M sodium thiosulfate solution.

In the following experiments it would be convenient for two students to work together.

A. Qualitative Rate Comparisons

Prepare a water bath using a 1-L beaker, and adjust the temperature to about 35°. During the experiment, keep the temperature at 35 ± 2°. Place 2.0 mL of 0.2 M solutions of each of the following substrates (contained in 15 M acetic acid) in separate small test tubes: (a) phenol, (b) anisole, (c) diphenyl ether, (d) acetanilide, (e) p-bromophenol, and (f) α-naphthol. If stock solutions are provided, use special care to ensure that one solution is not contaminated by another. Suspend the carefully labeled test tubes partially in the water bath by looping a piece of copper wire around the neck of the test tube and over the rim of the beaker.

Prepare some droppers with a capacity of 2.0 mL. Calibrate these by comparison with known volumes, and mark the droppers by scratching the glass with a file. These will be used to introduce the bromine solution into the test tubes containing the substrates. It is essential that the addition be done as rapidly as possible and that the volumes be fairly accurate. Transfer to an Erlenmeyer flask about 45 mL of a solution containing 0.05 M bromine in 15 M acetic acid and allow the solution to equilibrate in the water bath for a few minutes. Add 2.0 mL of the bromine solution to one of the test tubes containing a substrate. Make the addition rapidly, mix the solution quickly, and note the exact time of addition. Observe the reaction mixture and note how long it takes for the bromine color to become faint yellow or to disappear. Repeat this procedure with each of the substrates, making sure that you use the same end-point color for each one.

When the reaction is slow—that is, no decolorization occurs within 5 min—go on to another compound while waiting for the end-point to be reached. Record the reaction times, and on the basis of these observations arrange the compounds in order of increasing reactivity toward bromine.

If it is impossible to determine the relative rates accurately at 35°, repeat the experiment with the compounds in doubt at 0°, using an ice-water bath.

B. Semiquantitative Rate Measurements

Obtain stock solutions of each of the following prepared with 15 M acetic acid: 0.05 M bromine, 0.5 M p-nitrophenol, 0.5 M acetanilide, and 0.5 M phenyl ether.

Prepare standard color comparison solutions by diluting the 0.05 M stock bromine solution according to the directions in Table 7.4. Fill small test tubes with about 4 mL of each of these solutions. Adjust the levels of solution in each tube so that all are filled to a common depth. In a test tube rack arrange the tubes in order of decreasing concentration of bromine; place a white background beneath them as an aid in color discernment.

TABLE 7.4 PREPARATION OF STANDARD BROMINE SOLUTIONS FOR COLOR COMPARISON

	MIX TOGETHER:		
Solution	Bromine Solution	Solvent (mL)[a]	Concentration (M)
A	3.0 mL of stock[b]	3.0	2.5×10^{-2}
B	3.0 mL of stock	9.0	1.3×10^{-2}
C	6.0 mL of B	6.0	6.3×10^{-3}
D	6.0 mL of C	6.0	3.1×10^{-3}
E	6.0 mL of D	6.0	1.6×10^{-3}
F	6.0 mL of E	6.0	7.8×10^{-4}
G	3.0 mL of F	3.0	3.9×10^{-4}

[a] 15 M acetic acid in water.
[b] Stock bromine solution (0.05 M).

Calibrate an additional test tube of the same size, measuring the volume to the nearest 0.1 mL necessary to fill it to the same depth as the comparison tubes. Add to this tube one-half the calibrated volume of one of the stock substrate solutions. In a second tube place this same volume of the 0.05 M bromine solution. Working with the aid of a partner, pour the measured bromine solution into the calibrated test tube containing the substrate solution and stir with a small stirring rod to ensure homogeneity; note and record the time of mixing to the nearest second. Continue, with one partner marking the time that the color of the reaction mixture most closely matches each standard bromine solution, and the second partner reading and recording the times to the nearest second. The colors of the solutions should be compared by viewing the tubes vertically through the length of the tubes. Repeat, using each of the other substrate solutions and any others that the instructor may assign. As time allows, perform duplicate determinations on each substrate; to minimize temperature effects carry out all measurements during the same laboratory period.

Treatment of Data. **1.** From the color comparison data and the bromine concentrations in each standard solution given in Table 7.4, plot for each substrate the molar concentration of bromine remaining (vertically) versus time (seconds, horizontally). Include as an additional point the initial concentration of bromine, 2.5×10^{-2} M, at time, $t = 0$.

2. Draw the best curve that you can through the points which you have drawn. Draw a line tangent to the curve where it crosses the $t = 0$ point. Determine the slope of this line. These slopes, when compared for the various substrates, give an estimate of the *initial relative rates* and thus of the relative reactivities.

3. The rate expression for bromination (equation 16) may be rewritten as follows:

$$\text{Rate of reaction} = \frac{\Delta[Br_2]}{\Delta t} = k_1[Br_2]$$

where $\Delta[Br_2]/\Delta t$ = the rate of disappearance of bromine.

When the reaction has just started, the percentage change in the bromine concentration is very small compared to the total amount present. Under these conditions it is possible to substitute the *initial rate* determined in step 2 for $\Delta[Br_2]/\Delta t$. Since we know the initial rate and we know the initial concentration of bromine, it is possible to substitute these values into the above rate equation and solve for k_1, the rate constant. Do this for each substrate.

C. Quantitative Measurements

Obtain three sample tubes for use with the spectrophotometer. These tubes resemble small test tubes, but the dimensions and glass thickness are carefully controlled in manufacture; they are called *cuvettes*.

Have at your disposal stock solutions (0.5 M) of acetylsalicylic acid (aspirin), diphenyl ether, and acetanilide in 16 M acetic acid, as well as a 0.02 M solution of bromine in 16 M acetic acid. You will also need a supply of two or three 2-mL pipets for accurate sample measurement.

Since all solutions absorb light to some extent, though they may appear transparent to the naked eye, it is necessary to calibrate the spectrophotometer to zero absorbance using a solution containing all components of the solution to be measured *except the absorbing species* (bromine in this case). Into a clean and dry cuvette place 2 mL of 16 M acetic acid and 2 mL of a 0.5 M solution in 16 M acetic acid of the substrate whose rate of substitution is to be measured. Use 2-mL pipets to ensure accurate measurement of these small volumes. Stir the solution to make sure that it is homogeneous and place the cuvette in the spectrophotometer. Note that the cuvette has an alignment mark to ensure that it is always placed in the spectrophotometer in the same orientation. Align this mark with the corresponding engraved mark on the front of the sample holder. Close the door to the sample holder and adjust the absorbance reading to zero. This calibration and all future readings are done at 400 nm, the wavelength at which elemental bromine absorbs light.

In preparing for and executing the following experiments, you must be well organized so that you can work quickly. It may take several tries before you are able to work rapidly enough to get acceptable results. Once you have zeroed the spectrophotometer for a given substrate, clean and dry the *same* cuvette used in that step, and continue with the following operations: (1) to the cuvette add 2 mL of 0.02 M bromine solution; (2) add 2 mL of the 0.5 M solution of substrate; (3) stir quickly, one time only; (4) a second person should record the exact time of mixing; (5) place the cuvette in the spectrophotometer with the correct orientation and close the door to the sample compartment; and (6) record a series of absorbance readings and the times at which these readings were taken. Obtain as many readings as possible before the absorbance drops below about 0.5, although readings may be taken below 0.5. It is easier to record the time at which the needle on the meter crosses a line on the absorbance scale rather than to attempt interpolation between those lines. Record the laboratory temperature.

Repeat the above for each of the other substrates. For those runs which are particularly fast it is advisable to perform duplicate runs.

Treatment of Data.   1.   The data recorded above constitute a series of values A_t (see equation 21). To obtain A_0, it is necessary to extrapolate these data to the initial time of mixing, t_0. Since the absorbance decreases exponentially during the course of the run, it is easiest to do this by plotting log A_t (vertically) versus time (seconds, horizontally). To determine log A_0 extrapolate the straight line obtained to zero time. Take the antilog of this intercept to obtain A_0. This must be done separately for each substrate.

2.   For each substrate, using the value of A_0 just obtained and the recorded values of A_t, calculate a series of values of 2.303 log A_0/A_t. On graph paper plot these values (vertically) versus time (seconds, horizontally). Draw the best straight line possible through these points. Calculate the slope of that line to obtain k_1 (sec^{-1}) for that substrate.

3.   After obtaining k_1 for each substrate, divide all values of k_1 by the *smallest* value to obtain the *relative* reactivities of the substrates toward electrophilic aromatic bromination.

8

DIENES
The Diels-Alder Reaction

One of the more useful synthetic methods available to the organic chemist is the reaction between a 1,3-diene and an alkene, sometimes referred to as the dienophile (Gr., *philos,* loving), to produce a derivative of cyclohexene (equation 1). If an alkyne is used as the dienophile instead of an alkene, a derivative of 1,4-cyclohexadiene will be formed (equation 2). The reaction, termed the Diels-Alder reaction in honor of its primary developers, Otto Diels and Kurt Alder, results in the formation of new carbon-carbon bonds between the two π-bonded carbons of the dienophile and the 1- and 4-carbons of the diene. Thus there is an overall 1,4-addition of the dienophile to the diene.

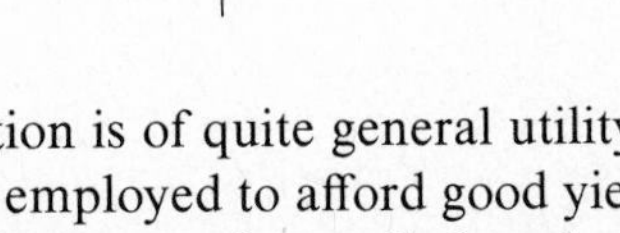

$$\tag{1}$$

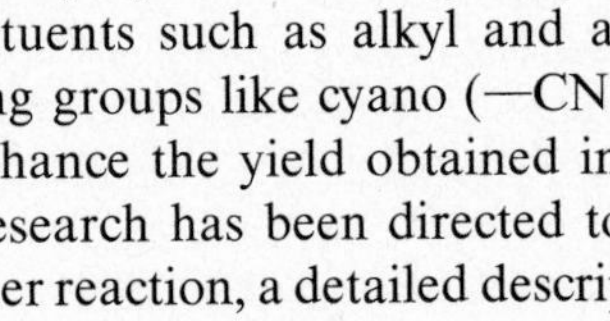

$$\tag{2}$$

The Diels-Alder reaction is of quite general utility, and many different dienes and dienophiles have been employed to afford good yields of adducts. The presence of electron-releasing substituents such as alkyl and alkoxy (—OR) on the diene and/or electron-withdrawing groups like cyano (—CN) and carbonyl (—CO—) on the dienophile seems to enhance the yield obtained in the reaction.

Even though much research has been directed toward the elucidation of the mechanism of the Diels-Alder reaction, a detailed description of the reaction pathway is still not available. In deriving a mechanism for the reaction, the following facts must be considered. The reaction exhibits first-order dependency upon the concentration of both the diene and the dienophile so that the process is second order overall. Kinetic measurements show that the rate of the reaction is normally not significantly changed by the addition of catalysts such as acids,[1] bases, or free radicals; by irradiation with light of various wavelengths;[2] or by the phase (gas or liquid) in which the reaction is performed. The evidence suggests that the transition state of the reaction is constructed from a single molecule of each of the reactants and that neither highly polar intermediates, for example, carbocations, nor free radicals are involved in the mechanistic pathway linking starting materials and products. A mechanism consistent with these observations is one in which bond breaking and bond making occur simultaneously in the transition state, so that little polarity or free radical character is developed (equation 3).[3]

$$\text{(3)}$$

Reactants   Transition state   Product

There is one important limitation to the Diels-Alder reaction, which is a result of the concerted nature of the process, and it is the requirement that the diene must be capable of attaining an *s-cis* conformation,[4] which is the geometry needed to give a *cis* double bond in the cyclic product. Reaction of a dienophile with a diene in an *s-trans* conformation would lead to a *trans* double bond, which is not possible in a six-membered ring for geometric reasons (models are helpful here).

s-cis   *s-trans*

---

[1] See reference 2 at the end of this chapter for a discussion of the catalytic effect of Lewis acids on some types of Diels-Alder reactions.

[2] There are important exceptions to this, particularly with compounds in which the diene grouping is part of an aromatic system. As an example, it is possible to accomplish a Diels-Alder reaction between maleic anhydride and benzene by *irradiation* of a mixture of the two; note that it is generally not possible to cause benzene to participate in thermally induced Diels-Alder reactions, because its aromatic stabilization is lost if a dienophile adds 1,4- to the ring. Students interested in investigation of the photochemical Diels-Alder reaction should consult reference 4.

[3] Theoretical analysis of the nature of the molecular orbitals in the diene and the dienophile and in the product resulting from their condensation has shown that a continuous transformation of the ground-state orbitals of the reagents into those of the Diels-Alder adduct is possible. Thus the reaction is said to be allowed as a concerted thermal process, and its mechanism is said to be controlled by orbital symmetry. Most organic lecture textbooks now contain a discussion of orbital symmetry. If yours does not and you wish a more complete discussion of the relationship between the symmetry of molecular orbitals and the mechanisms of organic reactions, see reference 5 or 6.

[4] The *s-cis* conformation of a 1,3-diene has the two conjugated double bonds on the same side of the single bond linking them. They are on opposite sides in the *s-trans* conformation.

The strict orientation in the transition state of the dienophile relative to the diene might lead one to expect high stereoselectivity in the Diels-Alder reaction, and this expectation is fulfilled. As an example, the adduct resulting from reaction of *trans, trans*-2,4-hexadiene and a dienophile is exclusively *cis*-3,6-dimethylcyclohexene (equation 4), whereas the product obtained when *cis, trans*-2,4-hexadiene is utilized is *trans*-3,6-dimethylcyclohexene.

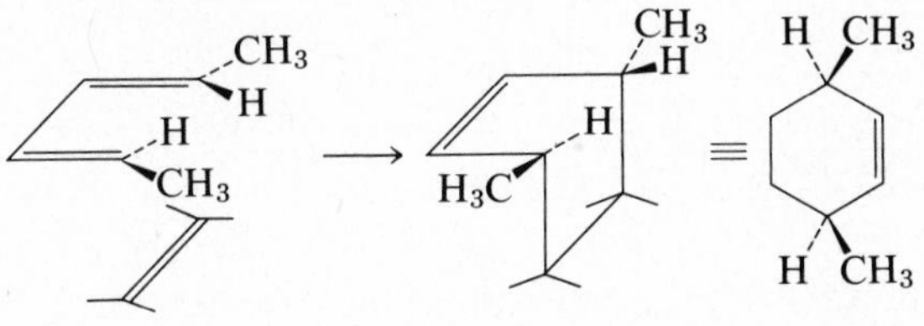

(4)

A second type of stereoselectivity is observed for the reaction. The addition of a dienophile such as maleic anhydride (**1**) to a *cyclic* diene such as cyclopentadiene (**2**) could in principle provide two products, **3** and **4** (equation 5). However, only a single adduct, **3**, is observed experimentally. The explanation for this stereoselectivity is not entirely clear, but one attractive theory is that stabilization, that is, a lowering in energy, of the transition state **5** leading to formation of **3** is provided by interaction between the *p*-orbitals of the diene and those of the dienophile. Analogous stabilization is not possible in the transition state **6** required for production of **4**.

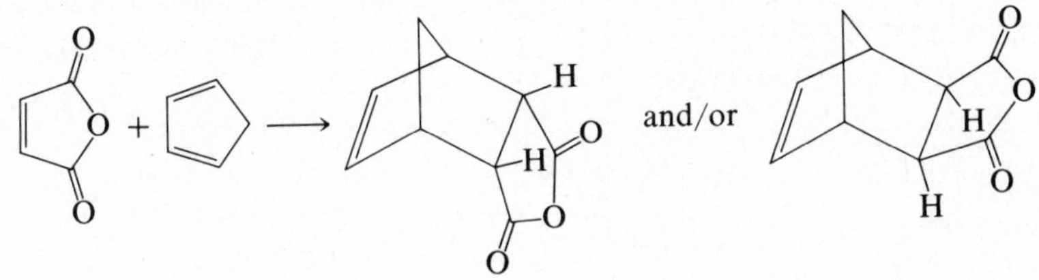

(5)

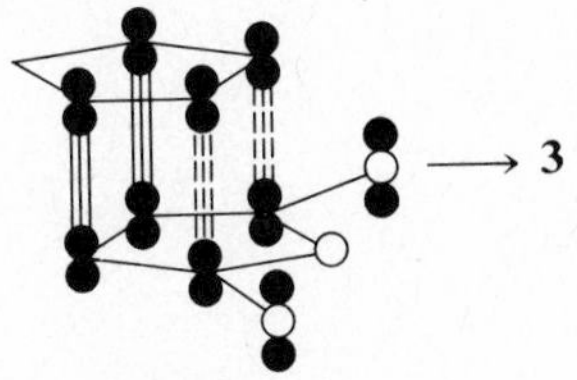

It should be noted that all Diels-Alder reactions are not as stereoselective as that between cyclopentadiene and maleic anhydride, so that mixtures of products are obtained. However, the product resulting from interaction of the diene and the dienophile in a transition state analogous to **5** generally predominates over that arising from a transition state similar to **6**.

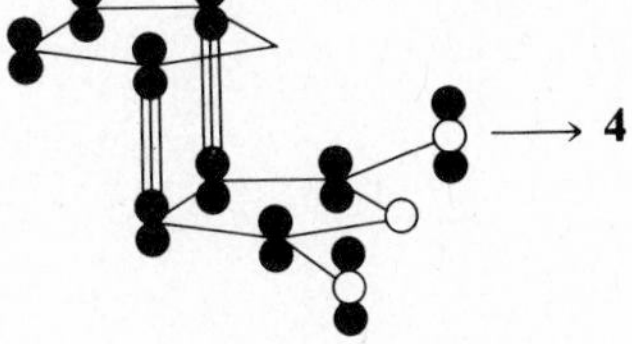

The Diels-Alder synthesis is remarkably free of complicating side reactions, and the yields of the desired product are often nearly quantitative. Probably the single most important side reaction observed is dimerization of the diene used, since one molecule of the diene can serve as the dienophile and another as the diene (equation 6). Such a reaction is usually important only in those instances in which a poor dienophile such as ethylene is employed.

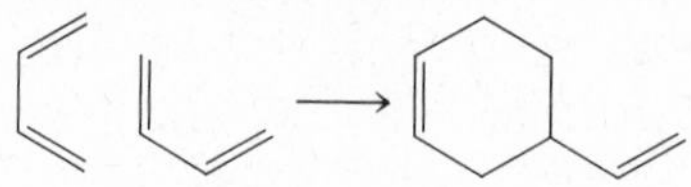

$$\text{image} \qquad (6)$$

EXPERIMENTAL PROCEDURE

A.   Preparation of *endo*-Norbornene-*cis*-5,6-dicarboxylic Anhydride

---

DO IT SAFELY

---

1.   Cyclopentadiene is a mildly toxic, volatile substance; prepare and use it in a hood if possible. In any event keep the diene cold at all times to minimize vaporization and possible inhalation of its vapors.

2.   Use open flames *only* in those steps in which you are directed to do so. The organic solvents and the cyclopentadiene used in this experiment are highly flammable. Be certain that all joints in the apparatus are tight before heating the dicyclopentadiene to produce the monomer.

The cyclopentadiene required for this experiment is not commercially available as such, because it readily dimerizes at room temperature to produce dicyclopentadiene (equation 7). Fortunately the equilibrium between the monomer and the dimer can be established at the boiling point of the dimer (170°), so that pure cyclopentadiene can be isolated by fractional distillation. The diene must be kept cold in order to prevent extensive dimerization before it can be used for a Diels-Alder reaction.

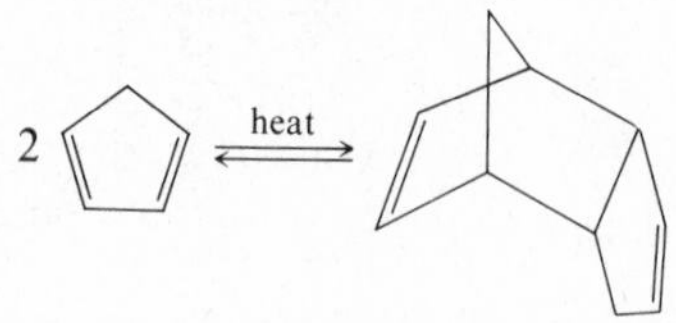

$$2 \; \text{image} \; \underset{}{\overset{heat}{\rightleftharpoons}} \; \text{image} \qquad (7)$$

Place 20 mL of dicyclopentadiene in a 100-mL round-bottomed flask, and attach the flask to an apparatus set for fractional distillation into an *ice-cooled* receiver. Using either a small burner or a heating mantle, gently heat the dimer until brisk refluxing occurs and the monomer begins to distil in the range 40–42°. Distil the monomer as rapidly as possible, but do not permit the temperature of the vapor to exceed 43–45°. Approximately 6 g of monomer should be obtained after distillation for about 30 min. The distillation should be terminated when approximately 5 g of residue remain in the flask. If the distilled cyclopentadiene is cloudy because of condensation of moisture in the cold receiver, add about 1 g of *anhydrous* calcium chloride to dry the diene.

While the distillation is in progress, place 6.0 g (0.061 mol) of maleic anhydride in a 125-mL Erlenmeyer flask, and dissolve it in 20 mL of ethyl acetate by heating on a steam bath. Add 20 mL of petroleum ether (bp 60–80°), and cool the solution thoroughly in an ice-water bath. To this cooled solution add 4.8 g (6.0 mL, 0.073 mol) of dry cyclopentadiene, and swirl the resulting solution until the exothermic reaction is complete and the adduct separates as a white solid.★ Heat the mixture on the steam bath until the solid has dissolved, and then allow the solution to cool slowly to room temperature.★ Filter the solution and determine the yield and melting point of the solid anhydride obtained. The reported melting point is 164–165°. The yield should be about 80% of theoretical. Test the product for unsaturation (Chapter 25, p. 507).

The anhydride can be converted to the corresponding diacid in the following manner. Place 4 g of the anhydride and 25 mL of distilled water in a 125-mL Erlenmeyer flask, and swirl the flask over a Bunsen burner until boiling occurs and all the oil which initially forms has dissolved. Allow the solution to cool to room temperature, and then scratch the inner wall of the flask at the air-liquid interface to induce crystallization.★ After crystallization has begun, cool the flask in ice to complete the process, and filter the solution. Determine the yield and melting point of the product. The reported melting point of the expected dicarboxylic acid is 180–182°. Perform appropriate tests to show whether the hydrolysis has destroyed the carbon-carbon double bond. Also test a saturated aqueous solution of the diacid with litmus paper and record the result.

B. Reaction of Cyclopentadiene with *p*-Benzoquinone

p-Benzoquinone (**7**) has two carbon-carbon double bonds which can serve as dienophiles. In this experiment the Diels-Alder adduct resulting from reaction of 2 mol of cyclopentadiene with 1 mol of *p*-benzoquinone will be prepared (equation 8).

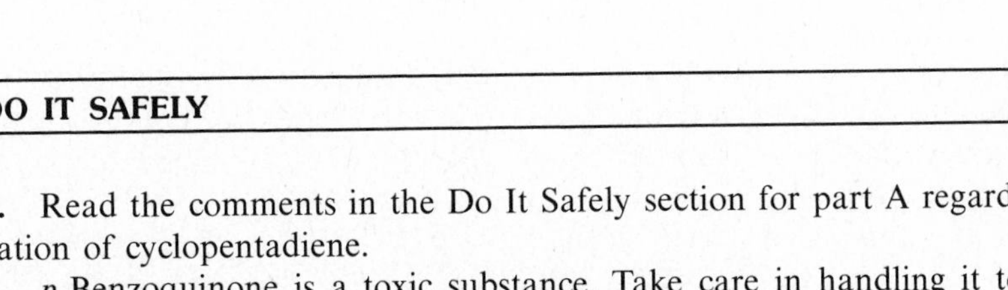

(8)

DO IT SAFELY

---

1. Read the comments in the Do It Safely section for part A regarding the preparation of cyclopentadiene.

2. *p*-Benzoquinone is a toxic substance. Take care in handling it to avoid contact with your skin. If contact should occur, wash your skin with soap and warm water, not solvents such as acetone or alcohol.

Prepare dry cyclopentadiene as described in part A. Add a solution of 3.2 g (4.0 mL, 0.049 mol) of cyclopentadiene and 10 mL of toluene to 2.7 g (0.025 mol) of *p*-benzoquinone[5] dissolved in 20 mL of toluene, and swirl the resulting solution until the mildly exothermic reaction has subsided.★ After 1 hr cool the reaction mixture in an ice-water bath and filter to isolate the solid adduct.

In order to collect a second crop of product, concentrate the mother liquor to one-third of its original volume with distillation apparatus, cool the solution, collect the precipitate, and wash it with a few milliliters of ice-cold toluene. Combine the two crops of the diadduct and recrystallize by dissolving the product in a minimum amount of boiling acetone (*Caution:* No flames!) and then adding water until turbidity begins to appear in the boiling mixture. Clarify the solution by addition of a *small* amount of acetone and allow the solution to cool slowly to room temperature. Complete the recrystallization by cooling in an ice-water bath.

The pure adduct forms beautiful iridescent white needles or platelets and has a reported melting point of 157–158°. The yield should be 60–70%.

C. Reaction of 1,3-Butadiene and Maleic Anhydride

1,3-Butadiene is a gas at room temperature (bp −4.4°). Diels-Alder reactions in which this diene is utilized are normally performed in closed steel pressure vessels (autoclaves) into which the diene is introduced under pressure. However, it has been discovered that 1,3-butadiene may be generated conveniently by the thermal decomposition of 3-sulfolene (equation 9), and that the diene will then react with a dienophile present in the reaction vessel. Such an *in situ* preparation of 1,3-butadiene will be used in this experiment to prepare 4-cyclohexene-*cis*-1,2-dicarboxylic anhydride (equation 10).

$$\text{[structure]} \quad SO_2 \xrightarrow{\text{heat}} \text{[structure]} + SO_2 \tag{9}$$

$$\text{[structure]} + \text{[structure]} \longrightarrow \text{[structure]} \tag{10}$$

DO IT SAFELY

Be certain that all joints in the apparatus used in this experiment are tight before heating the reaction mixture. Not only are the organic solvents flammable, but the sulfur dioxide which is evolved is toxic. It is important to be sure that the gas trap

[5]Commercial *p*-benzoquinone is generally rather impure and may require purification before being used in this experiment. However, a grade of this chemical having a melting point of 113–115° can be used successfully without further purification. *p*-Benzoquinone can be purified, if necessary, by sublimation in an apparatus such as that pictured in Figure 2.20. Consult your instructor for details of this method of purification.

is functioning properly before heating of the reaction mixture is initiated in order to avoid filling the room with toxic sulfur dioxide gas.

Place 10.0 g (0.085 mol) of 3-sulfolene, 6.0 g (0.061 mol) of finely pulverized maleic anhydride, and 4 mL of dry xylene in a 100-mL round-bottomed flask equipped with a water-cooled reflux condenser. Fit the condenser with the gas trap described in Chapter 5 and shown in Figure 5.1. Warm the flask gently while swirling to effect solution, then heat the solution to a gentle reflux, using either a small burner or a heating mantle; continue the heating for about 30 min.★ Cool the solution, add an additional 30 mL of xylene or toluene and about 1 g of decolorizing carbon; heat the mixture on a steam bath with constant swirling for about 5 min; during this period the condenser should be equipped with a drying tube rather than the gas trap to prevent introduction of water into the reaction mixture. Filter the hot mixture (*Caution:* Check for flames in your vicinity) using a short-stemmed funnel and a fluted filter (see Figure 2.16), and carefully add petroleum ether to the filtrate until cloudiness develops. Set the solution aside to cool to room temperature.★ Collect the crystals by vacuum filtration and dry them thoroughly. Record their weight and determine the melting point of the product. The reported melting point is 103–104°. Perform qualitative tests for unsaturation and hydrolyze 4 g of the anhydride to the diacid according to the procedure described in part A. Determine the yield and melting point of the diacid and test it for unsaturation. The reported melting point of 4-cyclohexene-*cis*-1,2-dicarboxylic acid is 164–166°.

If isolation of the anhydride is not desired, direct hydrolysis of it to the diacid can be accomplished in the following way. After the initial 30-min period of reflux, pour the reaction mixture into about 25 mL of water, and heat the resulting initially heterogeneous mixture for at least 30 min. Allow the mixture to stand until the next laboratory period. The crude diacid that results has mp 160–164° and can be purified by recrystallization from water.

EXERCISES

1. Why should 3-sulfolene and maleic anhydride be completely dissolved in the xylene before decomposition of the sulfolene is attempted?
2. Suggest a reason for the observation that cyclopentadiene dimerizes much more readily than do acyclic dienes.
3. In the reaction of butadiene with maleic anhydride, a small amount (4 mL) of xylene was used initially, but an additional and larger amount (30 mL) of xylene or toluene was subsequently required. Why was the total amount of organic solvent needed not added at the beginning of the reaction? Why was an additional quantity required in the procedure?
4. Why is it important to prevent introduction of water into the reaction mixture in the decolorizing step that precedes the isolation of the anhydride?
5. Explain why cyclopentadiene reacts more rapidly with *p*-benzoquinone than with another molecule of itself.
6. Write the structure of the product resulting from the reaction of *cis,trans*-2,4-hexadiene and 2,3-dimethyl-2-butene.

7. Write the structures of the products expected in the following reactions. If no reaction is to be anticipated, write N.R.

(a)  + CH_2=CH_2

(e) + CH_3O_2C—C≡C—CO_2CH_3

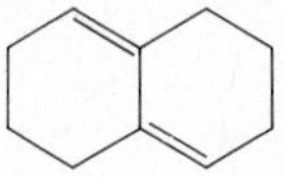

(b) + CH_3O_2C—C≡C—CO_2CH_3

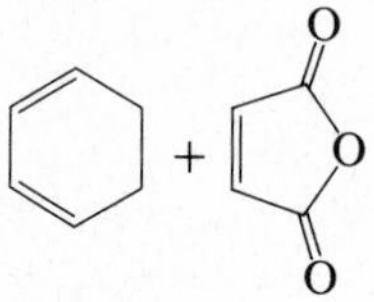

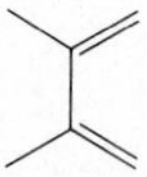

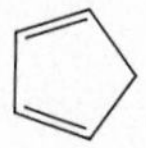

(f) 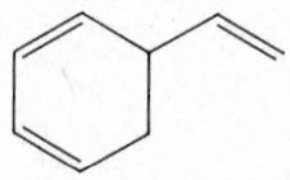 $\longrightarrow$ C_8H_{10}

(c) +

(g) 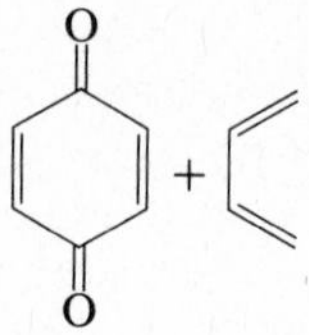 + (2 mol)

(d) + CH_2=$CHCHO$

8. The "cracking" of dicyclopentadiene to two moles of cyclopentadiene (equation 7) is an example of a reverse Diels-Alder reaction. Predict the products to be anticipated from an analogous reaction with the compounds shown below.

(a)

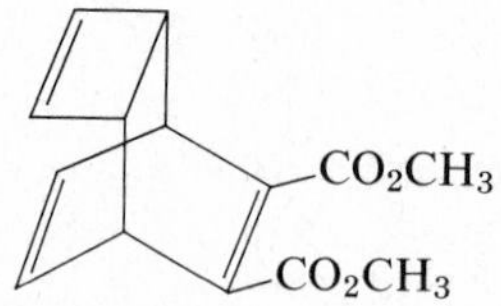

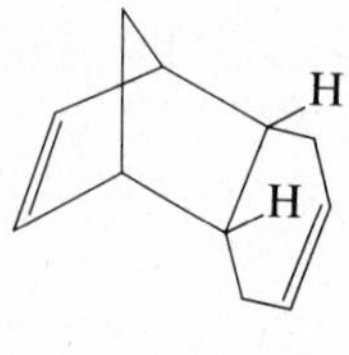

(c)

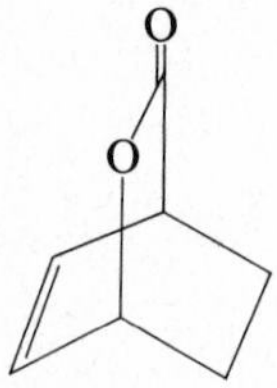

(b)

(d) 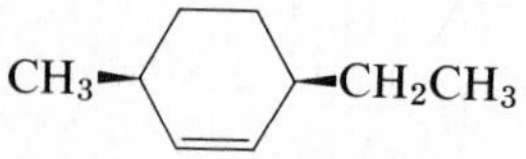 CH_3——CH_2CH_3

(e)

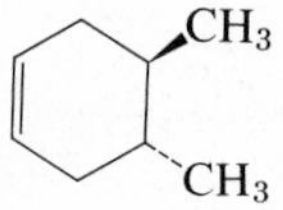

REFERENCES

1. J. Sauer, *Angewandte Chemie, International Edition,* **5,** 211 (1966). A general review.
2. P. Beltrame, in *Comprehensive Chemical Kinetics,* **9,** 94 (1973). A review emphasizing mechanistic aspects.
3. T. E. Sample, Jr., and L. F. Hatch, *Journal of Chemical Education,* **45,** 55 (1968).
4. R. E. Bozak and V. E. Alvarez, *Journal of Chemical Education,* **47,** 589 (1970).
5. R. B. Woodward and R. Hoffmann, *The Conservation of Orbital Symmetry,* Academic Press, New York, 1970.
6. R. Lehr and A. Marchand, *Orbital Symmetry,* Academic Press, New York, 1972.

SPECTRA OF STARTING MATERIALS AND PRODUCTS

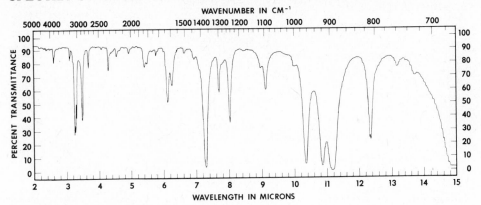

FIGURE 8.1 IR spectrum of 1,3-cyclopentadiene.

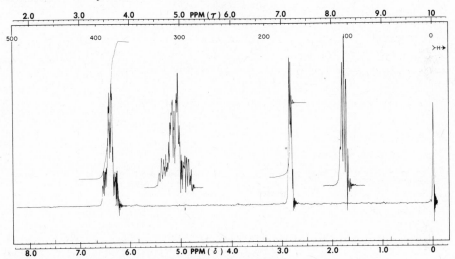

FIGURE 8.2 PMR spectrum of 1,3-cyclopentadiene.

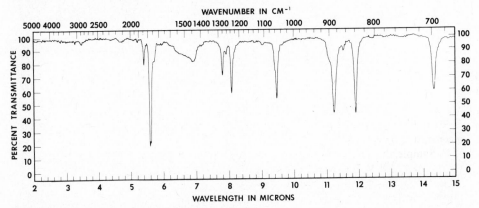

FIGURE 8.3 IR spectrum of maleic anhydride.

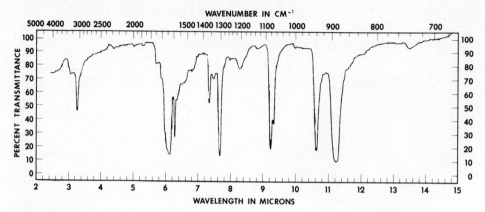

FIGURE 8.4 IR spectrum of *p*-benzoquinone.

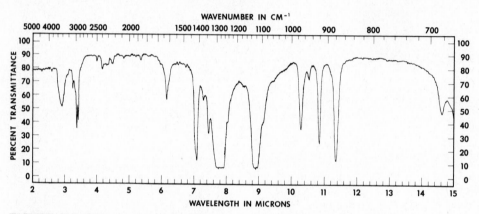

FIGURE 8.5 IR spectrum of 3-sulfolene.

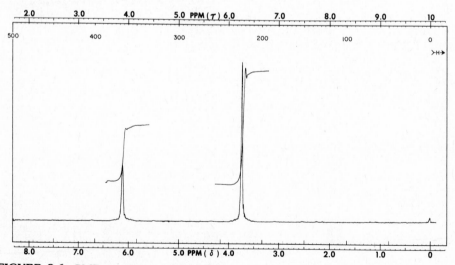

FIGURE 8.6 PMR spectrum of 3-sulfolene.

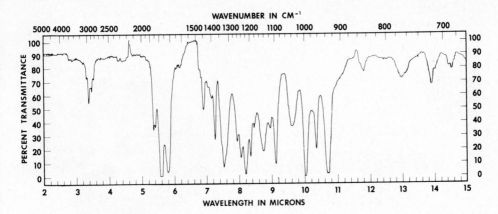

FIGURE 8.7 IR spectrum of 4-cyclohexene-*cis*-1,2-dicarboxylic anhydride.

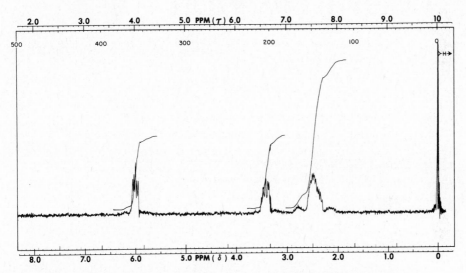

FIGURE 8.8 PMR spectrum of 4-cyclohexene-*cis*-1,2-dicarboxylic anhydride.

9

KINETIC AND EQUILIBRIUM CONTROL OF A REACTION

In predicting which of two competing reactions will predominate, a useful rule of thumb is to choose the reaction which is more exothermic. This rule is based on the fact that *most commonly* the more exothermic reaction will have the smaller heat of activation, and thus will have the greater rate of reaction. For example, Figure 9.1 depicts the potential energy changes involved in the conversion of compound X into

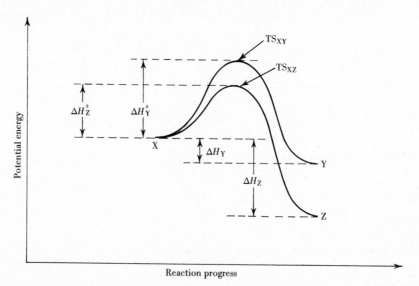

FIGURE 9.1 Typical reaction profile for competing reactions.

the products Y and Z by two competing reactions (equations 1 and 2). The more

$$X \rightleftharpoons Y \tag{1}$$

$$X \rightleftharpoons Z \tag{2}$$

exothermic reaction (equation 2) produces the more stable product, Z (the one lying at the lower energy level). Note that the energy of the transition state leading to Z (TS_{XZ}) is lower than the energy of the transition state leading to Y (TS_{XY}); that is, the relative energies of the transition states are in the same order as the relative energies of the products to which they lead. Hence the heat of activation required for production of Z ($\Delta H_Z^\ddagger$) is lower than the heat of activation required for production of Y ($\Delta H_Y^\ddagger$), so that the rate of the reaction producing Z is greater than the rate of the reaction producing Y.

Most organic reactions either are practically irreversible or they are carried out under conditions such that equilibrium between products and starting materials is not attained, so the yields of products are determined by the relative rates of competing reactions as described. The major product of such a reaction is said to be the product of *kinetic control*. However, when the experimental conditions are favorable for equilibrium to be reached between starting materials and products, the product that predominates initially because of kinetic control *may not* be the major product after equilibrium has been attained. In some reactions the product of kinetic control is less stable than another product formed at a lower reaction rate. An example of such a system (equations 3 and 4) is represented by Figure 9.2, in which product B is shown to

$$X \rightleftharpoons A \tag{3}$$

$$X \rightleftharpoons B \tag{4}$$

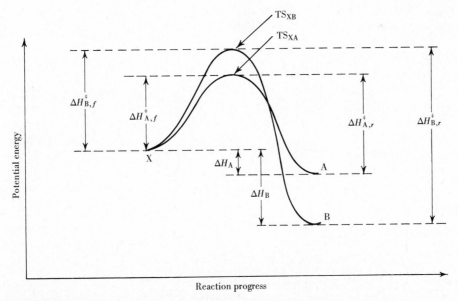

FIGURE 9.2 Reaction profile which predicts different products from kinetic and equilibrium control of competing reactions.

be more thermodynamically stable than product A. It should be noted that in contrast to the more usual relationships depicted in Figure 9.1, the relative energies of the transition states are not in the same order as the relative energies of the products to which they lead. Thus, although product A will predominate initially in the reaction mixture,[1] if the reactions are allowed to come to equilibrium, the more stable product B will be found to be the major product. For this reason it is called the product of *equilibrium control* (or thermodynamic control), whereas A is called the product of *kinetic control.*

There are many well-known reactions of various mechanistic types in which kinetic and equilibrium control lead to different major products under different experimental conditions. For example, the addition of hydrogen bromide to 1,3-butadiene at low temperatures gives the 1,2-adduct, 3-bromo-1-butene, as the main product (equation 5a), whereas at room temperature the main product is the 1,4-

$$CH_2=CH-CH-CH_3$$
$$|$$
$$Br \quad (80\%)$$
$$+ \quad (5a)$$
$$BrCH_2-CH=CH-CH_3$$
(*cis* and *trans*) $\quad$ (20%)

$-80°$

$$CH_2=CH-CH=CH_2$$
$$+$$
$$HBr$$

$40°$

$$CH_2=CH-CH-CH_3$$
$$|$$
$$Br \quad (20\%)$$
$$+ \quad (5b)$$
$$BrCH_2-CH=CH-CH_3$$
(*cis* and *trans*) $\quad$ (80%)

adduct, 1-bromo-2-butene (equation 5b). The 1,2-adduct is the product of kinetic control, since equilibrium is not attained at low temperatures. The 1,4-adduct, although formed more slowly, is the more stable product, so it predominates after equilibrium has been established at higher temperatures.

Sulfonation of naphthalene at 120° gives mainly 1-naphthalenesulfonic acid (equation 6a). However, at temperatures of 160° or above, or with prolonged heating at lower temperatures, the major product is 2-naphthalenesulfonic acid, the more stable isomer (equation 6b). A similar behavior is observed in the sulfonation of phenol, which gives mainly the *ortho*-isomer at lower temperatures (equation 7a) and mainly the *para*-isomer at higher temperatures (equation 7b).

Another aromatic substitution reaction which is reversible, and hence subject to

---

[1] It is interesting to note that although A is produced initially more rapidly than B because of the relative heats of activation for the competing forward reactions ($\Delta H^{\ddagger}_{A,f} < \Delta H^{\ddagger}_{B,f}$), A is also reconverted to starting material (X) more rapidly than B, because the heats of activation for the reverse reactions have the same relationship ($\Delta H^{\ddagger}_{A,r} < \Delta H^{\ddagger}_{B,r}$), owing to the greater stability of B than A. This is not true for competing reactions that have the more usual energy relationships represented by Figure 9.1, so that even when equilibrium conditions are attained in these systems, the same product (Z) will predominate; that is, the product of kinetic control is also the product of equilibrium control, although the proportion of products is not necessarily the same under all experimental conditions.

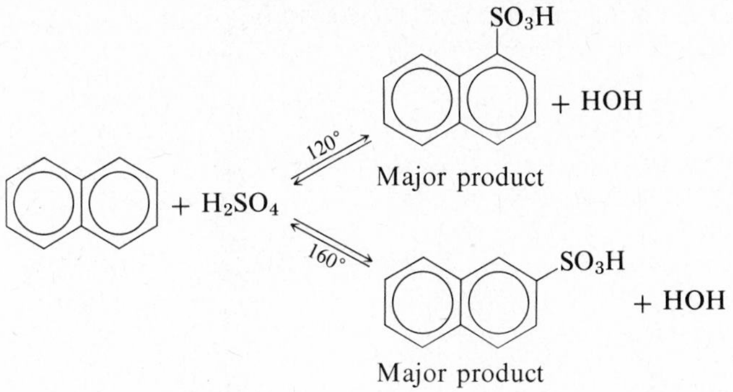

Major product (6a)

+ HOH

Major product (6b)

+ HOH

either kinetic or equilibrium control, is Friedel-Crafts alkylation. Reaction of toluene with methyl chloride and aluminum chloride at low temperatures gives chiefly o- and p-xylene (equation 8a); at higher temperatures the chief product is m-xylene (equation 8b).

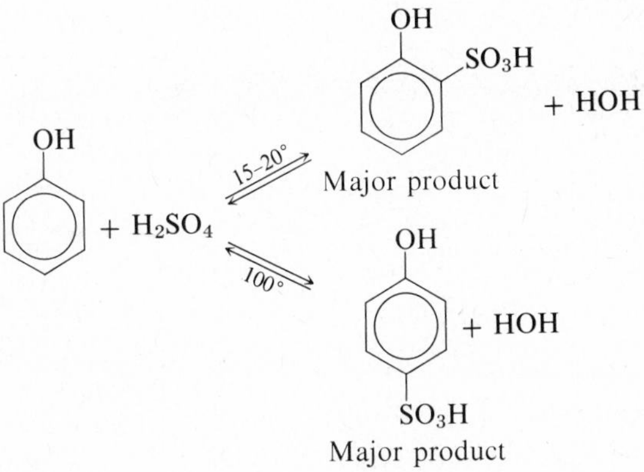

Major product (7a)

+ HOH

Major product (7b)

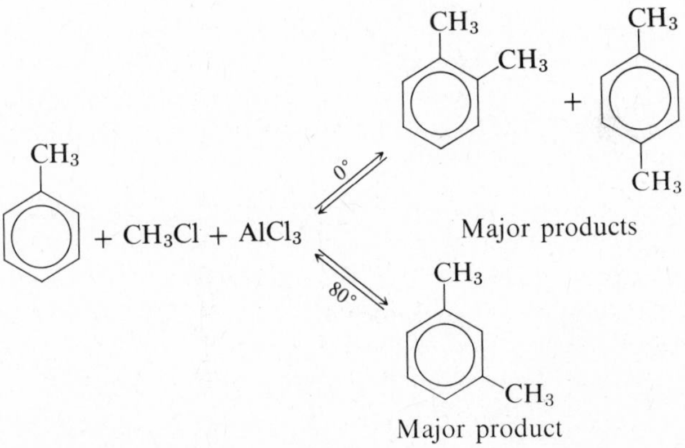

+ HOH

Major products (8a)

Major product (8b)

The elimination reaction that occurs when 2-phenyl-2-butyl acetate is heated in acetic acid gives a very different proportion of products when a small amount of a strong acid is added to the reaction mixture (equations 9a and 9b). Apparently the strong acid catalyzes the reaction so that equilibrium is attained, and the equilibrium product mixture contains a much higher proportion of the *cis*- and *trans*-isomers of the internal alkene, which are more thermodynamically stable than the terminal alkene.

The particular example chosen here to illustrate the principles of kinetic and equilibrium control of products involves the competing reactions of two carbonyl compounds, cyclohexanone (**2**, equation 10) and 2-furaldehyde (**4**, equation 11),

$$
\underset{\underset{CH_3}{|}}{\underset{C_6H_5C-CH_2CH_3}{\overset{OCOCH_3}{|}}}
\quad
\underset{50°}{\overset{CH_3CO_2H}{\nearrow}}
\quad
\underset{\substack{CH_3 \\ cis\ (53\%),\ trans\ (2\%)}}{C_6H_5C=CHCH_3} + \underset{\substack{CH_2 \\ (45\%)}}{C_6H_5C-CH_2CH_3}
\qquad (9a)
$$

$$
\underset{\substack{+\ C_7H_7SO_3H}}{\overset{CH_3CO_2H}{\underset{50°}{\searrow}}}
\quad
\underset{\substack{CH_3 \\ cis\ (81\%),\ trans\ (16\%)}}{C_6H_5C=CHCH_3} + \underset{\substack{CH_2 \\ (3\%)}}{C_6H_5C-CH_2CH_3}
\qquad (9b)
$$

$$
\underset{1}{\overset{O}{\overset{\|}{H_2NCNHNH_2}}} + \underset{2}{O=\bigcirc} \rightleftharpoons \underset{3}{\overset{O}{\overset{\|}{H_2NCNHN}}=\bigcirc} + H_2O \qquad (10)
$$

$$
\underset{1}{\overset{O}{\overset{\|}{H_2NCNHNH_2}}} + \underset{4}{O=CH-\!\!\bigcirc\!\!} \rightleftharpoons \underset{5}{\overset{O}{\overset{\|}{H_2NCNHN}}=CH-\!\!\bigcirc\!\!} + H_2O \qquad (11)
$$

with semicarbazide (**1**). The organic products of these reactions (**3** and **5**), which are known as the semicarbazones of the respective carbonyl compounds, are crystalline solids and have distinctive melting points by which they may be identified easily. For the purposes of the experiment, semicarbazide may be taken to be X of Figure 9.2, and the problem to be solved is identification of the semicarbazones of cyclohexanone and 2-furaldehyde with either A or B, that is, the experimental determination of which compound is the kinetically controlled product and which is the equilibrium product.

The reactions of carbonyl compounds with nucleophiles such as semicarbazide are subject to significant effects of pH on rates and equilibrium constants.[2] This is because of the different way in which the reactants are affected by reaction with hydrogen ions. The addition of a hydrogen ion to a carbonyl compound (equation 12) makes the carbonyl carbon atom *more* electrophilic because of the partial positive charge, as shown in the resonance hybrid of the conjugate acid **6**. On the other hand, addition of a hydrogen ion to the nucleophilic semicarbazide (**1**, equation 13)

[2]Actually, these reactions are affected not only by the concentration of hydrogen ions (of which pH is a measure) but also by the concentration (and nature) of any weak acids that may be present. A discussion of these effects is beyond the scope of this book; further information may be found in textbooks on physical-organic chemistry, in connection with specific and general acid catalysis.

$$R_2C{=}O + H^{\oplus} \rightleftharpoons \left[R_2C{=}\overset{\oplus}{O}H \longleftrightarrow R_2\overset{\oplus}{C}{-}OH\right] \qquad (12)$$
$$\underset{6}{}$$

$$\underset{1}{H_2N\overset{\overset{O}{\|}}{C}NHNH_2} + H^{\oplus} \rightleftharpoons \underset{7}{H_2N\overset{\overset{O}{\|}}{C}NH\overset{\oplus}{N}H_3} \qquad (13)$$

converts it to its conjugate acid (7), which is *not* nucleophilic. At very low pH (high H$^+$ concentration) the concentration of the nucleophile (1) is reduced, and at high pH (low H$^+$ concentration) the concentration of the activated electrophile (6) is reduced. Accordingly, for the reaction of each carbonyl compound with each nucleophilic reagent there is an optimum pH at which the product of the concentrations of the conjugate acid of the carbonyl compound and the nucleophilic reagent is maximized. In order to produce and maintain this optimum pH for reactions of aldehydes and ketones with reagents such as semicarbazide, phenylhydrazine, and hydroxylamine, these reactions are carried out in *buffered* solutions. With regard to the experiments of this chapter, the maximum rates of the reactions of both cyclohexanone and 2-furaldehyde with semicarbazide occur in the pH range 4.5–5.0, and the rates of both decrease at pH values either above or below this range.

In parts A–C of the experimental procedure, the use of a phosphate buffer, which provides reaction rates that are most satisfactory for demonstrating the principles of kinetic and equilibrium control, is recommended. In part F a bicarbonate buffer is employed, providing a higher pH than the phosphate buffer. The results from the experiments of part F should be compared with those obtained in the experiments of part C.

EXPERIMENTAL PROCEDURE

A. In a 50-mL Erlenmeyer flask, dissolve 1.0 g of semicarbazide hydrochloride and 2.0 g of dibasic potassium phosphate (K_2HPO_4) in 25 mL of water. This should give a solution having a pH of about 6.1–6.2.[3] Using a 1-mL graduated pipet, deliver 1.0 mL of cyclohexanone into a test tube and dissolve it in 5 mL of 95% ethanol. Pour the ethanolic solution into the aqueous semicarbazide solution, and swirl or stir the mixture immediately. Allow 5 or 10 min for crystallization of the semicarbazone to reach completion, then collect the crystals by suction filtration and wash them on the filter with a little cold water. Dry the crystals in air and determine their weight and melting point. The reported melting point of cyclohexanone semicarbazone is 166°.[*]

B. Prepare the semicarbazone of 2-furaldehyde by following the procedure of

---

[3]Sodium acetate (2.0 g) may be substituted for dibasic potassium phosphate in parts A and B of this experiment and may be used for the general preparation of semicarbazones of aldehydes and ketones as needed for qualitative organic analysis (Chapters 14 and 25). The solution prepared from semicarbazide hydrochloride and sodium acetate will have a pH of about 4.9–5.0. Sodium acetate *must not* be substituted for dibasic potassium phosphate in part C of this experiment.

part A exactly, except use 0.8 mL of 2-furaldehyde[4] instead of 1.0 mL of cyclohexanone. The reported melting point of 2-furaldehyde semicarbazone is 202°.★

C. Dissolve 3.0 g of semicarbazide hydrochloride and 6.0 g of dibasic potassium phosphate in 75 mL of water.[3] This solution will be referred to as *solution W.*

Prepare a solution of 3.0 mL of cyclohexanone and 2.5 mL of 2-furaldehyde in 15 mL of 95% ethanol. This solution will be referred to as *solution E.*★

1. Cool a 25-mL portion of solution W and a 5-mL portion of solution E *separately* in an ice bath to 0–2°, then add solution E to solution W and swirl them together; crystals should form almost immediately. Place the mixture in an ice bath for about 3–5 min, and then collect the crystals by suction filtration and wash them on the filter with a little cold water. Dry the crystals and determine their weight and melting point.

2. Add a 5-mL portion of solution E to a 25-mL portion of solution W at room temperature; crystals should be observed in about 1–2 min. Allow the mixture to stand at room temperature for 5 min, cool it in an ice bath for about 5 min, and then collect the crystals by suction filtration and wash them on the filter with a little cold water. Dry the crystals and determine their weight and melting point.

3. Warm a 25-mL portion of solution W and a 5-mL portion of solution E *separately* on a steam bath or in a water bath to 80–85°, then add solution E to solution W, and swirl them together. Continue to heat the mixture for 10–15 min, cool it to room temperature, and then place it in an ice bath for about 5–10 min. Collect the crystals by suction filtration, and wash them on the filter with a little cold water. Dry the crystals and determine their weight and melting point.★

D. In a 25-mL Erlenmeyer flask place 0.3 g of cyclohexanone semicarbazone (prepared in part A), 0.3 mL of 2-furaldehyde, 2 mL of 95% ethanol, and 10 mL of water. Warm the mixture until a homogeneous solution is obtained (about 1 or 2 min should suffice) and then an additional 3 min. Cool the mixture to room temperature and then in an ice bath. Collect the crystals on a filter, and wash them with a little cold water. Dry the crystals and determine their melting point.★

E. Repeat the experiment in part D, but use 0.3 g of 2-furaldehyde semicarbazone (prepared in part B) and 0.3 mL of cyclohexanone in place of the cyclohexanone semicarbazone and 2-furaldehyde.★

F. Dissolve 2.0 g of semicarbazide hydrochloride and 4.0 g of sodium bicarbonate in 50 mL of water. (The pH of this solution should be about 7.1–7.2.) Prepare a solution of 2.0 mL of cyclohexanone and 1.6 mL of 2-furaldehyde in 10 mL of 95% ethanol. Divide each of these solutions into two equal portions.

Mix one-half the aqueous solution and one-half the ethanolic solution at room temperature. Allow the mixture to stand at room temperature for 5 min, cool it in an ice bath for 5 min, and then collect the crystals by suction filtration and wash them on the filter with a little cold water. Dry the crystals and determine their weight and melting point.

Warm the other portions of the aqueous and ethanolic solutions *separately* on a steam bath or in a water bath to 80–85°, then combine them and continue heating the mixture for 10–15 min. Cool the solution to room temperature, and then place it in an

---

[4]For best results the 2-furaldehyde should be redistilled just before use.

ice bath for 5 to 10 min. Collect the crystals by suction filtration, and wash them on the filter with a little cold water. Dry the crystals and determine their weight and melting point.

On the basis of your results from experiments C, D, and E, deduce which semicarbazone is the product of kinetic control and which is the product of equilibrium control. To do this, you must first determine from the melting point of the crystals produced in parts C1, C2, and C3 whether the product in each part is the semicarbazone of cyclohexanone, 2-furaldehyde, or a mixture of the two. Note that in C1 the crystals of product separate almost immediately, in C2 after 1 or 2 min, and in C3 only after 10–15 min at a higher temperature; that is, the reaction time is shortest in C1, intermediate in C2, and longest in C3.

In considering experiments D and E, remember that the equilibrium product, being the thermodynamically more stable, is not easily converted into the less stable kinetic product. However, the kinetic product can more easily be converted into the more stable equilibrium product. This follows because the reverse reaction of the less stable kinetic product has a lower activation energy than the reverse reaction of the equilibrium product.

Your completed laboratory report should include the diagram called for in Exercise 1 and answers to Exercises 2 and 3 as well, unless you were instructed to omit some parts of the experiment.

EXERCISES

1. On the basis of the results from the experiments C, D, and E, draw a diagram similar to Figure 9.2, and clearly label the products corresponding to A and B.
2. On the basis of the results from the experiments of part F, explain the effect of the higher pH on the reactions between semicarbazide and the two carbonyl compounds.
3. What results might be expected if sodium acetate buffer (which provides a pH of approximately 5) were used in experiments analogous to those of part C? Explain.
4. Figure 9.5 is the uv spectrum of 2-furaldehyde.
   a. Calculate ε_{max} for the absorption bands present in the spectrum.
   b. Assuming that the uv spectrum of 2-furaldehyde is the same in 95% ethanol as it is in methanol, and that cyclohexanone does not absorb significantly in the uv spectrum, show how the concentration of 2-furaldehyde in solution E could accurately be determined.

SPECTRA OF STARTING MATERIALS

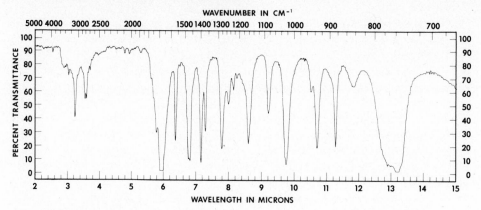

FIGURE 9.3 IR spectrum of 2-furaldehyde.

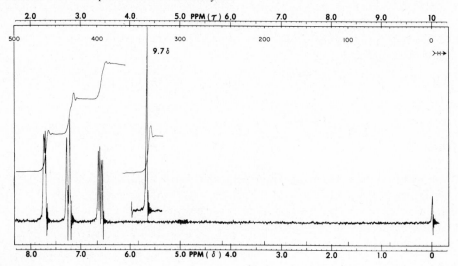

FIGURE 9.4 PMR spectrum of 2-furaldehyde.

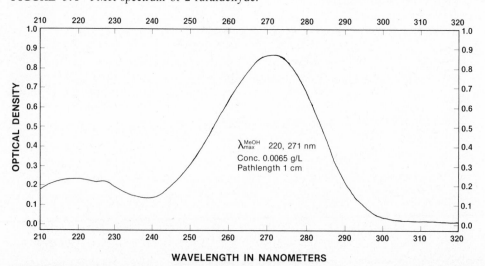

FIGURE 9.5 UV spectrum of 2-furaldehyde.

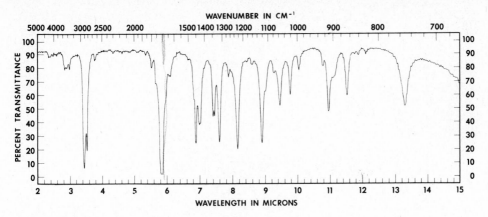

FIGURE 9.6   IR spectrum of cyclohexanone.

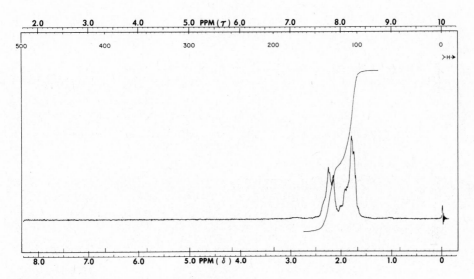

FIGURE 9.7   PMR spectrum of cyclohexanone.

10
NUCLEOPHILIC SUBSTITUTION

10.1 Nucleophilic Substitution at Saturated Carbon

The substitution of one group for another on a saturated carbon atom is a commonly utilized process in effecting organic molecular transformations. The reaction is exemplified in equation 1, where $Nu:$ represents a molecule or ion that has *nucleophilic* character (for example, Cl^-, Br^-, HO^-, NC^-, H_2O, NH_3), and L represents the *leaving group*. The nucleophilic atom in all nucleophiles bears at least one nonbonding pair of electrons that may be donated in the formation of a new covalent bond. Effective leaving groups are those which are able to accept readily the pair of bonding electrons from the alkyl group R as that bond breaks, and hence are usually groups that lead to the *conjugate bases* of fairly strong acids. Thus Cl^- is an efficient leaving group as the conjugate base of H—Cl, whereas HO^- is a poor leaving group owing to the relatively weak acidity of H—OH.

$$R—L + Nu: \longrightarrow R—Nu + L: \tag{1}$$

Types of Nucleophilic Substitution. Although nucleophilic substitution is a very general reaction for aliphatic compounds (R—L), the mechanism followed in a given transformation is strongly dependent on the nature of the alkyl group, R. There are two distinctly different mechanisms concerning which a rather considerable amount of information has been accumulated over the years. These are designated by the symbols S_N1 (substitution, nucleophilic, and 1 for unimolecular) and S_N2 (where 2 stands for bimolecular). The two types of mechanistic pathways are depicted in equations 2 and 3, respectively.

The S_N1 pathway involves two successive steps. In the first step the leaving group heterolytically fragments from the remainder of the molecule, taking its pair of bonding electrons with it. Although equation 2a does not show the interactions, this step is aided considerably by polar interactions between solvent molecules and the incipient ions, a phenomenon that will be investigated kinetically in the experiments of Section 10.2. The carbon atom takes on a positive charge, becoming a carbocation (or carbonium ion). In the second step the nucleophile combines with the positively charged carbon atom to form the product. Normally the concentration of the nucleophile is large compared to that of $L:$, so that the reverse of reaction (2a) is relatively unimportant. Clearly one should anticipate that the first of the two steps will be the slower, since it involves net bond breaking and must be endothermic, whereas the second involves bond formation and must be exothermic. The experimental observation that the rates of many substitution reactions depend only on the concentration of the substrate, R—L, and are independent of the concentration of the nucleophile is consistent with a mechanism of this type.

$S_N 1$ mechanism:

$$R_3C-L \xrightleftharpoons{\text{slow}} R_3C^{\oplus} + L:^{\ominus} \qquad (2a)$$

$$R_3C^{\oplus} + Nu:^{\ominus} \xrightarrow{\text{fast}} R_3C-Nu \qquad (2b)$$

$S_N 2$ mechanism:

$$Nu: \longrightarrow \underset{H}{\overset{R}{\underset{|}{H-C-L}}} \longrightarrow \left[Nu\cdots\underset{H\ H}{\overset{R}{\underset{|}{C}}}\cdots L \right] \longrightarrow Nu-\underset{H}{\overset{R}{C}}{\diagdown}H + L: \quad (3)$$

Transition state

The $S_N 2$ mechanism, on the other hand, is represented as a direct attack of the nucleophile on the backside (with respect to L) of the carbon atom. The C—L bond is broken concurrently with the formation of the C—Nu bond. As the substrate (RCH_2—L) and nucleophile are both involved in the transition state, this is a bimolecular reaction. Therefore the rate of an $S_N 2$ reaction is dependent on the natures and concentrations of both substrate and nucleophile.

In light of the two basic mechanisms, $S_N 1$ and $S_N 2$, outlined for nucleophilic substitution, it is not surprising to find that the following two factors play an important role in dictating the preferred mode of reaction for a particular substrate. (1) In going from CH_3—L to R_3C—L the number of alkyl groups attached to the central carbon atom is increased, making it more difficult for the backside attack by the nucleophile because of steric hindrance; this serves to decrease the ease of $S_N 2$ reaction. (2) As alkyl groups are added to the central carbon, the incipient carbocation of the $S_N 1$ reaction becomes more stable, facilitating the reaction along the $S_N 1$ pathway. (The relative stability of substituted carbocations has been discussed in Section 6.2.) These two effects reinforce one another in producing the general trends shown in the following diagram:

increasing ease of $S_N 2$ reaction

$$\overleftarrow{}$$

$$CH_3-L \quad RCH_2-L \quad R_2CH-L \quad R_3C-L$$

$$\overrightarrow{}$$

increasing ease of $S_N 1$ reaction

Competition between Substitution and Elimination.   In all reactions in which nucleophilic substitution occurs, there is the possibility of competing elimination reactions which produce alkenes. For the most part bimolecular elimination reactions (E2) compete with S_N2 substitution, and unimolecular elimination reactions (E1) compete with S_N1 substitution. E1 reactions have been discussed in detail in Section 6.2 and E2 reactions in Section 6.1.

The competition in the two cases is represented by equations 4 and 5. Generally, elimination is favored when strongly basic and only slightly polarizable nucleophiles are employed (for example, RO^-, NH_2^-, H^-, HO^-), and substitution is favored when weakly basic and highly polarizable nucleophiles are used (for example, I^-, Br^-, Cl^-, H_2O, $CH_3CO_2^-$). Polarizability is a measure of the ease of distortion of the electron cloud of the species by a nearby charged center (positively charged in this case).

S_N2 *versus* E2:

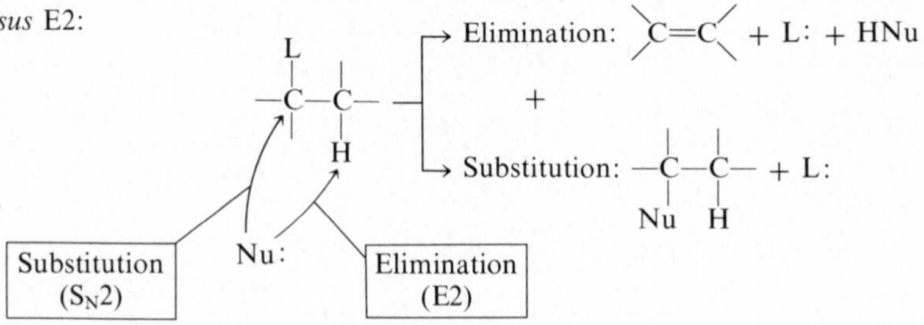

$$\text{(4)}$$

S_N1 *versus* E1:

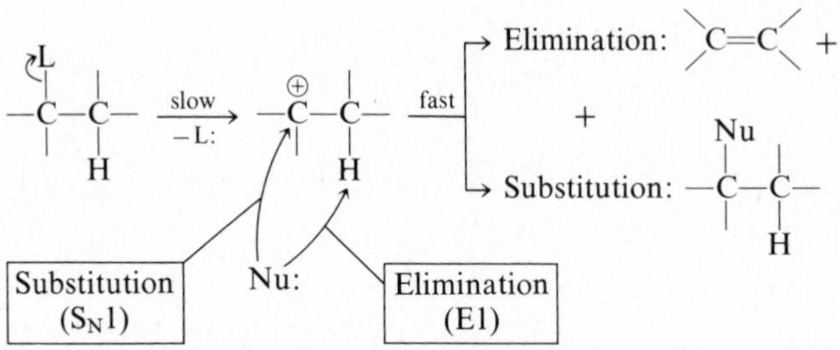

$$\text{(5)}$$

Preparation of 1-Bromobutane: an S_N2 Reaction.   A common method of converting a primary alcohol to an alkyl halide is shown in equation 6. This method involves the reversible reaction between an alcohol and a mineral acid (HX = HCl, HBr, HI). The

$$HX + R\text{—}OH \xrightleftharpoons[\text{heat}]{H_2SO_4} R\text{—}X + H_2O \qquad (6)$$

best yield of alkyl halide is of course obtained when the position of equilibrium lies well to the right.

The preparation of 1-bromobutane may be accomplished by heating 1-butanol with concentrated hydrobromic acid in the presence of concentrated sulfuric acid (equation 7). The mechanism is best written in two steps, which are given in equa-

$$CH_3CH_2CH_2CH_2\text{—}OH + HBr \xrightleftharpoons[\text{heat}]{H_2SO_4} CH_3CH_2CH_2CH_2\text{—}Br + H_2O \qquad (7)$$

tion 8. The first step involves the protonation of the alcohol to give oxonium ion **1**,

$$n\text{-}C_3H_7\text{-}CH_2\ddot{O}H + H^{\oplus} \rightleftharpoons n\text{-}C_3H_7\text{-}CH_2\text{-}\overset{\oplus}{O}\overset{H}{\underset{H}{}} \overset{Br^{\ominus}}{\rightleftharpoons} \begin{matrix} n\text{-}C_3H_7\text{-}CH_2\text{-}Br \\ + \\ H_2O \end{matrix} \qquad (8)$$

$$\mathbf{1}$$

an example of Lewis acid-Lewis base complex formation. Once formed, the oxonium ion undergoes displacement by the bromide ion to form the alkyl bromide and water (equation 8). In this S_N2 reaction *water is the leaving group* and *bromide ion is the nucleophile.* The sulfuric acid serves two distinct and important purposes: (1) as a dehydrating agent to reduce the activity of water and shift the position of equilibrium to the right and (2) as an added source of hydrogen ions to increase the concentration of oxonium ion **1**. The use of *concentrated* hydrobromic acid also helps to establish a favorable equilibrium.

The mixture of hydrobromic acid and sulfuric acid may be prepared in two ways. One method is to add concentrated sulfuric acid to concentrated hydrobromic acid. The second is to generate the hydrobromic acid *in situ* by adding concentrated sulfuric acid to sodium bromide (equation 9). These methods work well and give high yields of the bromide when low molecular weight alcohols are used. With higher

$$NaBr + H_2SO_4 \rightleftharpoons HBr + NaHSO_4 \qquad (9)$$

molecular weight alcohols, the *in situ* generation of HBr is not effective because of the low solubility of these alcohols in concentrated salt solutions. In these instances, concentrated hydrobromic acid is used. With very high molecular weight alcohols, the addition of hydrogen bromide gas to the alcohol at elevated temperatures appears to give the best results.

Although the added sulfuric acid is desirable as discussed, it may also give rise to two important side reactions. The alcohol may react with sulfuric acid to form a hydrogen sulfate ester (**2**, equation 10). This ester formation is reversible and the

$$ROH + H_2SO_4 \rightleftharpoons RO\text{-}SO_3H + H_2O \qquad (10)$$

$$\mathbf{2}$$

alcohol is regenerated (position of equilibrium shifted to the left) as its concentration is reduced in the production of alkyl bromide. The formation of the hydrogen sulfate ester would not directly decrease the yield, except that it undergoes reaction in two other ways to give undesired products. (1) On heating, it can undergo elimination to

$$RO\text{-}SO_3H \xrightarrow{heat} \text{alkenes} + H_2SO_4 \qquad (11)$$

$$RO\text{-}SO_3H + ROH \xrightarrow{heat} R_2O + H_2SO_4 \qquad (12)$$

give alkenes (equation 11). (2) It can react with another molecule of alcohol to give a dialkyl ether by an S_N2 reaction in which the nucleophile is ROH (equation 12). Both of these side reactions consume alcohol and result in a decreased yield of alkyl bromide. For primary alcohols the temperatures used for the substitution reaction are generally not high enough to cause these side reactions to be of great importance.

In the preparation of secondary bromides from secondary alcohols the method involving the use of concentrated sulfuric acid cannot be employed. This is because secondary alcohols are far more easily dehydrated by sulfuric acid to give the corresponding alkenes (see Section 6.2 and equation 11) than are primary alcohols. One means of circumventing this problem is to use 48% hydrobromic acid without the sulfuric acid. However, the secondary halides are generally best prepared by use of phosphorus trihalides (PX_3, X = Cl or Br, equation 13) or thionyl halides (SOX_2, X = Cl or Br, equation 14). The latter reagents are required when the alcohol is susceptible to *cationic rearrangements* (Section 6.2) so that the presence of acid is to be avoided. In these cases pyridine is used as the solvent, since this base reacts with the acid (HX) as it is formed in the reaction and serves to neutralize it (equation 14).

$$3\ ROH + PX_3 \longrightarrow 3\ RX + H_3PO_3 \tag{13}$$

$$ROH + SOX_2 \xrightarrow{C_5H_5N} RX + SO_2 + \underset{\underset{H}{\overset{\oplus}{N}}}{\bigcirc} X^{\ominus} \tag{14}$$

Preparation of Tertiary Alkyl Chlorides: an S_N1 Reaction.   Although the same general approaches as just described can be used for conversion of tertiary alcohols to their corresponding halides, only one method is found to give acceptable yields. This method involves the use of a mineral acid with the alcohol in a reaction which is typically S_N1. Phosphorus trihalides or thionyl halides are not applicable for the preparation of *t*-alkyl halides from the alcohols because elimination reactions predominate.

The reaction of 2-methyl-2-propanol with hydrochloric acid (equation 15) is illustrative of the S_N1 reaction. The mechanism, shown in equation 16, involves three

$$\underset{\underset{CH_3}{|}}{\overset{\overset{CH_3}{|}}{CH_3-C-OH}} + HCl \rightleftharpoons \underset{\underset{CH_3}{|}}{\overset{\overset{CH_3}{|}}{CH_3-C-Cl}} + H_2O \tag{15}$$

distinct steps. The second step (equation 16b) is the slowest step in this sequence, although it is faster than the corresponding S_N2 attack of Cl^- on oxonium ion **3**. The reverse is true for the oxonium ions of primary and secondary alcohols. This reactivity

$$\underset{\underset{CH_3}{|}}{\overset{\overset{CH_3}{|}}{CH_3-C-\overset{..}{O}H}} + H^{\oplus} \rightleftharpoons \underset{\underset{CH_3}{|}}{\overset{\overset{CH_3}{|}}{CH_3-\overset{\oplus}{C}-\overset{..}{O}}}\overset{H}{\diagdown}_{H} \tag{16a}$$
$$\text{3}$$

$$\underset{\underset{CH_3}{|}}{\overset{\overset{CH_3}{|}}{CH_3-C-\overset{\overset{\oplus}{O}}{}}}\overset{H}{\diagdown}_{H} \underset{\longleftarrow}{\overset{slow}{\longrightarrow}} \underset{\underset{CH_3}{}}{\overset{\overset{CH_3}{}}{CH_3-\overset{\oplus}{C}}} + H_2\overset{..}{\overset{..}{O}} \tag{16b}$$
$$\text{Relatively stable}$$
$$\text{tertiary carbocation}$$

$$CH_3-\overset{\oplus}{\underset{\underset{CH_3}{|}}{C}}\overset{CH_3}{\overset{|}{\diagup}} + Cl^{\ominus} \underset{}{\overset{fast}{\rightleftharpoons}} CH_3-\underset{\underset{CH_3}{|}}{\overset{\overset{CH_3}{|}}{C}}-Cl \qquad (16c)$$

difference between tertiary oxonium ions and primary and secondary oxonium ions is a reflection of the relative stabilities of the three types of carbocations that result from the loss of a water molecule. Although *relatively* slow compared to the other steps in the sequence, reaction 16b is quite a rapid one from an *absolute* viewpoint. Thus with *concentrated* hydrochloric acid a favorable equilibrium giving high yields of 2-methyl-2-chloropropane is established in a few minutes at room temperature. Clearly, with dilute hydrochloric acid the equilibrium is less favorable; refer to equation 15.

The principal side reaction is the E1 elimination, resulting from the loss of a proton from the incipient carbocation to give 2-methylpropene (equation 17). Under the reaction conditions, however, the reaction is readily reversible through the addition of HCl to the alkene to give 2-methyl-2-chloropropane (according to Markownikoff's rule, equation 18). Therefore this is not a serious complication.

$$CH_3-\overset{\oplus}{\underset{\underset{CH_3}{|}}{C}}\overset{CH_2-H}{\overset{|}{\diagup}} \longrightarrow CH_3-\underset{\underset{CH_3}{|}}{C}\overset{CH_2}{\diagdown} + H^{\oplus} \qquad (17)$$

$$CH_3-\underset{\underset{CH_3}{|}}{C}\overset{CH_2}{\diagup} + H^{\oplus} \longrightarrow CH_3-\overset{\oplus}{\underset{\underset{CH_3}{|}}{C}}\overset{CH_3}{\diagdown} \overset{Cl^{\ominus}}{\longrightarrow} CH_3-\underset{\underset{CH_3}{|}}{\overset{\overset{CH_3}{|}}{C}}-Cl \qquad (18)$$

EXPERIMENTAL PROCEDURE

DO IT SAFELY

1. Be sure to examine carefully your glassware for cracks or chips. Use of such glassware may lead to an unfortunate accident if the piece should completely break during the course of a reaction, spilling chemicals and reagents. All too often such breakage is the direct result of careless storage of glassware and apparatus in the laboratory drawer. Take the time to store your equipment properly because it may save you from an accident caused by an unnoticed crack in a flask.

2. Concentrated sulfuric acid and water mix with the evolution of substantial quantities of heat. Always add the acid *to* the water in order to disperse the heat through warming of the water. Add the acid slowly and with swirling to ensure continuous mixing.

3. During the extractions performed in the work-up procedure of this experiment (*particularly* for the one which uses concentrated sulfuric acid), the wearing of rubber gloves is advisable to avoid chemicals on your skin.

1-Bromobutane.   Place 0.30 mol of sodium bromide in a 250-mL round-bottomed flask, and add about 35 mL of water and 0.30 mol of 1-butanol. Mix thoroughly and cool the flask in an ice bath. *Slowly* add 35 mL of *concentrated* sulfuric acid to the cold mixture with swirling and continuous cooling. Remove the flask from the ice bath. Fit it with a reflux condenser, and warm the flask gently until most of the salts have dissolved. Add boiling stones and heat the mixture to gentle reflux. A noticeable reaction occurs and two layers form, the upper layer being the alkyl bromide; the inorganic layer has a high concentration of salts and thus is much more dense than 1-bromobutane. Continue heating at reflux for 45 min.★

Equip the flask for simple distillation. Distil the mixture rapidly, and collect the distillate in an ice-cooled receiver. Codistillation of 1-bromobutane and water occurs. Continue the distillation until the distillate is clear. The head temperature should be around 115° at this point. (The increased boiling point is caused by codistillation of sulfuric acid and hydrobromic acid with water.)★

Transfer the distillate to a separatory funnel. Add about 30 mL of water and shake gently, with venting. Note that two layers are formed; decide which of these is the organic layer by withdrawing a little of the upper layer with a dropper and adding it to about 1 mL of water in a small test tube. Your observation should allow you to identify the layers; the contents of the test tube may then be poured into the separatory funnel. Separate the layers and discard the *aqueous* layer. (Be sure you know which layer is which.) Place the organic layer back into the separatory funnel and to it add an equal volume of *cold, concentrated* sulfuric acid. Shake the mixture gently. Remove and discard the sulfuric acid layer, using the test just described if necessary to identify the layers. (The sulfuric acid layer may be discarded in the sink by pouring *slowly* into a stream of running cold water.) In succession wash the organic layer first with about 20 mL of water, then with 10 mL of 2 *M* aqueous sodium hydroxide solution, and finally with about 20 mL of water.

Transfer the cloudy 1-bromobutane layer to a small Erlenmeyer flask, and dry it over a little *anhydrous* magnesium sulfate.★ Swirl occasionally for a period of 10–15 min. By gravity filtration, filter the mixture into a clean and dry 50-mL round-bottomed flask. To avoid the loss of 1-bromobutane, which is soaked into the drying agent on the filter paper, pour about 5 mL of dichloromethane through the filter and allow it to drain into the flask along with the remainder of the 1-bromobutane. Add two or three boiling chips, and equip the flask for simple distillation. Carefully remove the dichloromethane as a forerun and collect the product, boiling at 99–103°. Weigh the product and compute the yield. Yields of 70–80% may be obtained in this experiment. Analyze your product by glpc on a nonpolar column (SF-96 or SE-30 are satisfactory). Obtain the retention times of standard samples of 1-bromobutane and of 2-bromobutane. In your report discuss the relative percentages of 1- and 2-bromobutanes in your product in terms of the mechanism(s) pertaining during the course of the reaction performed. Obtain ir spectra of your product and of 1-butanol and discuss the differences observed.

2-Chloro-2-methylpropane.   Place 0.50 mol of 2-methyl-2-propanol and 1.5 mol of hydrogen chloride provided by reagent grade *concentrated* (12 *M*) hydrochloric acid in a 250-mL separatory funnel. After mixing, swirl the contents of the separatory funnel gently, without the stopper on the funnel. After swirling about 1 min, stopper and invert the funnel; after inverting, vent to release excess pressure by carefully opening the stopcock (be careful to avoid shaking until this is done). Shake the funnel for several minutes, with intermittent venting. Allow the contents to stand until the mixture has separated into two distinct layers; these should be completely clear.

Separate the layers and wash the organic layer with 50 mL of *saturated* sodium chloride solution and then with 50 mL of *saturated* sodium bicarbonate solution. On initial addition of the bicarbonate solution, vigorous gas evolution will normally occur; gently swirl the *unstoppered* separatory funnel until the vigorous gas effervescence ceases. Stopper the funnel and invert gently; vent immediately to release gas pressure. Shake the separatory funnel gently, with *frequent* venting; then shake vigorously, again with frequent venting. Separate the organic layer (the density of a saturated sodium bicarbonate solution is about 1.1), wash it with about 40 mL of water, and then with 40 mL of *saturated* sodium chloride solution. Carefully remove the aqueous layer.

Transfer the 2-chloro-2-methylpropane to a small Erlenmeyer flask, and dry with *anhydrous* magnesium sulfate.* Swirl occasionally for a period of 10–15 min. Filter the mixture into a small round-bottomed flask, and equip the flask for simple distillation. Collect the fraction boiling from 49–52° in a receiver cooled with ice water. The yield of 2-chloro-2-methylpropane should be 75–85%.

2-Chloro-2-methylbutane. 2-Chloro-2-methylbutane may be prepared from 2-methyl-2-butanol, using the same general procedure as that described for 2-chloro-2-methylpropane, using 0.50 mol of 2-methyl-2-butanol. The 2-chloro-2-methylbutane will be obtained as the product on final distillation; material boiling from 83–85° should be collected. The yield should be about 75%.

EXERCISES

1. Observe that some water is added to the initial reaction mixture in the preparation of 1-bromobutane. How might the yield of 1-bromobutane be affected by the failure on the part of the student to add the water? What products would be obtained? How might the yield of 1-bromobutane be affected by adding twice as much water as is called for?
2. In the purification process the impure 1-bromobutane is "washed" with concentrated sulfuric acid. What impurities would a wash of this sort remove? Why?
3. Let us suppose that a student added 0.3 mol of 1,3-dihydroxypropane, $HOCH_2CH_2CH_2OH$, instead of 0.3 mol of 1-butanol. In principle, what product would you expect to be produced from the glycol? Write an equation for its formation. Do you think that it would be produced in good yield under the conditions of this experiment (and using the same quantities of reagents called for in this exercise)? Why or why not?
4. Draw the structures of the other alcohols which are isomeric with 2-methyl-2-propanol. Arrange these alcohols in order of *increasing* reactivity toward concentrated hydrochloric acid. Which, if any, of these alcohols would you expect to give a reasonable yield of the corresponding alkyl chloride under such conditions?

5. The work-up procedure in the 2-chloro-2-methylpropane preparation calls for using sodium bicarbonate. From your experience in the laboratory, you know that this work-up is accompanied by vigorous gas evolution, which increases the difficulty of handling and requires considerable caution. Alternatively, you might consider using a dilute solution of sodium hydroxide instead of the sodium bicarbonate. Comment on the relative advantages and disadvantages of using these two basic solutions in the work-up. On the basis of these considerations, why were you instructed to use sodium bicarbonate, even though it is more difficult to handle?

SPECTRA OF STARTING MATERIALS AND PRODUCTS

The ir and pmr spectra of 1-chlorobutane (Figures 5.5 and 5.6, respectively) are quite similar to those of 1-bromobutane, so that the spectra of the latter compound are not included. The ir and pmr spectra of 2-chloro-2-methylbutane are provided as Figures 6.2 and 6.3, respectively. Also the pmr spectrum of 1-butanol is included as Figure 4.14.

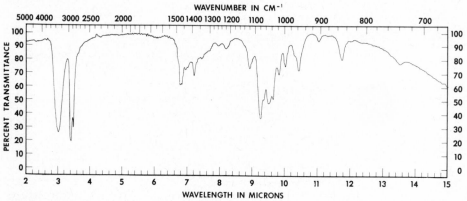

FIGURE 10.1   IR spectrum of 1-butanol.

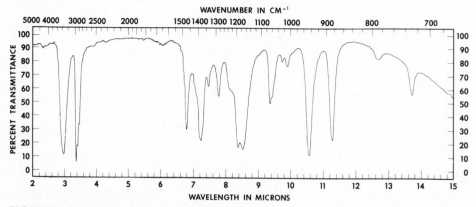

FIGURE 10.2   IR spectrum of 2-methyl-2-butanol.

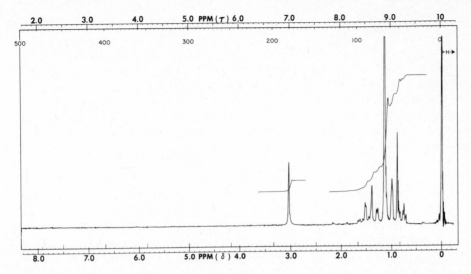

FIGURE 10.3   PMR spectrum of 2-methyl-2-butanol.

10.2   Chemical Kinetics: Evidence for Nucleophilic Substitution Mechanisms

During the course of chemical reactions, continual changes in the concentrations of reactants and products are both expected and observed. Thus as a reaction proceeds, the concentration of each reactant decreases until that of the limiting reagent becomes zero, at which point the reaction is complete. The concentration of the product concomitantly increases from zero at the beginning of the reaction to its maximum value at the end of the reaction. Conceptually, the rate or velocity of a reaction refers to the question: How fast do these concentrations change *as a function of time*? The concept of reactivity is also relevant here because of two compounds the more reactive will reach complete reaction in shorter time and will consequently exhibit a greater *rate* of change of concentration. As a branch of chemical science it would be difficult to overstate the central importance of the study of chemical kinetics. This approach to the study of chemical reactions involves the investigation of the interplay—sometimes subtle—of factors and variables that influence the velocity of a reaction. From this type of information considerable insight into the nature and details of the mechanism of a reaction may be gained.

Some of the most important criteria in support of the S_N1 and S_N2 mechanisms have been obtained from kinetic studies. For reaction at primary and most secondary carbon atoms it is found that the *rate* of substitution is proportional to the concentrations of the substrate *and* the nucleophile. This situation is represented by the following rate law:

$$\text{rate} = k_2[\text{R—L}][\text{Nu:}] \tag{19}$$

The proportionality constant, k_2, relating rate and concentrations is known as the *rate constant*. This equation indicates that the transition state involves both the substrate and the nucleophile, and that the reaction can be represented mechanistically as shown in equation 3, that is, as an S_N2 reaction.

On the other hand, the rates of substitution at tertiary and some secondary carbon atoms are found to be proportional to the concentration of the substrate but *independent* of the concentration of the nucleophile. Therefore the transition state of the slow step in these cases involves only the substrate and not the nucleophile.[1] This is interpreted to indicate an initial slow ionization of the substrate to yield a carbocation as represented in the S_N1 mechanism (equation 2). For this reaction, then, the rate law is as given in equation 20:

$$\text{rate} = k_1[\text{R--L}] \tag{20}$$

In addition to the concentration dependence, rates of substitution are also dependent on such factors as the temperature and the nature of the reactants and solvent. For the sake of simplicity the dependency of the rate on these factors may be considered as being reflected in the value of the rate constant; for example, if a temperature change results in a doubling of the rate (at the same concentration of reactants), then the rate constant is larger by a factor of 2.

S_N1 Reaction and Its Rate Law. S_N1 reactions are first-order reactions in that the rate of reaction is proportional to the first power of the concentration of the substrate (equation 20). It should be evident that a graphical plot of *rate* versus *concentration* in such a case would yield a straight line of slope k_1 and that doubling the concentration would double the rate. The rate constant is independent of the concentration.

When a substance is being consumed in a first-order reaction, its concentration decreases exponentially with time. If C_0 is the *initial* concentration of the substrate (at time $t = 0$, or t_0) and C_t is its concentration at any elapsed time t (where t is measured in any unit of time, for example, seconds, minutes, or hours) after the reaction is started, the relationship between these variables is

$$C_t = C_0 e^{-k_1 t} \tag{21}$$

The rate constant, k_1, has the units of $(\text{time})^{-1}$, for example $\sec^{-1}$. Equation 21 may be rewritten as

$$\ln (C_0/C_t) = k_1 t \tag{22}$$

or

$$2.303 \log (C_0/C_t) = k_1 t \tag{23}$$

If the initial concentration of the reactant (C_0) is known and if one measures the concentration of the reactant (C_t) at a series of measured time intervals while the reaction is proceeding, then the rate constant may be determined in either of two ways. (1) The values of C_t and t measured at each point during the reaction may be substituted into equation 23, which is then solved for k_1. Thus several values of k_1 are determined, and these values are then averaged. The rate constant (the *correct* one) is not easily obtained by this method because the average value will be affected without

---

[1] Molecules of solvent are involved in these reactions through solvation of ions. However, the concentration of the solvent remains *constant* within experimental error during the course of the reaction. Thus the solvent concentration is included as part of the rate constant, k_1.

bias, that is, without mathematical compensation for any measurements that may be "bad" (incorrect owing to experimental error). (2) An alternative and better method of calculation of k_1 from the experimental data consists of preparing a plot of $\log (C_0/C_t)$ versus the time t. Assuming the reaction to be first order and the data to be fairly accurate, a reasonably straight line should be obtained. This line is drawn on the graph in such a way as to lie closest to the largest number of points. The line may be drawn *with bias*, purposely giving more "weight" to the majority of points that will lie close to the line and ignoring or giving less credence to those points which appear to be in error. The slope of this line ("the best straight line")[2] is the rate constant k_1 if equation 22 is used (natural logarithms) or $k_1/2.303$ if equation 23 is used. In the latter case the slope must be multiplied by 2.303 to obtain k_1.

Both S_N1 and S_N2 reactions can be studied from a kinetic standpoint, but bimolecular reactions are generally more difficult to examine experimentally and the calculations for them are more tedious. Therefore we shall limit our studies to a detailed examination of the kinetics of S_N1 reactions, the rate law for which we have just discussed. The experiments presented here are designed to illustrate methods of studying chemical kinetics and measuring the effects of structure on reactivity, as exemplified by the solvolysis of tertiary alkyl halides.

Solvolysis refers to a substitution reaction in which the solvent (SOH) functions as the nucleophile (equation 24). In principle, solvolysis reactions can be carried out in any nucleophilic solvent, such as water (hydrolysis), alcohols (alcoholysis), car-

$$\text{R—X} + \text{SOH} \longrightarrow \text{R—OS} + \text{H}^\oplus + \text{X}^\ominus \tag{24}$$

boxylic acids (acetolysis with acetic acid and formolysis with formic acid), or liquid ammonia (ammonolysis). A major limitation in choosing a solvent is dictated by the solubility of the substrate in the solvent. It is important that the reaction mixture be homogeneous; if it is not, surface effects at the interface of the phases will make the kinetic results difficult to interpret and probably nonreproducible as well. In the experiments described here, the solvolyses will be carried out in mixed solvents consisting of 2-propanol and water.

In order to determine the rate constants (k_1) for these solvolysis reactions, one must make use of the rate expression given in equation 23. The quantities to be measured in addition to time are C_0 and C_t. The best procedure for measuring C_t is based on the fact that for every molecule of alkyl halide that reacts, one molecule of HX is produced (equation 24). Thus the progress of the reaction may be followed by determining the concentration of hydrogen ion, $[\text{H}^+]$, produced as a function of time. Note, however, that $[\text{H}^+]$ gives the amount of alkyl halide that has *reacted* and not the amount which remains, that is, $[\text{H}^+]_t = C_0 - C_t$. Then C_t is determined by subtracting $[\text{H}^+]_t$ from C_0 (equation 25). Experimentally, $[\text{H}^+]_t$ is determined by withdrawing

$$C_t = C_0 - [\text{H}^+]_t \tag{25}$$

---

[2] An alternative method of obtaining the best straight line is the application of a least-squares treatment to the data. This mathematical treatment will also yield a "biased" line, but will do so without the subjective wishful thinking that may influence human judgment. The student who wishes to apply this method may consult the following reference for details: G. H. Brown and C. M. Sallee, *Quantitative Chemistry*, Prentice-Hall, Englewood Cliffs, N. J., 1963, pp. 123 ff.

an aliquot from the reaction mixture with a pipet (the volume must be accurately measured) and adding it to a quantity of 98% 2-propanol sufficient to "quench" the reaction so that it no longer proceeds at a measureable rate. The elapsed time is noted, and the quenched sample is titrated with dilute sodium hydroxide solution.

The most reliable procedure for determining C_0 is to allow the reaction to go to completion, at which time all the alkyl halide will have been converted to product. Titration of an aliquot of the reaction mixture taken at this time (the "infinity point") will provide $[H^+]_\infty$, which must necessarily be equal to the initial concentration of the alkyl halide, C_0. Thus equation 25 becomes

$$C_t = [H^+]_\infty - [H^+]_t \tag{26}$$

So long as the *same* sodium hydroxide solution is used for all titrations, including the infinity titration, it is not necessary to use standardized base in the titrations. Furthermore it will not be necessary actually to calculate the $[H^+]$ values. These concentrations are directly proportional to the *volume* of the sodium hydroxide solution delivered from the buret in reaching the titrimetric end point. With this consideration and using equation 26, equation 23 becomes

$$2.303 \log \frac{(\text{mL of NaOH})_\infty}{(\text{mL of NaOH})_\infty - (\text{mL of NaOH})_t} = k_1 t \tag{27}$$

It is to be emphasized that application of equation 27 requires that the volumes of all aliquots be the same (including the infinity titration) and that the *same* solution of sodium hydroxide be used for all titrations.

Factors Influencing the Rate of S_N1 Reactions.   Some of the particular factors that influence the rate of S_N1 reactions are described in the following paragraphs; experiments are provided which will allow some of them to be investigated.

   1.   *Solvent composition.* The nature of the solvent greatly influences the rate of S_N1 reactions. The solvent effect may be considered as a dependence of the rate on the *polarity* of the solvent. The polarity influences the rate of the initial ionization (equation 2a), the slow step of the S_N1 reaction. The more polar the solvent, the more rapid the ionization. This is because of the greater ability of the more polar solvent to stabilize charged species through solvation.

   As mentioned, the experiments make use of mixed solvents consisting of 2-propanol and water. The effects of solvent composition on rate can be studied using mixtures containing 60:40, 55:45, and 50:50 (v:v) 2-propanol:water.

   2.   *Effect of concentration.* Equation 20 indicates that the rate of solvolysis is proportional to the concentration of alkyl halide. This equation assumes that the rate constant is independent of the concentrations of alkyl halide.

   3.   *Effect of temperature.* Temperature changes also influence the rate of a chemical reaction. A *rough* rule of thumb is that the rate doubles for each 10° rise in temperature. Thus, in going from 25 to 45° (a 20° rise in temperature), the rate of reaction increases by a factor of *about 4*. To examine the effects of temperature on solvolysis reactions, two temperatures can be utilized conveniently in the laboratory: one is room temperature, and the other is 0°, which can be obtained with the use of an ice-water bath.

4. *Effect of alkyl group.* In solvolysis reactions tertiary halides may be used with complete confidence that the substitutions are proceeding by the S_N1 mechanism. Two compounds suitable for the study are 2-chloro-2-methylpropane and 2-chloro-2-methylbutane, whose preparations are described in Section 10.1. A determination of the rate of solvolysis of these compounds will provide data which can be used to illustrate the reactivity difference between two tertiary alkyl groups, the *t*-butyl group, $(CH_3)_3C-$, and the *t*-pentyl group, $(CH_3)_2(C_2H_5)C-$.

5. *Effect of leaving group.* Although no experiments are included here to evaluate the effect of different leaving groups on the rate of a solvolysis reaction, it should be apparent that such effects should be manifested: The bond to the leaving group is being broken in the rate-determining step of the reaction. For example, alkyl bromides solvolyze somewhat faster than the corresponding alkyl chlorides, since bromide is a better leaving group than chloride. These reactivity differences are reflected in the values of the rate constants.

It should be noted that the composition of the final product mixture in an S_N1 reaction has no influence on the rate of the reaction or therefore on the value of the rate constant. This follows since it is the *first* step of an S_N1 reaction that is rate determining. The product-determining steps follow this step. For example, in the solvolysis of 2-chloro-2-methylpropane in aqueous 2-propanol (equations 28–31), there are three possible products, depending on the fate of the intermediate carbocation. Note, however, that in their formation from 2-chloro-2-methylpropane all three products share the same rate-determining step (equation 28).

$$CH_3-\underset{\underset{CH_3}{|}}{\overset{\overset{CH_3}{|}}{C}}-Cl \xrightarrow{\text{slow}} CH_3-\overset{\oplus}{C}\underset{CH_3}{\overset{CH_3}{<}} + Cl^{\ominus} \tag{28}$$

$$CH_3-\overset{\oplus}{C}\underset{CH_3}{\overset{CH_3}{<}} \xrightarrow{\text{fast}} CH_3-C\underset{CH_3}{\overset{CH_2}{<}} + H^{\oplus} \tag{29}$$

$$CH_3-\overset{\oplus}{C}\underset{CH_3}{\overset{CH_3}{<}} + H-OH \xrightarrow{\text{fast}} CH_3-\underset{\underset{CH_3}{|}}{\overset{\overset{CH_3}{|}}{C}}-OH + H^{\oplus} \tag{30}$$

$$CH_3-\overset{\oplus}{C}\underset{CH_3}{\overset{CH_3}{<}} + (CH_3)_2CHOH \xrightarrow{\text{fast}} CH_3-\underset{\underset{CH_3}{|}}{\overset{\overset{CH_3}{|}}{C}}-OCH(CH_3)_2 + H^{\oplus} \tag{31}$$

EXPERIMENTAL PROCEDURE

In experiments of the type given here, you should strive for consistency and accuracy. Prepare yourself ahead of time, and establish a systematic approach to the collection and recording of the necessary data. Before coming to class, prepare in your notebook a table for recording the following: (1) which solvent you are using and the volume used in preparing the reaction mixture; (2) the weight or volume of alkyl halide used in preparing the reaction mixture; (3) the time, t_0, at which the kinetic run is initiated; (4) the temperature of the reaction mixture; (5) the results of a "blank" titrimetric determination; (6) a series of times at which aliquots are withdrawn; and (7) the initial and final buret readings observed in the titration of each aliquot.

This experiment involves a series of quantitative measurements carried out in a relatively short time. You should be prepared to work *rapidly*, although *carefully*, in order to maintain a high standard of accuracy. Buret readings should be made to the nearest 0.02 mL if possible, although precision within 0.05 mL will normally be satisfactory. Time measurements should be made *at least* to the nearest minute.

General Kinetic Procedure.   Throughout this experiment, use the same pipet and buret, as well as the same sodium hydroxide solution.

Using a graduated cylinder, accurately measure into a 250-mL Erlenmeyer flask equipped with a well-fitting rubber stopper 100 mL of the solvent that has been assigned to you. Using a thermometer, measure and record the temperature of this solution. In a second flask obtain 80 mL of 98% 2-propanol[3] to use for quenching purposes.

Obtain in a third flask 125–150 mL of approximately 0.04 M sodium hydroxide solution. Stopper the flask with a well-fitting rubber stopper. Set up a 50-mL buret, rinse it with a small amount of the sodium hydroxide solution, fill it, see that all air bubbles are out of the tip, and cover the top of the buret with a test tube or small beaker to minimize absorption of carbon dioxide from the air.

Put about 2 mL of phenolphthalein indicator solution in a test tube and have it available, with a dropper, for use in each titration. Connect a short length of rubber tubing to the nearest aspirator or vacuum line for use in drawing air through the pipet for a minute or two after each sampling in order to dry it before taking the next sample. Have available a watch that may be read at least to the nearest minute.

To initiate a kinetic run, add the alkyl halide to the solvent mixture. A sample size of approximately 1 g is satisfactory, either weighed out or measured with a 1-mL pipet. Swirl the mixture gently to obtain homogeneity. Note and record as t_0 the time of addition. Keep the flask tightly stoppered to avoid evaporation and as a consequence a change of concentration.

While waiting to make the first measurement, determine a "blank" correction for the solvent. Using a graduated cylinder, measure into a 125-mL Erlenmeyer flask a separate 10-mL portion of the 2-propanol-water mixture being used. (*Note:* Do *not* use the mixture containing the alkyl halide.) Next add 10-mL of 98% 2-propanol and 4–5 drops of phenolphthalein to the blank, and titrate with base to a faint pink color that persists for 30 sec. In all titrations use a white background (paper or a towel) below the titration flask, and accentuate the lower edge of the meniscus in the buret by holding dark paper or some other dark object just below it to make it easier to read. The blank correction will probably be no more than 0.05–0.15 mL.

At regular intervals take a 10-mL sample from the reaction mixture with a 10-mL pipet, and add it to 10-mL (measured with a graduated cylinder) of 98%

---

[3]Commercial 2-propanol is 98% pure and is satisfactory for quenching the solvolysis reactions.

2-propanol contained in a 125-mL flask. Be sure to note the time of addition of the aliquot, probably best taken as the time at which one-half the aliquot has been added to the alcohol used to quench the reaction. Titrate with base to the phenolphtholein end point, as in the blank determination.

The suggested *approximate* times of taking aliquots using various solvents under various conditions are listed below:

(1) 50% 2-propanol-water and 2-chloro-2-methylpropane: 10, 20, 35, 50, 75, and 100 min

(2) 55% 2-propanol-water and 2-chloro-2-methylpropane: 15, 30, 50, 75, 100, and 135 min

(3) 60% 2-propanol-water and 2-chloro-2-methylpropane: 20, 40, 70, 100, 130, and 170 min

(4) 50% 2-propanol-water and 2-chloro-2-methylbutane: 10, 20, 30, 40, 50, and 60 min

(5) 55% 2-propanol-water and 2-chloro-2-methylbutane: 15, 30, 45, 60, 80, 110, and 140 min

(6) 60% 2-propanol-water and 2-chloro-2-methylbutane: 20, 40, 60, 80, 100, and 120 min

The *fastest* of the above reactions will require about 4 hr to reach 99.5% completion (the effective infinity point). The slowest requires over 12 hr. Therefore it is most convenient to wait until the next laboratory period to perform the infinity titration necessary to obtain C_0. Stopper the reaction flask *tightly* to avoid evaporation, and store the flask in your desk. Be sure also to save at least 30 mL of the sodium hydroxide solution in a *tightly* stoppered flask so that it will be available to you for the infinity titration in the following laboratory period.

Treatment of Data.   1.   Using the buret readings, determine by difference the number of mL of sodium hydroxide solution used in each titration. Apply the blank correction to all values by subtracting it from each volume. Use the corrected volumes in your calculations. Using the recorded times for each aliquot, determine in each case the elapsed time from t_0. Apply equation 27 by calculating the log term, multiplying that value by 2.303 and *plotting* the result (vertically) versus the elapsed time t (in hours) for each kinetic point. Draw the best straight line through the points. Determine the slope of the line; this slope is the rate constant k_1. It should be about 0.2 hr^{-1} for 60% 2-propanol-water and 2-chloro-2-methylpropane and about 0.8 hr^{-1} for 50% 2-propanol-water and 2-chloro-2-methylpropane.

2.   Using the same data, calculate the value of k_1 separately for each kinetic point using equation 27. Compare the *average* of these values with the rate constant obtained graphically. Also compare this average with each of the values that were averaged. Which procedure, graphical or averaging, allows you most easily to spot a point that is likely in error?

3.   From equation 27 it can be seen that the half-life ($t_{1/2}$, the time necessary for one-half of the original alkyl halide to react) is given by the relation:

$$t_{1/2} = \frac{0.69}{k_1} \tag{32}$$

Calculate the half-life of your reaction using the value of k_1 obtained from the graph. Go back and examine your experimental data. If about one-half of the total volume of NaOH used in the infinity titration would not have been consumed in a titration done at this time, an error has been made in the calculations. Recheck them carefully.

    4. Consider the magnitudes and types of errors involved in the various measurements made in this experiment. Using the unavoidable errors in measurement which would have been expected in volume, time, and titration measurements, calculate the *maximum* error in k_1 which can be expected, for example, $k_1 = 0.80 \text{ hr}^{-1} \pm 0.05 \text{ hr}^{-1}$.

EXERCISES

1. Give as many advantages as possible for using an infinity titration and equation 27 instead of calculating alkyl halide concentrations and using equation 23. Be sure to consider what the alternatives are in obtaining C_0.
2. Explain clearly how equation 27 derives from equation 23.
3. Show how equation 32 is derived from equation 23.
4. Suppose the flask had not been stoppered until the following laboratory period, when the infinity titration was done. Would the calculated rate constant have been too large or too small if some evaporation had occurred?
5. Why does the titration end-point color fade after 30–60 sec?
6. List the possible errors involved in the determination of rate constants by the procedure used. State the relative importance of each.

11

CARBENES AND ARYNES
Highly Reactive Intermediates

For a more complete understanding of any organic reaction, recognition and characterization of possible intermediates formed during the conversion of starting materials to products are required. The types of intermediates encountered in organic chemistry are varied, both in their relative reactivities and in their electronic structures. Charged intermediates such as carbocations and carbanions are usually very reactive, and their preparation and isolation commonly require special experimental precautions such as the use of inert atmospheres and the rigorous exclusion of moisture from the reaction. For instance, the preparation of a Grignard reagent, $RMgX$ (X = halogen), an organometallic substance having the properties anticipated for a carbanion, is successful only if the reagents and apparatus used are dry (Chapter 12). Some charged intermediates, however, are sufficiently stable so that they can be isolated as salts without the use of special experimental techniques, as is the case in the preparation of tropylium iodide, $C_7H_7^+I^-$, a salt of a carbocation (Chapter 18).

Uncharged intermediates quite often are stable molecules which can be readily isolated by use of appropriate reaction conditions. As an example, benzophenone (**1**), a stable ketone, undoubtedly intervenes as an intermediate in the preparation of triphenylmethanol by reaction of methyl benzoate with two moles of phenylmagnesium bromide (equation 1). Although benzophenone is not isolated in the procedure

$$C_6H_5CO_2CH_3 + C_6H_5MgBr \longrightarrow \underset{\textbf{1}}{C_6H_5COC_6H_5} \xrightarrow[\text{(2) } H_3O^{\oplus}]{\text{(1) } C_6H_5MgBr} (C_6H_5)_3COH \qquad (1)$$

described in Section 12.2, it could be if the conditions of the reaction were modified. Many neutral intermediates cannot be easily isolated, however, because of their high reactivity. Typical of these are free radicals such as those generated in the free radical-chain halogenation of alkanes (Section 5.1).

In this chapter we consider the generation and reactions of two very reactive *neutral* intermediates which owe their high reactivity to two quite different molecular properties. The first of these substances has the general formula **2** (R = H, alkyl, aryl, halogen, cyano, and so forth) and is called a *carbene*. The presence in this divalent form of carbon of only six rather than eight electrons in the outer valence shell imparts high reactivity to such intermediates and makes them *electrophilic* in character. The experiment described in Section 11.1 involves the generation and chemistry of *dichlorocarbene* (**2**, R = Cl).

The other intermediate to be considered in this chapter is *benzyne* (**3**), a member of a class of compounds designated as *arynes*. The high reactivity of benzyne, or dehydrobenzene as this molecule is sometimes called, can be considered a result of the large amount of strain resulting from introduction of a triple bond into a six-membered ring.

$$R-\overset{..}{C}-R$$

2   **3**

11.1 Carbenes

Carbenes can be generated in a variety of ways, two of which are given here. One method involves decomposition of a diazo compound (**4**) by irradiation or heat or with a metallic catalyst (equation 2). This method is generally much too hazardous to be attempted in the introductory organic laboratory because the diazo compounds are generally unstable and potentially explosive.

$$\left[R_2C=N=N \longleftrightarrow R_2\overset{\ominus}{C}-\overset{\oplus}{N}\equiv N: \right] \xrightarrow[\text{or metallic catalyst}]{\text{light or heat}} R_2C: + :N\equiv N: \tag{2}$$

4   **2**

An alternative approach to producing carbenes is that termed *α-elimination*. This technique fundamentally involves initial formation of a carbanion of the type **5**, followed by decomposition of the carbanion to a carbene by loss of an anion such as halide ion (equation 3), M = metal, X = halogen). This route to carbenes is quite

$$\overset{\displaystyle R}{\underset{\displaystyle X}{R-\overset{|}{\underset{|}{C}}:\ominus M\oplus}} \longrightarrow R_2C: + X\ominus + M\oplus \tag{3}$$

5   **2**

safe experimentally but does have the disadvantage that yields of products may be lower than when diazo compounds are utilized as sources of carbenes.[1]

---

[1] It should be noted here that current investigations indicate that in many instances α-elimination may not lead to a "free" carbene, as does photochemical or thermal decomposition of a diazo compound, but rather that the reactive intermediate in the former procedure is either **5** or **2** complexed with the salt MX; for the specific case of dichlorocarbene, however, α-elimination appears to produce the "free" species. Whatever the true nature of the active intermediate resulting from an α-elimination, it is observed to react in much the same way in the addition reaction as does a carbene generated from a diazo compound.

Two types of reactions characteristic of carbenes are their *insertion* into carbon-hydrogen bonds (equation 4) and their *addition* to carbon-carbon double and triple bonds (equation 5). The insertion reaction is of great interest mechanistically but has relatively little synthetic utility. The addition reaction, however, has received wide application in the synthesis of molecules containing three-membered rings. Of importance in terms of the mechanism of the reaction is the observation that the addition of carbenes usually occurs in a stereospecifically *cis* fashion. As an example,

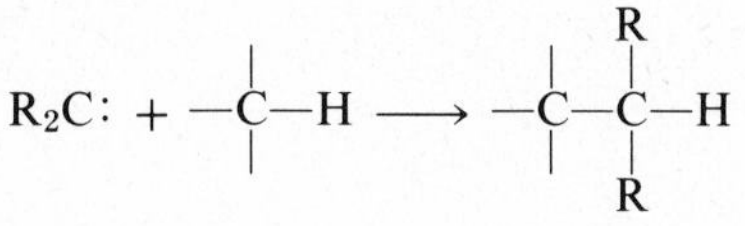

$$R_2C: + \quad \overset{|}{\underset{|}{C}}-H \quad \longrightarrow \quad -\overset{|}{\underset{\underset{R}{|}}{C}}-\overset{R}{\underset{|}{C}}-H \tag{4}$$

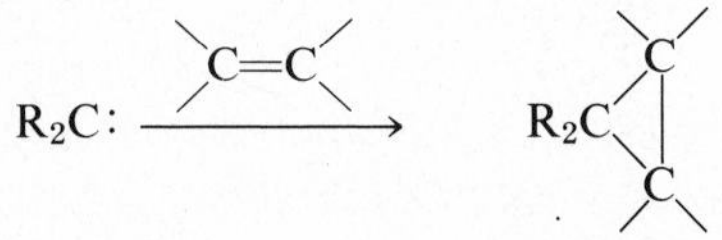

$$R_2C: \quad \xrightarrow{\quad \diagdown C = C \diagup \quad} \quad R_2C\diagup\overset{C}{\underset{C}{\diagdown}} \tag{5a}$$

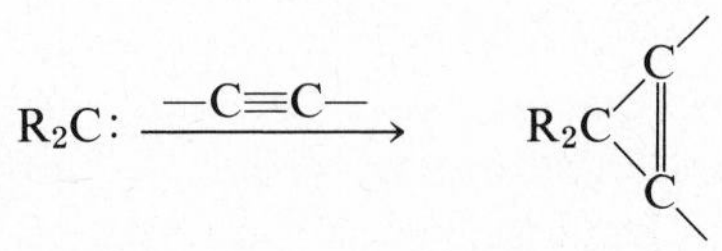

$$R_2C: \quad \xrightarrow{\quad -C \equiv C- \quad} \quad R_2C\diagup\overset{C}{\underset{C}{\diagdown}} \tag{5b}$$

the reaction of *trans*-2-butene with a carbene, R_2C:, produces only the adduct **6**, whereas the corresponding process using the *cis*-alkene gives **7** exclusively.

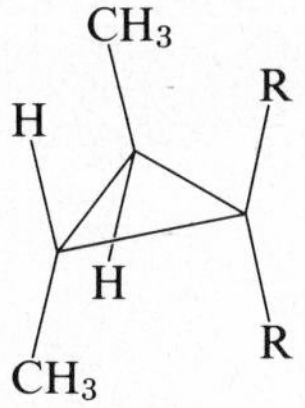

6	**7**	**8**

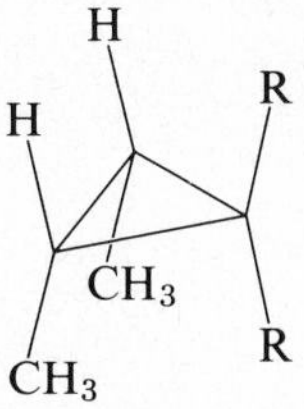

It should be emphasized that the high chemical reactivity of carbenes requires that they be generated *in the presence* of the substrate with which reaction is to occur. Thus success of the addition reaction described depends upon having the 2-butene admixed with the precursor to the carbene.

Precursors **5**, required for generation of carbenes by the method of α-elimination, are normally prepared by the reaction of a strong base with a halogenated substrate. Thus treatment of a mixture of chloroform and an alkene with potassium *t*-butoxide yields **8** (M = K), which can subsequently decompose, with loss of potassium chloride, to dichlorocarbene (**2**, R = Cl); the carbene may then immediately add to the alkene to produce a 1,1-dichlorocyclopropane (equation 6). Alternatively, the salt **8** (M = Na), can be produced by the thermal decomposition of sodium trichloroacetate (equation 7). Experiments using both sodium trichloroacetate and chloroform as sources of dichlorocarbene for the addition reaction with alkenes are described in this section.

$$t\text{-BuO}^{\ominus}\text{K}^{\oplus} + \text{CHCl}_3 \longrightarrow t\text{-BuOH} + \text{Cl}_3\text{C}^{\ominus}\text{K}^{\oplus} \xrightarrow{-\text{KCl}} \text{Cl}_2\text{C}: \qquad (6a)$$

$$\text{Cl}_2\text{C}: + \ \overset{}{\underset{}{\text{C}}}=\overset{}{\underset{}{\text{C}}} \longrightarrow \qquad (6b)$$

$$\text{Cl}_3\overset{\frown}{\text{C}}\overset{}{\underset{\underset{\text{O}}{\|}}{\text{C}}}\overset{\frown}{\text{O}}^{\ominus}\text{Na}^{\oplus} \xrightarrow[\text{heat}]{-\text{CO}_2} \text{Cl}_3\text{C}^{\ominus}\text{Na}^{\oplus} \xrightarrow{-\text{NaCl}} \text{Cl}_2\text{C}: \qquad (7)$$

Dichlorocarbene from Sodium Trichloroacetate.   The generation of dichlorocarbene by thermal decarboxylation of sodium trichloroacetate requires a thoughtful choice of solvent. First of all, the boiling point of the solvent should be sufficiently high to induce a suitable rate of decomposition of the trichloroacetate. Second, the solubility of the salt in the solvent chosen should be *low.* This somewhat unusual criterion for an appropriate solvent is dictated by knowledge that carbenes in general are susceptible to attack by nucleophiles so that the higher the concentration of trichloroacetate *in solution,* the greater the possibility of occurrence of competing reactions of the type shown in equation 8; these would act to decrease the yield of the desired addition product. Dimerization of the carbene to give tetrachloroethylene would also be expected to become more important with increasing solubility of its precursor salt in the reaction medium, and this would have a similar effect on the yield of the reaction. Finally, the solvent, of course, must not itself react with the carbene or any of its precursors. All the desired criteria are nicely met by a mixture of tetrachloroethylene and diglyme,[2] $(\text{CH}_3\text{OCH}_2\text{CH}_2)_2\text{O}$; the boiling point of the mixture used in the procedure is approximately 110°.

$$\text{CCl}_3^{\ominus}\text{Na}^{\oplus} + :\text{CCl}_2 \longrightarrow \text{Cl}_2\overset{\overset{\text{Cl}}{|}}{\text{C}}\overset{\overset{\text{Na}^{\oplus}}{}}{\overset{\frown}{\text{C}}\text{Cl}_2} \longrightarrow \overset{\text{Cl}}{\underset{\text{Cl}}{}}\text{C}=\text{C}\overset{\text{Cl}}{\underset{\text{Cl}}{}} + \text{NaCl} \qquad (8)$$

Indene (**9**) will be used as the unsaturated substrate with which the dichlorocarbene is to react. In principle a number of products resulting from both the insertion and addition reactions are possible. Fortunately the reaction mixture is not as complex as it might be, since carbenes formed by a α-elimination preferentially give products resulting from the addition, rather than from the insertion, of the carbene. However, even if the reaction is limited to addition of dichlorocarbene to indene, there is still possible a variety of products, all of which are shown in Figure 11.1.

A prediction concerning which of the possible products might be favored can be made on the basis that a fundamental difference exists between compound **10**, which bears the imposing name, 1,1-dichloro-1*a*,6*a*-dihydrocycloprop[a]indene, and all the

---

[2] Diglyme is an abbreviation of the term *di*ethylene*gly*col *di*methyl ether.

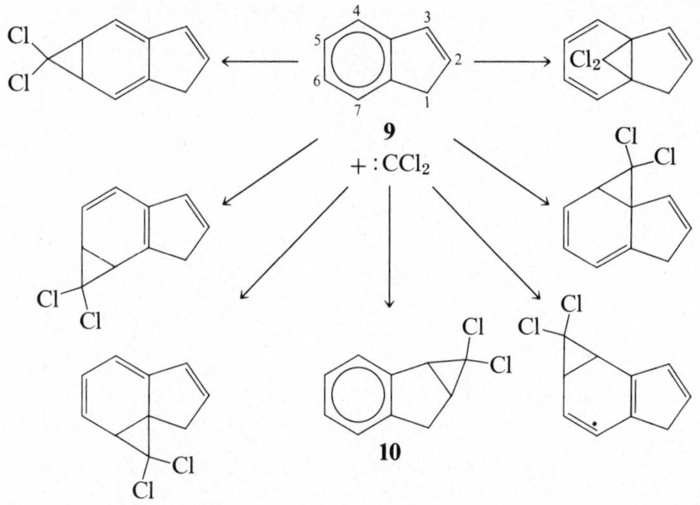

FIGURE 11.1 Possible addition products of dichlorocarbene and indene.

other potential products of addition. The formation of **10**, by attack of dichloro-carbene on the double bond between carbon atoms 2 and 3 of indene, permits preservation of the aromaticity of the benzene ring, whereas the production of any of the other adducts results in the loss of the aromaticity in the molecule. Therefore a preference for addition of dichlorocarbene across the 2,3-double bond might be anticipated. In fact this preference is realized experimentally, since generation of dichlorocarbene in the presence of indene, followed by work-up of the reaction mixture under mild conditions, affords **10** in 70% yield as the only isolable product.

Although the attack of dichlorocarbene on indene turns out to be selective, a possible further complication should be noted. Three-membered rings bearing

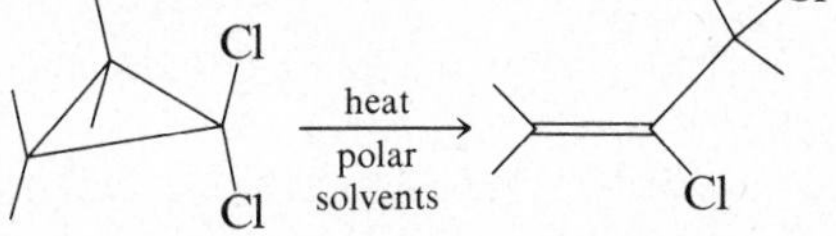

$$(9)$$

geminal halogens as in **10** are relatively unstable and rearrange in polar media and/or with heating to ring-opened isomers (equation 9). In the case of **10**, rearrangement is known to occur readily at temperatures greater than 50°. Compound **10**, by analogy to the general reaction given in equation 9, would be expected to rearrange to a mixture of **11** and **12** (equation 10), but it has been observed experimentally that on treatment with an ethanolic solution of potassium hydroxide, **10** yields 2-chloro-naphthalene **(13)** in high yield. Compounds **11** and **12** may serve as intermediates in the formation of **13**, but they readily lose the elements of hydrogen chloride to produce the more highly stable naphthalene system.

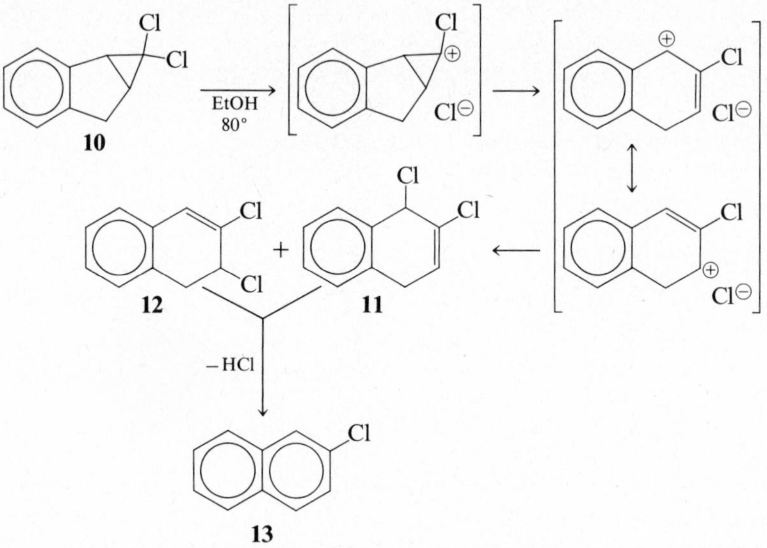

(10)

In the Experimental Procedure the addition product (10) is produced by heating indene and sodium trichloroacetate at a temperature greater than 100° in a polar medium. As a consequence 13, the rearrangement product from 10, is isolated in this experiment.[3] The overall equation for the reaction is therefore the following:

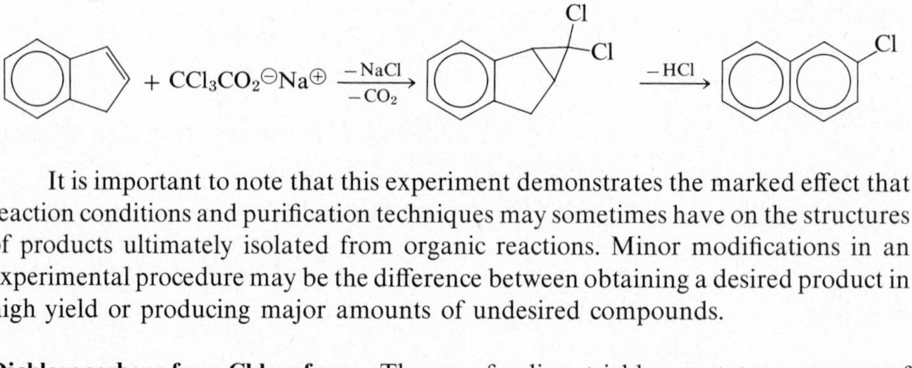

It is important to note that this experiment demonstrates the marked effect that reaction conditions and purification techniques may sometimes have on the structures of products ultimately isolated from organic reactions. Minor modifications in an experimental procedure may be the difference between obtaining a desired product in high yield or producing major amounts of undesired compounds.

Dichlorocarbene from Chloroform.   The use of sodium trichloroacetate as a source of dichlorocarbene has limitations in its applicability, the most obvious of which would be the ability of substrate alkenes and product dichlorocyclopropanes to withstand temperatures in excess of 100°. Synthetic methods for generation of dichlorocarbene at lower temperatures have been developed, and the most widely applied of these involves base-promoted decomposition of chloroform. Before 1970 the base of choice was potassium *t*-butoxide, a strong base in which the steric hindrance made reactions of it or of *t*-butyl alcohol, its conjugate acid, with dichlorocarbene unimportant. However, potassium *t*-butoxide is reasonably expensive and difficult to handle in the laboratory, primarily because of its extremely hygroscopic character. Bases less

---

[3]Students interested in preparing compound 10 should consult reference 2 given at the end of this section.

expensive than *t*-butoxide—for example, hydroxide or methoxide—either were much too insoluble in the reaction medium (chloroform and the alkene) to promote deprotonation of the chloroform at a reasonable rate and/or themselves reacted with the dichlorocarbene under the reaction conditions. Thus aqueous hydroxide transforms chloroform to carbon monoxide, presumably by way of the carbene, as shown in equation 11. Unsatisfactory yields of addition products were the net result of the use of such bases.

$$\text{HO}^\ominus + \text{H}{-}\text{CCl}_3 \longrightarrow \text{HOH} + \text{Cl}_2\text{C:}^\ominus \xrightarrow{-\text{Cl}^\ominus} \text{Cl}_2\text{C:} \tag{11}$$

$$\text{CO} \xleftarrow{-\text{HCl}} \underset{\overset{\|}{\text{O}}}{\text{Cl}{-}\text{C}{-}\text{H}} \xleftarrow{-\text{HCl}} \underset{\overset{|}{\text{H}}}{\text{Cl}_2\text{C}{-}\text{OH}}$$

The desired possibility of using an inexpensive and easily handled base, specifically sodium hydroxide, to produce dichlorocarbene from chloroform became a reality in the early 1970s as a result of research designed to solve an important and classical problem of organic chemistry. This was to achieve an efficient reaction between two reagents that are insoluble in one another, without having to resort to the use of solvents or mixtures of solvents that codissolve the reagents. The technique is called *phase transfer catalysis,* and although our concern will be its use for the generation of dichlorocarbene from a two-phase mixture of aqueous sodium hydroxide and chloroform, it has applicability to a variety of organic reactions.[4]

The concept underlying the technique is simple. Combination of aqueous sodium hydroxide with a mixture of chloroform and an alkene would of course result in formation of two layers, one of which would contain most if not all of the organic reagents and the other the hydroxide. The desired overall reaction to give a dichlorocyclopropane (equation 6) would be inefficient for two reasons: (1) sodium hydroxide is quite insoluble in the organic phase and therefore not much dichlorocarbene would be produced in the phase containing the substrate alkene; (2) chloroform and most simple alkenes are poorly soluble in aqueous media so that little reaction would occur in this phase. Furthermore, hydrolysis (equation 11) would be expected to predominate in the aqueous layer, owing to the enormous difference in the concentrations of alkene and hydroxide in this phase.

Now suppose it were possible to devise a species that would make the hydroxide ion more soluble in the organic medium without at the same time greatly increasing the solubility of water in this phase. Such a result would be very helpful, since it not only would place hydroxide ions in the presence of high concentrations of chloroform and alkene but also would suppress hydrolysis of any dichlorocarbene that is produced. The likelihood of addition of the carbene to the alkene would thus be enhanced. The characteristics of the "phase transfer catalyst" necessary to "drag" hydroxide out of an aqueous into an organic environment are described in the next paragraph.

---

[4] See later discussion and also references 3 and 4 at the end of this section.

Consider a salt such as trioctylmethylammonium chloride (14), known commercially as Aliquat 336. This substance is quite soluble in nonpolar (organic) media,

$$[CH_3(CH_2)_6CH_2]_3NCH_3^{\oplus}Cl^{\ominus}$$
$$\mathbf{14}$$

owing to the presence of the long hydrocarbon chains. It has limited solubility in water, despite its ionic character. The equilibria shown in Figure 11.2 then merit consideration. Basically, in the aqueous phase the sodium cation of the ion pair (15) is replaced by the tetraalkylammonium ion to generate a new ion pair (16); this pair has a more favorable partition coefficient with the organic phase than does (15), thereby increasing the concentration of hydroxide ions in the organic phase. This ion subsequently initiates the production of dichlorocarbene by deprotonation of chloroform. As shown in the figure, the catalyst (14) is ultimately re-formed in the organic phase and is free to repartition between it and the aqueous phase, thus returning the tetraalkylammonium ion to the aqueous layer for production of ion pair (16). It is clear from the equilibria shown that a catalytic amount of 14 can serve repeatedly to transfer hydroxide ions from the aqueous to the organic phase.

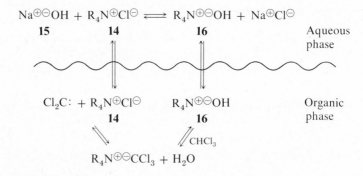

FIGURE 11.2 Partitioning with phase transfer catalyst.

In summary, phase transfer catalysts function by generation, in a mixture of two immiscible reagents, of a new ion pair that promotes partitioning of one of the reagents into the layer containing the other. The result is large enhancement in the rate of bimolecular reaction between the two reagents; accelerations of 10^4–10^9 (!) are not uncommon when a phase transfer catalyst is added to a heterogeneous liquid mixture of reagents. Note that the technique currently emphasizes the use of *cationic* organic catalysts which therefore form ion pairs with *anionic* reagents; a sampling of the scope of the method is given in equations 12–14. It is much more difficult and synthetically less useful to prepare stable anionic organic species that would complex with cationic reagents.

$$C_7H_{15}CH_2Br \text{ (in } CH_2Cl_2) + NaCN \text{ (in } H_2O) \xrightarrow[\text{transfer catalyst}]{\text{phase}} C_7H_{15}CH_2CN \qquad (12)$$
$$\text{(Organic phase)} \qquad \text{(Aqueous phase)}$$

$$C_6H_5CH_2OH \text{ (in benzene)} + KMnO_4 \text{ (in } H_2O) \xrightarrow[\text{catalyst}]{\text{phase transfer}} C_6H_5CO_2H \qquad (13)$$
$$\text{(Organic phase)} \qquad \text{(Aqueous phase)}$$

$$C_6H_5CH_2CH_2Br \text{ (in } CH_2Cl_2) + NaOH \text{ (in } H_2O) \xrightarrow[\substack{\text{transfer} \\ \text{catalyst}}]{\text{phase}} C_6H_5CH{=}CH_2 \quad (14)$$

(Organic phase)　　　　　(Aqueous phase)

Given the general principles that form the basis for phase transfer catalysis, it is clear that one factor that will determine the overall rate of a reaction will be the efficiency of partitioning of reagents between phases. This will be a function of, among other things, the total surface area of the two immiscible reagents in contact with each other. To increase this area the reaction mixture must be agitated vigorously in order to produce emulsification, wherein tiny droplets of the immiscible layers develop. This is generally accomplished by rapid mechanical stirring. However, if the appropriate apparatus is unavailable, vigorous shaking will sometimes produce the desired result, as is the case in the experimental procedure given here for the preparation of 7,7-dichlorobicyclo[4.1.0]heptane (equation 15) from cyclohexene. Magnetic stirring is normally insufficient to produce and maintain the necessary emulsification.

$$\text{(cyclohexene)} + CHCl_3 \xrightarrow[R_4N^+Cl^-]{\substack{NaOH \\ H_2O}} \text{(7,7-dichlorobicyclo[4.1.0]heptane)} + NaCl \quad (15)$$

EXPERIMENTAL PROCEDURES

A. Preparation of 1-Chloronaphthalene

DO IT SAFELY

1. Do not expose yourself unnecessarily to vapors of tetrachloroethylene; as a rule chlorinated solvents are more toxic than many of the organic solvents commonly used in the laboratory.
2. Use great care during the steam distillation to ensure that a blockage does not develop in your apparatus; blockage could lead to development of excessive pressure in the system and an explosive release of steam and hot water.
3. Use a steam bath to remove diethyl ether if it is used as a solvent in this experiment.

Equip a 250-mL round-bottomed flask with a reflux condenser fitted with a thermometer adapter, or 1-hole rubber stopper, through which is passed a glass tube. Attach to this tube a length of rubber tubing bearing a short piece of glass tubing. Place 0.081 mol of sodium trichloroacetate,[5,6] which has previously been dried for at

---

[5] If commercial sodium trichloroacetate (Columbia Organic Chemicals) is used, take into account the fact that the product is 95% pure, in making the calculation of the number of grams required.

[6] If desired, sodium trichloroacetate can be prepared in the following manner. Prepare a solution of 0.090 mol of sodium hydroxide in 10 mL of water, and cool this solution thoroughly in an ice-water bath. Place 0.086 mol of trichloroacetic acid in a 125-mL filter flask, and, while swirling this flask in the ice water

least 1 hr at 90°, 20 mL of tetrachloroethylene, 5 mL of diglyme, and 0.086 mol of indene[7] in the flask. Bring this mixture to a gentle reflux with the aid of a microburner, swirling the flask occasionally to ensure good mixing. Continue to heat until the evolution of carbon dioxide has ended, as indicated by the absence of bubbling when the tube leading from the top of the condenser is inserted into a few milliliters of tetrachloroethylene contained in a test tube. At this time the granular sodium trichloroacetate should have disappeared, and finely divided sodium chloride should have separated.*

Before removing the burner from beneath the flask, disconnect the bubbler in order to prevent suction of tetrachloroethylene into the reaction flask. Detach the flask from the cendenser, add 25 mL of water to the reaction mixture, and attach the flask to an apparatus set for steam distillation, using an external source of steam (see Figure 2.12).

Steam-distil the reaction mixture as rapidly as the ability of the condenser to cool the distillate will permit, collecting the distillate in a graduated cylinder. When approximately 26 mL of an organic layer has been obtained, which will require that a total volume of 60 to 65 mL of distillate be collected, replace the graduated cylinder with a 250-mL Erlenmeyer flask and discard the liquid in the graduated cylinder.* Continue the distillation until no more oily material can be detected in the distillate. Collection of about 200 mL of distillate will be required to obtain all the crude 2-chloronaphthalene.* During the course of the distillation check to make sure that solid does not collect in the condenser and block it. If solid does begin to form, shut off the flow of water through the condenser, and allow the uncondensed steam to sweep the solid out of the condenser.

With the aid of a water bath, cool the distillate to room temperature, and then transfer it to a separatory funnel. Some crystalline 2-chloronaphthalene may remain in the Erlenmeyer flask, but it will be recovered later. Extract the aqueous solution in the funnel three times with separate 25-mL portions of either dichloromethane or technical diethyl ether, and combine the extracts in the 250-mL Erlenmeyer flask used to collect the steam distillate.*

Remove the solvent from the crude product by heating the solution on a steam bath (preferably in the hood).* Alternatively, the solvent can be removed by performing a simple distillation (*No flames* if ether was used!). Transfer the crude 2-chloronaphthalene to a smaller Erlenmeyer flask, rinse the larger flask with 15 mL of 95% ethanol, and add this rinse along with some decolorizing carbon to the product. Boil this solution for a few minutes on the steam cone, and then filter to

---

bath, add to it about 9 mL of the cold solution of base. Now add 1 drop of 0.04% Bromocresol Green solution so that a yellow color is visible. Complete the titration of trichloroacetic acid by adding the hydroxide solution dropwise to a point where a single drop produces a color change from yellow to blue. If the end point is overshot, add some more acid and repeat the titration.* Stopper the flask and attach it to a water aspirator by means of vacuum tubing. Place the flask *within* the rings of a steam cone, and concentrate the solution by evacuating the flask and heating with steam. After the solution has been evaporated to dryness,* which will require about 20 min, transfer the solid to a watch glass, break up any large lumps, and dry the sodium trichloroacetate in an oven at 90°.

[7] Indene is not stable for long periods and therefore should be purified immediately before use by means of a simple distillation.

remove the carbon.★ Adjust the temperature of the filtrate to about 45°, and, while maintaining this temperature, add water dropwise until the solution becomes turbid. At this point allow the solution to stand undisturbed until it cools to room temperature.★ If the 2-chloronaphthalene fails to crystallize but rather begins to oil out, a problem characteristic of many low-melting solids, scratch the walls of the flask at the interface between the solution and the air in an attempt to induce crystallization. In some cases it may be necessary to evaporate a small sample of the solution to obtain the oily crude product which, on scratching, may provide a few seed crystals that can be added to the solution to encourage crystallization.

When no more solid crystallizes at room temperature, cool the flask in an ice-water bath to complete the crystallization, and collect the product. A second crop of product can be obtained if the filtrate is reheated to 45° and water added dropwise until the solution again becomes turbid. Cooling and scratching as before should provide additional 2-chloronaphthalene of somewhat lower purity than the first crop.

Pure 2-chloronaphthalene is an iridescent white solid[8] (mp 58–59°). The yield of product in this experiment should be 20–30%.

B. Preparation of 7,7-Dichlorobicyclo[4.1.0]-heptane

DO IT SAFELY

1.   The 25 *M* sodium hydroxide solution used in this experiment is particularly caustic. If during the course of carrying out the reaction you become aware of a slight stinging sensation on your hands, wash them well with warm water, and rinse the area with 1% acetic acid solution. If you prepare the hydroxide solution yourself, cool it to room temperature before using.

2.   During the shaking process in this procedure, pressure may build up in the stoppered flask; be sure to vent the flask frequently while shaking in order to relieve that pressure.

Prepare an ice-water bath for cooling purposes. Weigh 1 g of trioctylmethyl-ammonium chloride[9] into a 100-mL round-bottomed flask (a 125-mL Erlenmeyer flask with a tight-fitting rubber stopper may be used). Add to the flask 0.10 mol of cyclohexene, 0.10 mol of chloroform, and 20 mL of 25 *M* aqueous sodium hydroxide solution. Stopper the flask and shake it vigorously for several seconds. A mildly exothermic reaction will result as a thick emulsion forms. Unstopper the flask and insert a thermometer. Use the ice-water bath to maintain the reaction mixture at a temperature of 50–60°. To maintain the emulsion, frequently restopper the flask and shake vigorously for a moment. Between the shaking operations continue to monitor

[8] For a discussion of the reason for the blue color often observed in the crude 2-chloronaphthalene and even in the recrystallized product obtained in this experiment, see Exercise 6 at the end of this section.

[9] Trioctylmethylammonium chloride for use as a phase transfer catalyst is available under the trade name Aliquat 336 from the McKerson Corporation, 1219 East 27th Street, Minneapolis, Minnesota 55407. Aliquat 336 is a mixture of trioctylmethylammonium and tridecylammonium chlorides, with the former predominating in the mixture.

the temperature and control it in the 50–60° range. The reaction will be complete in about 20 min, and the reaction mixture will begin to cool on its own. When the mixture has cooled to about 30°, use the ice-water bath to cool it to room temperature, and then transfer the mixture to a separatory funnel. Rinse the reaction flask with 50 mL of water, and add the rinse to the funnel. Separate the organic layer. Extract the aqueous layer with 10–15 mL of diethyl ether that has first been used to rinse the original reaction flask. Separate the ether layer, and combine it with the original organic layer. Wash these combined layers with 15–20 mL of water, and then dry the organic layer over anhydrous sodium sulfate. Filter the solution into a 50-mL round-bottomed flask, and equip the flask for simple distillation. Using either a steam bath or a heating mantle, distil to remove the ether. When the ether is removed, continue the distillation with either a heating mantle or a burner, collecting the product between 195–200°. Use an air-cooled rather than a water-cooled condenser for the final stage of the distillation to minimize the risk of cracking the condenser at the high temperatures.

EXERCISES

1-CHLORONAPHTHALENE
1. Why is tetrachloroethylene rather than some other liquid such as water used to test for the evolution of CO_2?
2. Predict which double bond in phenanthrene would be most reactive toward a carbene, and give an explanation for your choice.

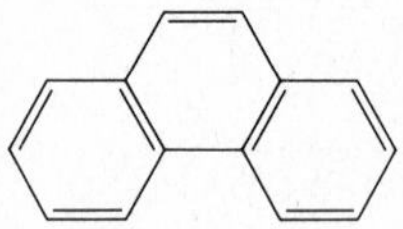

Phenanthrene

3. What properties of the diglyme-tetrachloroethylene mixture make it suitable as a solvent in the reaction?
4. What organic compounds distil in the first stage of the steam distillation?
5. When attempting a mixed-solvent recrystallization of the type performed in this experiment, the solvent in which the solute is sparingly soluble is normally added to a *boiling* solution of the solute and the other solvent. What special circumstances prohibit the addition of water to a boiling solution of 2-chloronaphthalene in 95% ethanol in order to effect saturation? Why is it more efficient to saturate the ethanolic solution at about 45° than at room temperature, for example?
6. The 2-chloronaphthalene isolated by the procedure described in this experiment sometimes has a bluish tinge caused by the presence of a small amount of a chloroazulene. The chlorine substituent is in the seven- rather than the five-membered ring but its exact position is unknown. Propose a mechanism for the

formation of the chloroazulene from indene and dichlorocarbene. Suggest a reason why it is difficult to remove the azulene from 2-chloronaphthalene even after repeated recrystallization.

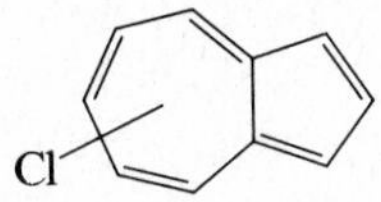

7. Write out the expected products of the reaction between dichlorocarbene and the following alkenes under mild conditions.

(a) CH_2=C
$\begin{array}{c} CH_3 \\ \diagup \\ \diagdown \\ CH_3 \end{array}$ + $:CCl_2$ (1 mol)

(b) 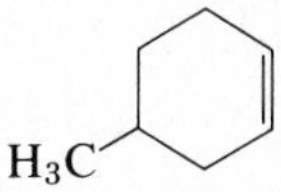 + $:CCl_2$ (1 mol)

(c) 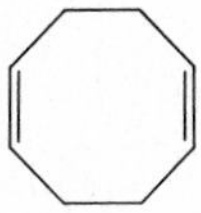 + $:CCl_2$ (2 mol)

8. Suggest why sodium trichloroacetate decarboxylates much more readily than sodium acetate.

7,7-DICHLOROBICYCLO[4.1.0]-HEPTANE

1. Explain why vigorous agitation or stirring is required during a reaction utilizing phase transfer catalysis.
2. In this experiment trioctylmethylammonium chloride is employed as a phase transfer catalyst. Explain whether or not the use of *tetra*octylammonium chloride would be as efficient as the catalyst actually used. What about tetra*methyl*-ammonium chloride?

REFERENCES

1. W. Kirmse, *Carbene Chemistry*, 2d ed., Academic Press, New York, 1971.
2. W. E. Parham, H. E. Rieff, and P. Swartzentruber, *Journal of the American Chemical Society*, **78**, 1437 (1956).
3. R. A. Jones, *Aldrichimica Acta*, **9**, 35 (1976).
4. G. Gokel and W. Weber, *Journal of Chemical Education*, **55**, 350, 429 (1978).

SPECTRA OF STARTING MATERIAL AND PRODUCTS

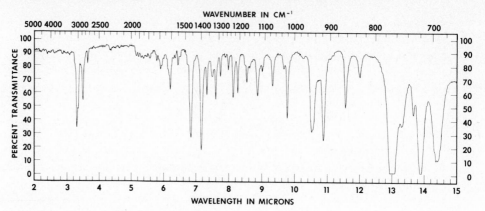

FIGURE 11.3 IR spectrum of indene.

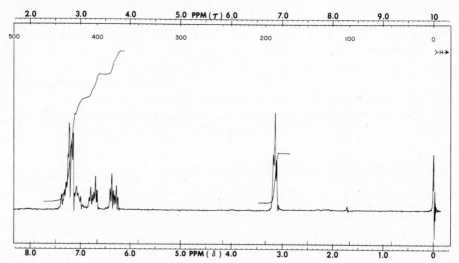

FIGURE 11.4 PMR spectrum of indene.

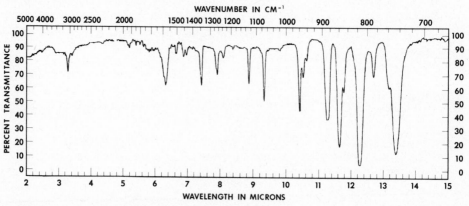

FIGURE 11.5 IR spectrum of 2-chloronaphthalene.

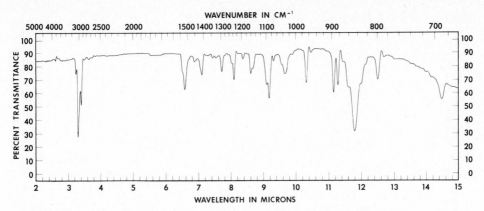

FIGURE 11.6 IR spectrum of 7,7-dichlorobicyclo[4.1.0]-heptane.

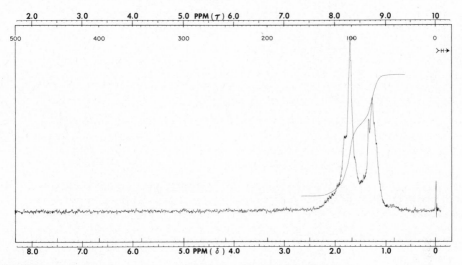

FIGURE 11.7 PMR spectrum of 7,7-dichlorobicyclo[4.1.0]-heptane.

11.2 Arynes

Arynes, which can be considered to be cyclohexadienynes, have been recognized as reaction intermediates only since the early 1950s. Evidence which prompts postulation of their existence includes the observations that treatment of **17** with amide ion produces an equimolar mixture of **18** and **19** (equation 16), that reaction of **20** with hydroxide ion generates **21** and **22** (equation 17), and that pyrolysis of **23** in

$$\text{(16)}$$

* denotes ^{14}C

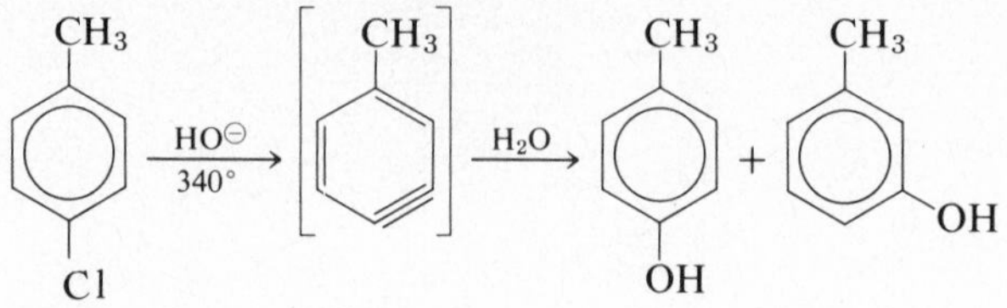

$$\text{(17)}$$

the presence of anthracene yields triptycene (**26**, equation 18). Equations 16 and 17

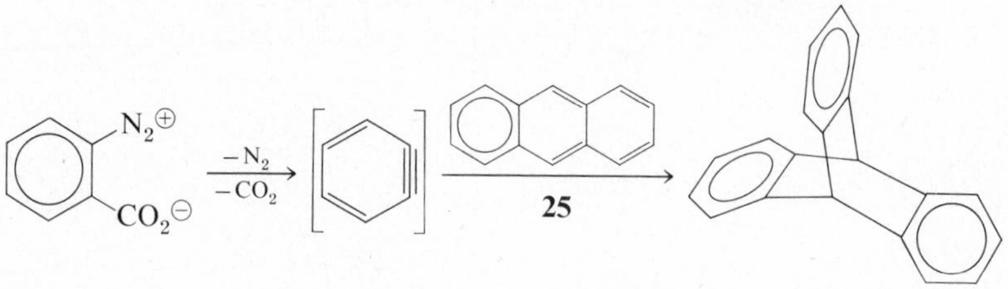

$$\text{(18)}$$

show the production of an aryne by an elimination reaction followed by reaction of the aryne with a nucleophile. Equation 18 exemplifies the generation of the aryne, benzyne (**24**), by thermal decomposition of an unstable "zwitterionic" species, (**23**), with subsequent reaction of the benzyne, acting as a dienophile, with a diene.

Although benzyne itself is often generated by reaction of an aryl halide with a strong base (equations 16 and 17), a more convenient procedure involving synthesis and decomposition of benzenediazonium-2-carboxylate (**23**) will be used here. As noted, **23** is unstable and can be violently explosive, particularly when dry, if it is subjected to shock. The *isolation* of **23** therefore represents a hazardous and, at the least, an unwise endeavor. It is possible to circumvent the hazards in this instance, and similarly in many other cases where preparation and reaction of unstable intermediates are involved, by generation *and* decomposition of **23** *in situ,* the goal being to maintain a *low* concentration of benzenediazonium-2-carboxylate (**23**).

Because it is the desire to decompose **23** to benzyne about as fast as **23** is produced, it is necessary to have present from the very beginning any reagent that is to react with the benzyne. Ideally, then, the reagent for trapping the benzyne should not itself react with the reagents involved in the production of **23**. This is the case in the procedure outlined below; anthracene (**25**), an aromatic hydrocarbon, does not react to any significant extent with either isopentyl nitrite (**27**) or anthranilic acid (**28**, Figure 11.8) under the reaction conditions used.

Benzenediazonium-2-carboxylate (**23**) is the product of a reaction between isopentyl nitrite (**27**), a diazotizing reagent, and anthranilic acid (**28**). A possible reaction mechanism is outlined in Figure 11.8. Intermediate **23** is prepared at a temperature sufficiently high to induce its decomposition to benzyne, which is subsequently trapped by anthracene (**25**) to provide the Diels-Alder adduct triptycene (**26**, equation 18). Anthracene is functioning as a diene in this reaction.

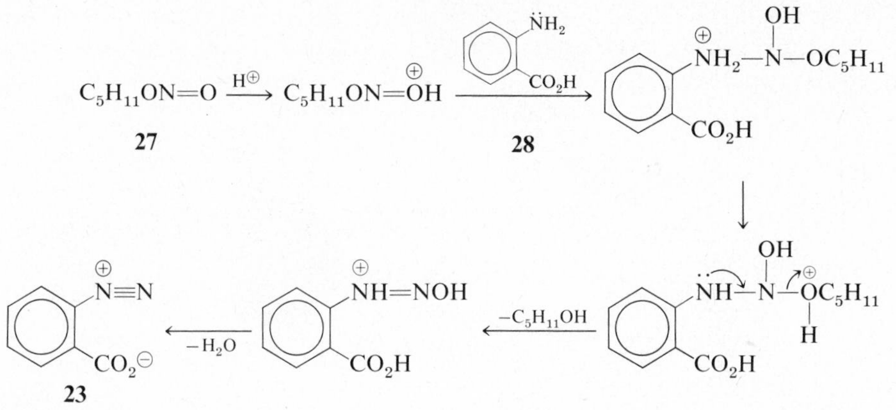

FIGURE 11.8 The preparation of benzenediazonium-2-carboxylate.

The formation of triptycene indicates that benzyne adds preferentially across the 9- and 10-positions of anthracene. The fundamental reason for this selectivity is that **26** has the greatest amount of aromatic stabilization of all the possible products.[10]

The choice of solvents for this reaction is crucial. Clearly, it is important that all starting reagents and the intermediate **23** be soluble in the medium. It is particularly crucial that **23** not be insoluble, because if it were it would precipitate from the reaction mixture and accumulate, an occurrence that *must* be avoided. Furthermore, the boiling point of the solvent must be sufficiently high so that **23** is formed and decomposed at a reasonable rate. A final requirement is that the solvent be *aprotic* in order to minimize side reactions involving nucleophilic attack of solvent on benzyne; for example, *protic* solvents such as alcohols would react with benzyne by adding across the highly reactive "triple bond" (equation 19). All these requirements are met

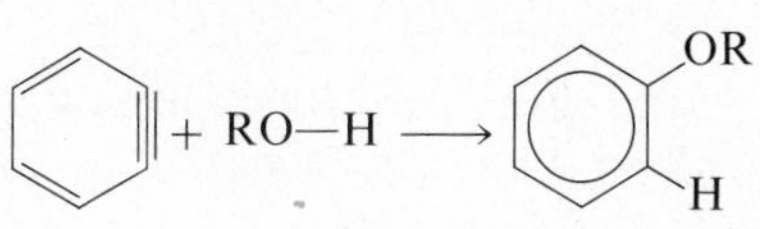

$$(19)$$

by 1,2-dichloroethane, an aprotic solvent of modest polarity. The anthranilic acid is added to the reaction mixture as a solution in diglyme (see footnote 2), another aprotic solvent of moderate polarity. The high-boiling point and water solubility of this ether facilitate isolation of the final product (see Exercise 2).

An important technical problem encountered in the purification of the reaction mixture is the removal of unreacted anthracene which, as might be anticipated on structural grounds, has solubility characteristics similar to those of triptycene. This problem is solved by preparing a derivative of anthracene that is soluble in aqueous base. Thus treatment of the crude reaction mixture with maleic anhydride will result

---

[10] A similar explanation was used to rationalize the preferred attack of dichlorocarbene on the 2,3-carbon-carbon double bond of indene (Section 11.1).

in the destruction of excess anthracene by the formation of the Diels-Alder adduct, **29**, which affords the water-soluble salt **30** on hydrolysis with aqueous potassium hydroxide (equation 20).

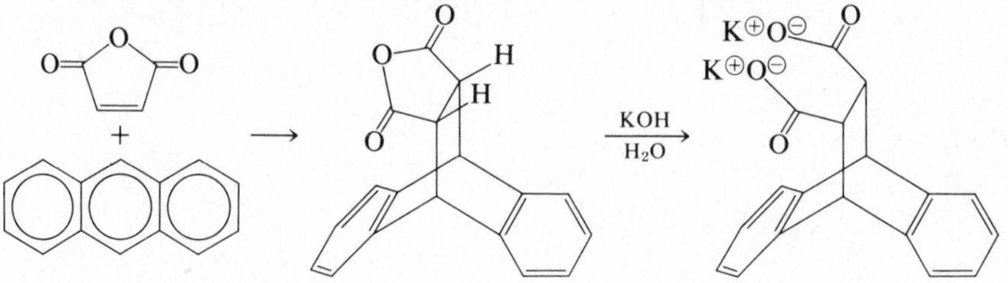

(20)

29   30

EXPERIMENTAL PROCEDURE

DO IT SAFELY

1.   If your apparatus is clamped to a ring stand, agitate the reaction mixture by picking up the ring stand and shaking it gently; be certain all clamps are securely fastened before doing this! If your apparatus is attached to immovable supports, however, loosen the clamps and carefully swirl the intact apparatus.

2.   Be certain to control the rate of addition of the solution of anthranilic acid so as to maintain gentle reflux. If the solution is not refluxing, the lower temperature may slow the decomposition of benzenediazonium-2-carboxylate sufficiently to allow it to accumulate in the reaction mixture. If the rate of addition is initially too slow to achieve reflux, increase the rate slightly. You may also need to return the apparatus briefly to the steam bath to reinitiate reflux. In the unlikely event that solid benzene-diazonium-2-carboxylate begins to appear in the reaction flask during the diazotiza-tion step, *stop the addition of anthranilic acid at once and return the apparatus to the steam bath,* heating until the solid disappears.

3.   Under no circumstances should you remove any part of the reaction mixture during the diazotization step and allow it to evaporate to dryness! When *dry,* even small quantities of **23** are dangerously explosive!

Place 5.4 g (0.030 mol) of anthracene, 2.7 g (0.023 mol) of isopentyl nitrite, and 40 mL of 1,2-dichloroethane in a 100-mL round-bottomed flask equipped with a Claisen connecting tube bearing an addition funnel and a reflux condenser; the funnel should be on the straight arm of the connecting tube. Heat the mixture to reflux over a steam bath. After the mixture has started to reflux, remove the appara-tus from the steam bath. While the mixture is still warm, add a solution of 3.0 g (0.022 mol) of anthranilic acid in 15 mL of diethylene glycol dimethyl ether (diglyme) dropwise from the addition funnel over a period of about 10 min. Agitate the reaction mixture frequently to ensure good mixing.[11] The exothermicity of the reaction should maintain a moderate rate of reflux. After the addition is complete, heat the mixture at reflux over a steam bath for 15 min. Refit the flask for simple distillation, and distil until a head temperature of 150° is reached. Allow the reaction mixture to cool somewhat, add 3.0 g (0.030 mol) of maleic anhydride, and heat this mixture at reflux for 5 min.★ Cool the flask in an ice-water bath and slowly add, while stirring, a solution of 6.0 g of potassium hydroxide in 20 mL of methanol and 10 mL of water.★ Collect the crude triptycene by vacuum filtration of the cold mixture, and wash it with 4:1 methanol-water solution until the washings are colorless (15 to 20 mL).★

Purify the crude product by dissolving it in 2-butanone (10 mL per gram of triptycene), heating the solution on a steam bath, adding a small amount of decolorizing carbon, and filtering the solution.★ Concentrate the filtrate to about two-thirds of its original volume, add an equal amount of methanol, and cool the solution in an ice-water bath. Collect the product by suction filtration, and wash it with a small amount of cold methanol. The reported melting point of triptycene is 254–255.5°. Do not attempt to determine this melting point in a mineral oil bath.

EXERCISES

1. What substances are removed by the simple distillation?
2. Outline the reasons why the high-boiling point and the water solubility of the particular ether chosen as one of the solvents in this experiment facilitate isolation of the triptycene formed in the reaction.
3. Why is the simple distillation necessary?
4. What is the purpose of adding methanol during the recrystallization of triptycene?
5. Predict the structure of the product resulting from addition of 1 mol of bromine to anthracene.
6. Explain, on the basis of stabilization (delocalization) energies, why the product of reaction between benzyne and anthracene is triptycene, rather than the compound shown below.

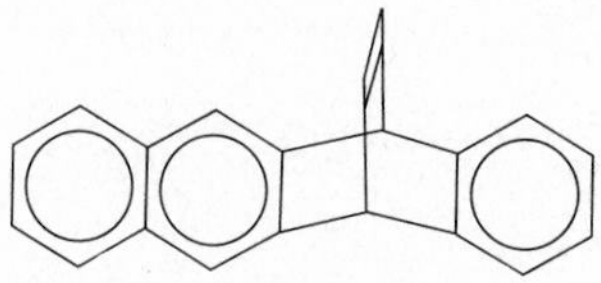

REFERENCES

1. L. Friedman and F. Logullo, *Journal of Organic Chemistry,* **34,** 3091 (1969).
2. H. Heaney, *Chemical Reviews,* **62,** 81 (1962).
3. R. W. Hoffmann, *Dehydrobenzene and Cycloalkynes,* Academic Press, New York, 1967.

[11] If magnetic stirrers are available, equip your flask with a stirring bar, and accomplish the mixing with the aid of the stirrer.

SPECTRA OF STARTING MATERIAL AND PRODUCT

The IR spectrum of triptycene is given in Figure 4.8.

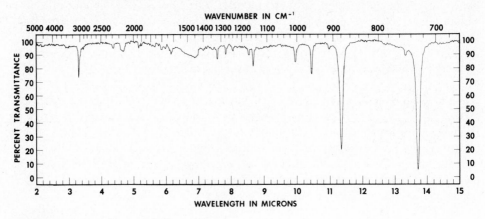

FIGURE 11.9 IR spectrum of anthracene.

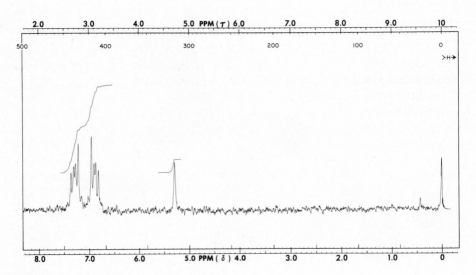

FIGURE 11.10 PMR spectrum of triptycene.

12

ORGANOMETALLIC CHEMISTRY

12.1 Introduction

Over the past 10 to 15 years the use of organometallic reagents as organic synthetic tools has expanded dramatically beyond the use of the organomagnesium compounds introduced by Victor Grignard in the early part of the twentieth century. One important synthetic advantage of these substances rests in the polarity of the carbon-metal bond, which is *reversed* relative to the bonds of carbon to oxygen and halogen, for example (compare structure **1** with **2** and **3**). It is instructive to recognize

$$(\overset{\delta}{R}-\overset{\delta}{CH_2})_n M \qquad \overset{\delta}{R}-\overset{\delta}{CH_2}-X \qquad \overset{R}{\underset{R}{>}}\overset{\delta}{C}=\overset{\delta}{O}$$

$$\textbf{1} \qquad\qquad\qquad \textbf{2} \qquad\qquad \textbf{3}$$

that the metal-bound carbon in **1** is *nucleophilic,* whereas the designated carbon atoms of **2** and **3** are *electrophilic.* This complementary reactivity provides for the types of reactions shown in equations 1 and 2; they are particularly valuable reactions in organic synthesis, since they lead to the formation of new carbon-carbon bonds.

$$R-CH_2-M + R'-CH_2-X \longrightarrow R-CH_2-CH_2-R' + MX \tag{1}$$

$$R-CH_2-M + \overset{R'}{\underset{R'}{>}}C=O \longrightarrow R-\overset{R'}{\underset{R'}{\overset{|}{C}}}-O^{\ominus}M^{\oplus} \tag{2}$$

The discussion in the following sections concerns the use of organometallics involving magnesium, copper, and lithium, and in the cases of magnesium and copper, experiments are included to demonstrate the applicability of these substances to organic synthesis.

12.2   The Grignard Reagent: Its Preparation and Reactions

The preparation of the Grignard reagent, RMgX, is represented in equation 3,

$$RX + Mg \xrightarrow[\text{(solvent)}]{\text{dry ether}} RMgX \tag{3}$$

where RX represents an organic halide. The exact nature of the Grignard reagent in solution is not known. It is believed to be a mixture of numerous species, but some of the principal species are undoubtedly R_2Mg and MgX_2 in equilibrium with 2 RMgX. These species are highly solvated by the ether solvent and are complexed with one another. It is customary to represent the Grignard reagent by the formula RMgX when writing chemical equations, but it should be kept in mind that the species in solution are of a much more complex nature.

The ether solvent is an essential part of the Grignard reagent, because it is known to form a complex with RMgX. Several cases are known where Grignard reagents have been prepared in the absence of ethers, but the yields are not good. The most common ether solvent is diethyl ether, $(C_2H_5)_2O$, owing to its low cost and ease of removal (its boiling point is 36°). Other ethers, for example, dibutyl ether, bp 142°, have been used successfully when higher boiling solvents are required. Some cyclic ethers such as tetrahydrofuran have been used with excellent success in the preparation of certain Grignard reagents.

The organic halide may, in general, be of any organic moiety (alkyl or aryl) and the halide may be bromide, chloride, or iodide. Alkyl iodides, bromides, or chlorides and aryl iodides or bromides can be converted to the corresponding Grignard reagents using diethyl ether as the solvent. Aryl chlorides, on the other hand, do not react with magnesium in diethyl ether, but they react readily when tetrahydrofuran is used as the solvent.

The initiation of the reaction between the halide and magnesium is sometimes difficult, with the iodides being more reactive than the bromides, which are in turn more reactive than chlorides. Aryl halides are less reactive than their alkyl counterparts, and it has been found that the aryl bromides and the alkyl chlorides react about equally well. In cases where one is dealing with a particularly unreactive halide, the initiation of the reaction may sometimes be accomplished by the addition of a small amount of iodine to the reaction mixture. This is believed to convert a small amount of the halide to the iodide, which is more reactive.

The preparation of the Grignard reagent *must* be carried out under *anhydrous* conditions and, if possible, in the absence of oxygen. It is very important to maintain completely dry conditions throughout, since the presence of water inhibits the initiation of the reaction as well as destroys the reagent once it is formed.

The reaction that occurs when the Grignard reagent comes in contact with water is shown in equation 4. The reagent is a strong base because one of the carbon atoms

$$RMgX + H_2O \longrightarrow RH + HOMgX \tag{4}$$

bears substantial negative charge ($R^-Mg^{++}X^-$). The Grignard reagent, acting as a base, removes a proton from water, which acts as an acid. The overall effect is the destruction of the reagent, with the formation of a hydrocarbon (RH) and a basic magnesium salt. The hydrocarbon is inert, so that the net effect is the production of hydroxide ion in solution. In fact it is possible to standardize a solution of a Grignard reagent by adding a known volume of it to a known volume of standardized mineral acid and backtitrating with standardized base.

In addition to the reaction with water (equation 4), there are other side reactions that may occur during formation of the Grignard reagent, as shown in equations 5–7.

Reaction with oxygen:

$$2 \text{ RMgX} + \text{O}_2 \longrightarrow 2 \text{ ROMgX} \tag{5}$$

Reaction with carbon dioxide:

$$\text{RMgX} + \text{CO}_2 \longrightarrow \text{RCO}_2\text{MgX} \tag{6}$$

Coupling:

$$\text{.RMgX} + \text{RX} \longrightarrow \text{R—R} + \text{MgX}_2 \tag{7}$$

It is possible to minimize these reactions by taking certain precautions when carrying out the experimental work. The reaction with oxygen and carbon dioxide may be avoided by carrying out the reaction under an inert atmosphere (such as nitrogen or helium gas). In research work this is routinely done; however, when diethyl ether is used as a solvent, it excludes a certain amount of air from the reaction vessel owing to its very high vapor pressure. The coupling reaction (equation 7) is an example of a Wurtz reaction. It is not possible to eliminate this coupling reaction completely, but it may be minimized by using dilute solutions so as to avoid localized high concentrations of halide. This can be done by very efficient stirring and by slowly adding the halide to the magnesium in ether. Normally the rate of addition of halide (dissolved in ether) and the rate of reflux (when diethyl ether is used) should be adjusted so that they are about equal. Alkyl iodides are much more prone to coupling reactions than are the bromides and chlorides, so that the latter are preferable for preparing Grignard reagents even though they are less reactive.

The experiments appearing in this chapter involve the preparation of two Grignard reagents from organic bromides and three typical reactions of these reagents, to produce a tertiary alcohol, a carboxylic acid, and a secondary alcohol.

Preparation of Phenylmagnesium Bromide and Its Reaction with Methyl Benzoate.

The equation for preparation of phenylmagnesium bromide indicates that 1 mol of magnesium must be added for each mole of bromobenzene used; the main reaction is

$$\text{C}_6\text{H}_5\text{Br} + \text{Mg} \xrightarrow{\text{(C}_2\text{H}_5)_2\text{O}} \text{C}_6\text{H}_5\text{MgBr} \tag{8}$$

The most important side reaction is the coupling process

$$\text{C}_6\text{H}_5\text{MgBr} + \text{C}_6\text{H}_5\text{Br} \longrightarrow \text{C}_6\text{H}_5—\text{C}_6\text{H}_5 + \text{MgBr}_2 \tag{9}$$

but it is not a significant problem. Once the reagent is prepared, it is used directly in subsequent reactions; it is not necessary to separate the small amount of the by-product biphenyl from the reagent before using it.

When phenylmagnesium bromide is allowed to react with an ester such as methyl or ethyl benzoate, the chief product of the reaction is triphenylmethanol. The formation of this compound can be envisioned as arising from the two-step reaction shown by equations 10 and 11. The intermediate benzophenone, whose formation is shown in equation 10 as arising from the reaction of 1 mol of methyl benzoate with 1 mol of phenylmagnesium bromide, then reacts with 1 more mol of phenylmagnesium bromide to give the salt of triphenylmethanol, as shown in equation 11.

$$
C_6H_5\overset{\displaystyle O}{\underset{\displaystyle \|}{C}}\!-OCH_3 + C_6H_5\!:\!^{\ominus}MgBr^{\oplus} \xrightarrow{\text{ether}} \left[C_6H_5\overset{\displaystyle O^{\ominus}MgBr^{\oplus}}{\underset{\displaystyle C_6H_5}{\overset{\displaystyle |}{\underset{\displaystyle |}{C}}}}\!-OCH_3 \right] \longrightarrow C_6H_5\overset{\displaystyle O}{\underset{\displaystyle \|}{C}}\!-C_6H_5 \quad (10)
$$
$$
+ CH_3OMgBr
$$

$$
C_6H_5\overset{\displaystyle O}{\underset{\displaystyle \|}{C}}\!-C_6H_5 + C_6H_5\!:\!^{\ominus}MgBr^{\oplus} \longrightarrow C_6H_5\overset{\displaystyle O^{\ominus}MgBr^{\oplus}}{\underset{\displaystyle C_6H_5}{\overset{\displaystyle |}{\underset{\displaystyle |}{C}}}}C_6H_5 \xrightarrow{H_2O} (C_6H_5)_3COH \quad (11)
$$
$$
+ HOMgBr
$$

Subsequent hydrolysis of the salt gives triphenylmethanol. In order to avoid precipitation of a basic magnesium salt (HOMgX), the hydrolysis is carried out with an acid solution (H_2SO_4 or NH_4Cl), which keeps the magnesium in solution in the form of its normal salts.

The synthesis of triphenylmethanol is accomplished experimentally by first preparing the Grignard reagent and then adding the methyl benzoate to it with stirring and cooling (the reaction is highly exothermic). In working up the reaction mixture, water is added to hydrolyze the magnesium salts to the final organic products; a little mineral acid is also commonly added to dissolve the basic magnesium salt which is formed. If this is not done, the magnesium hydroxide will produce an unworkable emulsion. The organic products (triphenylmethanol, benzophenone, biphenyl, unchanged methyl benzoate, and benzene—the benzene coming from action of water on unchanged phenylmagnesium bromide) all remain in the organic layer (consisting primarily of diethyl ether). The principal organic products present are triphenylmethanol and biphenyl, the former being present in the major amount. It is possible to separate these two compounds from one another, owing to the solubility of the hydrocarbon product (biphenyl) in hydrocarbon solvents (such as petroleum ether and hexane). Recrystallization of the crude product mixture from these solvents leaves pure triphenylmethanol as the solid. This is because the polar, hydroxylic triphenylmethanol is much less soluble in hydrocarbon solvents than is biphenyl.

Reaction of Phenylmagnesium Bromide with Carbon Dioxide. The reaction that occurs when phenylmagnesium bromide is allowed to react with carbon dioxide and the resulting magnesium salt hydrolyzed is shown in equation 12; the major product is a carboxylic acid (benzoic acid). The principal side reactions encountered in this

$$
C_6H_5\!:\!^{\ominus}MgBr^{\oplus} + O\!=\!C\!=\!O \longrightarrow C_6H_5\!-\!\overset{\displaystyle O}{\underset{\displaystyle \|}{C}}\!-O^{\ominus}MgBr^{\oplus} \xrightarrow{H_2O} C_6H_5\overset{\displaystyle O}{\underset{\displaystyle \|}{C}}\!-OH \quad (12)
$$
$$
+ HOMgBr
$$

process result in the formation of a ketone or a tertiary alcohol, as shown by the steps in equation 13. These side reactions can, however, be minimized in several ways. First

$$C_6H_5CO_2MgBr + C_6H_5MgBr \longrightarrow (C_6H_5)_2CO + MgBr_2 + MgO$$

$$\downarrow C_6H_5MgBr$$

$$(C_6H_5)_3COMgBr \xrightarrow{H_2O} (C_6H_5)_3COH \qquad (13)$$
$$+ \text{HOMgBr}$$

of all, the bromomagnesium salt of the carboxylic acid, because it is moderately insoluble in ether, precipitates from solution and is thus effectively prevented from undergoing further reaction. Second, dry ice is used as the source of carbon dioxide, so that the temperature is maintained at $-78°$ throughout. Finally, a large excess of carbon dioxide is always used, thus enhancing the probability that phenylmagnesium bromide will react with carbon dioxide rather than with the magnesium salt (to produce the ketone) or with the ketone (to yield the magnesium salt of the alcohol), both of which are present in lower concentrations than carbon dioxide and are also less reactive toward the Grignard reagent.

Experimentally this reaction is carried out by preparing the Grignard reagent, which is then poured slowly and with swirling onto an excess of finely powdered dry ice. The excess dry ice is then allowed to evaporate, and on treatment of the resulting mixture (which contains diethyl ether as the solvent) with dilute mineral acid, the carboxylic acid is liberated and dissolves in the diethyl ether. The crude product mixture, which includes benzoic acid, benzophenone, triphenylmethanol, benzene, and biphenyl, is extracted into the diethyl ether layer. However, the desired product, benzoic acid, is separated quite readily from the by-products by extracting it into a dilute solution of sodium hydroxide. The acid is converted to its soluble sodium salt, whereas the other compounds remain in the organic layer. Careful acidification of the aqueous layer regenerates the benzoic acid, which can be purified further by recrystallization.

Preparation of *n*-Butylmagnesium Bromide and Reaction with 2-Methylpropanal. *n*-Butylmagnesium bromide is prepared from 1-bromobutane using a procedure entirely analogous to the preparation of phenylmagnesium bromide. It is then allowed to react with 2-methylpropanal to give the salt of 2-methyl-3-heptanol, as shown in equation 14, and the salt is hydrolyzed with dilute sulfuric acid.

$$
\overset{\displaystyle O}{\underset{\displaystyle \underset{CH_3}{|}}{CH_3CH_2CH_2CH_2MgBr + CH_3CHCH}} \longrightarrow \overset{\displaystyle O^{\ominus}MgBr^{\oplus}}{\underset{\displaystyle \underset{CH_3}{|}}{CH_3CH_2CH_2CH_2CHCHCH_3}} \qquad (14)
$$

$$H_2O \Big| H_2SO_4$$

$$
\overset{\displaystyle OH}{\underset{\displaystyle \underset{CH_3}{|}}{CH_3CH_2CH_2CH_2CHCHCH_3}}
$$

It should be pointed out that the experiments described here represent two possible ways in which organometallic compounds can be allowed to react with substrates. In the preparation of triphenylmethanol and 2-methyl-3-heptanol, the substrates (methyl benzoate and 2-methylpropanal, respectively) are added to the Grignard reagent; this mode of addition is designated *normal addition*. On the other hand, in the preparation of benzoic acid the Grignard reagent is added to the substrate (carbon dioxide); this procedure is designated *inverse addition*. The method of addition is dictated by the chemical reactions involved. For example, in the preparation of benzoic acid, if carbon dioxide gas were bubbled into the Grignard reagent (normal addition), the reaction illustrated in equation 13 would be the predominant one. There would be an excess of the Grignard reagent, which would react further with the bromomagnesium salt of benzoic acid once it formed. To obtain a good yield of benzoic acid, it is necessary to minimize this reaction, so the Grignard reagent should be added to a large excess of carbon dioxide. On the other hand, in the preparation of triphenylmethanol the Grignard reagent could equally well be added to the methyl benzoate. However, to do this would involve transferring the Grignard reagent to a dropping funnel and adding it to another flask containing the substrate.

EXPERIMENTAL PROCEDURE

DO IT SAFELY

1. During the course of this experiment each person in the room will be using diethyl ether as a solvent. Owing to the extreme flammability of diethyl ether, each of you must remain aware of the potential problems its use entails. Open containers of "ether" should not be kept at your bench. The total volume of ether needed for the experiment may be anticipated and obtained in one trip to the hood; however, it should be kept in a *stoppered* container in order to prevent accidental fires, evaporation, and its absorption of moisture from the air. We recommend that *no* student in the room be allowed to obtain ether until *all* students have completed the flame-drying of their glassware and have put away their Bunsen burners.

2. When flame-drying your apparatus before the preparation of the Grignard reagent, take care not to heat too strongly in the direct vicinity of the magnesium metal in the flask. The resulting surface oxidation of the metal may make the reaction more difficult to initiate. Moreover, you should not directly heat the "ring seals" of the condenser (where the inner and outer tubes of the condenser are joined). All too frequently these seals crack from stresses created by heating. If your dropping funnel is equipped with a Teflon stopcock, avoid heating in the area of the stopcock because it will soften and flow.

3. The conduct of a Grignard experiment should provide you with an appreciation of the problems and techniques of controlling a highly exothermic reaction. In all such instances the key to safe performance is the control over the rate at which the reaction is *allowed* to proceed. Typically, one of the reagents necessary in the rate-determining step of the reaction is introduced into the reaction mixture by *slow* dropwise addition in order to maintain that reagent at a low concentration. The rate of the reaction, and thus the rate of heat evolution, is therefore controlled by the rate of addition. The heat generated in the reaction is carried away to the reflux condenser by the boiling solvent. The rate of the reaction must be controlled within the capacity of the condenser to condense the reluxing vapors. In these experiments keep an ice-water bath at hand, and be prepared to use it if the reaction appears to be getting out of control.

A. Assembly and Drying of Apparatus

In a 250-mL round-bottomed flask place 0.10 g atom of magnesium turnings and a magnetic stirring bar, if this stirring option is available. Equip the flask with a Claisen connecting tube to which are attached a reflux condenser and a dropping funnel. Position the dropping funnel above the flask and the condenser on the side arm of the connecting tube. The joint connecting the flask to the Claisen tube should be well lubricated. Attach calcium chloride drying tubes to the top of the condenser and to the dropping funnel.

When flaming the glassware with a burner, heat the apparatus thoroughly in order to evaporate water from the *inside* surface of the apparatus. Begin heating the glassware at the bottom, working the flame in succession up each of the two side extensions of the apparatus. In this way the evaporated moisture is forced upward ahead of the flame, where it is eventually absorbed by the desiccant in the drying tubes. *Allow the apparatus to cool to room temperature before continuing.*

B. Preparation of the Grignard Reagent

Once the glassware has cooled to room temperature, add 10 mL of *anhydrous* diethyl ether to the flask in one portion through the dropping funnel. Prepare in 25 mL of anhydrous diethyl ether a solution of 0.12 mol of whichever halide has been assigned. Make sure this solution is homogeneous, and add it to the dropping funnel (stopcock closed). Prepare an ice-water bath, and have it ready for use if needed. If you are using magnetic stirring, turn on the stirrer motor. Be sure water is running through the condenser.

Add a 2-to-3-mL portion of the halide-ether solution from the dropping funnel onto the magnesium turnings. If you are not using stirring, loosen the clamp to the flask, and manually swirl the flask in order to mix the contents. A change in the appearance of the reaction mixture, as evidenced by the presence of a slightly cloudy (chalky) solution and by the formation of small bubbles at the surface of the magnesium turnings, indicates that the reaction has started. Once the reaction has started, the ether will be observed to reflux, and the flask will become slightly warm. If the reaction has started, disregard the instructions in the next paragraph.

If the reaction does not start on its own, obtain two or three additional turnings of magnesium, and crush them thoroughly with a heavy spatula or the end of a clamp. Remove the dropping funnel just long enough to add these broken pieces of magnesium to the flask, and then replace the funnel. (The clean, unoxidized surfaces of magnesium which are exposed should aid in the initiation of the reaction.) If, after an additional 3 to 5 min, the reaction still has not started, consult your instructor. The best remedy at this point is to warm the flask and add 2 or 3 drops of 1,2-dibromoethane to the mixture.

Once the reaction has started, either on its own or as a result of using the remedies discussed above, and the ether is observed to be refluxing smoothly, add an extra 20-mL portion of anhydrous diethyl ether to the reaction mixture through the condenser. (This serves to dilute the reaction mixture and to minimize the coupling reaction.) The rest of the halide-ether solution should now be added *dropwise* to the reaction mixture at a rate that is just fast enough to maintain a gentle reflux. If it is added too fast, not only may the reaction get out of control, but the yield will also be reduced, owing to increased coupling. If the reaction becomes too vigorous, cool the flask a bit with the ice-water bath, and reduce the rate of addition. This total addition should take about 15–30 min. If the spontaneous boiling of the mixture becomes too slow, increase slightly the rate of addition. If this does not serve to accelerate the rate of reflux, use a steam bath or heating mantle to gently heat the mixture as necessary during the remainder of addition.

At the end of the reaction the solution will normally have a tan to brown, chalky appearance, and most of the magnesium will have disappeared, although residual bits of metal usually remain.

Use the Grignard reagent as soon as possible after its preparation, following one of the procedures given in part C. Either C1 or C2 applies if phenylmagnesium bromide was prepared; C3 applies if *n*-butylmagnesium bromide was prepared.

C.   Reactions of the Grignard Reagent

1.   Preparation of Triphenylmethanol.   Once the reaction mixture for the preparation of phenylmagnesium bromide (part B) has cooled to room temperature, dissolve 0.045 mol of methyl (or ethyl) benzoate in about 20 mL of anhydrous diethyl ether, and place this solution in the dropping funnel (stopcock closed). Cool the reaction flask with an ice-water bath, and then begin slow, dropwise addition of the solution of methyl benzoate to the phenylmagnesium bromide solution. This reaction is exothermic; control the rate of reaction by adjustment of the addition rate and by occasional cooling of the reaction flask with the ice-water bath as necessary. If you are not using magnetic stirring, swirl the flask from time to time during the addition. Frequently a white solid forms during the reaction; this is a sign that the reaction is proceeding normally. After the addition is complete and the exothermic reaction has subsided, either of the following may be done in order to complete the reaction. (1) Heat the reaction mixture at reflux for 30 min, using either a heating mantle or a steam bath. (2) Stopper the flask after cooling to room temperature, and allow the mixture to stand until the next laboratory period (no reflux required).[*]

In a 250-mL Erlenmeyer flask place about 50 mL of 6 *M* sulfuric acid and about 30 g of ice; pour the entire reaction mixture into this flask with swirling. Continue swirling until the heterogeneous mixture is completely free of undissolved solids. It may be necessary to add more diethyl ether (technical) to dissolve all the organic material. Transfer the entire mixture to a separatory funnel, shake it vigorously but carefully, venting the funnel often to relieve pressure, and remove the aqueous layer.★ Wash the organic layer with 3 M sulfuric acid and then with saturated sodium chloride solution. Repeat the washings with salt solution until the resulting aqueous layer is no longer acidic. Dry the organic layer with anhydrous sodium sulfate, and filter the solution into a round-bottomed flask of suitable size. Remove the ether by simple distillation (no flames). After the crude solid residue has dried, determine its melting range.★ It should weigh 11–12 g and may melt over a wide range.

The crude solid may be purified by recrystallization from a 2:1 mixture of cyclohexane:absolute ethanol. Use a minimum amount of the boiling solvent, carrying out the operation in the hood (or use a funnel inverted over the flask which is attached to an aspirator). Once all the material is in solution, evaporate the solvent slowly until small crystals of triphenylmethanol start to form. Remove the heat and allow crystallization to continue at room temperature; complete the process by cooling the solution to 0° until no more crystals appear to form. Isolate the product by filtration; it should be colorless.[1] Determine the melting point and yield of the product. The yield should be 5–6 g. The reported melting point of triphenylmethanol is 164°.

2. Preparation of Benzoic Acid. In a 500-mL Erlenmeyer flask place 25–30 g of solid carbon dioxide (dry ice) which has been coarsely crushed and protected from moisture as much as possible. Add the phenylmagnesium bromide solution to the dry ice slowly with gentle swirling. The mixture normally becomes somewhat viscous, and after the addition is complete, an extra 50- to 75-mL portion of technical diethyl ether should in turn be added to the mixture. Allow the excess carbon dioxide to evaporate. This is perhaps most easily done by allowing the flask to stand overnight and reach room temperature on its own accord; it would be preferable to leave the flask in the hood, properly labeled with your name, rather than to leave it in a locker.★[2]

After the excess dry ice is gone, treat the mixture with about 50 mL of 3 M sulfuric acid which has been mixed with 25–30 g of ice. The addition of the acid to the crude reaction mixture should be done with care in order to avoid foaming; if the ether has evaporated appreciably, more technical diethyl ether may be added so that the total volume of ether is about 100 mL. Transfer the entire mixture to a separatory funnel after mixing it thoroughly by swirling. Rinse the flask with a small portion of

---

[1]An alternative method of recovering the triphenylmethanol from the crude mixture containing it and biphenyl is as follows. Add about 150 mL of petroleum ether and about 50 mL of technical-grade diethyl ether to the crude product. Heat the mixture on a steam bath (preferably in the hood) until the crystals dissolve; add more ether if needed. Boil off the solvent *until* the first crystals appear, and then stop heating. Allow the flask to cool, and collect the crystals which form. Proceed to determine the melting point of the crystals, and recrystallize if needed.

[2]The process of removing the excess CO_2 can be expedited in one of several ways: (1) shake or swirl the flask while warming it *very slightly* in a warm water bath; (2) stir the viscous mixture that forms during the reaction; or (3) add small amounts of warm water to the reaction mixture. All these methods may cause a sudden loss of CO_2 gas, and great care should be taken in hastening this process. At no time should the flask be stoppered. Any warming that is done also causes the ether to evaporate, and care should be taken to replace the ether before going to the next step.

technical diethyl ether, and add the rinse to the separatory funnel. Shake the funnel cautiously (with venting), and separate the layers. Extract the aqueous layer two additional times with 15-mL portions of diethyl ether, and combine all the ether extracts.

Extract the ethereal solution with three 20-mL portions of a 1 *M* solution of sodium hydroxide. Treat the alkaline extract with decolorizing carbon and filter. Add 6 *M* hydrochloric acid until precipitation is complete and filter the solid.★ The solid (benzoic acid) may be recrystallized from water if needed. After drying the product, determine the melting point and yield; the yield should be about 6 g.

3.   Preparation of 2-Methyl-3-heptanol.   After the reaction involving the preparation of *n*-butylmagnesium bromide has subsided and the reaction mixture has cooled to room temperature, continue with the preparation of 2-methyl-3-heptanol as follows. Dissolve 0.090 mol of freshly distilled 2-methylpropanal in 15 mL of anhydrous diethyl ether, and place this solution in the dropping funnel (stopcock closed). Begin the dropwise addition of the solution of 2-methylpropanal to that of the *n*-butylmagnesium bromide. This reaction is highly exothermic; control the rate of the reaction by judicious control of the drop rate and by cooling the reaction mixture with an ice-water bath as necessary. Either stir or swirl the reaction mixture occasionally. The addition may require 15–20 min. After the addition is complete, allow the reaction mixture to stand about 15 min.★

Place about 150 mL of crushed ice in a 500-mL beaker, and add 9 mL of concentrated sulfuric acid. Pour the reaction mixture slowly, and with stirring, into the ice-acid mixture. After the addition is complete, transfer the cold mixture, which may contain some precipitate, to a separatory funnel and shake it gently. The precipitate should dissolve. Separate the layers and extract the aqueous layer with two 25-mL portions of diethyl ether; add these extracts to the main ether layer. Wash the combined ether solutions (in the separatory funnel) with 30 mL of saturated sodium bisulfite solution, venting the funnel to relieve pressure, then with two 30-mL portions of saturated sodium bicarbonate solution, and finally with 30 mL of saturated sodium chloride solution. Dry the ether solution over anhydrous magnesium sulfate.★

Filter the solution from the drying agent into a 250-mL round-bottomed flask, and remove most of the ether from the product by simple distillation, using a steam bath or heating mantle, taking the usual precautions against the flammability of ether. Transfer the residue (the product with some remaining ether) to a 50-mL round-bottomed flask and prepare for fractional distillation.

Insulate the top of the distilling flask and the still head with glass wool or aluminum foil to ensure steady distillation of the rather high boiling product. Collect separately any forerun boiling below 165°, and the product, 2-methyl-3-heptanol, boiling at 165–168°. Weigh the product and calculate the yield.

EXERCISES

1.   Comment on the use of steam distillation as a possible alternative procedure for the purification of the crude triphenylmethanol. Consider the possible unchanged

starting materials and the products which are formed, and indicate which of these would steam-distil and which would not. Explain in detail how this method of purification would yield pure triphenylmethanol (if it would at all).

2. Ethanol is quite often present in technical-grade diethyl ether. If this grade were used, what effect, if any, would the ethanol have on the formation of the Grignard reagent? Explain.

3. In the work-up of the reaction mixture from *n*-butylmagnesium bromide and 2-methylpropanal, successive washes with solutions of sodium bisulfite, sodium bicarbonate, and sodium chloride are used. What is the purpose of each of these steps?

4. The ir spectrum of 2-methyl-3-heptanol prepared by the procedure of this chapter sometimes shows an absorption at 1720 cm^{-1}. Give an explanation for this absorption.

5. Give a plausible structure for $C_6H_5MgBr \cdot 2(C_2H_5)_2O$. How do you think the ether molecules are bound to C_6H_5MgBr?

6. Arrange the following compounds in increasing order of reactivity toward a Grignard reagent: methyl benzoate, benzoic acid, benzaldehyde, acetophenone, benzoyl chloride. Explain the basis for your decision, making use of mechanisms where needed. What is (are) the product(s) of reaction of each of the above carbonyl-containing compounds with *excess* phenylmagnesium bromide?

7. How might primary, secondary, and tertiary alcohols be prepared from an aldehyde or a ketone and an organometallic reagent? Suggest suitable carbonyl-containing compounds, as well as suitable organometallic reagents. Write chemical reactions for these preparations, and indicate stoichiometry where important.

SPECTRA OF STARTING MATERIALS AND PRODUCTS

The spectra of 2-methylpropanal are given in Figures 13.1 and 13.2.

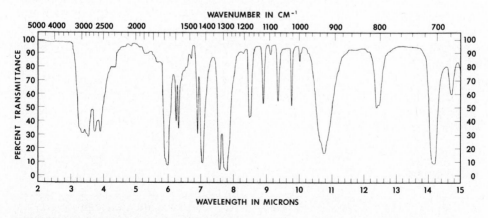

FIGURE 12.1 IR spectrum of benzoic acid.

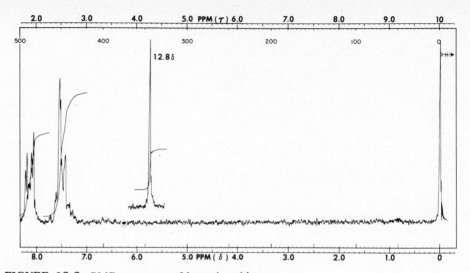

12.8δ

FIGURE 12.2 PMR spectrum of benzoic acid.

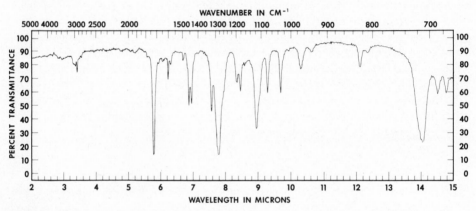

WAVENUMBER IN CM⁻¹

FIGURE 12.3 IR spectrum of methyl benzoate.

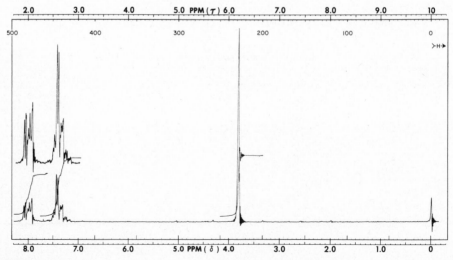

FIGURE 12.4 PMR spectrum of methyl benzoate.

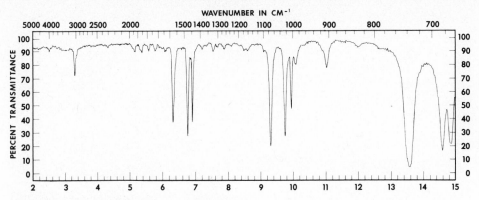

FIGURE 12.5   IR spectrum of bromobenzene.

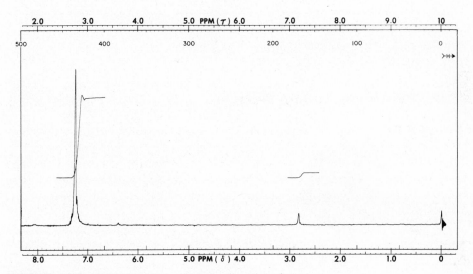

FIGURE 12.6   PMR spectrum of triphenylmethanol.

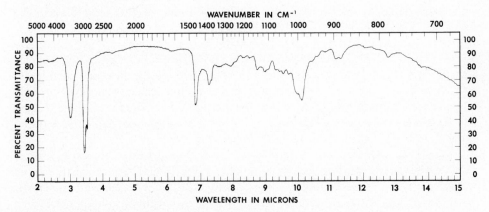

FIGURE 12.7   IR spectrum of 2-methyl-3-heptanol.

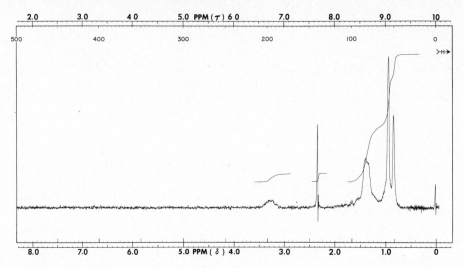

FIGURE 12.8   PMR spectrum of 2-methyl-3-heptanol.

12.3   The Organolithium Reagent

Alkyl and aryl lithium compounds play a role in organic synthesis second only to that of the Grignard reagents discussed in Section 12.2. Because of their importance we feel it is desirable to include a brief discussion of their preparation and reactivity; however, owing to possible difficulties in working with the reactive lithium metal, we have opted not to include any experiments involving the preparation of organolithium reagents.

The preparation of the lithium reagent, RLi, is indicated in equation 15, where RX represents an alkyl or aryl halide. The solvent does not appear to play as

$$RX + 2\,Li \xrightarrow[\substack{\text{or} \\ \text{hydrocarbon solvent}}]{\text{dry ether solvent}} RLi + LiX \tag{15}$$

important a role in the preparation of organolithium reagents as it does in the preparation of Grignard reagents. This has been shown by the fact that it is possible to prepare organolithium reagents in both hydrocarbon solvents (such as pentane and hexane) and ether solvents. However, there is a difference in reactivity of the reagents toward the solvent. Ethereal solutions of the Grignard reagent may be stored under anhydrous conditions for some time, but organolithium reagents may be kept only when a hydrocarbon solvent is used. The latter cleave ethers, even though the reagent can be prepared in ether solution. When ether is used, the reagent must be used immediately after its preparation and may not be stored.

It should be noted that 2 g atoms of lithium are required for each mole of organic halide, in contrast to the one-to-one mole ratio of magnesium to halide. Pure lithium metal can be used effectively in preparing the lithium reagent, although it has been found that the use of lithium containing 0.8% by weight of sodium metal often increases the ease with which the reaction starts.

The reactivity of the alkyl and aryl halides toward lithium metal is the same as that for the Grignard reagent. The major exception is the necessity of using tetrahydrofuran as the solvent for the preparation of the Grignard reagent from aryl chlorides. Most alkyl halides form lithium reagents by reaction with lithium metal either in diethyl ether or in a hydrocarbon solvent.

The side reactions arising from the preparation of the lithium reagent are the same as those found for the Grignard reagents (see equations 5, 6, and 7). The same conditions of extreme dryness must be utilized; the side reactions may be minimized by use of an inert atmosphere and dilute solutions.

The lithium reagents usually react in a manner similar to Grignard reagents. In some instances lithium reagents are preferable because they are generally more reactive than their halomagnesium counterparts. Furthermore, once the lithium reagent has reacted with a substrate, the resulting salt that forms is generally more soluble in the solvent than the corresponding salt resulting from the Grignard reagent.

An advantage of the greater reactivity of the lithium reagent is realized in *metalation reactions,* in which the lithium reagent, RLi, abstracts a proton from a hydrocarbon, R′H, as shown in equation 16. This reaction has been used to determine

$$RLi + R'H \longrightarrow RH + R'Li \qquad (16)$$

the acidity of hydrocarbons and can also be used to introduce substituents into various aromatic compounds. *n*-Butyllithium (equation 17) has been used extensively for this purpose; some typical examples are shown in equations 18–20.

$$n\text{-}C_4H_9Br + 2\ Li \xrightarrow[-10°]{\text{ether}} n\text{-}C_4H_9Li + LiBr \qquad (17)$$
$$85\% \text{ yield}$$

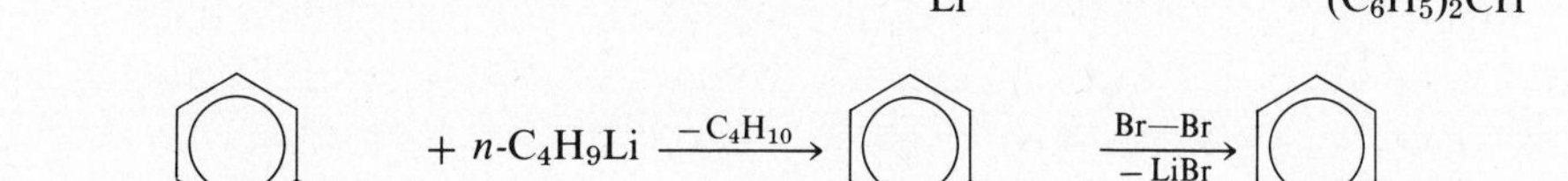

(18)

(19)

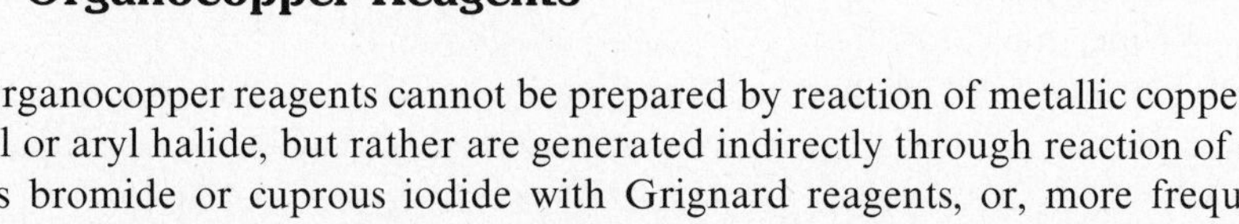

(20)

12.4   Organocopper Reagents

Organocopper reagents cannot be prepared by reaction of metallic copper with an alkyl or aryl halide, but rather are generated indirectly through reaction of either cuprous bromide or cuprous iodide with Grignard reagents, or, more frequently, organolithium reagents. For example, reaction of methyl lithium with one molar equivalent of cuprous iodide yields the polymeric methyl copper, an *organocopper* (**4**, equation 21). Addition of a second molar equivalent of methyl lithium provides lithium dimethylcuprate, an *organocuprate* (**5**, equation 22). The summation of these two reactions is shown in equation 23. The synthetic utility of these two types of

$$(x)\ CH_3Li + (x)\ CuI \longrightarrow (CH_3\!-\!Cu)_x + (x)\ LiI \qquad (21)$$
$$\mathbf{4}$$

$$(CH_3\!-\!Cu)_x + (x)\ CH_3Li \longrightarrow (x)\ (CH_3)_2Cu^{\ominus}Li^{\oplus} \qquad (22)$$
$$\mathbf{5}$$

$$2\ CH_3Li + CuI \longrightarrow (CH_3)_2Cu^{\ominus}Li^{\oplus} + LiI \qquad (23)$$

copper reagents relative to the organometallic reagents described in the preceding two sections is exemplified in the following discussion.

When alkylcuprates react with primary alkyl halides, coupling products arising from S_N2-type reactions may be isolated in good yield, as shown in equation 24, for example. The yields of coupling products are much higher than those obtained with Grignard reagents or alkyl lithium compounds, so that this is the chosen procedure for preparing hydrocarbons by this type of approach. Alkyl copper reagents react similarly, although normally alkylcuprates are used.

$$2\ (CH_3)_2CH\!-\!Li \xrightarrow{\ CuI\ } [(CH_3)_2CH]_2Cu^{\ominus}Li^{\oplus} \xrightarrow{\ 2\ CH_3CH_2Br\ } 2\ (CH_3)_2CH\!-\!CH_2CH_3 \quad (24)$$

Organocopper reagents may also be employed effectively with acid chlorides in the preparation of ketones (equation 25). Such reactions provide good yields because organocopper reagents react only sluggishly with ketones. Reaction of acid chlorides with Grignard reagents and alkyl lithium compounds does not often provide acceptable yields of ketonic products because of the tendency of ketones to undergo *further* reaction with the organometallic reagent to produce alcohols. In this application copper provides the same advantage as does cadmium in the similar use of organocadmium reagents, a reaction discussed in many organic chemistry textbooks.

$$2\ R\overset{\displaystyle O}{\underset{\displaystyle \|}{-}}C\!-\!Cl + (CH_3)_2Cu^{\ominus}Li^{\oplus} \longrightarrow 2\ R\overset{\displaystyle O}{\underset{\displaystyle \|}{-}}C\!-\!CH_3 + CuCl + LiCl \qquad (25)$$

Usually Grignard reagents react with α, β-unsaturated ketones to produce both **6** and **7**, products of 1,2- (normal) and 1,4- (conjugate) addition, respectively (equation 26). The reaction is very sensitive to steric effects, and examples range from those

$$R\!-\!CH\!=\!CH\!-\!\overset{\displaystyle O}{\underset{\displaystyle \|}{C}}\!-\!R' \xrightarrow[\text{(2) } H_3O^{\oplus}]{\text{(1) } CH_3MgX} R\!-\!CH\!=\!CH\!-\!\underset{\displaystyle \underset{\textstyle CH_3}{|}}{\overset{\displaystyle \overset{\textstyle OH}{|}}{C}}\!-\!R' + R\!-\!\underset{\displaystyle \underset{\textstyle CH_3}{|}}{CH}\!-\!CH_2\!-\!\overset{\displaystyle O}{\underset{\displaystyle \|}{C}}\!-\!R' \quad (26)$$
$$\qquad\qquad\qquad\qquad\qquad\qquad\qquad\qquad \mathbf{6}\qquad\qquad\qquad\qquad\qquad\qquad \mathbf{7}$$

in which the conjugate addition product is the major product to those in which it is either the minor product or is not produced at all.

In contrast, use of alkyl lithium reagents in similar reactions provides only products of addition to the carbonyl group (equation 27). Conjugate addition is not observed.

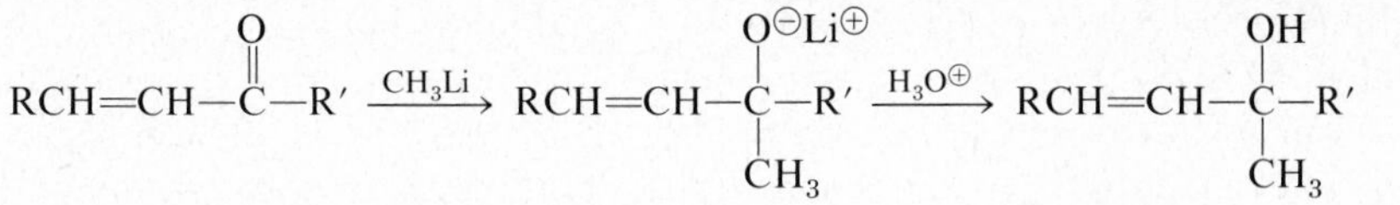

$$RCH{=}CH{-}\overset{\overset{\textstyle O}{\|}}{C}{-}R' \xrightarrow{CH_3Li} RCH{=}CH{-}\underset{\underset{\textstyle CH_3}{|}}{\overset{\overset{\textstyle O^{\ominus}Li^{\oplus}}{|}}{C}}{-}R' \xrightarrow{H_3O^{\oplus}} RCH{=}CH{-}\underset{\underset{\textstyle CH_3}{|}}{\overset{\overset{\textstyle OH}{|}}{C}}{-}R' \quad (27)$$

If the product of conjugate addition is desired, alkyl lithium reagents clearly cannot be used, and the unpredictability of the Grignard addition is a nuisance. However, both alkyl copper and alkylcuprate reagents react with α, β-unsaturated ketones to provide either totally or predominately the conjugate addition product. Thus reaction of isophorone (**8**, 3,5,5-trimethylcyclohex-2-enone) with lithium dimethylcuprate or methyl copper provides 3,3,5,5-tetramethylcyclohexanone in excellent yield (equation 28). The corresponding reaction of isophorone with methyl-

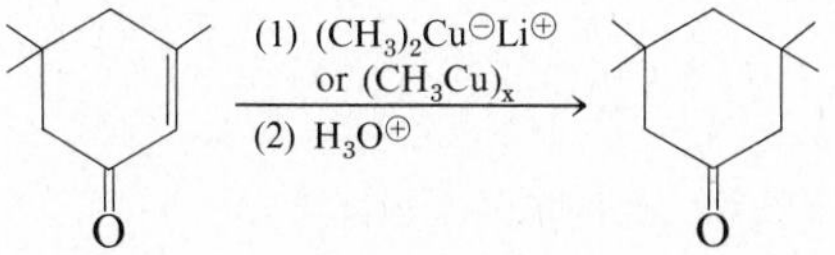

$$(28)$$

magnesium iodide, in contrast, produces the tertiary alcohol **9** in 90% yield (equation 29).

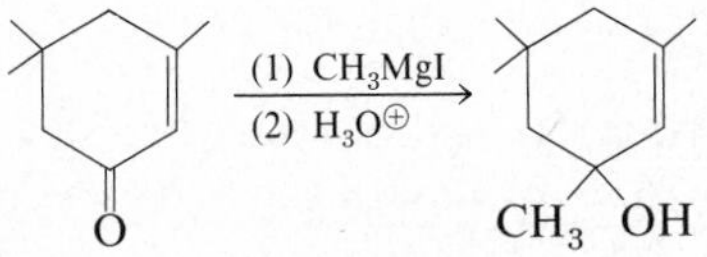

$$(29)$$

In the following experiment, isophorone (**8**) will be employed in reaction with a solution of methylmagnesium iodide to which has been added a *catalytic* amount of cuprous bromide. In this way methyl copper is produced *in situ* (equation 30). Methyl copper then reacts with **8** to produce the enolate ion **10** (equation 31), regenerating

$$(x)\ CH_3MgI + (x)\ CuBr \longrightarrow (CH_3{-}Cu)_x + (x)\ MgIBr \quad (30)$$

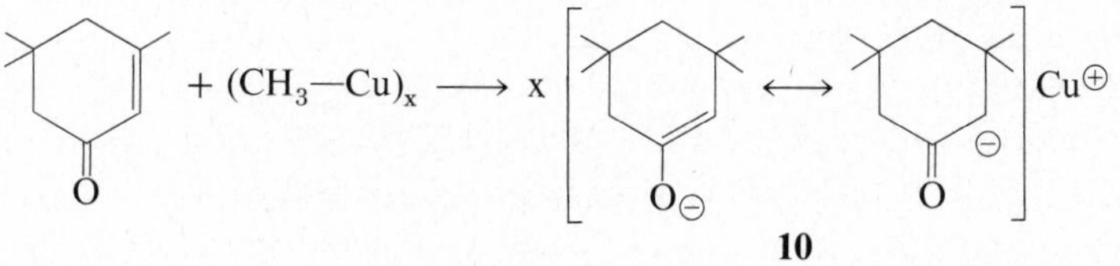

$$(31)$$

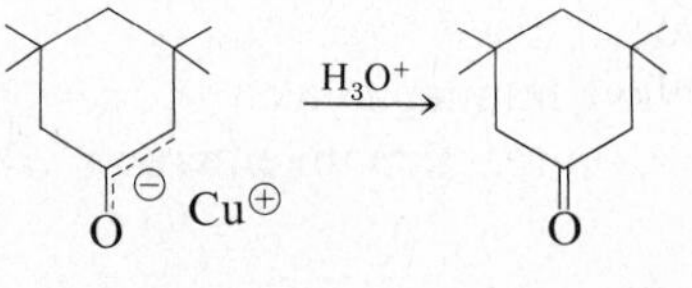

$$(32)$$

the cuprous ion for further reaction with additional methylmagnesium iodide, according to equation 30. Following neutralization (equation 32), 3,3,5,5-tetra-methylcyclohexanone may ultimately be isolated in up to 80% yield. Methyl copper is apparently somewhat more reactive toward **8** than methylmagnesium iodide, since the reaction shown in equation 29 does not compete effectively. *Small* amounts of 1,3,5,5-tetramethyl-1,3-cyclohexadiene **(11)** may be isolated during distillation of the final product; the diene presumably arises by dehydration of the tertiary alcohol **9** (equation 33).

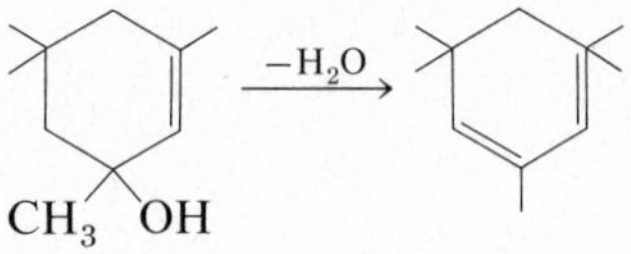

$$\xrightarrow{-H_2O}\qquad\qquad (33)$$

CH$_3$ OH

9   11

EXPERIMENTAL PROCEDURE

DO IT SAFELY

1. Before preparation of methylmagnesium iodide, read the Do It Safely directions in Section 12.2.

2. Methyl iodide is an alkylating agent. As with all alkylating agents of moderate or greater reactivity, care should be taken to avoid breathing its vapors or allowing contact with the skin. Although methyl iodide has *not* been implicated, several other alkylating agents are either suspected or proven carcinogens.

According to the procedures given in parts A and B of the Experimental Procedure in Section 12.2, prepare a solution of methylmagnesium iodide. Use 0.10 g-atom of magnesium turnings and 0.12 mol of iodomethane. If small amounts of magnesium metal remain at the end of the preparation, add a few more drops of methyl iodide, and allow the reaction to proceed a little longer. When the magnesium has completely dissolved, add 0.2 g of cuprous bromide to the solution of methyl-magnesium iodide, and cool the flask in an ice-water bath to 5°.

Prepare a solution of 0.08 mol of isophorone in 15 mL of anhydrous diethyl ether, and place this solution in the dropping funnel (stopcock closed) of the appara-tus used to make the Grignard reagent. While continuing to cool the flask, begin dropwise addition of the isophorone solution in the reaction mixture; the drop rate should be adjusted to require about 20 min to complete the addition. After the addition is complete, remove the cooling bath, and allow the mixture to stand (or stir) at room temperature for about 15 min. Then heat the mixture at reflux for a period of about 30 min, using either a steam bath or a heating mantle. Cool the mixture to room temperature with the ice-water bath.

Place 60 g of ice and 6 g of glacial acetic acid in a 500-mL beaker, and with swirling slowly pour the cooled reaction mixture over the ice. When most of the ice has melted, decant the liquid into a separatory funnel, and separate the ether layer. Extract the aqueous layer with an additional 25 mL of technical diethyl ether, and combine this extract with the original ether layer.★ Extract the ethereal solution twice with 15-mL portions of 0.5 *M* sodium carbonate solution, followed by two washings with 15-mL portions of water and finally with 15 mL of brine. Dry the organic solution over *anhydrous* sodium sulfate for about 15 min with occasional swirling and filter.★

Remove the ether by simple distillation, using a steam bath or a heating mantle, and then transfer the crude product to a smaller flask.★ Owing to the high-boiling nature of the product, the final distillation will need to be carried out with either a heating mantle or a burner. If heating mantles are not available so that a burner must be used, first connect the flask to either a house vacuum line or an aspirator, and evacuate the flask with swirling for a few minutes. It will help to warm the flask. This will ensure *complete* removal of flammable ether before using the burner.

Distil the residual oil, collecting 3,3,5,5-tetramethylcyclohexanone in the boiling range of 193–197°. If care is taken, a small forerun of 1,3,5,5-tetramethyl-1,3-cyclo-hexadiene **(11)** may be collected in the range 151–155°. Weigh your product and calculate the yield. Obtain an ir spectrum of the product, and compare it with that of isophorone (Figure 12.9).

EXERCISES

1. At what point in the work-up procedure is the crude product isolated from at least most of the copper and magnesium salts present in the reaction mixture?
2. Explain the role of the acetic acid used in the procedure.
3. Explain how 0.08 mol of isophorone can react with methyl copper when only about 0.02 mol of cuprous ion was added to the flask.

SPECTRA OF STARTING MATERIAL AND PRODUCT

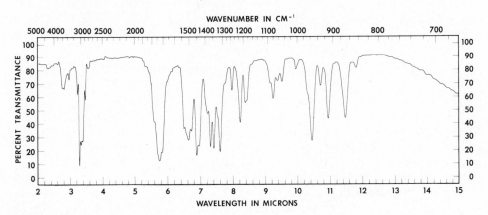

FIGURE 12.9 IR spectrum of isophorone.

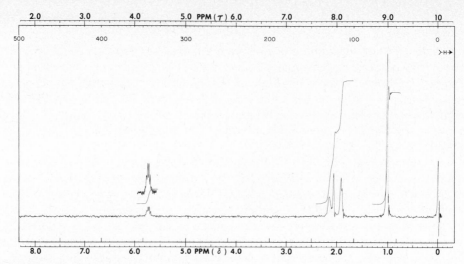

FIGURE 12.10 PMR spectrum of isophorone.

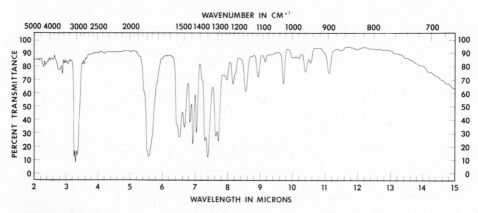

FIGURE 12.11 IR spectrum of 3,3,5,5-tetramethylcyclohexanone.

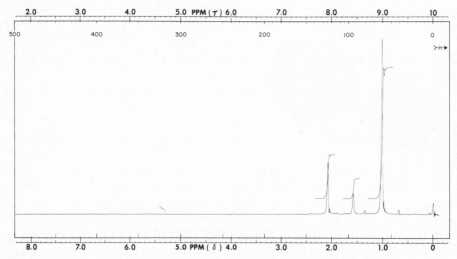

FIGURE 12.12 PMR spectrum of 3,3,5,5-tetramethylcyclohexanone.

13

OXIDATION REACTIONS OF ALCOHOLS AND CARBONYL COMPOUNDS

13.1 Preparation of Aldehydes and Ketones by Oxidation of Alcohols

Aldehydes and ketones play a central role in organic synthesis; accordingly, efficient methods for their preparation are of great importance. These substances can be synthesized from alkynes by acid-catalyzed hydration (equation 1) or by hydro-

$$R-C{\equiv}C-R(H) \xrightarrow[\substack{H_2SO_4 \\ H_2O}]{HgSO_4} R-\overset{\overset{\displaystyle O}{\|}}{C}-CH_2-R(H) \tag{1}$$

boration followed by oxidation (equation 2), and from carboxylic acids or their derivatives by reaction with organometallics or a variety of reducing agents (equations 3–6). One of the most common synthetic schemes, however, is the oxidation of

$$R-C{\equiv}C-R(H) \xrightarrow[\substack{(2)\ H_2O_2/HO^{\ominus}}]{(1)\ B_2H_6} R-CH_2-\overset{\overset{\displaystyle O}{\|}}{C}-R(H) \tag{2}$$

$$R-\overset{\overset{\displaystyle O}{\|}}{C}-OH \xrightarrow[\substack{(2)\ H_3O^{\oplus}}]{(1)\ R'Li} R-\overset{\overset{\displaystyle O}{\|}}{C}-R' \tag{3}$$

$$R-\overset{\overset{\displaystyle O}{\|}}{C}-Cl + R'_2Cd \longrightarrow R-\overset{\overset{\displaystyle O}{\|}}{C}-R' \tag{4}$$

$$R-C{\equiv}N \xrightarrow[\substack{(2)\ H_3O^{\oplus}}]{(1)\ R'MgX} R-\overset{\overset{\displaystyle O}{\|}}{C}-R' \tag{5}$$

$$R-C\equiv N \xrightarrow[\text{(2) } H_3O^\oplus]{\text{(1) } LiAlH_4/-80°} R-\overset{\displaystyle O}{\overset{\|}{C}}-H \tag{6}$$

primary and secondary alcohols with chromic acid, H_2CrO_4, or with potassium permanganate (equation 7). A description of the use of the former oxidizing agent for conversion of alcohols to aldehydes and ketones follows; procedures for the use of both reagents are included in the Experimental Procedure.

$$R-\overset{\displaystyle OH}{\underset{}{\overset{|}{C}H}}-R'(H) \xrightarrow[\text{or } KMnO_4]{H_2CrO_4} R-\overset{\displaystyle O}{\overset{\|}{C}}-R'(H) \tag{7}$$

Chromic acid is not stable for long periods and is therefore produced when required by reaction of sodium or potassium dichromate with an excess of an acid such as sulfuric or acetic (equation 8) or by dissolution of chromic anhydride in water (equation 9). In the latter preparation either sulfuric or acetic acid is also added, since

$$Na_2Cr_2O_7 + 2\ H_2SO_4 \longrightarrow [H_2Cr_2O_7] \xrightarrow{H_2O} 2\ H_2CrO_4 + 2\ NaHSO_4 \tag{8}$$

$$CrO_3 + H_2O \longrightarrow H_2CrO_4 \tag{9}$$

the rate of oxidation of alcohols by chromic acid is much greater in acidic solutions. For preparation or oxidation of substances that would decompose under strongly acidic conditions, either chromic anhydride dissolved in pyridine or basic potassium permanganate can be used as the oxidizing agent.

The stoichiometry of the chromic acid oxidation of alcohols is such that *two* equivalents of chromic acid *oxidize* three equivalents of an alcohol to the corresponding carbonyl compound (equation 10). Examination of equation 8 shows that

$$3\ R\overset{\displaystyle OH}{\underset{}{\overset{|}{C}H}}R'(H) + 2\ H_2CrO_4 + 3\ H_2SO_4 \longrightarrow 3\ R\overset{\displaystyle O}{\overset{\|}{C}}R'(H) + Cr_2(SO_4)_3 + 8\ H_2O \tag{10}$$

two equivalents of chromic acid are produced from one equivalent of dichromate. Thus if either sodium or potassium dichromate is used to generate chromic acid, only *one* equivalent of *dichromate* is required to oxidize *three* equivalents of alcohol. However, according to equation 9, only one equivalent of chromic acid is provided per equivalent of chromic anhydride; thus *two equivalents of anhydride* per *three* equivalents of alcohol are necessary.

The general mechanism of oxidation is well understood. Alcohols form chromate esters (**1**) in the presence of chromic acid, just as they do when allowed to react with carboxylic acids (see Section 15.2). The ester **1** is relatively unstable and decomposes by an elimination of the E2 type to produce the carbonyl compound (equation 11); the elimination is normally the rate-limiting step in the reaction.

$$\underset{R}{\overset{R'(H)}{\diagup}}CH-OH + HO\overset{\displaystyle O}{\underset{\displaystyle O.}{\overset{\|}{\underset{\|}{Cr}}}}OH \xrightarrow{-H_2O} \underset{R}{\overset{R'(H)}{\diagup}}CH-O\overset{\displaystyle O}{\underset{\displaystyle O}{\overset{\|}{\underset{\|}{Cr}}}}OH \longrightarrow \underset{R}{\overset{R'(H)}{\diagup}}C{=}O + H_2CrO_3 \tag{11}$$

1

During the course of the oxidation of the alcohol, chromium undergoes reduction from the $+6$ to the unstable $+4$ valence state. A disproportionation rapidly occurs between chromium (VI) and chromium (IV) to produce chromium (V), a new oxidizing agent which we shall write as $HCr^{V}O_3$ (equation 12). The 2 mol of chromium (V) will oxidize 2 more moles of alcohol to the carbonyl compound, and chromium (III), a stable oxidation state of this metal, will be generated (equation 13). The sum of equations 11–13 corresponds to equation 10.

$$H_2Cr^{IV}O_3 + H_2Cr^{VI}O_4 \longrightarrow 2\ HCr^{V}O_3 + H_2O \tag{12}$$

$$2\ HCrVO_3 + 2\ R\!-\!\overset{\overset{\displaystyle R'(H)}{|}}{C}HOH + 3\ H_2SO_4 \longrightarrow 2\ \underset{R}{\overset{R'(H)}{\diagup}}C\!=\!O + Cr_2(SO_4)_3 + 6\ H_2O \tag{13}$$

Some major side reactions complicate the oxidation of a primary alcohol to an aldehyde. Perhaps the most important of these is the ready oxidation of aldehydes to carboxylic acids by chromic acid (equation 14). This further, undesired oxidation can be minimized by adding the chromic acid *to* the primary alcohol, so that an excess of the oxidizing agent is *not* present in the reaction mixture, and by distilling the aldehyde from the reaction mixture as it is formed. As a consequence the desired aldehyde must be quite volatile, having a boiling point of less than about $150°$ if it is to be prepared in high yield by chromic acid oxidation.

$$3\ R\!-\!\overset{\overset{\displaystyle O}{\|}}{C}\!-\!H + 2\ H_2CrO_4 + 3\ H_2SO_4 \longrightarrow 3\ R\!-\!\overset{\overset{\displaystyle O}{\|}}{C}\!-\!OH + Cr_2(SO_4)_3 + 5\ H_2O \tag{14}$$

Even in cases in which the aldehyde is volatile, a poor yield of product is sometimes obtained. This is due to the facile formation from the aldehyde of the hemiacetal, **2**, which is subsequently oxidized to the ester of a carboxylic acid (equation 15). Occasionally this "side" reaction is turned into a useful synthesis of the ester, as is the case in the preparation of butyl butanoate by the chromic acid oxidation of 1-butanol.

$$RCH_2OH \xrightarrow{H_2CrO_4} R\!-\!\overset{\overset{\displaystyle O}{\|}}{C}\!-\!H \xrightarrow[RCH_2OH]{H^{\oplus}} R\!-\!\overset{\overset{\displaystyle OH}{|}}{\underset{\underset{\displaystyle 2}{OCH_2R}}{C}}\!-\!H \xrightarrow{H_2CrO_4} R\!-\!\overset{\overset{\displaystyle O}{\|}}{C}\!-\!OCH_2R \tag{15}$$

Ketones are much more stable toward oxidizing agents in mildly acidic media than are aldehydes, so that the side reactions mentioned in the oxidation of primary alcohols do not occur to a significant extent in the conversion of secondary alcohols to ketones. Under alkaline or *strongly* acidic conditions, however, enolizable ketones will undergo oxidation with cleavage to give two carbonyl fragments. For example, cyclohexanone, which can be obtained in good yield by the chromic acid oxidation of cyclohexanol, can be converted to hexanedioic acid by treatment with potassium permanganate under mild alkaline conditions (equation 16). The reaction undoubtedly requires initial conversion of the ketone to the enol, **3**, which is then oxidized by the permanganate (equation 17).

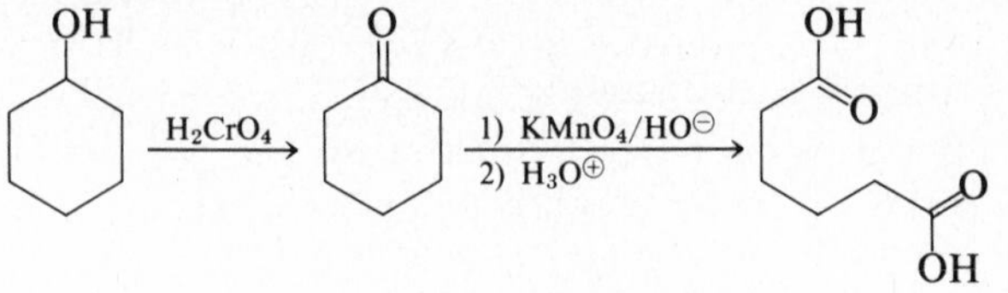

$$\text{(16)}$$

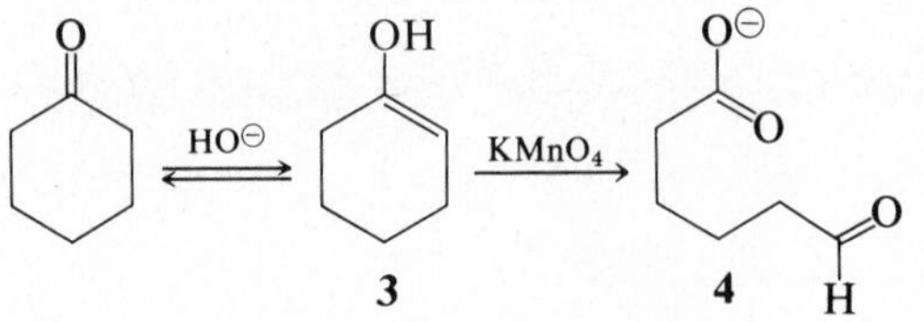

$$\text{(17a)}$$

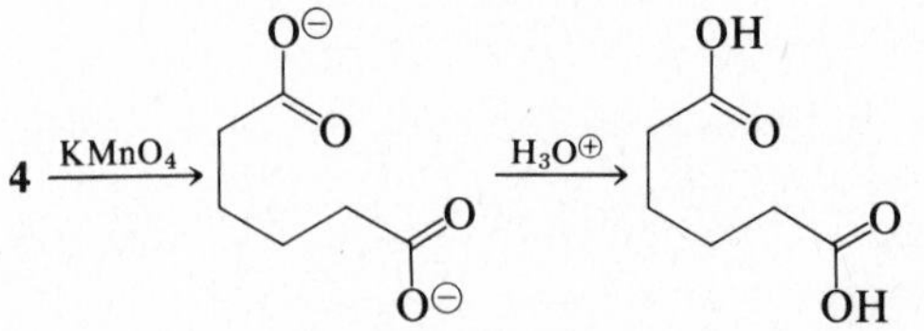

$$\text{(17b)}$$

Cyclohexanone is a symmetrical ketone and can give only the single enol, **3**. If a ketone is not symmetrical, it can produce two different enols (equation 18), each of which will be oxidized by permanganate to different products. The formation of complex mixtures of products when unsymmetrical ketones are oxidized is a complication which detracts from the synthetic utility of this reaction.

$$\text{(18)}$$

EXPERIMENTAL PROCEDURE

DO IT SAFELY

1. The extra 2 or 3 min taken at the end of a laboratory period to store the equipment and glassware safely and correctly in your drawer is well worth the effort. Carelessly stored equipment leads to breakage, particularly of glassware. Not only may unnecessary expenditures be required, but unnoticed hairline or star cracks in flasks and other glassware may lead to accidents and possible injury. Examine your glassware carefully for these types of cracks; cracked glassware being used to heat reaction mixtures or for distillations may suddenly break during heating, resulting in the spillage of large amounts of dangerous or flammable chemicals.

2. When mixing sulfuric acid and water, *always* be sure to add the acid *to* the water and to swirl the container to ensure continuous mixing. The dissolution of sulfuric acid in water generates much heat; when acid is added to water, the heat is dispersed through warming of the water. Swirling prevents the layering of the denser sulfuric acid at the bottom of the flask and the attendant possibility that hot acid will be splattered by the steam generated when the two layers are later unexpectedly mixed by agitation.

   3. When preparing and handling solutions of chromic acid or potassium permanganate, it is advisable to wear rubber gloves or to take proper care to avoid contact of the acids with your skin. Each solution will cause unsightly stains on your hands for several days, and the chromic acid-sulfuric acid solution in particular may cause rather severe chemical burns. Wash your hands thoroughly with soap and warm water if necessary, and in the case of chromic acid rinse the affected area with sodium bicarbonate solution.

A. Preparation of 2-Methylpropanal

   Prepare a solution of chromic acid by dissolving 0.10 mol of potassium dichromate in 150 mL of water and then *slowly* adding with swirling 22 mL of *concentrated* sulfuric acid. Allow this solution to cool to room temperature, using an ice-water bath if it is desired to hasten the process. Fit a 250-mL, round-bottomed flask with a Claisen adapter, and equip the adapter with a dropping funnel above the flask and a Hempel column on the parallel side arm. Attach to the top of the Hempel column a still head bearing a thermometer and a water-cooled condenser. Position a 50- or 100-mL graduated cylinder to receive distillate. Place 0.30 mol of 2-methyl-1-propanol and 25 mL of water in the flask along with two or three boiling stones, and heat the mixture until gentle boiling begins. Discontinue heating and immediately begin to add the red-orange solution of chromic acid from the dropping funnel to the hot mixture of alcohol and water at a rate such that all the acid will have been added in about 15 min. This rate will cause the reaction mixture to boil vigorously, and a mixture of alcohol, aldehyde, and water will steam-distil, giving a head temperature of 80–85°. This temperature should be maintained by appropriate adjustment of the rate of addition of acid. After all the chromic acid has been added and the mixture has stopped distilling on its own, heat and distil the dark green [chromium (III)] reaction mixture for an additional 15 min.★ Note the amount of water contained in the two-phase distillate, transfer the mixture to a separatory funnel, and add 0.5 g of sodium carbonate. Shake the funnel thoroughly, with venting to relieve any pressure, and then saturate the aqueous layer by adding about 0.2 g of sodium chloride for each milliliter of water to salt out any dissolved product. Shake the funnel again to effect solution of the salt, remove the organic layer, and dry it over anhydrous sodium sulfate.★ Filter the crude organic product into a 50-mL distilling flask, and isolate the 2-methylpropanal, bp 63–66°, by fractional distillation. If the fractionation is continued after the aldehyde distils, unchanged alcohol, bp 107–108°, and some isobutyl isobutyrate, bp 148–149°, can be isolated.

   Prepare the 2,4-dinitrophenylhydrazone of 2-methylpropanal, mp 186–188°, by the procedure given in Section 14.1. Apply to the 2-methylpropanal the chromic acid in the acetone test described in Section 13.2.

   The preparation of the sodium bisulfite addition product of the aldehyde can be accomplished by adding 6 mL of saturated sodium bisulfite solution to 1 mL of the aldehyde contained in a 125-mL Erlenmeyer flask. Swirl the flask, and then allow the mixture to stand for 10 min. Reaction will be indicated by warming of the solution. Add 25 mL of 95% ethanol, swirl well, and then cool the solution in an ice-salt bath. Collect the sodium bisulfite addition product by vacuum filtration, washing the filter cake once with 95% ethanol and once with diethyl ether. You should *not* attempt to determine the melting point of the solid. The 2-methylpropanal can be regenerated by adding 5–10 mL of either 1 *M* sodium carbonate solution or dilute hydrochloric acid to the addition compound and gently warming the mixture.

Although sodium bisulfite addition compounds are generally not good derivatives for characterizing carbonyl compounds, they are often used to purify aldehydes, methyl ketones, and cyclic ketones such as cyclohexanone and cyclopentanone.

B.   Preparation of Cyclohexanone

Prepare a solution of chromic acid by dissolving 0.081 mol of potassium dichromate in 125 mL of water contained in a 250-mL Erlenmeyer flask and *carefully* adding, with swirling, 19 mL of concentrated sulfuric acid. Cool the resulting orange-red solution of chromic acid to room temperature, and then, in one portion, *add it to* a mixture of 0.20 mol of cyclohexanol and 75 mL of water in a 500-mL Erlenmeyer flask. Thoroughly mix the solutions by swirling, and determine the temperature of the reaction mixture. The mixture should quickly become warm; when it reaches a temperature of 55°, cool the flask in an ice-water bath or a pan of cold water so that a temperature of 55–60° is maintained. When the temperature of the solution no longer exceeds 60° on removal of the flask from the cooling bath, allow the flask to stand for 1 hr with occasional swirling.★

Transfer the reaction mixture to a 500-mL round-bottomed flask, add 100 mL of water and two or three boiling stones, and attach the flask to an apparatus set for fractional distillation. Distil the mixture until approximately 100 mL of distillate, which consists of an aqueous and an organic layer, has been obtained.★ Place the distillate in a separatory funnel, saturate the aqueous layer by adding sodium chloride (about 0.2 g of salt per mL of water will be required) and swirling to effect solution. Separate the layers, and extract the aqueous layer with 15 mL of dichloromethane. Combine this extract with the organic layer, and dry the solution over *anhydrous* magnesium sulfate.★ Filter the solution into a 50- or 100-mL flask, and equip the flask for simple distillation (Figure 2.2). Remove the low-boiling dichloromethane, and then continue the distillation, collecting cyclohexanone as a colorless liquid in the boiling range 152–155°. Weigh the product, and calculate the percent yield.

Following the procedure outlined in Section 14.1, prepare the 2,4-dinitrophenylhydrazone of cyclohexanone, mp 162–163°. Also apply to cyclohexanone the chromic acid test described in Section 13.2.

C.   Oxidation of Cyclohexanone to Hexanedioic Acid (Adipic Acid)

In a 500-mL Erlenmeyer flask combine 0.10 mol of cyclohexanone and a solution of 0.20 mol of potassium permanganate in 250 mL of water. To this mixture add 2 mL of 3 *M* sodium hydroxide solution, and note the temperature of the reaction mixture. The mixture will warm to 45° and should be held at that temperature by intermittent cooling with an ice- or cold-water bath. When the temperature no longer rises above 45° on removal of the flask from the cooling bath, allow the flask to stand for an additional 5–10 min (a drop in the temperature should occur), and then complete the reaction by heating the mixture at gentle reflux for a few minutes with a hot plate or a burner. Test for the presence of permanganate by placing a drop of the reaction mixture on a piece of filter paper; any unchanged permanganate will appear as a purple ring around the brown manganese dioxide. If permanganate remains, add *small* portions of *solid* sodium bisulfite to the reaction mixture until the spot test is negative.* Vacuum-filter the mixture through a pad of filter-aid,[1] thoroughly wash the brown filter cake with water,* and then concentrate the filtrate to a volume of about 65 mL by heating with a hot plate or a burner.* If the concentrate is colored, add decolorizing carbon, reheat the solution to boiling for a few minutes, and then filter. Carefully add concentrated hydrochloric acid to the filtrate until the solution tests acidic to litmus paper, and then add an additional 15 mL of acid.* Allow the solution to cool to room temperature, and isolate the precipitated hexanedioic acid by vacuum filtration. The acid is a white solid, mp 152–153°, and can be recrystallized from ethanol-water if necessary.

EXERCISES

1. What is the purpose of washing the steam distillate with sodium carbonate in the preparation of 2-methylpropanal?
2. Write the structure of the sodium bisulfite addition product of 2-methylpropanal.
3. Why is the chromic acid solution added *to* the alcohol, rather than the reverse, when the preparation of an aldehyde is being attempted?
4. Why are the two-phase distillates obtained in the oxidation of 2-methyl-1-propanol and of cyclohexanol first saturated with sodium chloride before separation of the organic layer?
5. Point out what modifications in procedure might be made in the oxidation of 2-methyl-1-propanol with chromic acid to *maximize* the yield of isobutyl isobutyrate.
6. Write out the balanced equation for the oxidation of cyclohexanol to cyclohexanone by potassium dichromate and aqueous sulfuric acid.
7. The costs of sodium dichromate dihydrate and potassium dichromate are $11.75/lb and $10.00/lb, respectively. Determine which reagent would be more economical to use for oxidation of 1 mol of cyclohexanol to cyclohexanone.
8. Why is the reaction mixture made alkaline in the oxidation of cyclohexanone to hexanedioic acid?
9. Write the balanced equation for the oxidation of cyclohexanone to hexanedioic acid.
10. Give the products to be expected on oxidation of 2-methylcyclohexanone with alkaline permanganate.

[1] The pad can be prepared by first making a slurry of 0.5 to 1 g of filter-aid in a few milliliters of water. The slurry is poured onto a filter paper contained in a Büchner funnel attached to a filter flask; vacuum is then slowly applied to remove the water from the slurry to leave a thin, even pad of filter-aid.

REFERENCES

1. *Oxidation in Organic Chemistry,* K. B. Wiberg, editor, Academic Press, New York, 1965, Vol. 5-A, Chapters 1 and 2.
2. *Oxidation,* R. L. Augustine, editor, M. Dekker, New York, 1969, Vol. 1, Chapters 1 and 2.

SPECTRA OF STARTING MATERIAL AND PRODUCTS

The ir and pmr spectra of cyclohexanone are given in Figures 9.6 and 9.7, respectively; those of cyclohexanol are given in Figures 6.23 and 6.24. Additionally, the ir and pmr spectra of 2-methyl-1-propanol are provided in Figure 4.13.

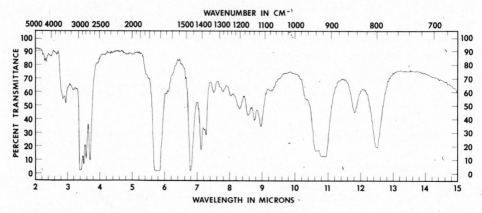

FIGURE 13.1   IR spectrum of 2-methylpropanal.

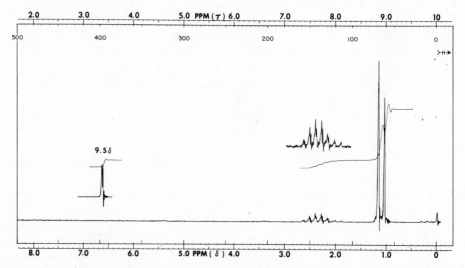

FIGURE 13.2   PMR spectrum of 2-methylpropanal.

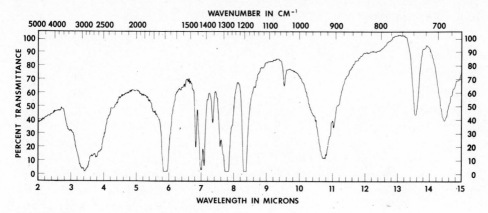

FIGURE 13.3 IR spectrum of hexanedioic acid.

13.2 Classification of Alcohols and Carbonyl Compounds by Means of Chromic Acid

The greater ease of oxidation of aldehydes than of ketones (Section 13.1) is the basis of several classical qualitative tests. Fehling's and Benedict's reagents contain complex salts (tartrate and citrate, respectively) of cupric ion as oxidizing agents, and Tollens' reagent consists of a silver diammine salt. The first two have been used as tests for aliphatic aldehydes (although they are more often used to detect reducing sugars), and the Tollens test has been widely employed to distinguish between aldehydes (both aliphatic and aromatic) and ketones.

Tertiary alcohols are much more resistant toward oxidation than primary and secondary alcohols. Potassium permanganate and certain other oxidizing agents have been used to distinguish tertiary alcohols from those of the other two types.

There are some disadvantages to all of these tests. A reagent that gives quick and definitive results is chromic acid, prepared by dissolving chromic anhydride in sulfuric acid (Section 13.1) and used in acetone solution. This reagent oxidizes primary and secondary alcohols and all aldehydes with a distinctive color change, and it gives no visible reaction with tertiary alcohols and ketones under the conditions of the test.

$$RCH_2OH \xrightarrow{H_2CrO_4} RCHO \xrightarrow{H_2CrO_4} RCO_2H$$

$$R_2CHOH \xrightarrow{H_2CrO_4} R_2CO \xrightarrow{H_2CrO_4}\!\!\!\!/\!/ \text{ no visible reaction}$$

$$R_3COH \xrightarrow{H_2CrO_4}\!\!\!\!/\!/ \text{ no visible reaction}$$

Thus the chromic acid reagent gives a clear-cut distinction between primary and secondary alcohols and aldehydes on the one hand and tertiary alcohols and ketones on the other. Aldehydes may be distinguished from primary and secondary alcohols by means of Tollens', Benedict's, or Fehling's test, and lower molecular weight primary and secondary alcohols may be differentiated on the basis of their rates of reaction with concentrated hydrochloric acid containing zinc chloride—the Lucas reagent (see Chapter 25, Alcohols, part A.2).

EXPERIMENTAL PROCEDURE

Place in a test tube 1 mL of *reagent grade* acetone (or solvent grade which has been distilled from potassium permanganate), and dissolve in it 1 drop of a liquid or about 10 mg of a solid alcohol or carbonyl compound. Add 1 drop of the acidic chromic anhydride reagent[2] to the acetone solution, and shake the tube to mix the contents. A positive oxidation reaction is indicated by disappearance of the orange color of the reagent and the formation of a green or blue-green precipitate or emulsion.

Primary and secondary alcohols and aliphatic aldehydes give a positive test within 5 sec. Aromatic aldehydes require 30 to 45 sec. Color changes occurring after about 1 min should not be interpreted as positive tests; other functional groups such as ethers and esters may slowly hydrolyze under the conditions of the test, releasing alcohols which in turn provide "false positive" tests. Tertiary alcohols and ketones produce no visible change in several minutes. Phenols and aromatic amines give dark precipitates, as do aromatic aldehydes having hydroxyl or amino groups on the aromatic ring.

Test a number of compounds of each type chosen from the following classes of compounds: primary alcohols, secondary alcohols, tertiary alcohols, aldehydes, ketones.

EXERCISES

1. Write the formula of the expected oxidation product from each alcohol or aldehyde tested.
2. What is the green precipitate formed in a positive test?
3. What structural feature do tertiary alcohols and ketones have in common, and what is the relationship of this feature to their resistance toward oxidation?
4. Why can the Lucas reagent *not* be used to distinguish between 1-octanol and 2-octanol? (Refer to Chapter 25, Alcohols, part A.2)

REFERENCES

1. F. G. Bordwell and K. M. Wellman, *Journal of Chemical Education,* **39,** 308 (1962).
2. J. D. Morrison, *Journal of Chemical Education,* **42,** 554 (1965).

[2]The chromic acid reagent is prepared as follows. Add 25 g of chromic anhydride (CrO_3) to 25 mL of concentrated sulfuric acid and stir until a smooth paste is obtained. Dilute the paste *cautiously* with 75 mL of distilled water, and stir until a clear orange solution is obtained.

13.3 Base-catalyzed Oxidation-Reduction of Aldehydes: The Cannizzaro Reaction

Aldehydes that have *no* hydrogen atoms on the carbon atom adjacent to the carbonyl group (the α-carbon atom) undergo mutual oxidation and reduction in the presence of strong alkali. Those which have hydrogens on the α-carbon atom undergo other reactions preferentially, as described in Section 14.2. Since aldehydes are intermediate in oxidation state between alcohols and carboxylic acids, it is not too surprising to find that this reaction, called the Cannizzaro reaction, occurs. The mechanism, which has been supported by considerable evidence, follows quite logically in view of the well-known ease of addition of nucleophiles to the carbonyl group. The first step explains the function of the strong basic catalyst:

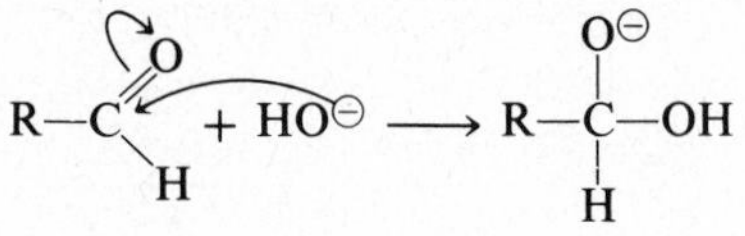

$$\text{R--C}\overset{O}{\underset{H}{}} + HO^{\ominus} \longrightarrow \text{R--}\overset{O^{\ominus}}{\underset{H}{\text{C}}}\text{--OH} \tag{19}$$

$$\text{R--}\overset{O^{\ominus}}{\underset{(H)}{\text{C}}}\text{--OH} + \text{R--C}\overset{O}{\underset{H}{}} \longrightarrow \text{R--C}\overset{O}{\underset{OH}{}} + \text{R--}\overset{O^{\ominus}}{\underset{H}{\text{C}}}\text{--H} \tag{20}$$

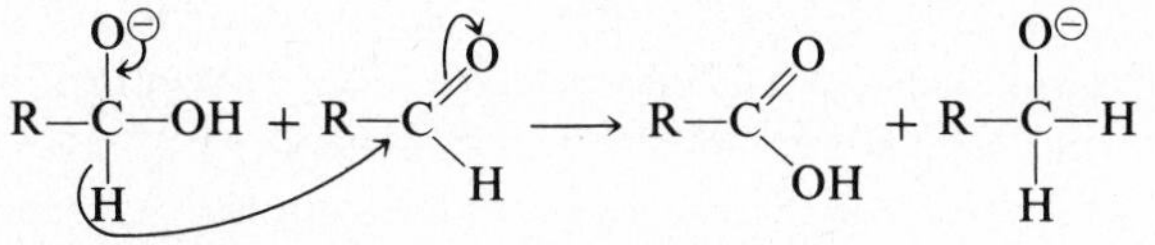

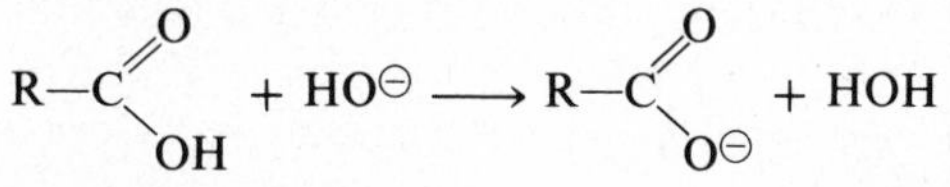

$$\text{R--C}\overset{O}{\underset{OH}{}} + HO^{\ominus} \longrightarrow \text{R--C}\overset{O}{\underset{O^{\ominus}}{}} + HOH \tag{21}$$

$$\text{R--}\overset{O^{\ominus}}{\underset{H}{\text{C}}}\text{--H} + HOH \longrightarrow \text{R--}\overset{OH}{\underset{H}{\text{C}}}\text{--H} + HO^{\ominus} \tag{22}$$

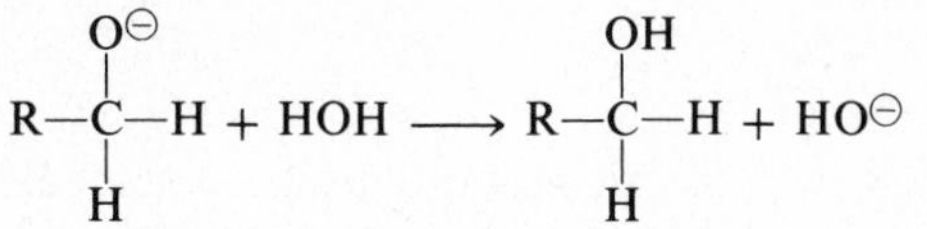

Summation of equations 19 through 22 gives the equation for the overall reaction:

$$2\,\text{RCHO} + HO^{\ominus} \longrightarrow RCO_2^{\ominus} + RCH_2OH \tag{23}$$

Aromatic aldehydes are the most common type that undergoes the Cannizzaro reaction, but formaldehyde and trisubstituted acetaldehydes also react in this way. Reaction between two *different* aldehydes of these types may occur, producing the acids and alcohols corresponding to each of the two aldehydes. Such reactions are referred to as *crossed-Cannizzaro reactions*.

EXPERIMENTAL PROCEDURE

Dissolve 2 g of solid potassium hydroxide in 2 mL of distilled water by swirling in a 100-mL beaker; cool the mixture to room temperature in a water bath. Put 2 mL of benzaldehyde in an 18 × 150-mm test tube, and add the concentrated potassium hydroxide solution to it. Cork the tube securely, and shake the mixture vigorously until an emulsion is formed. Allow the stoppered tube to stand in your desk until the next laboratory period. Crystallization should occur in the interim.

Add about 1 mL of water to the mixture, stopper the tube, and shake it. If all the crystals obtained do not dissolve, add a little water, break up the solid mass with a glass rod, stopper, and shake again. Repeat this procedure until all the solid is in solution. Pour the solution into a separatory funnel, and extract it three times with 10-mL portions of diethyl ether (shake the mixture *gently* to avoid forming an emulsion).

The ether solution may be dried over anhydrous magnesium sulfate and examined by gas chromatography[3] to determine the proportion of benzyl alcohol and unchanged benzaldehyde present. If the reaction is carried out on a larger scale, the ether solution may be shaken with aqueous sodium bisulfite solution to remove the benzaldehyde, and subsequently washed, dried, and distilled to yield benzyl alcohol as a high-boiling fraction, bp 205°. Alternatively, a crystalline derivative of benzyl alcohol may be prepared. The phenyl urethan (mp 78°) or 3,5-dinitrobenzoate (mp 113°) are suitable (see Chapter 25, Alcohols, parts B.1 and B.2).

Following its extraction with ether, pour the alkaline aqueous solution into a mixture of 5 mL of concentrated hydrochloric acid, 4 mL of water, and about 5 g of crushed ice, stirring vigorously. Cool the resultant mixture by placing its container in an ice-water bath, and then collect the crystals of benzoic acid.★ Dry the crystals and determine their melting point. Save some of this product for comparison with a product of an experiment in Chapter 14, page 318.

EXERCISES

1. When the Cannizzaro reaction is carried out in D_2O solution, no deuterium becomes attached to carbon in the alcohol produced. How does this support the mechanism given in equations 19–22?
2. By what means would aqueous sodium bisulfite remove unchanged benzaldehyde from the reaction mixture?
3. Write an equation for the reaction of benzaldehyde and formaldehyde with concentrated potassium hydroxide solution. Show all products.
4. How would propanal react with potassium hydroxide solution under the conditions of this experiment? Answer the same question for the reaction of 2,2-dimethylpropanal.

SPECTRA OF STARTING MATERIAL AND PRODUCTS

The ir and pmr spectra of benzoic acid are given in Figures 12.1 and 12.2.

---

[3]A 1.5-m column with 5% silicone gum rubber as the stationary phase is satisfactory. In one experiment about 20% of the original benzaldehyde was found present after 24 hr at room temperature.

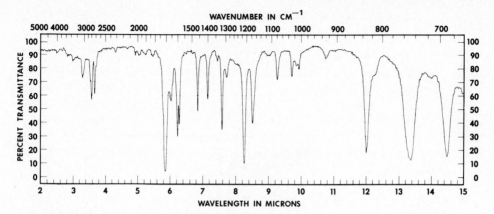

FIGURE 13.4 IR spectrum of benzaldehyde.

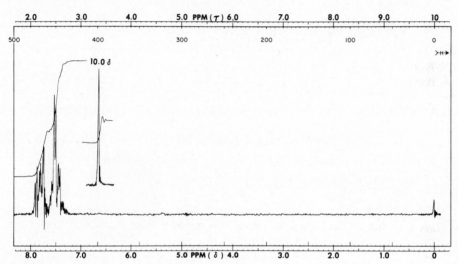

FIGURE 13.5 PMR spectrum of benzaldehyde.

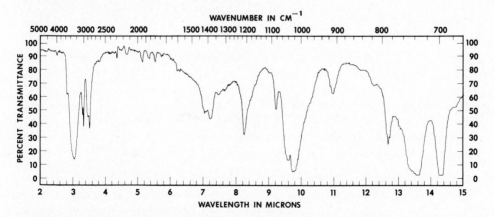

FIGURE 13.6 IR spectrum of benzyl alcohol.

14

SOME TYPICAL REACTIONS OF CARBONYL COMPOUNDS

14.1 Reactions Involving Nucleophilic Addition to the Carbonyl Group

The addition of a nucleophilic reagent (Nu—E) to a carbonyl compound may be written in the general form

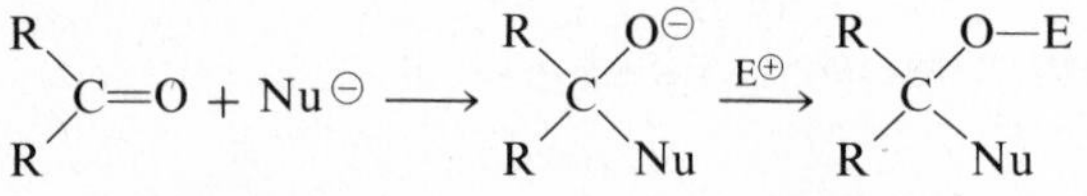

$$
\begin{array}{c}
R \\

\end{array}
C=O + Nu^{\ominus} \longrightarrow
\begin{array}{c}
R \quad O^{\ominus} \\
C \\
R \quad Nu
\end{array}
\xrightarrow{E^{\oplus}}
\begin{array}{c}
R \quad O-E \\
C \\
R \quad Nu
\end{array}
\tag{1}
$$

where Nu is a nucleophile and E is an electrophile. One of the most important examples of this reaction is the addition of Grignard reagents in which Nu is an alkyl group and E is MgX. In Section 12.2 experiments are described in which phenyl-magnesium bromide adds to carbon dioxide and methyl benzoate; the reactions of Grignard reagents with aldehydes, ketones, and various other carbonyl compounds also provide extremely valuable synthetic procedures.

Addition of Derivatives of Ammonia. Nitrogen compounds that may be considered to be derivatives of ammonia constitute other important types of nucleophilic reagents that add to carbonyl compounds (equation 2). In the case of *primary amines*

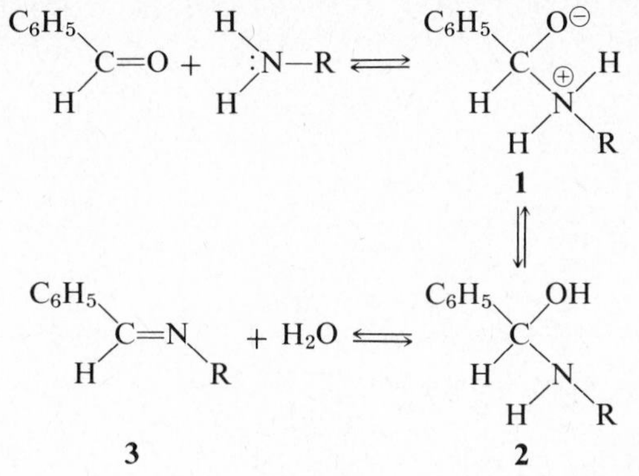

$$
\begin{array}{c}
\underset{H}{\overset{C_6H_5}{>}}C{=}O \;+\; \underset{H}{\overset{H}{>}}\!:\!N{-}R \;\rightleftharpoons\; \underset{H}{\overset{C_6H_5}{>}}\!\!\underset{\underset{H}{\overset{\oplus}{N}}\diagdown R}{C}\overset{O^{\ominus}}{\diagdown H} \\
\mathbf{1}
\end{array}
$$

$$\big\Updownarrow$$ (2)

$$
\underset{H}{\overset{C_6H_5}{>}}C{=}N\diagdown R \;+\; H_2O \;\rightleftharpoons\; \underset{H}{\overset{C_6H_5}{>}}\!\!\underset{\underset{H}{\overset{}{N}}\diagdown R}{C}\overset{OH}{\diagdown}
$$

$$\qquad \mathbf{3} \qquad\qquad\qquad\qquad \mathbf{2}$$

(R = alkyl or aryl) the addition product, **2**, readily loses a molecule of water to produce an *imine*, **3**. Although aliphatic amines give imines that are unstable and polymerize, aromatic amines give stable imines, also called Schiff bases, that can be isolated. These imines are readily hydrolyzed to regenerate the carbonyl compound; that is, the reaction of equation 2 is reversible.

Imines may undergo addition of a mole of hydrogen in the presence of a nickel catalyst to produce secondary amines (**4**, equation 3), a process that is analogous to hydrogenation of alkenes to alkanes and of carbonyl compounds to alcohols. It is not

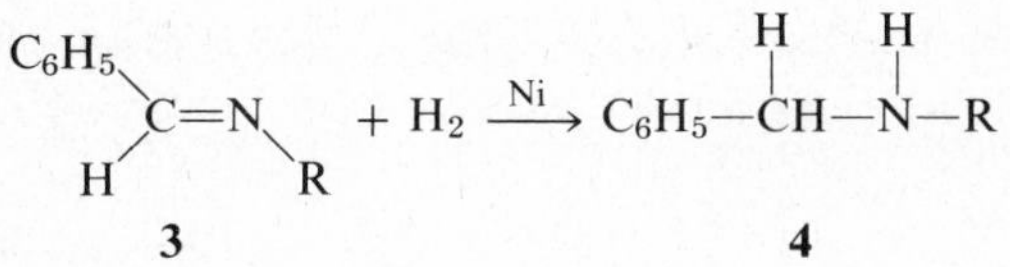

$$
\underset{H}{\overset{C_6H_5}{>}}C{=}N\diagdown R \;+\; H_2 \;\xrightarrow{\;Ni\;}\; \underset{}{C_6H_5{-}\overset{\overset{H}{|}}{C}H{-}\overset{\overset{H}{|}}{N}{-}R} \tag{3}
$$

$$\qquad \mathbf{3} \qquad\qquad\qquad\qquad\qquad \mathbf{4}$$

necessary to isolate the imine. In fact reaction mixtures from an aldehyde or ketone and an amine (either aromatic or aliphatic) or even ammonia may be subjected directly to catalytic hydrogenation to produce primary or secondary amines in good yields. Presumably imines are transitory intermediates in these reactions.

One of the limitations to this procedure for the synthesis of amines is that any other functional group in the imine that is sensitive to catalytic hydrogenation such as C=C, N=O (as in NO$_2$), or C—X (X = Cl, Br) may be reduced. This limitation has been removed by the discovery of the alkali metal hydrides, which are more selective reducing agents. For example, N-cinnamyl-*m*-nitroaniline (**9**) may be synthesized starting with cinnamaldehyde (**5**) and *m*-nitroaniline (**6**), using sodium borohydride as the reducing agent. Thus the carbon-nitrogen double bond of the intermediate imine (**7**) is reduced, but neither the carbon-carbon double bond nor the nitro group is affected (equations 4–6).

The reaction of sodium borohydride with the imine (equation 5) is analogous to the addition of a Grignard reagent to a carbonyl compound. A hydride ion (H$:^-$) is

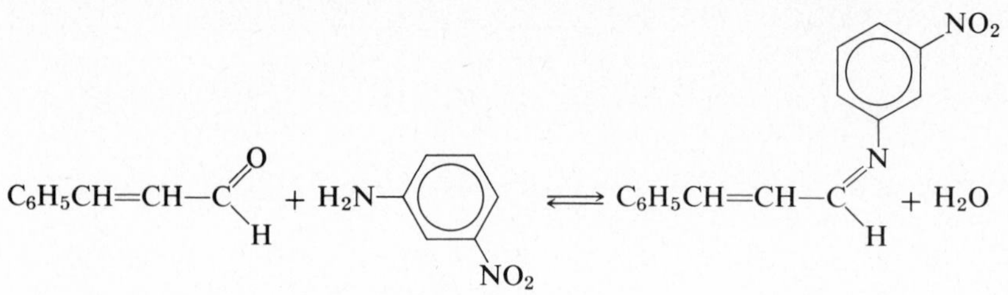

$$C_6H_5CH{=}CH{-}C{\overset{O}{\underset{H}{\diagup}}} + H_2N{-}\underset{NO_2}{\underset{\bigcirc}{}}{-}NO_2 \rightleftharpoons C_6H_5CH{=}CH{-}C{\overset{N{-}}{\underset{H}{\diagup}}} + H_2O \quad (4)$$

5   **6**   **7**

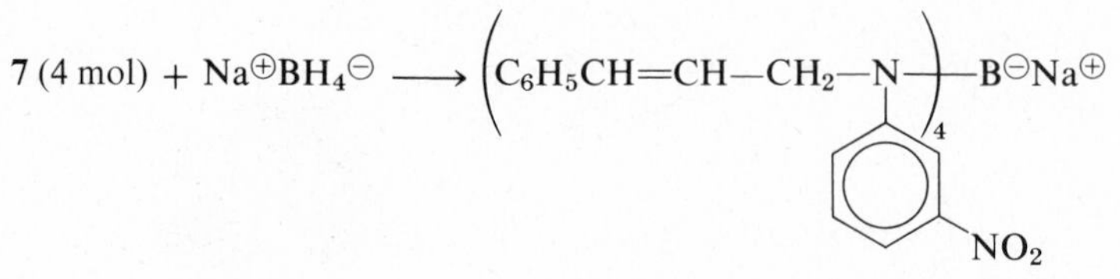

$$7\,(4\text{ mol}) + Na^{\oplus}BH_4^{\ominus} \longrightarrow \left(C_6H_5CH{=}CH{-}CH_2{-}N{-}\right)_4 B^{\ominus}Na^{\oplus} \quad (5)$$

8

$$8 + 3\,H_2O \longrightarrow 4\,C_6H_5CH{=}CH{-}CH_2{-}NH{-} + NaH_2BO_3 \quad (6)$$

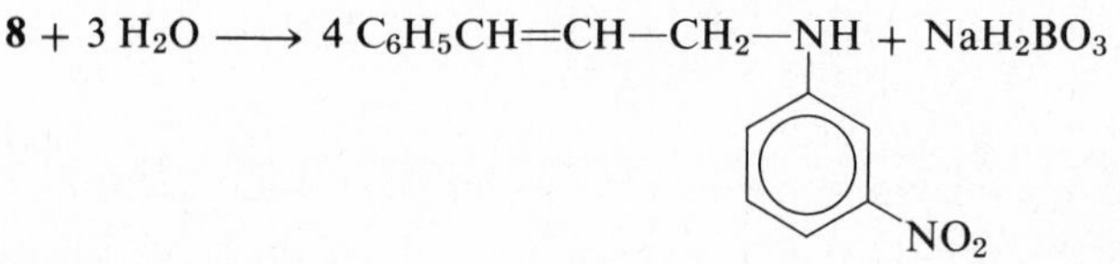

9

transferred from the borohydride anion (BH_4^-) to the electrophilic carbon of the carbon-nitrogen double bond, and the electron-deficient boron adds to nitrogen. All four hydrogens of the borohydride anion are transferred to imine carbons in this way, producing an organoboron anion, **8**, which is subsequently decomposed with water to yield the secondary amine (**9**, equation 6).

Although the formation of the imine is a reversible reaction (equation 4), it can be brought effectively to completion by removing the water from the reaction mixture as it is produced.[1] A convenient way to do this is to heat the carbonyl compound and primary amine in cyclohexane solution, allowing the cyclohexane to distil continuously. The cyclohexane forms a minimum-boiling azeotrope with the water and removes it as it is formed.

Sodium borohydride is conveniently used in methanol solution. Although it reacts slowly with the solvent, its rate of reaction with the imine is much faster. In small-scale reactions it is more convenient to use a sufficient excess of the borohydride to allow for its partial decomposition by reaction with the solvent than to use a less reactive solvent in which the borohydride is less soluble.

Other useful derivatives of ammonia that react with carbonyl compounds according to equation 2 are hydroxylamine (R = OH), semicarbazide (R = NHCONH$_2$), and various arylhydrazines (R = NHAr). The reaction of semicarbazide with carbonyl compounds to produce semicarbazones was presented in

[1]Compare with the similar problem encountered in esterification reactions, which are also equilibrium processes, Section 15.2.

Chapter 9 to illustrate the principle of kinetic and equilibrium control. However, the main importance of the reaction of carbonyl compounds with semicarbazide, hydroxylamine, and the arylhydrazines is that the products (semicarbazones, oximes, and arylhydrazones) are almost invariably crystalline solids, whereas the carbonyl compounds from which they are derived are often liquids. Arylhydrazines which are commonly employed in making crystalline derivatives of carbonyl compounds include phenylhydrazine, p-nitrophenylhydrazine, and 2,4-dinitrophenylhydrazine. The formation of a 2,4-dinitrophenylhydrazone, **10**, is represented by equation 7.

$$\text{R}_2\text{C}=\text{O} + \text{H}_2\text{NNH}-\text{C}_6\text{H}_3(\text{NO}_2)_2 \rightleftharpoons \text{R}_2\text{C}\underset{\text{NH}}{\overset{\text{OH}}{\Big|}} \rightleftharpoons \text{R}_2\text{C}=\text{N} + \text{H}_2\text{O} \qquad (7)$$

As the oximes, semicarbazones, and arylhydrazones are generally very insoluble in the reaction medium, they precipitate from solution, and the reversibility of the reaction therefore does not prevent their isolation in high yields.[2]

Directions for the preparation of semicarbazones, oximes, and 2,4-dinitro-phenylhydrazones are given in the Experimental Procedure section. The directions for semicarbazones are more generally applicable than those given in Chapter 9.

Addition of Wittig Reagents.   In 1953 G. Wittig discovered that certain organophosphorus compounds, called ylides, **13**, add to carbonyl compounds to form unstable intermediates called betaines, **14**, which decompose to produce an alkene and a phosphine oxide, **15**. The highly reactive ylide is generated in the presence of the carbonyl compound by the action of a strong base, usually phenyllithium, on a phosphonium halide, **12**. The latter compound is produced most commonly by reaction of a primary or secondary organic halide with triphenylphosphine (**11**, equation 8; R may be alkyl, aryl, CN, or $CO_2C_2H_5$, among others).

$$(\text{C}_6\text{H}_5)_3\text{P}: + \underset{\text{R}_2}{\overset{\text{R}_1}{\text{CH}-\text{Br}}} \longrightarrow (\text{C}_6\text{H}_5)_3\overset{\oplus}{\text{P}}-\underset{\text{R}_2}{\overset{\text{R}_1}{\text{CH}}} \text{ Br}^{\ominus} \qquad (8)$$

$$\text{11} \qquad\qquad\qquad\qquad \text{12}$$

$$\text{12} \xrightarrow{\text{C}_6\text{H}_5\text{Li}} \left[(\text{C}_6\text{H}_5)_3\text{P}=\underset{\text{R}_2}{\overset{\text{R}_1}{\text{C}}} \longleftrightarrow (\text{C}_6\text{H}_5)_3\overset{\oplus}{\text{P}}-\underset{\text{R}_2}{\overset{\ominus\quad\text{R}_1}{\text{C}}} \right] + \text{C}_6\text{H}_6 + \text{LiBr} \qquad (9)$$

$$\text{13}$$

[2]These reactions exhibit an interesting and important dependence of rate on the pH of the medium. Consult Chapter 9 for a discussion of this subject.

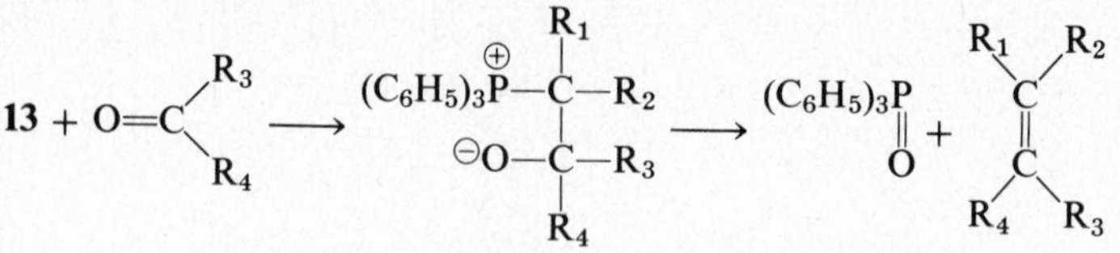

$$13 + O=C\begin{smallmatrix} R_3 \\ \\ R_4 \end{smallmatrix} \longrightarrow (C_6H_5)_3\overset{\oplus}{P}\begin{matrix} R_1 \\ | \\ -C-R_2 \\ | \\ \ominus O-C-R_3 \\ | \\ R_4 \end{matrix} \longrightarrow (C_6H_5)_3\underset{O}{\overset{\|}{P}} + \underset{R_4 \quad R_3}{\overset{R_1 \quad R_2}{C}} \qquad (10)$$

$$\qquad\qquad\qquad\qquad\qquad\qquad\qquad\quad \textbf{14} \qquad\qquad\qquad \textbf{15}$$

The electron-rich carbon is nucleophilic, as can be seen from the polarized resonance structure of the ylide, and it adds to a carbonyl group in the expected way to give the betaine. The decomposition of the betaine is thought to involve a four-center transition state; as the second C—C bond and the P—O bond are forming, the P—C and O—C bonds are breaking.

The overall effect of reactions 8–10 is conversion of a carbon-oxygen double bond to a carbon-carbon double bond, that is, $C=O \rightarrow C=C$. The Wittig synthesis thus represents a very general method of preparation of alkenes, one that has two important advantages over older procedures: (1) the carbonyl group is replaced *specifically* by a carbon-carbon double bond, without the formation of isomeric alkenes; and (2) the reactions are carried out in basic rather than in acidic media under very mild conditions.

Various modifications of the original Wittig synthesis have been made. One modification that is convenient for an introductory experiment (because it does not require an inert atmosphere and the use of the somewhat dangerous organolithium reagents) involves the use of triethyl phosphite (16) in place of triphenylphosphine. Activated organic halides (such as benzyl and allyl chlorides and α-bromo esters) will

$$(C_2H_5O)_3P + C_6H_5CH_2-Cl \longrightarrow (C_2H_5O)_3\overset{\oplus}{P}-CH_2C_6H_5 \quad Cl^{\ominus} \qquad (11)$$

$$\qquad\quad \textbf{16} \qquad\qquad\qquad\qquad\qquad\qquad \textbf{17}$$

$$17 \xrightarrow[\text{heat}]{} (C_2H_5O)_2\overset{\overset{O}{\|}}{P}-CH_2C_6H_5 + C_2H_5Cl \qquad (12)$$

$$\qquad\qquad\qquad\qquad \textbf{18}$$

$$18 + CH_3O^{\ominus}Na^{\oplus} \longrightarrow (C_2H_5O)_2\overset{\overset{O}{\|}}{P}-\overset{\ominus}{C}HC_6H_5 \; Na^{\oplus} + CH_3OH \qquad (13)$$

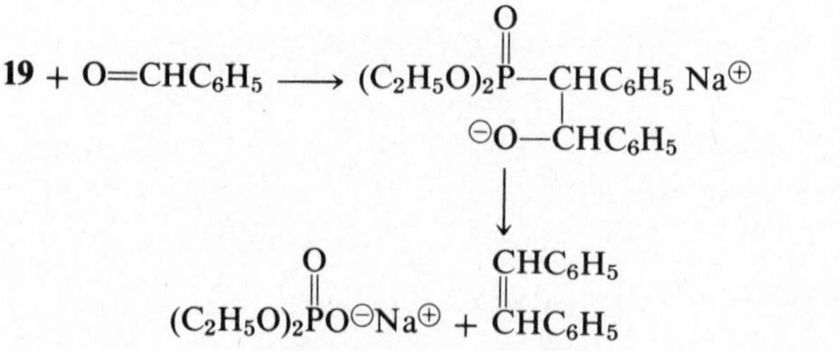

$$\qquad\qquad\qquad\qquad\qquad\qquad \textbf{19}$$

$$19 + O=CHC_6H_5 \longrightarrow (C_2H_5O)_2\overset{\overset{O}{\|}}{P}-CHC_6H_5 \; Na^{\oplus}$$

$$\qquad\qquad\qquad\qquad\qquad\qquad\qquad \overset{|}{\ominus}O-CHC_6H_5 \qquad\qquad (14)$$

$$\downarrow$$

$$(C_2H_5O)_2\overset{\overset{O}{\|}}{P}O^{\ominus}Na^{\oplus} + \overset{\overset{CHC_6H_5}{\|}}{C}HC_6H_5$$

$$\qquad\qquad \textbf{21} \qquad\qquad\qquad \textbf{20}$$

react satisfactorily with esters of phosphorous acid (equation 11). The phosphonium salt, **17**, produced is unstable toward heat; for example, the product from benzyl chloride evolves ethyl chloride and yields the phosphonate ester **18**. Treatment of this ester with sodium methoxide produces a nucleophile analogous to an ylide, and in the presence of benzaldehyde the reactions of equations 13 and 14 take place to yield an alkene (**20**, stilbene) and sodium diethylphosphate **(21)**. The latter is water-soluble and hence easily separated from the alkene.

There are two geometric isomers of stilbene: the *trans* isomer is a solid at room temperature, mp 126–127°, whereas the *cis* isomer is a liquid, mp 6°. The phosphonate ester modification of the Wittig reaction is reported to give the pure *trans* isomer, even though the original Wittig procedure (using triphenylphosphine as starting material) is reported to give a mixture of 30% *cis*-, 70% *trans*-stilbene. The ir, pmr, and uv spectra of the two isomers are distinctly different and may be used in both qualitative and quantitative analysis of mixtures.

Optional or Additional Experiments.   Other aldehydes may be used in place of benzaldehyde in the phosphonate modification of the Wittig reaction. With *trans*-cinnamaldehyde **(22)**, the product is *trans,trans*-1,4-diphenyl-1,3-butadiene **(23)**. If

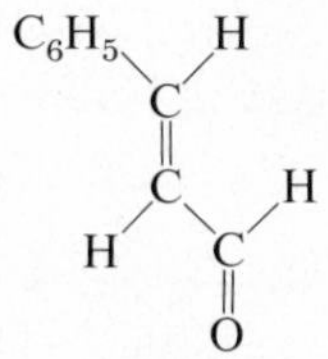

22 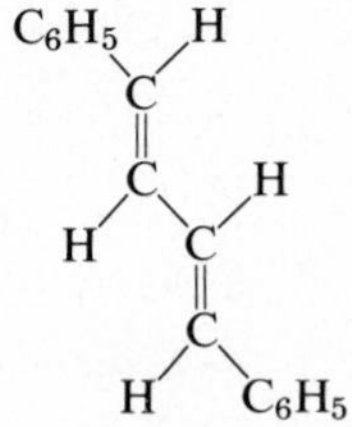   **23**

some students prepare this diene, it will be interesting to compare its uv absorption spectrum with that of stilbene to see the effect of the additional conjugated double bond. The uv spectra of *cis*- and *trans*-stilbene are provided as Figure 14.17; if it is not practical to record the uv spectrum of **23** in your laboratory, consult references 3 and 4 at the end of this section.

When 2-furaldehyde **(24)** is substituted for benzaldehyde, the product is *trans*-2-styrylfuran, **25**.

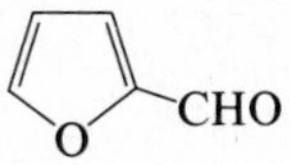

24 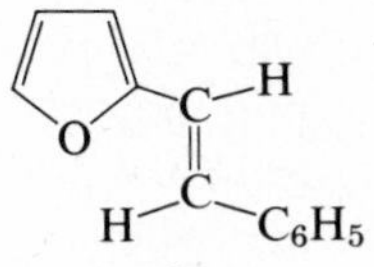   **25**

EXPERIMENTAL PROCEDURE

A. Addition of Derivatives of Ammonia

DO IT SAFELY
---

1. In those experiments in which methanolic sodium borohydride solutions are used, you should take care to avoid contact of the solution with your skin because it is highly caustic. Should you accidentally allow contact to occur, wash the area with copious quantities of cold water.

2. Flasks which contain methanolic sodium borohydride solution must *not* be stoppered. The solution evolves hydrogen gas, and dangerous buildup of pressure could occur in a stoppered flask.

1. Addition of Primary Amines to Produce Imines; Synthesis of Secondary Amines by Sodium Borohydride Reduction of Imines

a. Cinnamaldehyde and *m*-Nitroaniline. In a 100-mL round-bottomed flask place 0.022 mol of cinnamaldehyde, 0.020 mol of *m*-nitroaniline, 25 mL of cyclohexane, and three or four boiling stones. Set up the apparatus for simple distillation using a graduated cylinder as a receiver, and heat the reaction mixture on a steam bath or with a heating mantle. (If a burner must be used, take the proper precautions against the flammability of cyclohexane.) Distil until nearly all the cyclohexane has been removed; 22–24 mL of distillate should be obtained in 25–30 min.

During the distillation prepare a solution of 0.020 mol of sodium borohydride in 15 mL of methanol. This solution should be prepared in an *unstoppered* vessel no more than 5 to 10 min before it is used because of the reaction between sodium borohydride and methanol. When the distillation has been completed, remove the heat source, and either pour out about 0.5 mL of the residual liquid or, preferably, insert a micropipet and take a 0.5-mL sample of the liquid. Add 3–4 mL of methanol to this sample in a test tube, swirl to dissolve it, and then place the test tube in ice. (Read the directions of the next paragraph while the solution is cooling.) Collect any crystals which separate, dry them, and determine their melting point. The reported melting point of the imine, N-cinnamylidene-*m*-nitroaniline, is 92–93°. The imine may be purified by recrystallization from methanol.

Add 20 mL of methanol to the remainder of the residue from the distillation (crude imine). Attach a Claisen connecting tube equipped with a water-cooled reflux condenser and an addition funnel.[3] Pour the methanolic solution of sodium borohydride into the addition funnel (stopcock closed), and then add this solution to the imine solution dropwise, or in several portions, at a rate such that the addition is completed within 5 min. Swirl the reaction mixture while the addition is being made. After all of the borohydride solution has been added, heat the reaction mixture at reflux for 15 min.

---

[3] If a Claisen adapter is not available, the condenser may be attached directly to the flask and an addition funnel inserted in the top of the condenser through a *slotted* cork or rubber stopper. (The slot is to prevent a closed system.)

Cool the reaction mixture to room temperature, and pour it into 50 mL of water. Stir the mixture, and allow it to stand with occasional stirring for 10–15 min. Collect the orange crystals by suction filtration, and wash them with water.* Weigh the dry product, and calculate the yield. Determine the melting point of the product, N-cinnamyl-*m*-nitroaniline. It may be recrystallized from 95% ethanol; the melting point of the pure secondary amine is 106–107°.

b. Cinnamaldehyde and Aniline. Aniline may be substituted for *m*-nitroaniline, and the previous procedure followed with the following modification: after the reduction step has been completed, pour the reaction mixture into 50 mL of 3 *M* hydrochloric acid *instead of water*. The free secondary amine, N-cinnamylaniline, is a liquid at room temperature; for ease of handling, the product is converted by hydrochloric acid to the salt, which is only slightly soluble in cold water.

The melting point of the intermediate imine, N-cinnamylideneaniline (which may be isolated as described above for the corresponding imine from *m*-nitroaniline) is reported to be 109° (yellow leaflets from ethanol). The reported melting point of N-cinnamylanilinium chloride is 185°.

The picric acid salt (picrate) of N-cinnamylaniline (mp 137°) may be prepared as follows. Add 0.3 g of N-cinnamylanilinium chloride to 3 mL of 1.5 *M* sodium hydroxide solution and 2 mL of chloroform in a test tube. Shake the mixture, allow it to settle, and then remove the aqueous layer with a pipet. Add 2 mL of water, shake, and again remove the aqueous layer. Add a small amount of anhydrous magnesium sulfate to the chloroform solution, shake the mixture intermittently during 5–10 min, and then filter the solution from the magnesium sulfate.* Evaporate the chloroform solution, and dissolve the residue (the free secondary amine) in 2 mL of ethanol. Add 3 mL of a cold, saturated solution of picric acid, prepared by shaking excess picric acid with ethanol at 0° and filtering the mixture. Warm the ethanolic solution in a water bath at 100° for 5 min and then cool it in an ice-water bath. Collect the yellow crystals, and determine their melting point (reported 137°).

c. Benzaldehyde and *m*-Nitroaniline. The imine and the amine may be prepared from these reactants by the procedure used with cinnamaldehyde and *m*-nitroaniline. The melting point of the imine, N-benzylidene-*m*-nitroaniline (crystallized from methanol), is 70°. The melting point of N-benzyl-*m*-nitroaniline is 106°.

d. Benzaldehyde and Aniline. The imine and the amine–hydrochloric acid salt may be prepared from these reactants by the procedure used with cinnamaldehyde and aniline. The imine, N-benzylideneaniline, is more soluble in ethanol than the corresponding *m*-nitroaniline derivative, and its melting point is lower (54°), so it may be more difficult to crystallize. The reported melting point of N-benzylanilinium chloride is 214–216°.

2. The Preparation of Crystalline Derivatives of Carbonyl Compounds
a. Semicarbazones. Dissolve 0.5 g of semicarbazide hydrochloride[4] and 0.8 g of

---

[4] Neither semicarbazide nor hydroxylamine is stable as the free base, so both are usually stored in the form of the hydrochloric acid salt, or "hydrochloride."

sodium acetate in 5 mL of water in a test tube, and then add about 0.5 mL of the carbonyl compound. Stopper and shake the tube vigorously, remove the stopper, and place the test tube in a beaker of boiling water. Discontinue heating the water, and allow the test tube to cool to room temperature in the beaker of water. Remove the test tube to an ice-water bath, and scratch the side of the tube with a glass rod at the interface between the liquid and air. The semicarbazone may be recrystallized from water or aqueous ethanol.

If the carbonyl compound is insoluble in water, dissolve it in 5 mL of ethanol. Add water until the solution becomes turbid, then add a little ethanol until the turbidity disappears. Add the semicarbazide hydrochloride and sodium acetate, and continue as above from this point.

b. Oximes. Dissolve 0.5 g of hydroxylamine hydrochloride in 5 mL of water and 3 mL of 3 M sodium hydroxide solution, and then add 0.5 g of the aldehyde or ketone. If the carbonyl compound is insoluble in water, add just enough ethanol to give a clear solution. Warm the mixture on a steam bath (or boiling-water bath) for 10 min, and then cool it in an ice-water bath. If crystals do not form immediately, scratch with a glass rod the side of the container at and below the liquid level. The oxime may be recrystallized from water or aqueous ethanol.

In some cases the use of 3 mL of pyridine and 3 mL of absolute ethanol in place of the 3 mL of 3 M sodium hydroxide solution and 5 mL of water will be found to be more effective. A longer heating period is often necessary. After the heating is finished, pour the mixture into an evaporating dish, and remove the solvent with a current of air in a hood. Triturate (grind) the solid residue with 3–4 mL of cold water, and filter the mixture. Recrystallize the oxime from water or aqueous ethanol.

c. 2,4-Dinitrophenylhydrazones. If the reagent is not supplied, it is prepared by dissolving 1 g of 2,4-dinitrophenylhydrazine in 5 mL of concentrated sulfuric acid. This solution is then added, with stirring, to 7 mL of water and 25 mL of 95% ethanol. After being stirred vigorously, the solution is filtered from any undissolved solid.

Dissolve 1 or 2 drops of a liquid (or about 100 mg of a solid) in 2 mL of 95% ethanol, and add this solution to 2 mL of the 2,4-dinitrophenylhydrazine reagent. Shake the mixture vigorously; if a precipitate does not form immediately, let the solution stand for 15 min.

If more crystals are desired for a melting point, dissolve 200–500 mg of the carbonyl compound in 20 mL of 95% ethanol, and add this solution to 15 mL of the reagent. The product may be recrystallized from aqueous ethanol.

B. Addition of Wittig Reagents

DO IT SAFELY

---

1. Organophosphorus compounds are toxic, and benzyl chloride is irritating and lachrymatory. You should avoid contact of these substances with your skin and inhalation of their vapors. We recommend that experiments using these reagents be performed in the hood if possible and that rubber gloves be worn, particularly when measuring out the reagents. Should these substances come into contact with your skin, wash the area thoroughly with soap and warm water, and rinse with large quantities of water.

    2.   Carbon disulfide is highly volatile and extremely flammable. Avoid its use near open flames or hot plates.

1. *trans*-**Stilbene.** Weigh 0.045 mol of triethyl phosphite into a 25- or 50-mL round-bottomed flask, and add 0.045 mol of benzyl chloride and several small boiling stones. Connect a water-cooled condenser to the flask, and heat the mixture at gentle reflux for 1 hr. While this reaction mixture is cooling to room temperature, place 0.045 mol of sodium methoxide in a 125-mL Erlenmeyer flask, and immediately add 20 mL of N,N-dimethylformamide. Pour the *cool* phosphonate reaction mixture into the Erlenmeyer flask, and rinse the round-bottomed flask with 20 mL of N,N-dimethyl-formamide, adding the rinse solution to the rest of the reaction mixture. Cool the Erlenmeyer flask and its contents in an ice-water bath, stirring with a thermometer until the temperature of the solution is below 20°. While continuing to stir the cooled solution, slowly add 0.045 mol of benzaldehyde, cooling as necessary so that the temperature of the mixture does not rise above 35°. Remove the flask from the cooling bath, and allow it to stand at room temperature for about 10 min. Add 15 mL of water with stirring, collect the crystals on a Büchner funnel, and wash them with cold 1:1 methanol-water. Determine the weight and the melting point of the crystalline product. A mixture of ethanol and ethyl acetate may be used to recrystallize impure stilbene. Submit a sample of the product in carbon disulfide or carbon tetrachloride solution for infrared analysis, and compare the spectrum with those of Figures 14.13 and 14.14.

2. **1,4-Diphenyl-1,3-butadiene.** This diene may be prepared by following the procedure above up to the point of addition of benzaldehyde. Use 0.04 mol of cinnamaldehyde in place of benzaldehyde, and continue to follow the procedure up to the point of collecting the crystals. Wash the crystals first with water and then with methanol until the filtrate is colorless. Weigh the product and determine its melting point. The melting point of *trans,trans*-1,4-diphenyl-1,3-butadiene is reported to be 150–151°.

3. *trans*-**2-Styrylfuran.** Follow the procedure for *trans*-stilbene exactly except substitute an equimolar amount of freshly distilled 2-furaldehyde for benzaldehyde. The melting point of *trans*-2-styrylfuran is reported to be 54–55°. This product may be recrystallized from methanol.

EXERCISES

ADDITION OF DERIVATIVES OF AMMONIA

  1. What causes the turbidity in the distillate collected during the heating of the carbonyl compound and primary amine in cyclohexane solution?

2. Why was the use of *dry* cyclohexane not specified?
3. Although 95% ethanol is a satisfactory solvent for recrystallization of N-cinna-myl-*m*-nitroaniline, it is not suitable for recrystallization of the corresponding imine. Why?
4. Note the molar proportion of $NaBH_4$ to imine used in your experiment. How does this compare with the stoichiometric requirement? Account for the proportion used.
5. Sodium borohydride may be "stabilized" toward reaction with solvent methanol by addition of sodium hydroxide. Explain.
6. In Figures 14.3 and 14.4, assign as many of the ir absorption peaks to structural components of the molecules as you can. Which peaks give evidence of the conversion of $-\overset{|}{C}=N-$ to $-\overset{|}{\underset{H}{C}}-\overset{|}{\underset{H}{N}}-$ and of retention of the carbon-carbon double bond in the secondary amine?
7. Can you identify the ir absorptions of the NO_2 group by comparing Figures 14.3 and 14.4 with Figure 14.6? Explain the basis of your identification.
8. By comparing the pmr spectra of Figures 14.5 and 14.11, identify the vinyl hydrogen $(-CH=CH-)$ signals in Figure 14.5 and the methylene $(-CH_2-)$ signals in both Figures 14.5 and 14.11.
9. By calculating the integrations in Figures 14.5, 14.8, and 14.11 and, particularly, by considering the signals of Figure 14.8, see if you can discover the "hiding place" of the elusive N—H signal.
10. There are two NH_2 groups in the semicarbazide molecule, yet only one of them is nucleophilic. Note which one this is, and explain.
11. The melting points of 2,4-dinitrophenylhydrazones are generally higher than those of the corresponding phenylhydrazones. Suggest an explanation for this behavior.
12. How does scratching the side of a container with a glass rod help to induce crystallization?

ADDITION OF WITTIG REAGENTS

1. Compare the mechanism of the addition of Grignard and Wittig reagents to the carbonyl group, pointing out similarities and differences.
2. Write equations for the preparation of the following alkenes by the original Wittig synthesis from triphenylphosphine and the necessary organic halides and carbonyl compounds:
    (a) $C_6H_5CH=C(CH_3)_2$
    (b) $C_6H_5C(CH_3)=CHCH_3$
    (c) $CH_3CH_2C(CH_3)=CH_2$
3. Write equations for Wittig syntheses of the compounds of Exercise 2 using alternative organic halides and carbonyl compounds.
4. Write equations for the preparation of the following alkenes by the phosphonate ester modification (starting with triethyl phosphite):

(a) $C_6H_5C(CH_3)\!\!=\!\!C(CH_3)C_6H_5$

(b) $CH_2\!\!=\!\!CH\!-\!CH\!\!=\!\!CH\!-\!C_6H_5$

(c) $(CH_3)_2C\!\!=\!\!CH\!-\!CO_2C_2H_5$

5. Would you expect an ylide, **13**, in which R = CN, to be more or less stable than one in which R = alkyl?

6. Why should the sodium methoxide not be exposed to the atmosphere for more than a few minutes?

7. Why should the aldehydes used as starting materials in Wittig syntheses be free of carboxylic acids?

8. What peaks in the ir spectra of the stilbenes (Figures 14.13 and 14.14) would be most useful for quantitative analysis of a mixture of *cis* and *trans* isomers?

REFERENCES

1. E. J. Seus and C. V. Wilson, *Journal of Organic Chemistry,* **26,** 5243 (1961).

2. G. Wittig and U. Schollkopf, *Chemische Berichte,* **87,** 1318 (1954).

3. J. L. Bills and C. R. Noller, *Journal of the American Chemical Society,* **70,** 958 (1948).

4. J. H. Pinckard, B. Willie, and L. Zechmeister, *Journal of the American Chemical Society,* **70,** 1939 (1948).

5. A. Maercker, in *Organic Reactions,* A. C. Cope, editor, John Wiley & Sons, New York, 1965, Vol. 14, Chapter 3. A good general review.

SPECTRA OF STARTING MATERIALS AND PRODUCTS

The ir, pmr, and uv spectra of 2-furaldehyde are given in Figures 9.3, 9.4, and 9.5. The ir and pmr spectra of benzaldehyde are provided in Figures 13.4 and 13.5. The ir and pmr spectra of aniline are given in Figures 15.4 and 15.5.

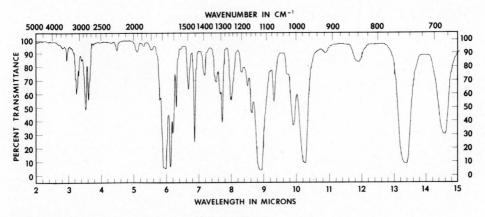

FIGURE 14.1 IR spectrum of cinnamaldehyde.

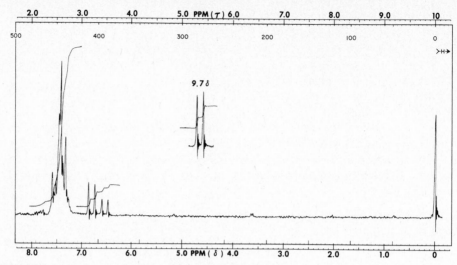

FIGURE 14.2   PMR spectrum of cinnamaldehyde.

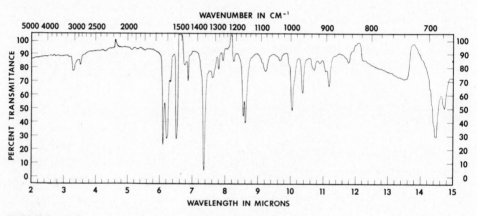

FIGURE 14.3   IR spectrum of N-cinnamylidene-*m*-nitroaniline (in CCl₄ solution).

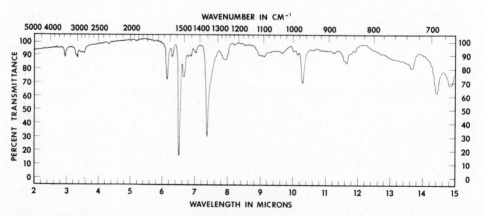

FIGURE 14.4   IR spectrum of N-cinnamyl-*m*-nitroaniline (in CCl₄ solution).

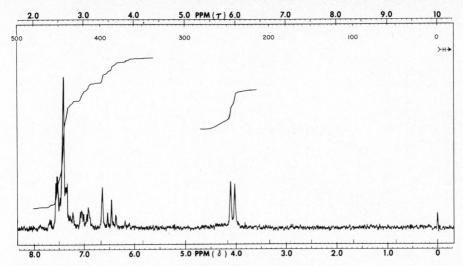

FIGURE 14.5 PMR spectrum of N-cinnamyl-*m*-nitroaniline.

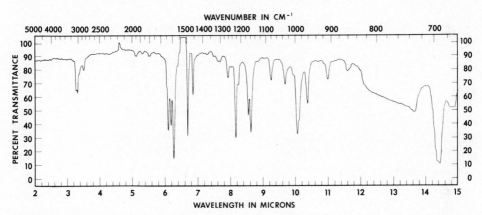

FIGURE 14.6 IR spectrum of N-cinnamylideneaniline (in CCl_4 solution).

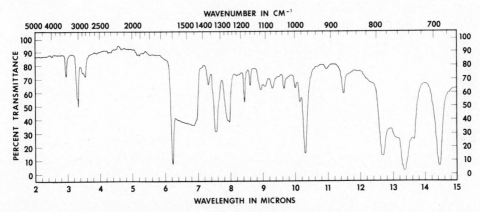

FIGURE 14.7 IR spectrum of N-cinnamylaniline (in CS_2 solution).

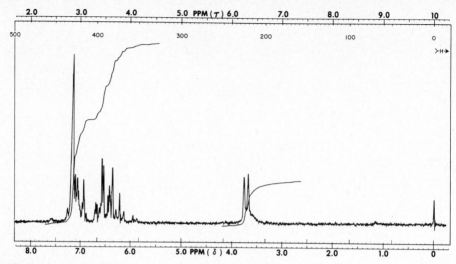

FIGURE 14.8   PMR spectrum of N-cinnamylaniline.

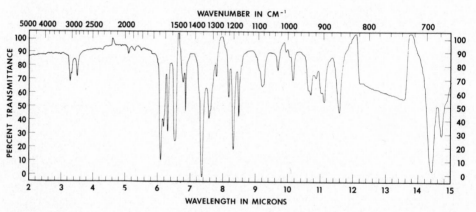

FIGURE 14.9   IR spectrum of N-benzylidene-*m*-nitroaniline (in CCl_4 solution).

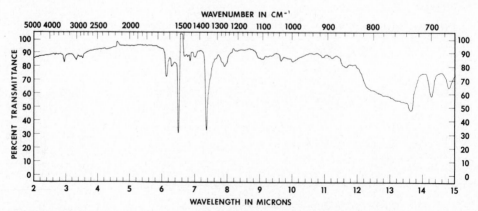

FIGURE 14.10   IR spectrum of N-benzyl-*m*-nitroaniline (in CCl_4 solution).

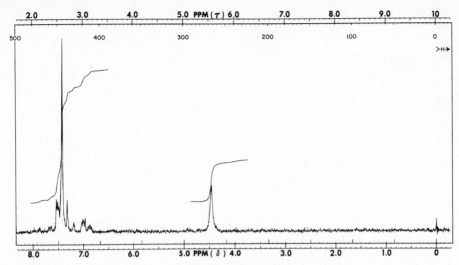

FIGURE 14.11 PMR spectrum of N-benzyl-*m*-nitroaniline.

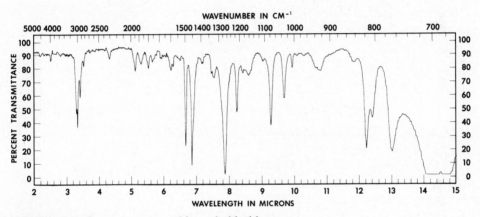

FIGURE 14.12 IR spectrum of benzyl chloride.

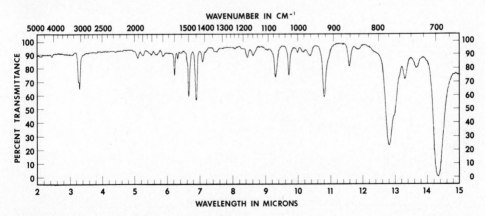

FIGURE 14.13 IR spectrum of *cis*-stilbene.

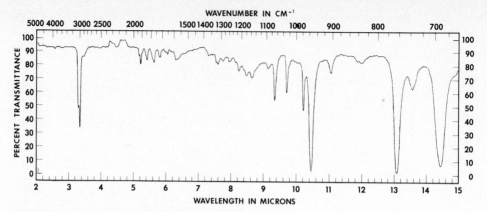

FIGURE 14.14 IR spectrum of *trans*-stilbene.

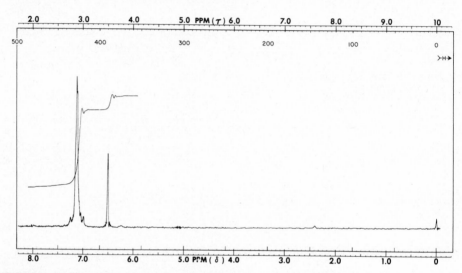

FIGURE 14.15 PMR spectrum of *cis*-stilbene.

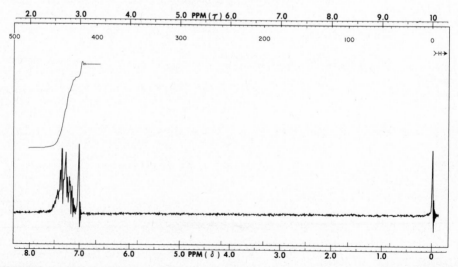

FIGURE 14.16 PMR spectrum of *trans*-stilbene.

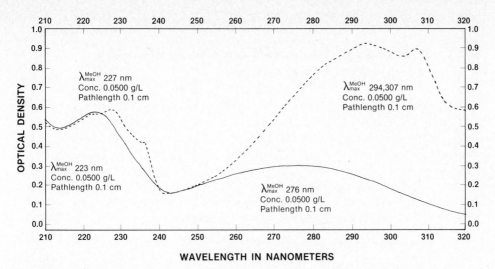

FIGURE 14.17   UV spectra of *cis*- (solid line) and *trans*- (dashed line) stilbene.

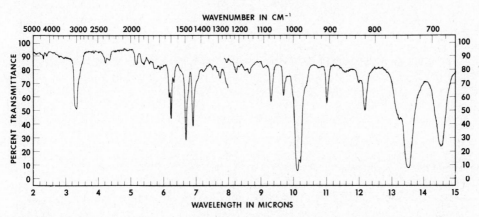

FIGURE 14.18   IR spectrum of *trans,trans*-1,4-diphenyl-1,3-butadiene.

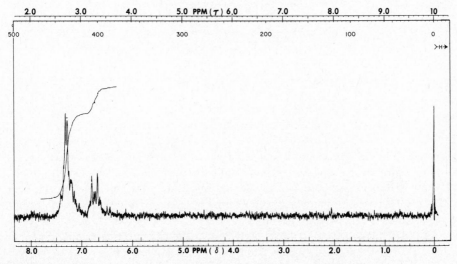

FIGURE 14.19   PMR spectrum of *trans,trans*-1,4-diphenyl-1,3-butadiene.

14.2 Reactions of Carbonyl Compounds Involving the α- and β-Carbon Atoms

The Haloform Reaction. The reactions studied in Chapter 13 and in the preceding sections of this chapter all take place at the carbonyl group, that is, oxidation to a carboxyl group, reduction to an alcohol group, or addition of various reactants across the carbon-oxygen π-bond. The nature of the carbonyl group also affects its neighboring α- and β-carbon atoms. The partial positive charge on the carbonyl carbon tends to make the hydrogens on the α-carbons (the "α-hydrogens") easily removed as protons, and the resulting carbanion is stabilized by resonance with the carbonyl π-electron system to produce what is termed an *enolate anion* (**26**, equation 15).

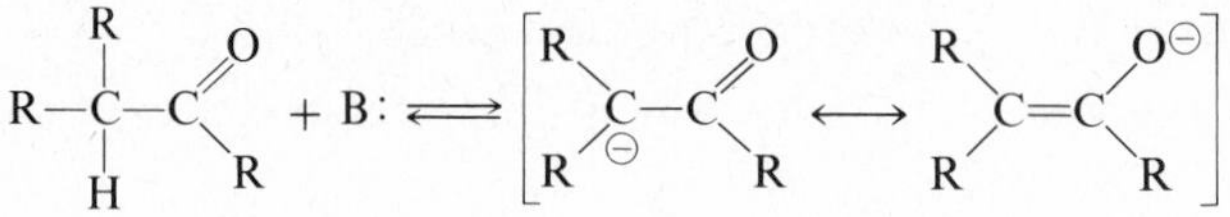

$$
R-\underset{\underset{H}{|}}{\overset{\overset{R}{|}}{C}}-\overset{O}{\underset{R}{C}} + B: \rightleftharpoons \left[\underset{R}{\overset{R}{C^{\ominus}}}-\overset{O}{\underset{R}{C}} \longleftrightarrow \underset{R}{\overset{R}{C}}=\overset{O^{\ominus}}{\underset{R}{C}} \right] + BH^{\oplus} \qquad (15)
$$

26

When an aldehyde or ketone that has α-hydrogens is treated with a halogen in basic medium, the enolate anion reacts rapidly with the halogen (equation 16). In the

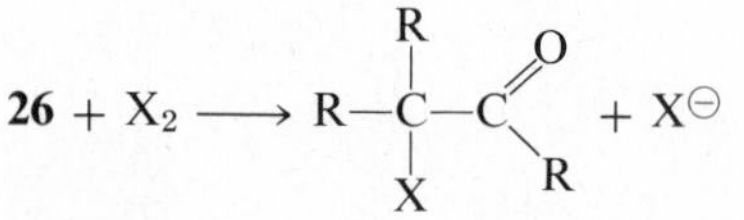

$$
26 + X_2 \longrightarrow R-\underset{\underset{X}{|}}{\overset{\overset{R}{|}}{C}}-\overset{O}{\underset{R}{C}} + X^{\ominus} \qquad (16)
$$

case of acetaldehyde or a methyl ketone, all three of the α-hydrogens of the methyl group are replaced by halogen to give **27** (equation 17, R = H or alkyl).

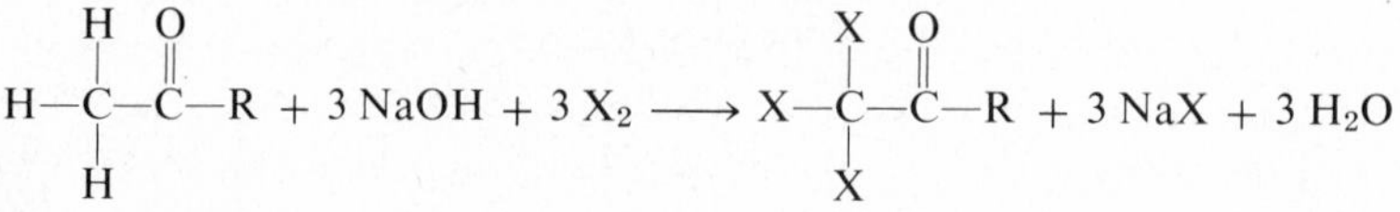

$$
H-\underset{\underset{H}{|}}{\overset{\overset{H}{|}}{C}}-\overset{O}{\overset{\|}{C}}-R + 3\,NaOH + 3\,X_2 \longrightarrow X-\underset{\underset{X}{|}}{\overset{\overset{X}{|}}{C}}-\overset{O}{\overset{\|}{C}}-R + 3\,NaX + 3\,H_2O \qquad (17)
$$

27

When the first α-hydrogen is replaced by halogen, the remaining hydrogens attached to this α-carbon atom become more acidic because of the inductive effect of the halogen, so that further substitution by halogen at this site occurs more rapidly than at the other α-carbon atom. The inductive effect of the three halogens makes the carbon atom of the carbonyl group particularly susceptible to nucleophilic addition of hydroxyl. The intermediate adduct **28** readily undergoes cleavage, as shown in equation 18, and the fragments **29** and **30** are immediately converted to the observed products, a haloform, **31**, and a carboxylic acid salt, **32**, as shown in equations 19 and 20. (**31** and **32** may also be formed simply by transfer of a proton from **30** to **29**). The

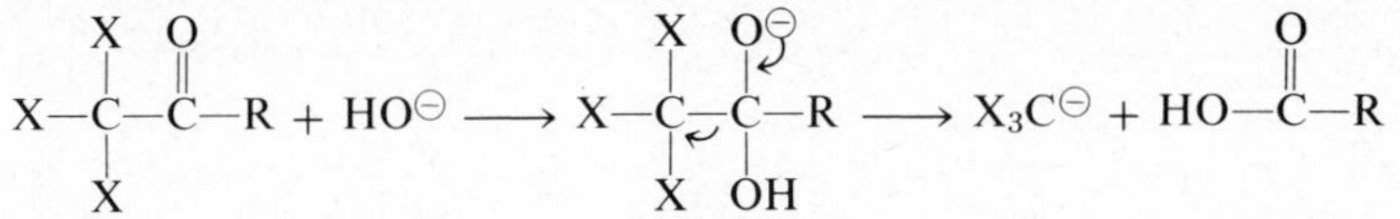

$$X-\overset{\overset{\displaystyle X}{|}}{\underset{\underset{\displaystyle X}{|}}{C}}-\overset{\overset{\displaystyle O}{\|}}{C}-R + HO^{\ominus} \longrightarrow X-\overset{\overset{\displaystyle X}{|}}{\underset{\underset{\displaystyle X}{|}}{C}}-\overset{\overset{\displaystyle O^{\ominus}}{|}}{\underset{\underset{\displaystyle OH}{|}}{C}}-R \longrightarrow X_3C^{\ominus} + HO-\overset{\overset{\displaystyle O}{\|}}{C}-R \qquad (18)$$

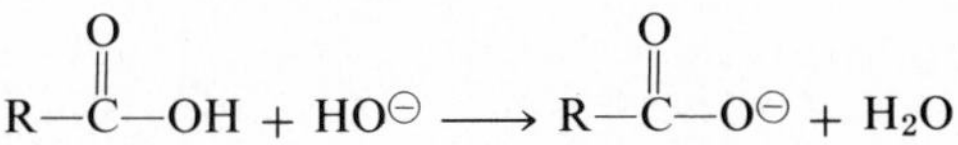

$$\mathbf{28}\mathbf{29}\mathbf{30}$$

$$X_3C^{\ominus} + H_2O \longrightarrow HCX_3 + HO^{\ominus} \qquad (19)$$
$$\mathbf{31}$$

$$R-\overset{\overset{\displaystyle O}{\|}}{C}-OH + HO^{\ominus} \longrightarrow R-\overset{\overset{\displaystyle O}{\|}}{C}-O^{\ominus} + H_2O \qquad (20)$$
$$\mathbf{32}$$

overall equation for the reaction of a methyl ketone, or acetaldehyde (R = H), is given by equation 21.

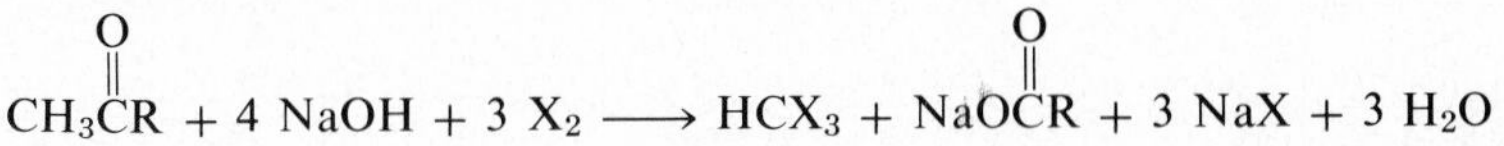

$$CH_3\overset{\overset{\displaystyle O}{\|}}{C}R + 4\ NaOH + 3\ X_2 \longrightarrow HCX_3 + Na O\overset{\overset{\displaystyle O}{\|}}{C}R + 3\ NaX + 3\ H_2O \qquad (21)$$

Although all carbonyl compounds having hydrogens on the α-carbons undergo halogenation at these positions, only those with *methyl* groups adjacent to the carbonyl group undergo carbon-carbon cleavage, presumably because *three* halogens attached to one carbon atom are required to weaken the bond to the point of rupture. Thus, although nearly all carbonyl compounds react with halogens in basic media, only those with α-methyl groups produce haloforms and carboxylic acid salts. This fact is exploited in two important ways. (1) Since iodoform (HCI$_3$) is a highly insoluble crystalline yellow solid with a characteristic odor, it is readily detected, and its formation is used as a *qualitative test* for the structural moieties shown here

$$CH_3-\overset{\overset{\displaystyle O}{\|}}{C}-R \qquad CH_3-\overset{\overset{\displaystyle OH}{|}}{C}H-R$$

(R = H, alkyl, or aryl). *Alcohols* of this structural type give a positive test because of the oxidizing power of the reagent, which partially converts them to the corresponding carbonyl compounds. (2) The conversion of a methyl ketone to a carboxylic acid with one less carbon atom is often useful in synthesis. In this case chlorine is the halogen of choice because it is the least expensive and most readily available.

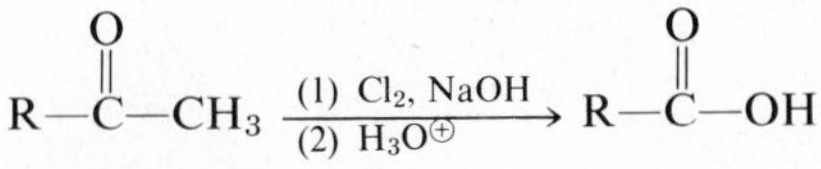

$$R-\overset{\overset{\displaystyle O}{\|}}{C}-CH_3 \xrightarrow[\text{(2) }H_3O^{\oplus}]{\text{(1) }Cl_2,\ NaOH} R-\overset{\overset{\displaystyle O}{\|}}{C}-OH \qquad (22)$$

Aldol Additions and Condensations. Another important and general reaction of enolate anions (besides reaction with halogens) is addition to the carbonyl group of the aldehyde or ketone from which the enolate anion was derived. In this way a dimeric anion, **33**, is produced (equation 23), which stabilizes itself by abstracting a proton from the solvent (water or alcohol), as shown in equation 24. The β-hydroxy

$$\left[\begin{array}{c} \diagup \\ C=C \\ \diagup^{\ominus} \end{array} \begin{array}{c} O \\ \parallel \\ C \\ \diagdown H \end{array} \longleftrightarrow \begin{array}{c} \diagup \\ C=C \\ \diagdown \end{array} \begin{array}{c} O^{\ominus} \\ \diagup \\ C \\ \diagdown H \end{array} \right] + \begin{array}{c} | \\ -C-C \\ | \end{array} \begin{array}{c} O \\ \parallel \\ \diagdown H \end{array} \longrightarrow \begin{array}{c} O^{\ominus} \\ | \\ -C-C-C-C \\ | \quad | \end{array} \begin{array}{c} O \\ \parallel \\ \diagdown H \end{array} \qquad (23)$$

26 \qquad\qquad\qquad\qquad 33

$$33 + H_2O \longrightarrow \begin{array}{c} OH \\ | \\ -C-C-C-C \\ | \quad | \end{array} \begin{array}{c} O \\ \diagup \\ \diagdown H \end{array} + HO^{\ominus} \qquad (24)$$

34

carbonyl compound of type **34** is called an "aldol," since it is both an aldehyde and an alcohol. The term "aldol *addition*" is also applied generally to the base-catalyzed self-addition of ketones as well as of aldehydes. The overall change for the reaction of acetaldehyde is given in equation 25. Most β-hydroxy aldehydes and ketones undergo dehydration readily to α,β-unsaturated aldehydes and ketones (equation 26). In this

$$2 \; CH_3CHO \xrightarrow{HO^{\ominus}} CH_3\overset{\displaystyle OH}{\underset{\displaystyle |}{CH}}-CH_2CHO \qquad (25)$$

$$CH_3\overset{\displaystyle OH}{\underset{\displaystyle |}{CH}}-CH_2CHO \longrightarrow CH_3CH=CHCHO + H_2O \qquad (26)$$

case the overall reaction may be referred to as an "aldol *condensation*," since a molecule of water is eliminated from the adduct.

In general ketones do not undergo self-addition as readily as aldehydes; in fact special conditions must usually be employed to obtain good yields in such reactions. (Refer to a textbook for a discussion of the reaction of acetone, for example). "Mixed (or crossed) aldol condensations" of two different aldehydes or of an aldehyde with a ketone are possible. Such mixed condensations are practical for synthesis in the case of (1) two aldehydes *if* one of the aldehydes has no α-hydrogens (and hence only serves as a carbonyl acceptor of the enolate anion derived from the other aldehyde) and (2) a ketone and aldehyde under conditions such that the ketone undergoes no appreciable self-condensation. A good example of the latter case is the reaction of benzaldehyde with acetophenone in the presence of dilute sodium hydroxide solution (equations 27–30). The product, **38**, is called benzalacetophenone. Under the condi-

$$C_6H_5\overset{\displaystyle O}{\overset{\displaystyle \parallel}{C}}-CH_3 + HO^{\ominus} \xrightarrow{-H_2O} \left[C_6H_5\overset{\displaystyle O}{\overset{\displaystyle \parallel}{C}}-\overset{\displaystyle \ominus}{C}H_2 \longleftrightarrow C_6H_5\overset{\displaystyle O^{\ominus}}{\overset{\displaystyle \parallel}{C}}=CH_2 \right] \qquad (27)$$

35

$$35 + C_6H_5\overset{\displaystyle O}{\overset{\displaystyle \parallel}{C}}H \longrightarrow C_6H_5\overset{\displaystyle O^{\ominus}}{\underset{\displaystyle |}{C}}H-CH_2\overset{\displaystyle O}{\overset{\displaystyle \parallel}{C}}C_6H_5 \qquad (28)$$

36

$$\textbf{36} + H_2O \longrightarrow \underset{\textbf{37}}{C_6H_5\overset{\underset{|}{OH}}{C}H-CH_2\overset{\underset{\|}{O}}{C}C_6H_5} + HO^{\ominus} \qquad (29)$$

$$\textbf{37} \longrightarrow \underset{\textbf{38}}{C_6H_5CH{=}CH\overset{\underset{\|}{O}}{C}C_6H_5} + H_2O \qquad (30)$$

tions of the experiment, the dehydration of the "aldol," **37**, is spontaneous (equation 30).

Reactions of Benzalacetophenone. As expected, benzalacetophenone gives reactions typical of both the ketone function and the alkenic double bond. For example, a semicarbazone and a 2,4-dinitrophenylhydrazone may be obtained by the procedures of Section 14.1.

Bromine adds readily to the carbon-carbon double bond. Since two asymmetric carbon atoms are produced in this reaction, there should be four possible stereoisomers of the dibromide, two enantiomeric pairs (see Chapter 19).

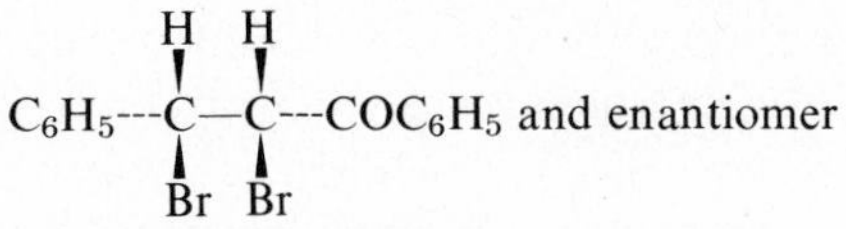

$$C_6H_5{-}{-}{-}\overset{\overset{\displaystyle H}{|}}{\underset{\underset{\displaystyle Br}{|}}{C}}{-}\overset{\overset{\displaystyle H}{|}}{\underset{\underset{\displaystyle Br}{|}}{C}}{-}{-}{-}COC_6H_5 \text{ and enantiomer}$$

erythro

$$C_6H_5{-}{-}{-}\overset{\overset{\displaystyle H}{|}}{\underset{\underset{\displaystyle Br}{|}}{C}}{-}\overset{\overset{\displaystyle Br}{|}}{\underset{\underset{\displaystyle H}{|}}{C}}{-}{-}{-}COC_6H_5 \text{ and enantiomer}$$

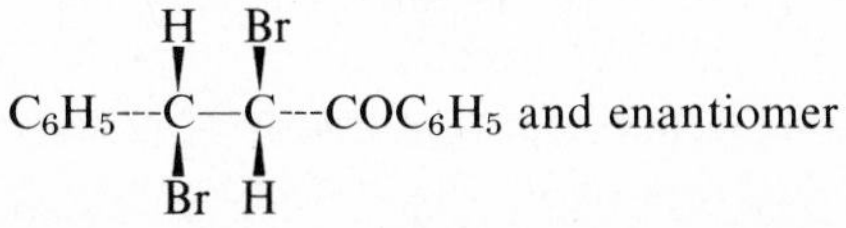

threo

The higher-melting pair (mp 159–160°) is thought to have the *erythro* configuration, and the lower-melting pair (mp 123–124°) the *threo* configuration. It should be noted that benzalacetophenone is capable of geometric (*cis-trans*) isomerism; the mixed aldol condensation gives predominantly the more stable *trans* isomer. If the addition of bromine to benzalacetophenone were stereospecific, only one of the pairs of enantiomers should result from each geometric isomer (see Section 6.3); however, the higher-melting dibromide is the major product of addition of bromine to both *cis-* and *trans*-benzalacetophenone.

As mentioned in Section 6.3, nucleophiles will add to compounds having carbon-carbon double bonds conjugated with carbonyl groups, although they do not add to simple alkenes. The addition of aniline to benzalacetophenone represents a good example of the addition of a nucleophile to an α,β-unsaturated ketone (equation 31). The driving force for this addition is the formation of the resonance-

$$\underset{\textbf{38}}{C_6H_5CH{=}CH\overset{\underset{\|}{O}}{C}C_6H_5} + C_6H_5NH_2 \longrightarrow \underset{\textbf{40}}{C_6H_5\underset{\underset{\displaystyle NHC_6H_5}{|}}{C}H-CH_2\overset{\underset{\|}{O}}{C}C_6H_5} \qquad (31)$$

stabilized intermediate, **39** (or the anion formed from it by loss of one of the protons from the nitrogen atom). The final product, β-anilino-β-phenylpropiophenone (**40**),

$$
\begin{array}{c}
\underset{\text{O}}{\underset{\|}{C_6H_5CH=CH-\overset{\|}{C}C_6H_5}} \\
+ \\
C_6H_5NH_2
\end{array}
\longrightarrow
\left[
\begin{array}{c}
C_6H_5CH-\overset{\ominus}{C}H-\overset{\text{O}}{\overset{\|}{C}}C_6H_5 \\
\overset{\oplus}{N}H_2C_6H_5 \\
\updownarrow \\
\overset{\text{O}^{\ominus}}{C_6H_5CH-CH=\overset{|}{C}C_6H_5} \\
\overset{\oplus}{N}H_2C_6H_5 \\
\mathbf{39}
\end{array}
\right]
$$

$$
\begin{array}{c}
\overset{\text{O}}{C_6H_5CH-CH_2-\overset{\|}{C}C_6H_5} \\
NHC_6H_5 \\
\mathbf{40} \\
\updownarrow \\
\overset{\text{OH}}{C_6H_5CH-CH=\overset{|}{C}C_6H_5} \\
NHC_6H_5 \\
\mathbf{41}
\end{array}
$$

may be formed directly from **39** or by the 1,4-addition product **41**, the less stable enol form of **40**.

Alkylation of Enolate Ions.   The aldol addition (equations 23 and 24) is of great value because it represents a method of producing a new carbon-carbon double bond, either intra- or intermolecularly, and because the products contain functional groups on which a variety of further useful synthetic transformations can be performed. In a general way the crucial carbon-carbon bond-forming step (equation 23) in the process can be characterized as reaction of a nucleophile, the enolate ion, with an electrophile, the carbonyl carbon atom. It should not be surprising that organic chemists have studied the reaction of enolate ions with other potential carbon electrophiles in order to broaden the scope and utility of enolate chemistry; use of one such class of electrophiles is discussed in the following paragraphs.

That alkyl halides serve as electrophiles is evidenced by their tendency to undergo S_N1 and S_N2 reactions (Chapter 10). They do not function well for alkylation of enolate ions generated under the usual reaction conditions for the aldol addition (dilute aqueous or alcoholic base) for a variety of reasons, among which are two side reactions: (1) the competing self-addition of the carbonyl compound (the aldol addition) and (2) reaction of the base with the alkyl halide. Even if the enolate is first generated *irreversibly* and the alkyl halide is subsequently added, problems such as polyalkylation (equation 32) lower the yield of the reaction and detract from its

$$
\underset{}{\overset{\text{O}}{\bigcirc}}
\xrightarrow[-78^\circ]{[(CH_3)_2CH]_2N^\ominus Li^\oplus}
\underset{\mathbf{42}}{\overset{\text{O}^{\ominus}\ Li^\oplus}{\bigcirc}}
\xrightarrow{R-X}
\underset{\mathbf{43}}{\overset{\text{O}}{\bigcirc}}
+
\overset{\text{O}}{\bigcirc}_R
+
\underset{R}{\overset{\text{O}}{\bigcirc}}_R
\qquad (32)
$$

utility. From a mechanistic standpoint, the difficulty that causes polyalkylation is mainly that the alkylation step itself is a relatively sluggish process compared to proton transfer. Consequently the desired monoalkylation product, **43**, formed initially, undergoes deprotonation with **42** serving as the base, and the new enolate ion(s) that result can then be alkylated; a mixture of mono-, di-, and further alkylated products along with the unalkylated starting material can thus result as shown in equation 32.

Most of the experimental variations that have been developed to aid in overcoming the problems of self-addition and polyalkylation in the alkylation of ketones are too sophisticated to be used in the introductory course. However, the alkylation procedure in this section circumvents the difficulties mentioned by employing an enolate precursor that is less prone to self-addition than is a normal ketone or aldehyde and by using a specialized alkylating agent with which polyalkylation is a disfavored process.

The source of the enolate in our procedure is dimethyl malonate **(44)**, a carbonyl-containing compound that is expected to be reasonably acidic owing to the presence of *two* carbonyl groups that can assist in the delocalization of the negative charge in the conjugate base **(45, equation 33)**. This expectation is realized in that **44** has a pK_a of about 14 (the pK_a of acetone is about 25 and that of a *mono*ester such as methyl acetate is about 25). Consequently, if the diester **44** is treated with an appropriate base, ionization to the enolate ion **45** will occur. The addition of this ion to the starting diester **44**, a process leading to self-condensation products, is less

$$CH_3O-\overset{\overset{O}{\|}}{C}-\overset{\overset{H}{|}}{\underset{H}{C}}-\overset{\overset{O}{\|}}{C}-OCH_3 + B^{\ominus}M^{\oplus} \rightleftharpoons B-H + CH_3O-\overset{\overset{O_{\diagdown M^{\oplus}}O}{}}{C}\underset{\underset{H}{C}}{\overset{\ominus}{}}C-OCH_3 \quad (33)$$

44   **45**

important than the similar reaction with aldehydes and ketones because of the lesser tendency of the carbonyl group of esters to suffer nucleophilic attack. This is due to electron delocalization in the ester, a type of stabilization that is lost upon addition of a nucleophile (equation 34).

$$R-\overset{\overset{\cdot\cdot}{O}\cdot}{\underset{}{C}}\overset{\curvearrowleft}{\underset{}{}}\overset{\cdot\cdot}{\underset{}{O}}-R' \longleftrightarrow R-\overset{:\overset{\cdot\cdot}{O}:^{\ominus}}{\underset{}{C}}=\overset{\oplus}{O}-R' \xrightarrow{Nu:^{\ominus}} R-\overset{:\overset{\cdot\cdot}{O}:^{\ominus}}{\underset{\underset{Nu}{|}}{C}}-\overset{\cdot\cdot}{O}-R' \quad (34)$$

The alkylating agent used in the procedure is 1,3-dibromopropane **(46)**. It is chosen for two reasons. First, it circumvents the problems associated with polyalkylation of dimethyl malonate (difficultly separable mixtures of mono- and dialkylated products normally occur). Second, it leads to formation of a derivative of cyclobutane, illustrating the utility of enolate alkylation in producing cyclic compounds that might otherwise be difficult to synthesize. Thus the dibromide **46** can be considered as a *dialkylating* agent in its reaction with **44** to produce dimethyl cyclobutane-1,1-dicarboxylate **(47, equation 35)**. The *intramolecular* reaction that produces **47** by way of

$$CH_2(CO_2CH_3)_2 + Br-CH_2CH_2CH_2-Br \xrightarrow[CH_3OH]{NaOCH_3} \underset{}{\diamondsuit}\overset{CO_2CH_3}{\underset{CO_2CH_3}{}} \quad (35)$$

44   **46**   **47**

the enolate (49) of the monoalkylation product 48 (equation 36) is favored by entropy over the *intermolecular* process that would convert 48 to 50 (equation 37). In any event, 47, 48, and 50 are each separable by distillation because of the differences in their boiling points.

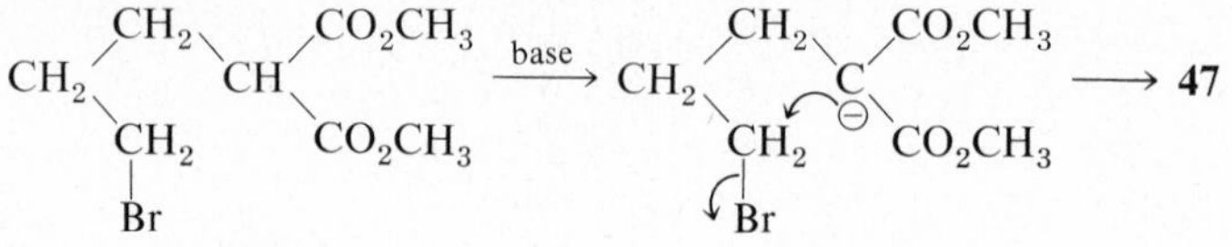

$$
\begin{array}{ccc}
\underset{\textbf{48}}{\overset{\displaystyle CH_2 \diagup CH_2 \diagdown CH \diagup CO_2CH_3}{}} & \xrightarrow{\text{base}} & \underset{\textbf{49}}{} \longrightarrow 47
\end{array} \tag{36}
$$

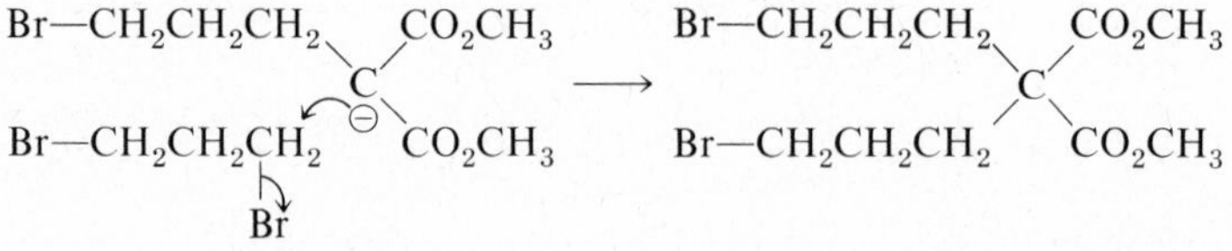

$$
\underset{\textbf{49}}{} \longrightarrow \underset{\textbf{50}}{} \tag{37}
$$

The choice of the specific base to be used in the reaction is important since there is more than one site in the malonate 44 that is susceptible to attack by base. Thus instead of deprotonating 44, (the desired reaction, equation 33), a base could *add* to the carbonyl carbon of an ester function, with subsequent elimination of methoxide ion (equation 38). This potential reaction excludes the use of bases such as hydroxide

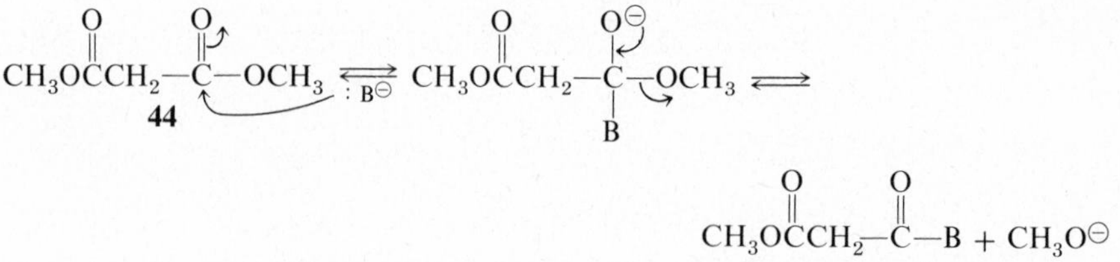

$$
\underset{\textbf{44}}{CH_3OCCH_2-\overset{\displaystyle O}{\overset{\|}{C}}-OCH_3} \underset{: B^{\ominus}}{\rightleftarrows} CH_3OCCH_2-\underset{B}{\overset{\displaystyle O^{\ominus}}{\underset{|}{\overset{|}{C}}}}-OCH_3 \rightleftarrows
$$

$$
CH_3OCCH_2-\overset{\displaystyle O}{\overset{\|}{C}}-B + CH_3O^{\ominus} \tag{38}
$$

and ethoxide since their reaction with 44, according to equation 38, would lead to hydrolysis and transesterification, respectively, of the ester functions of 44.[5] Selection of methoxide ion for the alkylation is an obvious choice not only because this base is sufficiently strong to deprotonate the diester (the pK_a of methanol is about 16) but also because addition of it to the carbonyl group is harmless (equation 38, $B = CH_3O$).

The undesired effect that the use of hydroxide ion would have on the reaction (equation 38, $B = HO$) should give a clue regarding an important experimental precaution: namely, all reagents, solvents, and apparatus employed in the alkylation should be *scrupulously dry*. Water would, of course, react with the methoxide to generate the unwanted hydroxide, which would in turn promote hydrolysis of the ester.

---

[5] The steric bulk of *t*-butoxide suppresses attack of this base at the carbonyl carbon atom. However, this reagent is too expensive for use in our procedure.

Intermolecular alkylation of **48** (equation 37) represents one side reaction that is possible, but others must also be considered. Two examples are the reactions of methoxide ion with 1,3-dibromopropane (**46**) and with the monoalkylated intermediate **48** to produce ethers (equation 39). Fortunately, this type of S_N2 process is not

$$CH_3O^{\ominus} + Br{-}CH_2CH_2CH_2{-}R \longrightarrow CH_3O{-}CH_2CH_2CH_2{-}R \qquad (39)$$

competitive with the desired deprotonation reactions of **44** and of **48**. The type of reaction represented by equation 39 may be further suppressed by *dropwise* addition of methoxide to the reaction mixture; this keeps the concentration of methoxide at a level throughout the reaction that is lower than it would be if the base were added to the reaction mixture in a single portion.

The malonate ion **45** can, of course, promote the same type of side reaction on the intermediate **48** as does methoxide (compare equations 39 and 40). Since the undesired reaction in equation 40 is enhanced by high concentrations of malonate

$$Br{-}CH_2CH_2CH_2CH\Big\langle {}^{CO_2CH_3}_{CO_2CH_3} \quad + \; {}^{\ominus}CH\Big\langle {}^{CO_2CH_3}_{CO_2CH_3} \quad \longrightarrow$$

$$\qquad\qquad \textbf{48} \qquad\qquad\qquad\qquad \textbf{45}$$

$$\begin{array}{c} CH_3O_2C \\ \\ CH_3O_2C \end{array}\!\!\Big\rangle CH{-}CH_2CH_2CH_2CH\Big\langle \begin{array}{c} CO_2CH_3 \\ \\ CO_2CH_3 \end{array} \qquad (40)$$

$$\textbf{51}$$

(**44**) whereas the desired reaction in equation 36 is not, the dimethyl malonate (**44**) should be added in a dropwise fashion to the reaction mixture just as is the methoxide solution. Ideally, then, the reaction would be effected by use of a three-necked flask equipped with *two* addition funnels for the simultaneous dropwise addition of each solution. Such apparatus is often unavailable in the beginning organic laboratory, so the procedure given utilizes a *single* addition funnel. Thus a methanolic solution of sodium methoxide and **44** is prepared, and this is added in a dropwise manner to the 1,3-dibromopropane (**46**). This procedure, which is only slightly less efficient than that using more complex apparatus, takes advantage of the reluctance of the malonate **44** to undergo self-condensation in the presence of sodium methoxide.

A final potential side reaction is worth noting. The anion **45** of dimethyl malonate is an *ambident* ion; that is, because of charge delocalization, **45** has two sites that may function as nucleophilic centers in reacting with electrophiles. Although only reactions involving the nucleophilic carbon atom of **45** have been discussed thus far, attack of an oxygen on **46** would produce **52**, a so-called O-alkylated product (equation 41). However, the reaction conditions chosen in our procedure suppress production of **52**, largely because the protic molecules of the solvent (methanol) and the counter ion (sodium) tend to cluster about the oxygen atoms of **45**. This is because there is greater charge density at oxygen than at carbon owing to the difference in their electronegativities. Thus the molecules of solvent and the counter ion tend to sterically inhibit addition of an electrophile at oxygen.

$$
CH_3O-\overset{O}{\underset{H}{\overset{\|}{C}}}\underset{C}{\overset{\oplus}{\overset{M}{\diagup}}}\overset{O}{\underset{\|}{C}}-OCH_3 + Br-CH_2CH_2CH_2Br \longrightarrow
$$

45   **46**

$$
CH_3O-\overset{O}{\overset{\|}{C}}\underset{\underset{H}{}}{\overset{}{C}}\overset{O-CH_2CH_2CH_2Br}{\overset{}{C}}-OCH_3 \qquad (41)
$$

52

The substituted malonic ester **47**, produced by the alkylation reaction discussed in the preceding paragraphs, can readily be converted to the corresponding cyclobutane-1,1-dicarboxylic acid **53** by hydrolysis under basic conditions, followed by careful acidification of the dibasic salt that results (equation 42). A procedure for isolation of cyclobutane-1,1-dicarboxylic acid is given in the Experimental Procedure.

$$
\textbf{47} \xrightarrow[\text{(2) } H_3O^\oplus]{\text{(1) } KOH/H_2O/\Delta} \quad
\begin{array}{c} CH_2 \diagdown \quad CO_2H \\ \qquad C \\ CH_2 \diagup \quad CO_2H \end{array} \qquad (42)
$$

53

Alternatively, **47** can be hydrolyzed under acidic conditions to provide **53**, which when further heated to about 180° suffers *decarboxylation* to produce cyclobutanecarboxylic acid (**54**, equation 43). Completion of synthesis of cyclobutanecarboxylic acid by carrying out the sequence of reactions shown in equations 35 and 43 represents an example of the "malonic ester synthesis" of substituted acetic acids. The synthesis of substituted acetic acids by the *direct* alkylation of the esters of acetic acid is much less efficient and more difficult than by the malonic ester synthesis. This is because the α-hydrogens of monoesters are much less acidic than those of malonic esters, and much stronger, more difficultly handled bases must be used.

$$
\textbf{47} \xrightarrow[\Delta\Delta]{H_3O^\oplus} \quad
\begin{array}{c} CH_2 \diagdown \quad CO_2H \\ \qquad C \\ CH_2 \diagup \quad H \end{array} \quad + \ 2\ CH_3OH + CO_2 \qquad (43)
$$

54

The decarboxylation shown in equation 43 is an example of a general reaction wherein carboxylic acids having unsaturation at the β-position lose carbon dioxide on heating, presumably by way of a six-centered transition state (equation 44). The enediol intermediate **55** rapidly tautomerizes to the final product **54**.

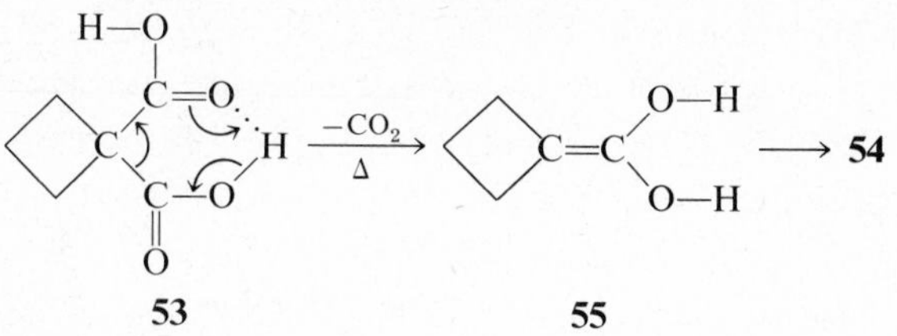

$$ \text{(44)} $$

EXPERIMENTAL PROCEDURE

A. The Haloform Reaction

DO IT SAFELY

Aqueous solutions of the sodium hypohalites are oxidizing agents. Avoid contact of them with your skin. Wash any affected areas with quantities of warm water.

1. Sodium Hypoiodite: The "Iodoform Test." Apply the following procedure to small samples of 2-propanol, acetophenone, cyclohexanone, and pinacolone.

If the substance is water-soluble, dissolve 2 to 3 drops of a liquid or an estimated 50 mg of a solid in 2 mL of water in a small test tube, add 2 mL of 3 M sodium hydroxide, and then slowly add 3 mL of iodine solution.[6] In a positive test the brown color disappears, and yellow iodoform separates. If the substance tested is insoluble in water, dissolve it in 2 mL of dioxane, proceed as above, and at the end dilute with 10 mL of water.

Iodoform can be recognized by its odor and yellow color and, more definitely, by its melting point, 119°. The substance can be isolated by suction filtration of the test mixture or by adding 2 mL of chloroform, shaking the stoppered test tube to extract the iodoform into the small lower layer, withdrawing the clear part of this layer with a capillary dropping tube, and evaporating it in a small tube on the steam bath. The crude solid is recrystallized from methanol-water.

2. Sodium Hypochlorite: Preparation of Benzoic Acid. To 40 mL of 5% sodium hypochlorite solution ("Clorox" and "Purex," for example) in a 150-mL beaker, add 1 mL of acetophenone. Swirl the mixture vigorously, and notice whether any heat of reaction may be detected. Place the beaker on a steam bath and heat gently, swirling from time to time, for about 10 min to volatilize the chloroform. Add about 10 drops of acetone to the warm solution to destroy any excess hypochlorite.

Add a small amount of decolorizing carbon to the warm solution, heat it to boiling while swirling gently but continuously, and filter the hot mixture. To the hot filtrate add 5 mL of concentrated hydrochloric acid with stirring, allow the mixture to

---

[6]The iodine reagent is prepared by dissolving 25 g of iodine in a solution of 50 g of potassium iodide in 200 mL of water.

cool to room temperature, and finally cool it in an ice-water bath. Collect the crystals, wash them with a little cold water, dry them, and determine their melting point.

Determine the melting point of a mixture of this product with the one from the Cannizzaro reaction (Section 13.3).

B. The Aldol Condensation

DO IT SAFELY

Benzalacetophenone and its dibromide derivative are skin irritants. If contact occurs, gently swab the affected area with a piece of cotton holding pure mineral oil, or wash any affected area thoroughly with soap and warm water.

1. Preparation of *trans*-Benzalacetophenone. In a 125-mL Erlenmeyer flask mix 25 mL of 3 *M* sodium hydroxide solution, 15 mL of 95% ethanol, and 0.05 mol of acetophenone. Cool the mixture in an ice-water bath, and, while swirling, add an equimolar amount of benzaldehyde. Allow the mixture to warm to room temperature, and occasionally shake it (stoppered) vigorously for 1 to 2 hr. When a yellow oil separates, it will be helpful to induce crystallization by adding a few seed crystals of benzalacetophenone and/or scratching the side of the flask with a stirring rod.

Cool the reaction mixture in an ice-water bath until crystallization appears to be complete, and collect the product by vacuum filtration. Wash the crystals with cold water and then with a little ice-cold 95% ethanol. Reserve a few crystals to be used for seeding, and recrystallize the crude product from 95% ethanol, using 4–5 mL of solvent per gram of crystals. Determine the melting point of the recrystallized material. The melting point of pure *trans*-benzalacetophenone is reported to be 58–59°.

2. Reactions of Benzalacetophenone; Benzalacetophenone Dibromide. Dissolve 1 g of dry *trans*-benzalacetophenone in 4 mL of carbon tetrachloride. Cool the solution in an ice-water bath, and add dropwise, while swirling, 4 mL of a solution of bromine in carbon tetrachloride, prepared by dissolving 1 g of bromine in 6 mL of carbon tetrachloride. Allow the mixture to stand for 5 to 10 min and collect the crystals. Wash them with a little cold 95% ethanol, dry them, and determine their melting point.

3. Addition of Aniline to Benzalacetophenone. Dissolve 1 g of benzalacetophenone in 20 mL of 95% ethanol, and add an equimolar amount of aniline. Stir or shake until the solution is homogeneous, and allow it to stand (stoppered) overnight. If crystals have not separated, scratch the side of the tube or flask with a glass rod, and cool the mixture in an ice-water bath. Collect the crystals, dry them, and determine their melting point (the reported melting point of β-anilino-β-phenylpropiophenone is 175°).

C. Alkylation of Dimethyl Malonate

DO IT SAFELY
---

1. Methanolic sodium methoxide and ethanolic potassium hydroxide are strongly caustic solutions. If they should come into contact with your skin, wash the affected area immediately with large quantities of cold water.

2. Strong heating of mixtures containing undissolved solids can cause superheating that results in severe bumping. Heat such mixtures carefully, and if possible, stir them to minimize any bumping.

3. When performing a vacuum distillation, take care not to use any glassware that is cracked or otherwise damaged. Damaged glassware may implode under reduced pressure, especially when heated, and may result in spillage of hot chemicals and danger of cuts from broken glass.

1. Preparation of Dimethyl Cyclobutane-1,1-dicarboxylate.[7] Prepare a solution of methanolic sodium methoxide by adding 0.15 mol of solid sodium methoxide *in 5 or 6 portions* to 50 mL of ice-cold anhydrous methanol contained in a dry Erlenmeyer flask fitted with a stopper. The heat of solution is quite high, so that the mixture should be chilled with swirling in an ice-water bath as the portions of methoxide are added. Once all the methoxide has been added and dissolved, further add to the *chilled* solution 0.080 mol of dimethyl malonate (see footnote 7) to provide a methanolic solution of sodio dimethyl malonate (**45**, M = Na). Note that this solution also contains the additional equivalent of sodium methoxide needed for the final cyclization to product (equation 36). Keep the flask stoppered as much as possible throughout the preparation of this solution to avoid exposure to atmospheric moisture.

Equip a 250-mL round-bottomed flask with a magnetic stirring bar (if this is available) and a Claisen connecting tube fitted with a condenser and an addition funnel. Protect both the condenser and the addition funnel with drying tubes filled with anhydrous calcium chloride. This glassware assembly must be thoroughly dried by heating with the flame of a burner before continuing. If you are uncertain of this operation, refer to part A of the Experimental Procedure in Section 13.2. Also read note 2 in that Do It Safely section. Once the glassware has dried and cooled to near room temperature, add to the flask through the addition funnel a solution of 0.075 mol of 1,3-dibromopropane in 15 mL of anhydrous methanol. Close the stopcock of the addition funnel, and add to the funnel the previously prepared methanolic solution of sodio dimethyl malonate. Commence stirring (if you are using magnetic stirring equipment), and heat the reaction mixture to reflux. Discontinue heating once the 1,3-dibromopropane solution is boiling, and begin dropwise addition of the sodio dimethyl malonate solution at such a rate that the total addition will be completed in

---

[7]The methanol used in this experiment must be anhydrous and should be dried over 3-Å molecular sieves for 24 hr before using. It is best to distil dimethyl malonate just prior to its use; however, if this is not done, drying of the malonate over 3 Å molecular sieves for 24 hr will normally suffice.

about 30 min. If you are not using magnetic stirring, the mixture must be manually swirled frequently during the addition. The reaction is mildly exothermic, and reflux *may* be maintained without external heating. If it is not, use either a steam bath or a heating mantle to maintain gentle reflux of the reaction mixture. On completion of the addition, heat the mixture at reflux until addition of 3 or 4 drops of the reaction mixture to about 0.5 mL of water in a test tube gives a solution that is no longer alkaline to pH indicator paper.★ If neutrality has not been attained, as evidenced by this test, after 1.5 hr at reflux, add sufficient glacial acetic acid to make the reaction mixture weakly acidic.★

Allow the mixture to cool slightly, and arrange the flask for simple distillation. Distil the methanol as rapidly as bumping caused by precipitated sodium bromide will allow; if possible, continue stirring during the distillation to minimize the force of bumping. Collect the methanol in a graduated cylinder. After 60–65 mL have been distilled, allow the residue to cool to room temperature.★ Add 75 mL of water, and swirl the mixture until all salts are dissolved. Transfer the two-phase mixture to a separatory funnel, and add 25 mL of technical diethyl ether to dissolve the organic products. Separate the layers, and extract the aqueous phase three additional times with 25-mL portions of diethyl ether. Wash the combined ethereal extracts once with about 20 mL of brine, and dry the solution over magnesium sulfate.★ Remove the ether and any residual methanol by simple distillation, terminating the distillation when the head temperature reaches 80° or so.★ Place the residue in an appropriately sized round-bottomed flask, and *vacuum*-distil the product with the aid of either a water aspirator or the house vacuum line. The desired product has bp 116–117° (20 mm) or 174–176° (165 mm) and should be colorless. Weigh the product and calculate the yield. Normally yields of 40–60% may be anticipated. The distillation residue is largely the undesired tetraester **50**, bp 185–189° (2 mm).

2.   Cyclobutanecarboxylic Acid.   Combine 3 mL of concentrated hydrochloric acid and 1.5 mL of water with each 0.006 mol of dimethyl cyclobutane-1,1-dicarboxylate that is to be hydrolyzed, and place this mixture in an appropriately sized flask that has previously been fitted with a reflux condenser and magnetic stirring bar (if available). Heat and stir (or occasionally swirl) the mixture at reflux until the solution becomes homogeneous,★ and then continue heating for an additional hour.★ Arrange the flask for simple distillation; once the methanol, water, and hydrochloric acid have been removed, continue heating the distillation residue until *gas* evolution ceases.★ If an oil bath has been used for heating, it must ultimately be heated to about 180° to complete the decarboxylation. Crude product can be isolated by simple distillation of the residue at atmospheric pressure, bp 195–196°. Redistillation under vacuum, bp 105–108° (25 mm), gives pure acid in an overall yield of 60–70%.

3.   Cyclobutane-1,1-dicarboxylic Acid.   Combine 6.1 g of potassium hydroxide and 5 mL of 95% ethanol with each 0.025 mol of dimethyl cyclobutane-1,1-dicarboxylate that is to be hydrolyzed, and place this mixture in an appropriately sized flask equipped with a reflux condenser. Heat the mixture at reflux for 2 hr,★ and then arrange the apparatus for simple distillation. Remove nearly all of the methanol and ethanol by distillation,★ and then evaporate the residue to dryness on a steam bath.★

Dissolve the solid residue in a *minimum* amount of water, and then make this solution strongly acidic by *cautious* addition of concentrated hydrochloric acid.★ Extract the aqueous mixture four times with 15-mL portions of technical diethyl ether, dry the combined ethereal extracts over magnesium sulfate,★ filter, and remove most of the ether by simple distillation (*no flames*). Completion of the evaporation to dryness may conveniently be accomplished by fitting the flask with a stoppered vacuum adapter and attaching this apparatus to a water aspirator or house vacuum line; gentle warming of the flask on a steam bath will hasten the removal of the ether. Alternatively, the flask may be allowed to sit unstoppered until the next laboratory period. The resulting beautifully crystalline solid is the desired product, mp 157–158°. It can be recrystallized from ethyl acetate if desired (no flames). The yield should be 80–85%.

EXERCISES

THE HALOFORM REACTION

1. Ethanol gives a clearly positive "iodoform test," yet when sodium hypoiodite is used to oxidize ethanol to acetaldehyde, the yield is found to be less than 10%. Explain how one is able to obtain a good qualitative test in spite of this fact. (*Hint:* Calculate the molecular weight of iodoform.)

2. Suggest a reasonable explanation for the failure of ethyl acetoacetate, $CH_3COCH_2CO_2C_2H_5$, to give iodoform on treatment with sodium hypoiodite. (*Hint:* Consider the mechanism of the base-catalyzed Claisen condensation, which is a reversible reaction.)

3. Dibenzoylmethane, $C_6H_5COCH_2COC_6H_5$, gives iodoform when treated with sodium hypoiodite, even though the compound is not a methyl ketone. Explain this behavior, using equations.

THE ALDOL CONDENSATION

1. Why is the mixed aldol condensation, which produces benzalacetophenone, the main reaction that occurs in this experiment rather than the two possible side reactions, self-condensation of acetophenone and Cannizzaro reaction of benzaldehyde?

2. Write equations to show the major products from reaction of the following carbonyl compounds and bases:
   (a) propionaldehyde, formaldehyde, dilute sodium hydroxide
   (b) benzaldehyde, concentrated potassium hydroxide
   (c) benzaldehyde, ethyl acetate, sodium ethoxide in ethanol
   (d) acetone, benzaldehyde (2 mol), dilute sodium hydroxide

3. Suggest a reason for the fact that the *trans* isomer of benzalacetophenone predominates in the reaction mixture from the aldol condensation.

4. How may *cis*-benzalacetophenone be isomerized to the *trans* isomer?

ALKYLATION OF DIMETHYL MALONATE

1. What is the expected consequence of using methanolic sodium *ethoxide* rather than sodium *methoxide* in the alkylation reaction?

2. Write reactions showing the by-products anticipated if the alkylation is conducted with reagents and/or solvents contaminated with water.

3. Why might the reaction mixture in the alkylation not become neutral after the 1.5 hr period of reflux?

4. What consequences might be expected if the reaction mixture in the alkylation were still basic during the work-up procedure? If acid is added to achieve neutrality, why is it important that the reaction mixture not be made *strongly* acidic?

5. Explain why dimethyl malonate (**44**) is expected to be somewhat more acidic than the monoalkylation product **48**.

6. Calculate the equilibrium concentration of sodio dimethyl malonate (**45**) present if 0.18 mol of dimethyl malonate (**44**) is added to 100 mL of 3 M sodium methoxide in methanol. The pK_a of dimethyl malonate is about 14 and that of methanol is about 16.

7. Draw the structure of the self-condensation product expected to arise from methoxide-promoted reaction of two molecules of dimethyl malonate (**44**). Would this substance be predicted to be more or less acidic than **44**? Why?

8. In the alkylation reaction, why is the bulk of the methanol removed by distillation *prior to* addition of water and subsequent extraction with diethyl ether? In other words, why does the work-up procedure not simply call for addition of water to the neutral reaction mixture followed by extraction of this solution with diethyl ether?

9. Why do you think the base-promoted rather than the acid-catalyzed hydrolysis is the preferred method for preparing cyclobutane-1,1-dicarboxylic acid?

10. Dipotassium cyclobutane-1,1-dicarboxylate does *not* decarboxylate when heated above 150°, whereas the corresponding diacid does. Provide an explanation for the difference in reactivity.

11. Why might calcium chloride *not* be a particularly good choice as a drying agent for the ethereal solution of cyclobutane-1,1-dicarboxylic acid obtained during the procedure for base-catalyzed hydrolysis of the diester?

SPECTRA OF STARTING MATERIALS AND PRODUCTS

The ir and pmr spectra of benzaldehyde are given in Figures 13.4 and 13.5, respectively. The ir spectra of cyclobutane-1,1-dicarboxylic acid in both mineral oil and perfluorokerosene are given in Figure 4.7.

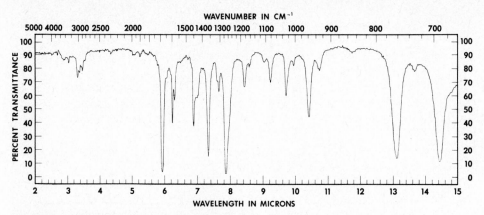

FIGURE 14.20 IR spectrum of acetophenone.

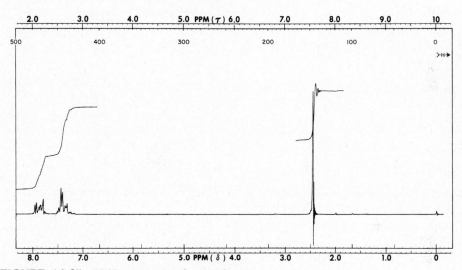

FIGURE 14.21 PMR spectrum of acetophenone.

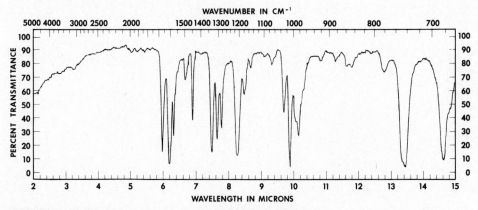

FIGURE 14.22 IR spectrum of *trans*-benzalacetophenone.

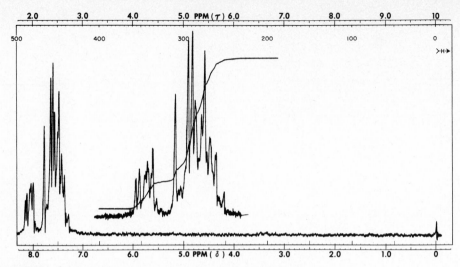

FIGURE 14.23 PMR spectrum of *trans*-benzalacetophenone.

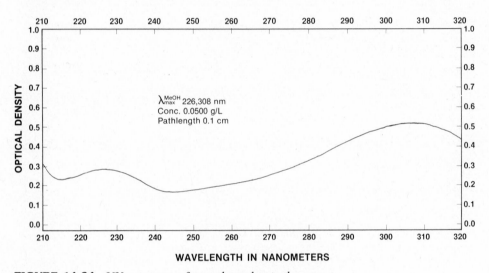

λ_{max}^{MeOH} 226,308 nm
Conc. 0.0500 g/L
Pathlength 0.1 cm

WAVELENGTH IN NANOMETERS

FIGURE 14.24 UV spectrum of *trans*-benzalacetophenone.

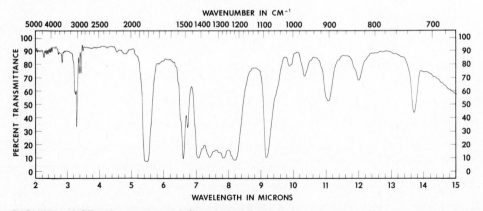

WAVENUMBER IN CM^{-1}

WAVELENGTH IN MICRONS

FIGURE 14.25 IR spectrum of dimethyl malonate.

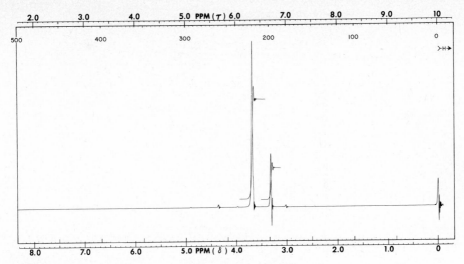

FIGURE 14.26 PMR spectrum of dimethyl malonate.

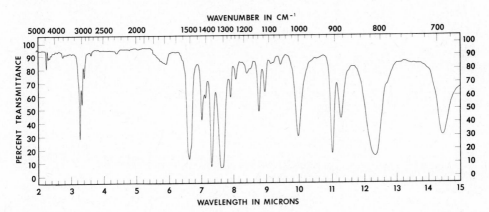

FIGURE 14.27 IR spectrum of 1,3-dibromopropane.

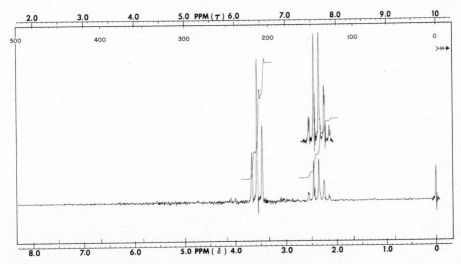

FIGURE 14.28 PMR spectrum of 1,3-dibromopropane.

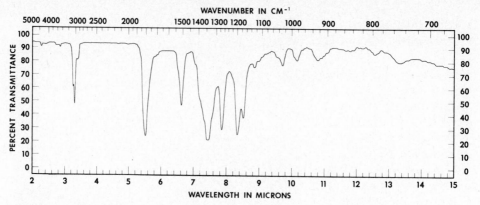

FIGURE 14.29 IR spectrum of dimethyl cyclobutane-1,1-dicarboxylate.

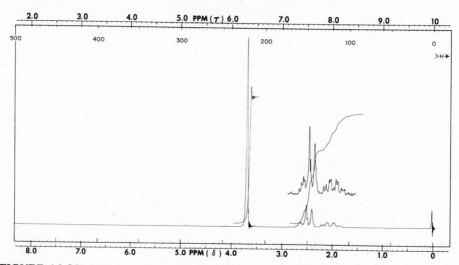

FIGURE 14.30 PMR spectrum of dimethyl cyclobutane-1,1-dicarboxylate.

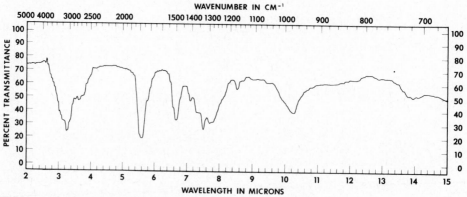

FIGURE 14.31 IR spectrum of cyclobutanecarboxylic acid.

14.3 Identification of an Alcohol or Carbonyl Compound by Means of Three Reagents

The reactions of alcohols, aldehydes, and ketones with the reagents, chromic acid in acetone, 2,4-dinitrophenylhydrazine, and sodium hypoiodite, have been discussed in Sections 13.2, 14.1, and 14.2. By use of only these three reagents, with proper interpretation of the results of their reactions, it is possible to identify an unknown alcohol or carbonyl compound if there is a known limitation on the number of possibilities for the unknown.

As an illustration, it is possible to decide on the identity of any one of the compounds in the following sets of five by using only the three reagents. In some instances only two tests may be required. Your instructor will provide you with an unknown compound labeled as belonging to one of the sets. Identify the compound by using a *minimum* number of tests.

Set A	*Set B*	*Set C*
1-Pentanol	2-Propanone	1-Octanol
2-Butanol	2-Methyl-1-propanol	Cinnamaldehyde
2-Butanone	2-Methylpropanal	Acetophenone
Benzaldehyde	2-Propanol	4-Methyl-2-pentanol
2-Methyl-2-butanol	2-Methyl-2-propanol	Cyclohexanone

EXERCISE

Write out balanced equations for all positive reactions which you observe, and give a careful explanation of their meaning in terms of the structure of your unknown compound.

15

CARBOXYLIC ACIDS
AND THEIR DERIVATIVES

Organic compounds having the general structural formula RCOX, in which R is an alkyl or aryl substituent and X represents a functional group with an atom other than carbon or hydrogen bonded to the carbonyl carbon, are classified as carboxylic acids or their derivatives. Thus X is OH in a carboxylic acid, OR′ in an ester, OCOR′ in an anhydride, and NR_2 in an amide. Nitriles, RCN, are also sometimes classified as derivatives of carboxylic acids because their hydrolysis produces these acids.

It is of interest that many naturally occurring materials are carboxylic acids or their derivatives. For example, fats are triesters of straight-chain carboxylic acids containing from 12 to 18 carbon atoms and glycerol (1,2,3-trihydroxypropane), and proteins are basically polyamides. In addition, volatile esters are partially responsible for the pleasant odors of many fruits, whereas carboxylic acids are often the cause of unpleasant odors, such as those of rancid butter and of goats. Soaps are the sodium or potassium salts of the same carboxylic acids contained in fats.

The chemistry of carboxylic acids and their derivatives in many ways parallels that of aldehydes and ketones (see Chapters 13 and 14) in that reactions at both the carbonyl carbon and the α-carbon are known. An example of attack at the α-carbon is the Hell-Volhard-Zelinsky reaction in which a carboxylic acid is converted to an α-haloacid (equation 1, X = Cl or Br). The Claisen ester condensation exemplifies reaction at the α-carbon of one molecule of an ester and at the carbonyl carbon of a second (equation 2). These two conversions are analogous to the base-catalyzed

$$
\underset{\substack{|\\ R'}}{\overset{\substack{H\\ |}}{R-C-COOH}} + X_2 \xrightarrow[\text{or } PX_3 \text{ (cat.)}]{P \text{ (cat.)}} \underset{\substack{|\\ R'}}{\overset{\substack{X\\ |}}{R-C-COOH}} + HX \tag{1}
$$

$$
2\ RCH_2\overset{\overset{\displaystyle O}{\|}}{C}OR' \xrightarrow{R'O^{\ominus}} RCH_2\overset{\overset{\displaystyle O}{\|}}{C}-\underset{\underset{\displaystyle R}{|}}{CH}\overset{\overset{\displaystyle O}{\|}}{C}OR' + R'OH \tag{2}
$$

halogenation of aldehydes and ketones and to the base-catalyzed aldol condensation, respectively. Of course, a third site of reaction, not present in aldehydes and ketones, is the functional group X, present in carboxylic acids and their derivatives.

The types of reactions to be considered in this chapter are limited to those occurring at the carbonyl carbon or the substituent X. Although the specific experiments to be described all involve benzoic acid or its derivatives, C_6H_5COX, the student should bear in mind that other carboxylic acids and their derivatives are prepared or react in a similar manner. As a rule, however, aliphatic derivatives of acids are more reactive than their aryl counterparts, undergoing hydrolysis much more readily, for example. This is due to the conjugative stabilization by the aromatic ring of the carbonyl group in the aromatic derivatives.

15.1 Carboxylic Acids

Carboxylic acids can be synthesized in a variety of ways, which include the hydrolysis of nitriles or of any derivative of the acid, and oxidation of an aldehyde or ketone. The carbonation of a Grignard reagent, described in Section 12.2, is another useful synthetic route to these acids.

A special method available for the preparation of aryl carboxylic acids involves the oxidation of the side-chain R of an alkyl-substituted benzene with alkaline permanganate or with chromic acid solution (equation 3). The carbon atom bonded

$$ArCH_2R \xrightarrow[\text{or } H_2CrO_4]{KMnO_4/\text{base}} ArCOOH \tag{3}$$

to the aromatic ring must bear *at least one* hydrogen for this method of preparation to be efficient. If more than one alkyl side chain is present, each will be oxidized to a carboxyl group. Identification of the resulting acid will allow determination of the positions of attachment of the side chains on the aromatic ring, if this is not known.

In the preparation of a carboxylic acid described in this experiment, ethylbenzene, which is readily available from the Friedel-Crafts alkylation of benzene, is oxidized with alkaline permanganate to benzoic acid (equation 4). The mechanism of this conversion is not well understood but probably involves the removal of the benzylic hydrogen in a free-radical process initiated by permanganate ion; subsequent reactions lead first to the alcohol and then to the ketone, which is oxidized further to the carboxylic acid. The mechanism of this final oxidation step parallels that given for the oxidation of cyclohexanone to hexanedioic acid (Section 13.1).

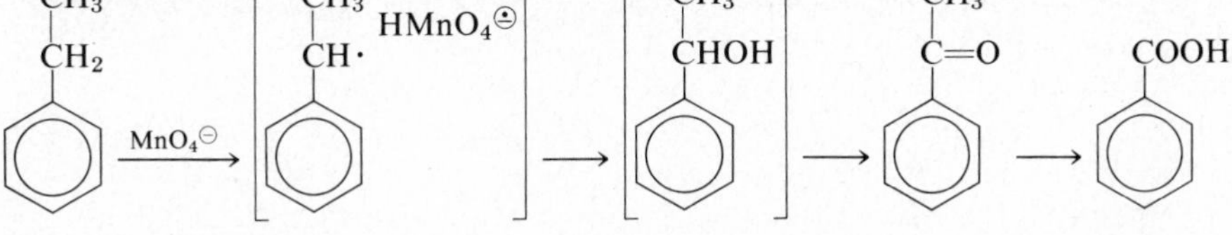

As a group, carboxylic acids are considered weak acids as compared to mineral acids such as hydrochloric and have K_a's of the order of 10^{-5}. Nevertheless, carboxylic acids are sufficiently acidic to give a test with litmus paper and readily to undergo reaction with bases to form salts.

Use is often made of the acidity of the carboxylic acids in their removal from a reaction mixture by extraction of the mixture with base. The organic acid can be regenerated from the basic solution by acidification with mineral acid. A good way to differentiate carboxylic acids from phenols (ArOH, another type of weak organic acid) depends on the ability of carboxylic acids to react with the weak base bicarbonate to produce the carboxylate salt and carbon dioxide, which bubbles out of the solution. Most phenols, being weaker acids ($K_a \approx 10^{-10}$), will not react with this base.

The relatively high acidity of carboxylic acids enables ready determination of the *equivalent weight* (sometimes called "neutralization equivalent") of the acid by titration with standard base. The equivalent weight of an acid is that weight, in grams, of acid which reacts with one equivalent of base. As an example, suppose that 0.1000 g of an unknown acid required 16.90 mL of 0.1000 N sodium hydroxide solution to be titrated to a phenolphthalein end point. This means that 0.1000 g of the acid corresponds to (16.90 mL)(0.1000 equivalent/1000 mL) or 0.0016901 equivalent of the acid, or that one equivalent of the acid weighs 0.1000/0.00169 or 59.201 g. Thus the following expression applies:

$$\text{equivalent weight} = \frac{\text{grams of acid}}{(\text{volume of base consumed in liters})(N)}$$

where N is the normality of the standard base.

Because each carboxylic acid function in a molecule will be titrated with base, the equivalent weight corresponds to the molecular weight of the acid divided by n, where n is the number of acid functions present in the molecule. Thus for the example given, the molecular weight may be 59.20 if a single acid function is present, 118.4 if two are present, and 177.6 if three are present. If the molecular weight of an unknown compound is known, then the number of acid groups in the molecule can be calculated by dividing the molecular weight by the equivalent weight. Hence, if the molecular weight of the unknown compound is 118 and its equivalent weight is 59.201, the unknown must have *two* titratable acid functions.

15.2   Carboxylic Acid Esters

The esters of carboxylic acids are most commonly obtained by allowing the acid and an alcohol to react in the presence of a mineral acid (equation 5). The role of the catalyst undoubtedly is to promote attack of the nucleophilic oxygen of the alcohol on the carbonyl carbon of the acid by protonation of the carbonyl oxygen to give **1**. The process is one of equilibria between a variety of compounds, and in order that the esterfication be a reaction of high yield, the equilibria must be shifted toward the products, the desired ester and water. This can be accomplished either by removing one or more of the products from the reaction mixture as formed or by employing a large excess of one of the starting reagents.

$$R-C{\overset{O}{\underset{OH}{}}} \underset{-H^\oplus}{\overset{+H^\oplus}{\rightleftharpoons}} \left[R-C{\overset{OH}{\underset{OH}{}}}\right]^\oplus \underset{-R'OH}{\overset{+R'OH}{\rightleftharpoons}} R-\underset{\underset{\underset{\oplus}{HOR'}}{|}}{\overset{\overset{OH}{|}}{C}}-OH$$

(5)

$$R-C{\overset{O}{\underset{OR'}{}}} \underset{+H^\oplus}{\overset{-H^\oplus}{\rightleftharpoons}} \left[R-C{\overset{OH}{\underset{OR'}{}}}\right]^\oplus \underset{+H_2O}{\overset{-H_2O}{\rightleftharpoons}} R-\underset{\underset{OR'}{|}}{\overset{\overset{OH}{|}}{C}}{\overset{\oplus}{=}}OH_2$$

The effect of the latter approach can be understood by consideration of the mass law relating starting materials and products (equation 6). Increasing the amount of either the alcohol or the carboxylic acid will result in an increase in the amount of products formed since the equilibrium constant, K, for the reaction must remain constant at a given temperature, no matter what quantity of either reagent is used.

$$RCOOH + R'OH \overset{K}{\rightleftharpoons} RCOOR' + H_2O$$

$$K = \frac{[RCOOR'][H_2O]}{[RCOOH][R'OH]}$$

(6)

Occasionally it is not economically feasible to use a large excess of either reagent, and other methods must be employed. These may involve actual removal of water by azeotropic distillation with the esterifying alcohol or the equivalent of removal of the water from the reaction mixture by use of either a large excess of acid catalyst, which converts water to its conjugate acid, H_3O^+, or by use of a solvent such as 1,2-dichloroethane in which the ester is soluble but water is not. The procedure adopted in the esterification of benzoic acid with methanol (equation 6, R = C_6H_5, R' = CH_3), a reaction described in the Experimental Procedure, is to use a large excess of the alcohol, which is inexpensive.

If the carboxylic acid has bulky substituents on the carbons α and/or β to the carboxyl group, the esterification either may be very slow or may not occur at all because of steric hindrance to the attack of the alcohol on the intermediate **1** (equation 5). An alternative method for the preparation of esters of such hindered acids involves initial conversion of the acid to its acid chloride (see below) and subsequent reaction of this compound with an alcohol to produce the ester (equation 7).

$$R-\overset{\overset{O}{\|}}{C}-OH \longrightarrow R-\overset{\overset{O}{\|}}{C}-Cl \xrightarrow{R'OH} R-\overset{\overset{O}{\|}}{C}-OR' + HCl$$

(7)

Hydrolysis of esters to carboxylic acids, the reverse of esterification, can be accomplished by heating the ester with aqueous acid or base, or in the case of esters which are insoluble in water, by heating with methanolic or ethanolic potassium hydroxide (equation 8). Promotion by base rather than acid is generally preferred because the rate of hydrolysis of an ester is greater in basic medium. Moreover, if at least a molar equivalent of base is used, the acid will be converted to its salt, which makes the reaction irreversible.

$$
\underset{\substack{O\\\parallel}}{R-C-OR'} + H_2O \xrightarrow{\text{H}_3\text{O}^\oplus \text{ or HO}^\ominus} \underset{\substack{O\\\parallel}}{R-C-OH} + R'OH \tag{8}
$$

It is possible to carry out the hydrolysis of an ester with alkali in a quantitative manner so that a value, termed the *saponification equivalent,* can be derived. This value is analogous to the equivalent weight of an acid in that it is the molecular weight of the ester divided by the number of ester functions in the molecule. Therefore the saponification equivalent is the number of grams of ester required to react with one gram-equivalent of alkali. This equivalent is determined by hydrolyzing a weighed amount of the ester with an excess of standardized alkali and then titrating the excess alkali to a phenolphthalein end point with standardized hydrochloric acid. The saponification equivalent is then given as follows:

$$
\text{Saponification equivalent} = \frac{\text{grams of ester}}{\text{equivalents of alkali consumed}}
$$

$$
= \frac{\text{grams of ester}}{(\text{volume of alkali in liters})(N) - (\text{volume of acid in liters})(N')}
$$

where N is the normality of the standard base and N' is the normality of the standard acid.

15.3   Carboxylic Acid Halides (Acyl Halides) and Anhydrides

Acid halides, substances in which the hydroxyl function of the carboxylic acid has been replaced by a halogen, are very useful and reactive compounds. Some of the more important reactions of acid halides are those with various nucleophiles, and a few of these are shown in Figure 15.1. Thus acid halides are converted to carboxylic acids by hydrolysis, to esters by reaction with alcohols, to amides when treated with ammonia or amines, and to anhydrides when allowed to react with carboxylic acids or their salts.

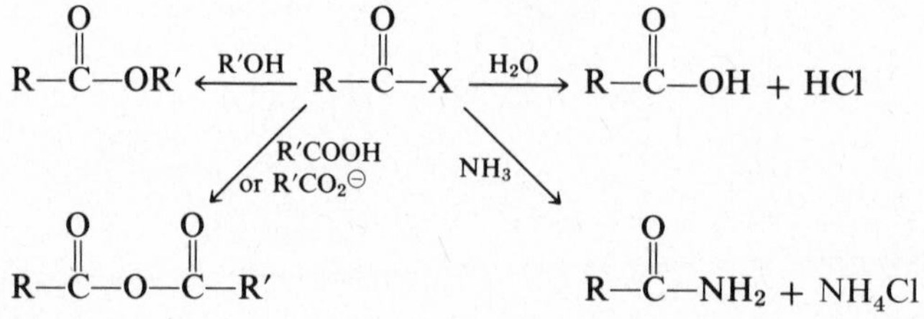

FIGURE 15.1   Some reactions of acid halides.

A generalized mechanism, which is common to almost all reactions of derivatives of carboxylic acids with nucleophiles, can be applied to the conversions shown in Figure 15.1. In this mechanism the nucleophile, NuH (or Nu⁻), adds to the carbonyl carbon of the derivative, RCOX, to produce the tetrahedral intermediate, **2**, which then decomposes with elimination of HX (or X⁻) to give the product (equation 9). Although the overall reaction is one of *substitution* of Nu for X, the mechanism of the conversion involves an addition step and an elimination step, so it is therefore markedly different from the familiar substitutions of the S_N1 and S_N2 types. The

$$R\overset{O}{\overset{\|}{-}}C\overset{}{-}X + \; :NuH \longrightarrow R\overset{O^\ominus}{\overset{|}{-}}\underset{\oplus NuH}{C}\overset{}{-}X \longrightarrow R\overset{O}{\overset{\|}{-}}C\overset{}{-}Nu + HX \qquad (9)$$

$$\mathbf{2}$$

success of such a substitution reaction might be expected to depend to a great extent upon the relative basicities of the reagent, NuH (or Nu⁻), and the leaving group, XH (or X⁻), and such a dependence is observed experimentally. Thus it is impossible to convert an ester (X = OR) to an acid chloride (X = Cl) by reaction of the ester with chloride ion because chloride ion is a much weaker base than is alkoxide, RO⁻.

The rate at which the substitution reaction occurs with acid halides is enhanced compared to other acid derivatives because the halogen is electron-withdrawing and therefore increases the electrophilicity of the carbonyl carbon. In addition, halide ions, being very weak bases, are good leaving groups.

Acid chlorides can be prepared by treatment of the carboxylic acid with thionyl chloride (equation 10), phosphorus trichloride (equation 11), or phosphorus penta-

$$RCOOH + SOCl_2 \rightarrow RCOCl + HCl + SO_2 \qquad (10)$$

$$3 \; RCOOH + PCl_3 \rightarrow 3 \; RCOCl + H_3PO_3 \qquad (11)$$

$$RCOOH + PCl_5 \rightarrow RCOCl + POCl_3 + HCl \qquad (12)$$

chloride (equation 12). Acid bromides may be produced by using the corresponding bromine-containing reagents but seldom are because, in sharp contrast to the relative reactivities of *alkyl* bromides and chlorides in S_N1 and S_N2 reactions, acid bromides are *less* reactive than the corresponding chlorides.

The phosphorus pentahalides are more reactive halogenating agents than the others but are more difficult to handle in the laboratory. Consequently they are usually employed only in instances in which the other reagents fail, as is the case in the conversion of some aryl carboxylic acids and sterically hindered acids to their acid halides. It should be noted that the most complete utilization of the halogens present in the halogenating agent is accomplished in the reaction of the carboxylic acid with phosphorus trichloride, but use of this reagent is sometimes precluded in cases where the boiling points of the two products (equation 11) are similar.

Thionyl chloride is used for the preparation of benzoyl chloride from benzoic acid (equation 10, R = C_6H_5), as described in the Experimental Procedure. It is used in preference to phosphorus trichloride because the latter reagent, although somewhat less expensive, leads to reaction mixtures containing phosphorus acid, H_3PO_3, from which it is difficult to separate the benzoyl chloride. In contrast, the by-products from the reaction of thionyl chloride with benzoic acid are gases (HCl and SO_2) and therefore are readily removed from the reaction mixture. Furthermore, thionyl chloride has a low boiling point (79°) so that it may be separated from the product by simple distillation.

Methods are also given for the conversion of benzoyl chloride to esters, amides, and an anhydride. Although the reactions are typical of acid chlorides in general, it is important to note once again that the reaction conditions are somewhat more vigorous than those required for aliphatic acid chlorides. Thus hydrolysis of benzoyl chloride requires gentle heating, whereas acetyl chloride, CH_3COCl, reacts almost violently with cold water.

The preparation of benzoic anhydride from benzoyl chloride deserves some special comment. The reaction of benzoic acid with benzoyl chloride, using pyridine, C_5H_5N, as a catalyst, produces a high yield of the anhydride (equation 13). The role

$$C_6H_5\overset{O}{\overset{\|}{C}}-Cl \xrightarrow{} C_6H_5\overset{O}{\overset{\|}{C}}-\overset{\oplus}{N} \quad \overset{C_6H_5CO_2^{\ominus} \;\; H\overset{\oplus}{N}}{\xrightarrow{}} \quad C_6H_5\overset{O}{\overset{\|}{C}}-OCOC_6H_5 \qquad (13)$$

$$Cl^{\ominus}$$

3

of the catalyst is twofold. (1) It activates the acid chloride by formation of a pyridinium salt, **3**, which is more susceptible to nucleophilic attack than is the acid chloride itself. (2) It converts benzoic acid to the more nucleophilic benzoate ion. It is possible to obtain a still higher yield of benzoic anhydride if the benzoic acid is generated *in situ* by reaction of the acid chloride-pyridine complex, **3**, with 0.5 equivalent of water. The benzoic acid resulting from hydrolysis of the complex reacts immediately with a second mole of the complex to produce the desired anhydride.

Because of their greater ease of handling in the laboratory, anhydrides are often used in place of the more reactive acid halides for the acylation of alcohols and of amines to produce esters and amides, respectively (equations 14 and 15). Use of acetic anhydride to acetylate an amine is described in Section 17.2.

$$R-\overset{O}{\overset{\|}{C}}-O-\overset{O}{\overset{\|}{C}}-R + R'OH \longrightarrow R-\overset{O}{\overset{\|}{C}}-OR' + R-\overset{O}{\overset{\|}{C}}-OH \qquad (14)$$

$$R-\overset{O}{\overset{\|}{C}}-O-\overset{O}{\overset{\|}{C}}-R + R'NH_2 \longrightarrow R-\overset{O}{\overset{\|}{C}}-NHR' + R-\overset{O}{\overset{\|}{C}}-OH \qquad (15)$$

15.4 Carboxylic Acid Amides and Nitriles

The amides of carboxylic acids can be prepared by reaction of an acid halide or anhydride with either anhydrous or aqueous ammonia, if an unsubstituted amide is desired, or with an amine, if an N-alkylated amide is required. Thermal decomposition of the ammonium salt of a carboxylic acid also generates the corresponding amide (equation 16).

$$RCOOH + NH_3 \longrightarrow RCOO^{\ominus}NH_4{}^{\oplus} \xrightarrow{\text{heat}} R-\overset{\overset{\displaystyle O}{\|}}{C}-NH_2 + H_2O \qquad (16)$$

Amides can also be prepared by treatment of esters with amines (equation 17). This conversion generally requires heat, and in the case of volatile amines or ammonia must be carried out under pressure.

$$R-\overset{\overset{\displaystyle O}{\|}}{C}-OR' \xrightarrow{HNR_2} R-\overset{\overset{\displaystyle O}{\|}}{C}-NR_2 + R'OH \qquad (17)$$

Although amides undergo a variety of reactions, two of the more important are their reduction to amines with lithium aluminum hydride (equation 18) and their conversion to nitriles with dehydrating agents such as phosphorus oxychloride, $POCl_3$, phosphorus pentoxide, P_2O_5, or thionyl chloride (equation 19). The fact that

$$R-\overset{\overset{\displaystyle O}{\|}}{C}-NH_2 \xrightarrow[\text{(2) } H_3O^{\oplus}]{\text{(1) } LiAlH_4} R-CH_2-\overset{\oplus}{N}H_3 \qquad (18)$$

$$R-\overset{\overset{\displaystyle O}{\|}}{C}-NH_2 \xrightarrow[\text{or } SOCl_2]{\overset{POCl_3}{\text{or } P_2O_5}} R-C{\equiv}N \qquad (19)$$

amides are reduced to amines is somewhat unexpected, since all other derivatives of carboxylic acids and in fact the acids themselves are reduced to primary alcohols on treatment with lithium aluminum hydride (equation 20). In short, the functional group X is retained on reduction of amides but is lost in the case of the other derivatives of acids.

$$R-\overset{\overset{\displaystyle O}{\|}}{C}-X \xrightarrow[\text{(2) } H_3O^{\oplus}]{\text{(1) } LiAlH_4} R-CH_2-OH \qquad (20)$$
$$(X \neq NR_2)$$

The mechanism of dehydration of an amide to a nitrile is interesting. As an example, the conversion of benzamide to benzonitrile with thionyl chloride probably involves attack of the amide oxygen on the sulfur atom of thionyl chloride to produce an intermediate salt, **4**, which then decomposes to the nitrile by loss of sulfur dioxide and two moles of hydrogen chloride (equation 21, $R = C_6H_5$).

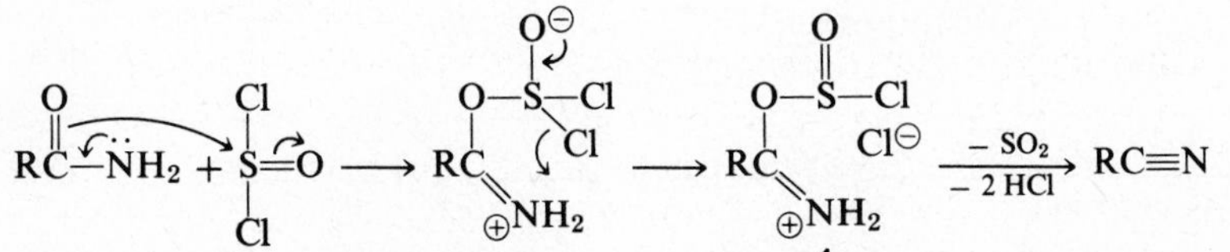

$$\xrightarrow[\text{- 2 HCl}]{\text{- SO}_2} RC{\equiv}N \qquad (21)$$

4

Nitriles can also be prepared by the S_N2 reaction between cyanide ion and a primary or secondary alkyl halide. These substances can be hydrolyzed to amides, which are isolable in some cases, and then to the corresponding carboxylic acids by treatment with aqueous acid or base (equation 22). They can also be converted into esters by heating with an alcohol in the presence of one molar equivalent of water and a catalytic amount of mineral acid (equation 23).

$$RC{\equiv}N \xrightarrow[\text{H}_2\text{O}]{\text{H}^{\oplus}\text{ or HO}^{\ominus}} \left[\underset{\displaystyle \text{RC}-\text{NH}_2}{\overset{\displaystyle O}{\|}} \right] \longrightarrow \underset{\displaystyle \text{RC}-\text{OH}}{\overset{\displaystyle O}{\|}} \qquad (22)$$

$$RC{\equiv}N \xrightarrow[\substack{\text{R'OH} \\ \text{H}_2\text{O}}]{\text{HCl}} \underset{\displaystyle \text{RC}-\text{OR'}}{\overset{\displaystyle O}{\|}} \qquad (23)$$

EXPERIMENTAL PROCEDURE

A. Oxidation of an Aromatic Side Chain: Preparation of Benzoic Acid from Ethylbenzene

Place 0.071 mol of potassium permanganate, 120 mL of water, 1.5 mL of 3 *M* aqueous sodium hydroxide, and 0.016 mol of ethylbenzene in a 500-mL round-bottomed flask fitted with a reflux condenser, and gently heat this mixture at reflux for 2 hr (or up to 3.5 hr if time permits).★ *Severe bumping will occur if strong heating is used.* Test the hot solution for unchanged permanganate by placing a drop of the reaction mixture on a piece of filter paper; if a purple ring appears around the brown spot of manganese dioxide, permanganate remains. Destroy any excess permanganate by adding small amounts of solid sodium bisulfite to the mixture until the spot test is negative. Do not add a large excess of bisulfite. Add 2 g of filter-aid to the hot mixture to promote its rapid filtration; manganese dioxide is so finely dispersed that it would tend to clog the pores of the filter paper otherwise. Filter the mixture by suction, and rinse the reaction flask and filter cake with two 10-mL portions of hot water.★ Concentrate the filtrate to about 30 mL by performing a simple distillation; if the concentrate is turbid, clarify it by a gravity filtration.★ Acidify the aqueous residue with concentrated hydrochloric acid, adding acid until no more benzoic acid separates (about 5–10 mL of acid will be required).★ Cool this mixture in an ice-water bath, and isolate the benzoic acid by suction filtration. The crude acid can be purified by recrystallization from hot water or by sublimation. If the latter procedure is to be used, the crude acid must be *dry*. Determine the melting point and yield of benzoic acid obtained.

B. Reactions of Benzoic Acid

DO IT SAFELY

---

1. Pay particular attention to the italicized cautionary notes included in the following procedures.

2. The acid chlorides used and prepared in some of the procedures are *lachrymators;* avoid exposure to their vapors. If the compounds come in contact with your skin, avoid burns by washing the area with dilute sodium bicarbonate solution and rinsing well with cold water.

3. Phenol is a moderately strong acid (pK$_a$ about 10) and can cause severe burns. If it comes in contact with your skin, wash the area thoroughly with dilute sodium bicarbonate solution and rinse well with warm water.

1. Acidity. Saturate 1 mL of hot water with benzoic acid, and test the solution with litmus paper, noting the result. Add a small amount of solid sodium bicarbonate or a few drops of 1.5 M aqueous sodium bicarbonate to the saturated solution, and note the result. Acidify the resulting solution with concentrated hydrochloric acid, and note the result. Compare the solubility of benzoic acid both in cold water and in cold 3 M aqueous sodium hydroxide.

2. Esterification: Preparation of Methyl Benzoate. Place 0.082 mol of benzoic acid and 0.62 mol of methanol in a 100-mL round-bottomed flask, and carefully pour 3 mL of concentrated sulfuric acid down the walls of the flask. Swirl the flask to mix the components thoroughly, attach a reflux condenser, and gently heat the mixture at reflux for 1 hr.★ Cool the solution and transfer it to a separatory funnel containing 50 mL of water. Rinse the flask with 40 mL of diethyl ether, and add the rinsings to the funnel. Shake the funnel thoroughly to facilitate extraction of methyl benzoate into the ether layer, *venting the funnel occasionally.* Drain off the aqueous layer, and wash the organic layer with a second 25-mL portion of water. Separate the layers, add 25 mL of 0.6 M aqueous sodium bicarbonate to the funnel (*Caution: Foaming may occur!*), and shake the mixture, *frequently* venting the funnel. Drain off the aqueous washes, and test to see that they are basic. If not, repeat the washing of the organic layer with aqueous bicarbonate until basic washes are obtained. After a final wash with saturated sodium chloride solution, dry the ether solution thoroughly over anhydrous magnesium sulfate.★

Filter the solution and remove the ether by distillation on the steam bath (*No flames!*). Then decant the methyl benzoate into a 100-mL round-bottomed flask, and attach the flask to an apparatus set up for simple distillation. Distil the ester using an air-cooled rather than a water-cooled condenser (which might crack because of the high boiling point, 199°, of the ester), collecting the material boiling above 190° in a tared Erlenmeyer flask.

Recover unchanged benzoic acid by acidifying the basic washes with concentrated hydrochloric acid and collecting any precipitate that results. Dry and weigh the precipitate so that recovered starting material can be allowed for when calculating the percentage yield in this reaction. Ascertain that the precipitate is benzoic acid by determining its melting point.

3. Preparation of Benzoyl Chloride. (Carry out this reaction in a hood if possible. If not, use the gas trap, which is shown in Figure 5.1.)[1] Equip a dry 100-mL round-bottomed flask with a water-cooled condenser connected to a gas trap as

---

[1]See paragraph 2 under Do It Safely.

shown in Figure 5.1, and place 0.20 mol of dry benzoic acid in the flask. Measure out 0.70 mol of thionyl chloride (see footnote 1) *in the hood,* pour it into the flask, and quickly attach a reflux condenser. Bring the mixture to a gentle reflux, and continue heating until the vigorous evolution of HCl and SO_2 ceases (about 30 min).[★] Quickly remove the flask from the condenser, attach it to an apparatus set for simple distillation, and add a fresh boiling stone. Distil the excess thionyl chloride using a burner or a heating mantle, collecting the distillate in a receiver *attached to a vacuum adapter* carrying a calcium chloride tube. This tube should be connected to a gas trap if the distillation is not performed in a hood. In this way the distillation can be carried out in an open system without exposure of the distillate to atmospheric moisture. After no more thionyl chloride distils, drain the water from the condenser and continue the distillation. Material boiling between 85–192° should be collected in a second flask, and the product, bp 192–198°, in a third. Return the low-boiling fraction to a bottle labeled "Recovered Thionyl Chloride," and pour the intermediate fraction down a sink *in the hood,* flushing with copious amounts of water. Store the benzoyl chloride (*Caution: Lachrymatory*) in an appropriately labeled *glass-stoppered* bottle. Determine the yield of product.

C.   Reactions of Benzoyl Chloride

(*Caution: Benzoyl chloride is a strong lachrymator. Perform the reactions in a hood if possible.*)

1.   Hydrolysis.   Add 5 drops of benzoyl chloride to 5 mL of water contained in an 18 × 150-mm test tube, and warm the mixture gently, with occasional shaking, until it is clear. Cool the solution and collect the precipitated benzoic acid. Compare the reactivity of this acid chloride to that of acetyl chloride by *cautiously* adding 2 to 3 drops of the latter to 5 mL of ice-cold water contained in a test tube, carefully shaking the tube, and observing the result.

2.   Ammonolysis.   Add 1 mL of benzoyl chloride dropwise to 5 mL of cold concentrated ammonium hydroxide in an 18 × 150-mm test tube, stopper the tube, and shake for about 2 to 3 min, venting the test tube periodically to relieve any pressure. Pour the mixture into 10 mL of water contained in an Erlenmeyer flask, and collect the precipitated benzamide. Determine the yield and melting point of this product. If you are to prepare enough benzamide to use in the preparation of benzonitrile described in part D, the amounts of reagents used in the ammonolysis must be increased by a factor of 15.

Benzanilide, $C_6H_5CONHC_6H_5$, can be prepared by combining 0.027 mol of aniline and 0.017 mol of benzoyl chloride in a 25-mL Erlenmeyer flask and adding, in portions and with vigorous shaking, 12 mL of 3 *M* aqueous sodium hydroxide. Isolate the product by vacuum filtration, and recrystallize the crude benzanilide from 95% ethanol. The reported melting point of this compound is 163°.

3.   Alcoholysis.   Methyl benzoate can be prepared by adding 0.01 mol of benzoyl chloride to 0.05 mol of methanol contained in an 18 × 150-mm test tube. Warm the tube gently on a steam cone for 2 to 3 min, and then add 2 mL of saturated sodium chloride solution. Note the odor of the layer which separates.

To prepare phenyl benzoate, add 1.5 mL of 90% aqueous phenol[2] to 10 mL of 3 M aqueous sodium hydroxide contained in a 25-mL Erlenmeyer flask, and to this solution add 1 mL of benzoyl chloride. Shake the mixture for 5 to 10 min, cool it in an ice-water bath, and isolate the product by vacuum filtration. Phenyl benzoate can be recrystallized from 95% ethanol-water and has a reported melting point of 72°.

4. Conversion to Benzoic Anhydride. Add 0.056 mol of benzoyl chloride to 15 mL of dry dioxane contained in a 50-mL Erlenmeyer flask, stopper the flask, and cool the solution to about 5°. Add 0.060 mol of pyridine, which has been dried over potassium hydroxide, and swirl the mixture for a few minutes, keeping the temperature below 10°. Introduce 0.028 mol of water with vigorous swirling, and then allow the solution to stand, with occasional swirling, in the ice-water bath for about 10 min. Pour the reaction mixture into a well-stirred mixture of 25 mL of concentrated hydrochloric acid, 40 g of cracked ice, and 100 mL of water, and collect the solid benzoic anhydride by vacuum filtration. (If the product should "oil out" rather than crystallize, cool the solution in an ice-water bath, and scratch to induce crystallization.) Shake the solid with 15 mL of cold 0.6 M aqueous sodium bicarbonate for a minute to remove any residual acid, and then filter the mixture. Wash the filter cake with cold water. The crude benzoic anhydride can be recrystallized by dissolving it in 10–15 mL of 95% ethanol, adjusting the temperature of the solution to 40°, and adding water until the mixture just becomes turbid. Allow the solution to cool to room temperature, and scratch to induce crystallization. It may be necessary to add a seed crystal and cool the solution in an ice-water bath to obtain the product. Complete the recrystallization by cooling the solution thoroughly in an ice-water bath, and isolate the product. Pure benzoic anhydride is colorless, mp 42–43°.

D. Preparation of Benzonitrile: Dehydration of an Amide

Dissolve 0.083 mol of finely pulverized benzamide in 50 mL of toluene contained in a 100-mL round-bottomed flask. Equip the flask with a reflux condenser, and connect the latter to a gas trap, as shown in Figure 5.1. Add 0.084 mol of thionyl chloride which has previously been measured out *in the hood* (see footnote 1). Observe the resulting solution to determine if some white crystals can be seen suspended in it. Bring the mixture to a gentle reflux, and continue the reflux until the evolution of HCl and SO_2 ceases; about 1 hr will be needed.★ Attach the flask to an apparatus set for simple distillation, and distil the toluene using a microburner or a heating mantle (*Caution: Toluene is flammable. All connections in the apparatus must be tight.*)★ Transfer the residue to a 25-mL flask, rinsing the larger flask with a little toluene and adding the rinse to the 25-mL flask, and continue the distillation. After all of the toluene has distilled, drain the water from the condenser and distil the benzonitrile (bp 188–192°). Record the yield of product.

[2]See paragraph 3 under Do It Safely.

EXERCISES

1. Write a balanced equation for conversion of ethylbenzene and permanganate to benzoic acid and manganese dioxide in the presence of hydroxide ion.
2. Write possible structures for a dicarboxylic acid which has the molecular formula $C_4H_6O_4$.
3. Calculate the equivalent weight of an acid, 0.2370 g of which could be titrated to a phenolphthalein end point with 27.50 ml of 0.0750 N sodium hydroxide solution.
4. Write a mechanism for the conversion of a carboxylic acid to an acid chloride with thionyl chloride.
5. Would it be feasible to attempt the preparation of benzaldehyde by permanganate oxidation of toluene? Why or why not?
6. Would you expect a carboxylic acid to be more or less soluble in water than in water containing mineral acids? Explain.
7. Write a balanced equation for reaction of benzoic acid with sodium bicarbonate.
8. Suggest a reason why phenols are less acidic than carboxylic acids.
9. Assuming that the equilibrium constant for esterification of benzoic acid with methanol is 3, calculate the yield of methyl benzoate expected using the molar amounts employed in this experiment.
10. Show how the expression for the saponification equivalent is derived.
11. Calculate the saponification equivalent that would be expected for dimethyl succinate, $CH_3O_2CCH_2CH_2CO_2CH_3$.
12. Give a mechanism for the hydrolysis of an ester in acidic and in basic media.
13. Why is it important that all apparatus used in the preparation of an acid chloride be as dry as possible?
14. Why is it possible to prepare unsubstituted amides by reaction of an acid chloride with ammonium hydroxide, which is nothing more than an *aqueous* solution of ammonia?
15. Suggest why acetyl chloride is significantly more reactive than benzoyl chloride.
16. Why must one carry out the reaction of benzoyl chloride with phenol in basic solution when no base is required for reaction of the acid halide with methanol?
17. Explain why anhydrides are less susceptible to nucleophilic attack than are acid chlorides. What evidence is there from experiments described in this chapter that this may be true? (*Hint:* Consider the method of purification of benzoic anhydride.)
18. Why might cyclohexane be less suitable than dioxane as the solvent for the preparation of benzoic anhydride described in this chapter?
19. What might the crystals be that can sometimes be observed when benzamide and thionyl chloride are combined?

SPECTRA OF STARTING MATERIALS AND PRODUCTS

For spectra of benzoic acid and of methyl benzoate, see Figures 12.1, 12.2, 12.3, and 12.4.

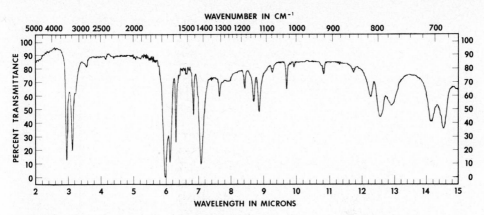

FIGURE 15.2 IR spectrum of benzamide.

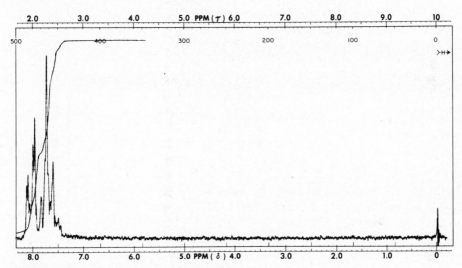

FIGURE 15.3 PMR spectrum of benzamide.

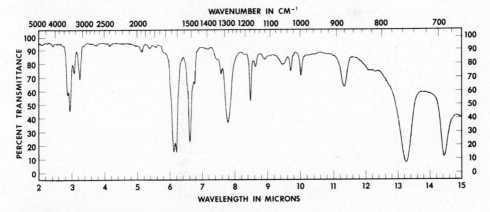

FIGURE 15.4 IR spectrum of aniline.

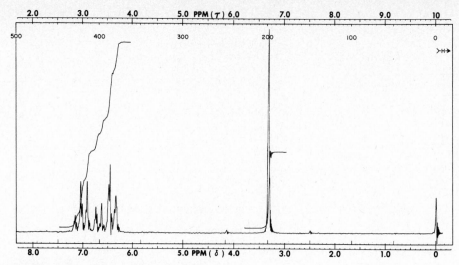

FIGURE 15.5 PMR spectrum of aniline.

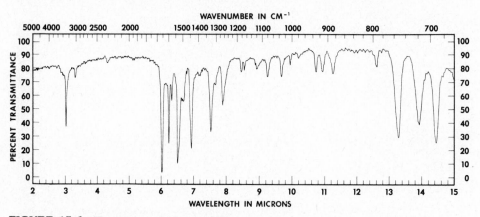

FIGURE 15.6 IR spectrum of benzanilide.

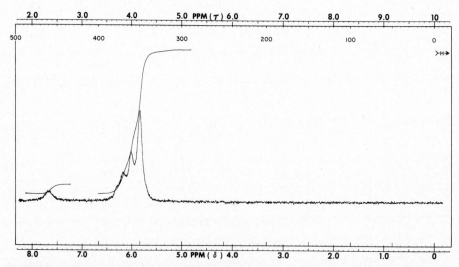

FIGURE 15.7 PMR spectrum of benzanilide.

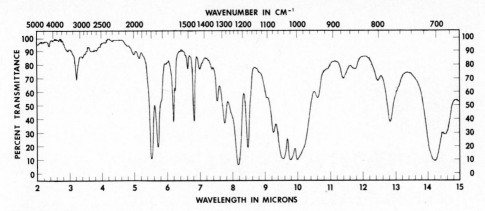

FIGURE 15.8   IR spectrum of benzoic anhydride.

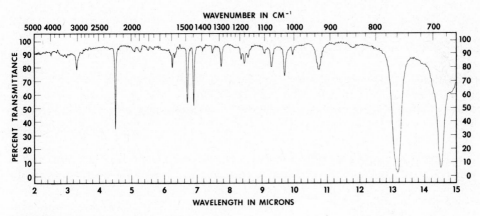

FIGURE 15.9   IR spectrum of benzonitrile.

16

HETEROCYCLIC SYNTHESIS

Heterocyclic compounds (heterocycles) are cyclic organic compounds which contain, in addition to carbon atoms, one or more atoms such as oxygen, nitrogen, or sulfur as part of a ring. These three heteroatoms are by far the most common. The importance of heterocycles in organic chemistry is emphasized by the fact that about one-third of the pages of *Beilstein* (see Chapter 26, under Class C heading) are devoted to them, and about one-fourth of the research articles in organic chemistry being published currently involve them. If we include the organic compounds in biochemical publications, the proportion of heterocyclic compounds is even much higher. This is because heterocyclic compounds occur widely in nature, and many others are synthesized because of the broad spectrum of biological activity they exhibit.

Oxygen-containing heterocycles are found in carbohydrates, flower pigments, and marijuana; nitrogen heterocycles occur in proteins, alkaloids, nucleic acids, vitamins, and coenzymes. Some of these latter compounds also contain sulfur heterocycles. Almost all medicinal compounds—whether natural or synthetic—contain one or more heterocyclic rings.

Heterocyclic compounds may usefully be divided into two classes in the same way as carbocyclic compounds, that is, aliphatic and aromatic. The chemistry of the aliphatic heterocyclic compounds requires little special attention because it is generally predictable on the basis of a knowledge of noncyclic compounds. Exceptions to this rule are the unusual properties of those heterocycles with very small rings (three- and four-members) and certain transannular reactions of medium-ring compounds.

Heterocyclic compounds classed as aromatic have properties in common with aromatic carbocyclic compounds, as might be expected; however, they may exhibit substantial differences. These differences may generally be attributed to the presence of polarized bonds resulting from differences in electronegativity between heteroatoms and carbon atoms or to the nonbonding electron pairs that the heteroatoms possess. Formally, aromatic heterocycles are related to aromatic hydrocarbons by (1) replacement of one or more carbon atoms by isoelectric heteroatoms (those providing an equal number of electrons for covalent bonding) and (2) replacement of a carbon-carbon double bond of a Kekulé structure by a heteroatom with at least one pair of unshared electrons. Some heterocycles contain both types of replacements. Examples of common parent ring systems which illustrate these relationships are shown here.

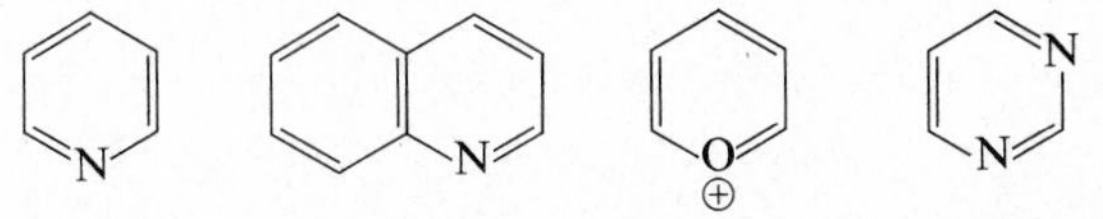

Pyridine   Quinoline   Pyrilium Ion   Pyrimidine

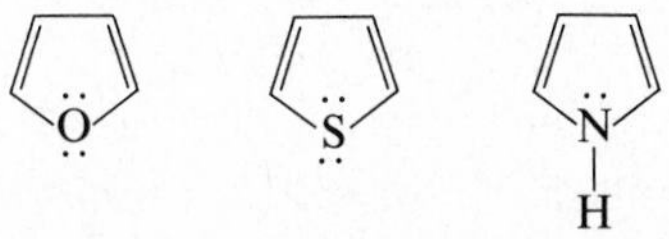

Furan   Thiophene   Pyrrole

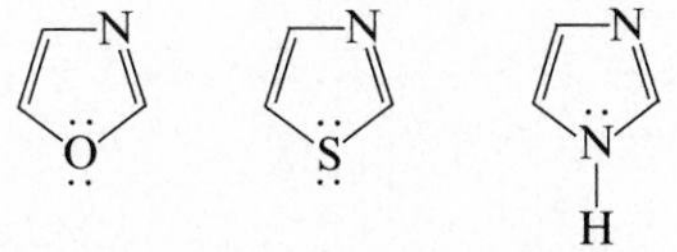

Oxazole   Thiazole   Imidazole

Four heterocyclic bases are involved in the celebrated genetic sequence code of DNA. Two of these (cytosine and thymine) are derivatives of pyrimidine, and the other two (adenine and guanine) are derivatives of purine, which may be considered to be a fusion of pyrimidine and imidazole.

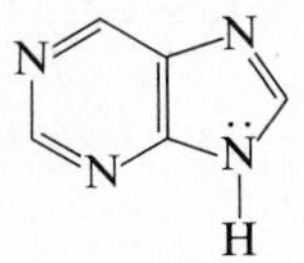

Purine

In this chapter we are not concerned primarily with the chemical properties of heterocyclic compounds but rather with the synthesis of two typical aromatic heterocycles. These experiments are illustrative of many similar synthetic procedures used to produce a wide variety of biologically active compounds.

16.1 2-Aminothiazole

The thiazole ring system is found in various naturally occurring compounds, the most important of which is vitamin B$_1$, or thiamine. This "vital amine" is one of the most important vitamins, and indeed the name *vitamin* resulted from this fact. It functions as a coenzyme in critical biochemical decarboxylations; its complete absence in the diet leads to beriberi in humans and to death in a week or two in experimental animals.

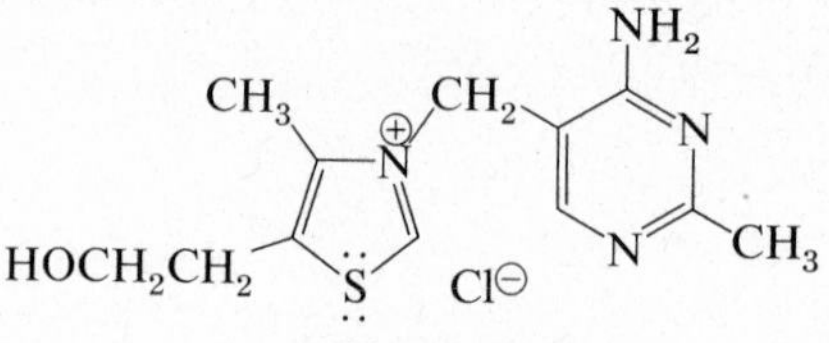

Thiamine

Sulfathiazole has been one of the more successful antibacterial drugs. A late step in its synthesis consists of reaction of 2-aminothiazole with *p*-acetamidobenzenesulfonyl chloride, with the ultimate formation of a heterocyclic derivative of sulfanilamide (see equation 13 in Chapter 17). In this chapter we are concerned with the preparation of the 2-aminothiazole used in this synthesis.

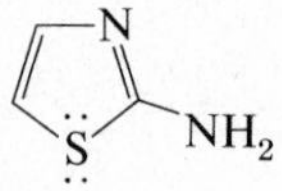

2-Aminothiazole

Because the thiazole ring system is not nearly as stable as the analogous carbon ring system, cyclopentadiene, it is not possible to substitute directly on the thiazole ring without destroying the molecule. Consequently most thiazole systems are made by simple condensation reactions. The preparation of 2-aminothiazole is typical of the general synthetic route to this type of compound, namely, condensation of an α-haloketone or aldehyde with a thioamide.

The main reaction leading to the preparation of 2-aminothiazole is shown in equation 1. Chloroacetaldehyde is condensed with thiourea by heating an aqueous solution of the two at reflux.

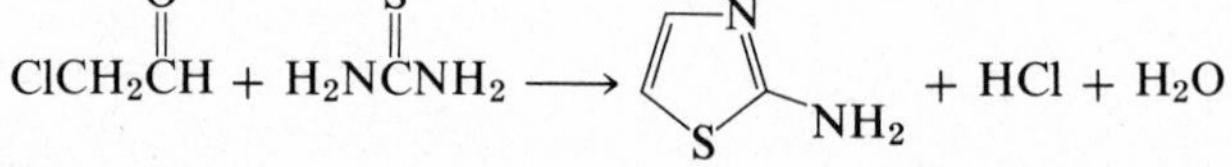

$$\text{ClCH}_2\overset{\overset{\text{O}}{\|}}{\text{CH}} + \text{H}_2\text{N}\overset{\overset{\text{S}}{\|}}{\text{C}}\text{NH}_2 \longrightarrow \quad + \text{HCl} + \text{H}_2\text{O} \qquad (1)$$

Chloroacetaldehyde is extensively hydrated in aqueous solution and can be obtained commercially as a 30–40% solution in water. Alternatively, it can be prepared *in situ* from acid hydrolysis of either an α,β-dichloroethyl alkyl ether (equation 2) or of a dialkyl chloroacetal (equation 3).

$$\text{ClCH}_2\overset{\overset{\text{OR}}{|}}{\text{CH}}\text{---Cl} \xrightarrow{\text{H}_3\text{O}^{\oplus}} \text{ClCH}_2\overset{\overset{\text{O}}{\|}}{\text{CH}} + \text{HCl} + \text{ROH} \qquad (2)$$

$$\text{ClCH}_2\overset{\overset{\text{OR}}{|}}{\text{CH}}\text{---OR} \xrightarrow{\text{H}_3\text{O}^{\oplus}} \text{ClCH}_2\overset{\overset{\text{O}}{\|}}{\text{CH}} + 2\,\text{ROH} \qquad (3)$$

Chloroacetaldehyde is believed to condense with thiourea through the enol form of the latter (perhaps more properly called an "iminethiol"). Nucleophilic addition of the imino group to the carbonyl function, followed by the loss of a molecule of water, gives 2-aminothiazole.[1] A plausible mechanism for these steps is shown in equation 4.

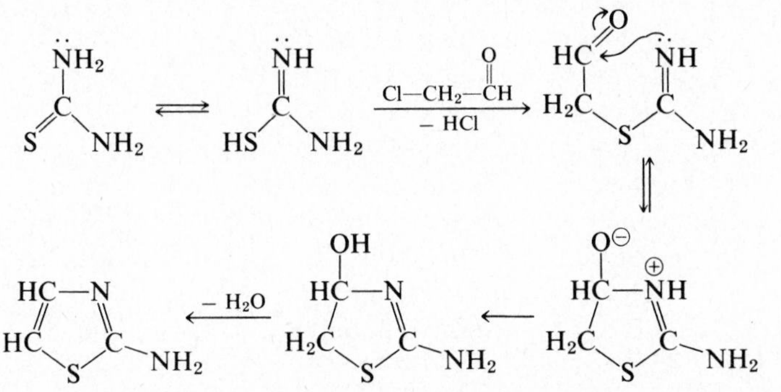

$$(4)$$

Important side reactions include the polymerization of chloroacetaldehyde, which produces materials that are water soluble and are removed easily during filtration.

EXPERIMENTAL PROCEDURE

DO IT SAFELY

This procedure calls for filtration of hot toluene and distillation of toluene, which is flammable. The boiling point of toluene is too high (110°) for it to be distilled from a steam bath, so a burner or electrical heating must be used. Take proper care to avoid a fire.

Prepare a solution of 0.2 mol of thiourea and 90 mL of water in a 500-mL round-bottomed flask; the water may need to be warmed to effect complete solution. Add 0.2 mol of chloroacetaldehyde dimethyl acetal and 1.5 mL of 85% phosphoric acid,[2] and equip the flask with an efficient reflux condenser. Heat the mixture under reflux with a heating mantle for 2.5 hr. Fit the flask for simple distillation, and slowly collect and then discard 60 mL of distillate. Cool the reaction mixture in an

[1] Hydrogen chloride is produced as the condensation reaction (equation 4) proceeds and converts the 2-aminothiazole into its hydrochloric acid salt.

[2] Alternative ways of obtaining chloroacetaldehyde are discussed in the text; if one of them is to be used, *substitute* the following reagents for the chloroacetaldehyde dimethyl acetal and phosphoric acid:

1. *Use of α,β-dichloroethyl ethyl ether:* Substitute 0.2 mol of α,β-dichloroethyl ethyl ether for the above reagents.

2. *Use of 40% aqueous chloroacetaldehyde:* Dissolve the thiourea in 44 mL of water and add 36 mL of a 40% aqueous solution of chloroacetaldehyde.

ice-water bath, and make it slightly basic to litmus by slow addition of 12 *M* sodium hydroxide solution. Avoid a large excess of base. After cooling for an additional 15 min, collect the precipitated solid by suction filtration. Concentrate the mother liquor to about two-thirds its initial volume, readjust the pH by adding more base until just basic to litmus, and recool. Collect the second crop of crystals and combine it with the first crop. Wash the product twice with 20-mL portions of ice-cold half-saturated sodium bisulfite solution, and press as dry as possible.★

For purification of this crude product, dissolve it in hot toluene (10 mL per gram of product), treat with decolorizing carbon, and filter by gravity while hot; filter-aid will help in the complete removal of the decolorizing carbon. Collect the filtrate in a distilling flask, and distil about one-half the toluene from the mixture. If the pressing of the initial precipitate has been thorough, the small amount of water retained is removed at this point by azeotropic distillation with toluene. When the remaining toluene solution is cooled, 2-aminothiazole (mp 89–90°) is obtained. Dry the solid, and determine its melting point and the yield.

EXERCISES

1. Explain why chloroacetaldehyde forms a much more stable hydrate than acetaldehyde.
2. What is the purpose of washing the crude 2-aminothiazole with sodium bisulfite solution?

SPECTRA OF PRODUCT

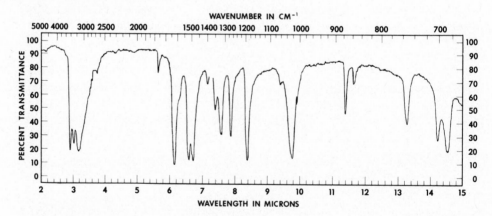

FIGURE 16.1 IR spectrum of 2-aminothiazole.

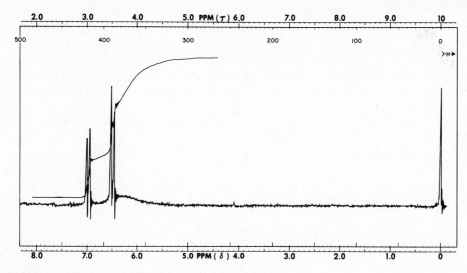

FIGURE 16.2   PMR spectrum of 2-aminothiazole.

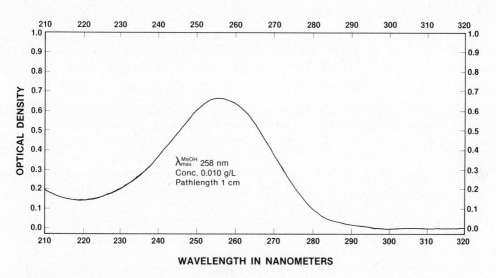

λ_{max}^{MeOH} 258 nm
Conc. 0.010 g/L
Pathlength 1 cm

WAVELENGTH IN NANOMETERS

FIGURE 16.3   UV spectrum of 2-aminothiazole.

16.2   4-Aminoquinolines

Quinine, which is obtained from the bark of the cinchona tree, has long been used as an antimalarial drug. One of the first successful synthetic antimalarial drugs

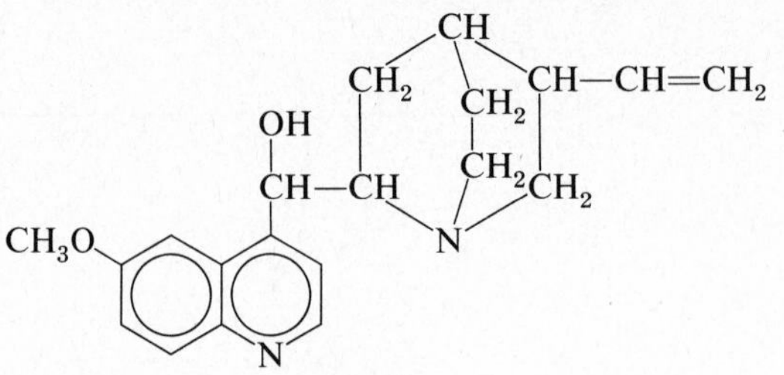

Quinine

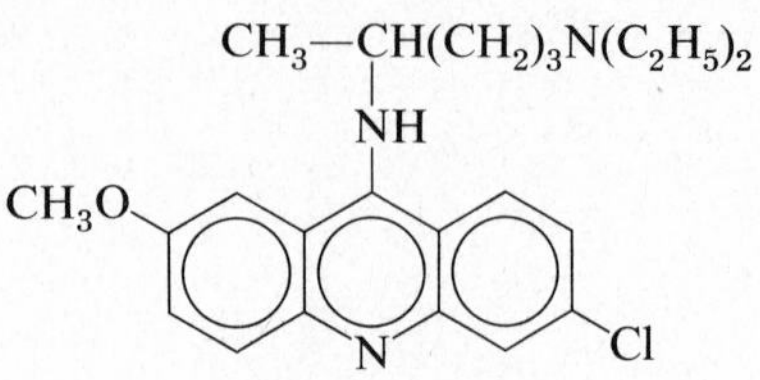

Quinacrine

was quinacrine (or Atebrin), first produced in Germany. During and after World War II thousands of heterocyclic compounds related structurally to quinine and quinacrine were synthesized by organic chemists and tested for antimalarial effectiveness. Many of these were derivatives of 4-aminoquinoline. One of the most successful was chloroquine (9a), which may be seen to be related structurally to both quinine and quinacrine.

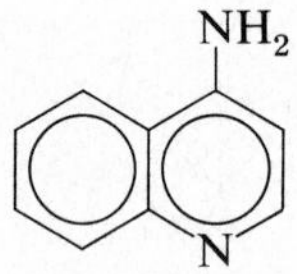

4-Aminoquinoline

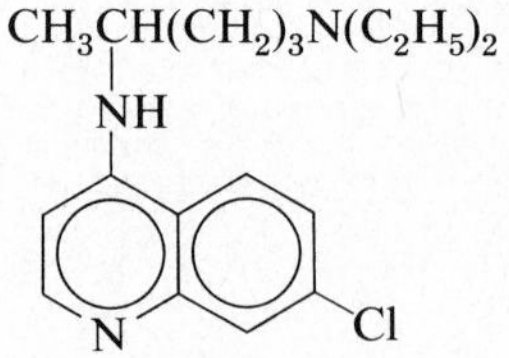

Chloroquine (9a)

A practical scheme for the synthesis of chloroquine and of other 4-aminoquinoline derivatives is outlined in Figure 16.4. This synthesis is typical of the majority of quinoline syntheses in that the cyclization is carried out to produce the pyridine ring rather than the benzene ring. Besides *m*-chloroaniline, many other substituted anilines may be used to produce quinolines having other substituents in the benzene ring, and the amine side chain may be varied by the reaction of different primary amines with 4,7-dichloroquinoline (7) in the last step.

In our experiment the primary amine used in the last step is *m*-chloroaniline, for several reasons. (1) It is relatively inexpensive. (2) It is the same compound needed for the first step of the synthesis. (3) It does not have the extremely disagreeable odor of the diamine required to produce chloroquine. (4) The product, 4-*m*-chloroanilino-7-chloroquinoline (9b), is a crystalline compound which is easily purified by recrystallization. It should be noted, however, that this product (9b) is *not* an effective antimalarial drug.

The first step in the synthesis is the condensation of *m*-chloroaniline (1) with diethyl ethoxymethylenemalonate (2) to produce 3, with the elimination of a molecule of ethanol (equation 5). A high-boiling solvent (diphenyl ether, bp 259°) is added to the crude product 3, which need not be isolated before carrying out the cyclization step (equation 6). The diphenyl ether solution is heated to reflux, where-

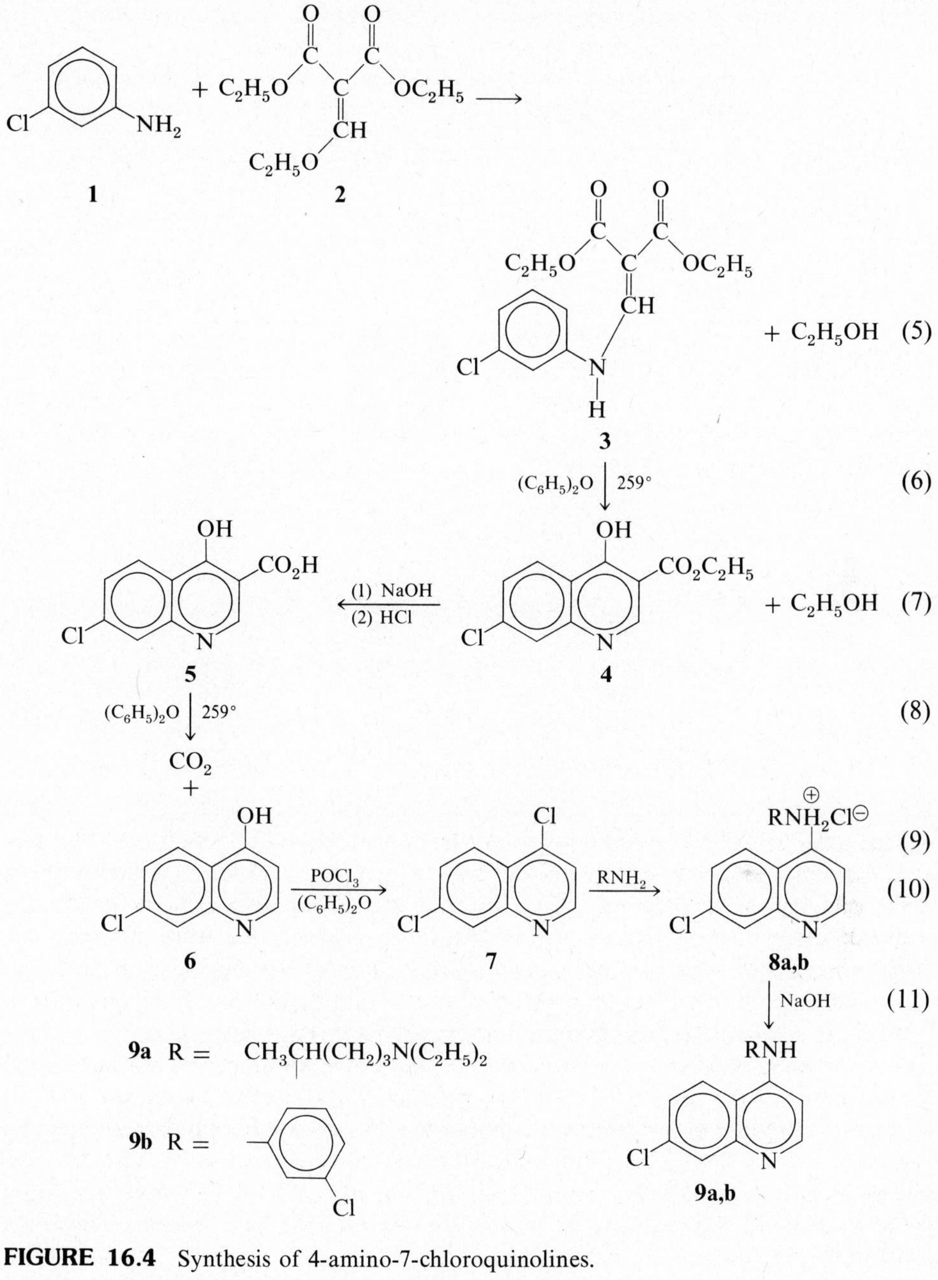

FIGURE 16.4 Synthesis of 4-amino-7-chloroquinolines.

upon the cyclization reaction begins, as is evidenced by the separation of crystals of 3-carbethoxy-7-chloro-4-hydroxyquinoline **(4)** from the boiling solution.

This cyclization step may be rationalized as involving an orbital symmetry allowed electrocyclic reaction, as shown in equation 12; see references 5 and 6 in

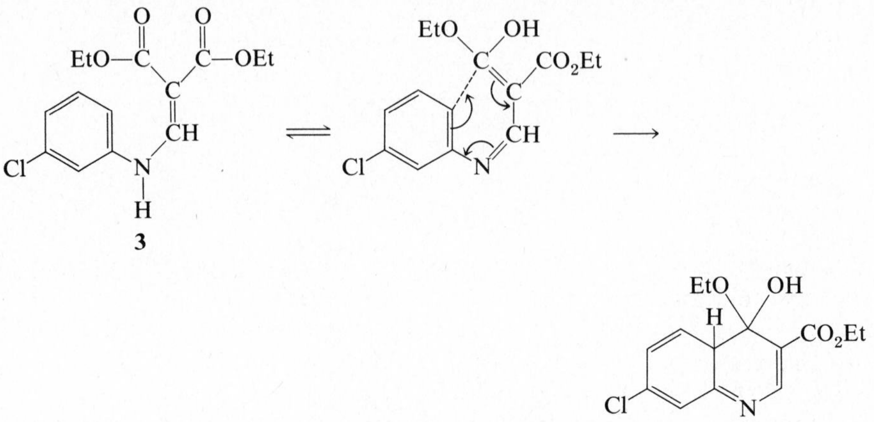

$$(12)$$

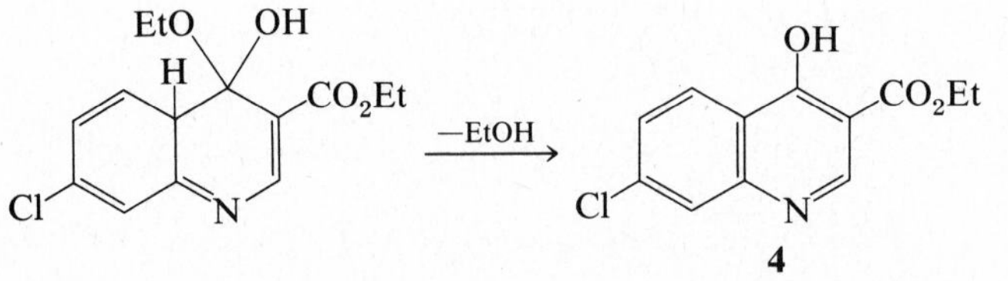

$$(13)$$

Chapter 8. The second carbethoxy group may appear to have no function in this step, and it is removed in subsequent steps. However, it apparently serves a useful purpose in influencing the *direction* of the cyclization. If ring closure occurred *ortho* to the chlorine atom of intermediate **3**, 5-chloroquinoline derivatives would result. The 5-chloroquinoline isomer of chloroquine is not an effective antimalarial; therefore any cyclization in this direction represents an undesirable side reaction. Fortunately, little of this occurs; the 5-chloroquinoline isomers have been detected only when the synthesis was carried out on a very large scale. It is interesting to speculate on the explanation for the selectivity of the ring closure (see Exercise 1).

The third step of the synthesis, hydrolysis of the ester **4**, is carried out by adding aqueous sodium hydroxide solution and heating to vigorous reflux. As hydrolysis occurs, the 7-chloro-4-hydroxyquinoline-3-carboxylic acid **(5)** produced goes into solution as the sodium salt. After cooling, the organic acid **(5)** is precipitated by the addition of hydrochloric acid.

The next step of the synthesis is the decarboxylation of **5**. This may be done successfully without drying or purification of the crude acid by heating it in diphenyl ether, the high-boiling solvent used in the previous steps. A clue to the ease of this decarboxylation may be gained from a consideration of a tautomeric form of the acid, **(5a)**, which may be seen to be a β-keto acid, a structure generally sensitive to decarboxylation.

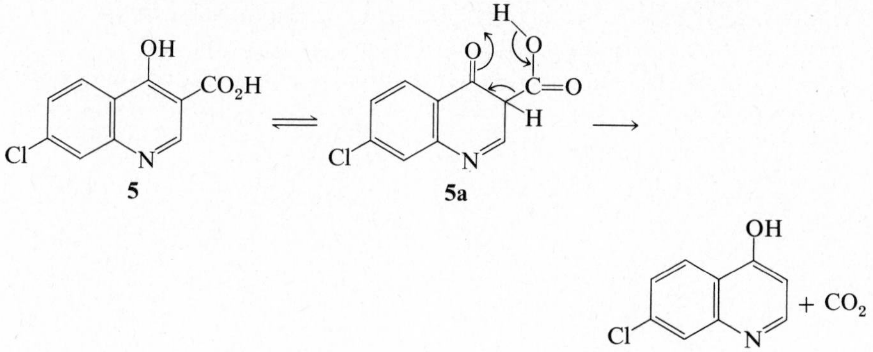

$$+ CO_2 \quad (14)$$

Subsequently, reaction with phosphorus oxychloride replaces the phenolic hydroxyl group of **6** with chlorine, producing 4,7-dichloroquinoline (**7**). The success of the final step depends on the wide difference in the ease of nucleophilic substitution of the two chlorine atoms in **7**. The chlorine in the 7-position has the inactivity characteristic of chlorobenzene, whereas the chlorine in the 4-position is quite reactive toward nucleophilic aromatic substitution by a primary amine. This is owing to the activating effect of the nitrogen in the same ring. The nitrogen is able to

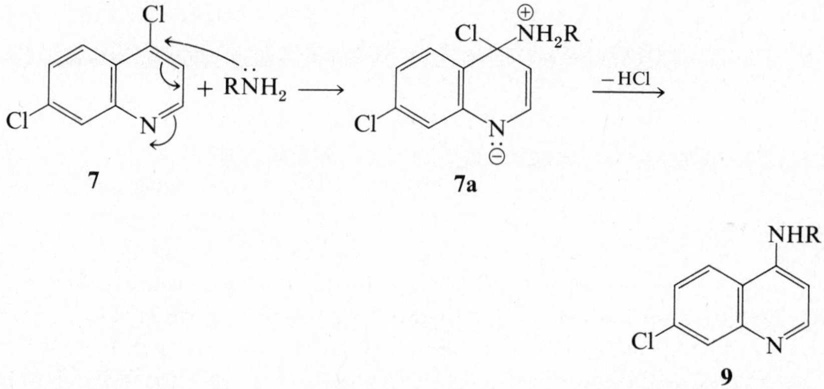

$$(15)$$

stabilize the intermediate ion **7a** (one of several resonance structures) by bearing the negative charge. Although the 7-chlorine may be considered also to be in conjugation with the nitrogen, the negative charge cannot be transmitted to nitrogen without disrupting the aromatic resonance of both rings (equation 16), whereas this is not true in the case of reaction at the 4-position (equation 15).

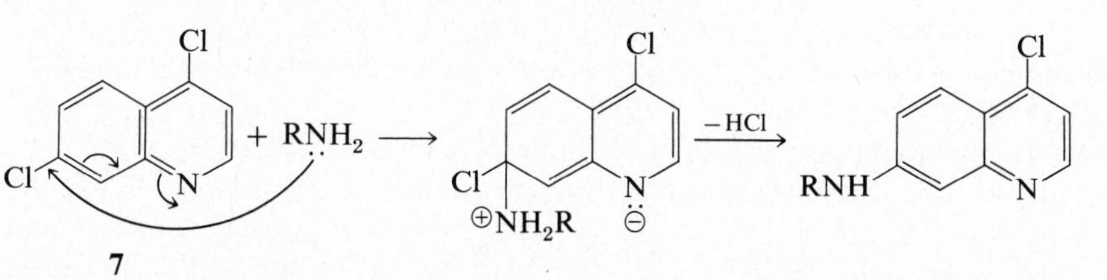

$$(16)$$

The last reaction is best carried out in the presence of a little hydrochloric acid, which acts as a catalyst, according to equation 17. Protonation by the acid of the basic nitrogen of the ring further enhances the reactivity of the ring toward nucleophilic aromatic substitution.

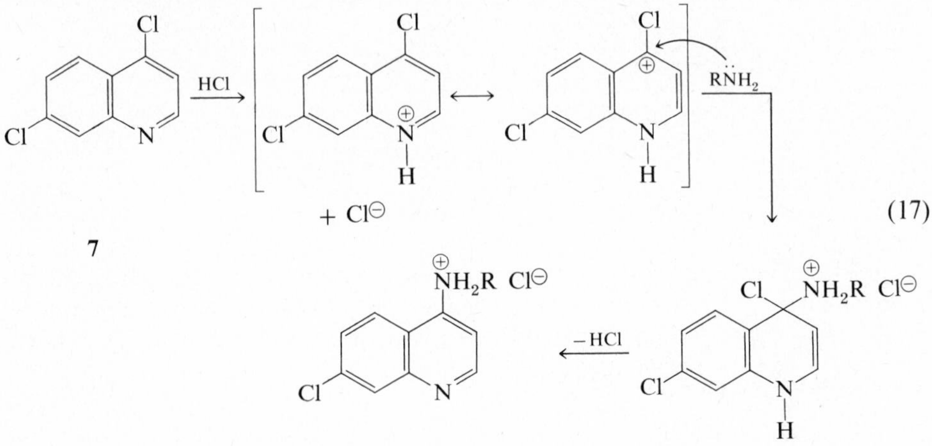

7

8a,b

(17)

EXPERIMENTAL PROCEDURE

A.   3-Carboethoxy-7-chloro-4-hydroxyquinoline (4)

DO IT SAFELY
---

1.   During a portion of this experiment it is necessary to use a burner to attain the required reaction temperature; in assembling the apparatus pictured in Figure 16.5, be sure that all ground-glass joints are greased and tight to avoid leakage of flammable vapors.

2.   Note the cautions given in the procedures about taking the melting points of the high-melting products. Do *not* use a liquid-bath apparatus for these determinations.

Equip a 250-mL round-bottomed flask with a still head to which a capillary ebullition tube is attached by means of a Neoprene thermometer adapter. Connect a vacuum adapter to the still head, and attach a small, round-bottomed receiving flask to the vacuum adapter, as shown in Figure 16.5. In the 250-mL flask place 0.050 mol of *m*-chloroaniline and 0.051 mol of diethyl ethoxymethylenemalonate; connect the vacuum adapter to a water aspirator and heat the reaction mixture on a steam cone under vacuum for 1 hr. During this time the ethanol produced according to equation 5 will be removed and will pass into the water stream of the aspirator before

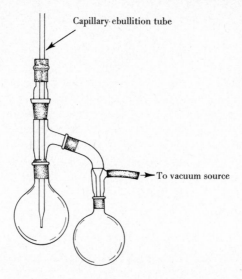

Capillary·ebullition tube

To vacuum source

FIGURE 16.5 Apparatus for preparation of 3-carboethoxy-7-chloro-4-hydroxyquino-line **(4)**.

condensing. Disconnect the aspirator hose from the adapter and then remove the still head from the reaction flask.[★3]

To the reaction mixture add 50 mL of diphenyl ether, and attach a reflux condenser *without water hoses connected* and heat the mixture to reflux, using a flame diffused by a wire gauze. Continue heating for 15 min after reflux temperature has been attained. During this period crystals of 3-carbethoxy-7-chloro-4-hydroxyquin-oline **(4)** will separate from the boiling solution. Remove the flame and allow the reaction mixture to cool to room temperature. Collect the crystalline ester **(4)** by vacuum filtration; press the filter cake with a clean cork or stopper to remove the solvent. Transfer the crystals to a 250-mL beaker, and triturate (stir) them well with about 75 mL of petroleum ether (bp 60–80°); collect the crystals again on a filter, and wash them with a little more petroleum ether. Resuspend the crystals once more in about 50 mL of petroleum ether, stir, and filter again.[★4]

B. 7-Chloro-4-Hydroxyquinoline-3-carboxylic Acid (5)

Place the crystals of 3-carbethoxy-7-chloro-4-hydroxyquinoline **(4)** in a 250-mL round-bottomed flask, and add 50 mL of 3 *M* sodium hydroxide solution and a few boiling stones. Attach a water-cooled reflux condenser, and heat the mixture to vigorous reflux for 1 hr.[★] Discontinue heating the reaction mixture, and allow it to cool for a few minutes, disconnect the condenser, and add about 0.5 g of decolorizing

[3]If desired, a small sample of **3** may be removed by pipet; upon cooling to 0°, it should crystallize; it may be purified by recrystallization from petroleum ether (bp 60–80°). The melting point of pure **3** is 55–56°.

[4]If desired, a sample of 3-carbethoxy-7-chloro-4-hydroxyquinoline **(4)** may be purified by recrystallization from pyridine. The reported mp is 295–297°. (*Caution:* Do not attempt to take a melting point over 200° with a liquid-bath apparatus; a metal block apparatus is appropriate for high-melting substances.)

charcoal. Replace the condenser and heat to reflux again for 5 min, swirling the flask if bumping occurs.★

Allow the reaction mixture to cool almost to room temperature and filter it, using a fluted filter. Wash the filter with 20 mL of hot water, adding the wash to the main filtrate. Acidify the filtrate to pH 4 with 3 *M* hydrochloric acid. Collect the 7-chloro-4-hydroxyquinoline-3-carboxylic acid **(5)** by suction filtration, and wash it on the filter with a little water.★ Take a small amount of the acid for purification by recrystallization from a large volume of ethanol. The reported melting point of the pure acid is 273–274° (dec). Owing to the high-melting behavior of **5**, only a melting-point apparatus utilizing a heated metal block (such as the Mel-temp apparatus shown in Figure 1.6) should be used to verify this melting point.

C.   4,7-Dichloroquinoline (7)

DO IT SAFELY
---

1.   Phosphorous oxychloride is lachrymatory and produces hydrogen chloride in contact with atmospheric water and moist mucous membranes; take appropriate precautions in handling it.

2.   4,7-Dichloroquinoline is also lachrymatory and is irritating to the skin. Be careful to avoid inhalation of its vapors and its contact with your skin. The use of rubber gloves is appropriate in this experiment.

Place the 7-chloro-4-hydroxyquinoline-3-carboxylic acid **(5)** from part B in a 250-mL round-bottomed flask with 50 mL of diphenyl ether and a few boiling stones. Attach a reflux condenser (without water cooling), and heat the mixture to reflux for 1 hr; the acid should dissolve in the hot solvent as the decarboxylation proceeds.

Allow the reaction mixture to cool to room temperature.★⁵ Add 5 mL of phosphorus oxychloride and some fresh boiling stones. Attach a water-cooled reflux condenser, and suspend a 360° thermometer from the top of the condenser by means of a copper wire so that the bulb of the thermometer is in the liquid reaction mixture. Heat the reaction mixture slowly to 135–140°, and swirl the flask from time to time during 1 hr of heating.★

When the reaction mixture has cooled to room temperature, transfer it to a separatory funnel with the aid of about an equal volume of diethyl ether and extract with four 50-mL portions of 3 *M* hydrochloric acid.⁶ Add a little crushed ice to the combined acid extracts, and neutralize with 3 *M* sodium hydroxide. Collect the precipitated 4,7-dichloroquinoline **(7)** on a filter. Resuspend the crystals in water, stir

---

[5]If isolation of pure 7-chloro-4-hydroxyquinoline is desired, pour out about 10 mL of the reaction mixture into a flask containing 10 mL of petroleum ether (bp 60–80°). Stir the mixture, and then collect the crystals and wash them with more petroleum ether. The crude product may be recrystallized from a large volume of water, using charcoal if necessary. The melting point of pure 7-chloro-4-hydroxyquinoline is 270–272°, sintering from 260° (use a metal-block apparatus).

[6]Take care to vent the separatory funnel immediately after each shaking operation because some heat may be produced in the extraction.

well, and collect them again on a filter. The product may be air-dried in a fume hood or, better, in a vacuum desiccator. The melting point of pure 4,7-dichloroquinoline is 83.5–84.5°.

D.   4-*m*-Chloroanilino-7-chloroquinoline (9b)

Place 2.0 g of 4,7-dichloroquinoline and 1.3 g of *m*-chloroaniline in 100 mL of water, and add a few drops of concentrated hydrochloric acid. Attach a reflux condenser, and heat the mixture to a gentle boil. Within a few minutes the hydrochloric acid salt of the product will begin to precipitate. Continue heating for an additional 10 min, allow the mixture to cool to room temperature, and collect the crystals by vacuum filtration. Resuspend the crystalline hydrochloride in about 20 mL of 3 *M* sodium hydroxide solution; heat the mixture to boiling for about 5 min while stirring. Collect the solid product, and recrystallize it from ethanol. Shining white plates of 4-*m*-chloroanilino-7-chloroquinoline should be obtained, mp 223–225°. (Read the last sentence of part B regarding the determination of melting point of high-melting substances.)

EXERCISES

1. 7-Chloro-4-hydroxyquinoline has also been synthesized starting with *m*-chloroaniline and diethyl oxaloacetate, which condense to give compound **10**. The sequence of reactions by which **10** is converted into 7-chloro-4-hydroxyquinoline

10

is parallel to that in the present synthesis. However, the decarboxylation of 7-chloro-4-hydroxyquinoline-2-carboxylic acid is more difficult than that of the isomer **5**, and a significant amount of 5-chloro-4-hydroxyquinoline is found, as well as the desired product, **6**.

   Outline all the steps in the synthesis of **6** from **10**, and suggest an explanation for the more difficult decarboxylation and for the production of 5-chloro-4-hydroxyquinoline.

2. Why is water *not* circulated through the reflux condenser during some of the heating periods in the procedures?

3. In taking the melting point of 7-chloro-4-hydroxyquinoline-3-carboxylic acid, a student noticed that the sample began to froth and bubble as it melted in the capillary tube. Explain.

4. (a) Identify the three major pmr absorptions in the spectrum of diethyl ethoxymethylenemalonate with the responsible hydrogen atoms. (b) Rationalize the very low field position of the absorption at 7.5 δ. (c) What does the pmr spectrum tell you about the equivalence or nonequivalence of the three ethyl moieties in this molecule?

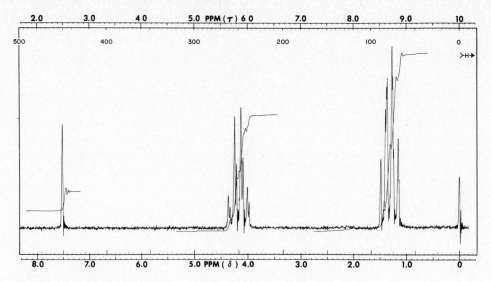

FIGURE 16.6 PMR spectrum of diethyl ethoxymethylenemalonate.

17

MULTISTEP ORGANIC SYNTHESES

17.1 Introduction

The synthesis of complex molecules from simpler starting materials is one of the most important aspects of organic chemistry. Some syntheses require as many as 20 or 30 sequential reactions or "steps." Multistep syntheses of such lengths are usually economically feasible only for end products that are biologically active and which cannot be obtained readily from natural sources (for example, cortisone). This is because of the cumulative effect on overall yield of less-than-quantitative yields of individual steps when there are many such steps. For example, even a five-step synthesis in which each step occurs in 80% yield gives an overall yield of only 33% [$(0.8)^5 \times 100 = 32.8$]. To offset this cumulative effect the synthetic chemist searches for reaction sequences capable of high yields and seeks to develop optimum experimental techniques to minimize losses in each step. A remarkably successful example of such an effort is the synthesis of the enzyme ribonuclease by R. B. Merrifield. This enzyme is a protein having 124 amino acid units (see Chapter 23). Its synthesis required putting the amino acid units together in the proper sequence by 369 chemical reactions, in which nearly 12,000 experimental operations such as additions of reagents, filtrations, and washings were involved. To accomplish this efficiently, Merrifield and his coworkers developed a "synthesis machine" which could be programmed to carry out the operations automatically. Although the overall yield of the biologically active enzyme was only 17%, because of the enormous number of steps, this represented an *average yield of over 99% for each step!*

Perhaps the most complicated natural product ever synthesized is vitamin B_{12} (Figure 17.1), whose total synthesis was announced in 1972 by R. B. Woodward and A. Eschenmoser as a result of collaborative work at Harvard and Zürich by 100 chemists from 19 different countries over a period of 11 years. Although the synthesis

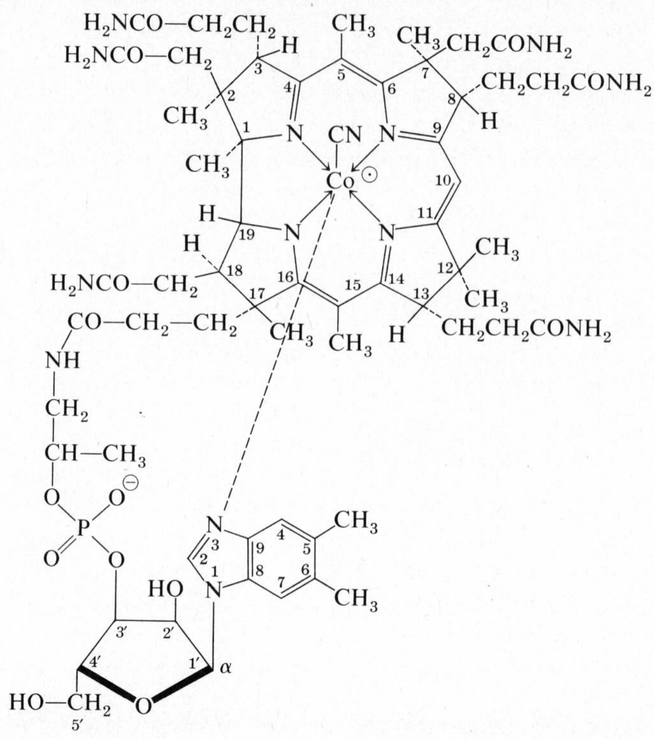

FIGURE 17.1 Structure of vitamin B_{12}.

will never be used as a practical source of the vitamin, it was a landmark in organic synthesis and was of significant value in that new reactions, techniques, and theories were developed in the course of the work, including the Woodward-Hoffman principle of conservation of orbital symmetry (see references 5 and 6 in Chapter 8).

The synthesis with the smallest number of steps is not invariably the best route to a target molecule. As a simple example, consider the synthesis of 2-chloropropane. It can be produced from propane in one step by direct chlorination, but it must be separated from 1-chloropropane which is formed concurrently. If the mixture of 1- and 2-chloropropane is treated with a base, both isomers undergo dehydrochlorination to propene, and subsequent addition of hydrogen chloride produces 2-chloropropane as the sole product. In this case the three-step process is the preferred way to obtain pure 2-chloropropane. Thus it may be seen that ease of separation and purification of desired product from by-products is an important consideration. Other considerations are availability and cost of starting materials, simplicity of equipment and instrumentation, energy costs (for example, low temperature reactions are desirable), activity of catalysts, regioselectivity of reactions in polyfunctional molecules, and stereochemical control.

Since so many variables must be considered in planning a synthesis for a complex molecule, it is not surprising that organic chemists are investigating the possibility of using computers to design multistep syntheses. The memory capacity and data-retrieval capabilities of computers can be put to use in handling an enormous amount of information of the kind mentioned, and it is hoped that optimum sequences of reaction steps can some day be predicted by computer programs. In considering this fascinating prospect, however, it is well to keep in mind that it is the information provided to the computer by humans (chemists) that will be vital to the success of such a project.

Some insight into the kind of information needed in the design of syntheses for complex molecules can be gained from carrying out a synthesis consisting of a fairly small number of steps. This chapter provides such an introduction by means of two examples, the synthesis of sulfathiazole (Section 17.2) and the synthesis of 1-bromo-3-chloro-5-iodobenzene (Section 17.3). There is also the possibility of putting together additional synthetic sequences by combining some of the separate experiments described in other chapters in this book. Suggestions for two such sequences are given in Section 17.4.

These syntheses provide experience in using the product of one reaction in subsequent steps, an experience which emphasizes the importance of good experimental technique and gives some insight into the excitement of producing a complex organic substance from simpler starting materials. In some of the procedures there are alternatives either of isolating intermediate products in pure form or of using crude reaction mixtures in the next step. This is the sort of choice the synthetic organic chemist faces constantly. It will be instructive for some students in a laboratory section to be assigned to one procedure and other students to the other; the two groups will then have an opportunity to compare the results of the alternative procedures.

17.2   The Synthesis of Sulfathiazole

Some of the most effective drugs belong to a general class called "sulfa" drugs, the members of which have a common structural feature in that they are derivatives of the parent compound, *p*-aminobenzenesulfonamide **(1)**, often called sulfanilamide. It should be noted that this compound contains two different —NH_2 moieties, one as the amino group and the other as the sulfonamido (—SO_2NH_2) group. Different substituents on the sulfonamido nitrogen give rise to the family of sulfa drugs.

$$H_2N—\langle\bigcirc\rangle—SO_2NH_2 \qquad H_2N—\langle\bigcirc\rangle—SO_2NH—\underset{S}{\overset{N}{\diagup\diagdown}}$$

$$\textbf{1} \qquad\qquad\qquad\qquad \textbf{2}$$

One of the more familiar examples of a sulfa drug is sulfathiazole **(2)**, whose total synthesis is outlined in Figure 17.2. Owing to the toxicity of benzene, we shall begin our experiment with the second step, using nitrobenzene as our starting material. (The theory of nitration of benzene and other aromatic hydrocarbons is described in Section 7.3.) The preparations of several interesting intermediate compounds, such as sulfanilamide itself, are included, and many of them can be isolated

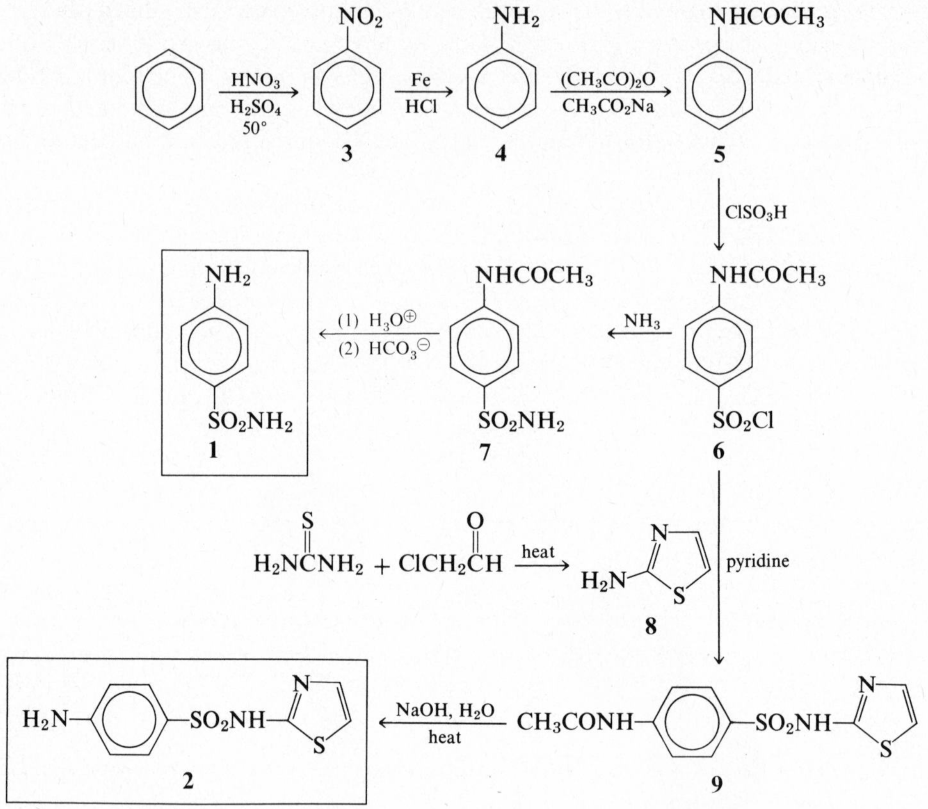

FIGURE 17.2 Laboratory synthesis of sulfanilamide (1) and sulfathiazole (2).

if desired or can be used in subsequent steps without extensive purification. A discussion of the reactions involved in each step is included.

Reduction of Aromatic Nitro Compounds: Aniline (4). A variety of methods is available for the reduction of an aromatic nitro compound to the corresponding amine. It should be noted that reduction of aliphatic nitro compounds is seldom found, owing to the difficulty in obtaining the starting compounds; on the other hand, aromatic nitro compounds are easily obtained. The general methods of reduction include catalytic hydrogenation and electrolytic and chemical reduction. In the laboratory the most commonly used method is chemical reduction, in which various metals are used in acidic solution; the most important commercial method is probably catalytic hydrogenation.

The mechanisms of these reductions are not well understood. The fact that various stable intermediate compounds have actually been isolated suggests a stepwise reaction. Some of the intermediates for the reduction are shown in equation 1. Differences in rates of reduction of the various intermediates may be important, but the nature of the reducing agent dictates the final product. For example, the reduction of nitrobenzene with tin metal and hydrochloric acid gives only aniline, whereas the use of zinc metal and ammonium chloride gives only N-phenylhydrox-ylamine **(11)**. Note that in equation 1, [H] means reduction by any of the possible reducing agents.

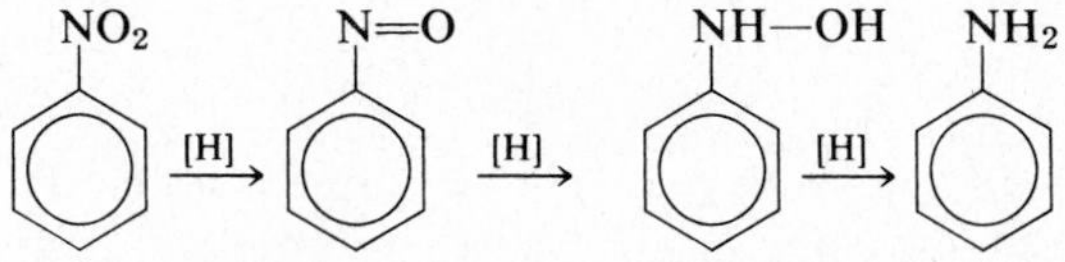

$$\underset{3}{\underset{NO_2}{\bigcirc}} \xrightarrow{[H]} \underset{10}{\underset{N=O}{\bigcirc}} \xrightarrow{[H]} \underset{11}{\underset{NH-OH}{\bigcirc}} \xrightarrow{[H]} \underset{4}{\underset{NH_2}{\bigcirc}} \tag{1}$$

In this experiment iron powder with a small amount of hydrochloric acid will be used to reduce nitrobenzene to aniline. The main reaction is shown in equation 2, in unbalanced form. This is a typical oxidation-reduction equation that can perhaps be

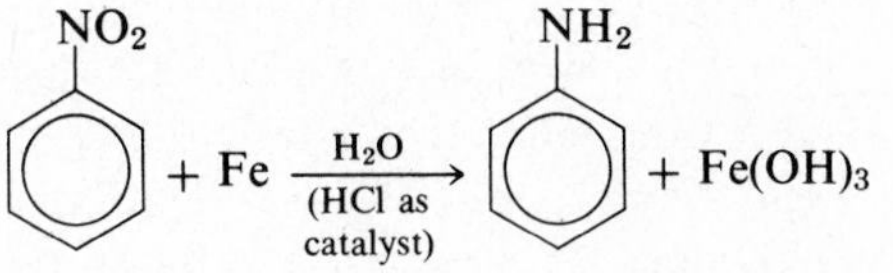

$$\underset{}{\underset{NO_2}{\bigcirc}} + Fe \xrightarrow[\substack{(HCl \ as \\ catalyst)}]{H_2O} \underset{}{\underset{NH_2}{\bigcirc}} + Fe(OH)_3 \tag{2}$$

most easily balanced by considering the two half-reactions given in equations 2a and 2b.

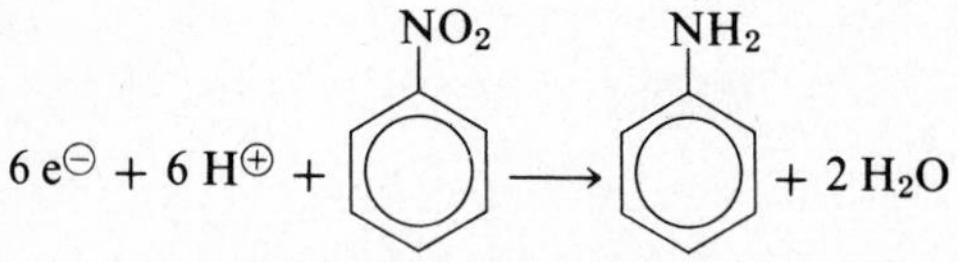

$$6\,e^{\ominus} + 6\,H^{\oplus} + \underset{}{\underset{NO_2}{\bigcirc}} \longrightarrow \underset{}{\underset{NH_2}{\bigcirc}} + 2\,H_2O \tag{2a}$$

$$3\,H_2O + Fe = Fe(OH)_3 + 3\,H^{\oplus} + 3\,e^{\ominus} \tag{2b}$$

Several possible side reactions occur but only to a minor extent under the reaction conditions used. These side reactions (equation 3) occur because of the presence of some of the intermediate products (equation 1), which then react with one another. Nitrosobenzene **(10)** can react with aniline to give azobenzene **(12)**, which can be reduced to hydrazobenzene **(13)** by iron. Hydrazobenzene can also undergo rearrangement in the presence of the acid catalyst to yield benzidine **(14)**; this rearrangement is often referred to as the *benzidine rearrangement*. N-Phenylhy-droxylamine **(11)** can undergo acid-catalyzed rearrangement to *p*-aminophenol **(15)**. Thus some of the anticipated side products are compounds **11–15**; their separation from aniline is discussed in the following paragraph.

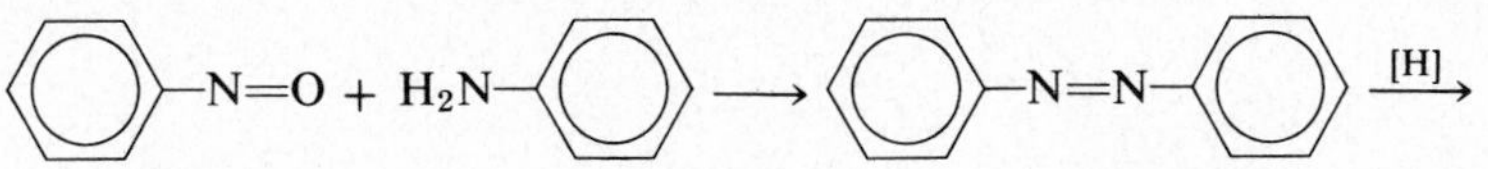

$$\underset{10}{\bigcirc\!-\!N=O} + \underset{4}{H_2N\!-\!\bigcirc} \longrightarrow \underset{12}{\bigcirc\!-\!N=N\!-\!\bigcirc} \xrightarrow{[H]}$$

$$\underset{13}{\bigcirc\!-\!NH\!-\!NH\!-\!\bigcirc} \xrightarrow{H^{\oplus}} \underset{14}{H_2N\!-\!\bigcirc\!-\!\bigcirc\!-\!NH_2} \tag{3a}$$

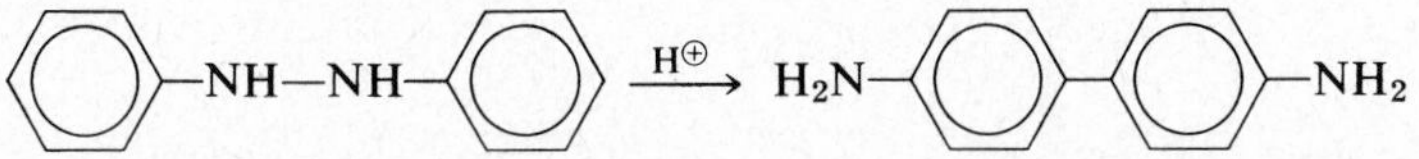

$$\text{(3b)}$$

(structure **11**: NH—OH on benzene ring) $\xrightarrow{H^{\oplus}}$ (structure **15**: NH$_2$ and OH on benzene ring)

11   15

Isolation of the aniline and its purification represents an interesting example of how solubility differences and steam distillation can be used. It is necessary to remove the aniline from its principal impurities, which are unchanged nitrobenzene, benzidine, and p-aminophenol. When the reaction is completed, some of the aniline will be present as its hydrochloric acid salt. The reaction mixture is made distinctly basic and then steam-distilled. The phenol, which will have been converted to its water-soluble, nonvolatile salt, will remain in the aqueous residue along with the high molecular weight benzidine. The aniline and nitrobenzene will steam-distil together. To separate these two from each other, it is necessary to convert the amine to its soluble hydrochloric acid salt and steam-distil again. The nitrobenzene will now steam-distil, whereas the salt will remain in the aqueous residue. The aniline may be recovered by making the acidic residue basic and extracting the amine into an organic solvent such as dichloromethane.

Alternatively, one could isolate the aniline from the unchanged nitrobenzene by treating the mixture of the two (obtained from steam distillation) with dilute acid and extracting the neutral nitrobenzene (the aniline will have been converted to the salt and will be in the aqueous layer) with an organic solvent, such as dichloromethane. Making the aqueous layer basic, followed by extraction with dichloromethane, will render the aniline in the second dichloromethane layer.

Acetanilide (5).   Aniline **(4)** is acetylated in aqueous solution, using a mixture of acetic anhydride and sodium acetate (equation 4). The water-insoluble aniline is first converted to the water-soluble hydrochloric acid salt; to this salt in water is added acetic anhydride, followed by the addition of sodium acetate to free the aniline from its salt. The free amine is then rapidly acetylated by the acetic anhydride. This provides an easy method of acetylation, and the overall yields are high. Under the conditions used, the acetanilide is produced as a solid, which can be filtered and recrystallized from water.

A side reaction which might occur is diacetylation, shown in equation 5. It appears that this is minimized by the procedure described, whereas acetylation in pure acetic anhydride (using the free amine) frequently leads to the production of the diacetyl compound.

(structure **4**: NH$_2$ on benzene ring) $\xrightarrow{HCl}$ (structure: NH$_3^{\oplus}$Cl$^{\ominus}$ on benzene ring) $\xrightarrow[CH_3CO_2Na]{(CH_3CO)_2O}$ (structure **5**: NH—$\overset{\overset{\textstyle O}{\|}}{C}CH_3$ on benzene ring) $+ \ 2\ CH_3CO_2H + NaCl$ $\qquad$ (4)

4   5

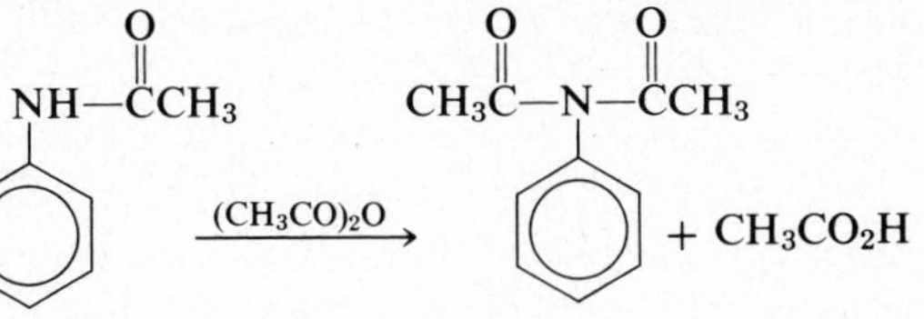

$$\underset{\textbf{5}}{\text{(acetanilide)}} \xrightarrow{(CH_3CO)_2O} \text{(diacetamidobenzene)} + CH_3CO_2H \qquad (5)$$

Chlorosulfonation Reactions: *p*-Acetamidobenzenesulfonyl Chloride (6). A simple, one-step reaction can be used to introduce the sulfonyl chloride group, —SO₂Cl, onto an aromatic ring. This reaction, known as chlorosulfonation, involves the use of chlorosulfonic acid (ClSO₃H) and an aromatic compound. As applied to acetanilide **(5)**, the major product is *p*-acetamidobenzenesulfonyl chloride (**6**, equation 6). Two moles of the chlorosulfonic acid are needed. The reaction is known to proceed through the sulfonic acid, which is converted to the sulfonyl chloride on further reaction with chlorosulfonic acid (equation 7).

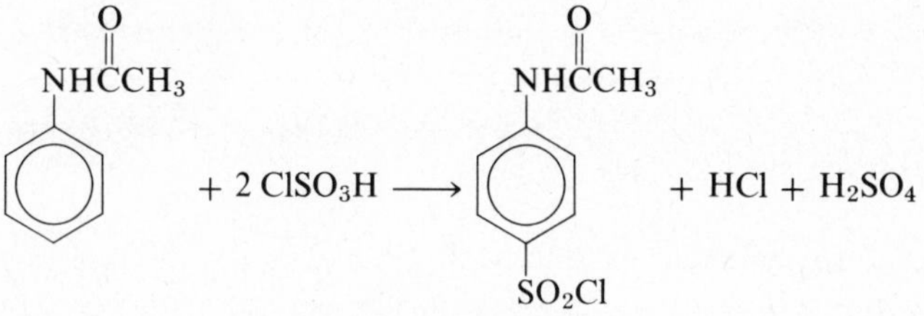

$$\underset{\textbf{5}}{\text{(acetanilide)}} + 2\,ClSO_3H \longrightarrow \underset{\textbf{6}}{\text{(p-acetamidobenzenesulfonyl chloride)}} + HCl + H_2SO_4 \qquad (6)$$

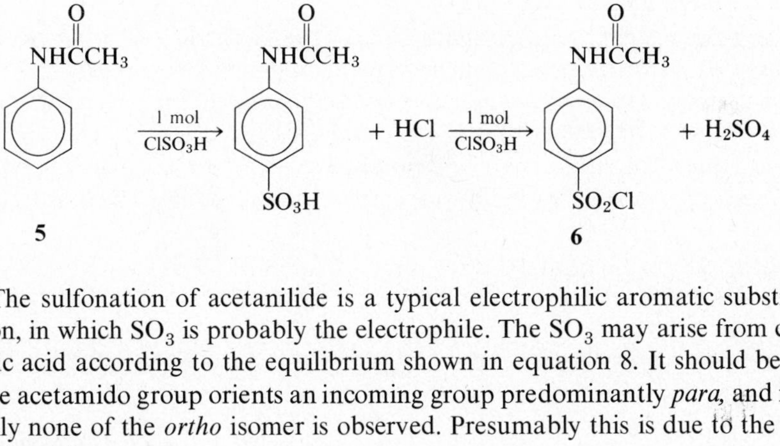

$$\underset{\textbf{5}}{\text{(acetanilide)}} \xrightarrow[ClSO_3H]{1\ mol} \text{(sulfonic acid)} + HCl \xrightarrow[ClSO_3H]{1\ mol} \underset{\textbf{6}}{\text{(sulfonyl chloride)}} + H_2SO_4 \qquad (7)$$

The sulfonation of acetanilide is a typical electrophilic aromatic substitution reaction, in which SO₃ is probably the electrophile. The SO₃ may arise from chlorosulfonic acid according to the equilibrium shown in equation 8. It should be noted that the acetamido group orients an incoming group predominantly *para*, and indeed virtually none of the *ortho* isomer is observed. Presumably this is due to the size of the acetamido group.

$$ClSO_3H \rightleftarrows SO_3 + HCl \qquad (8)$$

To isolate the product the reaction mixture is poured into ice water, and the product is obtained as a precipitate. The water serves to hydrolyze the excess chlorosulfonic acid (equation 9).

$$ClSO_3H + H_2O \rightarrow HCl + H_2SO_4 \qquad (9)$$

The preparation of sulfanilamide may be carried out without drying or purifying the *p*-acetamidobenzenesulfonyl chloride, since the latter will be treated with aqueous ammonia. However, if the sulfonyl chloride is to be converted to sulfathiazole, it must be dried, because this reaction requires anhydrous conditions. In general, sulfonyl chlorides are much less reactive toward water than are acid chlorides, but they do hydrolyze slowly to give the corresponding sulfonic acid (equation 10). In

$$
\underset{\underset{SO_2Cl}{\displaystyle\bigcirc}}{\overset{\overset{\displaystyle O}{\underset{\|}{NHCCH_3}}}{}} \;+\; H_2O \;\longrightarrow\; \underset{\underset{SO_3H}{\displaystyle\bigcirc}}{\overset{\overset{\displaystyle O}{\underset{\|}{NHCCH_3}}}{}} \;+\; HCl \tag{10}
$$

order to purify *p*-acetamidobenzenesulfonyl chloride, the wet product is dissolved in chloroform; the small amount of water which is present forms as a second layer, owing to its very low solubility in chloroform. The organic layer containing the desired product is then drawn off and the product is recrystallized from a suitable solvent. This represents a generally useful technique for purifying solids which contain small amounts of water.

Sulfanilamide [*p*-Aminobenzenesulfonamide (1)].   The preparation of sulfanilamide starts with *p*-acetamidobenzenesulfonyl chloride (**6**). The moist product can be used without purification, but it must be used *immediately*. The amide is produced by treatment of the sulfonyl chloride with an excess of aqueous ammonia, followed by removal of the acetyl group. This latter process can be carried out in acidic or basic solution without affecting the sulfonamido group, which hydrolyzes much less rapidly than the acetamido group. Acid hydrolysis will be used here, and the resulting amine forms a soluble hydrochloric acid salt under these conditions. To obtain free sulfanilamide the acidic solution must be neutralized with dilute base (sodium bicarbonate). These reactions are shown in equation 11. The sulfanilamide is purified further

$$
\underset{\underset{\underset{\mathbf{6}}{SO_2Cl}}{\displaystyle\bigcirc}}{\overset{\overset{\displaystyle O}{\underset{\|}{NHCCH_3}}}{}} \xrightarrow[\text{(NH}_3)]{\text{NH}_4\text{OH}} \underset{\underset{SO_2NH_2}{\displaystyle\bigcirc}}{\overset{\overset{\displaystyle O}{\underset{\|}{NHCCH_3}}}{}} \xrightarrow[\text{H}_2\text{O}]{\text{HCl}} \underset{\underset{SO_2NH_2}{\displaystyle\bigcirc}}{\overset{NH_3^{\oplus}Cl^{\ominus}}{}} \xrightarrow{\text{HCO}_3^{\ominus}} \underset{\underset{\underset{\mathbf{1}}{SO_2NH_2}}{\displaystyle\bigcirc}}{\overset{NH_2}{}} \tag{11}
$$

by recrystallization from water. A possible side reaction, though one which is not important under the conditions of the experiment, is the hydrolysis of the sulfonyl chloride during the first step shown in equation 11.

An important synthetic procedure has been introduced in the preparation of sulfanilamide—the use of a protective group (the acetyl group) which is subsequently removed at a later stage. It would appear that one reasonable method for preparing sulfanilamides would be to allow *p*-aminobenzenesulfonyl chloride (16) to react with ammonia or with some substituted amine, as is the case in the preparation of sulfathiazole. However, it is not practical to generate a sulfonyl chloride group in the presence of an amino group contained in the same molecule. The amino group of one molecule would react with the sulfonyl chloride group of another molecule and give a polymeric material containing sulfonamide linkages, as illustrated in equation 12.

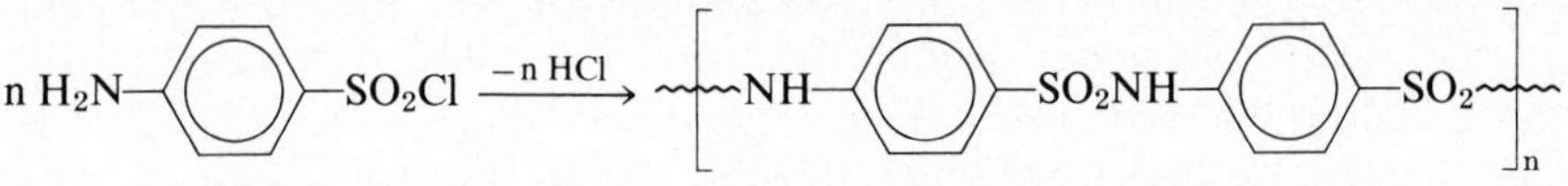

$$n \; H_2N\!-\!\langle\bigcirc\rangle\!-\!SO_2Cl \xrightarrow{-n \; HCl} \left[\!-\!NH\!-\!\langle\bigcirc\rangle\!-\!SO_2NH\!-\!\langle\bigcirc\rangle\!-\!SO_2\!-\!\right]_n \quad (12)$$

16

Thus to prepare the sulfanilamides, it is necessary first to "protect" the free amino group so as to allow the other functional group to be introduced. Such protection is a general requirement when a molecule contains functional groups which are reactive toward one another. In the present synthesis this is most easily done by acetylating the amine. Of course, the free amine can be regenerated by removing the acetyl group (called either the "protective group" or the "blocking group") after the sulfonamido group has been introduced. In using such a technique, care must be taken to ensure that the blocking group can be removed without affecting the second functional group; in the present case the acetyl group can easily be removed without hydrolyzing the sulfonamido group.

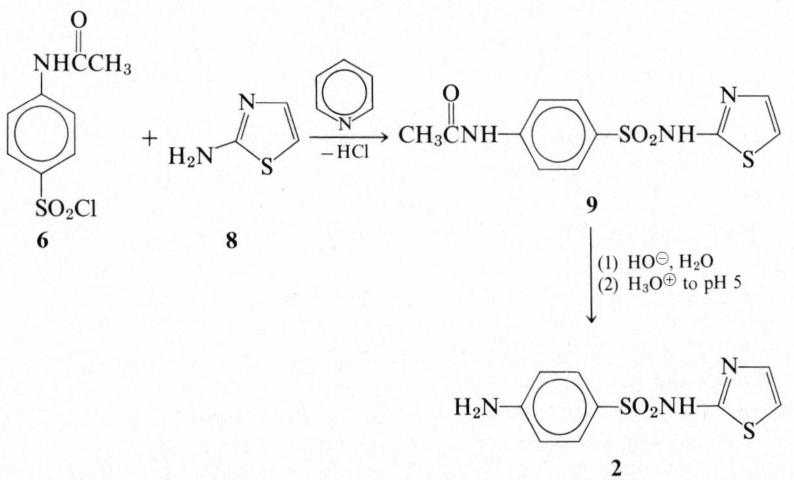

Sulfathiazole (2). The final reaction in the synthesis of sulfathiazole involves the condensation of *p*-acetamidobenzenesulfonyl chloride (6) with 2-aminothiazole (8), which occurs with the liberation of hydrogen chloride (see equation 13). (The preparation of 2-aminothiazole is described in Section 16.1.) As the reaction proceeds, the hydrogen chloride which is given off would react with 2-aminothiazole and convert it to its unreactive hydrochloric acid salt. Thus it is important to remove the hydrogen chloride as it is produced. Pyridine is most often used for this purpose because it is inexpensive and is easily removed from the reaction mixture. In this preparation pyridine will be used both to remove hydrogen chloride and to serve as a solvent for the reaction. Precautions must be taken to ensure that all equipment and reagents are dry to minimize hydrolysis of the sulfonyl chloride.

Another important principle of organic synthesis—the conservation of the more (or most) valuable reagent—is involved in this preparation. 2-Aminothiazole is more expensive than is p-acetamidobenzenesulfonyl chloride,[1] so the latter is used in slight excess to maximize conversion of the more expensive reagent to product.

Important side reactions occur when the sulfonyl chloride reacts with 2-amino-thiazole in alternative ways, and some of these reactions are shown in equation 14. None of the side products so obtained have any hydrogen atoms on a sulfonamide nitrogen, and they will therefore be insoluble in the alkaline solution used in the work-up procedure (see the Hinsberg reaction, Chapter 25, under Amines, part A.1).

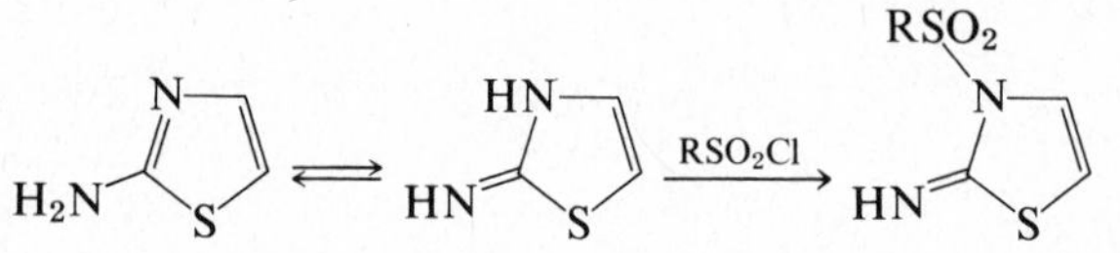

(14a)

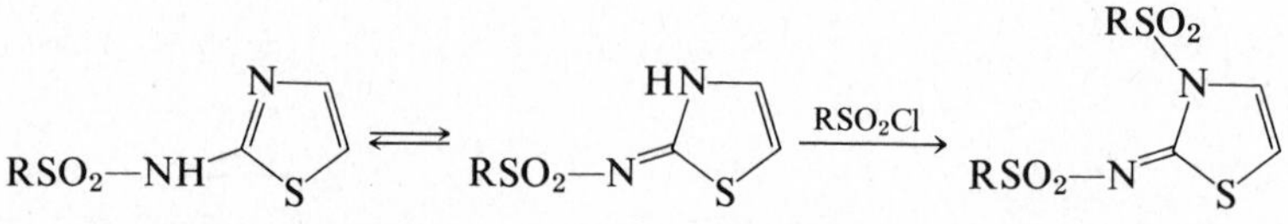

(14b)

Purification of the sulfathiazole is accomplished by treating the reaction mixture with water, collecting the solid which forms, and treating the solid with excess base and filtering. Compound **9** is soluble in base, whereas the side products indicated by equation 14 are not. After removal of the side products, the basic solution is heated to effect hydrolysis of the amide. Under the reaction conditions the sulfonamide group is not affected. The basic solution is acidified and then treated with sodium acetate until the pH of the solution is about 5. The acetate reacts with the mineral acid present to produce acetic acid, which in the presence of acetate ion yields a buffer solution. In strongly acidic solution sulfathiazole is protonated and is water-soluble, but after the solution is adjusted to pH 5, the sulfathiazole is liberated.

EXPERIMENTAL PROCEDURE

The experimental procedures start with nitrobenzene and carry through to sulfathiazole. The instructor will indicate at which point to start and how far along the synthesis to proceed. Remember to adjust quantities of reagents according to the amount of starting material which is available. Beginning with nitrobenzene, for

---

[1] This may not seem to be a valid statement to the student who started the preparation with nitrobenzene. However, acetanilide is available at very low cost, so that it is only one simple step to convert acetanilide to p-acetamidobenzenesulfonyl chloride. On the other hand, chloroacetaldehyde, which is needed for the preparation of 2-aminothiazole, is very expensive.

example, some of the reactions may give better yields than those indicated, whereas others may give poorer yields. This must be taken into account as you go to the next step; however, do not change the reaction times unless told to do so. It may be convenient to use different sized flasks.

A.   Aniline

DO IT SAFELY

---

1.   In all parts of this experiment be sure to examine your glassware carefully for "star cracks" or other weaknesses. Use of damaged glassware, particularly during heating operations, may lead to an accident because heat produces stresses which may cause such glassware to break, thus spilling the contents.

2.   Although nitrobenzene is not as dangerous as benzene, it is toxic, and care should be taken to avoid breathing its vapors or allowing contact with your skin. Develop respect for all the chemicals that you use in the laboratory, whether toxic or not. Many experienced chemists wear rubber gloves in the laboratory, particularly when handling liquids and solutions; this is a precautionary habit that you should try to develop.

Place 0.54 g-atom of iron powder[2] in a 500-mL round-bottomed flask; add to it 0.24 mol of nitrobenzene and 100 mL of water. Attach a reflux condenser, and add about 0.5 mL of concentrated hydrochloric acid through the top of the condenser. Shake the reaction mixture vigorously, and if the reaction does not start soon (as evidenced by the production of heat), heat it gently with a small flame (be prepared to cool the flask in a pan of water if it appears that the reaction is going to get out of control). After the reaction has started, add another 0.5 mL of concentrated hydrochloric acid, and shake the flask vigorously. When the reaction has subsided, bring the mixture to gentle reflux. After 15 min add 1 mL of concentrated hydrochloric acid, and then continue the heating under reflux for an additional 45 min.[3]

When the reflux period is complete, add 5 mL of 6 M sodium hydroxide solution directly to the reaction mixture, and equip the flask for steam distillation (Section 2.6 and Figures 2.12, 2.13, and 2.14). Steam-distil the mixture until the distillate dropping from the condenser no longer contains any visible amount of organic product. (The distillate may still be slightly cloudy, but if there is no visible amount of oil in it, it may be assumed that the distillation is complete.)★ Add 20 mL of concentrated hydrochloric acid to the distillate, and steam-distil this mixture until the *residue* in the distilling flask is clear and free from oily material.★

Make the acidic distillation *residue* basic with a minimum volume of 12 M sodium hydroxide solution. (*Use care:* Heat is evolved.) Saturate the basic solution

---

[2] Iron powder should be finely divided and should be free of an oxide coating. The reduction reaction is a heterogeneous surface reaction, so that a large area of metal surface is needed. Iron metal obtained by reduction of the oxide with hydrogen is termed "reduced with hydrogen" and is most suitable for this reduction reaction. Tin powder may be used in place of iron.

[3] If the reaction requires heating with a burner for initiation, heat under reflux for at least 90 additional min.

with sodium chloride (roughly 25 g of salt per 100 mL of solution), cool the mixture, and transfer it to a separatory funnel. Extract the organic product with two 50-mL portions of technical diethyl ether, using the first portion to rinse the flask in which the neutralization was done. Separate the aqueous layer from the organic layer as thoroughly as possible; transfer the combined organic extracts to a small Erlenmeyer flask, and dry with several sodium hydroxide pellets until the solution is clear.★ Transfer the solution by decantation into a distilling flask and distil it. Collect three fractions, one boiling between 35 and 90°, the second boiling between 90 and 180°, and the third boiling between 180 and 185°. Discard the first fraction, and if the second fraction is of significant volume, redistil it to obtain more product. Pure aniline is colorless but may darken immediately following distillation, owing to air oxidation. Aniline is often redistilled just before use to remove colored oxidation products. The yield of aniline should be 85–90%.

B.   Acetanilide

Dissolve 0.22 mol of aniline in 500 mL of water to which 18 mL of concentrated hydrochloric acid has been added; use a 1000-mL Erlenmeyer flask. Swirl the mixture to aid in dissolving the aniline. (If the solution is dark colored, add about 1 g of decolorizing carbon to it, swirl, and filter.) Prepare a solution containing 0.24 mol of sodium acetate trihydrate dissolved in 100 mL of water, and measure out 24 mL of acetic anhydride. Warm the solution containing the dissolved aniline to 50°, and add the acetic anhydride; swirl the flask to dissolve the anhydride. Then add the sodium acetate solution *immediately and in one portion.* Cool the reaction mixture in an ice-water bath, stirring while the product crystallizes. Collect the acetanilide by vacuum filtration, wash it with a small portion of cold water, and dry.★ Determine the melting point and yield of product; the yield should be between 65 and 75%. If impure or slightly colored acetanilide is obtained, it may be recrystallized from a minimum volume of hot water, using decolorizing carbon if necessary to give a colorless product. (See Section 2.7 for a discussion of the recrystallization of acetanilide from water.)

C.   *p*-Acetamidobenzenesulfonyl Chloride

DO IT SAFELY

---

1.   Be particularly careful in handling and transferring chlorosulfonic acid. It reacts vigorously with water; use *dry* glassware and avoid contact of the acid with your skin. Should this occur, wash the area *immediately* with large quantities of cold water, and then rinse with 0.6 *M* sodium bicarbonate solution. Because open containers of chlorosulfonic acid will fume from reaction with atmospheric moisture, it should be measured and transferred *only in the hood.*

2.   If you are to prepare sulfathiazole, the *p*-acetamidobenzenesulfonyl chloride is purified at the end of this procedure by use of boiling chloroform. In this step not only is a recrystallization accomplished but also water is removed from the product as a second layer from boiling chloroform solution. *This operation must be carried out in the hood* to avoid breathing and filling the room with chloroform vapors.

If possible, either carry out the following reaction in the hood or use the gas-removal apparatus described to prevent hydrogen chloride or sulfur dioxide gases from being vented in the laboratory. Equip a *dry* 250-mL round-bottomed flask with a Claisen connecting tube. Place a vacuum adapter filled with 4–6 mesh calcium chloride on the side arm of the Claisen tube, and connect the vacuum adapter to an aspirator trap by means of a length of tubing (see Figure 5.1). Grease all joints carefully because an airtight seal is required at all connections. Place 0.16 mol of *dry* acetanilide in the flask. *In the hood* measure 0.92 mol of chlorosulfonic acid into a 125-mL separatory funnel (be certain that the stopcock of the funnel is firmly sealed). *Use care in handling chlorosulfonic acid. Be careful to avoid contact with the skin and with moisture. Chlorosulfonic acid reacts vigorously with water; care should be used in washing any equipment that has contained the acid.* Stopper the funnel, and place it on the straight arm of the Claisen tube so that the chlorosulfonic acid will drop directly onto the acetanilide contained in the flask.

Cool the flask to 10–15° in a water bath containing a little ice, but do not cool below 10°. Turn on the water aspirator so that a slight flow of air occurs through the vacuum adapter; it is not necessary to turn on the aspirator full force. Open wide the stopcock of the funnel so that the chlorosulfonic acid is added as rapidly as possible to the flask; it may be necessary to lift the stopper on the funnel to equalize the pressure in the system if the flow of acid becomes slow. After completion of the addition, *gently* swirl the apparatus from time to time to speed the rate of dissolution of the acetanilide; maintain the temperature of the mixture below 20°. After most of the solid has dissolved, allow the reaction mixture to warm to room temperature, and then heat it on a steam bath until moderate agitation of the apparatus produces no increase in the rate of gas evolution; 10–20 min of heating will be required.

Cool the mixture to room temperature (or slightly lower), using an ice-water bath. Working in the hood, place 600 g of cracked ice and 100 mL of water in a large beaker or Erlenmeyer flask, and pour the reaction mixture slowly and with stirring onto the ice. *Use care in this step; do not add the mixture too quickly, and avoid splattering of the chlorosulfonic acid.* After the addition is complete, rinse the flask with a little cold water, and transfer this to the beaker. The precipitate that forms in the beaker is crude *p*-acetamidobenzenesulfonyl chloride, which may be white to pink in color. It may soon become a hard mass, and any lumps that form should be broken up with a stirring rod. Collect the crude material by vacuum filtration. Wash the solid with a small amount of cold water, and press it as dry as possible with a cork.

If sulfanilamide is to be prepared from the sulfonyl chloride, the crude product does not need to be purified, but it should be used immediately. *If sulfathiazole is to be made, the sulfonyl chloride must be purified.*

For purification, dissolve the crude product in a minimum amount of boiling chloroform (the approximate volume required is 135 to 185 mL). Preheat a separatory funnel on the steam bath, and transfer the boiling chloroform mixture to it. Remove the organic layer from the water layer as quickly as possible; be very careful in separating the layers. Rinse the funnel with an additional 40-mL portion of hot chloroform, and combine this rinse with the main organic layer. On cooling, colorless crystals of pure product are obtained.★ The yield should be 77–90%; the reported melting point of pure *p*-acetamidobenzenesulfonyl chloride is 149°. Air-dry the product and store it in a closed container.

D.   Sulfanilamide

Transfer the crude *p*-acetamidobenzenesulfonyl chloride to a 500-mL Erlenmeyer flask, and add 75 mL of concentrated ammonium hydroxide (28%). A very rapid exothermic reaction should occur. Break up with a stirring rod any lumps of solid which may remain; the reaction mixture should be thick but homogeneous. Heat the mixture on a steam bath for about 30 min. Because some ammonia vapors will be released into the air, heating should preferably be done in a hood; alternatively, invert a funnel over the flask, and attach the funnel to a water aspirator. Cool the flask in an ice-water bath, and add 6 *M* sulfuric acid to the cool reaction mixture until it is acidic to Congo red paper. Cool the reaction mixture again in an ice-water bath, and collect the product by vacuum filtration. Wash the crystals with cold water and dry.★ The yield should be 89–96%. The product is pure enough for the hydrolysis reaction, but if desired it may be purified as follows. Dissolve the crude *p*-acetamidobenzenesulfonamide in a minimum amount of hot water. If needed, use decolorizing carbon. After filtering, cool the aqueous solution in an ice-water bath, and collect and dry the crystals.★ The melting point of the pure product is 220°.

Weigh the *p*-acetamidobenzenesulfonamide, and transfer it to a 250-mL round-bottomed flask. Prepare a solution of dilute hydrochloric acid by mixing equal volumes of concentrated hydrochloric acid and water. Add to the amide an amount of dilute hydrochloric acid solution *twice* the weight of the amide. Attach a reflux condenser to the flask, and heat at a gentle reflux for 30 min. Boil gently and swirl the reaction mixture when first heating so as to aid in the dissolution of the organic material. To the homogeneous reaction mixture, add an equal quantity of water, and transfer the new mixture to a 600-mL beaker. Neutralize the excess acid which is present by the addition of small quantities of solid sodium carbonate; continue addition of the base until the solution is just alkaline to litmus paper. (*Caution:* Add the sodium carbonate in small quantities because foaming will occur!) A precipitate should form during neutralization. After making the solution just basic, cool the mixture in an ice-water bath to complete the precipitation of the product. Collect the crystals by vacuum filtration, wash with a small amount of cold water, and allow the product to air-dry.★ Purify the crude product by recrystallization from hot water (12 to 15 mL of hot water per gram of compound will be required). Decolorize the hot solution, if necessary, filter and cool the filtrate in an ice-water bath. It may be necessary to preheat the filter funnel so that the product will not crystallize in the funnel. Sulfanilamide will form on cooling to give long, white needles; collect the product by vacuum filtration. Determine the melting point (reported 163°) and the yield, which should be about 0.7 g per gram of starting *p*-acetamidobenzenesulfonamide.

Test the solubility of sulfanilamide in 1.5 *M* hydrochloric acid solution and in 1.5 *M* sodium hydroxide solution.

E.   Sulfathiazole

Dry about 40 mL of reagent grade pyridine over 2–3 g of potassium hydroxide pellets.[4] To a 250-mL round-bottomed flask equipped with a Claisen connecting tube, add 0.075 mol of 2-aminothiazole (Chapter 16) and about 30 mL of dry pyridine. Transfer the dry pyridine by use of a graduated pipet. Equip the flask with a calcium chloride drying tube and a thermometer. Add 0.082 mol of purified p-acetamido-benzenesulfonyl chloride to the reaction flask in small portions; swirl gently to ensure mixing, and make the addition at such a rate that the temperature inside the flask does not exceed 40°. Do not cool the reaction mixture at any time during the addition. After the addition is complete, heat the mixture on a steam bath for about 30 min. Cool the mixture, and then pour it into about 200 mL of warm water, stirring vigorously with a stirring rod. The oil which forms initially should solidify if a stirring rod is used to induce crystallization. Remove the solid by filtration,★ wash it with a small amount of water, and dissolve it in 38 mL of 2 M sodium hydroxide solution; warm on a steam bath, if needed, to bring the solid into solution. If any solid remains, filter it off and discard it. Make the resulting filtrate acidic to litmus by the addition of glacial acetic acid, collect the solid which forms by vacuum filtration, and press as dry as possible.★ Weigh the solid, add 10 times its weight of 2 M sodium hydroxide solution, and heat the mixture at reflux for 1 hr. After heating, cool the mixture, and slowly add concentrated hydrochloric acid; at first a solid will form, which will then redissolve as excess acid is added. Add only enough acid so that the solid redissolves; do not add excess acid. After decolorizing and filtering the solution, neutralize it with 2 M sodium hydroxide, and then add solid sodium acetate in small portions until the solution is basic to litmus paper. Bring the heterogeneous mixture to boiling and then cool in an ice-water bath; collect the solid by vacuum filtration,★ and recrystallize it from a minimum amount of boiling water. Collect the sulfathiazole by vacuum filtration and air-dry it. Determine the melting point (reported 201–202°) and the yield.

EXERCISES

1. Calculate the overall yield of cortisone, an important drug, produced in a 33-step synthesis, assuming an average yield of 90% in each step.

2. Outline a possible synthesis for the compound shown here, using benzene as the only source of an aromatic ring. Use any needed aliphatic or inorganic reagents.

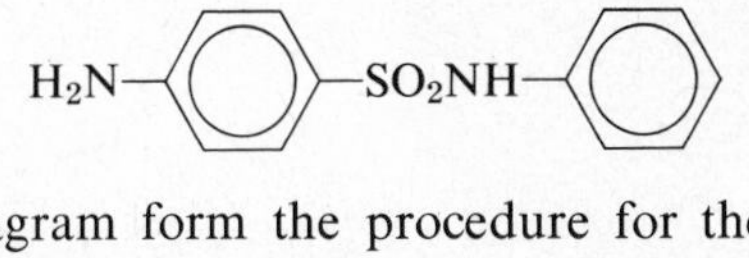

$$H_2N \text{—} \langle \bigcirc \rangle \text{—} SO_2NH \text{—} \langle \bigcirc \rangle$$

3. Outline in flow diagram form the procedure for the purification of aniline. Indicate the importance of each step in the procedure, and give reasons for doing the steam distillation first with a basic solution and then with an acidic solution. Write equation(s) for reactions which occur when base and then acid are added.

---

[4]This should be done during the laboratory period before the experiment is to be performed. Add potassium hydroxide pellets to the pyridine, and store in the laboratory desk until ready to use.

4. In the purification of aniline, sodium hydroxide pellets are used as the drying agent. Why is this compound used rather than magnesium sulfate, calcium chloride, or other common drying agents? Would potassium hydroxide be an acceptable substitute for sodium hydroxide?

5. In the preparation of sulfanilamide from *p*-acetamidobenzenesulfonamide, only the acetamido group is hydrolyzed. Give an explanation for this difference in reactivity of acetamido and sulfonamido groups.

6. Explain the results obtained when the solubility of sulfanilamide was determined in 1.5 *M* hydrochloric acid and in 1.5 *M* sodium hydroxide. Write equations for any reaction(s) occurring.

7. What would be observed if *p*-acetamidobenzenesulfonamide were subjected to vigorous hydrolysis conditions (that is, concentrated hydrochloric acid and heat for a long period)? Write an equation for the reaction which would occur.

8. With the aid of chemical equations explain the behavior of sulfathiazole toward acid and base. How do these reagents affect its solubility in water?

9. Outline the work-up of sulfathiazole in flow diagram form. Indicate the importance of each step in the procedure, and give reasons for each. Write equation(s) for the reactions which occur in the purification process.

10. The uv spectra of 2-aminothiazole and sulfathiazole are provided in Figures 16.3 and 17.10, respectively. Determine values of ε for each absorption maximum in each spectrum, and assign the chromophores responsible for them. Comment on what effect, if any, conversion of 2-aminothiazole to sulfathiazole has had on the nature of the uv absorption attributable to the chromophore of the former compound.

SPECTRA OF STARTING MATERIALS AND PRODUCTS

The ir and pmr spectra of aniline are given in Figures 15.4 and 15.5, respectively. The ir, pmr, and uv spectra of 2-aminothiazole are included as Figures 16.1, 16.2, and 16.3, respectively:

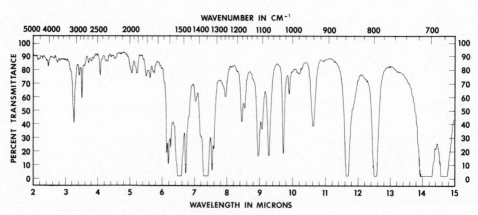

FIGURE 17.3 IR spectrum of nitrobenzene.

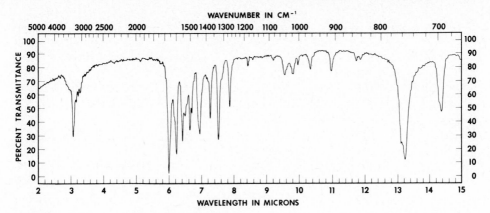

FIGURE 17.4   IR spectrum of acetanilide.

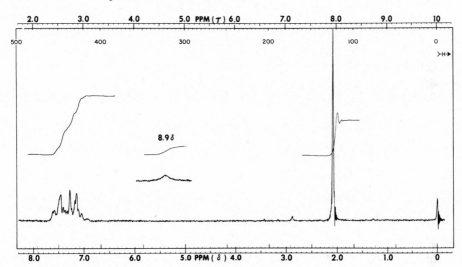

FIGURE 17.5   PMR spectrum of acetanilide.

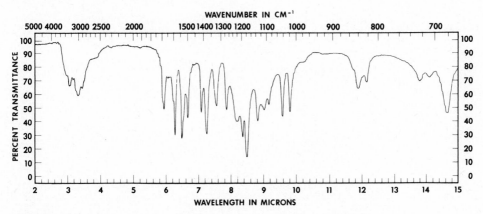

FIGURE 17.6   IR spectrum of *p*-acetamidobenzenesulfonyl chloride.

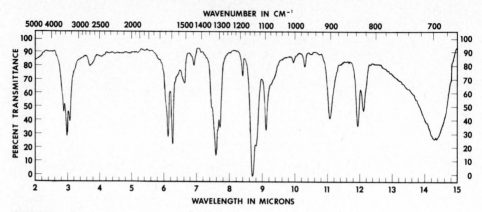

FIGURE 17.7 IR spectrum of sulfanilamide.

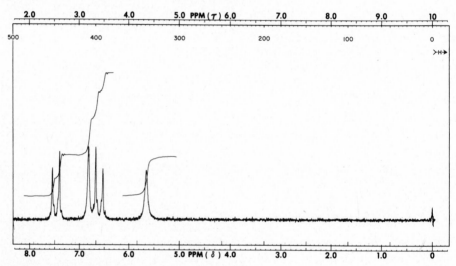

FIGURE 17.8 PMR spectrum of sulfanilamide.

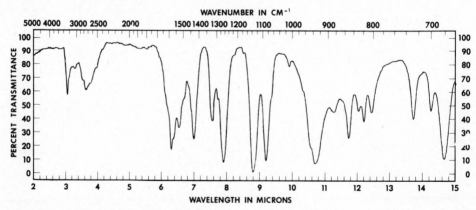

FIGURE 17.9 IR spectrum of sulfathiazole.

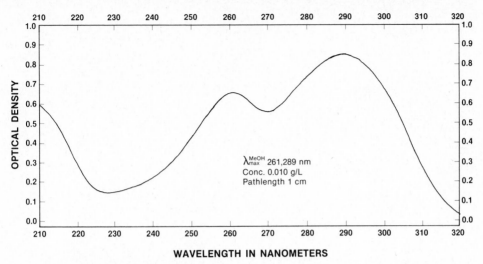

FIGURE 17.10   UV spectrum of sulfathiazole.

17.3   The Synthesis of 1-Bromo-3-chloro-5-iodobenzene

Another interesting multistep synthesis is the preparation of 1-bromo-3-chloro-5-iodobenzene **(17)**, which can be initiated from benzene. The complete sequence of this eight-step synthesis is outlined in Figure 17.11. Most of the reactions involve electrophilic aromatic substitution. Two important stratagems of aromatic chemistry are incorporated in the synthesis. (1) An amino group is "protected" by conversion to the amide and after several electrophilic substitution reactions are performed on the molecule, the free amine is regenerated. (2) An amino group, which was introduced into the molecule initially because of its directing influence and activating properties, is removed in the final step to yield the desired product. The usefulness of the amino group will be indicated as the various steps of the reaction sequence are discussed. It should be noted that the yields of each individual step range from 60 to 96%, which is always an important consideration in multistep syntheses. In general the reaction times are short enough so that one or more steps can be carried out in a single laboratory period.

The synthesis of **17** may be started anywhere along the reaction sequence, but owing to the toxicity of benzene, we recommend nitrobenzene as the first starting material. The reduction of nitrobenzene and conversion of aniline to acetanilide have already been discussed in Section 17.2 in connection with the synthesis of sulfathiazole. The experimental details of those reactions are found in that section. The discussion here will start with the synthesis of 4-bromoacetanilide **(18)** from acetanilide.

4-Bromoacetanilide (18).   The first halogen to be introduced into the molecule is bromine. The amino group activates the aromatic ring so greatly that any bromina-

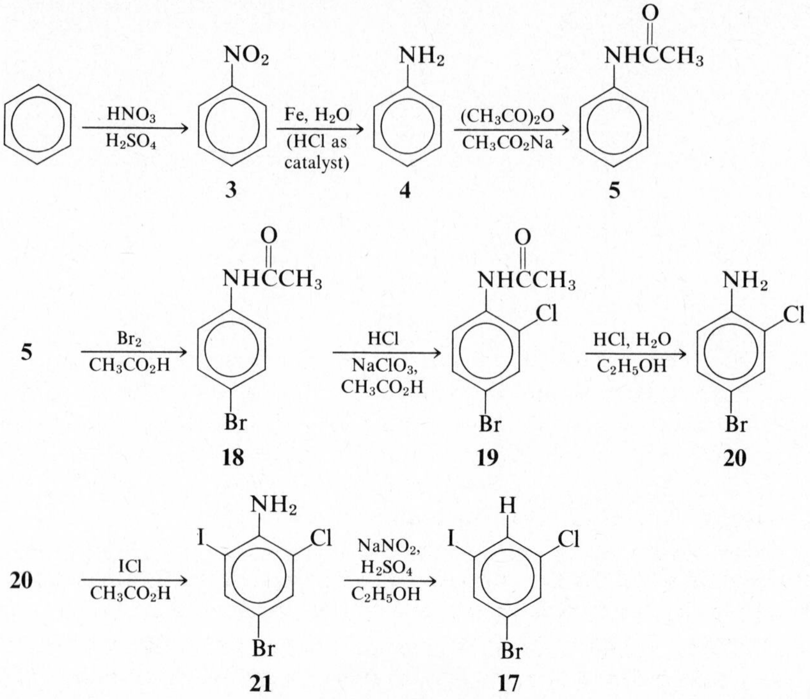

FIGURE 17.11 Laboratory synthesis of 1-bromo-3-chloro-5-iodobenzene (**17**).

tion conditions lead to trisubstitution on aniline, and indeed an aqueous solution of bromine is sufficient to do this (equation 15).

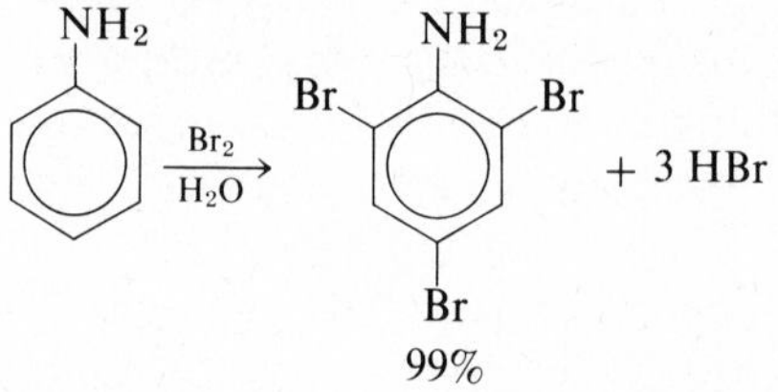

$$+ \ 3 \ HBr \qquad (15)$$

To decrease the reactivity of aniline so that monosubstitution can be realized, aniline is acetylated to give acetanilide. Under mild conditions (bromine in acetic acid) acetanilide can be monobrominated easily (note that no Lewis acid catalyst such as $FeBr_3$ is required since acetanilide is still quite reactive toward bromine).

As the acetamido group, CH_3CONH-, is *ortho,para*-directing, both 2-bromo- and 4-bromoacetanilide are produced; however, under the conditions of the reaction 95% of the 4-bromo isomer is obtained (equation 16). 4-Bromoacetanilide can be separated from the 2-bromo isomer by a single recrystallization of the crude product from methanol.

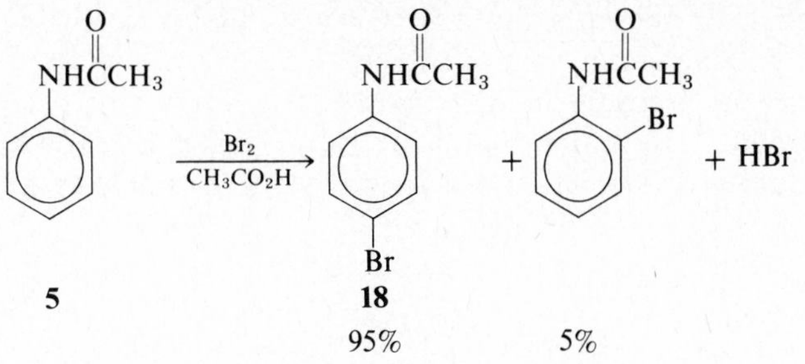

$$
\underset{\mathbf{5}}{\text{(NHCCH}_3\text{)}} \xrightarrow[\text{CH}_3\text{CO}_2\text{H}]{\text{Br}_2} \underset{\substack{\mathbf{18} \\ 95\%}}{\text{(NHCCH}_3\text{-Br)}} + \underset{5\%}{\text{(NHCCH}_3\text{-Br)}} + \text{HBr} \qquad (16)
$$

In general the separation and purification of *ortho* and *para* isomers from a mixture containing both of them is fairly simple, as illustrated by the ease with which 4-bromoacetanilide is purified. Even though different functional groups may be attached to an aromatic ring, the *para* isomer is more "symmetrical" than is the *ortho* isomer, and one generally observes the former to possess a much higher melting point than the latter. Also, the *para* isomer is normally less soluble in a given solvent than is the *ortho* isomer, so that fractional crystallization can be used to separate them from one another. For example, in this experiment the melting point of 4-bromoacetanilide is 167° and that of 2-bromoacetanilide is 99°. The 4-bromo isomer is much less soluble in methanol than is the 2-bromo isomer, so that recrystallization of a mixture of these compounds from methanol gives 4-bromoacetanilide as a solid whereas the 2-bromo isomer remains in solution.

2-Chloro-4-bromoacetanilide (19).   The next halogen to be introduced into the molecule is chlorine. The bromine atom in 4-bromoacetanilide deactivates the ring slightly toward further electrophilic substitution (as compared to acetanilide), but the compound is still reactive enough so that it can be chlorinated without using a catalyst. Chlorine in acetic acid will monochlorinate 4-bromoacetanilide to produce 2-chloro-4-bromoacetanilide (**19**, equation 17). Although it is possible to use chlorine

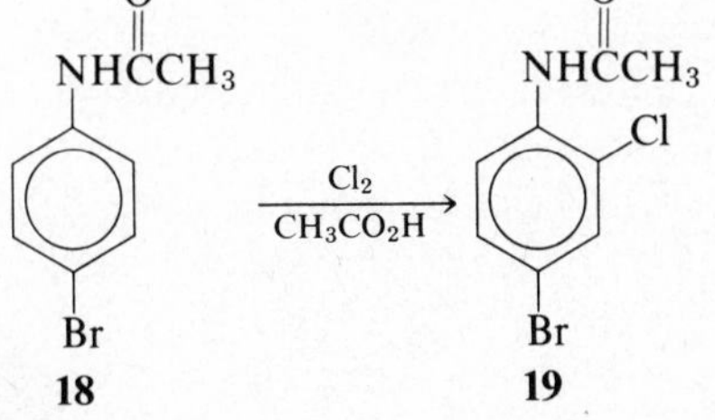

$$
\underset{\mathbf{18}}{\text{(NHCCH}_3\text{-Br)}} \xrightarrow[\text{CH}_3\text{CO}_2\text{H}]{\text{Cl}_2} \underset{\mathbf{19}}{\text{(NHCCH}_3\text{-Cl, Br)}} \qquad (17)
$$

gas dissolved in acetic acid, the difficulties and hazards of using this gas with large numbers of students make it desirable to generate chlorine gas *in situ*. One convenient method of *in situ* preparation of chlorine is to allow hydrochloric acid and sodium chlorate ($NaClO_3$) to react; an oxidation-reduction reaction occurs, in which chloride ion is oxidized to chlorine and chlorate ion is reduced to chlorine as shown in equation (18). In addition to ease and safety in handling, just the right amount of chlorine gas can be prepared by this method. The reaction between chlorine and 4-bromoacetanilide occurs very rapidly, and little if any dichlorination product is obtained. 4-Bromo-2-chloroacetanilide is purified by fractional crystallization from methanol.

$$
\begin{array}{ll}
\text{Oxidation:} & 2\,Cl^{\ominus} = Cl_2 + 2\,e^{\ominus} \\
\text{Reduction:} & 12\,H^{\oplus} + 2\,ClO_3^{\ominus} + 10\,e^{\ominus} = Cl_2 + 6\,H_2O \\
\hline
\text{Overall:} & 5\,Cl^{\ominus} + ClO_3^{\ominus} + 6\,H^{\oplus} = 3\,Cl_2 + 3\,H_2O
\end{array}
\tag{18}
$$

2-Chloro-4-bromoaniline (20). 2-Chloro-4-bromoacetanilide **(19)** can be converted to 2-chloro-4-bromoaniline in excellent yield using acid-catalyzed hydrolysis of the amide (equation 19). Concentrated hydrochloric acid in ethanol is employed, with the ethanol being used to increase the solubility of the amide.

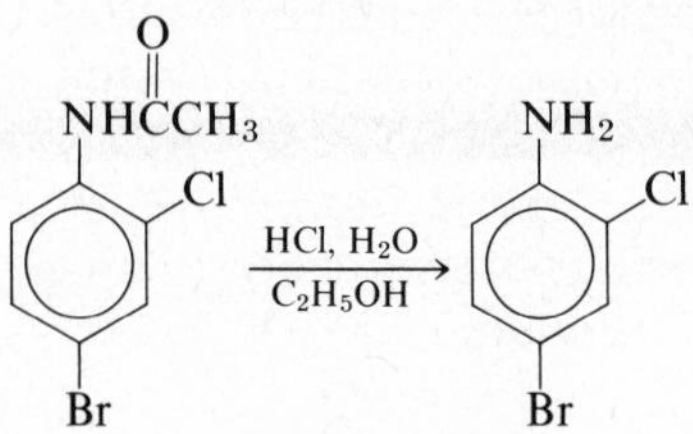

$$\tag{19}$$

2-Chloro-4-bromo-6-iodoaniline (21).   The 2-chloro-4-bromoaniline **(20)**, which was obtained in the previous reaction, is highly activated owing to the presence of the amino group. Of the positions *ortho* and *para* to the NH_2 group, only the 6-position is unsubstituted. Thus iodine may be added in that position by starting with the free amine and without need for the deactivating effect of the acetamido group. Any electrophilic substitution reaction which might be carried out on **20** would result in monosubstitution at the free 6-position.

Electrophilic substitution reactions involving iodine are generally more difficult to carry out than are chlorination and bromination reactions. One general method uses iodine and potassium carbonate, with the base being present to absorb the hydroiodic acid which forms. Another good method makes use of iodine monochloride, ICl, as the iodinating agent. Of the two halogens present in ICl, chlorine is more electronegative than is iodine. Thus when the molecule dissociates, the bonding electrons leave with chlorine to give Cl^- and I^+ (equation 20). I^+ serves as an electrophile which attacks the aromatic ring and replaces H^+ at the 6-position (equation 21).

$$
I\!-\!Cl \xrightleftharpoons{CH_3CO_2H} I^{\oplus} + Cl^{\ominus}
\tag{20}
$$

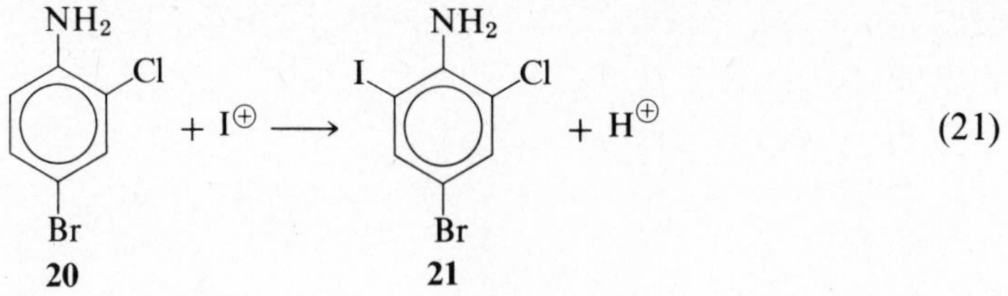

$$\text{20} + I^{\oplus} \longrightarrow \text{21} + H^{\oplus} \tag{21}$$

It should be pointed out that I^+ is a very weak electrophile compared to Br^+ or Cl^+. One of the main reasons for introducing iodine in the final step of the substitution sequence is that a very activated aromatic ring must be used to realize a good yield in any iodination substitution reaction. Thus bromine and chlorine are introduced with the acetamido group present, since electrophilic substitution with these elements requires a less activated ring than does iodine. Also, with the acetamido group present, the reaction can be controlled to give monosubstitution. Once the chlorine and bromine atoms are in the desired positions, the highly activating free amino group is regenerated and iodination is carried out with ease and without the possibility of disubstitution.

There are no important side reactions in this preparation; unchanged starting material can be separated from the product by fractional crystallization.

1-Bromo-3-chloro-5-iodobenzene (17).   The final step in the synthesis is the removal of the amino group from 2-chloro-4-bromo-6-iodoaniline (21). After the amino group has been diazotized with nitrous acid at $0°$, the diazo group can be removed by treatment with hypophosphorous acid (H_3PO_2) or by treatment with absolute ethanol. The former is the method most often encountered in textbooks, and indeed it is the best general reagent for replacing the diazo group by a hydrogen. When hypophosphorous acid is used, it is oxidized to phosphorous acid as the diazonium salt is reduced.

The use of ethanol has been found to be successful with certain types of aromatic compounds, in particular those which contain halogen. Because this method is easier to use in the laboratory, it will be employed here. Presumably an oxidation-reduction reaction occurs with ethanol as well; in this case ethanol is oxidized to acetaldehyde or acetic acid as the diazo compound is reduced (equation 22).

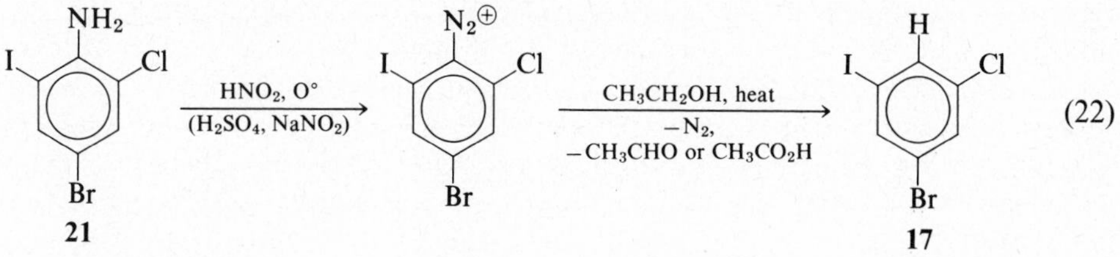

$$\text{21} \xrightarrow[\text{(H}_2\text{SO}_4, \text{NaNO}_2)]{\text{HNO}_2, \text{O}°} \text{N}_2^{\oplus} \xrightarrow[\substack{-\text{N}_2, \\ -\text{CH}_3\text{CHO or CH}_3\text{CO}_2\text{H}}]{\text{CH}_3\text{CH}_2\text{OH, heat}} \text{17} \tag{22}$$

One possibly important side reaction might occur when this procedure is used. The diazonium salt may dissociate to give a carbocation and nitrogen. The carbocation could then attack ethanol to give an aryl alkyl ether as a side product (equation 23). It is well known that this type of reaction occurs with certain types of

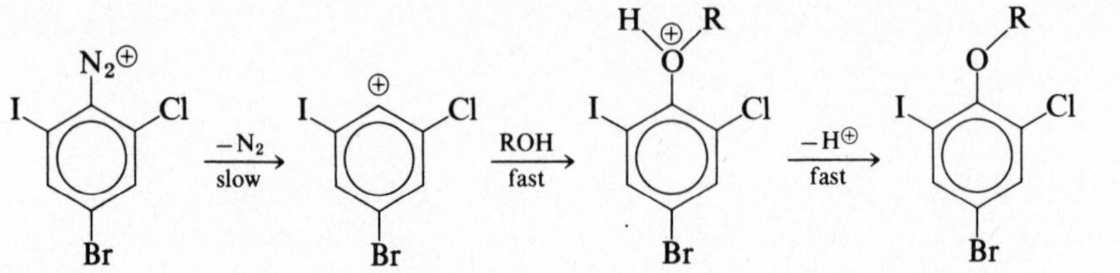

$$\text{(23)}$$

aromatic amines; for example, *o*- and *m*-toluidine (methylanilines) give high yields of the corresponding ethers when diazotized and treated with an alcohol. It appears, however, that ether formation is not important in the present reduction, and if any ether is produced, it is removed in the recrystallization of the product.

In diazotization reactions involving aromatic halo compounds, hydrohalic acids containing a *different* halogen should never be used. There are numerous examples of halogen interchange between the ring and the mineral acid during diazotization and subsequent reduction of the diazo group. This interchange occurs only with *ortho* and *para* halogens (of which there are three in compound **21**) but not with *meta* halogens. When a halogen is present in the ring, it is best to produce the nitrous acid from sodium nitrite and *sulfuric acid*, as is done here.

EXPERIMENTAL PROCEDURE

It is possible to start the synthesis anywhere along the sequence, but this normally will not extend beyond 4-bromoacetanilide, which is the last compound commercially available at a reasonable cost.

In each of the procedures except the final step, the crude product from the previous reaction can be used without further purification. However, methods of purification for each intermediate compound are given if the student wishes to purify the starting materials or to stop somewhere along the reaction sequence. The melting points reported are those for once-recrystallized material.

In carrying out a multistep synthesis, it may be necessary to increase or decrease the scale of a subsequent reaction, as dictated by the yield of a previous product. However, do not change reaction *times* unless told to do so by your instructor.

A. Aniline and Acetanilide

See Section 17.2 (pages 369–370) for experimental procedures for preparation of these substances.

B.   4-Bromoacetanilide

DO IT SAFELY
---

1.   *Bromine is a hazardous chemical.* Do not breathe its vapors or allow it to come into contact with your skin because it may cause *serious* chemical burns. All operations involving the transfer of the pure liquid or its solutions should be carried out in a ventilation hood; rubber gloves should be worn. If you get bromine on your skin, wash the area quickly with soap and warm water, and soak the skin in 0.6 *M* sodium thiosulfate solution (up to 3 hr if the burn is particularly serious).

2.   Glacial acetic acid is a vesicant which will cause severe blistering of the skin if it is allowed to remain there for long. If you get this substance on your skin, wash the area immediately with cold water and apply 0.6 *M* sodium bicarbonate solution.

Dissolve 0.060 mol of acetanilide in 30 mL of glacial acetic acid in a 250-mL round-bottomed flask, and add *to* this mixture slowly and with stirring or swirling a solution of 0.061 mol of bromine dissolved in 6 mL of glacial acetic acid. Wear rubber gloves while preparing and mixing these solutions. After the addition of the bromine-acetic acid solution is complete, stir for several minutes, and then add slowly, with stirring, 200 mL of water. On addition of water a solid will form. Prepare a saturated solution of sodium bisulfite in water, and add just enough of it to discharge the yellow color of the solution. Collect the product by vacuum filtration; wash the solid well with water, and allow it to air-dry.* The yield of crude product should be about 96%. Recrystallization of the crude product from methanol (about 3–4 mL of methanol per gram of compound) gives 4-bromoacetanilide, mp 171–172°.

C.   2-Chloro-4-bromoacetanilide

DO IT SAFELY

Because some chlorine gas may be vented from the reaction flask during this reaction, it would be preferable to carry out this reaction in the hood. If that is not possible, position an inverted funnel over the flask, and attach the funnel to a water aspirator by means of a length of rubber tubing. In this way you create your own miniature hood when the water aspirator is operating. Chlorine gas if inhaled may cause severe irritation to nasal membranes and bronchial tubes. Also read the Do It Safely section of part B regarding glacial acetic acid, a vesicant.

Suspend 0.050 mol of 4-bromoacetanilide in a mixture of 23 mL of concentrated hydrochloric acid and 28 mL of glacial acetic acid, using a 250-mL flask. Heat the mixture gently on a steam bath until it becomes homogeneous, and then cool the solution to 0°. To the cold mixture add 0.026 mol of sodium chlorate dissolved in about 7 mL of water. During the addition of the sodium chlorate solution, some chlorine gas is evolved. As the addition is carried out, a yellow precipitate forms and the solution turns yellow. Allow the reaction mixture to stand at room temperature for one hour, and collect the precipitate by vacuum filtration.* The material that is obtained is crude 2-chloro-4-bromoacetanilide; the yield should be about 97%. The crude product can be recrystallized from methanol (7–8 mL of methanol per gram of crude product) to give pure 2-chloro-4-bromoacetanilide, mp 153–154°.

D.   2-Chloro-4-bromoaniline

Mix 0.045 mol of crude 2-chloro-4-bromoacetanilide with 20 mL of 95% ethanol and 13 mL of concentrated hydrochloric acid in a 250-mL Erlenmeyer flask. Heat the mixture on the steam bath for about 30 min; during the heating the yellow precipitate should dissolve and should be replaced by a white precipitate. At the end of the heating add 90 mL of hot water. Swirl the flask to dissolve the white solid completely and pour the solution onto 50 g of ice. Add to the resulting mixture 12 mL of 14 *M* sodium hydroxide solution, stirring well during the addition. Light-brown crystals should precipitate during the addition of the base. Collect these by vacuum filtration, and dry them as thoroughly as possible.* The yield should be about 91%. Recrystallize the crude product from 30–60° petroleum ether (3–4 mL per gram of product) to give 2-chloro-4-bromoaniline, mp 65–66°.

E.   2-Chloro-4-bromo-6-iodoaniline

DO IT SAFELY

---

1.   Read the Do It Safely section of part B regarding the use of glacial acetic acid.

2.   Iodine is toxic. Care should be taken to avoid getting the solutions used in this experiment on your hands. The wearing of rubber gloves while preparing and transferring solutions in this experiment would be prudent.

Dissolve 0.024 mol of recrystallized 2-chloro-4-bromoaniline in 80 mL of glacial acetic acid, and add about 20 mL of water to the mixture. Prepare a solution of 0.030 mol of technical iodine monochloride in 20 mL of glacial acetic acid in an Erlenmeyer flask, and add this solution to the reaction mixture over a period of 8 min. Heat the resulting black mixture on the steam bath until its temperature is 90°, and then add just enough saturated sodium bisulfite solution to turn the color of the mixture bright yellow; note the volume of sodium bisulfite solution added. Dilute the reaction mixture with enough extra water such that the volume of the sodium bisulfite solution used *plus* the volume of added water is about 25 mL. Cool the reaction mixture in an ice-water bath; 2-chloro-4-bromo-6-iodoaniline will separate as light-brown crystals. Collect the solid by vacuum filtration, and wash the crystals with a small amount of 5 *M* acetic acid and then with water.*

It would be advisable to purify the product before going on to the final step of the synthesis. The crude product may be recrystallized from acetic acid-water as follows. Mix the product with glacial acetic acid in the ratio of about 20 mL of glacial acetic acid per gram of product. Heat the mixture on the steam bath, and slowly add to the solution as it is heating 5 mL of water per gram of product. On slow cooling, long colorless crystals of 2-chloro-4-bromo-6-iodoaniline will form; filter and dry the pure product. The recovery should be about 80% (based on crude product) and the melting point should be 96–98°.

F.   1-Bromo-3-chloro-5-iodobenzene

Suspend 0.006 mol of 2-chloro-4-bromo-6-iodoaniline in about 10 mL of absolute ethanol in a 250-mL round-bottomed flask. While stirring the mixture, add 4.0 mL of concentrated sulfuric acid dropwise. Equip the flask with a condenser, and add 0.010 mol of powdered sodium nitrite in small portions through the condenser. When the addition is complete, heat the mixture on a steam bath for about 10 min. Add 50 mL of hot water to the flask through the condenser, and steam-distil. Collect about 80 mL of distillate (which should be clear and should not contain any organic product). *The desired product will form as a solid in the condenser, so that care should be taken to ensure that the condenser does not plug up completely during the distillation.*[5] It is easiest to remove the solid product from the condenser by pouring diethyl ether through it; this dissolves and removes the product.★ Dry the ether solution over anhydrous magnesium sulfate, filter, and distil the ether. (*Caution:* Use a steam bath, and be sure that no flames are close by.) Recrystallize the residue from about 10 mL of methanol[6] to give 1-bromo-3-chloro-5-iodobenzene in about 40% yield. The long, nearly colorless needles should melt between 82 and 84°.

EXERCISES

1.   Outline synthetic procedures, starting from benzene, for each of the following compounds:
     (a)   1,3,5-tribromobenzene
     (b)   2-bromo-4-chloro-6-iodophenol
     (c)   2-bromo-4,6-dichloroaniline
2.   In the bromination of acetanilide, using bromine in acetic acid, the major product is 4-bromoacetanilide. Suggest a reason for this. Give another example of an electrophilic substitution reaction which gives the *para* isomer as the predominant product.
3.   Give the complete, stepwise mechanism for the reaction which occurs between iodine monochloride and 2-chloro-4-bromoaniline. Suppose that this same reaction is carried out using bromine monochloride, BrCl. What electrophilic

[5] If the condenser plugs up during the distillation, it can be opened up by running hot water or a slow stream of steam through the condenser jacket.

[6] It may be necessary to heat the filter funnel before filtering the solution because the solid will crystallize in the stem of a cold funnel before the filtrate reaches the flask. See the discussion in Section 2.7 under Hot Filtration and Figure 2.15.

substitution reaction might occur when this compound is allowed to react with 2-chloro-4-bromoaniline? Briefly explain.

4. Hydroxylic solvents such as water or low molecular weight alcohols are often used to purify amides. Briefly explain why these are used rather than hydrocarbon solvents, such as petroleum ether.

REFERENCE

A. Ault and R. Kraig, *Journal of Chemical Education,* **43,** 213 (1966).

SPECTRA OF STARTING MATERIALS AND PRODUCTS

The ir and pmr spectra of aniline are given in Figures 15.4 and 15.5, respectively, and the corresponding spectra of acetanilide in Figures 17.4 and 17.5.

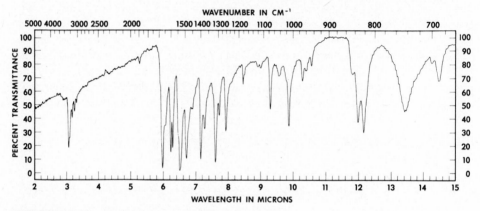

FIGURE 17.12 IR spectrum of 4-bromoacetanilide.

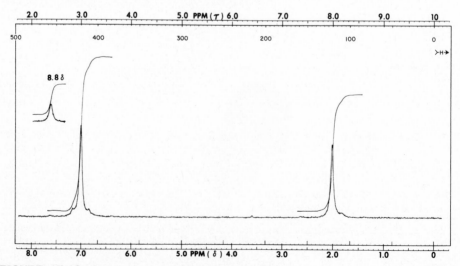

FIGURE 17.13 PMR spectrum of 4-bromoacetanilide.

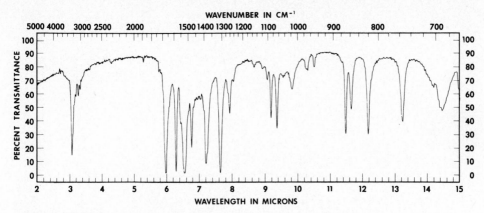

FIGURE 17.14 IR spectrum of 2-chloro-4-bromoacetanilide.

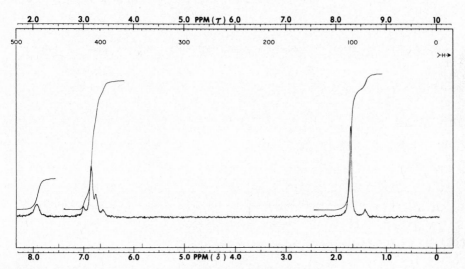

FIGURE 17.15 PMR spectrum of 2-chloro-4-bromoacetanilide.

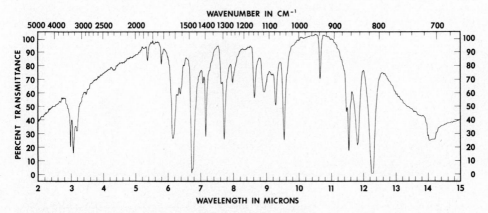

FIGURE 17.16 IR spectrum of 2-chloro-4-bromoaniline.

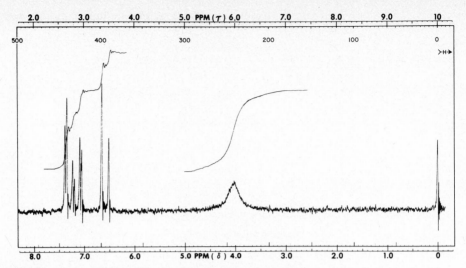

FIGURE 17.17 PMR spectrum of 2-chloro-4-bromoaniline.

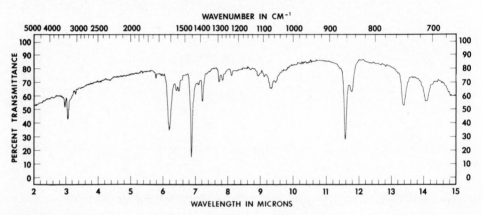

FIGURE 17.18 IR spectrum of 2-chloro-4-bromo-6-iodoaniline.

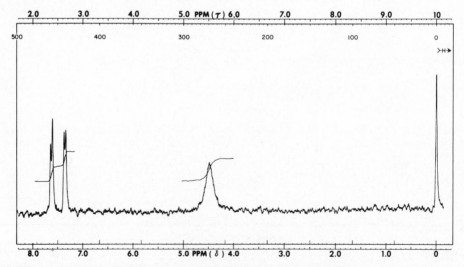

FIGURE 17.19 PMR spectrum of 2-chloro-4-bromo-6-iodoaniline.

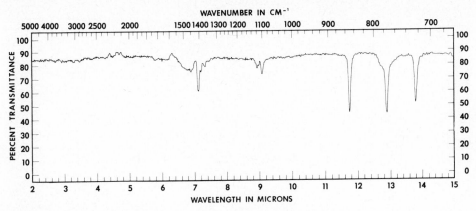

FIGURE 17.20   IR spectrum of 1-bromo-3-chloro-5-iodobenzene.

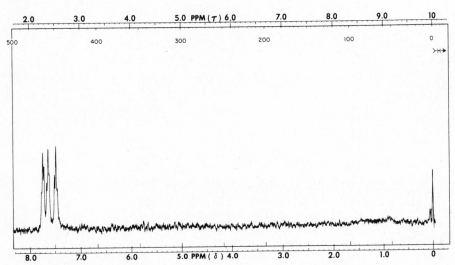

FIGURE 17.21   PMR spectrum of 1-bromo-3-chloro-5-iodobenzene.

17.4   Additional Multistep Synthetic Sequences

In Chapter 8 the preparation of 4-cyclohexene-*cis*-1,2-dicarboxylic anhydride (**22**) from maleic anhydride and 1,3-butadiene is described, as well as hydrolysis of **22** to the corresponding acid (**23**). In Chapter 6 directions are given for hydrogenation of **23** to *cis*-cyclohexane-1,2-dicarboxylic acid (**24**). These preparations may be combined as shown in Figure 17.22 to comprise a three-step sequence.

The preparation of 1-bromobutane (**25**) is described in Chapter 10 and the oxidation of 2-methyl-1-propanol (**26**) to 2-methylpropanal (**27**) is given in Chapter 13. 1-Bromobutane may be converted to the Grignard reagent (**28**) and its reaction with 2-methylpropanal carried out as described in Chapter 12 to produce 2-methyl-3-heptanol (**29**). These preparations may be combined as shown in Figure 17.23.

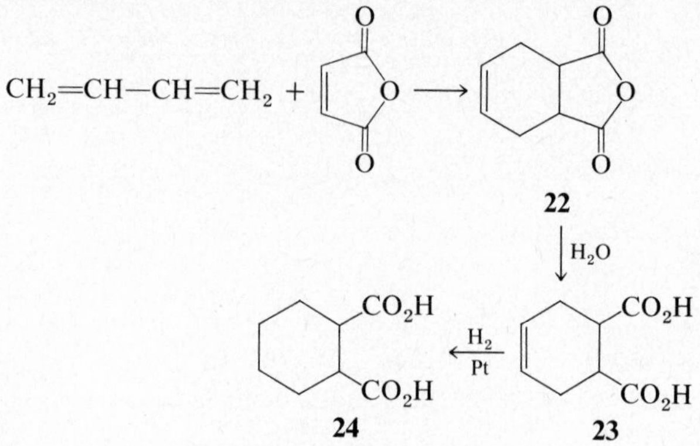

$$CH_2{=}CH{-}CH{=}CH_2 \; + \; \underset{\text{O}}{\overset{\text{O}}{\Big\|}}$$

22

$\downarrow H_2O$

24 $\xleftarrow[\text{Pt}]{H_2}$ **23**

FIGURE 17.22 Synthesis of cyclohexane-1,2-dicarboxylic acid.

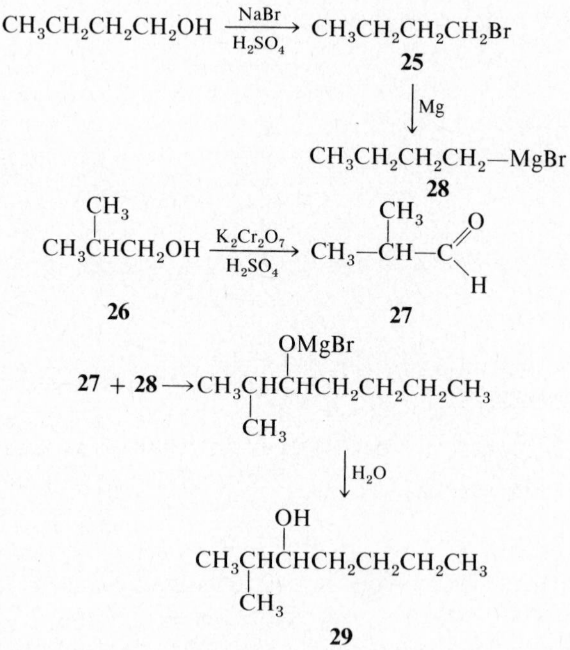

$$CH_3CH_2CH_2CH_2OH \xrightarrow[\text{H}_2\text{SO}_4]{\text{NaBr}} CH_3CH_2CH_2CH_2Br$$

25

$\downarrow Mg$

$$CH_3CH_2CH_2CH_2{-}MgBr$$

28

$$\underset{\textstyle \text{26}}{CH_3\overset{\textstyle CH_3}{\underset{\textstyle |}{C}}HCH_2OH} \xrightarrow[\text{H}_2\text{SO}_4]{K_2Cr_2O_7} \underset{\textstyle \text{27}}{CH_3{-}\overset{\textstyle CH_3}{\underset{\textstyle |}{C}}H{-}\overset{O}{C}{\diagdown}_H}$$

$$27 + 28 \longrightarrow CH_3\overset{\textstyle OMgBr}{\underset{\textstyle |}{C}}H\overset{}{\underset{\textstyle \underset{\textstyle CH_3}{|}}{C}}HCH_2CH_2CH_2CH_3$$

$\downarrow H_2O$

$$CH_3\overset{\textstyle OH}{\underset{\textstyle |}{C}}H\overset{}{\underset{\textstyle \underset{\textstyle CH_3}{|}}{C}}HCH_2CH_2CH_2CH_3$$

29

FIGURE 17.23 Synthesis of 2-methyl-3-heptanol.

18

NONBENZENOID AROMATIC COMPOUNDS

18.1 Aromaticity

In recent years many fascinating examples of compounds exhibiting chemical properties typical of benzene, yet having no apparent molecular similarity to it, have been discovered. These substances are said to have aromatic character, or "aromaticity," although there is not yet complete agreement among chemists as to an exact definition of this term.

The problem relates to which property or properties of benzene are considered to be most typical. The older concept of aromaticity involved resistance toward addition reactions and ease of substitution. This has been updated in terms of mechanisms; both ionic addition and free radical substitution are considered typical of aliphatic compounds, and electrophilic substitution is considered typical of aromatic compounds.

Another criterion that has been applied is *resonance energy,* which may be calculated from heats of combustion or hydrogenation. The resonance energy of an aromatic compound is a measure of the unusual stability of the molecule compared to some aliphatic model and is attributed to the cyclic conjugation of the aromatic compound. Naphthalene (1), phenanthrene (2), and other polycyclic aromatic hydrocarbons and their derivatives are obviously "benzenoid," since they essentially consist of two or more benzene rings fused together. The nitrogen-containing analogs of benzene, such as pyridine (3) and pyrimidine (4), are also usually considered benzenoid.

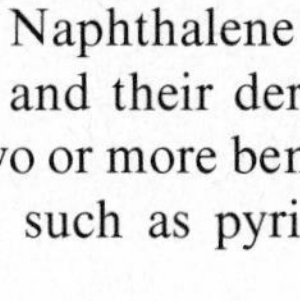

1 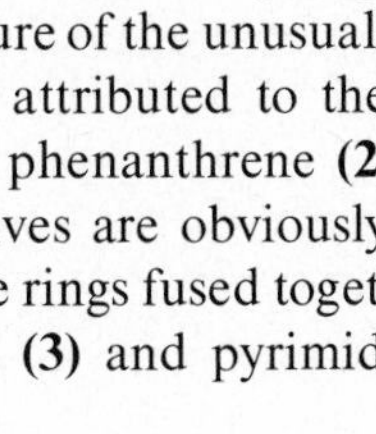   2 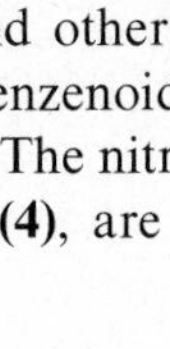   3 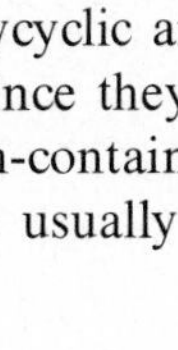   4

A different situation is encountered in azulene (**5**), which contains a cyclic conjugated system in fused five- and seven-membered rings. Azulene, then, must be considered nonbenzenoid. It is an isomer of naphthalene, to which it is converted quantitatively by heating above 350° in the absence of air (equation 1). Although less stable than naphthalene, azulene undergoes certain typical electrophilic substitution reactions.[1]

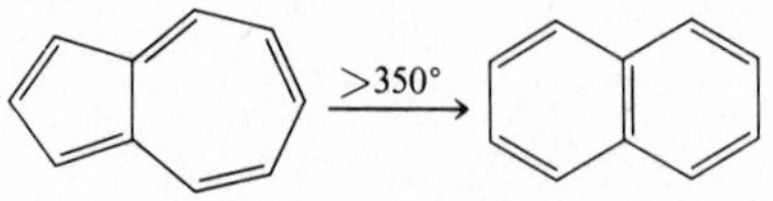

$$\text{5} \xrightarrow{\;>350°\;} \text{1} \tag{1}$$

5   **1**

The heterocyclic compounds with five-membered rings, such as furan (**6**), pyrrole (**7**), and thiophene (**8**), constitute another class of aromatic compounds.

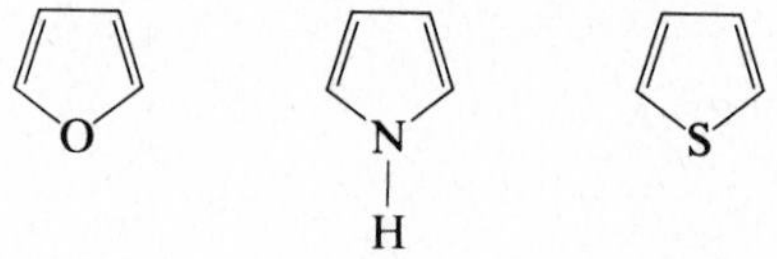

6   **7**   **8**

Strictly speaking they are not benzenoid, since they do not have six-membered rings. However, they are closely related to benzene and pyridine in that they also have an "aromatic sextet" of electrons. The hetero atoms provide a pair of electrons which enter into conjugation with the four π-electrons of the carbon-carbon bonds, as shown in the resonance structures **6**, **9**, and **10** (there are two additional resonance structures equivalent to **9** and **10**).

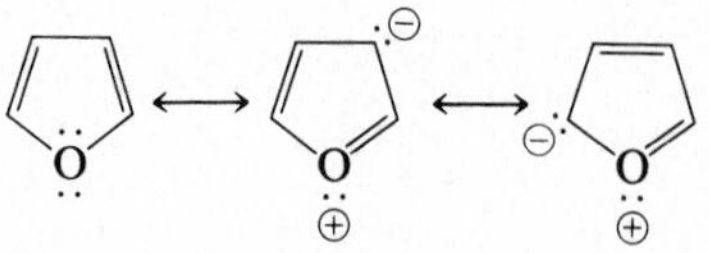

6   **9**   **10**   , and so on

A hydrocarbon system closely related to these heterocyclic compounds is the cyclopentadienyl anion, represented by resonance structures **12**, **13**, and **14**, and so

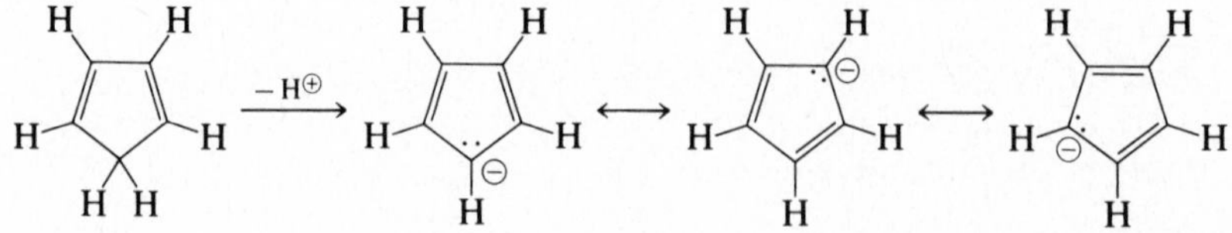

11   **12**   **13**   **14**   , and so on

forth, derived from cyclopentadiene (**11**). This compound is unusually acidic for a hydrocarbon (pK_a of about 20), presumably owing to stabilization of its anion by

[1]In connection with this discussion of azulene, see Exercise 6 in Section 11.1.

delocalization of negative charge as represented by resonance structures **12**, **13**, and **14**, and so forth.

One might infer from these examples that any cyclic conjugated system for which a number of reasonable contributing resonance structures may be written will have aromatic character. That this is not true is demonstrated by the fact that the cyclopentadienyl *cation* (**15**) is extremely unstable, even though the same number of contributing resonance structures (**15**, **16**, and **17**, and so forth) can be written as

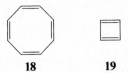

for the corresponding anion. Thus it may be seen that simple *valence bond theory* is inadequate to explain the difference in stability of cyclopentadienyl cation and anion, and certain other aspects of aromaticity.

A *molecular orbital theory* proposed by the German chemist E. Hückel in 1931 has provided an explanation for such problems. Hückel's theory, which was based on quantum mechanics, stated that conjugated cyclic systems in which the number of delocalized π-electrons is $4n + 2$, where $n = 0, 1, 2, 3$, and so forth, will exhibit special stability, that is, aromaticity. Benzene, pyridine, furan, and the cyclopentadienyl anion all fit "Hückel's rule" for $n = 1$; that is, the number of π-electrons is six $(4 \times 1 + 2 = 6)$. The cyclopentadienyl *cation* system involves *four* π-electrons, so according to Hückel's rule, it should not have special stability.

Hückel's rule also offers an explanation for the fact that cyclooctatetraene (**18**), with eight π-electrons, and cyclobutadiene (**19**), with four, are nonaromatic. On the

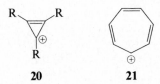

other hand, cyclopropenyl cations (**20**, $n = 0$), and cycloheptatrienyl cation (**21**, $n = 1$), are predicted to be unusually stable.[2] One of the most significant examples of

the success of prediction based on theory in organic chemistry is the realization of the preparation of these cations and the demonstration of their remarkable stability.

[2]It is interesting to note that the cycloheptatrienyl *anion*, which has eight π-electrons, is extremely unstable, even though the same number of equivalent contributing resonance structures can be written for the cycloheptatrienyl cation and anion.

A modern criterion of aromaticity may be described in connection with these cations. A carbocation may react with water reversibly to form an equilibrium mixture containing the corresponding alcohol and hydronium ion (equation 2). According to equation 5, the $pK_R^{\oplus}$ of a carbocation is equal to the pH of a solution in which the carbocation is half converted to the corresponding alcohol. An unconju-

$$R^{\oplus} + 2\,HOH \rightleftharpoons ROH + H_3O^{\oplus} \tag{2}$$

$$K_R^{\oplus} = \frac{[ROH][H_3O^{\oplus}]}{[R^{\oplus}]} \tag{3}$$

When $\dfrac{[ROH]}{[R^{\oplus}]} = 1$, then $K_R^{\oplus} = [H_3O^{\oplus}]$ $\tag{4}$

and therefore $pK_R^{\oplus} = \log\dfrac{1}{[H_3O^{\oplus}]} = pH$ $\tag{5}$

gated carbocation is extremely reactive toward water, so that the $pK_R^{\oplus}$ is a large negative number. The $pK_R^{\oplus}$ of an alkyl carbocation such as the *t*-butyl cation cannot be measured, and even the triphenylmethyl cation has a $pK_R^{\oplus}$ of -6.63. By contrast, the cycloheptatrienyl cation (**21**) has a $pK_R^{\oplus}$ of $+4.7$, and the cyclopropenyl cation (**20**, R = *n*-C$_3$H$_7$) has a $pK_R^{\oplus}$ of $+7.2$. The latter cation is thus stable even in neutral water solution; it is the most stable hydrocarbon cation yet known. Since the $pK_R^{\oplus}$ of a cation is a measure of its stability, it may also be used as a criterion of aromaticity in appropriate systems.

Another modern criterion of aromatic character is the existence of a *ring current,* which is induced in an aromatic compound when it is placed in a magnetic field. A pmr spectrometer makes use of a strong external magnetic field, and evidence for the ring current is provided by the pmr spectrum of an aromatic compound.[3] The ring current attributable to the circulation of π-electrons in benzene results in the deshielding of the six hydrogens (which are *outside* the aromatic ring) so that their pmr signals are shifted downfield (to larger δ values). According to the ring current theory, hydrogen atoms on the *inside* of an aromatic ring should experience a shielding effect, resulting in an upfield shift of their pmr signal. This theory has been confirmed by the pmr spectrum of the interesting hydrocarbon **22**, which is called [18]annulene. This compound, which has 18 π-electrons and thus should be aromatic

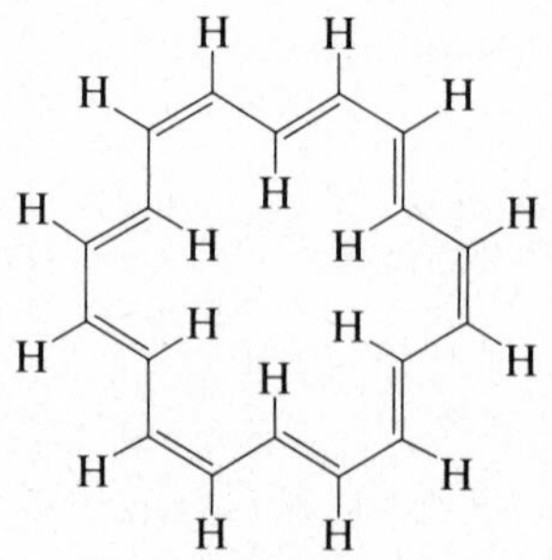

22

[3] Refer to Section 4.2.

according to Hückel's rule ($n = 4$), has 6 hydrogen atoms inside the conjugated ring system and 12 outside. The two pmr signals occur at 1.9 ppm *upfield* from TMS (integration = 6H) and 8.8 ppm *downfield* from TMS (integration = 12H); thus the inside hydrogen atoms are strongly shielded, and the outside hydrogen atoms are deshielded, as expected for a ring current effect.

18.2 Ferrocene

Ferrocene (23) was discovered accidentally in 1951. It is an orange, crystalline solid whose structure consists of two cyclopentadienyl anions bonded to a ferrous cation. The bonding between the metal and the organic anions involves the π-electrons of the two rings in such a way that all the carbon atoms are bonded equally to the central ferrous ion. Several other metals form "sandwich-type" compounds with cyclopentadienyl anions, called *metallocenes*. Among the most stable of these, besides ferrocene, are those from the dipositive ions of ruthenium and osmium, since in these compounds, as well as in ferrocene, the metal ion achieves the electronic structure of an inert gas atom.

Ferrocene has the classical aromatic properties of resistance toward reaction with acids and bases, even concentrated sulfuric acid. It is sensitive toward oxidizing acids, because the iron is oxidized to the ferric state and the resulting metallocene cation is much less stable (the iron no longer has the krypton electronic structure). Ferrocene does not undergo addition reactions typical of cyclopentadiene but readily undergoes electrophilic substitutions such as Friedel-Crafts acylation.

Depending on the catalyst and the conditions, either acetylferrocene (24) or 1,1'-diacetylferrocene (25) may be produced as the major product of acetylation. The

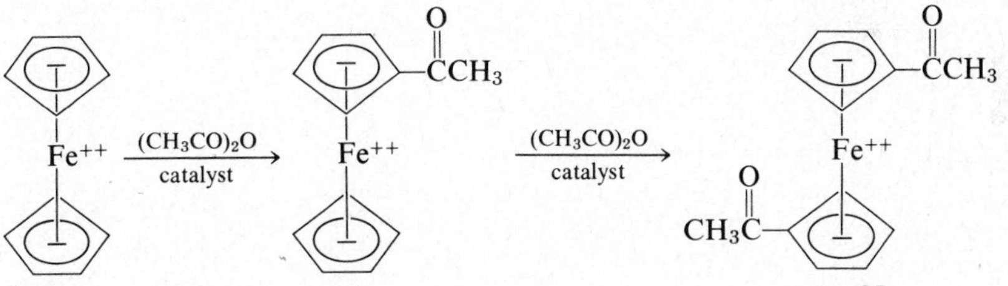

structure of the disubstituted product has been established by degradation studies, using reactions which removed the iron from the organic part of the molecule and which gave no products indicative of the presence of *two* acetyl groups on a cyclopentadienyl ring. This confirmed the expectation that the second acetyl group would not substitute on the same ring as the first, since the acetyl group would be expected to deactivate the ring toward a second electrophilic attack, by analogy to reactions of benzene derivatives. The fact that only one such diacetylferrocene has been found indicates that the cyclopentadienyl rings are able to rotate about the axis of the bonds to the metal, although the staggered conformation indicated in formula 23 is probably preferred.

The acetylation of ferrocene offers an excellent opportunity for a demonstration of the application of (1) thin-layer chromatography to monitor the course of a reaction and (2) column chromatography to effect separations of products from one another and from starting material on a preparative scale. The utility of these methods will be demonstrated in the experiments described.

EXPERIMENTAL PROCEDURE

A.   Acetylferrocene and 1,1'-Diacetylferrocene

DO IT SAFELY
---

1.   The acetylation procedure in part 1 calls for the use of acetic anhydride and of 85% phosphoric acid. The inhalation of vapors of acetic anhydride may cause irritation of mucous membranes. It should be measured and transferred in the hood. Each of these substances may cause acid burns of the skin. If either comes into contact with your skin, wash the area well with cold water, and apply a 0.6 M solution of sodium bicarbonate. If an acid burn seems particularly severe, prepare and apply a *paste* of sodium bicarbonate to the area.

2.   In each of the alternative chromatographic procedures of parts 2 and 3, it is necessary to evaporate volatile and flammable solvents. Take the usual precautions to avoid fire, either by using a steam bath *in the hood* or by setting up distillation apparatus to accomplish the evaporation.

1.   Preparation from Ferrocene.   Place 0.0054 mol of ferrocene and 0.11 mol of acetic anhydride in a 50-mL Erlenmeyer flask, and add 2 mL of 85% (8.7 M) phosphoric acid dropwise, with stirring. Attach a calcium chloride drying tube to the flask, and heat it gently for about 15–20 min on a steam bath. Pour the reaction mixture onto about 20 g of chipped ice (ignoring any tarry material that may cling to the bottom of the flask), and when the ice has melted, neutralize the solution by adding solid sodium bicarbonate. Add the bicarbonate with swirling until bubbles of carbon dioxide are no longer produced when small portions are added.[4] About 20–25 g of bicarbonate will be required. Approximately 20 g can be added as rapidly as foaming will allow; additional bicarbonate should be added more slowly because an excess of base is to be avoided. Cool the reaction mixture for about 30 min in an ice-water bath; then add a little solid sodium bicarbonate to the mixture. If reaction occurs, add more bicarbonate until no further bubbling occurs. (See Exercise 1.) Now collect by vacuum filtration the orange-brown solid that has precipitated, and wash it with water until the washings are pale orange. Dry the product in air.★

---

[4]The use of pH paper is a more exact way of determining when neutralization is complete. However, if the reaction mixture has darkened so much that the use of pH paper is impractical, the absence of bubbling on addition of bicarbonate is a satisfactory criterion of neutralization.

2. Purification by "Wet-Column" Chromatography.[5] Prepare a chromatographic column using a 50-mL buret filled about halfway with acid-washed alumina. Dissolve the dry crude product from part 1 in a minimum amount of toluene and introduce it onto the alumina column with a capillary pipet. Elute with petroleum ether (bp 60–80°) first, which should bring a band of yellow-orange material down the column much more rapidly than a red-orange material. After the yellow-orange band has been removed from the column with petroleum ether, the red-orange material may be removed more rapidly with petroleum ether–diethyl ether (1:1 by volume).

Evaporate the solvents from the fractions of eluate containing the yellow-orange and red-orange materials, and determine the melting point of the crystalline residues. The reported melting point of ferrocene is 173°, of acetylferrocene, 85°. Recrystallize the acetylferrocene from petroleum ether (bp 60–80°), and determine the yield.

Carry out a test with sodium hypoiodite on a small sample of the pure product to furnish evidence of its structure; see Section 14.2 for details of the iodoform test.

3. Purification by "Dry-Column" Chromatography. Dissolve the brown solid product from part 1 in a small quantity of acetone (about 5 mL), and add this solution to 2 g of activity III alumina.[6] Evaporate the solvent by stirring the resulting slurry under a *gentle* stream of air. When completely dry, the remaining solid should be granular and free flowing.

Prepare the column necessary for the separation in the following way. Place a loose plug of glass wool in one end of a 40-cm × 17-mm section of clean and dry Pyrex glass tubing. With the aid of a funnel, add 35 g (about 35 mL) of activity III alumina through the other end of the tubing. Hold the filled column in a vertical position, and tap the plugged end gently against a cushioned surface, for example, a towel on top of the lab bench, to pack the alumina firmly in the column. Tap the sides of the tubing to aid in compacting the alumina in the column. Clamp the column in a *vertical* position, and ascertain that the top of the alumina packing is level. If it is not, tap the column gently to achieve a level surface. (*Caution:* To avoid uneven development of solvent through the column, do *not* lay the column on its side after it has been packed.) To the top of the prepared column add the alumina on which the reaction mixture is adsorbed, and tap the column gently to provide a level surface. Complete the preparation of the column by adding enough fresh alumina to provide a level layer 6 mm in depth on top of the sample.

Develop the column in the following manner. *Carefully* pour dichloromethane onto the top of the column until a layer of about 5 cm is reached. As the solvent percolates down the column, add fresh solvent so as to maintain the 5-cm head on the column. When the solvent is within about 2 cm of the bottom of the column, remove the excess solvent from the top with a pipet. When the solvent has further advanced

---

[5]See Section 3.5 for details of preparing a chromatographic column. The column should be used the same day it is prepared, so it may be convenient to allow the crude acetylferrocene to dry in the desk until the next laboratory period when the column will be prepared and used.

[6]Activity III alumina suitable for dry-column chromatography can be prepared from commercially available activity I alumina by adding 6 wt % of water; or Woelm activity III alumina for dry-column chromatography can be obtained from Waters Associates, Inc., 61 Fountain Street, Framingham, Massachusetts 01701.

to within about 3 mm of the bottom, unclamp the column and lay it on its side to terminate development of the column. In the same manner as for thin-layer chromatography (see Figure 3.12), *immediately* determine the R_f values for ferrocene, acetylferrocene, and, if you can detect it, diacetylferrocene.[7]

Isolate the purified products by scraping the individual bands of products from the column with the aid of a spatula and extracting the alumina containing each band three times with 20- to 30-mL portions of technical diethyl ether. After each extraction decant the supernatant liquid into an appropriately sized flask; following the final extraction, evaporate the solvent to obtain the solid product. Determine the yield and melting point of the isolated acetylferrocene. The average yield of isolated acetylferrocene is about 30% of theoretical.

B. Acetylation of Ferrocene: Thin-Layer Chromatography

DO IT SAFELY

1. For part 1 read the Do It Safely notes included under part A.
2. For part 2 the use of acetyl chloride and aluminum chloride is required. Neither of these reagents should be exposed to air for any longer than the time necessary for their measurement because they each will react with atmospheric moisture to produce hydrochloric acid. Thus you should avoid allowing either chemical to come into contact with your skin and also avoid breathing the vapors of acetyl chloride.

1. **Phosphoric Acid Catalyst.** In a 125-mL Erlenmeyer flask fitted with a calcium chloride drying tube, dissolve 0.0054 mol of ferrocene in 0.21 mol of acetic anhydride by heating on a steam bath and swirling. Cool the solution to room temperature, and then add dropwise and with swirling 3 mL of 85% (8.7 M) phosphoric acid. Reattach the drying tube, note the time, and allow the mixture to stand at room temperature. After 2 hr[8] measure out a 5-mL aliquot of the reaction mixture, and pour it onto about 20 g of chipped ice. After the ice has melted, neutralize the mixture with solid sodium bicarbonate until effervescence ceases (see footnote 4). Extract the resulting mixture with diethyl ether, dry the ether extract with calcium chloride, and evaporate the ether.* Dissolve the solid residue in a small amount of toluene, and use the toluene solution to spot a thin-layer chromatographic (tlc) plate.[9] Dissolve some ferrocene in toluene, and place a spot alongside the spot from the reaction mixture. Use a mixture of toluene and absolute ethanol (30:1 by volume) to develop the tlc plate, and determine the R_f values of ferrocene and acetylferrocene.

[7]The characteristic colors of the individual components are ferrocene—yellow; acetylferrocene—reddish-orange; diacetylferrocene—tan.

[8]This period may be used for the preparation of tlc plates (see the reference by J. G. Kirchner at the end of Chapter 3), or if these are furnished ready for use (see footnote 9), part 2 of this experiment (using aluminum chloride catalyst) may be initiated.

[9]Commercial silica-gel plates (without fluorescent indicator) may be used conveniently. For details of the tlc technique, see Section 3.7.

Allow the remainder of the reaction mixture to stand with the drying tube attached and, after periods of 1 or 2 days and/or a week, measure out 5-mL aliquots, pour them onto 20 g of chipped ice, and repeat the work-up and tlc procedure described for the 2-hr reaction mixture. Compare the thin-layer chromatograms obtained from the aliquots of the reaction mixture after longer reaction times. Determine the R_f for 1,1'-diacetylferrocene, which should be found as a product from the longer reaction periods.

2. Aluminum Chloride Catalyst. Dissolve 1.0 g of ferrocene in about 15 mL of dichloromethane. Calculate amounts of acetyl chloride and powdered *anhydrous* aluminum chloride ($AlCl_3$) which will provide 1.5 mol of each of these compounds per mole of ferrocene. Weigh out the correct amounts of these reagents within ± 0.02 g, and dissolve both immediately in 25 mL of dichloromethane.

Add the solution of ferrocene to the solution of acetyl chloride and aluminum chloride, swirl to mix, note the time, and allow the mixture to stand in a flask protected by a calcium chloride drying tube. After intervals of 1, 15, 30, and 60 min, measure out a 5-mL aliquot of the reaction mixture, and pour it onto about 15 g of chipped ice. Neutralize the aliquot with solid sodium bicarbonate until effervescence ceases (see footnote 4), add about 10 mL of dichloromethane, and separate the layers. Wash the organic solution with distilled water, and dry it over calcium chloride.★ Evaporate the solvent from the extract, and dissolve the solid residue in a small amount of toluene.★ Carry out tlc analysis of the toluene solutions as described in part 1. Determine R_f values for ferrocene, acetylferrocene, and 1,1'-diacetylferrocene.

As an interesting extension of this experiment, some of the class members may use different molar ratios of ferrocene, acetyl chloride, and aluminum chloride, and the effect on the relative rates of production of acetylferrocene and 1,1'-diacetylferrocene may be determined. Other suggested molar ratios are ferrocene:acetyl chloride:aluminum chloride = 1:1:1 and 1:2:2.

EXERCISES

1. Why does more sodium bicarbonate react after the reaction mixture from ferrocene, acetic anhydride, and phosphoric acid has been neutralized once and then allowed to stand 30 min? (*Hint:* It is not necessary to wait and neutralize a second time when acetyl chloride is used as the acetylating agent.)
2. Judging from the procedure for the column chromatography of crude acetylferrocene, (a) which is more strongly adsorbed on alumina, ferrocene or acetylferrocene; (b) in which solvent, petroleum ether or diethyl ether, is acetylferrocene more soluble? why?
3. Ferrocene cannot be nitrated successfully by the usual mixed nitric acid-sulfuric acid medium. Why not?
4. If ferrocene were locked in the conformation shown in **23**, how many isomers of 1,1'-diacetylferrocene would there be? It may be helpful to draw the structure as it would appear from a position above the molecule as shown in **23**, that is, in line with the axis of the bonds between the rings and the iron.

SPECTRA OF STARTING MATERIAL AND PRODUCT

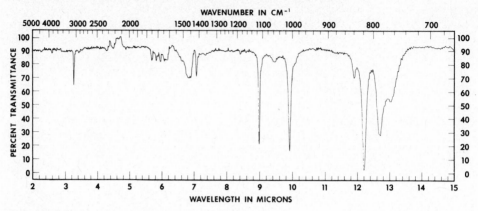

FIGURE 18.1 IR spectrum of ferrocene.

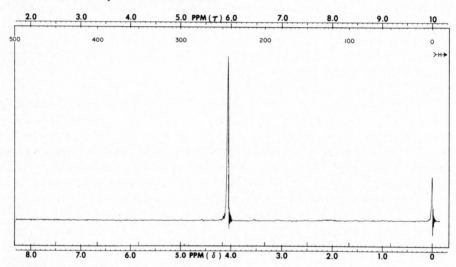

FIGURE 18.2 PMR spectrum of ferrocene.

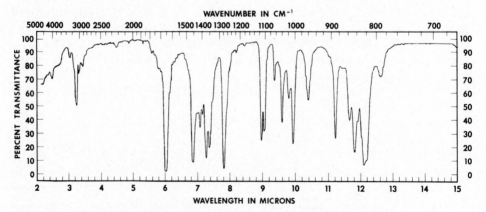

FIGURE 18.3 IR spectrum of acetylferrocene.

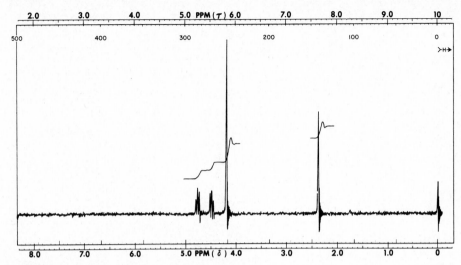

FIGURE 18.4 PMR spectrum of acetylferrocene.

18.3 Tropylium Iodide

In contrast to the accidental discovery of ferrocene, the successful synthesis of salts containing the cycloheptatrienyl cation moiety in 1954 was the result of intense competitive research efforts. As mentioned in Section 18.1, Hückel predicted in 1931 that the cycloheptatrienyl, or tropylium, cation should exhibit unusual stability indicative of aromaticity. Many chemists set out on syntheses designed to produce the cation with the intention of proving Hückel's theory to be either correct or incorrect. Ironically, the isolation of a salt containing the cation was delayed by initial failure to anticipate its water-solubility, which might have been expected if Hückel's prediction were correct! However, since no other carbocations had previously been found to be stable in water solution,[10] it is not surprising that this possibility was at first overlooked.

In this experiment a sequence of reactions is used which demonstrates the fact that the tropylium cation is much more stable than the triphenylmethyl cation (refer to Section 18.1 for the comparison in terms of $pK_R^{\oplus}$). Triphenylmethyl fluoborate **(27)** is first prepared from triphenylmethanol (**26**, equation 7) and then allowed to react with 1,3,5-cycloheptatriene (**28**, equation 8). Hydride ion ($H:^{\ominus}$) is transferred

$$(C_6H_5)_3COH + HBF_4 \xrightarrow{(CH_3CO)_2O} (C_6H_5)_3C^{\oplus}BF_4^{\ominus} + H_2O \qquad (7)$$
$$\quad\;\, \textbf{26} \qquad\qquad\qquad\qquad\qquad\qquad \textbf{27}$$

$$(C_6H_5)_3C^{\oplus}BF_4^{\ominus} + \;\; \underset{\textbf{28}}{\vcenter{\hbox{⬡}}}\!\!\begin{matrix}H\\H\end{matrix} \longrightarrow (C_6H_5)_3C-H + \underset{\textbf{30}}{\vcenter{\hbox{⬡}}}\!\!-H\;\; BF_4^{\ominus} \qquad (8)$$
$$\qquad \textbf{27} \qquad\qquad\qquad\qquad\qquad\qquad \textbf{29}$$

[10] Although the tripropylcyclopropenyl cation is more stable than the tropylium cation (see Section 18.1), it was not prepared until 1962.

$$\text{(cycloheptatrienyl)} + BF_4^\ominus + Na^\oplus I^\ominus \xrightarrow{H_2O} \text{(tropylium)} + I^\ominus + Na^\oplus BF_4^\ominus \qquad (9)$$

30   **31**

from the triene to the triphenylmethyl cation to produce the more stable tropylium cation, which precipitates from diethyl ether solution in the form of the fluoborate, **30**. Tropylium iodide (**31**) is much less soluble in water than the bromide (which was the first salt isolated) and can be precipitated by metathetical reaction of tropylium fluoborate with sodium iodide (equation 9).

EXPERIMENTAL PROCEDURE

DO IT SAFELY

1.   The 48% solution of fluoboric acid utilized in this experiment is an *extremely* corrosive liquid. Under no circumstances should you allow contact with your skin. Should this happen, *immediately* wash the area with large amounts of cold water, *and* apply a paste of sodium bicarbonate to the affected area.

2.   Fairly large quantities of diethyl ether are also used. Take the usual precautions against the presence of flames and hot plates in your area.

In a 250-mL beaker place 0.10 mol of acetic anhydride, and add 3.0 g of 48% fluoboric acid dropwise with stirring and cooling. Now add 0.015 mol of triphenyl-methanol to this mixture in small portions with stirring; a yellow precipitate should be present after the addition is complete.[11] Add to the mixture 100 mL of anhydrous diethyl ether, stir for a few minutes, and then permit the precipitate of triphenyl-methyl fluoborate to settle. Pour a few milliliters of ether carefully down the side of the beaker; if any turbidity appears, the precipitation was incomplete, and more ether should be added.

The next step is to collect the triphenylmethyl fluoborate on a filter. Since it is sensitive to moisture in the air, be prepared to dissolve it *immediately* in 10 mL of *anhydrous* nitromethane by rapidly carrying out the operations described. Use a 4-cm Büchner funnel and a well-fitting paper to collect the triphenylmethyl fluoborate by vacuum filtration. As soon as the filtrate has run through the funnel, release the vacuum, and insert the funnel stem into a clean, dry 25-mL Erlenmeyer flask. Pour about 7 or 8 mL of *anhydrous* nitromethane onto the triphenylmethyl fluoborate on the filter paper, and stir with a glass rod until solution is complete. Lift the paper slightly with the stirring rod so that the solution runs through the funnel and into the flask below. Add another 2 or 3 mL of anhydrous nitromethane to wash the filter paper and funnel, and allow the wash liquid to drain into the flask with the main part of the solution.

[11] It may appear that the triphenylmethanol does not dissolve or react; by careful observation it may be noticed that the white crystals of triphenylmethanol do dissolve, but they are replaced immediately by the yellow crystals of triphenylmethyl fluoborate.

To the nitromethane solution of triphenylmethyl fluoborate, add 0.010 mol of 1,3,5-cycloheptatriene. After stirring for 2 or 3 min, pour the solution into 100 mL of technical diethyl ether and stir. Allow the white precipitate of tropylium fluoborate to settle, and test for completeness of precipitation by adding a little ether.

Collect the precipitate by vacuum filtration, allowing time for removal of all the ether, and weigh the dry tropylium fluoborate. Prepare a solution of 5 g of sodium iodide in 20 mL of absolute ethanol by stirring and heating to boiling, then cool the solution to room temperature. Now dissolve the tropylium fluoborate in water, using about 4 mL of water per gram of tropylium fluoborate and warming gently (do not boil). To the warm aqueous solution of the tropylium fluoborate, add the alcoholic solution of sodium iodide. The solution will turn red immediately, and crystals of tropylium iodide may begin to appear. Cool the solution in an ice-water bath, collect the red crystals by vacuum filtration, and wash them with a few milliliters of *ice-cold absolute* ethanol. Air-dry the crystals and determine the yield.

EXERCISES

1. What is the function of the acetic anhydride that is added to the 48% fluoboric acid? Why does the mixture require cooling?
2. What would happen to triphenylmethyl fluoborate if it were allowed to stand in moist air for several hours? What *weight* of water is required to react with 0.015 mol of triphenylmethyl fluoborate?
3. Apparently $C_{10}H_{10}$ ([10]annulene) and $C_{14}H_{14}$ ([14]annulene) are much less stable than [18]annulene, even though these compounds fit Hückel's rule for $n = 2$ and $n = 3$. Suggest an explanation, considering possible geometries of [10]- and [14]annulene.

SPECTRA OF STARTING MATERIAL

For the pmr spectrum of triphenylmethanol, see Figure 12.6.

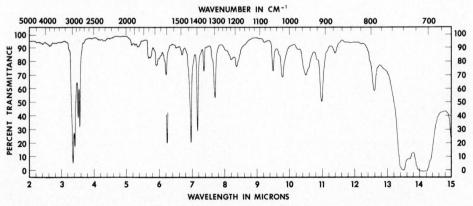

FIGURE 18.5 IR spectrum of 1,3,5-cycloheptatriene.

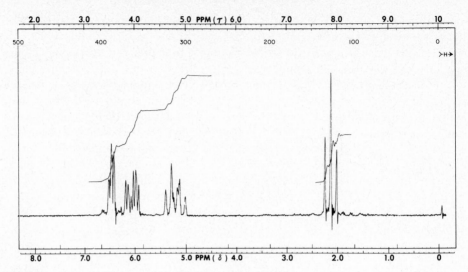

FIGURE 18.6 PMR spectrum of 1,3,5-cycloheptatriene.

19

ISOMERISM AND OPTICAL ACTIVITY
Resolution of Racemic α-Phenylethylamine

Isomers are compounds which have the same composition, that is, the same molecular formula, but have different properties. There are two main classes of isomers: *constitutional isomers* and *stereoisomers.* Constitutional isomers have sometimes been subdivided into three closely related types: *skeletal, positional,* and *functional.* All these have been referred to as "structural isomers," a term that is ambiguous, however, because the way in which stereoisomers differ from one another and from constitutional isomers is also a matter of structure.

As an example of *skeletal isomers,* consider butane and 2-methylpropane. Both

$$CH_3{-}CH_2{-}CH_2{-}CH_3 \qquad CH_3{-}CH \overset{\displaystyle CH_3}{\underset{\displaystyle CH_3}{\Big\langle}}$$

Butane   2-Methylpropane

have the molecular formula C_4H_{10}, but they are *constituted* differently in that butane has a continuous chain of four carbon atoms in its carbon skeleton, whereas 2-methylpropane has a skeletal structure that is branched. 1-Chloropropane and 2-chloropropane (both C_3H_7Cl) are examples of *positional isomers*. The position of

$$CH_3{-}CH_2{-}CH_2{-}Cl \qquad CH_3{-}\underset{\displaystyle Cl}{\overset{\displaystyle |}{CH}}{-}CH_3$$

the chlorine atom in the three-carbon chain is different in the two isomers. The similarity to skeletal isomerism is obvious; it is only the difference in the position of a chlorine atom, rather than a methyl group, which produces different structures. The distinction between skeletal and positional isomers becomes clear upon noting that the *carbon* skeletons of positional isomers are identical.

Functional isomers may be illustrated by ethanol and dimethyl ether (both

$$CH_3—CH_2—O—H \qquad CH_3—O—CH_3$$

C_2H_6O). Here it is the difference in the position of the oxygen atom that produces isomeric molecules with *different functional groups.* Thus skeletal and positional isomers will display similar chemical properties, whereas functional isomers may vary widely in their chemical behavior.

Stereoisomers are those which have the same composition and constitution but differ in the orientation of their atoms in space. One way in which the atoms of covalent molecules may assume different positions in space is by the rotation of parts of the molecule about single bonds, for example, the C—C bond in ethane. Two of the infinite number of *conformations* possible for ethane molecules are illustrated by perspective and Newman projection formulas. The differences in energy separating

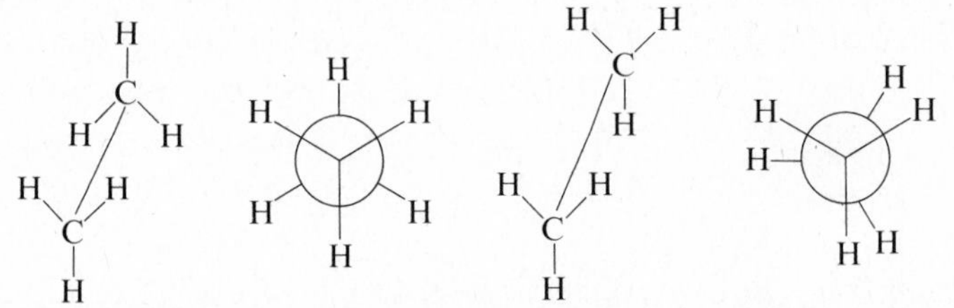

conformational isomers (conformers) are usually too small to allow for their isolation. Compounds capable of assuming two or more distinct conformations are typically found as equilibrium mixtures of these conformational isomers.

Configurational isomers are those in which the different spatial arrangements of the isomeric molecules cannot be produced by rotation about single bonds but require that bonds be broken and remade. The tetrahedral covalency of carbon leads to two different arrangements in space (configurations) of four different atoms (or groups of atoms) attached to a single carbon atom. Such a carbon atom is called *asymmetric* because the molecule lacks all elements of symmetry. The configurational isomers resulting from the two arrangements have the relationship of an object and its mirror image and are called *enantiomers.* Enantiomers have the unusual physical property of rotating the plane of polarized light in opposite directions. For this reason they have been called "optical isomers." The enantiomers of 2-chlorobutane illustrate this type of configurational isomerism.

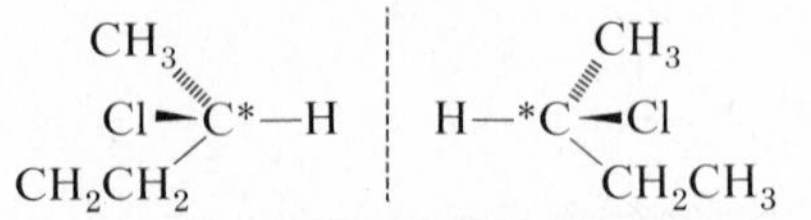

*Asymmetric carbon
atom

Enantiomers of 2-chlorobutane

Some organic molecules do not contain asymmetric carbon atoms, yet they may have two different configurations that are nonsuperimposable mirror images of one another. In all such cases the stereoisomers will be optically active. Any molecule whose mirror image is nonsuperimposable with itself is said to be *chiral* (from the Greek work for "hand") because the hands are perhaps the best-known illustration of mirror image dissymmetric objects. Molecular chirality is the necessary requirement for optical activity.

Achiral molecules may also have different configurations. In contrast to chiral molecules, some of these may be represented adequately in two dimensions. Ethylenic *cis-trans* isomers belong to this category. They are also often called *geometric*

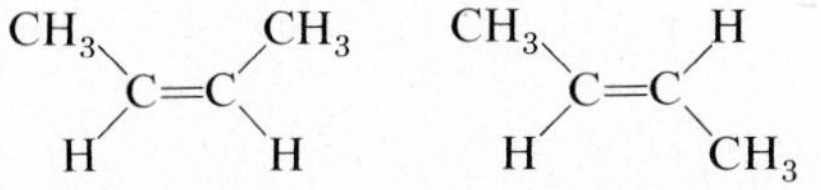

$$CH_3\diagdown_{}_{\diagup}CH_3$$

cis-2-Butene   trans-2-Butene

isomers, which is somewhat misleading because all stereoisomers differ in their geometry. Certain symmetrical cyclic compounds exhibit isomerism which is very similar to that of ethylenic *cis-trans* isomers, for example, 1,3-dimethylcyclobutane.

cis-1,3-Dimethylcyclobutane   trans-1,3-Dimethylcyclobutane

Of course cyclic molecules may also be chiral, resulting in an apparent overlap between *cis-trans* isomers and optical isomers.[1] For example, *cis*-1,2-dimethylcyclobutane is an achiral molecule having a plane of symmetry as shown, but *trans*-1,2-dimethylcyclobutane is chiral and exists in two mirror-image, optically active forms (enantiomers). Thus it may be seen that it is possible for molecules with

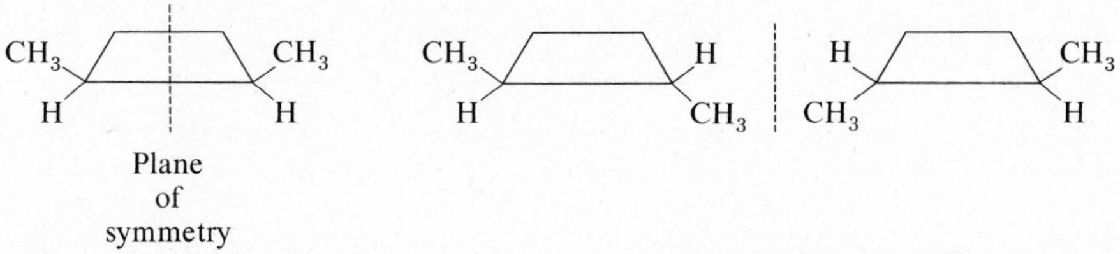

Plane
of
symmetry

cis-1,2-Dimethylcyclobutane (*meso*)   trans-1,2-Dimethylcyclobutane
(enantiomers)

---

[1] It is for this reason that we have not used optical isomers as a class name. The term "optical isomers," if it is used at all, should be reserved for those stereoisomers which are optically active; for example, *meso* compounds are often described ambiguously under the heading of "optical isomers" (see the later discussion of *meso* compounds).

multiple asymmetric carbon atoms to be achiral and hence not optically active. Such compounds are called "*meso.*" *cis*-1,2-Dimethylcyclobutane is an example. Tartaric acid provides the classic example; there are three stereoisomers of tartaric acid, a pair of optically active enantiomers and the achiral *meso* form.

$$
\begin{array}{ccc}
\text{COOH} & & \text{COOH} \quad\vdots\quad \text{COOH} \\
\text{H}-\text{C}-\text{OH} & \text{Plane of} & \text{H}-\text{C}-\text{OH} \quad\vdots\quad \text{HO}-\text{C}-\text{H} \\
\text{H}-\text{C}-\text{OH} & \text{symmetry} & \text{HO}-\text{C}-\text{H} \quad\vdots\quad \text{H}-\text{C}-\text{OH} \\
\text{COOH} & & \text{COOH} \quad\vdots\quad \text{COOH} \\
(\textit{meso}) & & (\text{Enantiomers})
\end{array}
$$

Tartaric acids

It should be noted that the relationship between *cis*-1,2-dimethylcyclobutane and the optically active *trans*-isomers is similar to that between *meso* and optically active tartaric acids. Such stereoisomers which are not enantiomers are called *diastereomers*. This definition includes *cis-trans* isomers of the ethylenic as well as the cyclic types. Thus *cis*- and *trans*-2-butene are diastereomers, as are *cis*-1,2-dimethyl-cyclobutane and either of the *trans*-1,2-dimethylcyclobutane enantiomers, and *meso*-tartaric acid and either of the two optically active tartaric acid enantiomers.

In the examples of diastereomers given in the preceding paragraph, one of the isomers is *meso*. In many cases, however, none of the diastereomers is *meso;* for example, there are four stereoisomers of 3-chloro-2-butanol, two pairs of enantiomers, and all are optically active. The *erythro*-isomers are diastereomers of the *threo*-isomers.[2]

$$
\begin{array}{cccc}
\text{CH}_3 \quad\vdots\quad \text{CH}_3 & & \text{CH}_3 \quad\vdots\quad \text{CH}_3 \\
\text{H}-\text{C}-\text{OH} \quad\vdots\quad \text{HO}-\text{C}-\text{H} & & \text{HO}-\text{C}-\text{OH} \quad\vdots\quad \text{HO}-\text{C}-\text{H} \\
\text{H}-\text{C}-\text{Cl} \quad\vdots\quad \text{Cl}-\text{C}-\text{H} & & \text{Cl}-\text{C}-\text{H} \quad\vdots\quad \text{H}-\text{C}-\text{Cl} \\
\text{CH}_3 \quad\vdots\quad \text{CH}_3 & & \text{CH}_3 \quad\vdots\quad \text{CH}_3 \\
\textit{erythro}\text{-3-Chloro-2-butanol} & & \textit{threo}\text{-3-Chloro-2-butanol}
\end{array}
$$

[2]The terms *erythro* and *threo* are derived from the structurally related four-carbon sugars erythrose and threose.

$$
\begin{array}{cccc}
\text{CHO} \quad\vdots\quad \text{CHO} & & \text{CHO} \quad\vdots\quad \text{CHO} \\
\text{H}-\text{C}-\text{OH} \quad\vdots\quad \text{HO}-\text{C}-\text{H} & & \text{HO}-\text{C}-\text{H} \quad\vdots\quad \text{H}-\text{C}-\text{OH} \\
\text{H}-\text{C}-\text{OH} \quad\vdots\quad \text{HO}-\text{C}-\text{H} & & \text{H}-\text{C}-\text{OH} \quad\vdots\quad \text{HO}-\text{C}-\text{H} \\
\text{CH}_2\text{OH} \quad\vdots\quad \text{CH}_2\text{OH} & & \text{CH}_2\text{OH} \quad\vdots\quad \text{CH}_2\text{OH} \\
\text{D- and L-Erythrose} & & \text{D- and L-Threose}
\end{array}
$$

19.1 Polarimetry

As noted in the preceding discussion, chiral molecules are optically active; that is, they have a rotational effect on the plane of polarized light[3] as it passes through the substance. The phenomenon may be explained as follows. Consider that a chiral molecule has a dissymmetric distribution of electron density. As the polarized light passes through the molecule, its electric component will necessarily be affected in a dissymmetric manner by the electrons of the molecule. The result is that the plane of the polarized light is twisted about the axis of propagation. Any substance having this effect on polarized light is termed optically active. An achiral molecule (one that is superimposable with its mirror image) will be optically inactive, since the light, as it passes through the substance, will on the average[4] encounter a symmetric distribution of electrons and will thus be unaffected.

In a symmetric environment enantiomers are identical with regard to all physical constants and chemical properties, with one exception: although each member of the enantiomeric pair rotates the plane of polarized light by angles of the same magnitude, they do so in *opposite* directions. By convention, when looking through the sample *into* the beam of light, a rotation observed to be clockwise, or to the right, is termed a *positive* rotation, and a counterclockwise rotation is termed a *negative* rotation. If the sample under consideration should be a mixture of exactly equal amounts of the two enantiomers, the rotation of one of the components will be exactly offset by the equal but opposite rotation of the other. The net rotation will be observed as zero. Such a mixture is called a *racemic modification.* In order that there be any net observed rotation in a mixture of enantiomers, one member of the pair must be present in amounts greater than the other.

The observed angle of rotation depends on a number of factors: (1) the nature of the compound, (2) its concentration (in solution) or density (neat liquid), (3) the length of the sample through which the light must pass (the path length), (4) temperature, (5) solvent, and (6) the wavelength of light. The concentration, or density, and path length of the sample are important considerations because they determine the average number of optically active molecules through which a beam of light will pass. The magnitude of rotation is the result of a cumulative effect, and if, on the average, the light passes through a greater number of active molecules, the observed magnitude of rotation will be greater. The dependence is essentially linear with respect to both concentration and path length.

---

[3] Electromagnetic radiation (including light) is composed of propagated electric and magnetic fields. From a wave-mechanical point of view, a ray of light is composed of two plane waves, one electric and one magnetic, oriented at right angles to each other. For simplicity, considering only the electric waves, ordinary light is completely unordered, with these waves oriented at all possible angles in the plane perpendicular to the direction of propagation. In plane-polarized light, each ray has the plane of its electric wave oriented parallel to the corresponding plane of all other rays.

[4] It must be realized that such an experiment is macroscopic rather than microscopic because even with a very small sample of the substance under consideration, a very large number of molecules are present. Any measurement must necessarily be an average observation of the effect on all rays within the polarized light by all molecules through which they pass.

It is certainly desirable to present the optical rotation as a physical constant of the active compound. Therefore it is common practice to report the *specific rotation* rather than the observed angle of rotation. The specific rotation is calculated by correcting the observed rotation to *unit* concentration and path length with the following equation:

$$[\alpha] = \frac{\alpha}{1 \times c} \quad \text{or} \quad \frac{\alpha}{1 \times d} \tag{1}$$

$[\alpha]$ = specific rotation (degrees)

α = observed rotation (degrees)

1 = path length (decimeters)

c = concentration (g/mL of solution)

d = density (g/mL, neat)

To specify the other variables on which the rotation depends, the temperature and wavelength (nm) employed are presented as superscript and subscript, respectively, on the symbol for specific rotation; the solvent used is denoted in parentheses following the numerical value and sign of the specific rotation, for example, $[\alpha]_{490}^{25°}$ +23.4° (CH_3OH). A sodium lamp emitting light at 589 nm, the sodium D line, is usually used as a light source. In this case it is common to use D in place of the numerical value of the wavelength in the expression, for example, $[\alpha]_D^{25°}$ −15.2° (H_2O).

Various types of commercial polarimetric apparatus are available. Some, such as the Rudolph polarimeter, are manually operated and require direct readings by the operator, whereas other, newer varieties are automatic and utilize photoelectric cells for measurements of greater accuracy and precision. Two simply constructed student-type polarimeters have been described,[5] and an inexpensive instrument similar to the one described in the Kapauan article (see footnote 5) is marketed by the Instruments for Research and Industry Company. One of these simple polarimeters may be used quite satisfactorily in the experiment on α-phenylethylamine described in the Experimental Procedure.

The basic principles of operation of all polarimeters are illustrated in Figure 19.1. Ordinary light passes through a polarizing element (the polarizer in the figure) which in a precise instrument is a Nicol prism but in a simple instrument may be a piece of Polaroid film. The beam of polarized light then passes through the sample tube, where rotation of the plane of the light will be effected by a neat optically active liquid or a solution of an optically active substance. To determine the angle of rotation, the viewer turns the analyzer, another Nicol prism or Polaroid film, until the light intensity observed is the same as that observed when the sample tube is empty or contains only pure solvent. The number of degrees through which the analyzer

[5]W. H. R. Shaw, "A Mailing-Tube Polarimeter," *Journal of Chemical Education,* **32,** 10 (1955); A. F. Kapauan, "A Simple Split-Field Polarimeter," *Journal of Chemical Education,* **50,** 376 (1973).

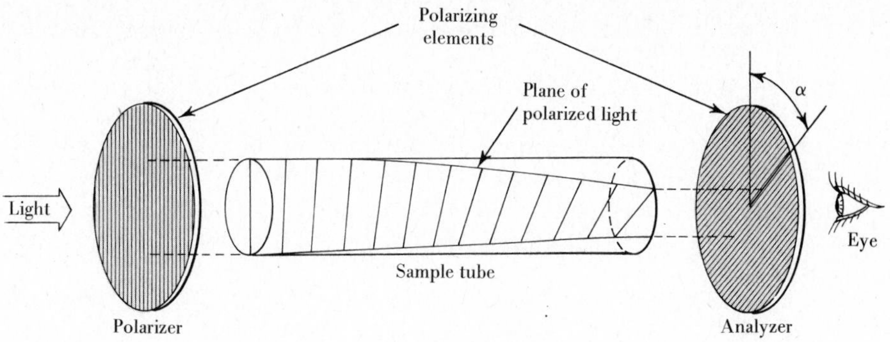

Polarizing
elements

Plane of
polarized light

α

Light

Eye

Sample tube

Polarizer

Analyzer

FIGURE 19.1 Schematic illustration of optical rotation (α = about $-45°$).

must be turned is the observed optical rotation, α, which can be converted to the specific rotation, $[\alpha]$, by applying equation 1.

19.2 Resolution of Racemic α-Phenylethylamine

α-Phenylethylamine (**3**) may be prepared from acetophenone (**1**) and ammonium formate (**2**).[6] The product obtained is not optically active, even though it contains an asymmetric carbon atom (marked in formula **3** with an asterisk), because the enantiomers are produced in exactly equal amounts; that is, a racemic modification is formed.

$$C_6H_5-\overset{\displaystyle O}{\overset{\|}{C}}-CH_3 + H-\overset{\displaystyle O}{\overset{\|}{C}}-O^{\ominus}NH_4^{\oplus} \longrightarrow C_6H_5-\overset{\displaystyle NH_2}{\underset{\displaystyle H}{\overset{|}{\underset{|}{C^*}}}}-CH_3 + H_2O + CO_2$$

1 **2** **3**

Enantiomers cannot be separated from one another by the standard methods of crystallization, distillation, or chromatography. This is because they have identical physical properties, with the exception of their rotation of polarized light in opposite directions, as noted in the preceding section. However, *diastereomeric isomers (diastereomers)* do differ in physical properties such as solubility, boiling point, and chromatographic adsorption characteristics. The most generally useful procedure for *resolving a racemic modification* of enantiomers involves converting the enantiomers to diastereomers, separating the diastereomers by one of the standard experimental procedures, and then regenerating the enantiomers from the separated diastereomers. This procedure requires the use of an optically active resolving agent; fortunately, nature provides an abundant source of these. Most of the naturally occurring organic compounds that have chiral centers are found in optically active form and hence can be used to resolve racemic modifications produced by synthesis.

[6]A. W. Ingersoll in *Organic Synthesis,* Collective Vol. II, A. H. Blatt, editor, John Wiley & Sons, New York, 1943, p. 503.

In our experiment we use optically active (+)-tartaric acid as the resolving agent for racemic α-phenylethylamine. The resolution scheme is outlined in Figure 19.2. By reaction with the optically active acid, the racemic amine is converted to a mixture of diastereomeric salts, which have different solubilities in methanol and hence can be separated by fractional crystallization. To obtain *both* enantiomeric isomers in a state of high optical purity may require a large number of careful and tedious crystallization steps, but it is usually possible to obtain by only one or two crystallization steps the enantiomer which gives the less soluble salt in reasonable optical purity.[7]

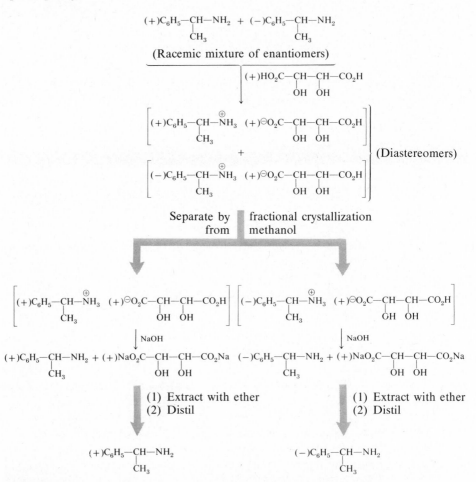

FIGURE 19.2 Resolution of (±)-α-phenylethylamine by means of (+)-tartaric acid.

[7]In some cases it is also possible to obtain the other enantiomer without resorting to many repeated crystallizations of the more soluble diastereomer. For example, in the research on which this experiment is based [W. Theilacker and H.-G. Winkler, *Chemische Berichte*, **87**, 690 (1954)], it was found that the *hydrogen sulfate salt* of the enantiomer which gave the more soluble tartrate salt was *less soluble* than the hydrogen sulfate salt of the racemic amine, so that both enantiomers could be obtained in optical purity quite conveniently.

By measuring the optical rotation of the α-phenylethylamine recovered from the less soluble salt in this experiment, it will be possible to determine not only whether it is the (+)- or (−)-enantiomer but also the extent of optical purity achieved in the resolution.

EXPERIMENTAL PROCEDURE

Dissolve an accurately weighed sample of approximately 5 g of racemic α-phenylethylamine in 35 mL of methanol, and determine its specific rotation, [α], using a polarimeter and equation 1.[8] Detailed instruction for the use of a polarimeter should be obtained from your instructor.

In a 1-L Erlenmeyer flask place 0.208 mol of (+)-tartaric acid and 415 mL of methanol and heat to boiling. To the hot solution add cautiously (to avoid foaming), with stirring, the 35 mL of solution recovered from the polarimeter and enough additional racemic α-phenylethylamine to make a total of 0.206 mol. Allow the solution to cool slowly to room temperature and to stand undisturbed for 24 hr or until the next laboratory period.* The amine hydrogen tartrate should separate in the form of white prismatic crystals. If the salt separates in the form of needlelike crystals, the mixture should be reheated until *all* the crystals have dissolved and then allowed to cool slowly. If any prismatic crystals of the salt are available, they should be used to seed the solution.

Collect the crystals of the amine hydrogen tartrate (17–19 g) on a filter, and wash them with a small volume of cold methanol.[9] Dissolve the crystals in about four times their weight of water, and add 15 mL of 14 *M* sodium hydroxide solution. Extract the resulting mixture with four 75-mL portions of diethyl ether,[10] wash the combined ether extracts with 50 mL of saturated sodium chloride solution, dry the ether solution over anhydrous magnesium sulfate, and then filter it from the desiccant.*

Remove the ether from the solution by distillation from a steam bath using an unpacked Hempel fractionating column, and then arrange to distil the residue (α-phenylethylamine) under aspirator pressure, using no fractionating column. It will be necessary to use a burner or an electric heater because the amine has a boiling point of 94–95° (28 mm). The yield of optically active α-phenylethylamine should be 5–6 g.

Weigh the distilled product accurately, and then transfer it quantitatively into

---

[8] The amounts of α-phenylethylamine and methanol specified here are those which give satisfactory results when a polarimeter sample tube with a light path of 2 dm or more and a volume of 35 mL or less is used. If a tube of shorter light path and/or larger volume is used, either the amount of α-phenylethylamine used as starting material must be increased, or two or more students will have to combine their yields of resolved product to give a solution of high enough concentration so that the observed optical rotation will be large enough for accuracy. Students should check with their instructor about the type of polarimeter sample tube that will be used.

[9] Do not discard the filtrate before you ask your instructor if you are to turn it in for recovery of the other enantiomer of α-phenylethylamine.

[10] Do not discard the aqueous solution; pour it into a bottle provided for recovery of tartaric acid.

about 35 mL of methanol. Measure the volume of the methanol *solution* accurately. It is the weight of the α-phenylethylamine (in grams) divided by the volume of the *solution* (in mL) that gives you the concentration (c) for equation 1. Transfer the solution to the polarimeter sample tube and determine the specific rotation from the observed rotation by using equation 1. The reported specific rotation of optically pure α-phenylethylamine is $[\alpha]_D^{25}$ 40.1° (neat).

EXERCISES

1. Report the specific rotation of your resolved product as (+) or (−). Suppose the observed rotation was found to be 180°. How could you determine whether the rotation was (+) or (−)?
2. How could you increase the optical purity of your product?
3. Describe clearly the point in the experimental procedure at which the major part of the other enantiomer was removed from your product.
4. How could the (+)-tartaric acid be recovered so that it can be used over again to resolve more racemic amine?
5. The absolute configuration of (+)-α-phenylethylamine has been shown to be R. Make a perspective drawing of this configuration.[11]
6. Suppose you had prepared a racemic organic acid and wished to obtain an optically active form of it. How could you do this?
7. Suggest a possible procedure for resolving a racemic alcohol.

REFERENCE

A. Ault, *Journal of Chemical Education,* **42,** 269 (1965); *Organic Syntheses,* **49,** 93 (1969).

19.3 Optional Projects

A. Preparation of a Crystalline Derivative of Racemic α-Phenylethylamine and of Optically Active α-Phenylethylamine

To demonstrate that resolution of a racemic modification does not change the chemical or physical properties of the compound (in a symmetric environment), other than the effect on polarized light, the crystalline benzoyl derivative, the benzamide **4,** may be prepared from a sample of racemic α-phenylethylamine and from a sample of amine which has been resolved. The reaction is shown in equation 2.

[11]The R,S or Cahn-Ingold-Prelog convention for designating absolute configuration is now being applied to all chiral compounds except carbohydrates, amino acids, and closely related compounds. The D,L system is still used for these latter types, but they too may eventually be denoted by the R,S system. See any modern organic textbook for a description of the R,S system.

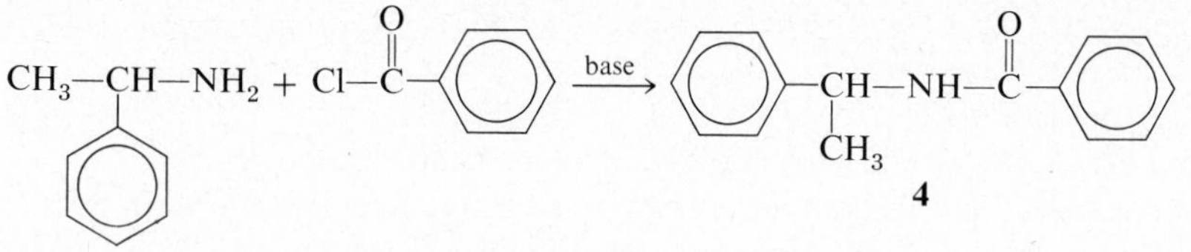

$$CH_3-CH-NH_2 + Cl-\overset{\overset{O}{\parallel}}{C}-\bigcirc \xrightarrow{\text{base}} \bigcirc-\underset{\underset{CH_3}{|}}{CH}-NH-\overset{\overset{O}{\parallel}}{C}-\bigcirc \qquad (2)$$

4

The experimental procedure for the preparation of benzamides from amines is described in Section 25.2. The melting point of this benzamide has been variously reported by different workers to be between 120° and 125°. It may be recrystallized from aqueous ethanol. Compare the melting points of derivatives prepared from racemic and optically active samples of α-phenylethylamine.

To determine if the formation of the benzamide occurs with retention of optical activity, the derivative produced from the optically active sample of α-phenylethylamine should be examined for optical rotation. This may be done by dissolving an accurately weighed sample of approximately 0.9 g of the benzamide in 38 mL of toluene and measuring the rotation in the polarimeter, calculating $[\alpha]$ by using equation 1. The $[\alpha]_D^{27}$ has been reported to be $+39.2°$ (benzene).

Determine not only the extent of optical purity of the benzamide but also whether the sign of rotation is the same as that of the resolved α-phenylethylamine.

B. Preparation of Di-α-phenylethylamine (α,α′-Dimethyldibenzylamine)

The reaction of optically active α-phenylethylamine with acetophenone to produce the imine **5** (equation 3) followed by catalytic hydrogenation might be

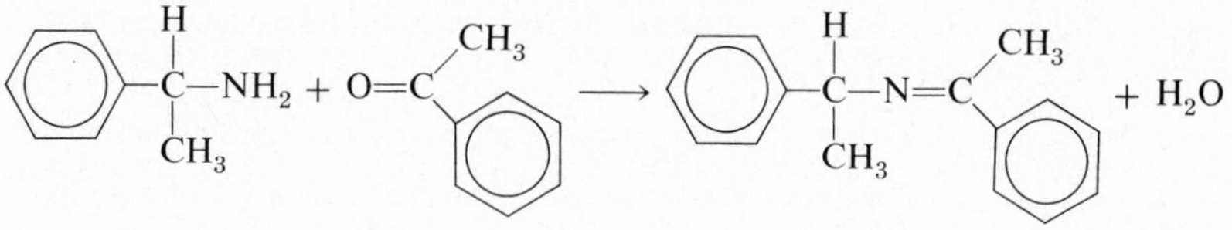

$$\bigcirc-\underset{\underset{CH_3}{|}}{\overset{\overset{H}{|}}{C}}-NH_2 + O=\underset{\underset{\bigcirc}{|}}{\overset{\overset{CH_3}{|}}{C}} \longrightarrow \bigcirc-\underset{\underset{CH_3}{|}}{\overset{\overset{H}{|}}{C}}-N=\underset{\underset{\bigcirc}{|}}{\overset{\overset{CH_3}{}}{C}} + H_2O \qquad (3)$$

5

$$5 + 2\,(H) \longrightarrow \begin{cases} \bigcirc-\underset{\underset{CH_3}{|}}{\overset{\overset{H}{|}}{C}}-NH-\underset{\underset{CH_3}{|}}{\overset{\overset{H}{|}}{C}}-\bigcirc \\ \textbf{6} \ (meso) \\ + \\ \bigcirc-\underset{\underset{CH_3}{|}}{\overset{\overset{H}{|}}{C}}-NH-\underset{\underset{H}{|}}{\overset{\overset{CH_3}{|}}{C}}-\bigcirc \\ \textbf{7} \ (\text{Optically active}) \end{cases} \qquad (4)$$

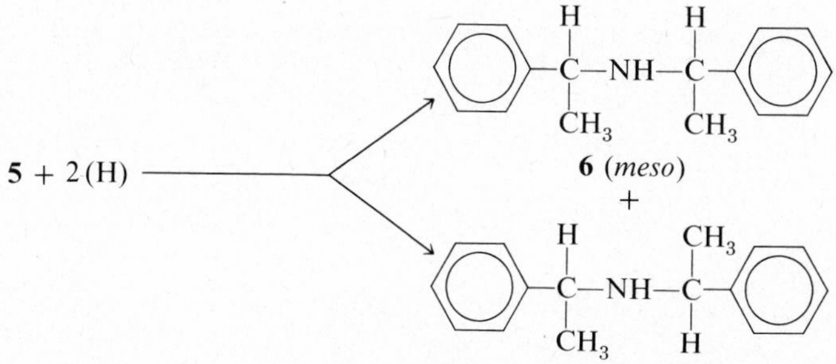

expected to produce the diastereomers of di-α-phenylethylamine, **6** and **7**. These should be capable of separation by ordinary means, since they should have significantly different physical properties. However, when these reactions were carried out and the di-α-phenylethylamine was converted to its hydrochloric acid salt, an attempted separation of the diastereomeric isomers by fractional crystallization failed; the high specific rotation of the salt indicated that the hydrogenation occurred stereoselectively, so that the optically active isomer, **7**, was almost exclusively the product formed.[12] When the reduction was carried out with lithium aluminum hydride instead of by catalytic hydrogenation, a lower specific rotation was found, but the optically active isomer was the major product.

The reactions of equations 3 and 4 are analogous to those of equations 4 to 6 in Chapter 14. The reaction of acetophenone, a ketone, however, is slower than the reactions of the aldehydes in Chapter 14, and it may not be possible to complete the reaction in one laboratory period. The reduction of **5** with lithium aluminum hydride was also carried out at a higher temperature and for a longer time than the reactions of sodium borohydride. *This reagent* ($LiAlH_4$) *is pyrophoric and must be handled with care.*

It is quite possible that $NaBH_4$ may be used in place of $LiAlH_4$ and that the conditions reported for the formation of the imine **(5)** are more severe than necessary. A trial with the procedures of Chapter 14 would be worth a try. In any case, with proper supervision the conversion of optically active α-phenylethylamine to di-α-phenylethylamine and the determination of the specific rotation of the secondary amine make an interesting project.

The specific rotation of the pure hydrochloric acid salt of optically active di-α-phenylethylamine, $[\alpha]_D^{20}$, is reported to be $+71.8°$ (ethanol). The specific rotation of the salt obtained by reduction of the imine **5**, followed by a work-up analogous to that described in the article of footnote 12, should be compared with $+71.8°$ to calculate the degree of optical purity of **7** obtained, and thus the proportion of **6** and **7** produced in the reduction of **5**.

[12]C. G. Overberger, N. P. Marullo, and R. G. Hiskey, *Journal of the American Chemical Society,* **83**, 1374 (1961).

20

PHOTOCHEMICAL DIMERIZATION OF 4-METHYLBENZOPHENONE; PINACOL-PINACOLONE REARRANGEMENT

20.1 Preparation and Photochemical Reaction of 4-Methylbenzophenone

The effect of light in producing chemical reactions has been recognized for many years. Indeed the role of sunlight in the photosynthesis of carbohydrates was known in the early 1800s. Nevertheless, organic photochemistry was slow to develop, and rapid progress occurred only after the development of spectroscopic techniques capable of elucidating complex organic structures and detecting transient species, such as those with lifetimes less than a millisecond. In addition to photosynthesis, which has now assumed an increased importance because of efforts to develop energy sources independent of petroleum and coal, many other significant photochemical reactions are recognized. Some are beneficial, such as those operating in the biochemical vision processes of all animals, and some are dangerous, such as those which alter DNA and induce skin cancer. There are many useful industrial applications of photochemistry, such as the light-induced chlorination of alkanes (Chapter 5) and the photochemical generation of free radicals in solution for initiation of addition polymerization (Chapter 21). Entirely aside from its practical aspects, photochemistry has fascinated organic chemists in recent years because of the avenues it has provided for the synthesis of exotic molecular structures such as basketene (pentacyclo-[4.2.2.0^{2,5}.0^{3,8}.0^{4,7}]deca-9-ene) and cubane (pentacyclo[4.2.0.0^{2,5}.0^{3,8}.0^{4,7}]octane).

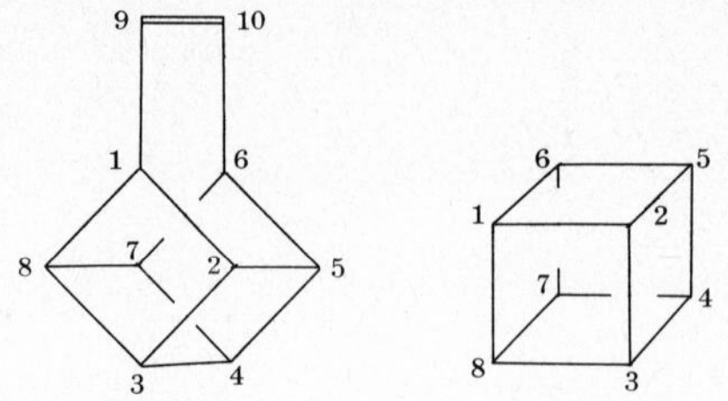

Basketene   Cubane

One of the classic photochemical reactions is the photoreduction of benzophenone (**1a**) to benzopinacol (**2a**) by the action of low-energy ultraviolet light in the presence of 2-propanol. To allow an additional interesting study, we utilize 4-methylbenzophenone (**1b**), which leads to 4,4′-dimethylbenzopinacol (**2b**) (equation 1), in our experiment.

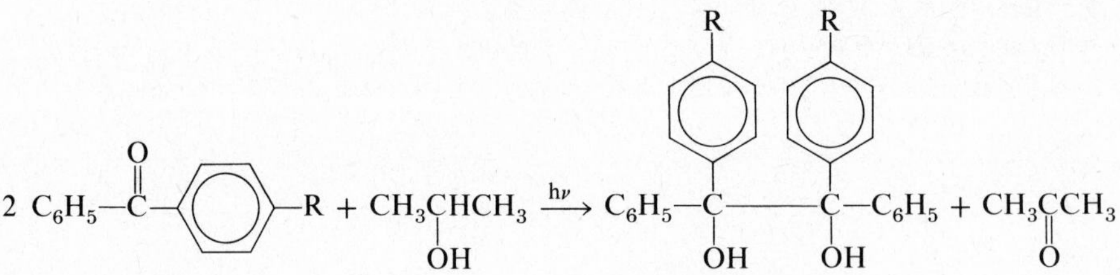

1a, R = H
1b, R = CH$_3$

2a, R = H
2b, R = CH$_3$

The 4-methylbenzophenone needed as starting material for the photochemical reaction may be prepared by a Friedel-Crafts *acylation* of toluene with benzoyl chloride (equation 2). In comparing this reaction with the Friedel-Crafts *alkylation* of

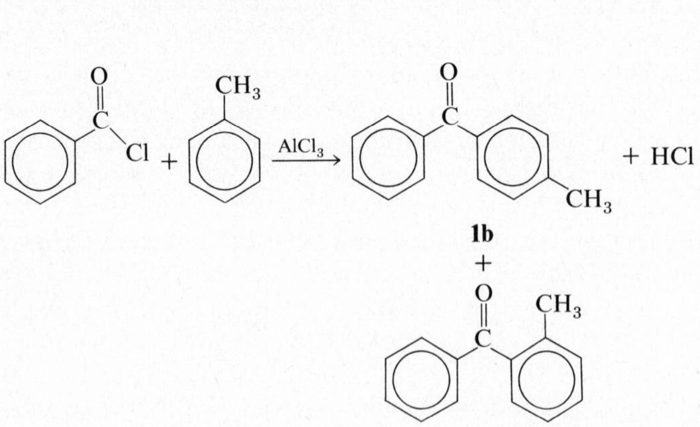

1b
+

3

(2)

p-xylene (Chapter 7), it is instructive to note that the amount of catalyst ($AlCl_3$) used in the acylation is much larger than in the alkylation. This is because the ketone formed in an acylation forms a 1:1 complex with aluminum chloride, such as **4** in the

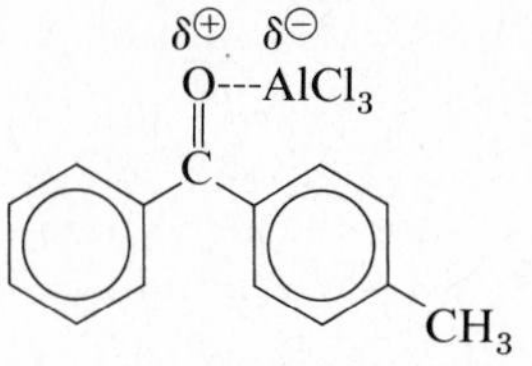

$$\delta^{\oplus} \quad \delta^{\ominus}$$
$$O\text{---}AlCl_3$$

4

present experiment. For this reason more than 1 mol of $AlCl_3$ per mole of benzoyl chloride is used in order to ensure an adequate supply of active catalyst throughout the acylation reaction.

A small amount of 2-methylbenzophenone (**3**) is produced, but it is easily removed from the desired 4-methylbenzophenone (a solid) by recrystallization because the 2-isomer is a liquid.

The sequence of events in the photochemical reduction-dimerization of 4-methylbenzophenone (**1b**) to 4,4′-dimethylbenzopinacol (**2b**) is outlined in equa-

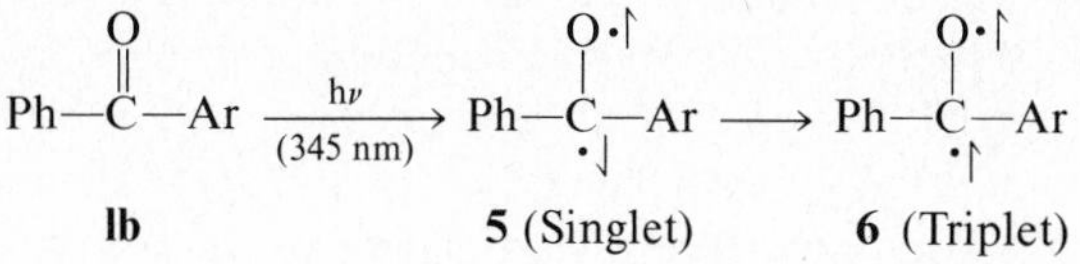

$$Ph\text{—}\overset{O}{\underset{}{C}}\text{—}Ar \xrightarrow[\text{(345 nm)}]{h\nu} Ph\text{—}\overset{O\cdot\uparrow}{\underset{\cdot\downarrow}{C}}\text{—}Ar \longrightarrow Ph\text{—}\overset{O\cdot\uparrow}{\underset{\cdot\uparrow}{C}}\text{—}Ar \qquad (3)$$

1b   **5** (Singlet)   **6** (Triplet)

$$Ph\text{—}\overset{O\cdot\uparrow}{\underset{\cdot\uparrow}{C}}\text{—}Ar + CH_3\text{—}\overset{OH}{\underset{}{CH}}\text{—}CH_3 \longrightarrow Ph\text{—}\overset{OH}{\underset{\cdot}{C}}\text{—}Ar + CH_3\text{—}\overset{OH}{\underset{\cdot}{C}}\text{—}CH_3 \qquad (4)$$

6   **7**   **8**

$$CH_3\text{—}\overset{OH}{\underset{\cdot}{C}}\text{—}CH_3 + Ph\text{—}\overset{O}{\underset{}{C}}\text{—}Ar \longrightarrow CH_3\text{—}\overset{O}{\underset{}{C}}\text{—}CH_3 + Ph\text{—}\overset{OH}{\underset{\cdot}{C}}\text{—}Ar \qquad (5)$$

8   **1b**   **7**

$$2\ Ph\text{—}\overset{OH}{\underset{\cdot}{C}}\text{—}Ar \longrightarrow Ph\text{—}\underset{\underset{Ar}{|}}{\overset{\overset{OH}{|}}{C}}\text{—}\underset{\underset{Ar}{|}}{\overset{\overset{OH}{|}}{C}}\text{—}Ph \qquad (6)$$

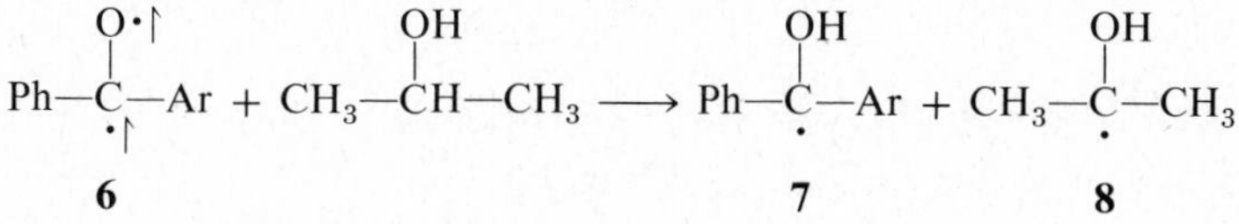

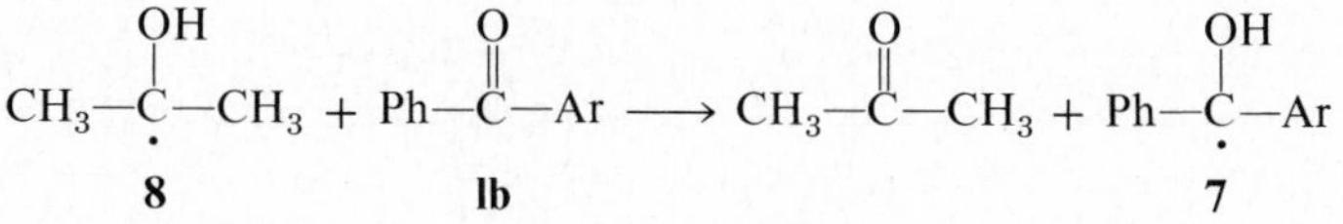

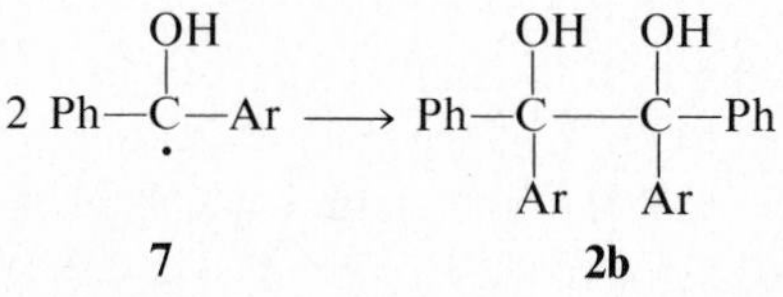

7   **2b**

tions 3–6, in which Ph is phenyl and Ar is *p*-tolyl. Note that the overall chemical change is just that written earlier in equation 1.

The first step is the absorption of low-energy ultraviolet light[1] (at about 345-nm wavelength) by **1b** to produce an excited molecule (**5**). The light energy absorbed effects promotion of a nonbonding (n) electron at oxygen to the lowest unoccupied antibonding π orbital of **1b**; hence this photochemical (electronic) excitation is called an $n \rightarrow \pi^*$ transition. The excited state **5** is a *singlet,* in which all electrons are still paired, since the excitation does not change the spin of the electron. Thus singlet excited states, like stable ground state molecules, do not have magnetic moments. The singlet state rapidly changes to the thermodynamically more stable *triplet* state whereby the spins of the electrons become unpaired. (Note the directions of the arrows in formulas **5** and **6**.) A triplet state has a magnetic moment that may be aligned in *three* different ways in a magnetic field—hence the name.

The triplet **6** has a long enough lifetime so that it can effectively abstract a hydrogen from the solvent, 2-propanol, producing the two free radicals **7** and **8** (equation 4). The radical **8** next donates a hydrogen atom to another molecule of 4-methylbenzophenone to give a second **7** radical and 2-propanone (equation 5). This reaction is energetically favorable because of the greater possibility for delocalization of the odd electron in **7** than in **8**. The reaction is completed by combination of two **7** radicals to form 4,4'-dimethylbenzopinacol (**2b**) (equation 6).

Although other routes may be formulated for conversion of the photoexcited molecules to the stable products, the mechanism outlined in equations 3–6 satisfies an important experimental observation, which is that the quantum yield of both 2-propanone and 4,4'-dimethylbenzopinacol are nearly unity when the light intensity is not high. (A quantum yield of unity means that one molecule of product is formed for each quantum of energy absorbed.) This suggests strongly that *two* of the radicals having structure **7** must be formed for each molecule of **1b** that becomes activated by light.

20.2 Pinacol-Pinacolone Rearrangement of 4,4'-Dimethylbenzopinacol

Many 1,2-diols undergo an acid-catalyzed rearrangement to carbonyl compounds. The prototype of this reaction is the conversion of 2,3-dimethyl-2,3-butanediol (pinacol) to 3,3-dimethyl-2-butanone (pinacolone) (Figure 20.1). The function of the acid is to assist in the loss of one of the OH groups by protonating it, thereby making it a better leaving group. The tertiary carbocation that results from this loss undergoes a 1,2-methyl shift; the driving force of this rearrangement is the formation of the thermodynamically more stable cation **9**. Note that in the resonance structure **9a** every atom has a full octet of electrons.

This type of molecular rearrangement has been the subject of a great number of mechanistic investigations. Interest has focused mainly on two aspects of the reaction: (1) in the case of unsymmetrically substituted ethanediols or pinacols such as **10**, to learn which OH group leaves and which becomes a carbonyl group; (2) in the case of

---

[1] A discussion of the molecular phenomenon of electronic excitation by absorption of ultraviolet light may be found in the introductory paragraphs of Chapter 4 and in Section 4.3.

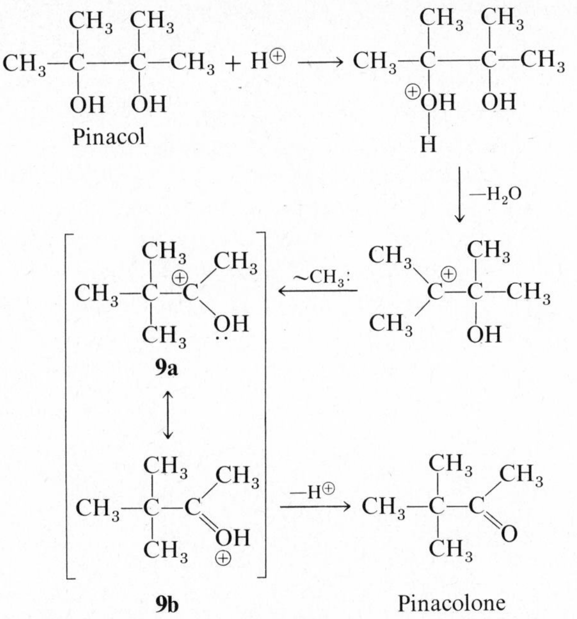

CH$_3$ CH$_3$

CH$_3$—C—C—CH$_3$ + H$^\oplus$ ⟶ CH$_3$—C—C—CH$_3$

OH OH

Pinacol

$\oplus$OH OH

H

↓ —H$_2$O

$$\left[\begin{array}{cc} \text{CH}_3\text{—C—C—CH}_3 & \xleftarrow{\sim\text{CH}_3\!:} \quad \text{CH}_3\text{—C—C—CH}_3 \\ \textbf{9a} & \\ \updownarrow & \\ \text{CH}_3\text{—C—C—CH}_3 & \xrightarrow{-\text{H}^\oplus} \quad \text{CH}_3\text{—C—C—CH}_3 \\ \textbf{9b} & \text{Pinacolone} \end{array} \right]$$

FIGURE 20.1 Rearrangement of pinacol to pinacolone.

symmetrical diols such as **11**, to learn which group, R or R', preferentially undergoes the 1,2-shift—that is, what are the *migratory aptitudes* of R and R'.

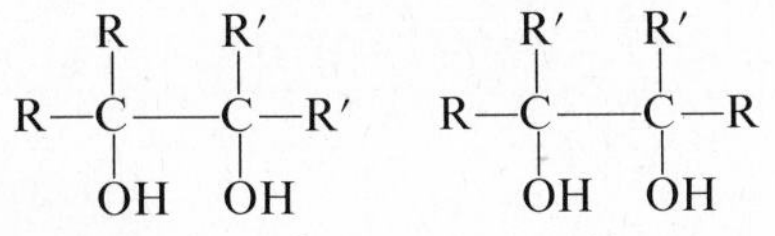

R R'

R—C——C—R'

OH OH

10

R' R'

R—C——C—R

OH OH

11

The answer to the first question is determined by the stability of the carbocation produced by the loss of OH in the form of H$_2$O. For example, compound **12** gives **15** rather than **16** as the major product because of the greater stabilization by delocalization of the charge by the phenyl groups in **13** than by the inductive effect of the methyl groups in **14**.

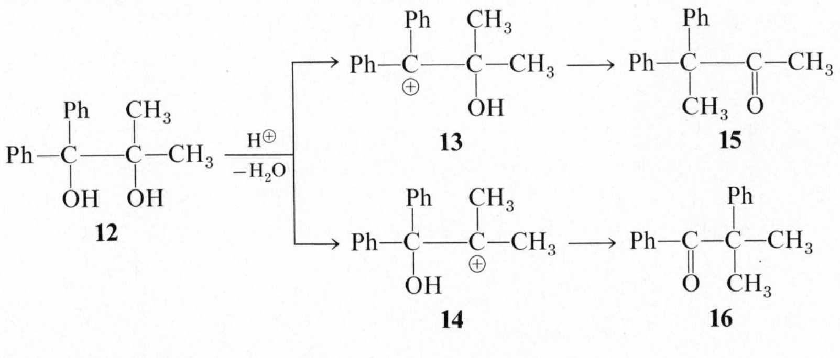

Ph CH$_3$

Ph—C——C—CH$_3$ ⟶ Ph—C——C—CH$_3$

$\oplus$ OH

13 CH$_3$ O

15

Ph CH$_3$

Ph—C——C—CH$_3$ $\xrightarrow[-\text{H}_2\text{O}]{\text{H}^\oplus}$

OH OH

12

Ph CH$_3$

Ph—C——C—CH$_3$ ⟶ Ph—C—C—CH$_3$

OH $\oplus$

14

Ph

O CH$_3$

16

An answer to the second question is not so simple. Much experimental evidence indicates that aryl groups "migrate" more readily than alkyl groups. The migratory aptitude of hydrogen relative to other groups cannot be assessed as successfully, however, because trisubstituted 1,2-ethanediols are subject to a number of competing side reactions. The most consistent correlations between structure and migratory aptitudes have been obtained from studies on diols of type **11** in which R = phenyl and R′ = *para*-substituted phenyl. Thus consider the changes in charge distribution that would be produced in a transition state such as that depicted by structure **17** by electron-withdrawal or electron-release by a *para*-substituent G. In this instance the migratory aptitudes of substituted phenyl groups relative to phenyl should be predictable on the basis of stabilization or destabilization of **17** by the group G.

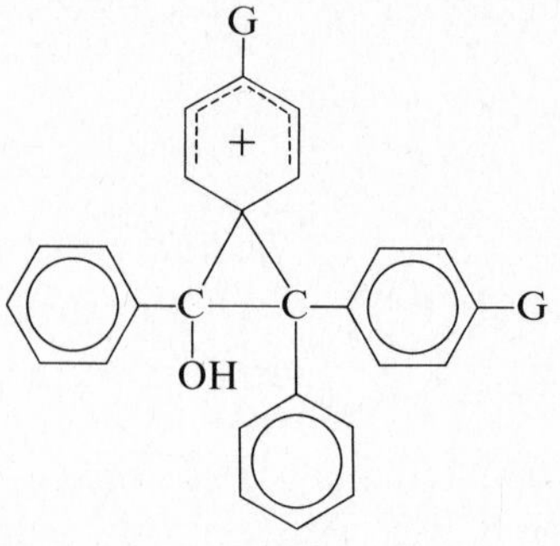

17

The unsymmetrical pinacol chosen for our experiment is 4,4′-dimethylbenzopinacol (**2b**), which on rearrangement has the potential of producing a mixture of the isomeric pinacolones **18** and **19**. The molar proportion of **18** and **19** in the reaction

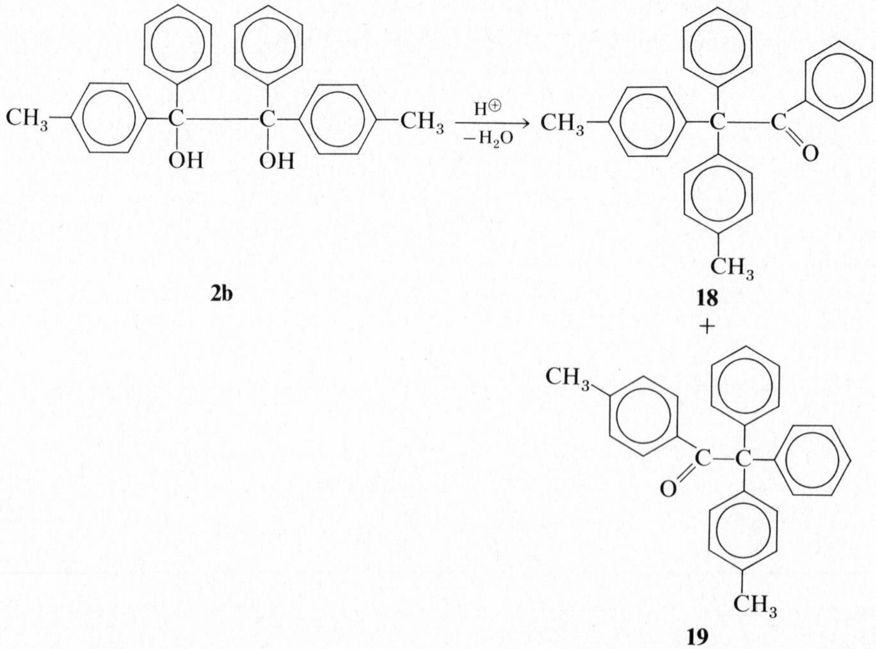

2b

18

+

19

mixture is a measure of the relative migratory aptitudes of the *p*-tolyl group and the phenyl group, respectively.

The rearrangement of **2b** is effected by heating it in acetic acid solution containing iodine. The catalytic effect of iodine is attributed to its function as a Lewis acid in aiding the removal of one of the OH groups, as shown in Figure 20.2. The water and iodine are removed by distillation along with the acetic acid solvent, leaving either or both of the two isomeric pinacolones **18** and **19** in the reaction mixture.

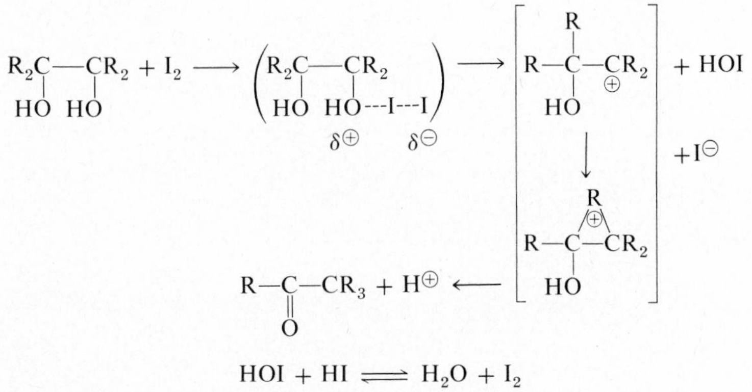

FIGURE 20.2 Catalysis of the pinacol-pinacolone rearrangement by iodine.

To simplify the determination of the proportion of **18** and **19** produced in the rearrangement reaction, the pinacolones are cleaved to the corresponding triaryl-methanes **20** and **22** and carboxylic acids **24** and **25**, as shown in Figure 20.3. The exact nature of the basic cleavage of the ketones is uncertain, but it has been demonstrated that the presence of each of the components of the reagent—potassium *t*-butoxide, water, and dimethylsulfoxide—and their proper proportions are critical to the success of the cleavage process. For simplicity, one may consider that the cleavage is effected by attack at the carbonyl group by hydroxide ion (produced from water and *t*-butoxide ion), which is an especially strong nucleophile in dimethylsulfoxide solution,[2] giving a mixture of the triarylmethanes and the potassium salts of the acids, **21** and **23**. The cleavage products can be separated by extraction of the alkaline aqueous solution with dichloromethane in which the triarylmethanes are soluble, leaving the acid salts in solution in the aqueous phase until concentrated hydrochloric acid is added.

The desired information about migratory aptitudes could be obtained by determining the molar ratio of either the triarylmethanes **20** and **22** or the carboxylic acids, benzoic acid (**24**), and *p*-toluic acid (**25**). It is experimentally simpler to analyze the mixture of the acids, however.

---

[2] Dimethylsulfoxide is a solvent which complexes certain cations (such as K^+) so effectively that it leaves the counter anion essentially "naked" and hence especially nucleophilic.

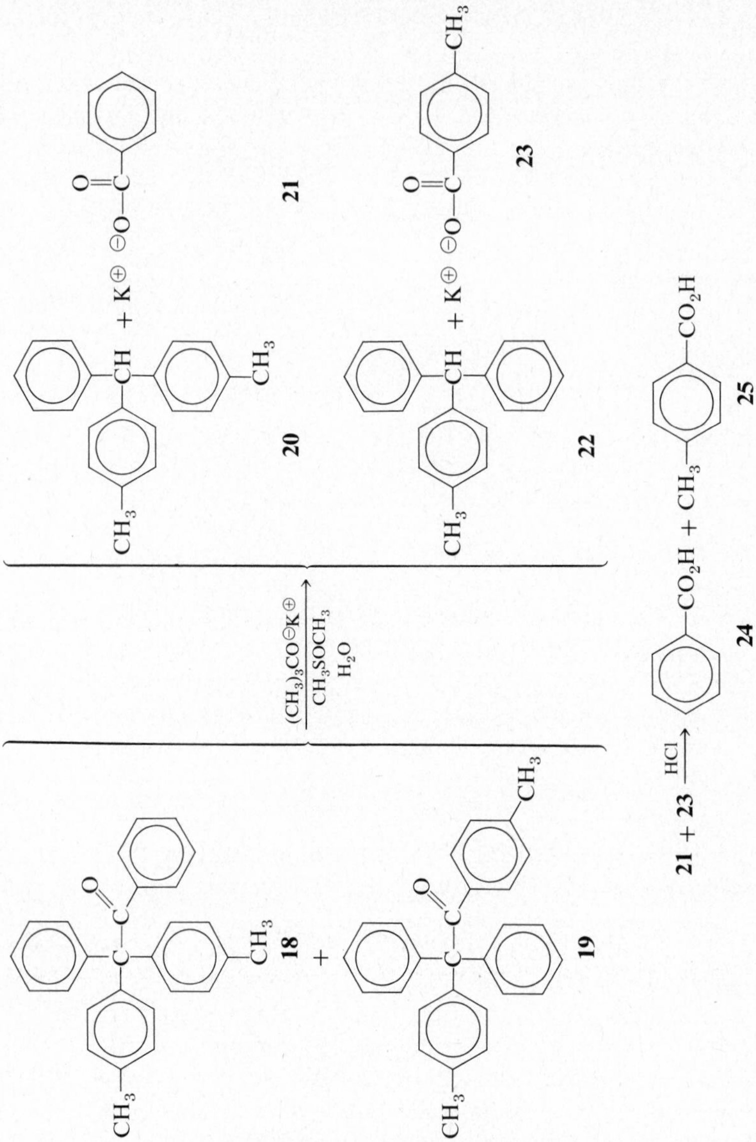

FIGURE 20.3 Cleavage of pinacolones.

The quantitative analysis of the mixture of acids may be carried out directly by pmr spectroscopy or by glpc analysis, and/or by converting the mixed acids to their methyl esters before glpc analysis. Your instructor may suggest that each of these methods of analysis be tested by different students and the results be compared.

EXPERIMENTAL PROCEDURE[3]

A. 4-Methylbenzophenone

DO IT SAFELY
---

1. As usual, examine your glassware for any cracks or other defects, particularly star cracks in round-bottomed flasks.

2. Both benzoyl chloride and aluminum chloride react with moisture to produce hydrogen chloride. When you handle these materials, be especially cautious to avoid their coming in contact with your skin and eyes. You may want to wear rubber gloves.

3. Before using a burner for the distillation at the end of this experiment, examine your work area for the presence of any flammable materials.

4. Do not use a water-cooled condenser if you distil 4-methylbenzophenone at atmospheric pressure; the large heat differential might crack the glass of the condenser.

Place 0.11 mol of *anhydrous* powdered aluminum chloride and 0.54 mol of toluene in a 250-mL round-bottomed flask. Attach a Claisen connecting tube, a water-cooled condenser, a separatory funnel, and a gas trap as shown in Figure 7.4. By means of a simple connector and rubber tube, connect the sidearm of the connecting tube to a gas trap (see Figure 5.1). Place 0.10 mol of benzoyl chloride in the separatory funnel, and begin its dropwise addition to the toluene-aluminum chloride mixture. Loosen the lower clamp holder so that by moving the clamp gently the mixture can be swirled during the addition of the benzoyl chloride. When the addition has been completed, tighten the clamp and heat the reaction mixture under reflux for 15 min.

Allow the mixture to cool until you can hold your hand against the flask, and then pour the contents into a 500-mL Erlenmeyer flask containing 100 g of crushed ice and 40 mL of concentrated hydrochloric acid. Swirl the mixture vigorously until the toluene layer turns to a yellow or orange color. If any solid is present, remove it by vacuum filtration; otherwise pour the liquid mixture into a separatory funnel, and remove and discard the water layer. Wash the toluene layer gently (swirling, not shaking, the separatory funnel so as to avoid producing an emulsion) with 20 mL of 1 M sodium hydroxide and then with 20 mL of water. Transfer the toluene layer to a 100-mL round-bottomed flask, and remove the solvent by distillation, using an

---

[3] Adapted from an article by N. M. Zaczek, J. C. Ruff, A. H. Jackewitz, and D. F. Roswell, *Journal of Chemical Education,* **48,** 257 (1971).

electric heater (or a small burner).★ (*Caution: Toluene is flammable. Be sure your connections are tight, and protect the distillate from flames.*) When all the toluene has distilled, remove the condenser, and connect a vacuum adapter and receiving flask directly to the still head and distil the product. 4-Methylbenzophenone is reported to have bp 327° (760 mm), 200° (25 mm), and 155° (3 mm). Stop heating when 1 or 2 mL of residue remain in the distillation flask.★

Cool the distillate in ice, and scratch the side of the flask with a glass rod at the surface of the liquid to induce crystallization of the product. A seed crystal would be helpful. Transfer the mushy crystals to a filter on a Büchner funnel, and press them to remove the liquid isomeric 2-methylbenzophenone. Recrystallize the 4-methyl-benzophenone from 80% ethanol; the pure, colorless needles have mp 57–58°. The yield should be about 8 g. A second crop of crystals may often be obtained by warming the filtrate from the first crop, adding water until cloudiness just persists and allowing the filtrate to cool after adding a seed crystal.

B.   4,4′-Dimethylbenzopinacol

DO IT SAFELY

The use of 275-watt sunlamp is recommended in this experiment to effect the photochemical dimerization. These lamps emit sufficient ultraviolet radiation to cause retinal damage, so take particular care *not to look directly at the lamp.* We recommend draping a hood opening with black cloth; a bank of two or three sunlamps may be used in the hood to irradiate all student samples simultaneously.

In a 24 × 200-mm test tube place 0.02 mol of 4-methylbenzophenone, 25 mL of 2-propanol, and 2 drops of glacial acetic acid. Warm the mixture to effect complete solution. Stopper the tube and either place it in a position to receive strong sunlight for about 1 week★ or, more reliably, expose it for 3 days to a 275-watt sunlamp at a distance of about 15 cm.★ Cool the mixture in ice, and collect the yellow-to-white solid on a filter by vacuum filtration. Transfer this product to a 100-mL round-bottomed flask, and dissolve it in 50 mL of 1-propanol by heating to reflux. Add water to the hot solution until faint cloudiness appears, and allow the solution to cool slowly to room temperature. The yield of amorphous white solid, mp 162–164°, should be about 2.4 g; it is a mixture of the diastereomers of 4,4′-dimethylbenzo-pinacol. Collect this product by vacuum filtration, dry it, and determine its melting point. It might be interesting to see if the diastereomers can be separated by tlc.★

C.   Pinacol-Pinacolone Rearrangement of 4,4′-Dimethylbenzopinacol

---
DO IT SAFELY

The glacial acetic acid used in this experiment may cause severe blistering of the skin if it is not washed off immediately. Its vapors may also cause irritation of the mucous membranes if inhaled. We recommend that you transfer this material in the hood and wear rubber gloves during the process. If you do get glacial acetic acid on your skin, wash the area thoroughly with cold water, and treat with 0.6 M sodium bicarbonate solution.

In a 100-mL round-bottomed flask place 0.0040 mol of 4,4′-dimethylbenzo-pinacol, 15 mL of glacial acetic acid, and sufficient 0.1 M solution of iodine in acetic acid to produce a pale-orange color. Add boiling chips and heat the mixture under reflux for 10 min; if the solution becomes colorless, add more iodine solution until the pale-orange color is restored. Cool the solution and evaporate the acetic acid under aspirator vacuum while heating on a steam bath. Add 5 mL of glacial acetic acid to the residue and evaporate again. Repeat this process twice, so that three 5-mL portions of glacial acetic acid have been added and then removed by heating under vacuum.

D.   Cleavage of the Isomeric Pinacolones

DO IT SAFELY
---

1.   Dimethylsulfoxide is rapidly absorbed through the skin. Avoid breathing its vapors or allowing contact of either it or its solutions with the skin.
2.   Potassium butoxide is strongly corrosive. To avoid caustic burns, if acciden-tal contact with either the solid or its solutions occurs, wash the affected area immediately with abundant quantities of cold water. Rubber gloves are recom-mended. (*Since potassium butoxide is very hygroscopic, the bottle should remain uncapped for only the minimally required time. The cap should be replaced tightly.*)

Dissolve the residual mixture of isomeric pinacolones obtained in the preceding section in 10 mL of dimethylsulfoxide in a small flask by warming, and then cool it to room temperature. In a separate 125-mL Erlenmeyer flask place 0.036 mol of potassium butoxide and 10 mL of dimethylsulfoxide, and then add 0.38 mL of water by means of a graduated pipet or syringe.

Pour the solution of the pinacolones into the potassium butoxide slurry, and swirl the intensely colored mixture for 2 min. Add 50 mL of cold water, and cool the mixture to room temperature. Extract the white slurry with two 20-mL portions of dichloromethane, which should remove the solid material. Discard the dichloro-methane extracts in a container provided for organic waste. Acidify the aqueous solution to pH of about 1 (use indicator paper) with concentrated hydrochloric acid, and then extract the acidic solution with three 15-mL portions of dichloromethane. Combine the dichloromethane extracts, and wash them in a separatory funnel with two 10-mL portions of water and one 10-mL portion of saturated sodium chloride solution. Dry the dichloromethane solution over anhydrous sodium or magnesium sulfate; filter and evaporate the solution until only a small solid or oily residue remains, which is a mixture of benzoic and p-toluic acids.

E.   Analysis of the Mixture of Pinacolone Degradation Products

The mixture of benzoic and p-toluic acids obtained by the preceding procedure may be analyzed quantitatively by one or more of the following three methods.

PMR Analysis.   Dissolve a small sample of the mixed acids in the minimum amount of deuteriochloroform containing 1% TMS. Record the 60 MHz pmr spectrum and integrate it. Calculate the proportion of benzoic and p-toluic acids from the ratio of the integrals of the methyl and aromatic hydrogens, using the following formulas:

$$\frac{I_A - \frac{4}{3}I_M}{5} \times 100 = \% \text{ benzoic acid}$$

$$\% \text{ } p\text{-toluic acid} = 100 - \% \text{ benzoic acid}$$

$$(I_A = \text{aromatic integral, } I_M = \text{methyl integral})$$

The formula is derived from the fact that both benzoic and p-toluic acid contribute to the aromatic integral, whereas only p-toluic acid contributes to the methyl integral.

GLPC Analysis.   1.   The mixed acids are dissolved in the minimum amount of 2-propanone, and a sample is injected onto a 1.5-m Chromosorb 101 column held at 250°, using a helium flow rate of 55 mL/min. Benzoic acid has a shorter retention time than p-toluic acid.

2.   The remainder of the 2-propanone solution of the mixed acids is evaporated, and 15 mL of a 1 M methanolic solution of hydrochloric acid is added to the residue. The methanolic solution is heated under reflux for 1.5 hr, and then all but about 1 mL of methanol is removed by evaporation. A sample of the residual solution is injected onto a 1.5-m QF-1 or SF-96 (20% on Gas Chrom-Q) column. A column temperature of 130° with a helium flow rate of 25 mL/min gives good separation. Methyl benzoate has a shorter retention time than methyl p-toluate.

EXERCISES

1.   Explain why little or no 3-methylbenzophenone is produced from benzoyl chloride and toluene.
2.   Why do you think the major product from benzoyl chloride and toluene is the 4-methyl isomer rather than the 2-isomer? What would you predict to be the proportion of isomeric products from acylation of t-butylbenzene by benzoyl chloride?
3.   Considering equation 4, why do you think the abstraction of H from 2-propanol by **6** occurs at C rather than O?
4.   Considering transition state **17**, what do you think the migratory aptitude of the substituted phenyl groups would be, relative to unsubstituted phenyl, for G = OCH$_3$ and G = Cl?
5.   Figure 20.9 provides the pmr spectrum of the mixture of acids obtained in part D in a typical experiment. Using the procedure discussed in part E, determine the percentage of benzoic acid and p-toluic acid obtained by degradation of the mixture of pinacolones.
6.   Explain the derivation of the equation relating the pmr integrals of aromatic and methyl signals to the proportion of benzoic and p-toluic acids in the product mixture.
7.   Calculate ε at the maxima in Figure 20.6, and assign the chromophores responsible for them.

SPECTRA OF STARTING MATERIALS AND PRODUCTS

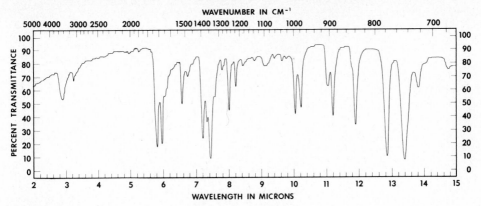

FIGURE 20.4 IR spectrum of 4-methylbenzophenone.

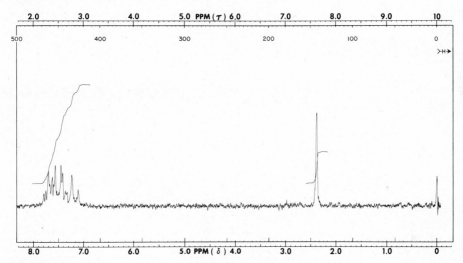

FIGURE 20.5 PMR spectrum of 4-methylbenzophenone.

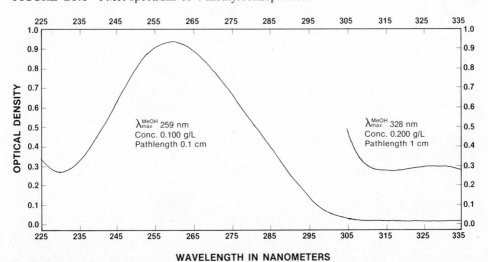

λ_{max}^{MeOH} 259 nm
Conc. 0.100 g/L
Pathlength 0.1 cm

λ_{max}^{MeOH} 328 nm
Conc. 0.200 g/L
Pathlength 1 cm

WAVELENGTH IN NANOMETERS

FIGURE 20.6 UV spectrum of 4-methylbenzophenone.

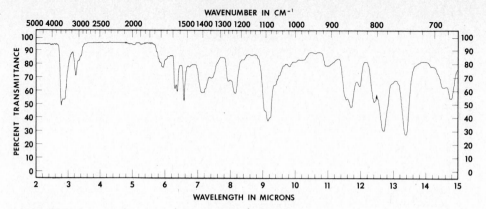

FIGURE 20.7 IR spectrum of 4,4'-dimethylbenzopinacol.

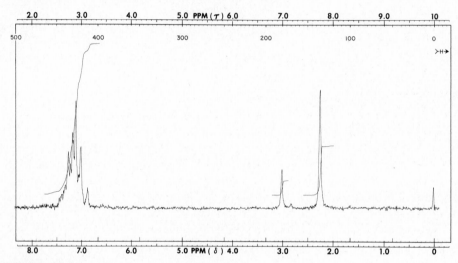

FIGURE 20.8 PMR spectrum of 4,4'-dimethylbenzopinacol.

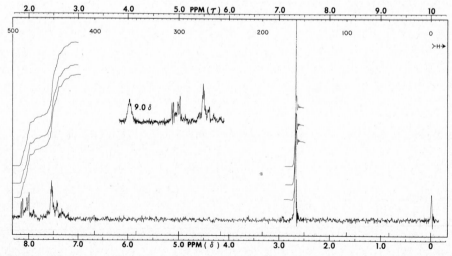

9.0 δ

FIGURE 20.9 PMR spectrum of mixture of benzoic acid and *p*-toluic acid obtained in part D.

21

POLYMERS

Polymer is the name given to a class of molecules characterized by great size (molecular weights in the range of thousands to hundreds of thousands) and by recurring structural units that are related to simpler molecules (monomers) from which the polymer may be considered to be derived. Because of the size of their molecules, polymers are also often referred to as *macromolecules*. Important examples of polymers are found in nature, for example, proteins, polysaccharides, rubber, and nucleic acids. Polymers may also be produced synthetically, as is attested by the myriad of synthetic plastics, elastomers, and fibers that have become commonplace in our contemporary society. Additional mention of some of the natural polymers may be found in Chapters 22 and 23; in this chapter we are concerned with typical examples of synthetic polymers.

21.1 Addition Polymers

Most synthetic polymers can be classified as of two major types, *addition polymers* and *condensation polymers*. This classification is based on the kinds of reactions (*polymerizations*) by which the polymers are produced. Addition polymers are produced by self-addition of a large number of small molecules (monomers). For example, as the name implies, polyethylene is produced by the catalyzed self-addition of thousands of ethylene molecules (equation 1). It may be noted that the molecular

$$(n)\ CH_2{=}CH_2 \xrightarrow[\text{and/or heat}]{\text{catalyst}} {+}(CH_2{-}CH_2)_n \tag{1}$$

$$n = 10,000\text{--}30,000$$

formula of the polymer is essentially the same as that of the monomer, since the material at the ends of the polymer chain is insignificant because n is such a large number. Polystyrene (2) is a polymer of styrene, or ethenylbenzene (1, equation 2).

$$\text{(n)} \quad \underset{\textbf{1}}{\overset{\displaystyle CH{=}CH_2}{\bigcirc\!\!\!\bigcirc}} \quad \xrightarrow{\text{catalyst}} \quad \underset{\textbf{2}}{\overset{\displaystyle \text{+}CH{-}CH_2\text{+}_{\overline{n}}}{\bigcirc\!\!\!\bigcirc}} \tag{2}$$

Some other common addition polymers have trade names which do not indicate their structure; for example, Teflon is a polymer of tetrafluoroethylene (**3**), and Plexiglas and Lucite are polymers of methyl 2-methylpropenoate (**4**, methyl methacrylate).

$$CF_2{=}CF_2 \qquad CH_2{=}\underset{\underset{\textstyle CH_3}{|}}{C}{-}CO_2CH_3$$

$$\textbf{3} \qquad\qquad\qquad \textbf{4}$$

Copolymers may be produced by the addition polymerization of a mixture of monomers. For example, a useful plastic film (such as Saran Wrap) is made by polymerizing a mixture of chloroethene (**5**) and 1,1-dichloroethene (**6**, equation 3).[1]

$$\text{(n) } CH_2{=}\underset{\underset{\textstyle Cl}{|}}{CH} + \text{(m) } CH_2{=}CCl_2 \longrightarrow \left(CH_2{-}\underset{\underset{\textstyle Cl}{|}}{CH}\right)_{\!n}\!\!\left(CH_2\underset{\underset{\textstyle Cl}{|}}{\overset{\overset{\textstyle Cl}{|}}{C}}\right)_{\!m} \tag{3}$$

$$\textbf{5} \qquad\qquad\qquad \textbf{6} \qquad\qquad\qquad \textbf{7}$$

Addition polymerization is usually a catalyzed *chain* reaction. The catalyst may be a cation, anion, or free radical, but the latter type is most commonly used. The mechanism of this type of polymerization as described below should be compared with that of the chlorination of hydrocarbons discussed in Section 5.1. The reaction is initiated by thermal decomposition of the catalyst, which in our experiment is *tert*-butyl peroxybenzoate (**8**); this compound produces the free radicals **9** and **10**

$$CH_3{-}\underset{\underset{\textstyle CH_3}{|}}{\overset{\overset{\textstyle CH_3}{|}}{C}}{-}O{-}O{-}\overset{\overset{\textstyle O}{\|}}{C}{-}\!\bigcirc \xrightarrow{\text{heat}} CH_3{-}\underset{\underset{\textstyle CH_3}{|}}{\overset{\overset{\textstyle CH_3}{|}}{C}}{-}O\cdot \;+\; \cdot O{-}\overset{\overset{\textstyle O}{\|}}{C}{-}\!\bigcirc \tag{4}$$

$$\textbf{8} \qquad\qquad\qquad\qquad \textbf{9} \qquad\qquad\qquad \textbf{10}$$

when heated (equation 4). If we let In· stand for either or both of these free radicals, we may indicate the course of the polymerization as shown in equations 5–8. Equation 5 indicates the function of the free radicals in *initiating* the polymerization.

---

[1] The abbreviated formula of **7** is not meant to imply that all of the two monomer units are bunched together in two blocks, although such "block copolymers" can be produced by special techniques. In the more common copolymers the different monomer units are distributed randomly in the chain.

$$\text{In} \cdot + \text{CH}_2\text{=CH} \longrightarrow \text{In}-\text{CH}_2-\text{CH} \cdot \qquad (5)$$
$$\underset{\text{C}_6\text{H}_5}{|} \qquad\qquad \underset{\text{C}_6\text{H}_5}{|}$$

$$\text{In}-\text{CH}_2-\underset{\underset{\text{C}_6\text{H}_5}{|}}{\text{CH}} \cdot + \text{n CH}_2\text{=}\underset{\underset{\text{C}_6\text{H}_5}{|}}{\text{CH}} \longrightarrow \text{In}-\text{CH}_2-\underset{\underset{\text{C}_6\text{H}_5}{|}}{\text{CH}}-(\text{CH}_2-\underset{\underset{\text{C}_6\text{H}_5}{|}}{\text{CH}})_{\overline{n}} \cdot \qquad (6)$$
$$\mathbf{11}$$

$$2 \quad \mathbf{11} \longrightarrow \text{In}-\text{CH}_2-\underset{\underset{\text{C}_6\text{H}_5}{|}}{\text{CH}}-(\text{CH}_2-\underset{\underset{\text{C}_6\text{H}_5}{|}}{\text{CH}})_{\overline{n}-1}\text{CH}_2-\underset{\underset{\text{C}_6\text{H}_5}{|}}{\text{CH}_2}$$
$$\mathbf{12}$$

$$+ \text{In}-\text{CH}_2-\underset{\underset{\text{C}_6\text{H}_5}{|}}{\text{CH}}-(\text{CH}_2-\underset{\underset{\text{C}_6\text{H}_5}{|}}{\text{CH}})_{\overline{n}-1}\underset{\underset{\text{C}_6\text{H}_5}{|}}{\text{CH}}\text{=CH} \qquad (7)$$
$$\mathbf{13}$$

$$\mathbf{11} + \text{R} \cdot \longrightarrow \text{In}-\text{CH}_2-\underset{\underset{\text{C}_6\text{H}_5}{|}}{\text{CH}}-(\text{CH}_2-\underset{\underset{\text{C}_6\text{H}_5}{|}}{\text{CH}})_{\overline{n}}\text{R} \qquad (8)$$

Equation 6 represents the *propagation* of the growing polymer chain. Equations 7 and 8 show possible *termination* processes. In equation 7 the free radical end of one growing polymer chain abstracts a hydrogen atom with its electron from the carbon atom next to the end of another polymer radical to produce one polymer molecule that is saturated at the end **(12)** and one polymer molecule that is unsaturated at the end **(13)**, a process termed *disproportionation*. In equation 8 an $\text{R} \cdot$ may be one of the initiating radicals, $\text{In} \cdot$, or another growing polymer chain.

An industrial process for the manufacture of styrene is outlined in Figure 21.1 and a laboratory synthesis in Figure 21.2. Although the industrial process involves one step less than the laboratory synthesis, it is not easily adapted for a student experiment for several reasons. It utilizes ethene and benzene as starting materials, and since ethylbenzene **(14)**, the initial product, is even more easily alkylated than benzene, a large ratio of benzene to ethene must be used to minimize the formation of

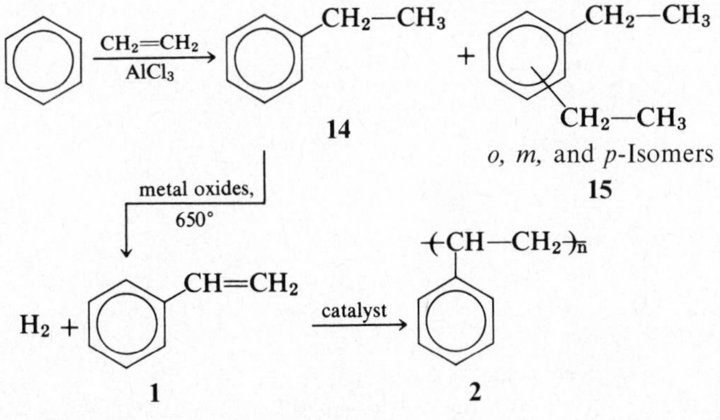

FIGURE 21.1 Industrial synthesis of polystyrene.

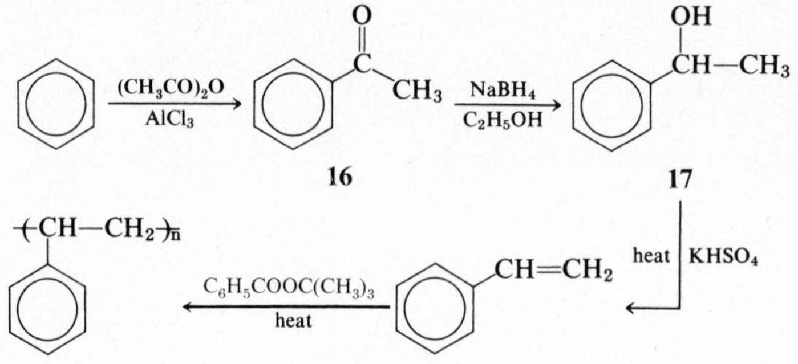

FIGURE 21.2 Laboratory synthesis of polystyrene.

diethylbenzenes, **15**. Even under these conditions some diethylbenzenes form and must be separated from the ethylbenzene. The catalytic dehydrogenation step also requires the difficult separation of the product, styrene **(1)**, from unchanged ethylbenzene.

The steps in our laboratory synthesis are now described in detail.

Acetophenone. Instead of the Friedel-Crafts *alkylation* of benzene employed in the industrial method, the laboratory synthesis utilizes *acylation* in the first step. The acylation of benzene is accomplished with acetic anhydride and aluminum chloride and is an example of a Friedel-Crafts synthesis. The product of this reaction is acetophenone **(16)**. Because of the toxicity of benzene, we shall begin our synthesis using the readily available (and nontoxic) acetophenone as starting material. We may note in passing, however, that Friedel-Crafts acylations are free of two undesirable features of alkylations: the tendency toward polysubstitution, and rearrangement of the alkylating agent. The introduction of an acyl group into an aromatic ring deactivates the ring toward further electrophilic substitution. The lack of rearrangement of acyl moieties makes acylation, followed by reduction, one of the most reliable synthetic procedures for the preparation of alkylbenzenes of authentic structures, for example, 1-propylbenzene (**18**, equation 9).

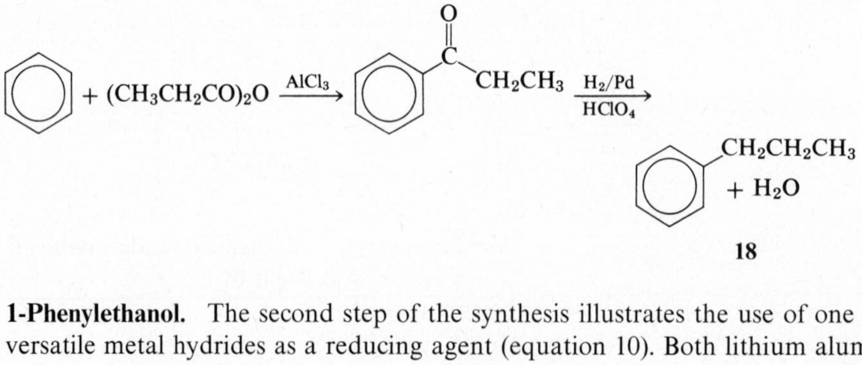

$$+ (CH_3CH_2CO)_2O \xrightarrow{AlCl_3} \quad \xrightarrow[HClO_4]{H_2/Pd} \quad + H_2O \qquad (9)$$

1-Phenylethanol. The second step of the synthesis illustrates the use of one of the versatile metal hydrides as a reducing agent (equation 10). Both lithium aluminum

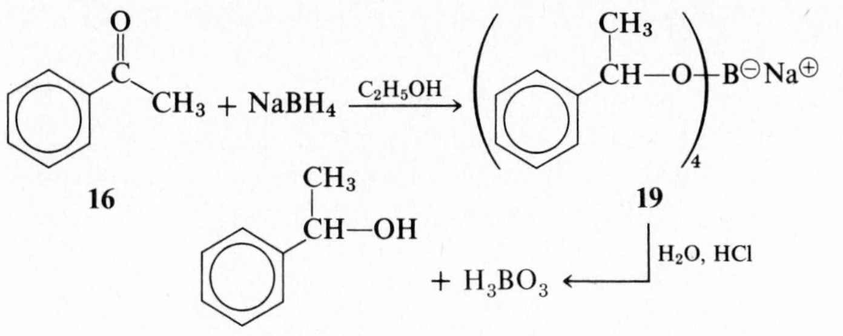

$$\text{16} + \text{NaBH}_4 \xrightarrow{\text{C}_2\text{H}_5\text{OH}} \left(\text{19} \right)_4 \text{B}^{\ominus}\text{Na}^{\oplus} \qquad (10a)$$

$$\text{17} + \text{H}_3\text{BO}_3 \xleftarrow{\text{H}_2\text{O, HCl}} \qquad (10b)$$

hydride and sodium borohydride reduce aldehydes and ketones to alcohols; the virtue of the latter (milder) reagent is that it may be used in alcoholic and even aqueous solution in contrast to lithium aluminum hydride. Although sodium borohydride reacts slowly with alcohols and water, it reacts much more rapidly with carbonyl compounds so that its reaction with these solvents is not troublesome.

The sodium borohydride transfers a hydride ion to the carbonyl carbon in the first step of the mechanism (equation 11). All the hydrogens attached to boron are

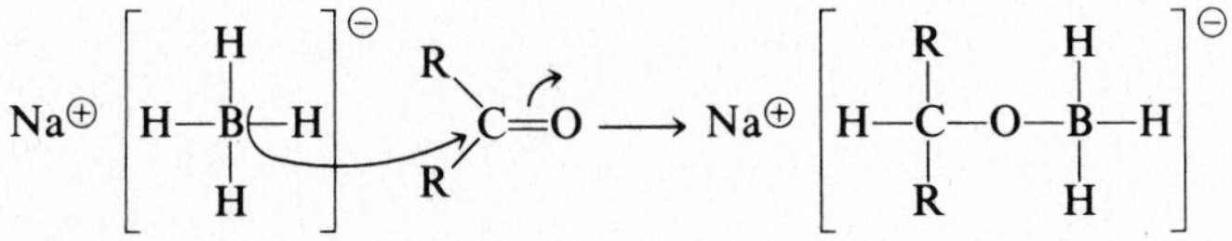

$$\text{Na}^{\oplus} \left[\text{H}-\text{B}(\text{-H})\text{H} \right]^{\ominus} \overset{R}{\underset{R}{\diagup}}\text{C}{=}\text{O} \longrightarrow \text{Na}^{\oplus} \left[\text{H}-\overset{R}{\underset{R}{\text{C}}}-\text{O}-\overset{H}{\underset{H}{\text{B}}}-\text{H} \right]^{\ominus} \qquad (11)$$

transferred in this way to produce the intermediate borate salt (**19**, equation 10a), which is decomposed upon the addition of water and acid to yield 1-phenylethanol (**17**, equation 10b). The 1-phenylethanol must then be separated from water, ethanol, inorganic acids, and salts. The separation is accomplished by evaporating the ethanol on a steam bath and then extracting the 1-phenylethanol with ether. The ether solution is dried by a dessicant, and the ether is evaporated. The 1-phenylethanol may be isolated by distillation under reduced pressure; the next step (dehydration) may be carried out on the crude product, however.

Styrene. The dehydration of 1-phenylethanol is accomplished by heating the alcohol with a mild dehydrating agent, potassium hydrogen sulfate, and collecting the water and styrene (**1**) as they distil at atmospheric pressure (equation 12). The crude styrene

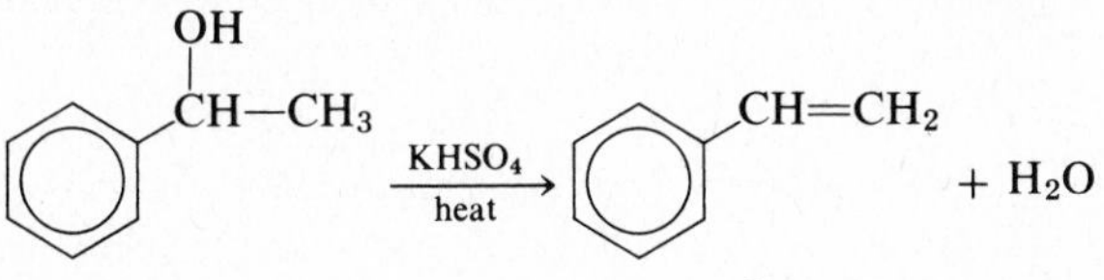

$$\text{17} \xrightarrow[\text{heat}]{\text{KHSO}_4} \text{1} + \text{H}_2\text{O} \qquad (12)$$

is then either extracted into diethyl ether, washed, and distilled under reduced pressure or it is diluted with xylene, and the solution is dried by azeotropic distillation. The choice depends on whether one wishes to carry out the fourth step, polymerization, on neat styrene or on a xylene solution of styrene.

Polystyrene. As mentioned, styrene is usually polymerized by free radical catalysis. One of the most commonly used catalysts in the industrial production of polystyrene is benzoyl peroxide, but its use requires extreme care in handling because of its tendency to decompose violently (explode!). The catalyst we use in our polymerization procedure, *tert*-butyl peroxybenzoate, is entirely safe to handle. In the following procedure, directions are given for producing the polymer in the form of an amorphous solid, a film, and a clear "glass."

EXPERIMENTAL PROCEDURE

A.   1-Phenylethanol

In a 150-mL beaker dissolve 0.032 mol of sodium borohydride (*Caution: Do not allow this strongly caustic reagent to come in contact with your skin*) in 25 mL of 95% ethanol, and add 0.100 mol of acetophenone dropwise. Stir the mixture with a thermometer during the addition, and keep the temperature below 50° by decreasing the rate of addition and by external cooling with an ice bath, if necessary. After the addition of the acetophenone is complete, allow the reaction mixture (which contains a white precipitate) to stand for 15 min at room temperature. Add about 10 mL of 3 *M* hydrochloric acid solution dropwise and with stirring; hydrogen will be evolved and most of the white solid will dissolve. Place the beaker containing the reaction mixture on a steam bath, and concentrate the solution by evaporation of the ethanol until two liquid layers separate.

Add 20 mL of diethyl ether, and transfer the mixture to a 125-mL separatory funnel. Shake the mixture gently, allow the layers to separate, and remove the ether layer.★ Extract the aqueous layer with a 10-mL portion of diethyl ether, and add this to the other ether solution. Dry the combined ether solutions over anhydrous magnesium sulfate; briefly swirl the solution with the drying agent, and then decant the solution into a dry flask. Add another portion of drying agent and repeat the swirling; filter the solution into a tared 100-mL round-bottomed flask.★

If pure 1-phenylethanol is to be isolated, add about 1 g of anhydrous potassium carbonate, and arrange for vacuum distillation.[2] (The potassium carbonate present in the distillation pot helps to avoid acid-catalyzed dehydration during the distillation.) Remove the diethyl ether at atmospheric pressure (steam bath), and then reduce the pressure to below 20 mm, if possible, and distil using an oil bath; the boiling point of 1-phenylethanol is 102.5–103.5° (19 mm), 97° (13 mm).

If pure 1-phenylethanol is not to be isolated by vacuum distillation, proceed in the following way. Connect the flask to an aspirator by means of an adapter or a one-hole rubber stopper fitted with a short glass tube. (The aspirator must be connected through a safety trap with a pressure release, such as that shown in Figure 2.17.) Gently apply suction to the flask by closing the screw clamp partially or completely, but momentarily; swirl the flask containing the ether solution so that the ether boils but does not bump. As the ether boils away, the pressure may be gradually

---

[2]See Section 2.5 and the references at the end of Chapter 2.

lowered until the full aspirator vacuum is applied. Warm the flask with your hand or with a water bath at room temperature. Calculate the yield of crude 1-phenylethanol.

B. Styrene

At this stage two or more students should combine their yields of 1-phenylethanol so that a total of at least 0.09 mol may be used in the dehydration step. This may be done by pouring the crude products together in one 100-mL round-bottomed flask with the aid of a little dry diethyl ether, which is then removed as described in the last paragraph of part A.

To about 0.09 mol of crude 1-phenylethanol, add 1.0 g of anhydrous potassium hydrogen sulfate and about 0.1 g of copper powder.[3] Attach a distilling head and a condenser to the 100-mL flask, and distil the mixture by heating in an oil bath whose temperature is gradually (during about 10 min) raised to 200–220°. The temperature of the distilling vapor should not exceed 130°. Collect the distillate and calculate the approximate yield of crude styrene from the volume of the organic layer. If the layers do not separate well, add a little sodium chloride to saturate the water layer.★

If you are to isolate pure styrene, add the mixture of styrene and water obtained as distillate to 30 mL of diethyl ether in a 125-mL separatory funnel. Shake gently, allow the layers to separate, and drain off the aqueous layer. Wash the ether solution first with 15 mL of saturated sodium carbonate solution and then with 10 mL of water. Dry the ether solution by swirling with about 1 g of anhydrous calcium chloride.★ Decant the ether solution into a 50-mL round-bottomed flask, wash the drying agent with 5 mL of diethyl ether, and combine the wash with the main ether solution. Attach a still head and condenser, add some boiling stones, and distil, using a steam bath. Remove the ether at atmospheric pressure, then arrange for vacuum distillation, using an aspirator (see footnote 2). The boiling point of styrene is 55° (30 mm), 48° (20 mm). Collect the styrene in a tared flask, and calculate the yield based on 1-phenylethanol (one step) as well as on acetophenone (two steps), considering the actual amount of material used in each step.

C. Polystyrene

Polymerization of Pure Styrene. Place about 5 g of pure, freshly distilled styrene in an 18 × 150-mm soft-glass test tube, and add 4 or 5 drops of *tert*-butyl peroxybenzoate. Clamp the test tube in a vertical position over a wire gauze, insert a 360° thermometer so that its bulb is in the liquid, and heat the styrene and catalyst with a small burner flame. When the temperature reaches 140°, remove the flame temporarily. If boiling stops, replace the flame to maintain gentle boiling. Since the polymerization is an exothermic reaction and since free radicals are produced by thermal decomposition of the catalyst, polymerization begins to occur rapidly. Check for a rapid increase in the rate of boiling, and be prepared to remove the flame if the refluxing liquid rises to the top of the test tube.

After the onset of polymerization the temperature may be seen to rise to 180° or

---

[3]Copper powder has been reported to reduce the tendency of the styrene (produced in this step) to polymerize during the distillation.

190°, much above the boiling point of styrene, 145°. The viscosity of the liquid may also be observed to increase rapidly during this time. As soon as the temperature begins to decrease, the thermometer should be removed, and the polystyrene may be poured onto a watch glass. (*Caution:* Do *not* touch the thermometer *before* the temperature decreases because movement of the thermometer in the boiling liquid might cause a sudden "bump," which would throw hot liquid out of the tube.) Note the formation of fibers as the thermometer is pulled out of the polymer.

The rate of solidification of the polystyrene will depend on the amount of catalyst used, the temperature, and the length of time the mixture was heated. It may be instructive for some students to use only 1 or 2 drops of catalyst per 5 g of styrene for comparison purposes. The properties of the polystyrene will also vary with the conditions of the polymerization process. (See Exercise 8.)

Solution Polymerization of Crude Styrene.   Add the mixture of styrene and water collected in step B to three times its volume of xylene (commercial mixture of isomers), and drain off the small water layer into a 125-mL separatory funnel. Distil the xylene solution until the distillate dropping from the end of the condenser is clear (all the water will have been removed by azeotropic distillation). Add 5 *drops* of *tert*-butyl peroxybenzoate for each 5 g of styrene (the amount estimated in the crude distillate from the dehydration reaction) in the solution, connect a reflux condenser, and heat to reflux with a burner for 30 min. Cool the solution to room temperature, and then pour about half of it into 150 mL of methanol. Collect the white precipitate of polystyrene by decantation or by vacuum filtration if decantation is not practical. Resuspend the polystyrene in fresh methanol, and stir it vigorously; then collect it on a filter, and allow it to dry.

Pour the remaining half of the polystyrene solution on a watch glass or the bottom of a large inverted beaker, and allow it to evaporate. A clear film of polystyrene should result.

EXERCISES

1. Suggest a formula for the white precipitate formed in the reaction of acetophenone with sodium borohydride, and write an equation for its reaction with water and hydrochloric acid.
2. When 1-phenylethanol is purified by distillation, reduced pressure is always used. Why?
3. Styrene is distilled under reduced pressure. Why?
4. Suggest an explanation for the fact that acylation occurs without rearrangement, whereas rearrangement often accompanies alkylation; for example, compare the reaction of CH_3CH_2COCl and $CH_3CH_2CH_2Cl$ with benzene and aluminum chloride.
5. What other reagents besides $NaBH_4$ and $LiAlH_4$ will reduce acetophenone to 1-phenylethanol?
6. What reagents will reduce acetophenone to ethylbenzene?

7. Suppose that in the first step of this synthesis an appreciable amount (about 10%) of diacetylation occurred and the other steps were carried out without purification of the intermediates. What difference in the physical properties of the ultimate polymer might result?

8. What effect on the average molecular weight of polystyrene would you expect to be produced by using a smaller proportion of catalyst to styrene?

9. Assume that the location and value of ε for styrene is the same in mixtures of methanol and water as it is in pure methanol (Figure 21.6). Describe a method for *accurately* determining the total amount of styrene present in the styrene-water mixture obtained upon dehydration of 1-phenylethanol. The method should not require any extractions. You may assume that there are no impurities present which absorb in the range of 210–320 nm.

REFERENCE

S. H. Wilen, C. B. Kremer, and I. Waltcher, *Journal of Chemical Education,* **38,** 304 (1961).

SPECTRA OF STARTING MATERIALS AND PRODUCTS

The ir and pmr spectra of acetophenone are given in Figures 14.20 and 14.21, respectively.

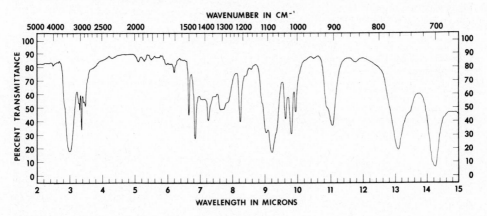

WAVENUMBER IN CM⁻¹

FIGURE 21.3 IR spectrum of 1-phenylethanol.

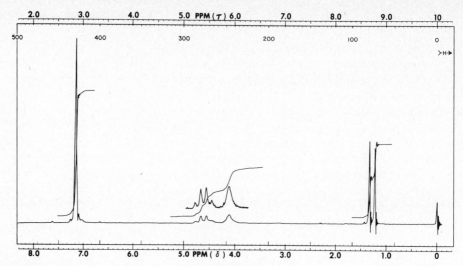

FIGURE 21.4 PMR spectrum of 1-phenylethanol.

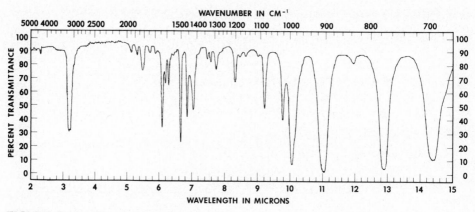

FIGURE 21.5 IR spectrum of styrene.

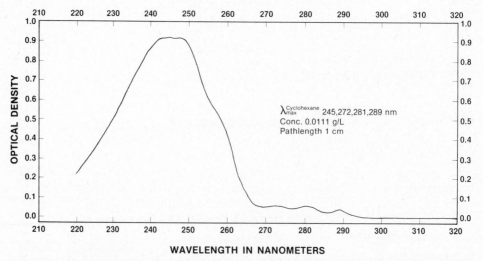

$\lambda_{max}^{Cyclohexane}$ 245,272,281,289 nm
Conc. 0.0111 g/L
Pathlength 1 cm

FIGURE 21.6 UV spectrum of styrene.

WAVENUMBER IN CM⁻¹

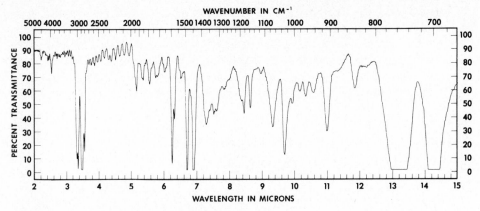

FIGURE 21.7 IR spectrum of polystyrene.

21.2 Condensation Polymers

The second major type of synthetic polymers consists of those produced, as the name implies, by condensation reactions—reactions in which the monomeric units are joined by intermolecular eliminations of small molecules such as water or methanol. Obviously, then, condensation polymers do not have the same molecular formula as their monomers, as is true in the case of addition polymers. For example, a polyester can be produced from 1,4-benzenedicarboxylic acid (terephthalic acid, **20**) and 1,2-ethanediol (ethylene glycol, **21**) as shown in equation 13. The reaction

$$(n) \quad \text{HO} \overset{\text{O}}{\underset{}{\text{C}}} - \bigcirc - \overset{\text{O}}{\underset{}{\text{C}}} \text{OH} \quad + \quad (n) \text{ HOCH}_2\text{CH}_2\text{OH} \xrightarrow{\text{H}^+}$$

$$\quad\quad\quad\quad\quad \textbf{20} \quad\quad\quad\quad\quad\quad\quad \textbf{21}$$

(13)

$$\text{H} \left(\text{O} - \overset{\text{O}}{\underset{}{\text{C}}} - \bigcirc - \overset{\text{O}}{\underset{}{\text{C}}} - \text{O} \text{CH}_2 - \text{CH}_2 \right)_n \text{OH} \quad + \quad (2n - 1)\,\text{H}_2\text{O}$$

mechanism is that of an ordinary acid-catalyzed esterification, but a polymer is formed because both the acid and the alcohol monomer molecules are difunctional. The polymer produced in this way does not have a very high molecular weight. To obtain higher molecular weight polyesters that have better properties for use in textile fibers, for example, the polymerization is carried out as a transesterification between dimethyl 1,4-benzenedicarboxylate and 1,2-ethanediol. In this condensation reaction the molecule eliminated in the monomer-linking process is methanol rather than water; methanol is removed more easily than water from a growing polymeric product because it is more volatile, and hence better yields of higher molecular weight polymer are obtained. Dacron, Terylene, and Mylar are trade names of commercial polyesters.

Polyamides are another type of useful condensation polymer, and a whole spectrum of such polymers has been produced from various diacids and diamines. Nylon-6,6 was the first commercially successful polyamide. (The numbers in the name derive from the six carbon atoms in each of the monomer molecules.) Nylon-6,6 is made from hexanedioic acid (adipic acid, **22**) and 1,6-hexanediamine (hexamethylenediamine, **23**), as shown in equation 14. In the industrial process equimolar

$$\text{(n) HO}-\overset{\overset{O}{\|}}{C}(CH_2)_4\overset{\overset{O}{\|}}{C}-OH + \text{(n) } H_2N(CH_2)_6NH_2 \longrightarrow$$

$$\quad\quad\quad \textbf{22} \quad\quad\quad\quad\quad\quad \textbf{23}$$

$$HO\left[\overset{\overset{O}{\|}}{C}(CH_2)_4\overset{\overset{O}{\|}}{C}-NH(CH_2)_6-NH\right]_n H + (2n-1)\,H_2O$$

$$\text{Nylon-6,6}$$

(14)

amounts of the diacid and diamine are mixed to give the salt, which is then heated to high temperature under vacuum to eliminate the water. The polymer so produced has a molecular weight of about 10,000 and a melting point of about 250°. Fibers can be spun from melted polymer, and if the fibers are stretched to several times their original length, they become very strong. This "cold drawing" serves to orient the polymer molecules parallel to one another so that hydrogen bonds form between C=O and N—H groups on adjacent polymer chains, greatly increasing the strength of the fibers. The strength of the fibers of silk, a well-known natural polymer of the protein type, is attributed to the same factor: the hydrogen bonds between the natural polyamide molecules.

The polyamide produced from decanedioic acid (sebacic acid, **24**) and 1,6-hexanediamine is called nylon-6,10 (equation 15).

$$HO-\overset{\overset{O}{\|}}{C}-(CH_2)_8\overset{\overset{O}{\|}}{C}-OH + H_2N-(CH_2)_6-NH_2 \longrightarrow$$

$$\quad\quad \textbf{24} \quad\quad\quad\quad\quad\quad \textbf{23}$$

$$HO\left(\overset{\overset{O}{\|}}{C}(CH_2)_8\overset{\overset{O}{\|}}{C}-NH-(CH_2)_6-NH\right)_n H + (2n-1)\,H_2O$$

$$\text{Nylon-6,10}$$

(15)

For this experiment the preparation of nylon-6,10 rather than nylon-6,6 is chosen. To produce a polyamide under simple laboratory conditions, the diacyl chloride derivative of the diacid is used because it is more reactive, and decanedioyl chloride is more stable toward hydrolysis than the corresponding six-carbon compound in contact with moist air or the aqueous phase of the reaction mixture. When the diacyl chloride is employed, the small molecule eliminated in the polymerization process is hydrogen chloride rather than water (equation 16). Sodium carbonate is

$$-CH_2-\overset{\overset{O}{\|}}{C}-Cl + H_2N-CH_2- \longrightarrow -CH_2\overset{\overset{O}{\|}}{C}-NH-CH_2- + HCl \quad\quad (16)$$

added to neutralize the acid formed by the reaction to avoid using an excess of the expensive diamine.

Using the reactive diacyl chloride makes it possible to carry out a polymerization under very mild conditions. When a solution of the diacyl chloride in a water-immiscible solvent is brought into contact with an aqueous solution of the aliphatic diamine at room temperature, a film of high molecular weight polymer forms at once where the two solutions meet. The film is thin but strongly coherent, and it can be pulled out of the interface between the two solutions, where it is immediately and continuously replaced. In this way a long cord or "rope" of polyamide can be produced in much the same way as a magician pulls a string of silk handkerchiefs out of a top hat. When this experiment was first described by two DuPont chemists, they characterized it as the "Nylon Rope Trick." [4] It does seem to be magic that a polymer formed in fractions of a second can attain an average molecular weight in the range 5,000 to 20,000!

To do justice to this experiment, equipment should be arranged to allow the polymer rope to be pulled out of the reaction zone as rapidly as it is formed. A convenient way to do this is illustrated in Figure 21.8. A can such as that in which coffee, fruit juice, or motor oil is packaged, preferably with a diameter of 10 cm or more, makes a good drum on which to wind the polymer. If the can has been emptied by punctures at the edges, it can then be punctured in the center of each end, and a

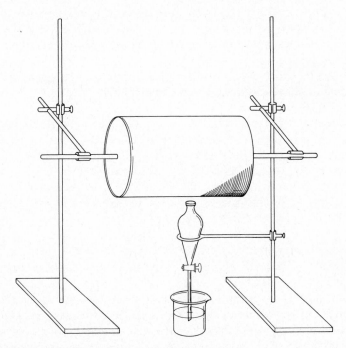

FIGURE 21.8 Apparatus for the "Nylon Rope Trick."

[4] The present experiment is adapted from the original article by P. W. Morgan and S. L. Kwolek, *Journal of Chemical Education,* **36,** 182 (1959).

wooden or metal rod can be passed through the center holes to make an axle for the drum. The rod can be supported horizontally by clamps attached to ring stands in the usual way. To be able to estimate the length of the nylon rope produced, a piece of string should be passed around the drum to measure its circumference; when the rope is being wound on the drum, the revolutions of the drum may be counted and the total length of the rope calculated. A length of 12 m or more can usually be obtained with the procedure described here.

EXPERIMENTAL PROCEDURE

DO IT SAFELY

1.   If pipets are used instead of syringes to measure the reactants, use a rubber bulb to draw up the liquid; never use your mouth with a pipet.

2.   Do not handle the polymer rope with your hands any more than is necessary until it has been washed free of solvent and reagents. Use rubber gloves, tongs, or forceps to manipulate it. If you touch the crude polymer, wash your hands with soap and warm water immediately thereafter.

3.   After all the rope has been drawn from the reaction mixture, stir the remaining mixture thoroughly until no more polymer forms. Any additional polymer so formed should then be separated from the liquid, washed well with water, and disposed of in the solid waste receptacle.

4.   If formic acid is used to form a film, take care not to get it on your skin; it will cause deep skin burns which are not immediately apparent.

Measure into a 250-mL beaker 2 mL (0.0093 mol) of decanedioyl chloride by means of a syringe or pipet. (The size of the beaker is important. In smaller beakers the polymer tends to stick to the walls, whereas in larger beakers poor "ropes" are obtained unless larger amounts of reagents are used.) Dissolve the decanedioyl chloride in 100 mL of dichloromethane. Place 1.1 g (0.0095 mol) of crystalline 1,6-hexanediamine (or 1.3 mL of a commercially available 80–95% aqueous solution) in a 125- or 250-mL separatory funnel, add 2.0 g of sodium carbonate, and dissolve both substances by adding 50 mL of water and shaking gently. Arrange the drum on which the polymer is to be wound at a height such that the beaker containing the decanedioyl chloride solution can be placed on the lab bench about 40 cm beneath and slightly in front of the drum.

Support the separatory funnel containing the other reagents in such a way that the lower tip of the funnel is centered no more than a centimeter above the surface of the dichloromethane solution of the decanedioyl chloride. Open the stopcock of the separatory funnel slightly so that the aqueous solution runs *slowly and gently* onto the surface of the organic solution. A film of polymer will form immediately at the interface of the two solutions. Use a long forceps or tongs to grasp the center of the polymer film and pull up to the front of the drum the rope that forms, loop it over the drum, and rotate the drum away from you so as to wind the rope onto the drum. (For the first turn or two it may be necessary for you to use your fingers to secure the rope to the drum. If it is, rinse your hands as soon as possible thereafter.) Continue to rotate the drum and wind the nylon rope onto the drum at a rapid rate until the reactants are used up, remembering to count the revolutions of the drum as you wind (counting may be facilitated by previously marking a spot on the drum with an adhesive label or a marking pen).

Replace the beaker with a large dish or pan containing about 200 mL of 50% aqueous ethanol, and unwind the nylon rope into the wash solution. After stirring the mixture gently, decant the wash solution, and transfer the polymer to a filter on a Büchner funnel. Press the polymer as dry as possible, and then place it in your desk to dry until the next laboratory period. Dispose of the residual reaction mixture as described in the Do It Safely section.

Examination of Dry Nylon-6,10.   You will probably encounter two surprises in this experiment. The first will be the apparently enormous amount of nylon rope obtained from about 3 g of starting material. The second surprise will be the decrease in bulk of the polymer on drying. You will learn that the "rope" was really a delicate *tube* that appeared much larger when it was swollen with solvents. When the nylon is thoroughly dry, weigh it and calculate the yield.

Film Formation.   The dry polymer may be dissolved in about 10 times its weight of 90–100% formic acid (*Caution:* see Do It Safely, paragraph 4) by stirring at room temperature. (Heating to dissolve will degrade the polymer.) If the viscous solution is spread on a glass plate, a film of nylon-6,10 will be left by evaporation of the formic acid. The plate should be left *in a hood* from one laboratory period to the following one.

Fiber Formation.   The dry polymer obtained in this experiment does not appear to have the properties expected of nylon; it is fragile and of low density. However, if the product is carefully melted by *gentle* heating in a metal spoon or spatula over a very small burner flame or an electric hot plate, fibers may be drawn from the melt with a small glass rod. It will probably be necessary for several students to combine their yields to provide enough polymer to be melted and drawn successfully. The polymer should not be heated much above the melting temperature, or it will become discolored and charred. When the molten polymer cools, it will be found to be much more dense, and its appearance will be more characteristic of a typical polyamide.

EXERCISES

1. Write an equation for the formation of a salt that might be produced from one molecule of hexanedioic acid and two molecules of 1,6-hexanediamine.
2. Using full structural formulas, draw a typical portion of a nylon-6,6 molecule; that is, expand a portion of the formula given in equation 13. Show at least two hexanedioic acid units and two 1,6-hexanediamine units.

3. Draw formulas that illustrate the hydrogen bonding that may exist between two polyamide molecules after fibers have been "cold drawn."

4. Nylon-6 is produced from caprolactam by adding a small amount of aqueous base and then heating to about 270°.

(a) Draw a representative portion of the polyamide molecule.

(b) Suggest a mechanism for the polymerization and decide whether it is of the condensation or addition type.

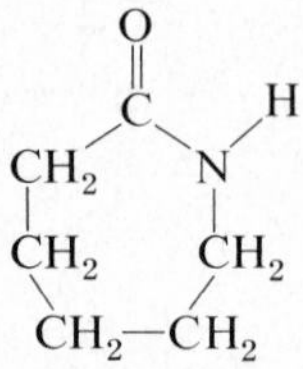

$$
\begin{array}{c}
\overset{O}{\overset{\|}{C}} \\
\text{CH}_2 \quad \text{N} \text{---H} \\
| \qquad | \\
\text{CH}_2 \qquad \text{CH}_2 \\
\text{CH}_2\text{---CH}_2
\end{array}
$$

Caprolactam

21.3 Polyurethanes

Polyurethanes are a type of synthetic polymer that does not fit neatly into either of the major classes of polymers described in Sections 21.1 and 21.2—addition polymers and condensation polymers. Polyurethanes are formed by an addition reaction, but the reaction is very different in mechanism from the usual free radical catalyzed addition polymerization. A typical polyurethane-forming reaction is the addition reaction between an aliphatic diol or triol and an aromatic diisocyanate, as in equation 17. The reaction involves nucleophilic addition of an alcohol function to

$$(n)\ \text{HO}-\boxed{\text{R}}-\text{OH} + (n)\ \text{O}{=}\text{C}{=}\text{N}-\boxed{\text{Ar}}-\text{N}{=}\text{C}{=}\text{O} \longrightarrow$$

$$
\text{HO}-\left(\boxed{\text{R}}-\text{O}-\overset{O}{\overset{\|}{\text{C}}}-\text{NH}-\boxed{\text{Ar}}\right)_n-\text{N}{=}\text{C}{=}\text{O} \tag{17}
$$

an electrophilic carbon, in this case the isocyanate carbon. Since no small molecules are eliminated in the polymer-forming step, however, the reaction cannot be considered a condensation. Despite the difficulty in categorizing polyurethanes according to type of reaction by which they are formed, they are interesting chemically and are extremely useful materials.

If a *small* amount of water is added to an alcohol-isocyanate starting mixture, the polyurethane may be produced in the form of a plastic *foam*. The water reacts with some of the isocyanate groups to form unstable carbamic acid groups which undergo decarboxylation, yielding carbon dioxide and amine groups (equation 18). The

$$
\text{R}-\text{N}{=}\text{C}{=}\text{O} + \text{H}_2\text{O} \longrightarrow \left(\text{R}-\underset{\underset{H}{|}}{\text{N}}-\overset{O}{\overset{\|}{\text{C}}}-\text{OH}\right) \longrightarrow \text{RNH}_2 + \text{CO}_2 \tag{18}
$$

carbon dioxide forms bubbles in the developing polymer, thus producing a foam; the newly created amine groups react with residual isocyanate groups so as to extend the polymer chains through urea linkages (equation 19) as well as urethane linkages.

$$R-N=C=O + R'NH_2 \longrightarrow R-\underset{H}{\overset{\overset{\displaystyle O}{\|}}{N}}-\underset{}{\overset{}{C}}-\underset{H}{\overset{}{N}}-R' \qquad (19)$$

The polyurethane foams may be made *flexible* or *rigid,* depending on the specific nature of the starting materials. Flexible foams are extensively used as cushioning materials for furniture, mattresses, and automobile seats. Rigid foams have more recently found wide application as strong, lightweight building and insulating materials.

Rigid Polyurethane Foams. The key to making a rigid polyurethane foam is to produce a polymer with a highly branched, cross-linked molecular structure.[5] How this can be done is illustrated in Figure 21.9, in which a symbolized *triol* and *triisocyanate* are the starting materials. After the initial reaction between two molecules, subsequent reactions indicated by dashed arrows will quickly lead to a highly cross-linked polymer. Although some foaming may be produced by carbon dioxide generated by reaction of water as described, an additional foaming agent is usually added. A halocarbon such as $CFCl_3$ is often used because it is just volatile enough to be vaporized by the heat generated by the exothermic polymerization reaction. Other desirable properties of the foaming agent are a low thermal conductivity, which increases the insulating value of the foam, and a slow rate of diffusion through the polymer cell walls.

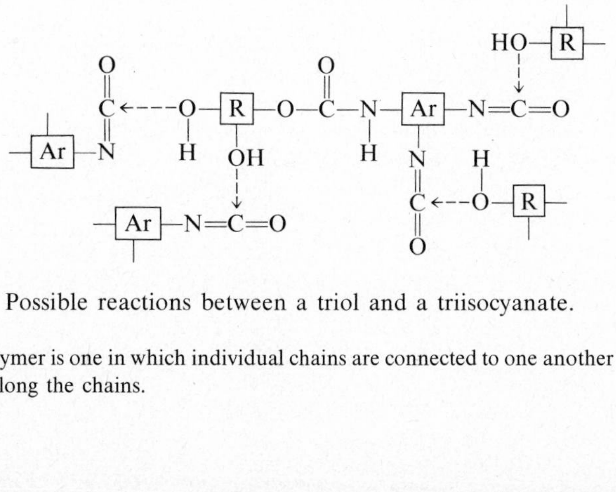

FIGURE 21.9 Possible reactions between a triol and a triisocyanate.

[5] A cross-linked polymer is one in which individual chains are connected to one another by covalent bonds at various points along the chains.

A great deal of chemistry and technology is involved in polyurethane foam production. For example, a *surfactant* (detergent) is usually added to the mixture of reactants and foaming agent to aid in the nucleation and stabilization of the bubbles in the foam because uniform small bubbles are critical to the structural and insulating properties of the foam. The rates of polymerization and bubble production must be precisely coordinated so that the bubbles are trapped by the hardening polymer at the optimum time. This is done by using a catalyst to control the rate of the polymerization reaction between the alcohol and the isocyanate functional groups.

In this experiment for the production of a rigid polyurethane foam, the alcohol and isocyanate components are not as simple as the triol and triisocyanate shown in Figure 21.9 but are actually a polyol and a polyisocyanate. The polyol is derived from *sucrose* (structure **11**, Chapter 22) by reaction of propylene oxide **(25)** with the hydroxyl groups of sucrose, as illustrated in equation 20. Although each sucrose

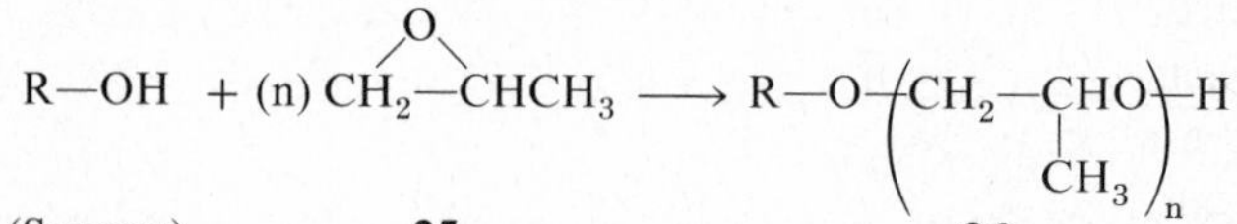

$$R\!-\!OH \;+\; (n)\, CH_2\!-\!CHCH_3 \;\longrightarrow\; R\!-\!O\!\!\left(\!CH_2\!-\!\underset{\underset{CH_3}{|}}{CH}O\!\right)_{\!n}\!\!\!H \qquad (20)$$

(Sucrose)   **25**   **26**

molecule has eight hydroxyl groups which might react according to equation 20, the reaction is controlled so that an *average* of four or five "propoxylated" groups are introduced per sucrose molecule to give a suitable polyol **(26)**. The polyol thus has the potential of reacting with isocyanate groups with its primary hydroxyl groups on the ends of the propoxyl chains, as well as the hydroxyl groups remaining on the sucrose moiety.

The polyisocyanate can be represented by a formula such as **27**. The reagent used

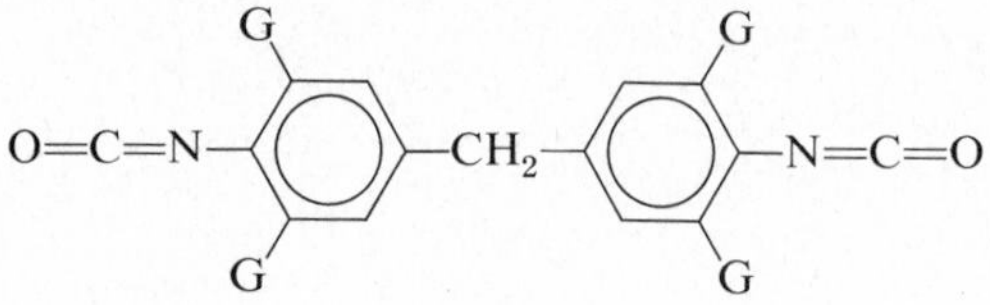

27

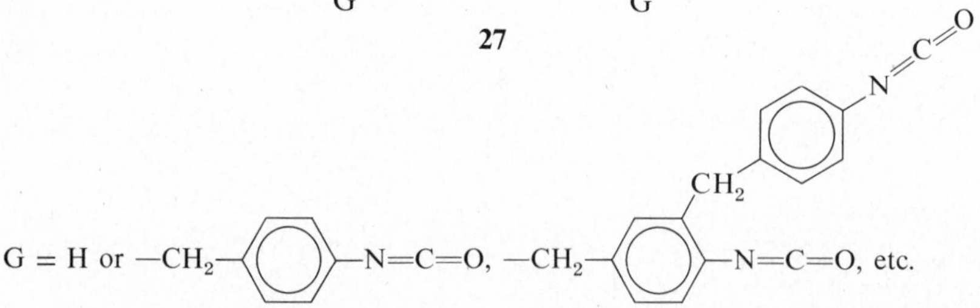

G = H or —CH₂—⟨○⟩—N=C=O, —CH₂—⟨○⟩—N=C=O, etc.

in this experiment is a complex one which consists mainly of molecules having three isocyanate groups per molecule (that is, in formula **27** three Gs may be Hs and one G

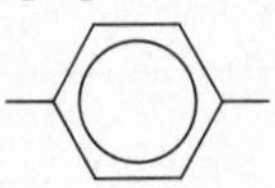

may be —CH₂—⟨○⟩—N=C=O) but which also contains molecules having two, four, or more isocyanate groups.[6] The polyfunctionality of both the alcohol and the isocyanate components of the reaction mixture leads to extensive cross-linking in the polyurethane product.

EXPERIMENTAL PROCEDURE

We recommend that this experiment be carried out by two or more students working as a team, for two reasons. First, the amounts of the major starting materials are rather large, and although the scale could be reduced, the amounts of the minor components are small and must be measured accurately and quickly, which would be more difficult on a smaller scale. Second, it will be helpful for the partners to cooperate in the weighings, observations, and timings involved. For the most efficient operation the partners should decide in advance what the duties of each will be.

After the reactants have been mixed at room temperature, four stages of the polymer foam formation are interesting to observe and time with a stopwatch or any watch with a second hand. In the industrial vernacular these are (1) *cream time,* the time of the first change in appearance of the complete reaction mixture to a light-brown creamy consistency as foaming begins; (2) *gel time,* the time when the rising foam changes from a bubble-filled liquid to an expanding gel and becomes sticky on the rising surface; (3) *tack-free time,* the time when the surface of the gel is no longer sticky or "tacky" to the touch; (4) *rise time,* the time when the foam has reached its maximum height.

The five components of the reaction mixture are to be weighed, in the exact sequence given, into a 400-mL beaker, using a top-loading balance which is accurate to 0.1 g. First record the tare weight of the beaker, then add 73.2 g of the polyol, POLY G-530 SA,[7] 1.0 g of the silicone surfactant, DC-193,[8] 1.0 g of the catalyst, THANCAT TD-33,[9] and 28.0 g of the fluorocarbon blowing agent, R-11B.[10] Record the total weight of the beaker and the first four components. Using a wooden tongue depressor, thoroughly mix the components, as in whipping cream, for 3–4 min until a creamy, emulsified mixture is obtained. Return the beaker to the balance, and add more fluorocarbon to bring the mixture back to the original weight (some of the volatile fluorocarbon will have been lost during the mixing); stir lightly. Now add 96.8 g of the polyisocyanate, PAPI.[11] *Note the time,* thoroughly stir the mixture for 10–15 sec, and then pour it into a 1-gal cardboard ice cream carton. Watch the surface

[6] Simple arylisocyanates are toxic, but the polyisocyanate used in this experiment is quite safe to handle.

[7] Olin Chemical Company. Other suitable propoxylated sucrose polyols are available from Union Carbide, Jefferson Chemical Company (a subsidiary of Texaco, Inc.), and Dow Chemical Company.

[8] Dow-Corning. A similar product is available from Union Carbide.

[9] Texaco, Inc. (a 33% solution of triethylene diamine in propylene glycol). A similar product is available from Air Products Company.

[10] DuPont's brand of $CFCl_3$.

[11] The Upjohn Company's polyarylpolyisocyanate.

of the liquid, and record the **cream time**. The foam will begin to rise almost immediately after this time; as the rate of rising decreases after about 70–80 sec, touch the surface of the rising foam with a clean tongue depressor and then draw it away. At **gel time** strings of foam will attach to the tongue depressor and stretch as it is pulled away. Repeat the touching of the surface until the foam fails to adhere to the tongue depressor; this is the **tack-free time**. Continue to observe the foam as its rate of rise decreases, and record the time that maximum height is finally attained; this is the **rise time**.

The formation of the rigid foam should be complete in an additional 2–5 min. After this time the firm foam can be cut with a razor blade to allow observation of the size, shape, and uniformity of the gas-filled cells.

EXERCISES

1. Write an equation for the reaction of glycerol with toluene diisocyanate (TDI), showing how a cross-linked polymer might be formed.

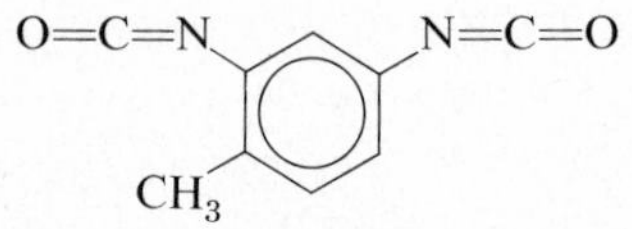

$$O=C=N \qquad N=C=O$$

$$CH_3$$

TDI

2. TDI is prepared from toluene, phosgene ($COCl_2$), and inorganic reagents. Indicate the steps in its synthesis.
3. A possible structure for a polyol made from sucrose and propylene oxide would be one produced by reaction of four moles of propylene oxide with each of the three primary alcohol groups in sucrose and with one of the secondary alcohol groups. Write a formula for this hypothetical polyol. The formula of sucrose is given in Section 22.3 (structure **11**).
4. A polyisocyanate such as the one used in this experiment is prepared from aniline, formaldehyde, phosgene ($COCl_2$), and inorganic reagents. Suggest possible steps in the synthesis.

22

CARBOHYDRATES

22.1 Introduction

Carbohydrates, also referred to as "sugars" or saccharides (Sanskrit, *sárkarā*, grit, gravel, sugar), are an extremely important class of naturally occurring polyhydroxy aldehydes (aldoses) and ketones (ketoses), or substances that yield aldoses or ketoses on hydrolysis. Many of the simplest carbohydrates, *monosaccharides,* display the general formula $C_nH_{2n}O_n$ [or $C_n(H_2O)_n$, a historical and misleading representation of the general formula, because it is hardly true that these substances are hydrates of carbon]. However, other monosaccharidic sugars, for example, the deoxy-sugars such as deoxyribose, and compounds containing heteroatoms such as nitrogen, sulfur, or phosphorus do not adhere to this formula.

More complex carbohydrates yield monosaccharides upon hydrolysis. Thus the disaccharide sucrose (table sugar) hydrolyzes to provide one molecule of D-glucose and one of D-fructose.[1] Polysaccharides, depending on their constitution, may be degraded to a mixture of monosaccharides or to only a single product; for example, starch is a mixture of the polymers amylose and amylopectin, each constituted only of D-glucose units. With particular reference to the structural and storage polysaccharides, it should be noted that polysaccharides are by far the most ubiquitous bioorganic compounds on earth. Although the experiments in this chapter necessarily utilize the very simplest of carbohydrates, the impression should be avoided that these may be the most abundant, or even the most significant, of the carbohydrates.

Carbohydrates provide the ultimate energy source in the food chain. D-glucose is synthesized in green leaves from carbon dioxide and water by the process of photosynthesis and the action of chlorophyll. This thermodynamically unfavorable process is made possible by the energy of sunlight. D-Glucose is combined in the plant to provide starch and cellulose. After ingestion, starch (and in some animals cellulose) is broken down again into D-glucose. One function of the liver is to recombine

---

[1] The D is a symbol used to designate the configuration of these sugars relative to that of D-glyceraldehyde, the standard of configuration for carbohydrates. See any modern organic textbook for further explanation.

D-glucose from the blood stream to form glycogen, which serves to store energy within the body. Glycogen as needed is reconverted into D-glucose, which is metabolized ultimately to carbon dioxide and water, providing the animal with the energy originally stored during photosynthesis. The other monosaccharides that are found, usually in combined form, in living systems are thought to be produced from D-glucose by the actions of various enzymes.

Carbohydrates are utilized within a living organism in many ways in addition to those of storage and transference of energy. In plants, for example, the polysaccharide cellulose is an important structural component providing rigidity and form. In animals, polysaccharides in combination with protein are important constituents of connective and other tissues. For example, the chondroitin sulfates are found in mammalian cartilages, tendons, heart valves, and cornea. A segment of the structure of chondroitin A, one of the three chondroitin sulfates that have been isolated, is shown here. Protein linkages are found at certain points along the chain of the polysaccharide.

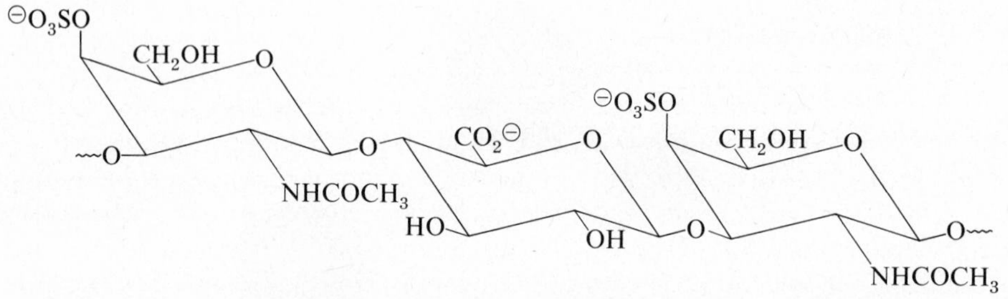

Moreover, carbohydrates serve as precursors in the biochemical formation of several other important bioorganic compounds. For example, they ultimately become involved in the biosynthesis of certain α-amino acids, the accepted pathway for which is shown in Figure 22.1. The pyridine derivative **3** is pyridoxine, or vitamin B_6. The Schiff bases **4** and **5** interconvert through enzyme-catalyzed tautomerization. Hydrolysis of either of these Schiff bases produces either the glycolysis product (**1** or **2**) or the nonessential amino acids alanine (**6**) or aspartic acid (**7**). Although this equation is vastly oversimplified because it omits the crucial role of enzymatic catalysis in all steps, it does serve to exemplify the significance of carbohydrates as precursors to a living system. Many other examples of the use of carbohydrates as biochemical "building blocks" may be cited; these include the occurrence of ribose and deoxyribose as structural constituents of nucleosides and deoxynucleosides necessary in the formation of RNA and DNA and the incorporation of ribitol in the biosynthesis of riboflavin, one of the B vitamins.

Although the foregoing discussion has been cursory, it provides an appreciation for the multiple functionality of carbohydrates in biochemistry and suggests that the proper and balanced "operation" of an organism as a chemical system is dependent on this class of compounds in many ways. In the following sections some of the chemical and physical properties of carbohydrates are investigated, and certain techniques used in the isolation and proof of structure of this interesting group of substances are described.

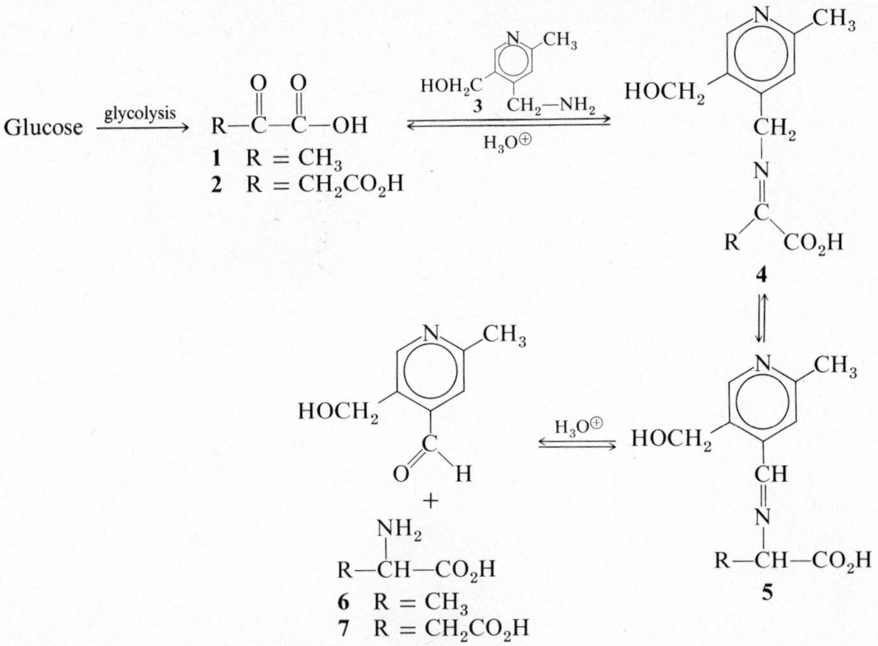

$$\text{Glucose} \xrightarrow{\text{glycolysis}} R-\overset{\overset{O}{\|}}{C}-\overset{\overset{O}{\|}}{C}-OH$$

1 R = CH$_3$
2 R = CH$_2$CO$_2$H

FIGURE 22.1 The chemical role of vitamin B$_6$ (**3**).

22.2 Mutarotation of Glucose

Figure 22.2 shows the open-chain structures in Fischer projection form of several of the more common monosaccharides.[2,3] Only a few of the monosaccharides shown are known to be naturally occurring; the remainder have been synthesized, however. By examination of the structures, it can be seen that monosaccharides generally contain more than one asymmetric carbon atom and are therefore subject to extensive stereoisomerism. An aldohexose, for example, has 4 asymmetric atoms and consequently may exist in 16 different stereoisomeric forms ($2^4 = 16$). Within this group of 16 isomers there are 8 enantiomeric pairs.

As is shown in Figure 22.2, D-glucose is one specific stereoisomer of the 16 aldohexoses. Consideration of the functional groups present in D-glucose and recollection that aldehydes react with alcohols to produce hemiacetals and acetals suggest that the open-chain form of this sugar might undergo an intramolecular reaction to produce a hemiacetal. Note that the conversion of an open-chain structure with its four asymmetric carbon atoms to a cyclic hemiacetal by, for example, reaction between the aldehyde group at C-1 and the C-5 hydroxyl group creates an additional

[2] Only the D isomers are shown. It may be noted that the sign of optical rotation, shown as ($+$) or ($-$), is *independent* of the relative configuration (D or L).

[3] Although the monosaccharides are shown in their *open-chain* form, those containing a chain of at least four carbon atoms exist predominantly in *cyclic* hemiacetal or hemiketal form.

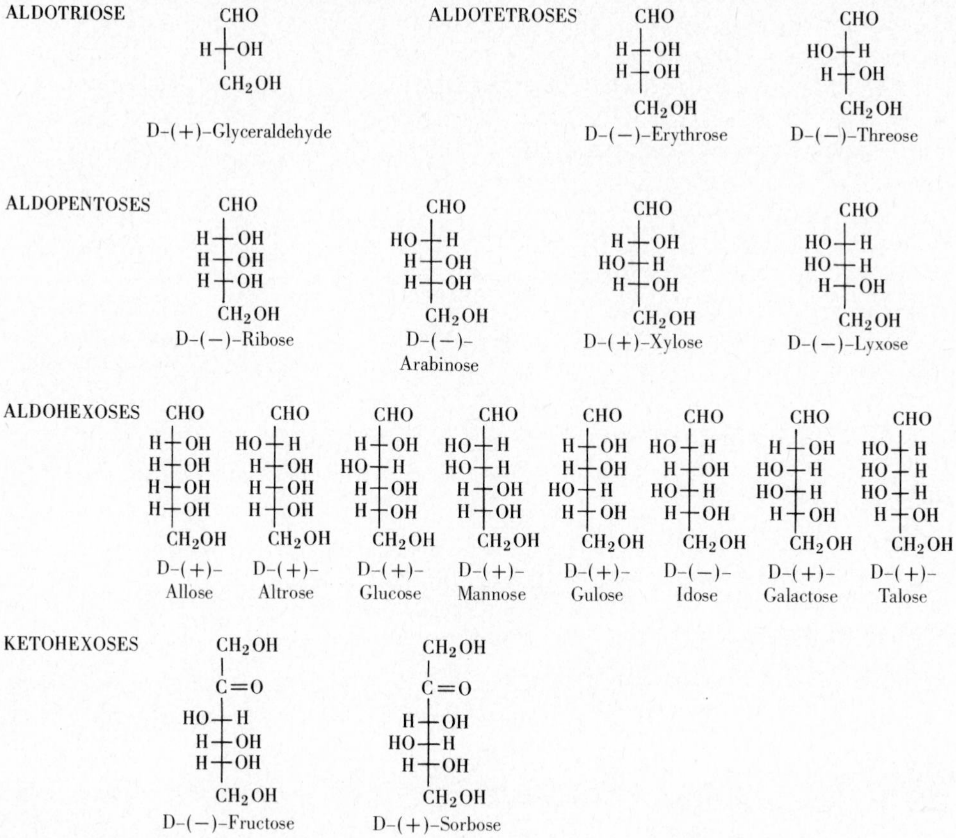

ALDOTRIOSE

```
        CHO
     H ─┼─ OH
       CH₂OH
```
D-(+)-Glyceraldehyde

ALDOTETROSES

```
        CHO              CHO
     H ─┼─ OH        HO ─┼─ H
     H ─┼─ OH         H ─┼─ OH
       CH₂OH            CH₂OH
```
D-(−)-Erythrose   D-(−)-Threose

ALDOPENTOSES

```
        CHO              CHO              CHO              CHO
     H ─┼─ OH        HO ─┼─ H         H ─┼─ OH        HO ─┼─ H
     H ─┼─ OH         H ─┼─ OH        HO ─┼─ H        HO ─┼─ H
     H ─┼─ OH         H ─┼─ OH         H ─┼─ OH         H ─┼─ OH
       CH₂OH            CH₂OH            CH₂OH            CH₂OH
```
D-(−)-Ribose   D-(−)-Arabinose   D-(+)-Xylose   D-(−)-Lyxose

ALDOHEXOSES

```
    CHO       CHO       CHO       CHO       CHO       CHO       CHO       CHO
 H ─┼─OH  HO ─┼─H   H ─┼─OH  HO ─┼─H   H ─┼─OH  HO ─┼─H   H ─┼─OH  HO ─┼─H
 H ─┼─OH   H ─┼─OH  HO ─┼─H  HO ─┼─H   H ─┼─OH   H ─┼─OH  HO ─┼─H  HO ─┼─H
 H ─┼─OH   H ─┼─OH   H ─┼─OH   H ─┼─OH  HO ─┼─H  HO ─┼─H  HO ─┼─H  HO ─┼─H
 H ─┼─OH   H ─┼─OH   H ─┼─OH   H ─┼─OH   H ─┼─OH   H ─┼─OH   H ─┼─OH   H ─┼─OH
   CH₂OH     CH₂OH     CH₂OH     CH₂OH     CH₂OH     CH₂OH     CH₂OH     CH₂OH
```
D-(+)-Allose   D-(+)-Altrose   D-(+)-Glucose   D-(+)-Mannose   D-(+)-Gulose   D-(−)-Idose   D-(+)-Galactose   D-(+)-Talose

KETOHEXOSES

```
      CH₂OH            CH₂OH
       |                |
       C=O              C=O
   HO ─┼─ H         H ─┼─ OH
    H ─┼─ OH       HO ─┼─ H
    H ─┼─ OH        H ─┼─ OH
      CH₂OH            CH₂OH
```
D-(−)-Fructose   D-(+)-Sorbose

FIGURE 22.2 Structures of several monosaccharides.

asymmetric center at C-1. Because there are two possible configurations about this new center, there must be possible two isomeric hemiacetals: **8**, the α-form, and **9**, the β-form (see Figure 22.3). Saccharides such as **8** and **9** that differ only in configuration at the carbon atom involved in the cyclization (C-1 for aldoses and C-2 for ketoses) are called *anomers*.

D-glucose does in fact exist in the two *diastereomeric* cyclic forms, **8** and **9**,

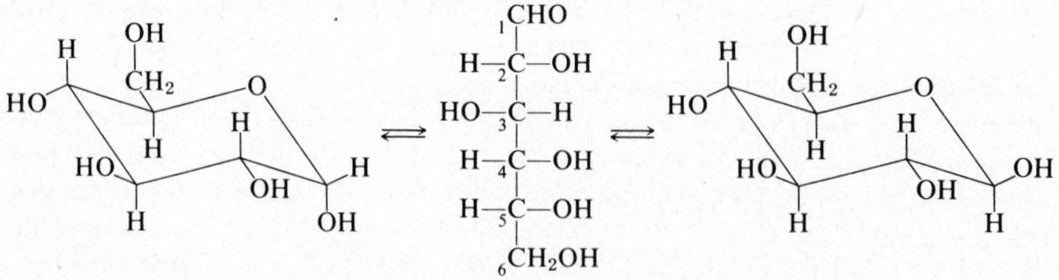

```
        ¹CHO
    H ─²C ─ OH
   HO ─³C ─ H
    H ─⁴C ─ OH
    H ─⁵C ─ OH
        ⁶CH₂OH
```

   8   10   9

FIGURE 22.3 Solution equilibria of D-glucose.

the two isomers being in equilibrium with one another in aqueous solution by way of the intermediacy of the open-chain structure **10** (Figure 22.3). The cyclic forms are the major components of the equilibrium mixture of **8, 9,** and **10**. Because compounds **8** and **9** are diastereomers rather than enantiomers, they have different physical properties and can be separated from one another by rather specific techniques of crystallization, as is done in the experiment.

The α-form has a specific rotation of $[\alpha]_D^{25°} + 112°$. When it is placed in water solution the specific rotation gradually changes until it reaches a constant $[\alpha]_D^{25°} + 52.7°$. The β-form, in water solution, undergoes a change of specific rotation from $[\alpha]_D^{25°} + 19°$ to $[\alpha]_D^{25°} + 52.7°$. Quite obviously, $+52.7°$ represents the specific rotation of an equilibrium mixture of **8** and **9**. The equilibration of diastereomeric isomers which gives rise to an equilibrium rotation is called *mutarotation*.

By following the rate of change of optical rotation of α- or β-D-glucose, information concerning the *rate* of mutarotation can be obtained. For a reaction of the type shown in equation 1, where A may be α-D-glucose **(8)** and B may be β-D-glucose,

$$A \underset{k_2}{\overset{k_1}{\rightleftarrows}} B \tag{1}$$

the following expression can be written.[4]

$$2.303 \log \frac{A_e - A_0}{A_e - A_t} = (k_1 + k_2)t \tag{2}$$

where A_e is the concentration of A at equilibrium, A_0 is the concentration of A at time $t = 0$, and A_t is the concentration at time t during the equilibration process. Note that the term $(A_e - A_0)$ represents the total extent of reaction from the beginning to the final establishment of equilibrium, that is, the total change in concentration of A, whereas the term $(A_e - A_t)$ represents the extent of reaction remaining at time t. We may substitute for these terms any other terms that also represent the ratio of the total extent of reaction and the extent remaining at time t. Any terms utilized must vary linearly during the course of reaction, as does concentration. Discussion in Chapter 19 indicates that optical rotation is linearly related to concentration; thus we may substitute the optical rotation of the solution measured at the beginning, end, and at time t for the concentrations in equation 2 to give equation 3:

$$2.303 \log \frac{\alpha_e - \alpha_0}{\alpha_e - \alpha_t} = (k_1 + k_2)t \tag{3}$$

Thus we see that measurement of the optical rotation of the initial solution of α-D-glucose, of the rotation of the same solution at equilibrium, and of the rotations at a series of times in between gives the data needed for calculation of $(k_1 + k_2)$. This is accomplished by plotting the left-hand term against t; the slope of the straight line obtained is $(k_1 + k_2)$. It is not possible to obtain k_1 and k_2 individually unless the value of the equilibrium constant is also known (see Experimental Procedure).

[4]The interested student can find the derivation of this expression in A. A. Frost and R. G. Pearson, *Kinetics and Mechanism,* 2d ed., John Wiley & Sons, New York, 1961, p. 186.

EXPERIMENTAL PROCEDURE

A.   Preparation of α-D-Glucose

DO IT SAFELY

The vapors of *glacial* acetic acid are extremely irritating to the nasal passages and the eyes. This material should be measured in the hood and brought to your desk only when you are ready for it. Also, glacial acetic acid will cause blistering of the skin. Wash off this material with cold water immediately, and apply a paste of sodium bicarbonate to the affected area for a few minutes if you accidentally allow any contact with the skin.

α-D-Glucose is the form obtained when D-glucose is crystallized slowly at room temperature from a mixture of acetic acid and water. Dissolve 50 g of anhydrous D-glucose (dextrose) in 25 mL of distilled water contained in a 500-mL Erlenmeyer flask. Heat the mixture on a steam bath to effect complete solution. The solution will be quite viscous and should be continuously stirred during this step. When solution is complete (*no* crystals remaining), remove the syrup from the steam bath, add 100 mL of *glacial* acetic acid that has been previously cooled in an ice-water bath to 17°,[5] and swirl the flask until the mixture becomes homogeneous. Stopper the flask with a cork, and allow it to remain undisturbed in the desk until the next laboratory period.

Collect the α-D-glucose by vacuum filtration, and wash the crystals thoroughly with 50–60 mL of 95% ethanol, followed by 50–60 mL of absolute ethanol. Either air-dry the crystals or dry them in an oven at approximately 80° for 1–2 hr.

B.   Rate of Mutarotation

Before coming to class, prepare in your laboratory notebook a format for recording the following information during the performance of this experiment: temperature, concentration, path length of sample tube, blank optical rotation, and the time of mixing of the solution. Also prepare a table for collecting a series of measurements of time and optical rotation.

Carefully fill with distilled water the sample tube for the polarimeter which is available for your use; be sure that no air bubbles remain trapped within. Place the sample tube in the polarimeter, and determine the blank reading of optical rotation for solvent. Record in your notebook the blank rotation and the length of the sample tube. Empty and carefully dry the sample tube.

Accurately weigh 10–15 g of α-D-glucose to the nearest 0.05 g, and transfer the sample *completely* to a dry 100-mL volumetric flask. Fill the flask with distilled water to within a few milliliters of the volumetric mark, tightly stopper the flask, and shake the contents to effect complete solution. The fine solid should dissolve completely within about 30 sec. When the solid is approximately one-half dissolved, note the time to the nearest minute and record it. As soon as the solid is dissolved, carefully fill

---

[5]Cooling below 17° may result in crystallization of acetic acid, whose freezing point is 16°.

the flask to the mark with distilled water, using a dropper. Again stopper and shake the flask until the solution is homogeneous (approximately 10–15 sec). Working carefully but rapidly, fill the polarimeter sample tube with this solution, leaving no air bubbles. Place a thermometer in the volumetric flask, and as soon as there is time, read and record the temperature. Insert the sample tube in the polarimeter, and determine the optical rotation of the solution. Record the rotation and the time of measurement to the nearest minute. During the next 6 or 7 min take a similar reading once each minute, recording the data obtained. Then take additional readings every 3 or 4 min for the next 40–50 min.

Approximately 4.5–5 hr will be required for the rotation to drop to within experimental error of the equilibrium value. It will be most convenient to store the remainder of the glucose solution in a tightly stoppered flask (to avoid evaporation) until the next laboratory period. At this time and using the same sample tube and polarimeter if possible, refill the sample tube with the equilibrated solution and take a final reading, which will be α_e.

Treatment of Data. **1.** Define the first recorded time at which a measurement was taken as time $t = 0$. The observed rotation at that time is then α_0. For each reading following the defined zero point, determine and record the elapsed time Δt (from $t = 0$) in minutes.

2. Perform the calculations needed for the graphical plot in the next operation.

3. Plot $2.303 \log (\alpha_e - \alpha_0)/(\alpha_e - \alpha_t)$ (vertically) versus elapsed time Δt (horizontally) on graph paper. Determine the slope of the best straight line that can be drawn through these points;[6] the value of the slope (in min^{-1}) is equivalent to the value $(k_1 + k_2)$.

4. From equation 4,[7] calculate the value of $(k_1 + k_2)$ to have been expected at the temperature at which the kinetic determination was made. Compare this with the value obtained experimentally.[8]

$$\log (k_1 + k_2) = 11.0198 - 3873/T(°\text{K}) \tag{4}$$

5. Plot α_t (vertically) versus Δt (horizontally) on a second sheet of graph paper. A curve should be obtained. Draw the best curved line through these points, and extrapolate the curve from time $t = 0$ back to the time of initial mixing. Using the extrapolated value of α obtained at that time, calculate the initial specific rotation, using equation 1 of Chapter 19, of the α-D-glucose used in the experiment. Remember, pure α-D-glucose has a specific rotation of $[\alpha]_D^{25°}$ $+112°$.

6. Using the value of the equilibrium constant, K_e, for the equilibrium between α- and β-D-glucose obtained by other workers, and the value of $(k_1 + k_2)$ obtained in this experiment, calculate the individual values of k_1 and k_2.

$$K_e = 1.762 = \frac{k_1}{k_2} \tag{5}$$

[6] See footnote 2 in Chapter 10.

[7] C. S. Hudson and J. K. Dale, *Journal of the American Chemical Society,* **39**, 320 (1917).

[8] From the authors' experience with this experiment, the experimental and calculated values of $(k_1 + k_2)$ should not differ by more than a factor of two.

C.   Specific Rotation of Saccharides

Prepare aqueous solutions of accurately known concentration (approximately 0.1 g/mL) of one or more pure sugars, as assigned by the instructor. Following the directions for the measurement of optical rotation given in part B, determine the specific rotations of each of these sugars.

EXERCISES

1. Would any different experimental results have been expected if one had started with β-D-glucose rather than with α-D-glucose? Explain.
2. From the specific rotation values of $+112°$ for α-D-glucose, $+19°$ for β-D-glucose, and $+52.7°$ for the equilibrium mixture, show by calculation that the value for the equilibrium constant given in equation 5 is correct.
3. Write a mechanism for the isomerization of α- to β-D-glucose.
4. An experimentally observed rotation appearing as $60°$ could just as correctly be interpreted as a rotation of $240°$, $420°$, $600°$, $-120°$, $-300°$, and so forth. How could it be determined which rotation is the correct one?

22.3   The Hydrolysis of Sucrose

Sucrose, a familiar foodstuff, is a carbohydrate having the structure **11**. As shown in the structure, the sucrose molecule is formed by a linkage between the hemiacetal form of α-D-glucose and one of the hemiketal forms of D-fructose (**12**). The obser-

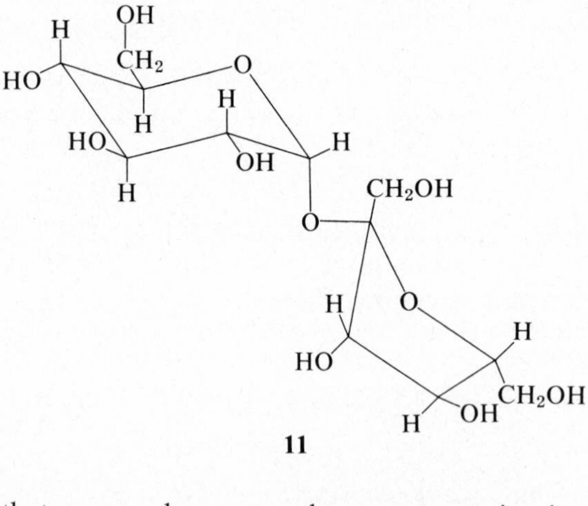

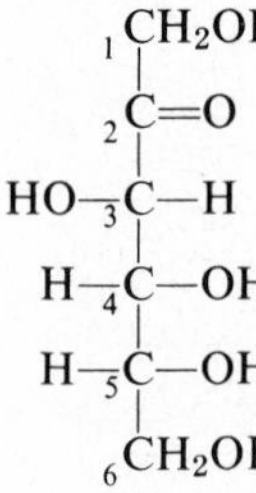

vation that sucrose does not undergo mutarotation is evidence that the linkage is through the glycosidic hydroxyl groups of each sugar.

Sucrose, $[\alpha]_D^{25°}$ $+66.5°$, undergoes acid-catalyzed hydrolysis to give a mixture of D-glucose and D-fructose (equation 6). Under the conditions of the hydrolysis the

$$\text{Sucrose} \xrightarrow{\text{H}_3\text{O}^\oplus} \text{D-Glucose} + \text{D-Fructose} \qquad (6)$$

glucose undergoes rapid mutarotation to give the equilibrium mixture of α- and β-forms, $[\alpha]_D^{25°}$ +52.7°, whereas the fructose is formed as an equilibrium mixture of the isomers shown in Figure 22.4 and has $[\alpha]_D^{25°}$ −92°. Isomer 13, a six-membered ring hemiketal, predominates in the equilibrium mixture. The observation that the sign of optical rotation changes on hydrolysis of sucrose has led to the name *invert sugar* for the product mixture. The enzyme *invertase* accomplishes the same chemical result as does the acid-catalyzed hydrolysis of sucrose.

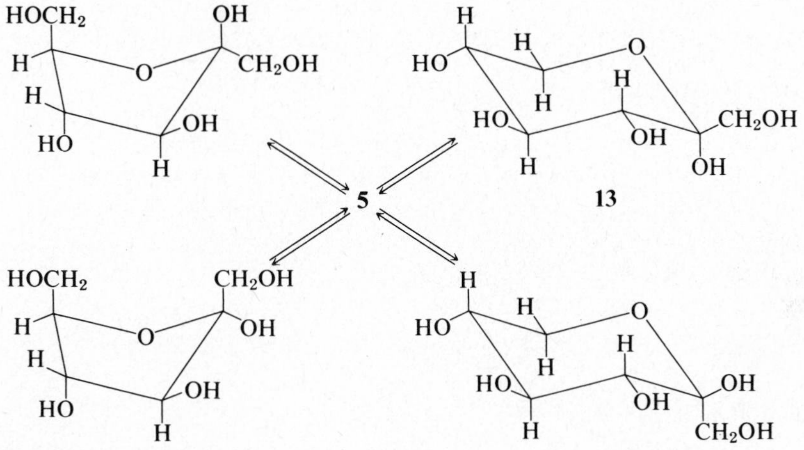

FIGURE 22.4 Solution equilibria of D-fructose.

EXPERIMENTAL PROCEDURE

Place 15 g of pure sucrose, whose weight has been accurately determined, in a 250-mL round-bottomed flask. Add about 80 mL of water, swirl the contents of the flask to effect solution, and then add about 0.5 mL of concentrated hydrochloric acid. Heat the solution at reflux for about 2 hr. During this time determine the specific rotation of sucrose following the general directions of parts B and C of the Experimental Procedure in Section 22.2 and the more specific directions of your instructor regarding the use of a polarimeter.

At the end of the period of reflux cool the reaction mixture to room temperature, and carefully transfer *all* the solution to a 100-mL volumetric flask. Use small amounts of water to rinse the round-bottomed flask, and add the rinses to the rest of the solution. Dilute the mixture to a volume of 100 mL. Using a polarimeter, determine the specific rotation of the product mixture from the hydrolysis of sucrose. Compare this value with the specific rotation of sucrose which you determined earlier.

EXERCISES

1. Explain the change in sign of the optical rotation of sucrose following hydrolysis. Calculate the specific rotation of *invert* sugar from the known equilibrium

rotations of D-glucose and D-fructose. How does this number compare with that determined experimentally?

2. In what way is the specific rotation of invert sugar analogous to the specific rotation of a racemic mixture?

22.4 Isolation of α,α-Trehalose

Trehalose is the name given to the D-glucosyl-D-glucosides. It derives from the isolation of α,α-trehalose, **14**, from the trehala manna, an oval shell built by certain insects, which has been shown to consist of 25–30% α,α-trehalose. The anomeric forms α,β- and β,β-trehalose have not been found in nature.

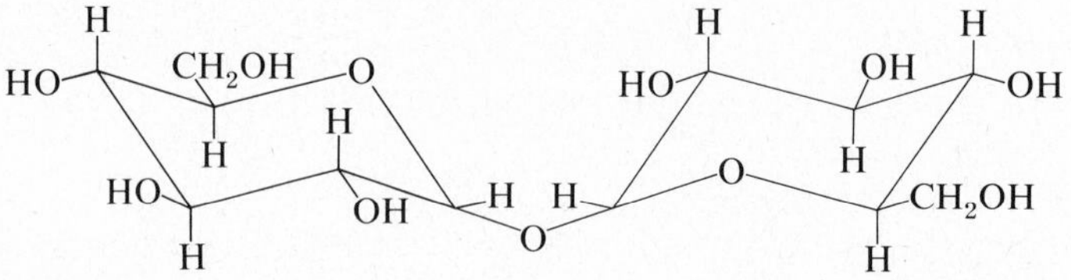

14

This disaccharide was probably first isolated from the ergot of rye, a fungus, in 1832. It has since been shown to occur in other fungi, bacteria, the blood of insects, certain algae and lichens, some of the higher plants, such as the resurrection plant, and yeast, as well as the trehala manna. α,α-Trehalose also occurs in combined form in human tubercle bacilli. The lipids from these bacilli may be separated into free fatty acids and natural fats, the fats containing no glycerol and being esters of fatty acids with α,α-trehalose. These types of fats are termed *microsides:* esters of sugars with fatty acids.

This experiment involves the isolation of α,α-trehalose from dried baker's yeast, in which its content may reach 10–15%. α,α-Trehalose is formed biosynthetically by yeast enzymes from D-glucose and stored within the cell, as is glycogen. These cells also contain an enzyme, trehalase, capable of enzymatically hydrolyzing the glycosidic linkage. Proliferating and active yeast cells utilize available D-glucose almost exclusively, producing carbon dioxide and water, and metabolize stored α,α-trehalose only after D-glucose has been consumed. Older yeast cells, on the other hand, ferment α,α-trehalose at least as fast as, and perhaps faster than, D-glucose. These observations probably offer at least a partial explanation for the observation that the α,α-trehalose content of baker's yeast decreases on storage.

Interestingly, baker's yeast has been shown to ferment α,α-trehalose, which has been added to the yeast, while leaving unaffected the α,α-trehalose stored within the yeast. Apparently there is a spatial separation in the cell between trehelase and its stored trehalose. There is some evidence that the enzyme may be at the cell surface.

The procedure for the isolation of trehalose from yeast offers interesting insight into the requirements for separating cellular components. Most of the materials extracted from the yeast are systematically removed before the trehalose is finally precipitated. Insoluble materials, such as polysaccharides, fibrous protein, and so forth, and various aromatic compounds, such as aromatic amino acids and heterocyclic compounds, are separated from the extract by filtration and treatment with activated charcoal. Globular proteins, of which the enzyme trehalase is one, are removed by heating to cause coagulation (denaturation) followed by precipitation as their insoluble zinc salts. Phosphorylated sugars are separated from the extract through the addition of barium hydroxide and filtration of their insoluble barium salts. Thus at the final precipitation α,α-trehalose is essentially the only ethanol-insoluble component in the extract.

EXPERIMENTAL PROCEDURE

A. Isolation of α,α-Trehalose

Prepare a paste from 32 g of dried baker's yeast and 68 mL of water. Add 250 mL of 95% ethanol and, with occasional stirring, allow the mixture to stand for about 30 min. Filter the mixture by vacuum filtration, and wash the filter cake with three 30-mL portions of 70% ethanol. Combine the washings with the main solution. To the filtrate add 20 mL of 1.2 M aqueous zinc sulfate, 1 mL of 1% phenolphthalein solution, and a sufficient quantity of saturated barium hydroxide solution to make the solution basic (about 50 mL will be required). Add 2 g of activated charcoal, and heat the mixture to 70° on a steam bath. Filter the hot solution through a Büchner funnel previously layered with filter-aid. (To form the layer of filter-aid, filter a slurry of filter-aid in 95% ethanol through the funnel, and then discard the ethanol.) Adjust the filtrate to about pH 7 with 0.1 M hydrochloric acid, and concentrate the solution to approximately 10 mL by *gentle* heating under vacuum. Slowly stir 80 mL of 95% ethanol into the resulting syrup, stopper the flask, and leave it undisturbed. Crystals will normally form within a couple of days; however, a week or more may occasionally be required. The crystals are frequently quite large, owing to their slow growth. If desired, crystallization may be hastened by the addition of a little more ethanol or by cooling in an ice-water bath.

Verify that you have successfully isolated α,α-trehalose by determination of its decomposition point (203°; dihydrate, 97°) and by measurement of its specific rotation (see Section 19.1 and parts B and C of the Experimental Procedure in Section 22.2); the $[\alpha]_D^{20}$ of α,α-trehalose is reported to be +178.3° (H_2O). Hydrolyze a portion of the trehalose by preparing a 0.5% solution in 1 M HCl and boiling for 20 min. That D-glucose is the sole monosaccharidic constituent of trehalose may be demonstrated by thin-layer chromatographic analysis (part B).

B. Thin-Layer Chromatography of Monosaccharides[9]

Obtain a 12-cm strip of cellulose chromatogram sheet (without fluorescent indicator). Spot the strip about 1 cm from the bottom with the dilute solution of an

---

[9] For an excellent compilation of techniques and procedures for the thin-layer chromatographic analysis of saccharides and their derivatives, consult B. A. Lewis and F. Smith, in *Thin-Layer Chromatography,* 2d ed., E. Stahl, editor, Springer-Verlag, New York, 1969, Chapter 10.

unknown sugar or sugar mixture and also with solutions of any desired known sugars for comparison purposes (use glucose for the trehalose hydrolysate). Spots should be separated by about 1 cm. Develop the plate, using as developing solvent a mixture of pyridine-ethyl acetate-acetic acid-water in the respective ratios 5:5:1:3. Development may require nearly 1.5 hr. The spots may be visualized by either spraying with *p*-anisidine phthalate reagent[10] or by leaving the plate in contact with iodine vapor. With the spray reagent, hexoses yield green spots and pentoses give red-violet spots after heating the plate at 100° for 10 min. Record the results by drawing a picture of the developed plate in your notebook.

---

[10] This reagent spray, for reducing sugars, is prepared as a solution of 1.23 g of *p*-anisidine and 1.66 g of phthalic acid in 100 mL of 95% ethanol.

23

AMINO ACIDS
AND PEPTIDES

23.1 Introduction

The amino acids constitute a highly important class of naturally occurring organic compounds. They are the monomeric units which are joined through amide linkages, called peptide bonds, to produce the important biopolymers on which every living system depends: the proteins **(1)**.

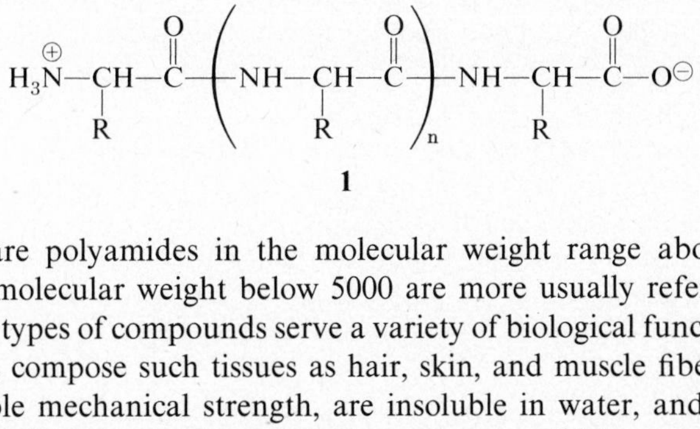

$$\overset{\oplus}{H_3N}-CH-\overset{\overset{\displaystyle O}{\|}}{C}\left(NH-CH-\overset{\overset{\displaystyle O}{\|}}{C}\right)_n NH-CH-\overset{\overset{\displaystyle O}{\|}}{C}-O^{\ominus}$$

$$\underset{R}{\qquad}\underset{R}{\qquad}\underset{R}{\qquad}$$

1

Proteins are polyamides in the molecular weight range above 5000; those polyamides of molecular weight below 5000 are more usually referred to as polypeptides. These types of compounds serve a variety of biological functions. Some, the *fibrous proteins,* compose such tissues as hair, skin, and muscle fiber. They possess quite appreciable mechanical strength, are insoluble in water, and chemically are relatively inert. Fibrous proteins generally possess very high, somewhat indefinite molecular weights and are sometimes polymerlike. Others, the *globular proteins* and smaller natural *peptides,* have much smaller molecular weights and exist as discrete chemical entities, often obtainable in crystalline form. They are water soluble and have characteristic reactivity. The globular proteins serve a variety of roles ranging from catalytic functions (enzymes) and overall regulatory functions (hormones) through immunological defense functions (antibodies). To a limited extent, differences in function are reflected in differences in molecular weight; for example, compare fibrous and globular proteins. However, particularly among the globular proteins, differences in biological properties are more completely determined by the exact sequence of different amino acids in the peptide chain: the *primary structure.*

The number of such possible arrangements is vast. For example, in a penta-peptide (**1**, n = 3) composed of five different amino acids, the number of different sequential arrangements of amino acids is 120. Each of these in principle would possess different biochemical reactivities. However, the biological properties of a peptide or protein is made more complex as a result of its three-dimensional structure.

Although each individual amide linkage is planar, owing to conjugation as shown in **2**, conformational differences in structure may arise through rotation about the remaining single bonds (those to C_α in **2**). These rotations allow the chain to coil

$$C_\alpha \diagdown \overset{\delta\oplus}{N}\diagup H$$
$$\underset{\delta\ominus}{\overset{}{O}}\diagdown C = N \diagdown C_\alpha$$

2

and to achieve stabilization through hydrogen bonding between amido hydrogens and carbonyls on separated peptide units. Such coiling constitutes the *secondary structure* of peptides.[1] Furthermore, owing to convolutions and gross foldings of the coiled chain, amino acid residues (peptide units) in widely separated positions of the chain may be brought into close proximity, acting together in concert and providing the peptide with its characteristic reactivity. This folding, which may be the result of a variety of structural influences, constitutes the *tertiary structure* of the peptide. Finally, the spatial relationship of one polypeptide chain to another results in the *quaternary structure* of the overall protein structure.[2]

A complete understanding of the biochemical behavior of a peptide, from a molecular and mechanistic point of view, must depend on the determination of its total structure. Of the various levels of structural complexity, determination of the primary structure (sequencing) must be deemed the most important. This is a logical consequence of the realization that the higher degrees of structural complexity are in the first instance dependent on the primary structure. It is this sequence which is genetically coded in DNA. Determination of the primary structure requires initial knowledge of the numbers of each kind of amino acid involved in the chain. This may be accomplished by the total hydrolysis of the peptide to provide a mixture of the amino acids constituting the structure (equation 1).

$$\overset{\oplus}{NH_3}-\underset{R_1}{CH}-\overset{O}{\overset{\|}{C}}-NH-\underset{R_2}{CH}-\overset{O}{\overset{\|}{C}}-NH-\underset{R_3}{CH}-\overset{O}{\overset{\|}{C}}-O^\ominus \xrightarrow{H_3O^\oplus}$$

$$\overset{\oplus}{NH_3}-\underset{R_1}{CH}-CO_2{}^\ominus + \overset{\oplus}{NH_3}-\underset{R_2}{CH}-CO_2{}^\ominus + \overset{\oplus}{NH_3}-\underset{R_3}{CH}-CO_2{}^\ominus \qquad (1)$$

[1] See any modern organic textbook for further information about the three-dimensional structural properties of peptides.

[2] For example, the tobacco mosaic virus, with an overall molecular weight of 41,000,000, is composed of many identical polypeptide subunits, each with a molecular weight of 17,500, held together by noncovalent interactions.

Qualitative and quantitative analysis of this mixture then determines the identities and relative numbers of each amino acid present. If the molecular weight is known, then the exact numbers of each amino acid in the chain may be determined. For example, if the hydrolysis of a peptide of unknown structure provided a mixture of amino acids analyzed to contain only alanine (ala) and glycine (gly, see Table 23.1) in a ratio of 2:1, respectively, the peptide could be a tripeptide (ala$_2$, gly),[3] a hexapeptide (ala$_4$, gly$_2$), and so on. If the unknown peptide was found to have a molecular weight of approximately 200, a general structure of unknown sequence would be established. Note that the molecular weight of (ala$_2$, gly) is 203 (2 ala + gly − 2 H$_2$O), whereas the molecular weight of (ala$_4$, gly$_2$) is 388 (4 ala + 2 gly − 5 H$_2$O). Experimental approaches to the determination of sequence will be discussed in Section 23.3.

Peptide bonds may be hydrolyzed under either acid- or base-catalyzed conditions. Although both procedures suffer from some disadvantages, the acid-catalyzed hydrolysis is preferable, primarily because alkaline conditions result in extensive racemization of the chiral center at the α-position as well as in degradation of some amino acid residues, for example, arginine (arg) and threonine (thr, see Table 23.1). Although certain amino acid residues are sensitive to acid and undergo partial destruction during acid-catalyzed hydrolysis (serine and tryptophan, for example), these effects are well understood, and quantitative corrections may be applied during careful work in the research laboratory. The mechanism of acid hydrolysis of the peptide linkage, which is simply that of an amide, is qualitatively similar to that for the hydrolysis of an ester (see equation 5, Section 15.2).

Total hydrolysis of a peptide is normally accomplished by treatment with 6 M HCl at 100–110° over a period of 16–20 hr. The hydrolysis is effected in a sealed tube in order to avoid evaporation and the charring that would result. An experiment involving peptide hydrolysis in the determination of structure of an unknown dipeptide is included in the experimental part of Section 23.3.

23.2 Analysis of Amino Acids

Although the number of conceivable structures containing both amino and carboxylic acid functional groups on the same carbon atom is vast indeed, fortunately only 20 or so are actually found in polypeptides from living sources. Nearly all these are α-amino acids of the type shown in **3**. The occasional exception contains an α-amino function as part of a ring, for example, proline. All have an α-hydrogen, so that the α-carbon atom is asymmetric except in the case of glycine, the simplest α-amino acid, in which there are two α-hydrogen atoms. With the exception of glycine and of a few D-amino acids derived from microorganisms, all the important amino acids found in polypeptides from living sources are of the L-configuration **(4)**. Table 23.1 includes many of the common, naturally occurring amino acids.

---

[3]The formula indicates that the peptide is composed of two alanine units and one glycine unit. The comma, by convention, indicates that the *sequence* of these units in the peptide is *unknown*.

TABLE 23.1   THE COMMON AMINO ACIDS

Name	Abbreviation	Formula	Isoelectric Point	Color from Pyridine-Isatin Reagent	Numerical Key to Figures 23.3 and 23.4
Alanine	ala	$CH_3CH(NH_2)CO_2H$	6.0	Pink-red	10
Arginine	arg	$H_2N-C(=NH)-NH(CH_2)_3CH(NH_2)CO_2H$	11.2	Deep pink	4
Asparagine	asn	$NH_2COCH_2CH(NH_2)CO_2H$	5.4	...	...
Aspartic acid	asp	$HO_2CCH_2CH(NH_2)CO_2H$	2.8	Bright red	6
Cysteine	cys	$HSCH_2CH(NH_2)CO_2H$	5.1	Yellow-brown	1
Glutamic acid	glu	$HO_2C(CH_2)_2CH(NH_2)CO_2H$	3.2	Bright red	9
Glutamine	gln	$NH_2CO(CH_2)_2CH(NH_2)CO_2H$	5.7	...	...
Glycine	gly	$H_2NCH_2CO_2H$	6.0	Orange-red	7
Histidine	his	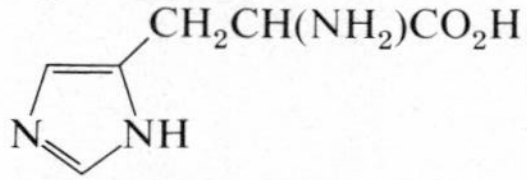 $CH_2CH(NH_2)CO_2H$ (imidazole ring)	7.5	Orange-red	3
Isoleucine	ile	$CH_3CH_2CH(CH_3)CH(NH_2)CO_2H$	6.0	Bright red	16
Leucine	leu	$(CH_3)_2CHCH_2CH(NH_2)CO_2H$	6.0	Orange-red	18
Lysine	lys	$NH_2(CH_2)_4CH(NH_2)CO_2H$	9.6	Red	2
Methionine	met	$CH_3S(CH_2)_2CH(NH_2)CO_2H$	5.7	Pink	14
Phenylalanine	phe	$C_6H_5CH_2CH(NH_2)CO_2H$	5.5	Red-brown	17
Proline	pro	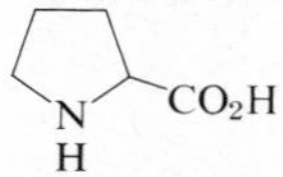 (pyrrolidine ring with CO_2H)	6.3	Intense blue	11
Serine	ser	$HOCH_2CH(NH_2)CO_2H$	5.7	Pink	5
Threonine	thr	$CH_3CH(OH)CH(NH_2)CO_2H$	5.6	Pink	8
Tryptophan	try	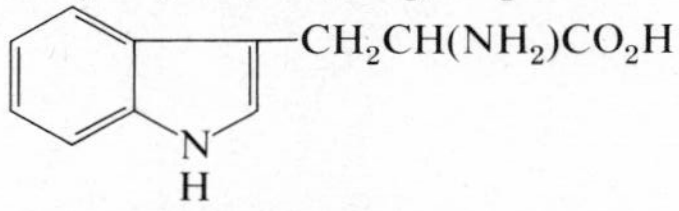 $CH_2CH(NH_2)CO_2H$ (indole ring)	5.9	Red-brown	15
Tyrosine	tyr	$p\text{-}HOC_6H_4CH_2CH(NH_2)CO_2H$	5.7	Light brown	12
Valine	val	$(CH_3)_2CHCH(NH_2)CO_2H$	6.0	Red	13

$$\underset{\textbf{3}}{H_3\overset{\oplus}{N}-\underset{\underset{R}{|}}{CH}-CO_2^{\ominus}} \qquad \underset{\textbf{4}}{\underset{R}{\overset{\overset{\oplus}{N}H_3}{\underset{|}{\overset{|}{H\cdots C}}}}-CO_2^{\ominus}}$$

Amino Acids as Acids and Bases. Note that the amino acids have both basic and acidic functional groups. As a result, in their crystalline forms they exist as zwitterions (internal salts, for example, **3**). Consequently they are high-melting solids that are generally insoluble in organic solvents but soluble in water.

Because of the acidic and basic character of an amino acid, there are established in aqueous solution pH-dependent equilibria among the forms shown in equation 2.

$$\underset{\textbf{5}}{\overset{\oplus}{N}H_3-\underset{\underset{R}{|}}{CH}-CO_2H} \rightleftharpoons \underset{\textbf{6}}{NH_2-\underset{\underset{R}{|}}{CH}-CO_2H} \rightleftharpoons$$

$$\underset{\textbf{7}}{\overset{\oplus}{N}H_3-\underset{\underset{R}{|}}{CH}-CO_2^{\ominus}} \rightleftharpoons \underset{\textbf{8}}{NH_2-\underset{\underset{R}{|}}{CH}-CO_2^{\ominus}} \tag{2}$$

The equilibria are displaced toward **8** in more alkaline solutions; in more acidic solutions **5** becomes more predominant. The equilibrium between **6** and **7** results in no change in hydrogen ion concentration; thus the ratio of **7** to **6** in solution is pH independent. The pH-dependent component of the equilibria in equation 2 is the relative concentrations of the species **5** and **8**. If electrodes are placed in a solution of an amino acid, there will be a net migration of the solute toward either the cathode or the anode, depending on whether **5** or **8**, respectively, is predominant.

At a certain pH, specific for each amino acid, the concentrations of **5** and **8** will be equal, and there will be no net migration of the solute toward the electrodes. This pH value is called the *isoelectric point*. Table 23.1 lists the isoelectric points for the common amino acids. Note that the values fall into three ranges: 2–3 for those amino acids containing additional acid groups as part of the side-chain R (the acidic amino acids), 5.5–6.5 for those amino acids containing neutral side chains, and 9–11 for those amino acids containing an additional basic site in the side chain (the basic amino acids). It should be noted that it is at the isoelectric point that amino acids have their *minimum* solubility in water.

Chromatographic techniques are utilized nearly universally for the analysis of mixtures of amino acids. In order of importance, as gauged from work carried out in research laboratories, these procedures involve ion exchange chromatography, paper chromatography, and thin-layer chromatography. Because amino acids are colorless, each of these techniques necessarily requires methods of detecting the separated amino acids. The most important detecting agent in use is ninhydrin (**9**).

The Ninhydrin Color-forming Reaction of Amino Acids. Ninhydrin reacts with amino acids of type **3** to produce characteristic blue-violet colors. The sensitivity and reliability of the test are such that 0.1 μmol of amino acid gives a color intensity that is reproducible to a few percent, so long as a reducing agent such as stannous chloride is present to prevent oxidation of the colored salt by dissolved oxygen. Although not all amino acids give the same color (for example, proline gives a pale-yellow color), most do, indicating that the colored product formed is the same in most cases, irrespective of the structure of the original amino acid. The sequence of steps involved in the color-forming reaction is shown in Figure 23.1.

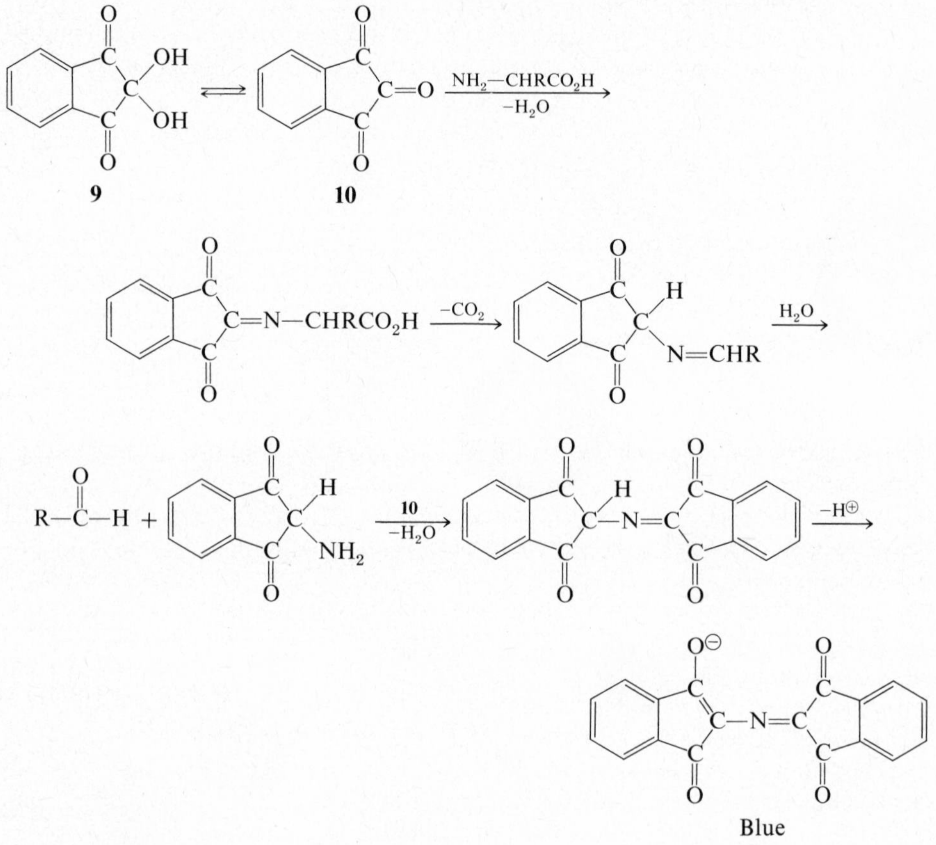

FIGURE 23.1 Chemistry of the ninhydrin color test.

The Automatic Amino Acid Analyzer. In the past, quantitative amino acid analyses were highly time-consuming, extremely tedious and required considerable amounts of peptide (about 25 g of protein for a full amino acid analysis). More recently, however, the use of ion exchange chromatography in conjunction with the *automatic amino acid analyzer* has revolutionized the practice of protein analysis. A complete analysis on automated equipment may be completed in 4–5 hr on no more than

0.1 μmol of protein! The amino acids are separated by elution ion exchange chromatography. The eluent is passed through the column at a constant rate. Because the amino acids elute at different rates, the flow of eluent at the base of the column contains different amino acids at different elapsed time intervals. As it leaves the column, the effluent is admixed with a solution of ninhydrin and then passed through a Teflon tube immersed in a boiling-water bath to speed up the color-forming reaction (Figure 23.1). The eluent stream then continues through a photoelectric colorimeter which continuously measures the color intensity; an electronic recorder is used to record the color intensity as a function of time. Because the color intensity in the effluent stream as a function of time is directly related to the elution times for the various amino acids, the recorder produces a chromatogram which is qualitatively similar to a gas chromatogram (Section 3.4). When correction factors are applied which relate the sensitivity of each amino acid to the ninhydrin color-forming reaction, the areas under the peaks are proportional to the molar ratios of the amino acids in the original mixture. Thus this procedure allows both qualitative and quantitative analyses of mixtures of amino acids.

The ion exchange resin employed consists of a sulfonated cross-linked polymer produced by copolymerization of styrene and *para*-divinylbenzene (Figure 23.2).

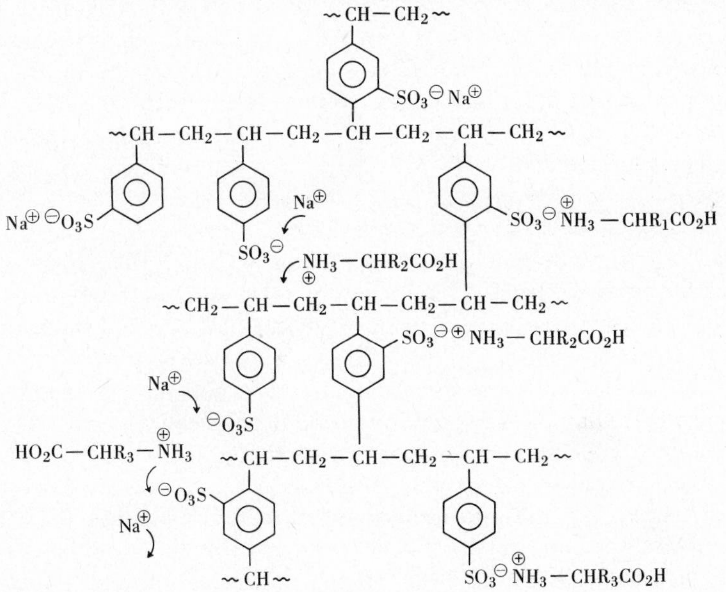

FIGURE 23.2 Cationic exchange chromatography of amino acids. The arrows represent displacement of one ion by another as the eluent passes down the column.

The sample is applied as a solution buffered at pH 2 to a column containing this resin. At this pH, below the isoelectric point of all amino acids, the amino acids are present in the conjugate acid form **5**. Acting as cations, they displace sodium ions and are held at the head of the column by the sulfonate groups of the resin. An empirical pattern of elution involving aqueous buffered solutions of sodium ion of increasing pH has been worked out that allows separation of all the common amino acids as a result of their individual abilities to be displaced from the resin by sodium ion. Although the factors that control resolution of the amino acids are complex, to a first approximation the differential elution pattern of two amino acids is controlled by the relative extent to which, at a given pH, they are present in the cationic form **5**. The more basic amino acids would then be expected to be eluted at higher pH values than the neutral or acidic amino acids.

Paper Chromatography of Amino Acids.   Paper chromatography is an especially valuable tool for the relatively rapid qualitative analysis of mixtures of amino acids. As noted in Chapter 3, paper chromatography is a type of partition chromatography in which the substrate is partitioned between the water, which is tightly bound within the cellulose fibers of paper (the stationary phase), and an organic solvent, which is allowed to migrate upward across the paper by capillary action (the mobile phase). Those acids having the highest solubility in the organic solvent relative to their solubility in water will have the greatest mobility and will migrate upward on the paper most rapidly. Because different amino acids will migrate at different rates, they will separate at different vertical displacements on the paper. Although the R_f values (see Figure 3.12) that may be used to identify the amino acid components present in the mixture are fairly reproducible, there is sufficient variation in the quality and type of paper used, in the temperature from run to run, and in the purity and exact composition of solvent as to make sole reliance on the values published by other workers somewhat risky. Consequently it is standard practice to run samples of known amino acids simultaneously with the unknown mixture to compare R_f values under identical conditions.

A very large variety of organic solvent systems and types of paper have been investigated for the purpose of separating mixtures of amino acids. The solvent and paper used are generally defined by the nature of the particular determination to be performed; that is, the experimental conditions are dependent on which amino acids are present in the mixture.

Because the separability of different amino acids depends greatly on the solvent system used, it is frequently found that some amino acids will separate into individual spots while others will remain unresolved in overlapping spots (see Figure 23.3a). The use of two-dimensional paper chromatography is standard practice to overcome the problem of overlapping. In this procedure the mixture is spotted on the paper in the lower left-hand corner, and the paper is developed by irrigation with one solvent. The paper is then removed from the developing chamber and allowed to dry. It is then turned 90° in orientation to the original direction of solvent flow and developed with a different solvent, chosen for its ability to separate the amino acids that did not separate with the first solvent. Thus the spots are displaced in two directions from the original spot rather than in one.

Figure 23.3 shows idealized reproductions of actual paper chromatograms. In Figure 23.3a a mixture of 18 amino acids has been developed in one-dimension using a mixture of 1-butanol-acetic acid-water as solvent in a ratio of 4:1:5 (by volume). Note that although there is significant separation of amino acids, several of the spots

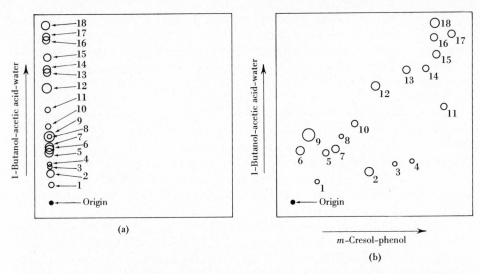

FIGURE 23.3 Paper chromatography of a mixture of 18 amino acids. (a) In one dimension. (b) In two dimensions using different solvents.

remain unresolved and overlapped [serine (5), aspartic acid (6), and glycine (7), for example]. In Figure 23.3b the same chromatogram is turned 90° and irrigated with *m*-cresol-phenol in a 1:1 (by weight) ratio buffered with a pH 9.3 borate solution. Observe that all the spots have now been resolved by the two-dimensional chromatography. The numbers that identify the amino acid responsible for each spot are keyed in Table 23.1.

Although two-dimensional paper chromatography provides greater resolution of the amino acids in a mixture, it suffers the disadvantage of allowing only one sample to be run at a time. Because it is desirable to run samples of known amino acids simultaneously with the unknown to facilitate identification, the following experiment involves only one-dimensional chromatography. The solvent system 1-butanol-acetic acid-water (4:1:5) is quite good for this purpose and is used.

EXPERIMENTAL PROCEDURE[4]

DO IT SAFELY

1. The solvent mixture used in the development of the paper chromatogram in this experiment may be somewhat irritating to nose and eyes. Avoid excessive exposure to these vapors or work in a hood if possible.

[4]The experimental directions provided are adapted in part from those given by L. Slaten, M. Willar, and Sister A. A. Green, *Journal of Chemical Education,* **33,** 140 (1956), which may be consulted for additional experimental details.

2. When spraying the chromatogram with either the ninhydrin or the isatin reagent, be cautious to neither breathe the aerosol mist nor allow contact with your skin. Spraying should be done in a hood behind an appropriate shield. It is advisable to wear rubber gloves. If these reagents do contact the skin, wash the affected area at once with soap and warm water.

Pour approximately 20 mL of solvent consisting of a 4:1:5 (by volume) mixture of 1-butanol, acetic acid, and water, respectively, into an 800-mL or 1-L beaker; avoid splashing the liquid onto the sides of the beaker. The depth of solvent in the beaker should be about 6 mm. Cover the beaker with a watch glass. If a piece of cotton is available, use it to close the opening at the lip of the beaker. Allow *at least 10 min* after preparing the chamber for the inside atmosphere to become reasonably saturated with the solvent vapor.

Standard aqueous amino acid solutions (approximately 0.07 M) will be provided in the laboratory. Among those amino acids which may be included are DL-aspartic acid, L-isoleucine, L-arginine, glycine, DL-serine, L-proline, DL-phenylalanine, DL-methionine, DL-threonine, L-lysine, L-cysteine, L-histidine, DL-alanine, L-leucine, DL-tryptophan, L-tyrosine, L-glutamic acid, and DL-valine. Two unknown amino acid mixtures will be provided, each of which will contain two or more amino acids chosen from those listed here. The instructor will indicate whether any additional amino acids, other than those listed, have been included.

Obtain a piece of chromatographic paper measuring 13 × 15 cm. The size may vary in length, depending on the number of samples to be run. Place a series of light pencil dots about 1–1.5 cm apart along a line parallel to a long edge of the paper and about 2 cm from this edge. The outermost dots should be about 1 cm from the short edges of the paper. Label each of the spots so that the solution spotted at each may readily be identified later from information recorded in your notebook. Avoid the common student error of losing track of the positions at which given solutions are spotted. Such errors eventually lead to confusion and probable misidentification of the components of the mixture. (*Caution:* Fingerprints contain significant amounts of amino acids, often enough to be easily detected by the methods used in this experiment. This is particularly true after the solutions of amino acids have been handled. Avoid touching the surface of the chromatographic paper. Handle it as little as possible, and then only along a thin strip on the edge *opposite* to that along which the samples are to be spotted.)

Spots are applied to the paper with a capillary pipet. If pipets are unavailable, the instructor will demonstrate how to draw them from capillary melting-point tubes. In spotting, a small drop of the solution is placed at the mark indicated and allowed to dry before development. If the drop is too small, the amount of sample may be insufficient for eventual detection of some of the components; if it is too large, it may lead to loss of resolution during development. The ideal size is 2–3 mm in diameter. Spots will increase in size by a factor of three or four times in the direction of mobility during development; they should not spread laterally very much. Practice spotting samples on scrap filter paper, and when you feel comfortable with your technique, spot the chromatographic paper with unknowns and any samples of knowns to be run. Be sure to use a *new* capillary pipet for each spot to avoid contamination of the samples being spotted.

After spotting, allow the spots to dry, and then coil the paper into a cylinder, fasten it with staples or paper clips, and insert it, sample edge down, into the developing chamber. The paper should not touch the sides of the beaker. When the solvent front has migrated to within about 2 cm from the upper edge of the paper, remove the paper and lightly mark the position of the solvent front. Stand the cylinder on a watch glass in the hood for a few minutes to dry. The drying should then be completed by spreading the paper out under a heat lamp, by placing it in an oven at about 105°, or perhaps by hanging it in a gentle stream of air in the hood.

When the paper is dry, spray it lightly and evenly with ninhydrin solution,[5] and redry it either under a heat lamp or in the oven (heat is necessary to the color-forming reaction). The colors should be readily visible after 20–30 min. It may be desirable to run additional chromatograms in order to obtain R_f values for additional standard samples, or to develop the chromatogram using different detecting agents. An excellent detecting agent that may be used in addition to ninhydrin is isatin.[6] This reagent produces colors by a reaction similar to that shown for ninhydrin in Figure 23.1. Isatin may be useful in order to distinguish between amino acids of similar R_f values because the different amino acids produce a greater variety of colors. The colors produced by isatin spray for the common amino acids are provided in Table 23.1. These colors develop after heating the chromatogram to 80–85° for 30 min.

Calculate R_f values for each of the standard samples and for each of the spots resolved in the unknowns (consult, if necessary, the legend to Figure 3.12 for the procedure of calculation). Systematically tabulate in your notebook the colors observed for all spots and the calculated R_f values. Identify the constituents of the unknowns, giving the justification for your conclusions. Fasten the chromatograms in your notebook as a permanent record.

EXERCISES

1. Rationalize the large observed difference in the isoelectric points of lysine (9.6) and glutamic acid (3.2).
2. Proline reacts with ninhydrin to produce a yellow color; the other amino acids in Table 23.1 produce a blue color. What structural feature of proline do you expect is responsible for this distinction in behavior? (*Hint:* Consult the mechanism of the color-forming reaction with ninhydrin as given in Figure 23.1.)
3. The empirical formula of the protein ribonuclease as obtained from cattle is $C_{566}H_{890}O_{168}N_{192}S_{13}$. It is obviously meaningless to use such a formula in the classical way as a basis for structural determinations. Suggest a more practical way of measuring the composition of a protein.
4. Why is the solubility of an α-amino acid at a minimum in a solution having a pH corresponding to the isoelectric point of the acid?

[5]The ninhydrin spray reagent is prepared as a 0.1% solution of ninhydrin in 95% ethanol.

[6]The isatin spray reagent is prepared as a solution of 1 g of isatin and 1.5 g of zinc acetate in 100 mL of 2-propanol and 1 mL of pyridine. The solution is effected by warming on a water bath at 80°, after which the solution must be kept cool.

23.3   Determination of Primary Structure

The most difficult aspect of the determination of primary structure of a poly-peptide is the establishment of the sequence of amino acid residues in the chain. The total hydrolysis and amino acid analyses discussed in the preceding section allow determination of the number and types of amino acids present; however, all infor-mation regarding sequence is lost at the hydrolysis step. The standardized approach to establishing sequence is discussed in the following paragraphs.

Terminal Residue Analyses.   Note that the amino acid residues at the termini of the polypeptide chain differ from the remainder of the residues: One, the *N-terminal residue,* is the only residue which contains a free *alpha* amino group; the other, the *C-terminal residue,* is the only residue which contains a free carboxyl group *alpha* to a peptide linkage (see **1**, for example). The special significance of these residues is that it is relatively simple to determine their identity. The importance of identifying these residues may be illustrated in the following example. There are 720 possible sequen-tial arrangements for a hexapeptide containing six *different* amino acid residues. If *either* of the terminal residues is determined, the remaining number of possible sequences is only 120; if both are known, this number is reduced to 24. Thus in this example terminal residue analyses would result in a reduction of the number of structures to be considered by a factor of 30! The results of terminal residue analysis, together with information gained by partial hydrolysis, will usually allow the se-quence of polypeptides to be determined.

A very successful method of identifying the N-terminal residue utilizes 2,4-dinitrofluorobenzene (DNFB). DNFB reacts by nucleophilic aromatic substitution in weakly alkaline aqueous solutions with free amino (N-terminal and lysyl), phenol (tyrosyl), and imidazole (histidyl) groups to provide dinitrophenyl (DNP) derivatized peptides (**11**, equation 3). Excess DNFB may be removed from the alkaline reaction mixture by extraction with diethyl ether. The DNP-peptide remains water soluble at

$$NH_2-CH-\overset{\overset{\displaystyle O}{\|}}{C}-NH-CH-CO_2H \underset{HO^{\ominus}}{\overset{DNFB}{\longrightarrow}}$$
$$\underset{R_1}{\quad} \qquad \underset{R_2}{\quad}$$

$$O_2N-\underset{NO_2}{\underbrace{\bigcirc}}-NH-\underset{R_1}{CH}-\overset{\overset{\displaystyle O}{\|}}{C}-NH-\underset{R_2}{CH}-CO_2H \;+\; \underset{NO_2}{\overset{OH}{\underset{}{\bigcirc}}}^{NO_2} \qquad (3)$$

11

alkaline pH, as does 2,4-dinitrophenol, which is formed by hydrolysis of DNFB during the reaction. The DNP-peptide (and, unfortunately, 2,4-dinitrophenol) is separated by adjustment of the reaction mixture to pH 1 and extraction with diethyl ether. The DNP-peptide no longer contains basic amino groups and consequently is not soluble in the acidic aqueous medium. Following removal of the ether, the

DNP-peptide is totally hydrolyzed with 6 M HCl at 100° for 24 hr to provide a mixture of both derivatized and underivatized amino acids. Table 23.2 portrays the

TABLE 23.2   CHEMICAL RESULTS OF 2,4-DINITROPHENYLATION AND HYDROLYSIS OF THE RESULTING DNP-PEPTIDE

Amino Acid Residue	As N-Terminal Residue	Other
Lysyl	DNP—NH—CH—CO$_2$H 　　　　　(CH$_2$)$_4$ 　　　　DNP—NH	$\overset{\oplus}{N}H_3$—CH—CO$_2$H 　　　　(CH$_2$)$_4$ 　　DNP—NH
Histidyl	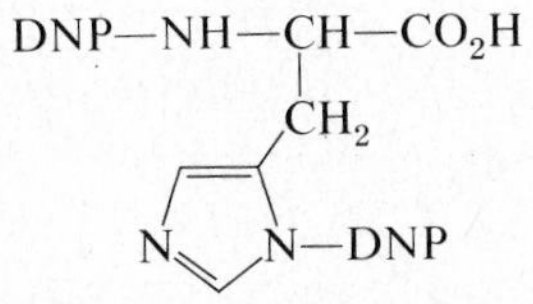 DNP—NH—CH—CO$_2$H 　　　　　CH$_2$ (imidazole ring) N—DNP	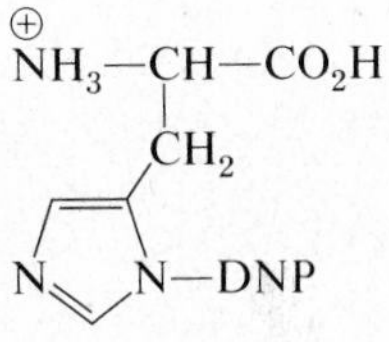 $\overset{\oplus}{N}H_3$—CH—CO$_2$H 　　　CH$_2$ (imidazole ring) N—DNP
Tyrosyl	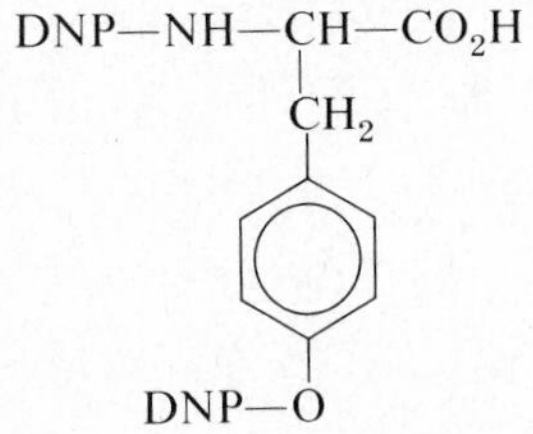 DNP—NH—CH—CO$_2$H 　　　　CH$_2$ (benzene ring) DNP—O	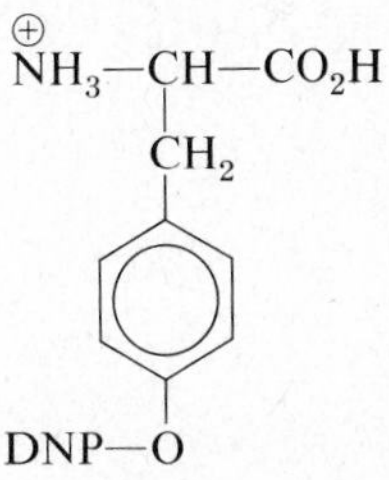 $\overset{\oplus}{N}H_3$—CH—CO$_2$H 　　　CH$_2$ (benzene ring) DNP—O
All other amino acids	DNP—NH—CH—CO$_2$H 　　　　　R	$\overset{\oplus}{N}H_3$—CH—CO$_2$H 　　　R

chemical results at this stage, with especial attention drawn to lysine, histidine, and tyrosine, those amino acids containing side-chain functional groups reactive to DNFB.

The N-terminal residue may now readily be distinguished from the other amino acid residues of the original peptide because it provides the only amino acid derivatized at the alpha position; the other acids bear free alpha amino groups. The N-terminal DNP-amino acid (with the exception of DNP-arginine) may be separated from the acidic hydrolysis mixture by extraction with ether and identified by standard chromatographic procedures.

C-Terminal residue analysis is normally accomplished in either one of two ways. Reaction of the peptide with anhydrous hydrazine results in hydrazinolysis of each of the peptide linkages, providing the C-terminal amino acid residue as the only free *alpha* amino acid in the product mixture (equation 4). As the only water-soluble fragment of the original peptide, it may be isolated and identified.

$$NH_2-\underset{R_1}{CH}-\overset{\overset{O}{\|}}{C}-NH-\underset{R_2}{CH}-CO_2H \xrightarrow{H_2NNH_2}$$

(4)

$$NH_2-\underset{R_1}{CH}-\overset{\overset{O}{\|}}{C}-NHNH_2 + \overset{\oplus}{N}H_3-\underset{R_2}{CH}-CO_2^{\ominus}$$

A second method of C-terminal residue determination makes use of the enzyme carboxypeptidase, a pancreatic enzyme whose characteristic reactivity is the hydrolysis of peptide bonds adjacent to free *alpha*-carboxyl groups. Because the peptide bond to the C-terminal residue is the only such bond in a peptide, carboxypeptidase selectively removes this residue as a free alpha amino acid, producing a shortened peptide chain:

$$\overset{\oplus}{N}H_3-\underset{R_1}{CH}-\overset{\overset{O}{\|}}{C}-NH-\underset{R_2}{CH}-\overset{\overset{O}{\|}}{C}-NH-\underset{R_3}{CH}-CO_2^{\ominus} \xrightarrow{carboxypeptidase}$$

(5)

$$\overset{\oplus}{N}H_3-\underset{R_1}{CH}-\overset{\overset{O}{\|}}{C}-NH-\underset{R_2}{CH}-CO_2^{\ominus} + \overset{\oplus}{N}H_3-\underset{R_3}{CH}-CO_2^{\ominus}$$

Carboxypeptidase will then remove the *new* C-terminal residue and so on. Analysis by paper chromatography may be used to identify the free amino acids present in the mixture. The amino acid corresponding to the original C-terminal residue will develop maximum concentration, as judged from spot color intensity on the chromatogram, at a shorter elapsed reaction time than acids from positions successively farther in from that end of the peptide chain, because it is the first residue removed.

Partial Hydrolysis and Sequence Determination.   In principle, the sequence of residues in a polypeptide might be determined by devising a procedure for selectively removing a terminal residue, identifying it, then removing and identifying the next, and so forth until all have been identified (a terminal sequence determination). A variety of chemical procedures are available which may be used in just this way. These procedures, however, are generally feasible only for relatively short peptides; for example, the primary structures of peptides containing up to 60 amino acid residues have been determined in this fashion with the aid of automated apparatus. Consequently the sequence of larger polypeptides and proteins is determined by effecting only their *partial* hydrolysis to produce a mixture of smaller peptides (dipeptides, tripeptides, and so on) which are separated and whose sequences are determined by terminal sequence analysis. When the sequences of a sufficient number of fragments are known, the primary structure of the original polypeptide may be logically deduced.[7]

The following experiment involves the determination of structure of an un-

[7]The interested student may consult any modern organic or biochemistry textbook for additional information, details, and examples of these and other procedures and their applications.

known dipeptide. It is apparent that the extent of accumulated information necessary to deduce the structure of a dipeptide is somewhat less than would be required for larger peptides, yet the procedures in the experiment demonstrate many of the techniques commonly used by protein chemists. The structural determination is accomplished with milligram quantities of the unknown peptide; thus experience is gained in the microtechniques of handling materials and solutions.

EXPERIMENTAL PROCEDURE

A.   Hydrolysis of an Unknown Dipeptide

Obtain a 10-cm length of soft glass tubing having an internal diameter of 1.0–1.5 mm. If this is not available, it may be cut from a drawn-out piece of larger diameter tubing. Seal one end by drawing it out, using a microburner. (*Caution:* The seal must be complete.) Place about 1 mg of an unknown dipeptide either on a porcelain spot plate or in a small test tube. Using either a syringe or a 0.1-mL pipet, add to the sample 30 μL (0.03 mL) of 6 M hydrochloric acid. Mix well and, using a disposable-type pipet, transfer the solution to the previously prepared hydrolysis tube. Seal the tube by drawing it out, affix an identification label, and heat the tube for 10–12 hr in an oven set at 110°.

After allowing the hydrolysis tube to cool, carefully open it, and with a disposable pipet transfer the solution to either a small watch glass or a spot plate. Evaporate the sample to dryness with a heat lamp, and to remove the last traces of hydrogen chloride add 20 μL of water and reevaporate. (*Caution:* In each of these evaporative steps, be careful not to char the sample.) Add 50 μL of water, and use 5–10 μL of this solution per spot in analyzing for the component amino acids of the dipeptide by paper chromatography, using the procedure provided in Section 23.2.

B.   N-Terminal Residue Analysis

DO IT SAFELY

Dinitrofluorobenzene (DNFB) is a vesicant; that is, it will cause blistering and burns when it comes in contact with the skin. Operations involving the transfer of DNFB or its solutions should be carried out in the hood if possible, and the substances themselves should be handled *only* by means of a small pipet or a syringe. *Do not pipet DNFB solutions by mouth.* If DNFB comes in contact with your skin, wash the area immediately with soap and warm water, and then rinse with 0.6 M sodium bicarbonate solution.

To a 12-mL conical centrifuge tube add 2 mg of an unknown dipeptide, 0.2 mL of water, 0.05 mL of 0.5 M aqueous sodium bicarbonate solution, and 0.4 mL of stock 2,4-dinitrofluorobenzene solution.[8] Stopper the tube and shake the mixture fre-

---
[8] The 2,4-dinitrofluorobenzene solution is prepared by dissolving 0.25 g of DNFB in 4.8 mL of absolute ethanol.

quently during 1 hr. The pH should be maintained at 8–9 by adding, as necessary, more 0.5 M sodium bicarbonate solution. Large amounts of precipitates indicate that the pH is too low.

After the 1-hr reaction period, add 1 mL of water and 0.05 mL of 0.5 M aqueous sodium bicarbonate solution to the reaction mixture in the centrifuge tube. Extract this solution three times with equal volumes of peroxide-free diethyl ether to remove unchanged DNFB. These extractions are carried out directly in the conical centrifuge tube by adding the portion of ether, stirring vigorously with a stirring rod, and by removing the ether layer with a pipet. If necessary, the solution may be centrifuged to hasten the separation of layers.

Using pH paper as a guide, adjust the pH of the aqueous layer to about pH 1 by adding approximately 0.1 mL of 6 M hydrochloric acid, and extract three times with 2-mL portions of diethyl ether. Combine the extracts in a test tube and evaporate the ether. The evaporation may be accomplished conveniently by placing the test tube in a beaker of warm water and blowing a gentle stream of air into the test tube.

All traces of ether must be removed from the DNP-peptide before its hydrolysis. To accomplish the removal of traces of ether simultaneously with the transfer of the DNP-peptide to a hydrolysis tube, which may be prepared by sealing one end of a 10-cm length of 5-mm glass tubing, add 0.2 mL of acetone to the dried DNP-peptide, and transfer the resulting solution to the tube. Evaporate the solution in the hydrolysis tube to dryness as before, using a disposable pipet to channel a gentle air stream into the tube. Add 0.5 mL of 6 M hydrochloric acid, seal and label the tube, and heat it at 100° in an oven for 10–12 hr.

Open the hydrolysis tube, and transfer the solution to a small test tube. After adding 1 mL of water, extract the hydrolysis solution three times with 2-mL portions of diethyl ether. Combine the ether extracts and evaporate to dryness as before. Dissolve the DNP-amino acid in 0.5 mL of acetone, and use this solution for chromatographic analysis.

If it is desired to confirm the identity of the C-terminal residue, evaporate the aqueous phase of the hydrolysate to dryness, using a heat lamp. After adding 0.1 mL of water and redrying, dissolve the residue, which contains the C-terminal amino acid, in 50 μL of water and identify by paper chromatography, following the procedure of Section 23.2. Note that if the C-terminal residue is lysine, histidine, or tyrosine (see Table 23.2), the procedure of Section 23.2 is not applicable.

C. Identification of N-Terminal DNP-Amino Acids by Thin-Layer Chromatography[9]

DO IT SAFELY

If the dinitrophenyl derivatives of the two amino acid residues of your unknown have similar R_f values with formic acid-water as a developing solvent, it will be necessary to use the benzene-acetic acid solvent system. Insofar as benzene has been

[9]For additional information, consult K. Wang and I. S. Y. Wang, "Chromatographic Identification of Dinitrophenylamino Acids on Polyester Film Supported Polyamide Layers," *Journal of Chromatography and Data,* **27,** 318 (1967).

implicated as a leukemia-causing carcinogenic substance, if you must use this solvent system, do both the chromatographic development and the subsequent drying of the plate *in the hood.* Do not remove any of the solvent from the hood. You should wear rubber gloves when handling the wet plate and should transfer the solution with a pipet. These precautions should be followed in order to avoid release of benzene vapors into the room and to prevent adsorption of benzene through the skin.

Obtain a 4 × 10-cm strip of polyamide chromatogram sheet for qualitative analysis of the N-terminal DNP-amino acid. On a line about 1.5 cm from a short side and with 1-cm spacings from each other and from the long sides, spot 5 μL each of the unknown DNP-amino acid-acetone solution and the two appropriate standard DNP-amino acid solutions provided in the laboratory. (Note that the appropriate solutions may be identified from the results of the amino acid analysis performed.) One of two solvents may be used to develop the chromatogram, according to the requirements of the analysis. Consult Figure 23.4 for information to aid in this decision. Use either benzene-glacial acetic acid (80:20, by volume) or 90% formic acid-water (50:50, by volume). The first solvent will require about 1.5 hr for development and the second about 1 hr. (*Caution:* DNP-Amino acids are light-sensitive; the chromatographic development should be performed in the dark by placing the chamber in the desk drawer or by covering the chamber with a cardboard box.) The DNP-amino acids produce yellow spots, as does 2,4-dinitrophenol. The phenol, in contrast to the DNP-amino acids, is colorless below pH 4. If you are uncertain which spot from the unknown is 2,4-dinitrophenol, add a drop of dilute hydrochloric acid to

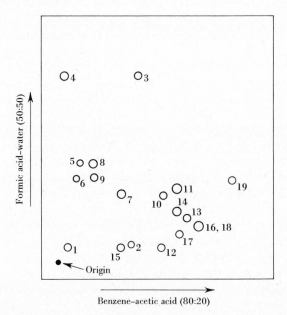

FIGURE 23.4 Two-dimensional chromatogram of 18 N-terminal DNP-amino acids (see second column of Table 23.2) and 2,4-dinitrophenol (circle 19). The numbers corresponding to each of the derivatized amino acids are keyed in Table 23.1.

each spot. If the DNP spots are hard to find because of low concentration, examination under ultraviolet light may be helpful. The colors of the spots will fade with time, so they should be outlined with a pencil after development. A drawing of the plate should be recorded in your notebook.

Using the information collected in these procedures, assign a structure for the unknown peptide.

EXERCISES

1. The rate of hydrolysis of gelatin, a protein, increases linearly with acid concentration over the range of 3.0–10.4 M hydrochloric acid. Explain, with reference to the mechanism of amide hydrolysis, why this should be the case.

2. Account for the observation that glycylvaline is hydrolyzed under acidic conditions much more rapidly than valylglycine.

3. When asparagine is present in peptides, an acid-catalyzed rearrangement is sometimes observed during hydrolysis to a mixture of alpha- (12) and beta-aspartyl (13) peptides. Account mechanistically for these transformations.

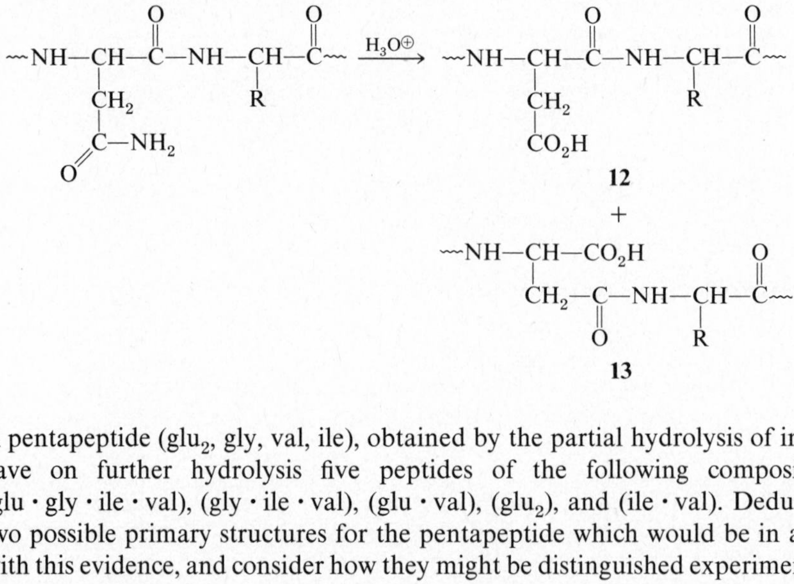

4. A pentapeptide (glu$_2$, gly, val, ile), obtained by the partial hydrolysis of insulin, gave on further hydrolysis five peptides of the following compositions: (glu · gly · ile · val), (gly · ile · val), (glu · val), (glu$_2$), and (ile · val). Deduce the two possible primary structures for the pentapeptide which would be in accord with this evidence, and consider how they might be distinguished experimentally.

5. Provide a mechanism for the reaction of 2,4-dinitrofluorobenzene with a peptide to produce an N-terminal DNP-peptide (11, equation 3). Would you expect 2,4-dinitrochlorobenzene to require more or less vigorous conditions if used in place of DNFB?

6. Provide a flow diagram which outlines the various separations that are effected on the basis of solubilities during the preparation of the N-terminal DNP-amino acid from a dipeptide.

7. What is the precipitate which forms at low pH during the 2,4-dinitrophenylation of a peptide?

REFERENCES

1. R. E. Dickerson and I. Geis, *The Structure and Actions of Proteins,* Harper & Row, New York, 1969.
2. S. Blackburn, *Protein Sequence Determination: Methods and Techniques,* Marcel Dekker, New York, 1970.
3. J. L. Bailey, *Techniques in Protein Chemistry,* Elsevier Publishing Company, New York, 1962.

24

NATURAL PRODUCTS
Classification, Isolation, and Characterization

24.1 Introduction

Naturally occurring organic compounds, that is, those substances found in and produced by living organisms, have been a source of fascination for centuries. The interest in these substances exists for a variety of reasons, ranging from the practical applications of such compounds in daily life to the scientific challenges presented by them. Thus people have used natural products to alleviate pain and to cure diseases, to provide colorful dyes for their bodies and their clothing, to flavor their foods, and to cause death, both of the animals on which they prey and of their enemies. From the standpoint of science it was the natural products, because of ready availability, that provided chemists with one of their first experimental challenges during the period when chemistry was developing from alchemy into a more exact science. It is noteworthy that natural products still present some of the greatest challenges to modern organic chemists.

As the scientific field now labeled "natural products chemistry" evolved, it was found convenient and profitable to place a given compound in one of four general categories defined on the basis of characteristic structural features found in most natural products. These categories are the carbohydrates (sugars), the acetogenins, the terpenes and steroids, and the alkaloids. Because the carbohydrates are discussed in some detail in Chapter 22, they are not further discussed here.

The *acetogenins* are a group of compounds that share the distinction that their *biosynthesis*[1] involves the head-to-tail polymerization of the two-carbon acetate unit to generate a *linear* polyacetyl chain, for example, **1**; as a class, then, the acetogenins are characterized by the absence of extensive branching in their carbon skeletons. Further transformations of the basic structure **1** can produce, among other sub-

---

[1] Biosynthesis is the synthesis of a chemical compound by a living organism.

stances, stearic acid (2), a fatty acid, and rhamnetin (3), a yellow pigment of the flavone type. The polymerization of acetate units to generate the general structure 1 is related mechanistically to the Claisen ester condensation, the active form of acetate in biological systems being a thioester, acetyl coenzyme A (CH_3COSEn, where En represents the coenzyme).

$$CH_3CO(CH_2CO)_nCH_2CO\text{---}$$
1

$$CH_3CH_2(CH_2CH_2)_{14}CH_2COOH$$
2

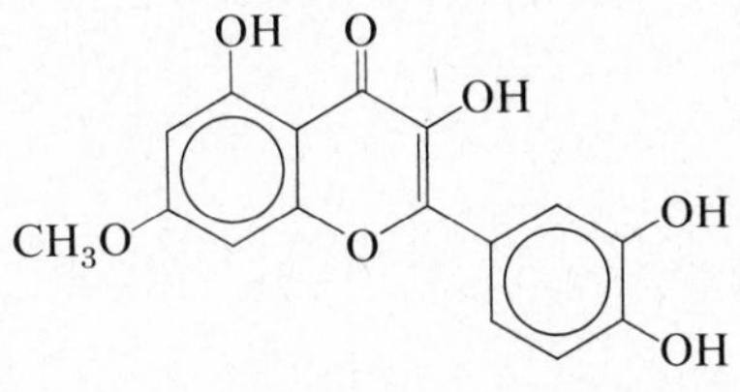

3

The substances that fall into the category designated *terpenes* and *steroids* can be viewed as having carbon skeletons constructed by 1,4-polymerization of the five-carbon, branched-chain isopentyl unit, 4. The active form of this unit in biosynthesis is isopentenyl pyrophosphate (5), which itself is produced in nature by condensation of three acetyl coenzyme A units, with one carbon atom being lost in the form of carbon dioxide. Thus in contrast to the generally linear character of the acetogenins, the terpenes and steroids possess carbon skeletons that are branched at regular intervals, as illustrated by the structure of vitamin A (6). It should be noted that substances in the terpene class contain two or more of the five-carbon units, two examples being 6 and camphor (7), whereas steroids such as cholesterol (8) consist of at least four isopentyl units.

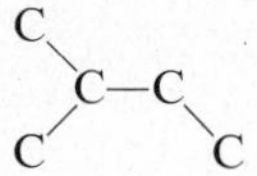

4

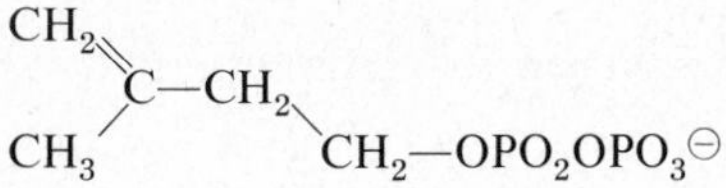

5

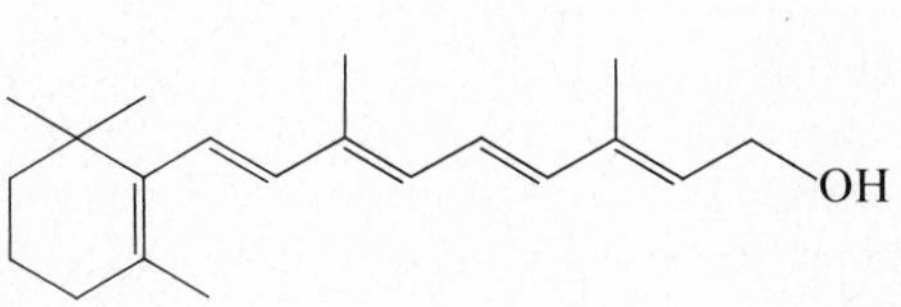

6

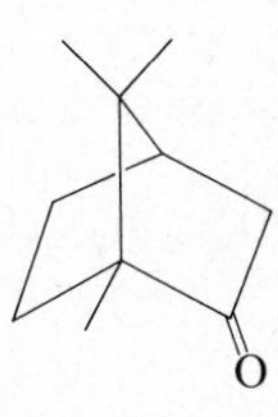

7

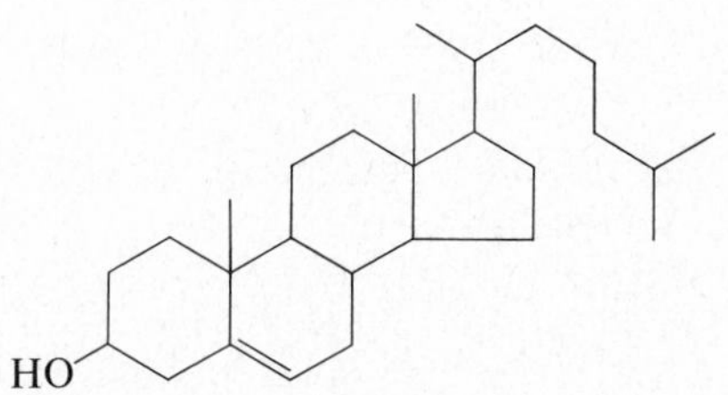

8

The final category of natural products to be considered here, the *alkaloids,* probably contains the most members and is the most diverse in terms of structural types. This is a consequence of the fact that all natural products which are basic (alkaline) are classed as alkaloids. The basicity of the substances in this category results from the presence in them of one or more nitrogen atoms; the nitrogen arises from incorporation of an α-amino acid unit, **9**, as one of the basic structural units during the biosynthesis of the alkaloid. Examples of alkaloids include the relatively simple nicotine **(10)** and the considerably more complex strychnine **(11)**.

$$R-CHCO_2H$$
$$\quad\ |$$
$$\ \ NH_2$$

9   **10**   **11**

Historically most natural products have been extracted from plants rather than from animals, owing to the generally greater chemical simplicity and availability of the former; microorganisms, which are "simple" in the same sense as are plants and also can be propagated rapidly and in large quantities, are assuming increasing importance as sources of natural products.

The isolation of a natural product in pure form normally presents a considerable challenge to the experimentalist, mainly because even the simplest plants and microorganisms represent mixtures of many organic compounds. The general approach taken in the isolation of a natural product can be summarized in the following way. The plant or microorganism is ground or homogenized into fine particles and the resulting material is then usually extracted with a solvent or mixture of solvents in which the desired natural product is expected to be soluble. If the natural product is volatile [as would be the case, for example, if an attempt were being made to characterize the substance(s) responsible for a particular odor], the volatile compounds in the extract can be detected and possibly isolated by application of glpc techniques. More often, however, the natural product is a relatively nonvolatile substance, so that removal of the solvent used in the extraction leaves an oil or a gum that requires further manipulation to achieve the resolution of the mixture into its various components. It should be noted that in rare instances some natural products, which when pure are crystalline compounds, will precipitate from solution during the removal of the solvent, and it was often just this type of fortuitous purification that facilitated the first investigations of natural products.

The more common situation with natural-products isolation is that the residual oil or gum might be treated with acids or bases in order to separate basic and acidic components, respectively, from neutral substances; slightly volatile compounds might be separated from nonvolatile ones by subjecting the residue to steam distillation.

One of the most powerful techniques developed as a result of attempts to purify natural products is that of chromatography of various types. Paper and column chromatography have been of dramatic importance to this branch of chemistry; more recently thin- and thick-layer, liquid-liquid, and gas-liquid chromatographic techniques have been increasingly used to aid in the resolution of a crude mixture of natural products into its various components.

The next stage facing the chemist working in the field of natural products is determination of the structure of the isolated product. Here again traditional procedures, such as qualitative tests for various functional groups, and chemical degradations to known substances were and are of great importance. More recently spectroscopic techniques such as mass spectrometry and ultraviolet, infrared, and nuclear magnetic resonance spectroscopy have greatly facilitated the determination of structure.

In many instances the final stage or goal for chemists working in this field is to develop a synthetic pathway that permits synthesis of the natural product. The synthesis of some natural products represents mainly an intellectual challenge and/or an opportunity to demonstrate the utility of new synthetic techniques. In some cases, particularly those in which the natural product has medicinal uses, the development of an efficient synthesis may be of importance because of the severely limited supply of the material from the natural source.

Some of the various techniques required for isolation of pure natural products are described in the experiments, which provide an introduction to this fascinating branch of organic chemistry.

24.2   Citral from Lemon Grass Oil

Terpenes are responsible for the characteristic flavors, odors, and colors of many substances encountered in nature. As an example, citral (12) is an aldehydic terpene that possesses a pleasant lemonlike odor and taste. It is amusing to note that whereas citral evokes pleasant odor and taste responses in humans, it apparently is less attractive to other organisms because certain insects such as ants are known to employ citral as one component of a secretion used to ward off potential predators. As might be expected from the nature of its odor, citral is of commercial importance as a constituent of perfumes in which a lemonlike essence is desired; it is also employed as an intermediate for the synthesis of vitamin A (6).

12

Owing to the commercial importance of citral, an extensive search for its presence in natural products has been made. Not too surprisingly, one source turns out to be oil from the skins of lemons and oranges, although it is only a minor component of the oil. Citral is the major component, however, of the oil that results from pressing of lemon grass; in fact 75–85% of this oil is the desired natural product.

Citral contains carbon-carbon double bonds, one of which is conjugated with a carbonyl (aldehyde) function. The presence of the double bond conjugated with the carbonyl group makes citral subject to polymerization, and the aldehyde function contained in it is readily oxidized to a carboxylic acid group, a reaction that is common with aldehydes. Thus citral is an extremely labile substance that reacts under conditions such as heat or light or the presence of reagents such as acids, bases, and oxygen that induce its polymerization and/or oxidation. The isolation of citral therefore potentially presents a significant challenge to the experimentalist. The task is greatly simplified, however, by the fact that citral is relatively volatile [bp 229° (760 mm)] and has a low solubility in water. These two properties make it a suitable candidate for steam distillation, a technique that allows distillation of citral from crude lemon grass oil at a temperature less than 100°, far below citral's normal boiling point, and in a neutral medium. It is worth noting that steam distillation is often the method of choice when reactive, volatile substances are to be separated from nonvolatile (or water-soluble) contaminants.

The citral isolated in this experiment is actually a mixture of the geometric isomers **12a**, geranial, and **12b**, neral. The separation of these two isomers is extremely difficult to achieve with the use of standard techniques and is not attempted in this experiment.

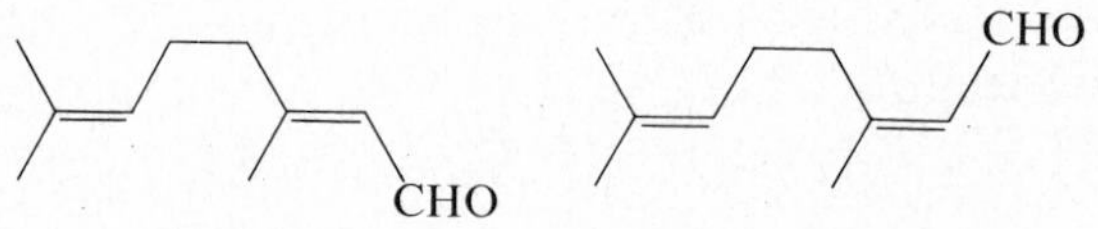

12a   **12b**

EXPERIMENTAL PROCEDURE

DO IT SAFELY
---

1. Be certain that the steam distillate is cool (below 30°) before attempting extraction of it with diethyl ether; otherwise excessive pressure may develop in the separatory funnel and blow out the stopper or stopcock.

2. Remember that diethyl ether is extremely flammable; take care that no flames are in your vicinity during its use or distillation.

Add 10 mL of lemon grass oil and 100 mL of water to a 250-mL round-bottomed flask, and attach the flask to an apparatus set for steam distillation, using an external source of steam (Figures 2.12 and 2.13). Steam-distil the mixture as rapidly as possible, continuing the distillation until droplets of oil no longer appear in the distillate; approximately 250 mL of distillate are required. Drain water from the steam trap (Figure 2.14) whenever necessary.* After allowing the distillate to cool to room temperature or below, add a portion of it to a separatory funnel containing 50 mL of technical diethyl ether, shake the funnel, and separate the aqueous and organic phases. Add another portion of the distillate to the organic phase in the separatory funnel, shake, and separate the layers as before. Repeat these steps until all the distillate has been shaken in portions with the ether, so that the citral has been removed from the aqueous phase. After the extraction is complete, the aqueous phases may be discarded.

Dry the organic phase over anhydrous calcium chloride,★ decant the dried organic solution into a 250-mL round-bottomed flask, and evaporate the solvent under aspirator vacuum. It may be advantageous to place the flask in a pan of water *at room temperature* during the evaporation of the ether. The residue is citral, bp 229°. Determine the percentage recovery of citral from the sample of lemon grass oil.

The isolated product can be characterized by obtaining ir, pmr, and/or uv spectra and comparing them with those in Figures 24.1, 24.2 and/or 24.3. Alterna-

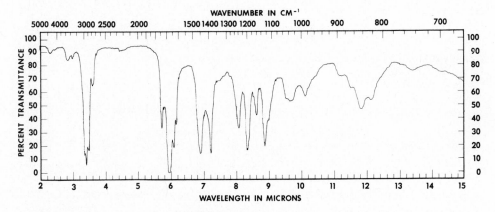

FIGURE 24.1 IR spectrum of citral.

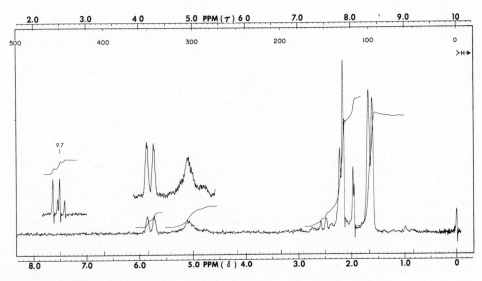

FIGURE 24.2 PMR spectrum of citral.

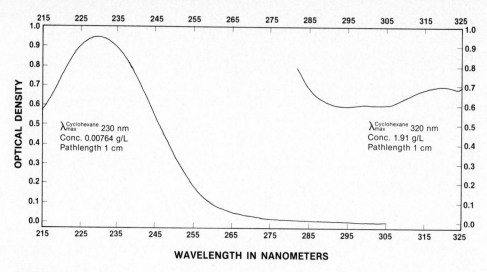

FIGURE 24.3 UV spectrum of citral.

tively, a glpc analysis can be performed, using an authentic sample of citral for comparison, to assess the nature and purity of the product that has been isolated. A typical glpc trace of the product is provided in Figure 24.4.

Chemical characterization of the product can be achieved by testing for unsaturation according to the procedures described in Chapter 25 (see p. 507) and for the presence of an aldehyde function by the chromic acid method outlined in Section 13.2. Solid derivatives of **12a** such as its 2,4-dinitrophenylhydrazone (mp 134–135°) and its semicarbazone (mp 164–165°) can be prepared by the procedures presented in Section 14.1.

Recrystallization normally removes any of the corresponding derivatives resulting from the minor amount of **12b** contained in the isolated citral. Note that the 2,4-dinitrophenylhydrazone of **12b** has a melting point of 171–172°, and the semicarbazone of this isomer melts at 125–126°.

EXERCISES

1. Calculate the relative amounts of geranial and neral present in the sample of citral as indicated by the glpc trace of Figure 24.4. By evaluating the relative areas of the aldehydic protons in the pmr spectrum (Figure 24.2) of citral, perform a similar calculation of the ratio of the two isomers present.
2. Why is the diethyl ether not removed from citral by distillation at atmospheric pressure, the more usual procedure?
3. Geranial (**12a**) is thermodynamically more stable than neral (**12b**). Suggest an explanation for this.

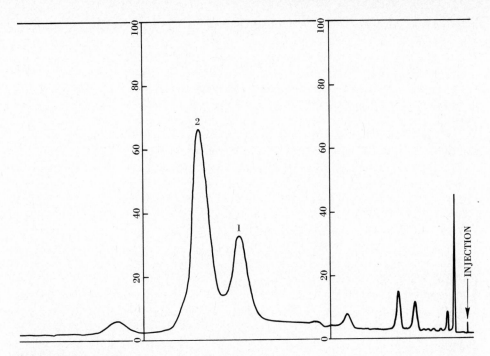

FIGURE 24.4 GLPC of steam distillate from lemon grass oil; peak 1 is neral and peak 2 is geranial.

24.3 Piperine from Black Pepper

A weakly basic substance that could be extracted from a variety of peppers was isolated and characterized in 1882 and given the name piperine, from the Latin name for pepper (*piper*). Piperine **(13)** is a 1,4-disubstituted butadiene having a specific geometry about the double bonds. This substance along with minor amounts of chavicine, a geometric isomer of piperine, constitutes about 10% of the weight of black pepper. Among other components of black pepper are starches (20–40%), volatile oils (1–3%), and water (8–13%).

13

Because both piperine and chavicine are relatively nonvolatile substances, they are not responsible for the aroma of black pepper. The taste of black pepper, however, is at least partially attributable to these substances; for example, although piperine is tasteless at first, it does ultimately produce a burning sensation and sharp aftertaste. The initial tastelessness of this substance may be a consequence of its extremely low solubility in water, so that it cannot penetrate the layer of saliva on the tongue and reach the taste buds; certainly, piperine that has been moistened with ethanol produces an immediate sharp taste when placed on the tongue. Some investigators have postulated that chavicine is responsible for the characteristic taste of pepper, but there still appears to be no general agreement regarding this point.

The isolation of piperine can be accomplished in an uncomplicated manner by extraction of ground pepper with 95% ethanol. Ideally the extraction would be performed using a Soxhlet apparatus such as that shown in Figure 3.3, so that only a relatively small volume of solvent would be required. Such apparatus is often not available in the undergraduate laboratory, so a round-bottomed flask fitted with a reflux condenser will be used instead; as a consequence, much more solvent will be required than would be with a Soxhlet apparatus.

The crude extract obtained by heating black pepper in ethanol contains, in addition to piperine and chavicine, some acidic, resinous materials that must not be allowed to precipitate with the piperine and thereby contaminate it. To prevent coprecipitation of piperine and the resin acids, dilute ethanolic potassium hydroxide is added to the concentrated extract to keep acidic materials in solution as their potassium salts.

Hydrolysis of Piperine.   When an attempt to determine the structure of an unknown natural product is being made, it is often found useful to cleave the unknown substance into smaller fragments by a chemical reaction; these fragments can generally be more readily identified than the original molecule. Once they have been identified, these smaller molecules represent pieces of a "jigsaw puzzle" which the chemist must put together in a rational way, the goal being, of course, to fit the pieces back together so that the structure of the original substance is duplicated.

In the case of piperine the gross structure is that shown in **13**; as noted, however, the stereochemistry about the double bonds has not been specified. In the original proof of structure of this substance, it was recognized that the single nitrogen atom present in the molecule was part of an amide linkage (see exercise 2), so that hydrolysis of this functional group should produce an acid and an amine, both of which might be of known structure. In fact base-catalyzed hydrolysis followed by appropriate work-up allowed isolation of piperidine **(14)**, a known cyclic amine, and

14

piperic acid, a substance that could be synthesized from the functionalized cinnamaldehyde **15** (equation 1). Consequently the gross structure **13** could be proposed for piperine.

$$R-CH=CH-CHO + (CH_3C)_2O \xrightarrow[\Delta]{NaOAc} R-CH=CH-CH=CH-CO_2H \quad (1)$$

15   Piperic acid

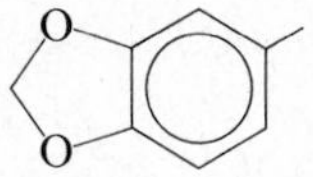

Assignment of the full structure to piperine of course requires that the geometry about the double bonds be specified. Given that piperic acid is one of the four geometric isomers shown here and that the melting point of each isomer is that given, isolation of piperic acid and determination of its melting point should allow assignment of the required stereochemistry (see Exercise 1).

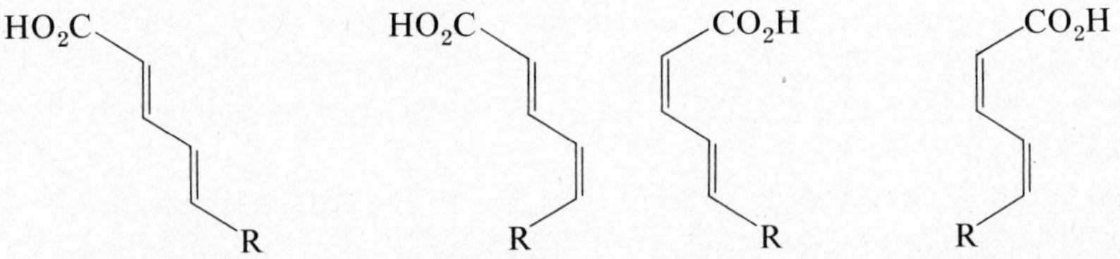

mp 215–217°   mp 134–136°   mp 154–156°   mp 200–202°

EXPERIMENTAL PROCEDURE

DO IT SAFELY

    1. Watch for excessive bumping during the initial step of the procedure; if care is not exercised in heating the ethanolic solution, it may erupt from the top of the condenser.

    2. Gaseous hydrogen chloride can cause severe burns in the respiratory tract if inhaled; *use this chemical only in the hood.*

A. Isolation of Piperine

    Place 30 g of finely ground black pepper in a suitably sized round-bottomed flask, add 300–350 mL of 95% ethanol, and gently heat the mixture under reflux for about 3 hr.[2]★ Because there is solid present in the boiling mixture, bumping may occur, particularly if heating is too vigorous. Filter the mixture by vacuum, and concentrate the filtrate to a volume of 20–30 mL by distillation. Add 30 mL of warm 2 *M* ethanolic potassium hydroxide solution to the residue from the distillation, stir the warm mixture well, and decant or filter the solution to remove any insoluble materials.★ While keeping the solution warm on a steam bath, add 15–20 mL of

---

[2]The period of reflux need not be done during a single laboratory session.

water; the solution should become turbid, and yellow needles may be discerned. Allow the resulting solution to stand until the next laboratory period, and then isolate the yellow precipitate of piperine that has formed. Recrystallize the crude piperine from acetone. The resulting crystals should be formed as fine yellow needles. Determine the yield and melting point of the product. The reported melting point of piperine is 129–131°.

B.   Hydrolysis of Piperine

(*Note:* Adjust quantities in the following procedure according to the amount of piperine that is available.) Heat a mixture of 1 g of piperine and 10 mL of 2 *M* ethanolic potassium hydroxide at reflux for 1.5 hr.* Evaporate the ethanolic solution to dryness by performing a vacuum distillation with the aid of a water aspirator and a steam bath.* Cool the receiver in an ice-salt bath during the distillation. Suspend the solid potassium piperate that remains in the stillpot in about 20 mL of hot water, and carefully acidify this suspension with 6 *M* hydrochloric acid. Collect the precipitate that results, wash it with cold water, and recrystallize the crude piperic acid from absolute ethanol. Determine the yield and melting point of the isolated product.

Qualitative detection of an amine (piperidine) in the distillate can be accomplished by dissolving a few drops of the distillate in a few milliliters of water and determining the pH of the solution. The distillate should also have an odor characteristic of an amine.

The piperidine can be isolated as its hydrogen chloride salt by saturating the distillate with *gaseous* hydrogen chloride (*in the hood!*), removing the ethanol by performing a vacuum distillation and recrystallizing the residue from absolute ethanol. Because salts of amines are generally very hygroscopic, the isolated salt should not be exposed to atmospheric moisture any more than is necessary. The reported melting point of piperidine hydrochloride is 242–244° (do not attempt to determine this melting point if the melting point apparatus being used contains a heating fluid such as mineral oil, which may ignite above 200°).

EXERCISES

1. On the basis of the melting point observed for piperic acid, assign a stereochemically complete structure to piperine. Given that chavicine has stereochemistry about both double bonds which is the opposite of that in piperine, write out the structure of that geometric isomer.
2. What approaches, both chemical and spectroscopic, might be taken to demonstrate the presence of an amide function in an unknown compound?
3. Is the piperine that is isolated in this experiment expected to be optically active? Why or why not?
4. Amides can be hydrolyzed with aqueous acid as well as with aqueous base. After considering the nature of the other functional groups present in piperine, suggest

a reason why base-catalyzed hydrolysis is the method of choice for hydrolysis of this particular substance.

5. What portion of the piperine molecule is responsible for its color? Would piperic acid be expected to be colored? Why or why not?

6. In the removal of ethanol that follows the hydrolysis of piperine, why is it important that the distillation receiver be cooled in an ice-salt bath?

7. In what class of natural products should piperine be placed? Why?

SPECTRA OF PRODUCTS

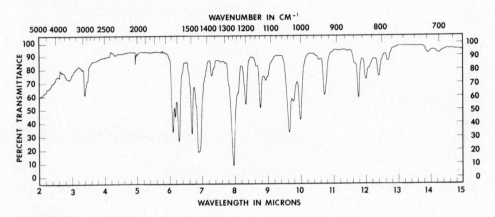

FIGURE 24.5 IR spectrum of piperine.

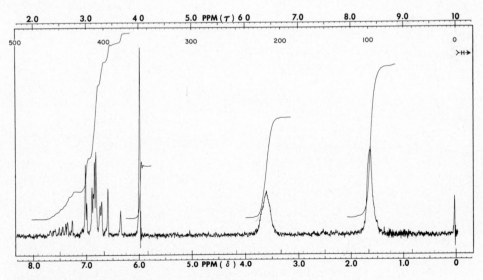

FIGURE 24.6 PMR spectrum of piperine.

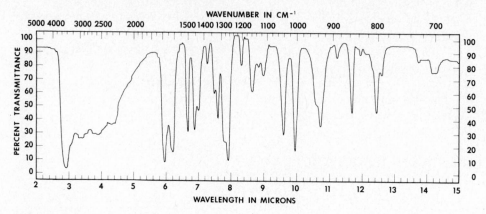

FIGURE 24.7   IR spectrum of piperic acid.

REFERENCES

1. J. B. Hendrickson, D. J. Cram, and G. S. Hammond, *Organic Chemistry,* 3d ed., McGraw-Hill Book Company, New York, 1970, Chapter 27.
2. J. B. Hendrickson, *The Molecules of Nature,* W. A. Benjamin, New York, 1965.
3. P. Yates, *Structure Determination,* W. A. Benjamin, New York, 1966.
4. R. Ikan, *Natural Products: A Laboratory Guide,* Academic Press, New York, 1969, p. 185.

25

IDENTIFICATION OF ORGANIC COMPOUNDS

Systematic procedures for the identification of organic compounds were developed much later than those for inorganic compounds and elements. The first successful scheme of organic qualitative analysis was developed by Professor Oliver Kamm and culminated in his textbook published in 1922. This scheme is the one on which most textbooks are still based,[1] and we shall refer to it, including the modifications and modernizations that have been made, as *classical qualitative organic analysis*.

As mentioned in Chapters 3 and 4, in recent years the development of instrumental methods of separation and analysis (particularly chromatographic and spectroscopic techniques) has revolutionized the laboratory practice of organic chemistry. However, the interest in classical qualitative organic analysis remains high because it is recognized by teachers and students alike as the most effective as well as the most interesting means of teaching fundamental organic chemistry. For this reason in this book we have retained an adequate outline of the classical scheme (Section 25.2) and have also provided an introduction to some of the more modern spectroscopic methods of identification and structure determination (Section 25.3).

25.1 Separation of Mixtures of Organic Compounds

In this chapter we are primarily concerned with the identification of a *pure* organic compound. It must be recognized, however, that when a chemist is faced with the problem of identifying an organic compound, it is seldom pure and is often mixed with by-products or starting materials. Modern methods of separation, particularly chromatographic procedures, make the isolation of a pure compound easier than it

---

[1] See references 1 and 2 at the end of Section 25.3.

used to be, but one must not lose sight of the importance of classical techniques of separation.[2] These have been utilized in many of the experimental procedures in the preceding chapters of this book.

The common basis of the procedures most often used to separate mixtures of organic compounds is the difference in *polarity* that exists or may be induced in the components of the mixture. This difference in polarity is exploited in nearly all the separation techniques, including distillation, recrystallization, extraction, and chromatography. The greatest differences in polarity, which make for the simplest separations, are those which exist between salts and nonpolar organic compounds. Whenever one or more of the components of a mixture can be converted to a salt, it can be separated easily and efficiently from the nonpolar components by extraction or distillation.

The procedures used to isolate and purify the products of some of the synthetic experiments described in this book serve as excellent illustrations of separations based on differences in polarity. For example, the procedure used for *aniline* (Section 17.1) is outlined in Figure 25.1. The first steam distillation is used to separate the

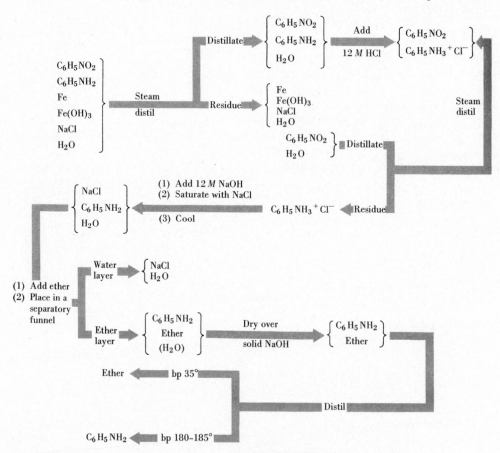

FIGURE 25.1 Separation of aniline from its reaction mixture (see Section 17.1).

[2] Methods of separation and purification of organic compounds are treated in detail in Chapters 2 and 3.

product, aniline, and unchanged starting material, nitrobenzene, from the nonvolatile inorganic compounds. Although the boiling points of aniline and nitrobenzene are far enough apart (24°) so that these compounds could be separated by careful fractional distillation, it is much simpler to convert aniline to its polar *nonvolatile* salt with hydrochloric acid and to remove nitrobenzene by a second steam distillation. The aniline is recovered from the salt by adding sodium hydroxide, and the free organic base is separated from the polar inorganic salt in the water solution by extraction into diethyl ether. A simple distillation then separates aniline from the volatile ether solvent.

In the work-up of the Grignard reaction in which *benzoic acid* is produced (Section 12.2), *extraction* rather than steam distillation is used to separate compounds which differ greatly in polarity (Figure 25.2). In the first partition between diethyl ether and water the inorganics are separated from the mixture of less polar organic

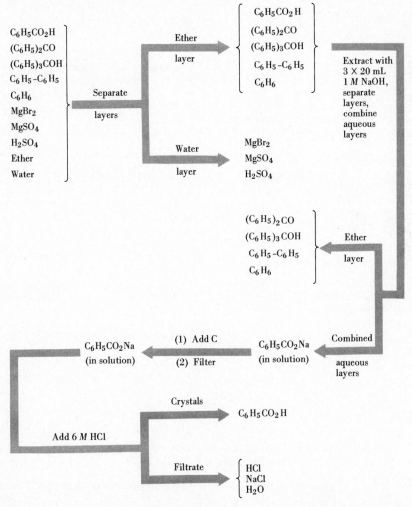

FIGURE 25.2 Separation of benzoic acid from its reaction mixture (see Section 12.2).

compounds. A large difference in polarity between the desired product, benzoic acid, and the by-products is then *induced* by converting the benzoic acid to its salt with sodium hydroxide. Ether extraction removes all the organic by-products from the alkaline water solution containing sodium benzoate. After charcoal treatment to remove traces of highly colored impurities, the benzoic acid is recovered from its salt by adding hydrochloric acid. Since benzoic acid is much less polar than its salt, it is only slightly soluble in water and precipitates from solution.

General Scheme for Separating Simple Mixtures of Water-insoluble Compounds.   A procedure that is adequate for separating mixtures of carboxylic acids, phenols, amines, and neutral compounds which have a low solubility in water and which do not undergo appreciable hydrolysis by reaction with dilute acids or bases at room temperature is outlined in Figure 25.3 and described later. The procedure is based primarily on the partition of compounds of significantly different polarities between diethyl ether and water and the separation of the liquid layers in a separatory funnel; refer to Section 3.1 for a discussion of the theory of extraction.

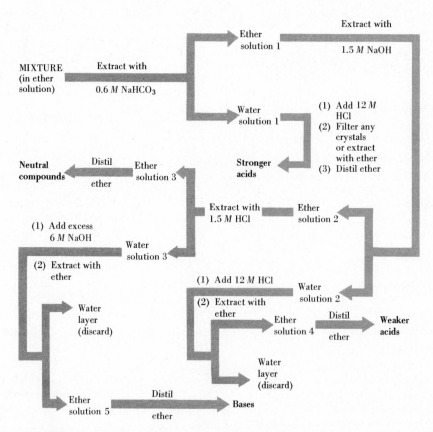

FIGURE 25.3   General scheme for separation of a simple mixture of water-insoluble compounds.

Experimental Procedure

DO IT SAFELY

Diethyl ether is removed from solutes after several extraction steps in this procedure. The ether should be dried over sodium or calcium sulfate, and the usual precautions against fire should be exercised, whether the ether is distilled using a safe heating source or is removed by using a water aspirator at room temperature.

The following experimental procedure, based on the scheme of Figure 25.3, may be used to separate the components of a mixture consisting of three or four compounds, 5 to 6 g of each, of the types described.

The mixture (15 to 25 g) is stirred with 50 mL of technical diethyl ether at room temperature. Any solid material which does not dissolve is collected on a filter and discarded. The ether filtrate is extracted with 20-mL portions of 0.6 M NaHCO$_3$ solution (*Caution:* The separatory funnel should be vented quickly after the first mixing because of possible buildup of pressure from carbon dioxide.) until the aqueous extract remains slightly basic. This solution, **water solution 1** of Figure 25.3, should contain the sodium salt of any carboxylic acid present in the mixture. The free organic acid is regenerated by careful acidification with 12 M hydrochloric acid. If a solid acid separates, it is collected by vacuum filtration; otherwise the aqueous acid solution is extracted with several 20-mL portions of diethyl ether, and the ether is distilled or evaporated.

The ether solution (**ether solution 1**) that was extracted with bicarbonate solution is extracted with two 20-mL portions of 1.5 M NaOH solution to remove a phenol or any other weak acid from the ether solution. The combined aqueous extracts (**water solution 2**) are acidified with 12 M hydrochloric acid, and the regenerated weak acid is extracted with several 20-mL portions of diethyl ether to yield **ether solution 4**.[3] The ether is removed from this solution as already described, leaving the weak acid as a residue.

Ether solution 2 is extracted with one or more 20-mL portions of 1.5 M hydrochloric acid until the aqueous extract (**water solution 3**) remains acidic. The ether layer, referred to in Figure 25.3 as **ether solution 3** should contain any neutral organic compounds which are recovered by removing the ether as described.

To **water solution 3** is added 5 M NaOH solution until the solution is strongly basic, and then it is extracted with several 20-mL portions of diethyl ether. The combined ether extracts (**ether solution 5**) should contain any organic base present in the original unknown mixture; the base may be isolated by removal of the ether.

It should be understood that this very generalized procedure may not give complete separation of all compounds of even the limited types for which it is intended. The products separated should be tested for purity by the usual methods, that is, by melting- or boiling-point determinations and, if possible, by gas chroma-

---

[3]If amphoteric compounds were included in the unknown mixture, they would be carried through to the water layer separated from **ether solution 4**.

tography or thin-layer chromatography. Before attempting identification of the individual compounds by any of the classical or modern instrumental methods described later, the samples should be purified by recrystallization, distillation, or chromatography.

25.2   Classical Qualitative Organic Analysis Procedure for Identification of a Pure Compound

The system consists of six fundamental steps, which are usually best carried out in the sequence listed.

1. **Preliminary examination** of physical and chemical characteristics
2. Determination of **physical constants**
3. **Elementary analysis**
4. **Solubility tests,** including acid-base reactions
5. **Classification tests;** functional reactivity other than acid-base reactions
6. **Preparation of derivatives**

It is a tribute to the effectiveness of the system that, although the number of organic compounds is many thousands of times the number of common inorganic ions, a known organic compound may usually be identified with more certainty than an inorganic compound. However, with the exception of a few guidelines, there is no rigid regimen of "cookbook" directions to be followed; a student must rely on her or his own judgment and initiative in choosing a course of attack on the unknown.

1.   The Preliminary Examination.   The preliminary examination may provide more information with less effort than any other part of the identification procedure, if it is carried out intelligently. The simple observation that the unknown is a *crystalline* solid, for example, eliminates from consideration a major fraction of all organic compounds, since most of them are liquids at room temperature. The *color* is also informative; most pure organic compounds are white (or colorless). A brown color is most often characteristic of small amounts of impurities; for example, aromatic amines and phenols quickly become discolored by the formation of trace amounts of highly colored air-oxidation products. Color in a pure organic compound is usually attributable to conjugated double bonds.

The *odor* of many organic compounds is highly distinctive, particularly among those of lower molecular weight. A conscious effort should be made to learn and recognize the odors which are characteristic of several classes of compounds such as the alcohols, esters, ketones, and aliphatic and aromatic hydrocarbons. The odors of certain compounds demand respect, even when they are encountered in small amounts and at considerable distance; for example, the unpleasant odors of thiols (mercaptans), isonitriles, and higher carboxylic acids and diamines cannot be described definitively, but they are recognizable once encountered. *Be cautious* in smelling unknowns, since some compounds are not only disagreeable but also irritating to the mucous membranes. Large amounts of organic vapors should never be inhaled because many compounds are toxic.

The *ignition test* is a highly informative procedure. Heat a small amount (1 drop of a liquid or about 50 mg of a solid) gently on a small spatula or crucible cover, at first above or to the side of a microburner flame. Make a note as to whether a solid melts at low temperature or only upon heating more strongly. Observe the flammability and the nature of any flame. A yellow, sooty flame is indicative of an aromatic or a highly unsaturated aliphatic compound; a yellow but nonsooty flame is characteristic of aliphatic hydrocarbons. Oxygen content in a substance makes its flame more colorless (or blue); extensive oxygen content lowers or prevents flammability, as does halogen content. The unmistakable odor of sulfur dioxide indicates the presence of sulfur in the compound.

If a white, nonvolatile residue is left after ignition, add a drop of water and test the solution with litmus or pH paper; a sodium (or other metal) salt is indicated by an alkaline test.

2. Physical Constants. If the unknown is a solid, determine its melting point by the capillary tube method (Section 1.4). If the melting range is more than 2°, recrystallize the sample (Section 2.7).

If the unknown is a liquid, determine its *boiling point* by the micro boiling-point procedure (Section 1.5). If the boiling point is indefinite or nonreproducible or if the unknown sample is discolored or inhomogeneous, distil it (Chapter 2) and determine the boiling point. Other physical constants which are useful for liquids, particularly in the case of hydrocarbons, ethers, and other less reactive compounds, are the *refractive index* and the *density* (Sections 1.6 and 1.7, respectively). Consult your instructor about the advisability of making these measurements; full directions are given in references 1 and 2 at the end of Section 25.3.

3. Elemental Analysis. A knowledge of the elements other than carbon and hydrogen present in an unknown organic compound is a great advantage in identification. The most commonly occurring elements are oxygen, nitrogen, sulfur, and the halogens. There are no simple tests for oxygen, and the other elements are most commonly held by covalent bonds, so that they do not respond directly to the usual ionic tests. If the organic unknown is fused with molten sodium, however, most compounds react so that N, S, and X are converted to the ions CN^-, $S^=$, CNS^-, and X^-. After the excess sodium is carefully decomposed, the aqueous solution containing these anions is analyzed by conventional methods of inorganic analysis.

Sodium Fusion. Support a small Pyrex test tube in a vertical position using a clamp with either an asbestos liner or no liner (*no rubber*). Weigh a 0.5 g sample of sodium-lead alloy ("dri-Na," 9:1 lead:sodium), and place it into the test tube. Heat the alloy with a flame until it melts and fumes of sodium are seen 1–2 cm up the walls of the test tube. *Do not* heat the test tube to redness. Add to the hot alloy 2–3 drops of a liquid sample or about 10 mg of a solid sample, being careful during the addition not to allow any of the sample to contact the sides of the hot test tube. If there is no visible reaction, heat the fusion mixture gently to initiate the reaction, then discontinue the heating, and allow the reaction to subside. Next, heat the test tube to redness for a minute or two, and then let it cool. Add 3 mL of *distilled water,* and heat gently for a few minutes to decompose the excess sodium with water. Filter the solution, washing the filter paper with about 2 mL of water. (If the filtration is not done, dilute the decanted solution with about 2 mL of water.) Discard in an appropriate container the lump of metallic lead which remains. Use this fusion solution in the following tests for sulfur, nitrogen, and the halogens.

a.   Sulfur.   Acidify a 1- to 2-mL sample of the solution with *acetic acid,* and add a few drops of 0.15 M lead acetate solution. A black precipitate of PbS indicates the presence of sulfur in the original organic compound.

b.   Nitrogen.   Check the pH of a 1-mL sample of the fusion solution with Hydrion E indicator paper. The pH should be about 13. If the pH is definitely above 13, add a *small* drop of 3 M sulfuric acid to bring the pH down to about 13. If the pH of the fusion solution is definitely below 13, add a *small* drop of 6 M NaOH to bring the pH up to about 13. Add 2 drops each of a saturated solution of ferrous ammonium sulfate and of 5 M potassium fluoride. Boil the mixture gently for about 30 sec, and then carefully add 3 M sulfuric acid to the mixture, 1 drop at a time until the precipitate of iron hydroxide *just* dissolves. Avoid an excess of acid. At this point the appearance of the deep-blue color of potassium ferric ferrocyanide (Prussian blue) indicates the presence of nitrogen in the original organic compound. If the solution is green or blue-green, filter it; a blue color remaining on the filter paper is a weak but positive test for nitrogen.

c.   Halogens.   Acidify about 2 mL of the fusion solution by dropwise addition of 6 M nitric acid. Follow the acidification with blue litmus paper. Boil the solution gently for 2–3 min to expel any hydrogen sulfide or cyanide that may be present. (Sulfide and cyanide, if present, will interfere with the test for the halogens.) Cool the solution and add several drops of 0.3 M aqueous silver nitrate solution. A *heavy precipitate* of silver halide indicates the presence of chlorine, bromine, or iodine in the original organic compound. A faint turbidity should not be interpreted as a positive test. *Tentative* identification of the particular halogen may be made on the basis of color: silver chloride is white, silver bromide is pale yellow, and silver iodide is yellow. *Positive* identification must be made by standard inorganic qualitative procedures,[4] or by means of thin-layer chromatography (Section 3.7), the procedure for which follows.

Obtain from the instructor a 2.5 × 7.5-cm strip of fluorescent silica gel chromatogram sheet. About 1 cm from one end, place four equivalently spaced spots as follows. At the left, using a capillary to provide the sample, spot the original test solution. Because this solution is likely to be relatively dilute in halide ion, it may need to be respotted several times; allow the spot to dry following each application. This may be hastened by blowing on the plate. Take care, however, to keep the spot as small as possible. Next, in order, spot samples of 1 M potassium chloride, 1 M potassium bromide, and 1 M potassium iodide. Develop the plate in a solvent mixture of 2-propanone, 1-butanol, concentrated ammonium hydroxide, and water in the volume ratio of 13:4:2:1 (see Section 3.7 for details). Following development, allow

[4]See references 1, 2, and 3 at the end of Section 25.3.

the plate to air-dry, and in a hood spray the plate lightly with an indicator spray prepared by dissolving 1 g of silver nitrate in 2 mL of water and adding this solution to 100 mL of methanol containing 0.1 g of fluorescein and 1 mL of concentrated ammonium hydroxide. Allow the yellow strip to dry, and then irradiate it for several minutes with a long wavelength ultraviolet lamp (366 nm). Compare the spots formed from the test solution with those formed from the solutions of known halides. (*Note:* Iodide gives two spots.)

4. Solubility Tests. Solubility tests, including acid-base reactions, were a fundamental part of the original Kamm system and still may be used to give indicative information. However, it should be recognized that definite assignment of an unknown to a formal solubility class is rather arbitrary because of the large number of compounds that exhibit borderline behavior, no matter where the dividing lines are drawn. The classification scheme is outlined in Figure 25.4.

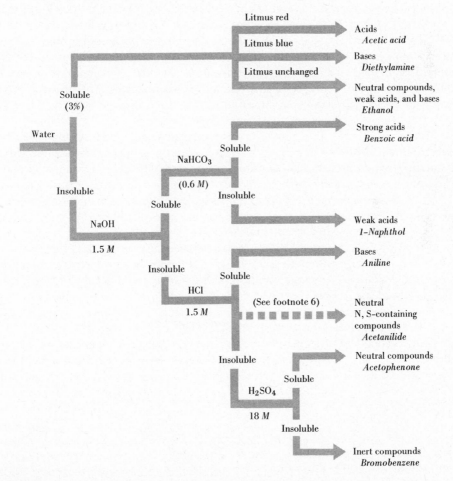

FIGURE 25.4 Classification by means of solubility and acid-base reactions, with typical examples of each class.

Make the tests with the solvents or reagents in left-to-right sequence; that is, determine the solubility in water first. In considering water solubility, for our purposes a compound is said to be soluble if it dissolves to the extent of 3 g in 100 mL of water or, more practically, 100 mg in 3 mL of water. If an unknown dissolves to this extent in water, do not test it in sodium hydroxide solution or any of the other reagents; instead, test its aqueous solution with litmus paper, and determine its subclassification as shown at the top right of Figure 25.4.

If an unknown is insoluble in water, test it in 1.5 M NaOH solution. In this and in the other reagent solutions, the definition of solubility is not 3% but any *greater solubility in the acid or base than in water*[5], which reflects the presence of an acidic or basic functional group in the water-insoluble unknown compound. Unknowns that are more soluble in dilute base than in water are tested for increased solubility (relative to water) in 0.6 M NaHCO$_3$ solution to distinguish between acids weaker or stronger than carbonic acid. Unknowns that are insoluble in both water and dilute base are next tested for solubility in 1.5 M hydrochloric acid; no attempt is made to distinguish between the weak and strong bases that dissolve in dilute acid solution.

Many compounds that are too weakly basic to dissolve in dilute aqueous acid will dissolve in or react with concentrated H$_2$SO$_4$. They are termed "neutral" on the basis of their behavior in aqueous solutions. However, there is no point in testing compounds that are soluble in dilute HCl or neutral water-insoluble compounds containing N or S in concentrated H$_2$SO$_4$, since they will invariably dissolve in or react with it.[6] Any detectable reaction such as evolution of a gas or formation of a precipitate is considered "solubility" in concentrated H$_2$SO$_4$.

The borderline for water solubility of monofunctional organic compounds, by our definition, is most commonly at or near the member of the homologous series containing five carbon atoms. Thus butanoic acid is soluble, pentanoic acid is borderline, and hexanoic acid is insoluble; 1-butanol is soluble, 1-pentanol is borderline, and 1-hexanol insoluble. The relation between molecular surface area and solubility is pointed up, however, by the observation that 2-methyl-2-butanol is clearly soluble (12.5 g/100 mL of water).

We strongly recommend that at this stage of the investigation each student present a *preliminary report* to the instructor of the results of his or her determination of physical constants, elementary analysis, and solubility tests. The instructor should advise the student of any errors (in the case of melting and boiling points a satisfactory limit of accuracy is ±5°) to prevent unnecessary loss of time.

5.   Classification Tests.   Under this heading are grouped all the reactions, other than acid-base reactions, which are indicative of other functional groups and allow assignment of the unknown to a structural class such as alkene, alkyne, alcohol,

---

[5] If the unknown does not completely dissolve in the reagent solution, do the following. After shaking the unknown with dilute acid or base, decant the liquid or filter it from undissolved sample, and carefully *neutralize* the filtrate; the formation of a precipitate or turbidity is indicative of greater solubility in the reagent than in water. It is important only to neutralize the filtrate because an unknown may show enhanced solubility in both acid and base solution if it contains *both* basic and acidic functional groups. To test for increased solubility in concentrated sulfuric acid relative to water, cautiously dilute the solution.

[6] This is the reason for the dashed arrow in Figure 25.4.

aldehyde, and so forth. Several reactions of this sort have been encountered previously in this book, and the remainder of these classification tests are presented here. The following structural classes (and reference pages) are included:

Neutral compounds
    Carbonyl compounds (aldehydes and
       ketones), p. 506
    Unsaturated compounds (alkenes), p. 507
    Alkyl halides, p. 509
    Aryl halides, p. 510
    Alcohols, p. 513
    Esters, p. 528
    Amides, p. 532

    Nitriles, p. 531
    Nitro compounds, p. 527
    Aromatic hydrocar-
       bons, p. 510
Acidic compounds
    Carboxylic acids, p. 519
    Phenols, p. 516
Basic compounds
    Amines, p. 520

6.   Preparation of Derivatives.   In the classical approach it is often useful to convert the suspected compound into another compound (a derivative) which is a solid. The melting points of two such solid derivatives, along with the melting point or boiling point of the unknown compound, often serve to identify the unknown completely. Some methods for preparing derivatives have been given, and other representative and dependable procedures for preparing solid derivatives of the other types of compounds are presented here. Short tables of some selected compounds, along with their melting or boiling points and melting points of several solid derivatives, are also presented.

Each of the classes of compounds tabulated will be considered in the following manner. *Classification tests* for the functional group characteristic of the type of compound are given (or in those cases where these tests have already been presented, reference is made to suitable places in the text), and these are followed by brief descriptions of methods of *preparation of several representative solid derivatives.* References are made to the tables of compounds and derivatives which follow.

As mentioned in connection with the outline of the six fundamental steps, it is generally wise to carry these out in the order listed. On the other hand, there is nothing necessarily sacred about the order of the first four steps. It usually makes little difference whether one carries out an elementary analysis before doing solubility tests or vice versa. For example, if a water-insoluble compound is found to be soluble in dilute hydrochloric acid, an amine is a possibility, and an immediate test for nitrogen is indicated. Conversely, if a positive nitrogen test had been found first, a solubility test for an amine would be a logical next step.

There is one warning that should be emphasized, however. Do not go directly to the preparation of a derivative on the basis of a hunch about the class of compound to which your unknown belongs or even its specific identity; rather, make certain of the type of functional group present by obtaining one or more positive classification tests before attempting the preparation of any derivative. Going directly from a melting point or odor to a derivative is the most common error made by beginners in qualitative organic analysis. For example, it is frustrating to try to make a ketone derivative when the unknown is actually an ester!

In deciding which of the classification tests to try first, one should use the information gained in the first four steps. To continue with the example used above, if the unknown has been found to contain nitrogen and to be soluble in dilute acidic solution, a classification test for an amine should be applied first. In the absence of clear indication of the type of compound, the most reliable classification tests should be applied first. Among the most reliable, certainly, are those for aldehydes and ketones.

ALDEHYDES AND KETONES (Tables 25.1 and 25.2)

A.   Classification Tests

1.   2,4-Dinitrophenylhydrazine.   This reagent will give a positive test for either an aldehyde or a ketone. Experimental details for this test are given in Section 13.1.

2.   Chromic Acid.   This reagent serves as a clear-cut method for distinguishing between aldehydes and ketones. The reactions are discussed in detail in Section 13.2, and the experimental procedures are given there.

3.   Iodoform Test Using Sodium Hypoiodite (NaOI).   This test is used to detect the presence of methyl ketones (compounds having a terminal $—COCH_3$ group), as compared to a ketone which does not contain this specific functional group. Hydroxymethyl compounds ($—CHOHCH_3$) also give iodoform with NaOI, so that it is always necessary to consider this alternative before making a final decision; the problem is solved by determining that a keto group is present or absent. This reaction is discussed in Section 14.2, and experimental procedures accompany that discussion.

4.   Tollens' Test.   Another method for distinguishing between aldehydes and ketones (an alternative to the chromic acid test) is the Tollens' test. A positive test determines the presence of an aldehyde function, whereas no reaction occurs with ketones. Tollens' reagent consists of silver ammonia complex, $Ag(NH_3)_2^{\oplus}$, in an ammonia solution. This reagent is reduced on reaction with an aliphatic or aromatic aldehyde, whereas the aldehyde is oxidized to the corresponding carboxylic acid; the silver is reduced from the $+1$ state to elemental silver and frequently is deposited as a silver mirror on the glass wall of the test tube. Thus the formation of the silver mirror is considered a positive test. Equation 1 shows the reaction that occurs.

$$RCHO + 2\,Ag(NH_3)_2^{\oplus} + 2\,HO^{\ominus} \longrightarrow 2\,\underline{Ag} + RCO_2^{\ominus}NH_4^{\oplus} + H_2O + 3\,NH_3 \quad (1)$$

Similar tests for aldehydes make use of Fehling's and Benedict's reagents, which contain complex salts (tartrate and citrate, respectively) of cupric ion as the oxidizing agents. With these reagents a positive test is the formation of a brick-red precipitate of cuprous oxide (Cu_2O), which forms when Cu^{++} is reduced to Cu^+ by the aldehyde. These two tests are more useful in distinguishing between aliphatic and aromatic aldehydes, since the aliphatic compounds give a fast reaction.

Experimental Procedure

Prepare Tollens' reagent by mixing solution A (2.5 g of silver nitrate in 43 mL of distilled water) and solution B (3 g of potassium hydroxide in 42 mL of distilled water) according to the following directions. Obtain 3 mL of both solutions A and B. To 3 mL of solution A, add concentrated ammonium hydroxide solution dropwise until the initial brown precipitate begins to clear. The solution should be grayish and almost clear. Add 3 mL of solution B, and again add concentrated ammonium hydroxide dropwise until the solution is almost clear.

To carry out the test, add 0.5 mL of the reagent to 3 drops or 50–100 mg of the unknown compound; the formation of a silver mirror or black precipitate constitutes a positive test.[7] If no reaction occurs at room temperature, warm the solution slightly in a beaker of warm water.

At the end of the laboratory period discard any unused Tollens' reagent; *do not store the solution because it decomposes on standing and yields an explosive precipitate.*

B. Derivatives

Two very suitable solid derivatives of aldehydes and ketones are the *2,4-dinitrophenylhydrazones* and the *semicarbazones*. Oximes are also sometimes useful. Methods for preparing and purifying each of these are given in Section 14.1; see Tables 25.1 and 25.2 for lists of possible compounds and their derivatives.

UNSATURATED COMPOUNDS

Classification Tests

Two common types of unsaturated compounds are alkenes and alkynes, characterized by the carbon-carbon double and triple bond, respectively, as the functional group. There are no simple direct ways to prepare solid derivatives of unsaturated aliphatic compounds having no other functional groups, but it is often useful to detect the presence of these two functional groups. The two common qualitative tests for unsaturation are the reaction of the compounds with *bromine in carbon tetrachloride* and with *potassium permanganate*. In both cases a positive test is denoted by decoloration of the reagent.

Bromine in Carbon Tetrachloride.   Bromine will add to the carbon-carbon double bond of alkenes (and alkynes) to produce dibromoalkanes (equation 2).[8] When this reaction occurs, molecular bromine is destroyed and its characteristic color disappears

---

[7] If the test tube is not scrupulously clean, the silver will not deposit on the wall as a mirror, but will appear as a black precipitate.

[8] More detailed descriptions of these qualitative tests may be found in Chapter 8 or in reference 1 at the end of Section 25.3.

(if bromine is not present in excess). The *rapid* disappearance of the color of bromine is a positive test for unsaturation. The test is not unequivocal, however, because some

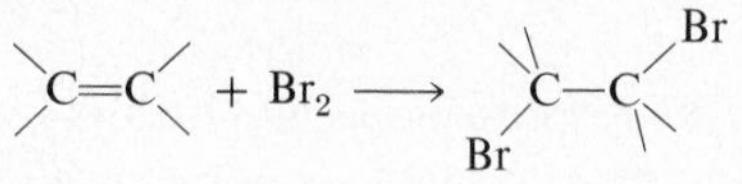

$$\ce{>C=C<} + Br_2 \longrightarrow \underset{\displaystyle Br}{\overset{\displaystyle Br}{\ce{>C-C<}}} \tag{2}$$

alkenes do not react with bromine and some react very slowly. In case of a negative test, moreover, it is usually good practice to carry out the permanganate test.

Experimental Procedure

To 1 or 2 mL of 0.1 *M* bromine in carbon tetrachloride solution, add 1 or 2 drops of the unknown. *Rapid* disappearance of the bromine color to give a colorless solution is a positive test for unsaturation.

Potassium Permanganate (the Baeyer Test).   A second qualitative test for unsaturation, the Baeyer test, depends on the ability of potassium permanganate to oxidize the carbon-carbon double bond to give alkanediols (equation 3) or the carbon-carbon triple bond to give carboxylic acids (equation 4). The permanganate is destroyed in

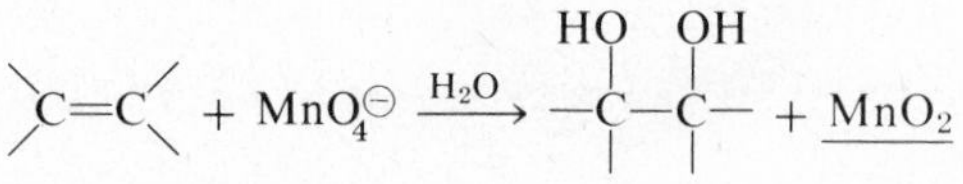

$$\ce{>C=C<} + MnO_4^{\ominus} \xrightarrow{H_2O} \underset{}{\overset{HO\ \ OH}{-C-C-}} + \underline{MnO_2} \tag{3}$$

$$R-C{\equiv}C-R' + MnO_4^{\ominus} \xrightarrow{H_2O} R-CO_2H + HO_2C-R' + \underline{MnO_2} \tag{4}$$

the reaction, and a brown precipitate of MnO_2 is produced. The disappearance of the characteristic color of the permanganate ion is a positive test for unsaturation. However, care must be taken, since compounds containing certain other types of functional groups (for example, aldehydes, containing the $-CH{=}O$ group) also decolorize permanganate ion.

Experimental Procedure

Dissolve 1 or 2 drops of the unknown in 2 mL of 95% ethanol, then add 0.1 *M* $KMnO_4$ solution dropwise, observing the results. Count the number of drops added before the permanganate color persists. For a *blank determination* count the number of drops that may be added to 2 mL of 95% ethanol before the color persists. A significant difference in the number of drops required in the two cases is a positive test for unsaturation.

ALKYL HALIDES

Classification Tests

Qualitative tests for alkyl halides are useful in deciding whether the compound in question is a primary, secondary, or tertiary halide. In general it is quite difficult to prepare solid derivatives of alkyl halides, so we limit this discussion to the two qualitative tests: (1) the reaction with *alcoholic silver nitrate* solution and (2) the reaction with *sodium iodide in acetone*.

1. Alcoholic Silver Nitrate. If a compound is known to contain a halogen (bromine, chlorine, or iodine), information concerning its environment may be obtained from observation of its reaction with alcoholic silver nitrate. The overall reaction is shown in equation 5. Such a reaction will be of the S_N1 type. As asserted in Section 10.1,

$$RX + AgNO_3 \xrightarrow{\text{ethanol}} \underline{AgX} + RONO_2 \tag{5}$$

tertiary halides are more reactive in an S_N1 reaction than secondary halides, which are in turn more reactive than primary halides. Differing rates of silver halide precipitation would be expected from halogen in each of these environments, namely, primary $<$ secondary $<$ tertiary. These differences are best determined by testing in separate test tubes authentic samples of primary, secondary, and tertiary halides with silver nitrate and observing the results.

Alkyl bromides and iodides react more rapidly than chlorides, and the latter may require warming to produce a reaction in a reasonable period.

Aryl halides are unreactive toward the test reagent, as are any vinyl or alkynyl halides generally. Allylic and benzylic halides, even when primary, show reactivities as great as or greater than tertiary halides because of resonance stabilization of the resulting allyl cation.

Experimental Procedure

Add 1 drop of the alkyl halide to 2 mL of a 0.1 M solution of silver nitrate in 95% ethanol. If no reaction is observed within 5 min at room temperature, warm the mixture in a beaker of boiling water and observe any change. Note the color of any precipitates; silver chloride is white, silver bromide is pale yellow, and silver iodide is yellow. If there is any precipitate, add several drops of 1 M nitric acid solution to it, and note any changes; the silver halides are insoluble in acid. To determine expected reactivities, test known primary, secondary, and tertiary halides in this manner. If possible, use alkyl iodides, bromides, and chlorides so that differences in halogen reactivity can also be observed.

2. Sodium Iodide in Acetone. Another method for distinguishing between primary, secondary, and tertiary halides makes use of sodium iodide dissolved in acetone. This test complements the alcoholic silver nitrate test, and when these two tests are used together, it is possible to determine fairly accurately the gross structure of the attached alkyl group.

The test depends on the fact that both sodium chloride and sodium bromide are not very soluble in acetone, whereas sodium iodide is. The reactions that occur (equations 6 and 7) are S_N2 substitutions in which iodide ion is the nucleophile; the order of reactivity is primary > secondary > tertiary.

$$RCl + NaI \xrightarrow{\text{acetone}} RI + \underline{NaCl} \qquad\qquad (6)$$

$$RBr + NaI \xrightarrow{\text{acetone}} RI + \underline{NaBr} \qquad\qquad (7)$$

With the reagent, primary bromides give a precipitate of sodium bromide in about 3 min at room temperature, whereas the primary and secondary chlorides must be heated to about 50° before reaction occurs. Secondary and tertiary bromides react at 50°, but the tertiary chlorides fail to react in a reasonable time. It should be noted that this test is necessarily limited to bromides and chlorides.

Experimental Procedure

Place 1 mL of the sodium iodide–acetone test solution in a test tube, and add 2 drops of the chloro or bromo compound. If the compound is a solid, dissolve about 50 mg of it in a minimum volume of acetone, and add this solution to the reagent. Shake the test tube, and allow it to stand for 3 min at room temperature. Note whether a precipitate forms; if no change occurs after 3 min, warm the mixture in a beaker of water at 50°. After 6 min of heating, cool to room temperature, and note whether any precipitate forms. Occasionally a precipitate forms immediately after combination of the reagents; this represents a positive test only if the precipitate remains after shaking the mixture and allowing it to stand for 3 min.

Carry out this reaction with a series of primary, secondary, and tertiary halides, both chlorides and bromides. Note in all cases the differences in reactivity as evidenced by the rate of formation of sodium bromide or chloride.

ARYL HALIDES AND AROMATIC HYDROCARBONS (Tables 25.3 and 25.4)

A. Classification Test

This test for the presence of an aromatic ring should be performed only on compounds that have been shown to be insoluble in concentrated sulfuric acid (see solubility tests). The test involves the reaction between an aromatic compound and chloroform in the presence of anhydrous aluminum chloride catalyst. The colors produced in this type of reaction are often quite characteristic for certain aromatic compounds, whereas aliphatic compounds give little or no color with this test. Some typical examples are tabulated below. Often these colors change with time and ultimately yield brown-colored solutions. Carbon tetrachloride may be used in place of chloroform; it yields similar colors.

Type of Compound	Color
Benzene and homologs	Orange to red
Aryl halides	Orange to red
Naphthalene	Blue
Biphenyl	Purple

The test is based upon a series of Friedel-Crafts alkylation reactions; for benzene, the ultimate product is triphenylmethane (equation 8).

$$3\ C_6H_6 + CHCl_3 \xrightarrow{\ AlCl_3\ } (C_6H_5)_3CH + 3\ HCl \tag{8}$$

The colors arise owing to formation of species such as triphenylmethyl cations $(C_6H_5)_3C^+$, which remain in the solution as $AlCl_4^-$ salts; ions of this sort are highly colored owing to the extensive delocalization of charge that is possible throughout the three aromatic rings.

The test is significant if positive, but a negative test does not rule out an aromatic structure; some compounds are so unreactive that they do not readily undergo Friedel-Crafts reactions.

Positive tests for aryl halides are difficult to obtain directly, and some of the best evidence for their presence involves indirect methods. Elemental analysis will indicate the presence of halogen. If *both* the silver nitrate and sodium iodide–acetone tests are negative, then the compound is most likely a vinyl or an aromatic halide, both of which are very unreactive toward silver nitrate and sodium iodide. Distinction between a vinyl and an aromatic halide can be made by means of the aluminum chloride-chloroform test.

Experimental Procedure

Heat about 100 mg of *anhydrous* aluminum chloride in a Pyrex test tube held almost horizontally until the material has sublimed to 3 or 4 cm above the bottom of the tube. Allow the tube to cool until it is almost comfortable to touch, and then add down the side of the tube about 20 mg of a solid, or 1 drop of a liquid, unknown, followed by 2 or 3 drops of chloroform. The appearance of a bright color ranging from red to blue where the sample and chloroform come in contact with the aluminum chloride is a positive indication of an aromatic ring.

B. Derivatives

Two types of derivatives can be used to characterize aromatic hydrocarbons and aryl halides. These are prepared by (1) nitration and (2) side-chain oxidation. The second method involves oxidation of a side chain to a carboxylic acid group. Since carboxylic acids are often solids, they themselves serve as suitable derivatives. The mechanism and equations for this oxidation reaction have been discussed (Section 15.1). The two common methods of oxidation use either chromic acid or potassium permanganate. Although the experimental details of a permanganate oxidation are given in Chapter 15, procedures for doing both types of oxidations are given here, since much smaller amounts of compound are normally used.

Experimental Procedure

1. Nitration. Some of the best solid derivatives of aryl halides are mono- and dinitration products. Two general procedures can be used for nitration. Whenever nitrating a compound, whether it is known or unknown, use care, since many of these compounds react vigorously under typical nitration conditions.

Method A. This method yields *m*-dinitrobenzene from benzene or nitrobenzene and the *p*-nitro derivative from chloro- or bromobenzene, benzyl chloride, or toluene. Dinitro derivatives are obtained from phenol, acetanilide, naphthalene, and biphenyl. Add about 1 g of the compound to 4 mL of concentrated sulfuric acid. Add dropwise to this mixture 4 mL of concentrated nitric acid; shake after each addition. Carry out the reaction in a large test tube or small Erlenmeyer flask, and heat at 45° for about 5 min, using a beaker of water to supply the heat. Then pour the reaction mixture onto 25 g of ice, and collect the precipitate on a filter. The solid may be recrystallized from aqueous ethanol if needed.

Method B. This method is the best one to use for nitrating halogenated benzenes, since dinitration occurs to give compounds which have higher melting points and are easier to purify than are mononitration products obtained from method A. The xylenes, mesitylene, and pseudocumene yield trinitro compounds. Follow the procedure used for method A, except that 4 mL of *fuming* nitric acid should be used in place of the *concentrated* nitric acid and the mixture should be warmed using a steam bath for 10 min. If little or no nitration occurs, substitute *fuming* sulfuric acid for the *concentrated* sulfuric acid. *Carry out this reaction in a hood.*

2. Side-Chain Oxidation. a. Permanganate Method. Add 1 g of the compound to a solution prepared from 80 mL of water and 4 g of potassium permanganate. Add 1 mL of 3 *M* sodium hydroxide solution, and heat the mixture at reflux until the purple color characteristic of the permanganate has disappeared; this will normally take from 30 min to 3 hr. At the end of the reflux period, cool the mixture and carefully acidify it with 3 *M* sulfuric acid. Now heat the mixture for an additional 30 min and cool it again; remove excess brown manganese dioxide (if any) by addition of sodium bisulfite solution. The bisulfite serves to reduce the manganese dioxide to manganous ion, which is water soluble. Collect the solid acid that remains by vacuum filtration. Recrystallize the acid from toluene or aqueous ethanol. If little or no solid acid is formed, this may be due to the fact that the acid is somewhat water soluble. In this case extract the aqueous layer with chloroform, diethyl ether, or dichloromethane, dry the organic extracts, and remove the organic solvent by means of a steam bath in the hood. Recrystallize the acid that remains. In this particular method the presence of base during the oxidation often means that some silicic acid will form on acidification; thus purification before determining the melting point is necessary.

b.   Chromic Acid Method.   Dissolve 7 g of sodium dichromate in 15 mL of water, and add 2 to 3 g of the compound to be oxidized. Add 10 mL of concentrated sulfuric acid to the mixture with mixing and cooling. Attach a reflux condenser to the flask, and heat gently until a reaction ensues; as soon as the reaction begins, remove the flame and cool the mixture if necessary. After spontaneous boiling subsides, heat the mixture at reflux for 2 hr. Pour the reaction mixture into 25 mL of water, and collect the precipitate by filtration. Transfer the solid to a flask, add 20 mL of 2 M sulfuric acid, and then warm the flask on a steam cone with stirring. Cool the mixture, collect the precipitate, and wash it with about 20 mL of cold water. Dissolve the residue in 20 mL of 1.5 M sodium hydroxide solution and filter the solution. Add the filtrate, with stirring, to 25 mL of 2 M sulfuric acid. Collect the new precipitate, wash with cold water, and recrystallize from either toluene or aqueous ethanol.

Lists of possible aromatic hydrocarbons and halides appear in Tables 25.3 and 25.4, along with derivatives.

ALCOHOLS (Table 25.5)

A.   Classification Tests

The tests for the presence of a hydroxy group not only detect the presence of the group but may also indicate whether it occupies a primary, secondary, or tertiary position.

1.   Chromic Acid in Acetone.   This test may be used to detect the presence of a hydroxy group, provided that it has been shown previously that the molecule does not contain an aldehyde function. The reactions and experimental procedures for this test are given in Section 13.2. It has been pointed out that chromic acid does not distinguish between primary and secondary alcohols, since primary and secondary alcohols *both* give a positive test, whereas tertiary alcohols do not.

2.   The Lucas Test.   This test is used to distinguish between primary, secondary, and tertiary alcohols. The reagent used is a mixture of concentrated hydrochloric acid and zinc chloride, which on reaction with alcohols converts them to the corresponding alkyl chlorides. With this reagent primary alcohols give no appreciable reaction, secondary alcohols react more rapidly, and tertiary alcohols react very rapidly. A positive test depends on the fact that the alcohol is soluble in the reagent, whereas the alkyl chloride is not; thus the formation of a second layer or an emulsion constitutes a positive test. The *solubility of the alcohol in the reagent places limitations on the utility of the test,* and in general only monofunctional alcohols with six or less carbon

Primary: $RCH_2OH + HCl \xrightarrow{ZnCl_2}$ No reaction   (9)

Secondary: $R_2CHOH + HCl \xrightarrow{ZnCl_2} R_2CHCl + H_2O$ (10)

Tertiary: $R_3COH + HCl \xrightarrow{ZnCl_2} R_3CCl + H_2O$ (11)

atoms, as well as polyfunctional alcohols, can be used. Note the similarity of this reaction with the nucleophilic displacement reactions between alcohols and hydrohalic acids which are discussed in Section 10.1. In the Lucas test the presence of zinc chloride, which is a Lewis acid, greatly increases the reactivity of alcohols toward hydrochloric acid.

Experimental Procedure

Add 10 mL of the hydrochloric acid–zinc chloride reagent (Lucas reagent) to about 1 mL of the compound in a test tube. Stopper the tube and shake; allow the mixture to stand at room temperature. Try this test with known primary, secondary, and tertiary alcohols, and note the *time* required for the formation of an alkyl chloride, which will appear either as a second layer or as an emulsion. Repeat the test with an unknown, and compare the result with the results from the knowns.

3.   The ceric nitrate reagent can also be used as a qualitative test for alcohols. Although this reagent has been used primarily for phenols, it does give a positive test with alcohols. Discussion about and experimental procedures for this test are given under Phenols in this chapter.

B.   Derivatives

Two common derivatives of alcohols are the urethanes and the benzoate esters; the former are best for primary and secondary alcohols, whereas the latter are useful for all types of alcohols.

1.   Urethanes.   When an alcohol is allowed to react with an aryl substituted isocyanate, $ArN=C=O$, addition of the alcohol occurs to give a urethane (equation 12).

$$ArN=C=O + R'OH \longrightarrow ArNH-\overset{\overset{\displaystyle O}{\|}}{C}-OR'$$ (12)

Some commonly used isocyanates are α-naphthyl, *p*-nitrophenyl, and phenyl isocyanate. A major side reaction is that of water with the isocyanate; water hydrolyzes the isocyanate to an amine, and the amine reacts with more isocyanate to give a disub-stituted urea (equations 13 and 14). Since the ureas are high-melting amides owing to

$$\text{ArN}{=}\text{C}{=}\text{O} + \text{H}_2\text{O} \longrightarrow (\text{ArNH}{-}\text{COOH}) \longrightarrow \text{ArNH}_2 + \text{CO}_2 \qquad (13)$$
A carbamic acid
(unstable)

$$\text{ArN}{=}\text{C}{=}\text{O} + \text{ArNH}_2 \longrightarrow \text{ArNH}{-}\overset{\overset{\textstyle O}{\|}}{\text{C}}{-}\text{NHAr} \qquad (14)$$
A disubstituted urea

their symmetry, their presence makes purification of the desired urethane quite difficult. In using this procedure, take precautions to ensure that the alcohol is *anhydrous*. The procedure works best for water-insoluble alcohols which can therefore be obtained easily in anhydrous form.

This type of derivative can also be useful for phenols; the procedure given here has been generalized so that it can be used for alcohols and phenols. Other deriva-tives of phenols are given later.

Experimental Procedure

DO IT SAFELY

Aryl isocyanates such as those used in the preparation of these derivatives are toxic. Take normal precautions in handling them. Quickly washing your hands with soap and warm water will remove the isocyanates from your skin.

Place 1 g of the *anhydrous* alcohol or phenol in a small round-bottomed flask, and add 0.5 mL of phenyl isocyanate or α-naphthyl isocyanate (recap the bottle of isocyanate tightly). If you are preparing the derivative of a phenol, also add 2 or 3 drops of dry pyridine as a catalyst. Affix a calcium chloride drying tube to the flask. Warm the reaction mixture with a heating mantle or a steam bath for 5 min. Cool the mixture in an ice-water bath, and scratch the mixture with a stirring rod to induce crystallization. Recrystallize the crude derivative from either carbon tetrachloride or petroleum ether. (*Note:* 1,3-Di-(α-naphthyl)-urea has mp 293° and 1,3-diphenylurea (carbanilide) has mp 237°; if your product shows one of these melting points, repeat the preparation, taking greater care to maintain anhydrous conditions.)

2. 3,5-Dinitrobenzoates. The reaction between 3,5-dinitrobenzoyl chloride and an alcohol gives the corresponding ester (equation 15). This method is useful for primary, secondary, and tertiary alcohols, especially those which are water soluble and which are likely to contain traces of water.

$$\text{ROH} + \text{Cl—C}\overset{\text{O}}{\underset{}{}}\!\!\bigcirc\!\!\overset{\text{NO}_2}{\underset{\text{NO}_2}{}} \longrightarrow \text{RO—C}\overset{\text{O}}{\underset{}{}}\!\!\bigcirc\!\!\overset{\text{NO}_2}{\underset{\text{NO}_2}{}} + \text{HCl} \qquad (15)$$

Experimental Procedure

Method A. (*Note:* 3,5-Dinitrobenzoyl chloride is reactive toward water; it should be used immediately after weighing. Take care to minimize its exposure to air and to keep the bottle tightly closed.) In a small flask bearing a reflux condenser mix 2 mL of the alcohol with about 0.5 g of 3,5-dinitrobenzoyl chloride and 0.5 mL of pyridine. Boil the mixture gently for 30 min (15 min is sufficient for a primary alcohol). Cool the solution and add about 10 mL of 0.6 M aqueous sodium bicarbonate solution. Cool this solution in an ice-water bath, and collect the crude crystalline product. Recrystallize the product from aqueous ethanol. A minimum volume of solvent should be used; adjust the composition of the solvent by adding just enough water to the alcohol so that the product dissolves in the hot solution but yields crystals when cooled.

Method B. Carry out this reaction in the hood, if possible, or use a gas trap such as that shown in Figure 5.1. To a 50-mL flask fitted with a reflux condenser (bearing a gas trap if a hood is not available) add 1 g of 3,5-dinitrobenzoic acid, 3 mL of thionyl chloride, and 1 drop of pyridine. Heat the mixture at reflux until the acid has dissolved and then for an additional 10 min. The total reflux time should be about 30 min. Equip the flask for simple distillation, as in Figure 2.2. Cool the receiving flask with an ice-salt bath, and attach the vacuum adapter to an aspirator by means of a safety trap such as that shown in Figure 2.17. Evacuate the system and distil the excess thionyl chloride by heating with a steam bath. When the thionyl chloride has been removed, cautiously release the vacuum. Discard the excess thionyl chloride by pouring slowly down a drain *in the hood* or by putting it in a container for recovered thionyl chloride. To the residue in the stillpot, which is 3,5-dinitrobenzoyl chloride, add in the same flask 2 mL of the alcohol and 0.5 mL of pyridine. Fit the flask with a reflux condenser bearing a drying tube. Proceed with the period of reflux following the procedure of method A.

A list of alcohols and their derivatives is given in Table 25.5.

PHENOLS (Table 25.6)

A. Classification Tests

Several tests can be used to detect the presence of a phenolic hydroxy group: (1) bromine water, (2) ceric nitrate reagent, and (3) ferric chloride solution. In addition to these, solubility tests give a preliminary indication of a phenol, since phenols are soluble in 1.5 M sodium hydroxide solution but generally insoluble in 0.6 M sodium bicarbonate solution. Care must be exercised here, however, because phenols containing highly electronegative groups are stronger acids and may be soluble in 0.6 M sodium bicarbonate. Examples are 2,4,6-tribromophenol and 2,4-dinitrophenol.

1. Bromine Water. Phenols are generally highly reactive toward electrophilic substitution and consequently are brominated readily by bromine water (see equation 16, for example). The rate of bromination is much greater in water than in

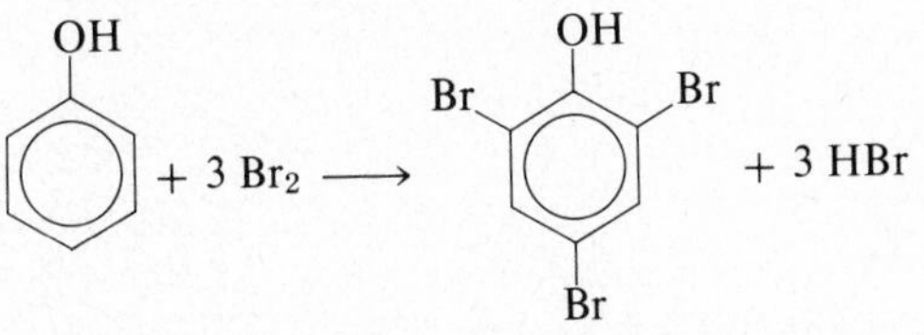

$$\text{(16)}$$

carbon tetrachloride solution. The water, being more polar than carbon tetrachloride, increases the ionization of bromine and thus enhances the ionic bromination mechanism. Although hydrogen bromide is liberated, it is not observed when using water as the solvent. Phenols are so reactive that all unsubstituted positions *ortho* and *para* to the hydroxy group are brominated. The brominated compounds so formed are often solids and can also be used as derivatives. Aniline and substituted anilines are also very reactive toward bromine and react analogously; however, solubility tests can be used to distinguish between anilines and phenols.

Experimental Procedure

Prepare a 1% aqueous solution of the unknown. If necessary, dilute sodium hydroxide solution may be added dropwise to bring the phenol into solution. Add dropwise to this solution a saturated solution of bromine in water; continue addition until the bromine color remains. Note how much bromine water was used, and try this experiment on phenol and aniline for purposes of comparison.

2. Ceric Nitrate Reagent. Alcohols and phenols are capable of replacing nitrate ions in complex cerate anions, resulting in a change from a yellow to a red solution

$$(\text{NH}_4^{\oplus})_2\text{Ce(NO}_3)_6^{\ominus} + \text{ROH} \longrightarrow (\text{NH}_4^{\oplus})_2\text{Ce(OR)(NO}_3)_5^{\ominus} + \text{HNO}_3 \qquad \text{(17)}$$
$$\underset{\text{Yellow}}{} \qquad \qquad \underset{\text{Red}}{}$$

(equation 17). Alcohols and phenols having no more than 10 carbon atoms give a positive test. Alcohols give a red solution. Phenols give a brown to greenish-brown precipitate in aqueous solution; in dioxane a red-to-brown solution is produced. Aromatic amines may be oxidized by the reagent and give a color indicating a positive test.

Experimental Procedure

Dissolve about 20 mg of a solid or 1 drop of a liquid unknown in 1–2 mL of water, and add 0.5 mL of the ceric ammonium nitrate reagent; shake and note the color. If the unknown in insoluble in water, dissolve it in 1 mL of dioxane, and add 0.5 mL of the reagent.

3. Ferric Chloride Test. Most phenols and enols react with ferric chloride to give colored complexes. The colors vary, depending not only on the nature of the phenol or enol but also on the solvent, concentration, and time of observation. Some phenols which do not give coloration in aqueous or alcoholic solution do so in chloroform solution, especially after addition of a drop of pyridine. The nature of the colored complexes is still uncertain; they may be ferric phenoxide salts that absorb visible light to give an excited state in which electrons are delocalized over both the iron atoms and the conjugated organic system. The production of a color is typical of phenols and enols; however, many of them do *not* give colors, so a negative ferric chloride test must not be taken as significant without supporting information (for example, the ceric nitrate and bromine water tests).

Experimental Procedure

Dissolve 30 to 50 mg of the unknown compound in 1–2 mL of water (or a mixture of water and 95% ethanol if the compound is not water soluble), and add several drops of a 0.2 M aqueous solution of ferric chloride. Most phenols produce red, blue, purple, or green coloration; enols give red, violet, or tan coloration.

B. Derivatives

Two useful solid derivatives of phenols are α-naphthyl urethane and bromo derivatives. The preparation of urethanes has already been discussed in this chapter under Alcohols. Although either substituted urethane could be prepared, the majority of the derivatives reported are the α-naphthyl urethanes. They are suggested as the urethanes of choice.

Experimental Procedure

The solution for bromination in this preparation should be supplied for you. (The solution is prepared by dissolving 10 g of potassium hydroxide in 60 mL of water and adding 6 g of bromine.) Dissolve 1 g of the phenolic compound in water or 95% ethanol, and add the brominating solution to it *dropwise*. Continue the addition until the reaction mixture begins to develop a yellow color, indicating excess bromine. Let the mixture stand for about 5 min; if the yellow coloration begins to fade, add another drop or two of the brominating solution. Add 50 mL of water to the mixture and then a few drops of 0.5 M sodium bisulfite solution to destroy the excess bromine.

Shake the mixture vigorously, and remove the solid derivative by vacuum filtration. It may prove necessary to neutralize the solution with concentrated hydrochloric acid to achieve precipitation of the solid derivative. This derivative may be purified by recrystallization from 95% ethanol or aqueous ethanol.

A listing of phenols and their derivatives is given in Table 25.6.

CARBOXYLIC ACIDS (Table 25.7)

A. Classification Test

One of the best qualitative tests for the carboxylic acid group is solubility in basic solutions. Carboxylic acids are soluble in both 1.5 M sodium hydroxide solution and in 0.6 M sodium bicarbonate solution, from which they can be regenerated by addition of acid. Solubility properties have already been discussed earlier in this section under Solubility Tests.

Information of a quantitative nature can be obtained about carboxylic acids, this being the *equivalent weight* or *neutralization equivalent*. This value can be obtained by titrating a known weight of the acid with a known volume of standardized base solution. Further discussion of the equivalent weight is given in Section 15.1.

Experimental Procedure

Dissolve an accurately weighed sample (about 0.2 g) of the acid in 50 to 100 mL of water or 95% ethanol or a mixture of the two. It may be necessary to warm the mixture to dissolve the compound completely. Titrate the solution with a *standardized* sodium hydroxide solution having a concentration of about 0.1 M. Use phenolphthalein as the indicator, and from these data calculate the equivalent weight as described in Section 15.1.

B. Derivatives

Three good solid derivatives of carboxylic acids are (1) amides, (2) anilides, and (3) p-toluidides. These derivatives are prepared from the corresponding acid chloride by treatment of the latter with either ammonia, aniline, or p-toluidine. The amides are generally less satisfactory than the other two because they tend to be more soluble in water and as a result are harder to isolate. The acid chlorides are most conveniently prepared from the acid, or its salt, and thionyl chloride.

$$\text{RCOOH (or RCOO}^{\ominus}\text{Na}^{\oplus}) + \text{SOCl}_2 \rightarrow \text{RCOCl} + \text{SO}_2 + \text{HCl (or NaCl)} \tag{18}$$

Amides: $\quad \text{RCOCl} + 2\,\text{NH}_3 \xrightarrow{\text{cold}} \text{RCONH}_2 + \text{NH}_4^{\oplus}\,\text{Cl}^{\ominus}$ \hfill (19)

Anilides: $\quad \text{RCOCl} + 2\,\text{C}_6\text{H}_5\text{NH}_2 \rightarrow \text{RCONHC}_6\text{H}_5 + \text{C}_6\text{H}_5\text{NH}_3^{\oplus}\,\text{Cl}^{\ominus}$ \hfill (20)

Toluidides: $\quad \text{RCOCl} + 2\,p\text{-CH}_3\text{C}_6\text{H}_4\text{NH}_2 \rightarrow \text{RCONHC}_6\text{H}_4\text{CH}_3\text{-}p$
$$+ p\text{-CH}_3\text{C}_6\text{H}_4\text{NH}_3^{\oplus}\,\text{Cl}^{\ominus} \tag{21}$$

Experimental Procedure

To prepare the acid chloride from either a carboxylic acid or its salt, place 1 g of the acid (or its sodium salt) in a small round-bottomed flask with 2 mL of thionyl chloride. Affix a water-cooled reflux condenser bearing a gas trap such as that shown in Figure 5.1. Using either a small burner or a heating mantle, heat the mixture at reflux for 15–30 min.

1. Amides. *In the hood* pour the mixture containing the acid chloride and unchanged thionyl chloride into 15 mL of *ice-cold,* concentrated ammonium hydroxide solution. Be very careful on this addition; the reaction is quite vigorous. Collect the precipitated amide derivative of the carboxylic acid on a filter, and recrystallize from water or aqueous ethanol.

2. Anilides and *p*-toluidides. Prepare the acid chloride of the carboxylic acid from 1 g of the acid and 2 mL of thionyl chloride. After the 30-min period of reflux, cool the mixture. Dissolve 1 g of either aniline or *p*-toluidine in 30 mL of cyclohexane; slight warming may be necessary to effect complete solution. Pour the cooled acid chloride into the cyclohexane solution of the amine, and heat the resulting mixture on a steam bath for 2–3 min. A heavy white precipitate of the amine hydrochloride will form; this precipitate should be removed by vacuum filtration and *set aside* (do not discard). In a separatory funnel wash the filtrate with 5 mL of water, followed by 5 mL of 1.5 *M* HCl, 5 mL of 1.5 *M* NaOH, and finally with 5 mL of water. In some cases precipitation may occur in the organic layer during one or more of these washings. If this occurs, warm the solution gently with a warm-water bath to redissolve the precipitate. Following the washings remove the cyclohexane from the organic layer either by distillation or by evaporation on a steam bath in the hood. Recrystallize the derivative of the carboxylic acid from aqueous ethanol. (*Note:* If little residue remains following evaporation of the cyclohexane, dissolve the precipitate which was removed earlier in about 10 mL of water. Stir and remove any undissolved solid by filtration, combining this solid with the residue obtained from the cyclohexane.)

AMINES (Tables 25.8 and 25.9)

A. Classification Tests

Two common qualitative tests for amines are the Hinsberg test and the nitrous acid test. We have not included the nitrous acid test in this edition because recent research indicates that the N-nitroso derivatives of secondary amines constitute one of the four or five major classifications of carcinogenic substances. The risk of producing an as-yet unrecognized carcinogenic material in this test outweighs any possible benefit of a test which can be misleading and difficult to interpret. The modified sodium nitroprusside test has been included as an alternative.

1. The Hinsberg Test. The reaction between primary or secondary amines and benzenesulfonyl chloride (equations 22 and 24, respectively) yields the corresponding substituted benzenesulfonamide. The reaction is carried out in excess base; if the amine is primary, the sulfonamide, which has an acidic amido hydrogen, is converted by base (equation 23) to the normally soluble potassium salt. Thus with few exceptions which are discussed in the next paragraph, primary amines react with benzenesulfonyl chloride to provide homogeneous reaction mixtures. Acidification of this solution regenerates the insoluble primary benzenesulfonamide. On the other hand, the benzenesulfonamides of secondary amines bear no acidic (amido) hydrogens: they typically are insoluble in both acid and base. Therefore secondary amines react to yield heterogeneous reaction mixtures, with production of either an oily organic layer or a solid precipitate.

Primary:

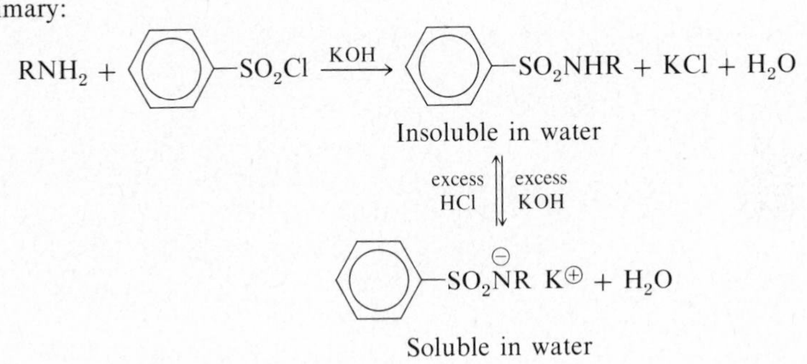

$$RNH_2 + \bigcirc\!\!-SO_2Cl \xrightarrow{\text{KOH}} \bigcirc\!\!-SO_2NHR + KCl + H_2O \qquad (22)$$

Insoluble in water

excess HCl $\updownarrow$ excess KOH

$$\bigcirc\!\!-SO_2\overset{\ominus}{N}R \; K^{\oplus} + H_2O \qquad (23)$$

Soluble in water

Secondary:

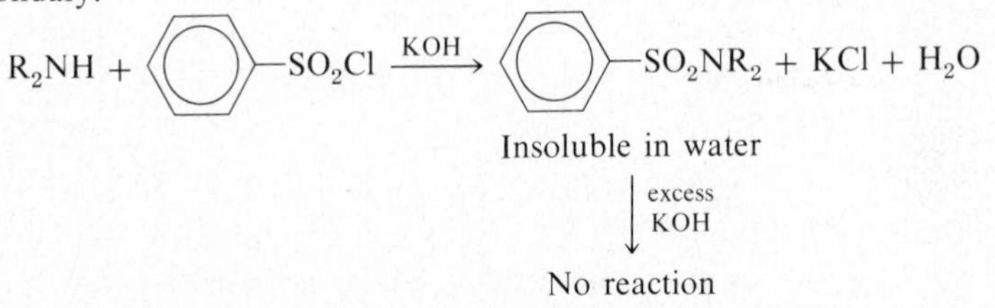

$$R_2NH + \bigcirc\!\!-SO_2Cl \xrightarrow{\text{KOH}} \bigcirc\!\!-SO_2NR_2 + KCl + H_2O \qquad (24)$$

Insoluble in water

$\downarrow$ excess KOH

No reaction

The distinction between primary and secondary amines then depends on the different solubility properties of their benzenesulfonamide derivatives. However, the potassium salts of *certain* primary sulfonamides are not completely soluble in basic solution. Examples are generally found among those primary amines of higher molecular weight and those having cyclic alkyl groups.[9] To avoid confusion and possible misassignment of a primary amine as secondary, the basic solution is separated from the oil or solid and acidified. The formation of an oil or a precipitate indicates that the derivative is partially soluble and that the amine is primary. It is important not to overacidify the solution because this may precipitate certain side products which may form, resulting in an ambiguous test. The original oil or solid

[9] P. E. Fanta and C. S. Wang, *Journal of Chemical Education,* **41,** 280 (1964).

should be tested for solubility in water and acid to substantiate the test for a primary or a secondary amine.

Tertiary amines behave somewhat differently.[10] Typically, under the conditions of the Hinsberg test the processes shown in equation 25 provide for the conversion of benzenesulfonyl chloride to potassium benzenesulfonate with recovery of the tertiary amine. Because tertiary amines are nearly always insoluble in the aqueous potassium hydroxide solution, the test mixture remains heterogeneous. It is worthwhile to note relative densities of the oil layer and of the test solution. Benzenesulfonamides are generally more dense than the solution, whereas the amines are less dense. The oil is separated and tested for solubility in aqueous acid; solubility usually indicates a tertiary amine.

Tertiary:

$$R_3N + \text{C}_6\text{H}_5\text{—}SO_2Cl \longrightarrow \text{C}_6\text{H}_5\text{—}SO_2\text{—}NR_3^{\oplus}Cl^{\ominus} \xrightarrow{KOH}$$

$$\text{C}_6\text{H}_5\text{—}SO_3^{\ominus}K^{\oplus} + NR_3 + H_2O \quad (25)$$

$$\xrightarrow{NR_3} \text{C}_6\text{H}_5\text{—}SO_2NR_2 + NR_4^{\oplus}Cl^{\ominus} \quad (26)$$

$$\text{C}_6\text{H}_5\text{—}SO_2Cl \xrightarrow{KOH} \text{C}_6\text{H}_5\text{—}SO_3^{\ominus}K^{\oplus} + KCl + H_2O \quad (27)$$

$$\xrightarrow[\text{ArNR}_2]{} \begin{array}{l} \text{Complex} \\ \text{mixture} \\ \text{including:} \end{array} \text{C}_6\text{H}_5\text{—}SO_2\text{—}NRAr \quad (28)$$

The test procedure should be followed as closely as possible. It is designed to minimize complications which may arise because of side reactions of tertiary amines with benzenesulfonyl chloride. As shown in equation 26, the initial adduct is subject to further reaction with another molecule of amine to produce the benzenesulfona-mide of a secondary amine. The relative competing rates of reaction of the adduct with hydroxide ion and with amine do not favor equation 26, particularly when excess amine is avoided, yet the formation of *small* amounts of an insoluble product may, through confusion, cause an amine to be incorrectly designated as secondary. Adduct formation such as shown in equation 25 is generally less of a problem with tertiary arylamines because they are normally much less soluble in the test solution and are less nucleophilic than trialkylamines. The competing hydrolysis of benzene-sulfonyl chloride by hydroxide ion (equation 27) allows the recovery of most of the

---

[10]Historically and almost invariably, sources of information have asserted that tertiary amines do not react with benzenesulfonyl chloride. For an interesting refutation of this widely accepted myth, see C. R. Gambill, T. D. Roberts, and H. Shechter, *Journal of Chemical Education,* **49,** 287 (1972).

amine. Moreover, tertiary arylamines are often subject to other side reactions producing a complex mixture of mainly insoluble products (equation 28). Because benzenesulfonyl chloride reacts more slowly with tertiary arylamines than with hydroxide ion, it is also possible to minimize the attendant ambiguity caused by these reactions of the tertiary amine by keeping the reaction time short and the temperature low.

To summarize the discussion, tertiary amines may produce small amounts of insoluble products if the concentration of the amine in the test solution is too high and if the reaction time is too long. By following the directions of the procedure and *taking care not to interpret small amounts of insoluble product as a positive test for secondary amines,*[11] the Hinsberg test may be used with confidence to designate an amine as primary, secondary, or tertiary.

Experimental Procedure

Mix 10 mL of 2 *M* aqueous potassium hydroxide, 0.2 mL or 0.2 g of the amine, and 0.7 mL of benzenesulfonyl chloride (*Caution:* It is a lachrymator) in a test tube. Stopper the tube and shake the mixture *vigorously,* with cooling if necessary, until the odor of benzenesulfonyl chloride is gone. In even the slowest case this should take no more than about 5 min. Test the solution to see that it is still basic; if it is not, add sufficient 2 *M* potassium hydroxide solution until it is.

If the mixture has formed two layers or a precipitate, note the relative densities, and separate the oil or solid by decantation or filtration. Test an oil for solubility in 0.6 *M* hydrochloric acid. The sulfonamide of a secondary amine is insoluble, whereas an amine is at least partially soluble (see footnote 5). If this solubility test indicates an amine, it may be either a tertiary amine or one of certain secondary amines that react with benzenesulfonyl chloride only very slowly because of steric bulk. Test a solid for solubility in water and in dilute acid. The potassium salt of a sulfonamide which is insoluble in base solution is usually soluble in water; the sulfonamide which forms from the potassium salt when placed in acid is insoluble in that medium. A solid sulfonamide of a secondary amine is insoluble in both water and acid. Acidify the solution from the original reaction mixture to pH 4 using pH indicator paper or a few drops of Congo red indicator solution; the formation of a precipitate or oil indicates a primary amine.

If the original mixture has not formed two layers, the test is indicative of a primary amine. Acidify the solution to pH 4; a sulfonamide of a primary amine will either separate as an oil or precipitate as a solid.

2. The Sodium Nitroprusside Tests. Two color tests to distinguish primary and secondary *aliphatic* amines have been available for many years,[12] although they have

---

[11] Tertiary amines often contain quantities of secondary amines as impurities. If it was not possible to obtain a reliable boiling point and the amine was not carefully distilled, small quantities of precipitate may form for this reason also, obscuring the test results.

[12] L. Simon, *Comptes Rendus,* **125,** 534 (1897); E. Ramini, *Chemisches Zentralblatt,* **11,** 132 (1898).

not been widely used. Recently, through a change in the solvent system and the introduction of a $ZnCl_2$ catalyst, these tests have been extended to primary, secondary, and tertiary *aromatic* amines.[13] Both the original and the modified tests are inconclusive for tertiary aliphatic amines. No attempt is made here to explain the complex color-forming reactions which occur. However, they most likely involve the reaction of the amine with either acetone (the Ramini test) or acetaldehyde (the Simon test) and the interaction of the products of these reactions with sodium nitroprusside to form colored complexes.

To apply these tests on an unknown amine, the conventional Ramini and/or the conventional Simon tests should first be performed. These will give positive results in the cases of primary and secondary aliphatic amines. If these tests are negative and an aromatic amine is suspected, then the modified versions of these tests may be performed. Figure 25.5 assists in the interpretation of the results of these tests.

Experimental Procedure

Ramini Test. To 1 mL of the sodium nitroprusside reagent[14] add 1 mL of water, 0.2 mL of acetone, and then about 30 mg of an amine. In most cases the characteristic colors given in Figure 25.5 appear in a few seconds, although in some instances up to about 2 min may be necessary.

Simon Test. To 1 mL of the sodium nitroprusside reagent[14] add 1 mL of water, 0.2 mL of 2.5 M aqueous acetaldehyde solution, and then about 30 mg of an amine.

	1° Aliphatic	2° Aliphatic	1° Aromatic	2° Aromatic	3° Aromatic
Ramini	Deep red	Deep red			
Simon	Pale yellow to red–brown	Deep blue			
Modified Ramini			Orange–red to red–brown	Orange–red to red–brown	Green
Modified Simon			Orange–red to red–brown	Purple	Usually green

FIGURE 25.5 Colors formed in the Ramini and the Simon tests.

[13] R. L. Baumgarten, C. M. Dougherty, and O. Nercessian, *Journal of Chemical Education,* **54,** 189 (1977).

[14] This reagent, for use in both the *conventional* Ramini and Simon tests, is prepared by dissolving 3.9 g of sodium nitroprusside ($Na_2[Fe(NO)(CN)_5] \cdot 2H_2O$) in 100 mL of 50% aqueous methanol.

As in the Ramini test, color formation will normally occur in a few seconds, although occasionally up to 2 min may be necessary.

Modified Ramini Test. To 1 mL of the *modified* sodium nitroprusside reagent[15] add in the following order: 1 mL of saturated aqueous zinc chloride solution, 0.2 mL of acetone, and then about 30 mg of an amine. Primary and secondary aromatic amines provide orange-red to red-brown colors within a period of a few seconds to 5 min. Tertiary aromatic amines give a color that changes from orange-red to green over a period of about 5 min.

Modified Simon Test. To 1 mL of the *modified* sodium nitroprusside reagent[15] add in the following order: 1 mL of saturated aqueous zinc chloride solution, 0.2 mL of 2.5 *M* aqueous acetaldehyde solution, and then about 30 mg of an amine. Primary aromatic amines give an orange-red to red-brown color within 5 min; secondary aromatic amines give a color changing from red to green within 5 min; tertiary aromatic amines give a color which changes from orange-red to green over a period of 5 min.

B. Derivatives

Suitable derivatives of primary and secondary amines are the benzamides and benzenesulfonamides (equations 29 and 30, respectively).

$$RNH_2 \text{ (or } R_2NH) + C_6H_5COCl \xrightarrow{\text{pyridine}} C_6H_5\overset{O}{\overset{\|}{C}}-NHR \text{ (or } C_6H_5\overset{O}{\overset{\|}{C}}-NR_2) \tag{29}$$

$$RNH_2 \text{ (or } R_2NH) + C_6H_5SO_2Cl \longrightarrow C_6H_5SO_2-NHR \text{ (or } C_6H_5SO_2NR_2) \tag{30}$$

Experimental Procedure

1. Benzenesulfonamides. The method of preparing the benzenesulfonamides has been discussed under the Hinsberg test. The derivatives can be prepared using that method, but sufficient amounts of material should be used so that the final product can be purified by recrystallization from 95% ethanol. If the derivative is obtained as an oil, it *may* crystallize by scratching in the presence of the mother liquor with a stirring rod. If the oil cannot be made to crystallize, separate it and dissolve it in a minimum quantity of hot ethanol and allow to cool. Note that some amines do not give *solid* benzenesulfonamide derivatives.

---

[15] This reagent, for use in both the *modified* Ramini and Simon tests, is prepared by dissolving 3.9 g of sodium nitroprusside in a solution containing 80 mL of dimethylsulfoxide and 20 mL of water. To avoid decomposition of the reagent, it should be stored in the refrigerator and dispensed in small dropper bottles as needed.

2. Benzamides. In a 50-mL round-bottomed flask dissolve 0.5 g of the amine in 5 mL of dry pyridine. *Slowly* add 0.5 mL of benzoyl chloride to this solution. Affix a drying tube to the flask and heat the reaction mixture to 60–70° for 30 min, using a water bath. Following the heating period, pour the mixture into 50 mL of water. If the solid derivative precipitates at this time, collect it on a filter, and when it is nearly dry, dissolve it into 20 mL of diethyl ether. If no precipitate forms, extract the aqueous mixture twice with 15-mL portions of diethyl ether. Combine the ether extracts. Wash the ether solution of the derivative in sequence with equal volumes of water, 1.5 M HCl, and 0.6 M sodium bicarbonate solution. Dry the ether layer over anhydrous magnesium sulfate, filter, and remove the ether by simple distillation. The solid residue, which constitutes the derivative, may be recrystallized from one of the following solvents: cyclohexane-hexane mixtures, cyclohexane–ethyl acetate mixtures, 95% ethanol, or aqueous ethanol.

Although one or both of the above methods are satisfactory with most primary and secondary amines, tertiary amines do not undergo the same reactions. In general one must take advantage of the fact that tertiary amines do form salts. Two useful crystalline salts are the ones from methyl iodide and picric acid (equations 31 and 32, respectively).

$$R_3N\colon + CH_3I \longrightarrow R_3\overset{\oplus}{N}-CH_3I^{\ominus} \tag{31}$$

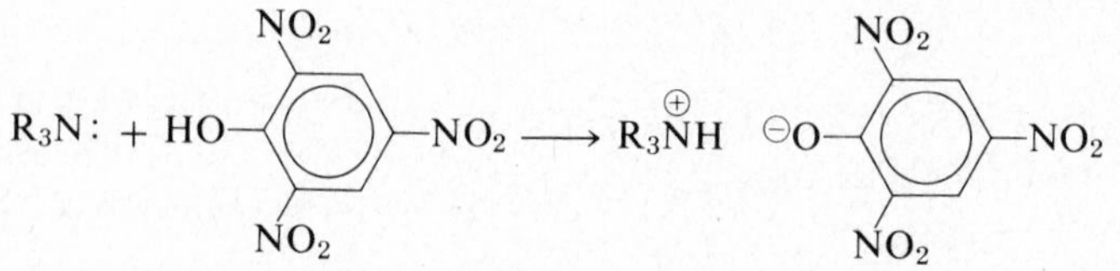

$$R_3N\colon + HO-\underset{NO_2}{\overset{NO_2}{\bigcirc}}-NO_2 \longrightarrow R_3\overset{\oplus}{N}H \quad {}^{\ominus}O-\underset{NO_2}{\overset{NO_2}{\bigcirc}}-NO_2 \tag{32}$$

3. Methiodides. To prepare the methyl iodide derivative mix 0.5 g of the amine with 0.5 mL of methyl iodide, and warm the test tube with a water bath for several minutes. Cool the test tube in an ice-water bath; the tube may be scratched with a rod to help induce crystallization. Purify the product by recrystallization from absolute ethanol or methanol or from ethyl acetate.

4. Picrates. The picric acid derivative (often called the picrate) may be prepared by mixing 0.3 to 0.5 g of the compound with 10 mL of 95% ethanol. If the solution is not complete, remove the excess solid by filtration. Add to the mixture 10 ml of a saturated solution of picric acid in 95% ethanol, and heat the mixture to boiling. Cool the solution slowly, and remove the yellow crystals of the picrate salt by filtration. Recrystallize the salt from 95% ethanol.

A list of primary and secondary amines and their derivatives can be found in Table 25.8; tertiary amines and derivatives appear in Table 25.9.

NITRO COMPOUNDS (Table 25.10)

A. Classification Test

Ferrous Hydroxide Test. Organic compounds that are oxidizing agents will oxidize ferrous hydroxide (blue) to ferric hydroxide (brown). The most common organic compounds that function in this way are the *nitro compounds,* both aliphatic and aromatic, which are in turn reduced to amines in the reaction (equation 33). Other

$$RNO_2 + 6\ \underset{\text{Blue}}{Fe(OH)_2} + 4\ H_2O \longrightarrow RNH_2 + 6\ \underset{\text{Brown}}{Fe(OH)_3} \tag{33}$$

less common types of compounds that give the same test are nitroso compounds, hydroxylamines, alkyl nitrates, alkyl nitrites, and quinones.

Experimental Procedure

In a 10×75-mm or smaller test tube,[16] mix about 20 mg of a solid or 1 drop of a liquid unknown with 1.5 mL of freshly prepared 5% ferrous ammonium sulfate solution. Add 1 drop of 3 *M* sulfuric acid and 1 mL of 2 *M* potassium hydroxide in methanol. Stopper the tube immediately and shake it. A positive test is indicated by the blue precipitate turning rust-brown within 1 min. (A slight darkening or greenish coloration of the blue precipitate should not be considered a positive test.)

B. Derivatives

Two different types of derivatives of nitro compounds can be prepared. Aromatic nitro compounds can be di- and trinitrated with nitric acid and sulfuric acid. Discussion of and procedures for nitration have been given under Aryl Halides. Refer to this section for additional information.

The other method for preparation of a derivative can be utilized for both aliphatic and aromatic nitro compounds. This involves the reduction of the nitro compound to the corresponding primary amine (equation 34), followed by conversion of the amine to a benzamide or benzenesulfonamide, as described under Amines. The reduction is most often carried out with tin and hydrochloric acid.

$$RNO_2 \text{ or } ArNO_2 \xrightarrow[\text{(2) NaOH}]{\text{(1) Sn, HCl}} RNH_2 \text{ or } ArNH_2 \tag{34}$$

Experimental Procedure

Carry out the reduction of the nitro compound by combining 1 g of the compound and 2 g of granulated tin in a small flask. Attach a reflux condenser, and

---

[16]A small test tube is used in order to lessen the exposure of the reagent to air, which may bring about some oxidation.

add, in small portions, 20 mL of 3 *M* hydrochloric acid. Shake after each addition. After addition is complete, warm the mixture for 10 min, using a steam bath. If the nitro compound is insoluble, add 5 mL of 95% ethanol to increase its solubility. Decant the warm, homogeneous solution into 10 mL of water, and add enough 12 *M* sodium hydroxide solution so that the tin hydroxide completely dissolves. Extract the basic solution with several 10-mL portions of diethyl ether. Dry the ether solution over potassium hydroxide pellets, and remove the ether by distillation.

The residue contains the primary amine. Convert it to one of the derivatives described under Amines.

A list of some nitro compounds and their derivatives appears in Table 25.10.

ESTERS (Table 25.11)

A.   Classification Test

A test for the presence of the ester group involves the use of hydroxylamine and ferric chloride. The former converts the ester to a hydroxamic acid, which then complexes with Fe(III) to give a colored species (equations 35 and 36).

$$
\underset{}{R-\overset{\overset{\displaystyle O}{\|}}{C}-OR'} + H_2NOH \longrightarrow \underset{\substack{\text{Hydroxamic} \\ \text{acid}}}{R-\overset{\overset{\displaystyle O}{\|}}{C}-NHOH} + R'OH \tag{35}
$$

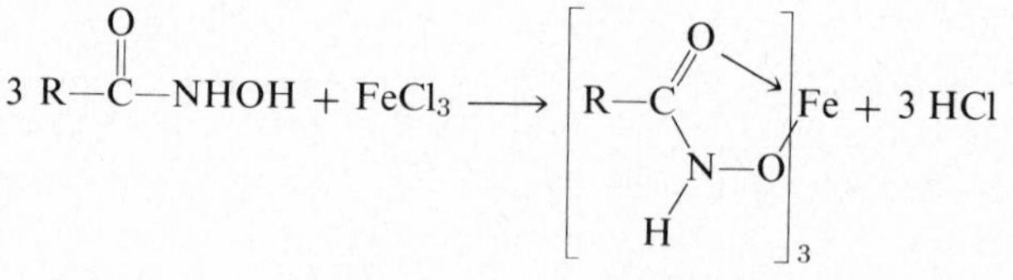

$$
3\ R-\overset{\overset{\displaystyle O}{\|}}{C}-NHOH + FeCl_3 \longrightarrow \left[R-\overset{\displaystyle O}{\underset{\underset{H}{N-O}}{C}}Fe \right]_3 + 3\ HCl \tag{36}
$$

Colored

All carboxylic acid esters (including polyesters and lactones) give magenta colors which vary in intensity depending on structural features in the molecule. Acid chlorides and anhydrides also give positive tests. Formic acid produces a red color, but other free acids give negative tests. Primary or secondary aliphatic nitro compounds give a positive test because ferric chloride reacts with the *aci* form (equivalent to the enol form of a ketone) which is present in basic solution. Most imides give positive tests. Some amides, but not all, give light magenta coloration, whereas most nitriles give a negative test. (A modification of the following procedure is given later which will yield a positive test for amides and nitriles.)

Experimental Procedure

Before performing the final test, it is necessary to run a preliminary (or "blank") test, since some compounds will give a positive test even though they do not contain an ester linkage.

1. Preliminary Test. Mix 1 mL of 95% ethanol and 50–100 mg of the compound to be tested, and add 1 mL of 1 M hydrochloric acid. Note the color which is produced when 1 drop of 0.6 M aqueous ferric chloride solution is added. If the color is orange, red, blue, or violet, the following test for the ester group does not apply and cannot be used.

2. Final Test. Mix 40–50 mg of the unknown, 1 mL of 0.5 M hydroxylamine hydrochloride in 95% ethanol, and 0.2 mL of 6 M sodium hydroxide. Heat the mixture to boiling, and after cooling it slightly, add 2 mL of 1 M hydrochloric acid. If the solution is cloudy, add more (about 2 mL) 95% ethanol. Add 1 drop of 0.6 M ferric chloride, and observe the color. Add more ferric chloride solution if the color does not persist, and continue to add it until it does. Compare the color obtained here with that from the preliminary test. If the color is burgundy or magenta, as compared to the yellow color in the preliminary experiment, the presence of an ester group is indicated.

B. Saponification Equivalent

Once an ester group has been detected, it is possible to obtain information about its equivalent weight. This value, termed the *saponification equivalent,* has already been discussed in Section 15.2; the principles and equations involved in this technique are presented there.

Experimental Procedure

Dissolve approximately 3 g of potassium hydroxide in 60 mL of 95% ethanol. Allow the small amount of insoluble material to settle to the bottom, and fill a 50-mL buret with the clear solution by decantation. Measure exactly 25.0 mL of the alcoholic solution into each of two flasks. Weigh *accurately* into one of the flasks a 0.3- to 0.4-g sample of pure, dry ester; the other basic solution will be used as a blank. Fit each flask with a reflux condenser.

Heat the solutions in both flasks at a gentle reflux for 1 hr. When the flasks have cooled, rinse each condenser with about 10 mL of distilled water, catching the rinse water in the flask. Add phenolphthalein, and separately titrate the solutions in each flask with *standardized* hydrochloric acid which is approximately 0.5 M.

The difference in the volumes of hydrochloric acid required to neutralize the base in the flask containing the sample and in the flask containing the blank corresponds to the amount of potassium hydroxide which reacted with the ester. The volume difference (in milliliters) multiplied by the molarity of the hydrochloric acid equals the number of *milli*moles of potassium hydroxide consumed. Using the titration data, calculate the saponification equivalent of the unknown ester.

If the ester does not completely saponify in the allotted time as evidenced by a *non*homogeneous solution, heat under reflux for longer periods (2–4 hr). In some cases higher temperatures may be required; if so, diethylene glycol must be used as a solvent *in place of* the original 60 mL of 95% ethanol.

C.   Derivatives

To characterize an ester completely, it is necessary to prepare solid derivatives of both the acid and the alcohol components. The problem here is to isolate both of these components in pure form so that suitable derivatives can be prepared. One such way is to carry out the ester hydrolysis in a high-boiling solvent. If the alcohol is low boiling, it can be distilled from the reaction mixture and characterized. The acid which remains in the mixture can also be isolated. Derivatives of acids and alcohols have already been discussed.

$$\text{RCOOR}' + \text{HO}^{\ominus} \longrightarrow \text{RCOO}^{\ominus} + \text{R}'\text{OH} \tag{37}$$

Experimental Procedure

In a small reaction vessel mix 3 mL of diethylene glycol, 0.6 g (2 pellets) of potassium hydroxide, and 10 drops of water. Heat until the solution is homogeneous, and cool to room temperature. Add 1 mL of the ester and equip the apparatus with a condenser. Heat to boiling again, with swirling, and after the ester layer dissolves (3–5 min), recool the solution. Equip for a simple distillation, and heat the flask strongly so that the alcohol distils; *all but high-boiling alcohols can be removed by direct distillation.* The distillate, which should be fairly pure and dry, can be used for the preparation of a solid derivative.

The residue which remains after distillation contains the salt of the carboxylic acid. Add 10 mL of water to the residue and mix thoroughly. Acidify the solution with 6 *M* sulfuric acid. Allow the mixture to stand, and collect any crystals by filtration. If crystals do not form, extract the aqueous acidic solution with diethyl ether or dichloromethane, dry the organic solution, and evaporate the solvent. Use the residual acid to prepare a derivative.

A list of esters and their boiling or melting points is given in Table 25.11. Alcohols and carboxylic acids and their derivatives are given in Tables 25.5 and 25.7, respectively.

NITRILES (Table 25.12)

A. Classification Test

A qualitative test which may be used for nitriles is similar to that for esters. Common nitriles (as well as amides) give a colored solution on treatment with hydroxylamine and ferric chloride (equation 38).

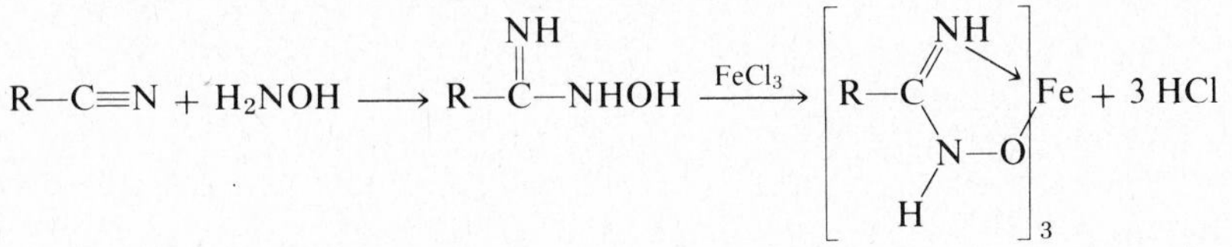

$$R-C\equiv N + H_2NOH \longrightarrow R-\overset{\overset{\displaystyle NH}{\|}}{C}-NHOH \xrightarrow{FeCl_3} \left[R-\overset{\overset{\displaystyle NH}{\diagdown}}{\underset{\underset{\displaystyle H}{|}}{C}}\overset{}{\underset{N-O}{}} Fe \right]_3 + 3\,HCl \quad (38)$$

Experimental Procedure

Prepare a mixture consisting of 2 mL of 1 M hydroxylamine hydrochloride in propylene glycol, 30–50 mg of the compound which has been dissolved in a minimum amount of propylene glycol, and 1 mL of 1 M potassium hydroxide. Heat the mixture to boiling for 2 min, and cool to room temperature; add 0.5–1.0 mL of a 0.5 M *alcoholic* ferric chloride solution. A red-to-violet color is a positive test. Yellow colors are negative, and brown colors and precipitates are neither positive nor negative.

B. Derivative

On hydrolysis, in either acidic or basic solution, nitriles are ultimately converted to the corresponding carboxylic acids. Using methods given previously, it is then possible to prepare a derivative of the acid.

Basic hydrolysis: $\quad RCN + NaOH \xrightarrow{H_2O} RCOO^{\ominus}Na^{\oplus} + NH_3 \quad\quad\quad (39)$

Acidic hydrolysis: $\quad RCN \xrightarrow[H_2SO_4]{H_2O} RCONH_2 \xrightarrow[H_2SO_4]{H_2O} RCOOH + NH_4^{\oplus} \quad\quad (40)$

Experimental Procedure

1. Basic Hydrolysis. Mix 10 mL of 3 M sodium hydroxide solution and 1 g of the nitrile. Heat the mixture to boiling, and note the odor of ammonia, or hold a piece of moist red litmus paper over the container, and note the color change. After the mixture is homogeneous, cool it and make it acidic to litmus. If the acid is a solid, collect the crystals by filtration. If it is a liquid, extract the acidic solution with diethyl ether; after drying the ether solution, remove the ether by distillation. The residue that remains is the acid. Prepare a suitable derivative of the acid, using procedures given previously.

2. Acidic Hydrolysis. Treat 1 g of the nitrile with 10 mL of concentrated sulfuric acid or concentrated hydrochloric acid, and warm the mixture to 50° for about 30 min. Dilute the mixture with water (*Caution:* Add the mixture slowly to water if sulfuric acid has been used), and heat the mixture at gentle reflux for 30 min to 2 hr. The organic layer will be the acid. Cool the mixture and either collect the crystals by filtration or extract the liquid with diethyl ether. Prepare derivatives of the acid.

A list of some nitriles is given in Table 25.12; carboxylic acids and their derivatives are given in Table 25.7.

AMIDES (Table 25.13)

A. Classification Test

A qualitative test for an amide group is the same as that given for a nitrile (equation 41). Follow exactly the procedure given for nitriles. The colors observed with amides are the same as those with nitriles.

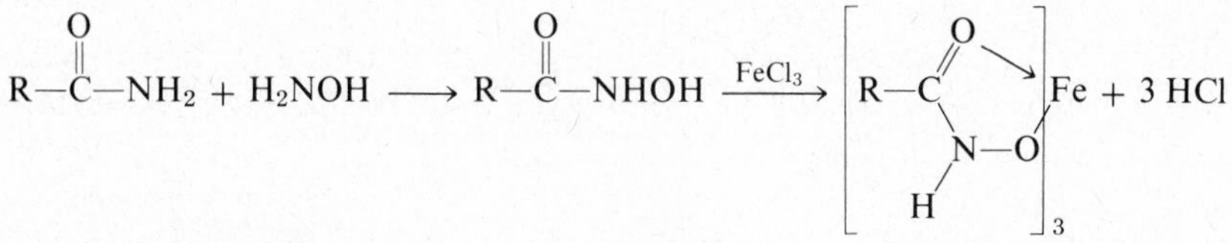

$$R-\overset{\overset{\displaystyle O}{\|}}{C}-NH_2 + H_2NOH \longrightarrow R-\overset{\overset{\displaystyle O}{\|}}{C}-NHOH \xrightarrow{FeCl_3} \left[R-\overset{O}{\underset{\underset{H}{N-O}}{C}} \right]_3 Fe + 3\,HCl \qquad (41)$$

B. Derivatives

Like nitriles, amides must be hydrolyzed (acidic or basic) to give an amine and a carboxylic acid (equation 42). In the case of unsubstituted amides, ammonia is liberated, but with substituted amides a substituted amine is obtained. In those cases it is necessary to classify the amine as being primary or secondary and to prepare derivatives of both the acid and the amine.

$$RCONR'_2 \xrightarrow[H_2O]{H^{\oplus} \text{ or } HO^{\ominus}} RCOOH + HNR'_2 \qquad (42)$$

$$(R' = \text{alkyl, aryl, or H})$$

Experimental Procedure

1. Basic Hydrolysis. Carry out the procedure described for the hydrolysis of nitriles. Distil the ammonia or volatile amine from the alkaline solution into a container of dilute hydrochloric acid. Neutralize this acidic solution, carry out the Hinsberg test, and prepare a derivative of the amine. If the amine is not volatile, it may be extracted from the aqueous layer with diethyl ether, the solution dried over potassium hydroxide pellets, and the ether removed to give the amine. After the amine has been obtained, either by distillation or extraction from the hydrolysis mixture, make the alkaline solution acidic, and isolate the acid (either by filtration if a solid or by extraction if a liquid). Characterize the acid by preparing a suitable solid derivative.

2. Acidic Hydrolysis. Carry out the hydrolysis, using the method described for nitriles. In this case the free acid is liberated and can be removed by filtration or extraction with diethyl ether. Prepare a derivative of the acid. Make the acidic hydrolysis mixture alkaline to liberate the amine. Collect the amine by distillation or extraction, characterize it by the Hinsberg test, and make a derivative.

A list of amides is given in Table 25.13. Lists of amines and acids and their derivatives are given in Tables 25.7 and 25.8.

25.3 Modern Spectroscopic Methods of Analysis

One major limitation to the classical system of qualitative organic analysis which was only mentioned in passing but which is inherent to the system is that only *known compounds* can be identified. The research chemist is constantly faced with the task of identifying new compounds. Although much information about the *type* of compound may be derived from the classical system, until recently the complete identification of an unknown organic structure required a combination of degradation and synthesis, which was usually a lengthy and laborious task.

The advent of modern spectroscopy has changed this picture dramatically; not only known compounds but also new and unknown compounds may be identified quickly and certainly by a combination of spectroscopic methods such as those described in Chapter 4. A number of examples might be cited of the structural identification in a matter of weeks or months of molecules of a complexity greater than those of other compounds which defied the lifework of several of the great nineteenth- and twentieth-century organic chemists. One of the significant early examples of this was the application of ir spectroscopy and X-ray diffraction to the determination of the structure of the penicillin-G molecule during World War II.

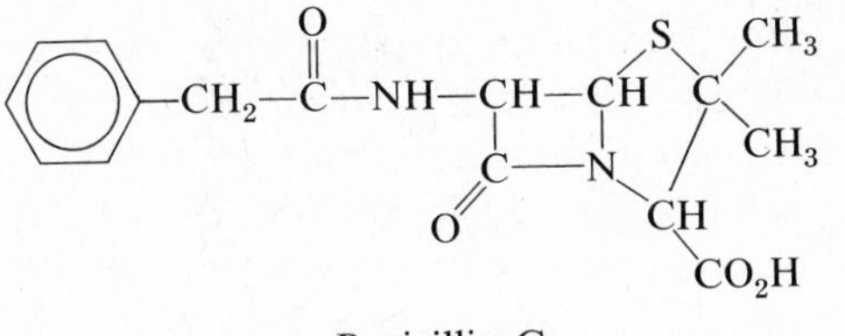

Penicillin-G

Modern procedure for identification of organic compounds usually involves a combination of classical and spectroscopic methods. Ideally a student should be introduced to applications of spectroscopy in organic chemistry through use of the instruments which produce the spectra. Because some of these instruments are very expensive, this is not feasible in many instances. The next best alternative is for students to have access to the spectra of typical known compounds for study and then to be provided with spectra of "unknowns" to identify.

In the preceding chapters, 210 ir, pmr, and uv spectra of starting materials and products have been presented. A careful study of these spectra, aided by the material in Chapter 4 and Appendixes 3 and 4 and by additional discussion on the part of the instructor, should enable a student to make meaningful use of ir, pmr, and uv spectra in the identification of unknown organic compounds. The spectral data may serve to complement or supplement the "wet" classification tests or in many cases may substitute for these tests. For example, a strong ir absorption in the 1690–1760 cm^{-1} region is certainly as indicative of a carbonyl as formation of a 2,4-dinitrophenyl-hydrazone, and pmr absorptions in the 6–8.5 δ region are a more reliable indication of any aromatic compound than a color test with $CHCl_3$ and $AlCl_3$.

It must be emphasized, however, that in spectroscopic analysis, just as in classical qualitative organic analysis, *careful interpretation of the data* and a certain amount of *chemical intuition* or common sense must be exercised. Students should be cautioned not to go overboard in their enthusiasm with modern spectroscopy. Although some problems can be resolved quickly and uniquely by spectroscopy, others can be resolved just as simply and much more economically by classical qualitative organic analytical procedures, and still others require the intelligent application of both the modern and the classical methods. Examples are given in the following paragraphs which should provide insight into the complementary application of classical testing procedures and of spectral analysis in the determination of structure of an unknown compound.

A student was given a liquid of unknown structure which had a boiling range of 143–145° and provided negative tests for halogen, nitrogen, and sulfur when subjected to elemental analysis by sodium fusion. The compound dissolved in water to give a neutral solution. Quite certainly, owing to its water solubility, the compound contains oxygen-bearing polar functional groups. As an aid to the functional group analysis, the student obtained ir and pmr spectra of the substance. These are shown in Figure 25.6. Immediately evident from the ir spectrum is the presence of an ester group because of the strong absorption at about 1750 cm^{-1}. Although this band could, for example, be indicative of a five-membered ring ketone, this and other possibilities are apparently negated by the presence of the peak at 2.0 δ, one of two peaks in the pmr spectrum integrating for three protons and apparently corresponding to methyl groups with no adjacent hydrogens. The higher field peak (2.0 δ) is characteristic of a methyl ketone or acetate, the latter being consistent with the 1750 and 1230 cm^{-1} bands in the ir spectrum. The second methyl peak at 3.3 δ is shifted to lower field, as would be expected if the methyl group were bonded to oxygen. This peak is apparently at too high a field for a methyl ester, since methyl esters more normally show the methyl absorption at about 3.7–4.1 δ. It is, however, within the range (3.3–4.0 δ) frequently observed for the aliphatic alpha-hydrogens of an alcohol or ether. Because neither of the spectra shows any evidence for an —OH group, the compound probably contains a CH_3O— group. The presence of an aliphatic ether is consistent with the C—O absorption observed at 1050 cm^{-1} in the ir spectrum. The multiplets centered at 4.1 and 3.5 δ each integrate for two hydrogens and show mutual spin-spin coupling. This pattern is almost certainly diagnostic of differently shielded, adjacent methylenes such as X—CH_2CH_2—Y. The low-field absorption for each of

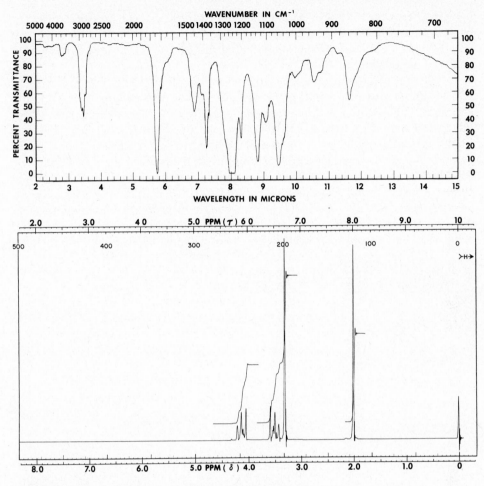

FIGURE 25.6 IR and PMR spectra of the first unknown.

the CH$_2$ groups indicates that each is bonded to oxygen. Only a single structure is seemingly compatible with all these spectral observations: 2-methoxyethyl acetate (**1**, bp 145°).

$$CH_3O-CH_2CH_2-O\overset{\overset{\displaystyle O}{\|}}{C}CH_3$$
1

The student (wisely) confirmed the structural assignment through the hydrolysis of the ester and formation of the 3,5-dinitrobenzoate derivative of the resulting alcohol; the derivative showed the correct melting point for 2′-methoxyethyl 3,5-dinitrobenzoate.

A second unknown, which had a boiling range of 227–230°, was obtained. The compound showed solubility in only concentrated sulfuric acid, according to the scheme of Figure 25.4. Sodium fusion analyses were negative for nitrogen, sulfur, and halogen. The compound provided the pmr and ir spectra shown in Figure 25.7. The uv spectrum in methanol had two maxima at 235 nm ($\varepsilon = 8630$) and at 304 nm ($\varepsilon = 579$). In an attempt to identify potential functional groups, these spectra were analyzed as follows.

The pmr spectrum shows four regions of absorption at 6.75, 4.78, 2.1–3.0, and 1.75 δ, with relative areas of $14:28:75:88$, respectively. This pattern of integration is consistent with a relative hydrogen abundance of $1:2:5:6$. The large singlet at 1.75 δ, integrating for six protons, is apparently caused by two methyl groups in closely similar environments. They are probably attached to carbon-carbon double bonds because of their chemical shift and the presence in the ir spectrum of absorptions at 3000–3100, 1645, and 895 cm^{-1}, all characteristic olefinic absorptions. The band at 895 cm^{-1} and the two-proton pmr absorption at 4.78 δ are at the positions expected for a terminal double bond (C=CH$_2$). The strong ir band at 1680 cm^{-1} indicates that the compound is probably either an α,β-unsaturated ketone or aldehyde, the latter being effectively eliminated by the absence of any additional evidence for —CHO in

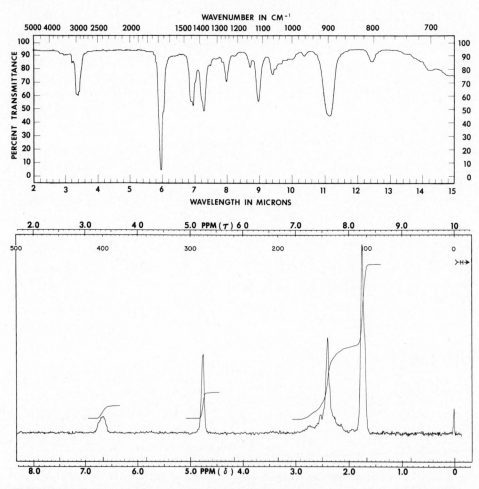

FIGURE 25.7 IR and PMR spectra of the second unknown.

either of the spectra. The conclusion that an α,β-unsaturated ketone is present is supported by the uv spectrum; the maximum occurring at the shorter wavelength is attributed to the $\pi \to \pi^*$ excitation, whereas the weaker absorption at longer wavelength is assigned to the $n \to \pi^*$ transition.

Although the one proton pmr peak at 6.75 δ is at an unusually low field, its position is quite in keeping with the possible presence of a conjugated ketone. Olefinic hydrogens beta to the carbonyl in these functions are strongly deshielded because of charge separation in contributing resonance structures:

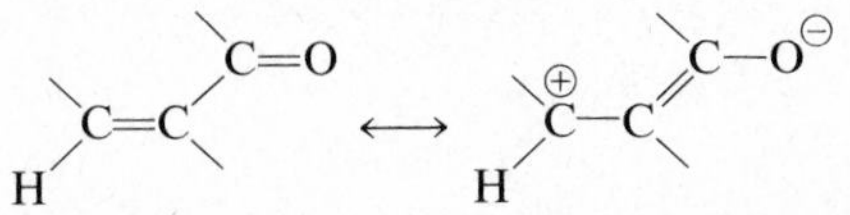

The four-proton multiplet from 2.1–3 δ represents aliphatic hydrogens (probably methylene and/or methine on the basis of the shift and general appearance of the multiplet) deshielded by adjacent groups such as vinyl or carbonyl.

The compound was chemically shown to contain *at least* one double bond through the observation that it decolorized both bromine and potassium permanganate solutions. That the compound was a ketone was verified through the formation of a 2,4-dinitrophenylhydrazone derivative (mp 190–191°). Reference to Table 25.2 led to the identification of the unknown as carvone (2), an essential oil isolated from caraway seed. (The reader should find it highly instructive to examine this structure carefully while reviewing the spectral analysis discussed above.)

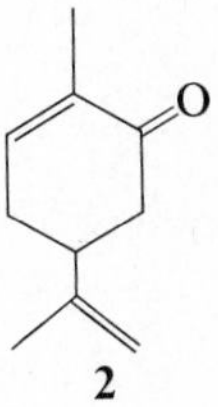

2

The preceding example should foster appreciation of the complementary aspects of spectral analysis and the "wet" classification scheme. Although the spectra were not in themselves specifically definitive in the assignment of structure to the unknown, they implicated the possible presence of certain functional groups and suggested appropriate classification tests. This allowed our imaginary student to avoid the effort and time that might conceivably have been spent in running through a series of negative tests.

REFERENCES

1. R. L. Shriner, R. C. Fuson, and D. Y. Curtin, *The Systematic Identification of Organic Compounds,* 5th ed., John Wiley & Sons, New York, 1964.
2. N. D. Cheronis, J. B. Entrikin, and E. M. Hodnett, *Semimicro Qualitative Organic Analysis,* 3d ed., Interscience Publishers, New York, 1965.
3. T. R. Hogness, W. C. Johnson, and A. R. Armstrong, *Qualitative Analysis and Chemical Equilibrium,* 5th ed., Holt, Rinehart and Winston, New York, 1966.
4. D. J. Pasto and C. R. Johnson, *Organic Structure Determination,* Prentice-Hall, Englewood Cliffs, N.J., 1969.

25.4 Tables of Derivatives

The following abbreviated tables of common organic compounds and their usual derivatives are arranged according to classes of compounds. The melting and boiling points listed represent the highest points in the range actually observed.

TABLE 25.1 ALDEHYDES

| | | | DERIVATIVES | |
Name of Compound	Boiling Point	Melting Point	Semi-carbazone	2,4-Dinitrophenyl-hydrazone
Propanal	49°		154°	156°
Glyoxal	50		270	328
Propenal	52		171	165
2-Methylpropanal	64		126	182
Butanal	75		106	123
Trimethylethanal	75		190	209
2-Methylbutanal	93		103	120
Trichloroethanal	98		90d	131
Pentanal	103			107
trans-2-Butenal	104		199	190
Hexanal	131		106	104
Heptanal	156		109	108
Furfural	161		202	230(214)
3-Cyclohexenecarboxaldehyde	165		155	
Benzaldehyde	179		222	237
Phenylacetaldehyde	194		156	121
Salicylaldehyde	197		231	248
o-Tolualdehyde	200		212	195
p-Tolualdehyde	204		215	234
Citronellal	207		92	78
o-Chlorobenzaldehyde	208		225	214
p-Chlorobenzaldehyde	214	47°	230	270
o-Anisaldehyde	246	39	215	254
p-Anisaldehyde	248		210	254d
Cinnamaldehyde	252		215	255d
p-Bromobenzaldehyde		57	228	
Vanillin		81	230	271d
p-Nitrobenzaldehyde		106	221	320d
p-Hydroxybenzaldehyde		115	224	280d

TABLE 25.2 KETONES

Name of Compound	Boiling Point	Melting Point	DERIVATIVES	
			Semi-carbazone	2,4-Dinitrophenyl-hydrazone
2-Propanone	56°		190°	126°
2-Butanone	80		136	117
3-Methyl-2-butanone	94		113	120
3-Pentanone	102		139	156
2-Pentanone	102		112	144
2,2-Dimethyl-3-butanone	106		158	125
4-Methyl-2-pentanone	117		132	95
1-Chloro-2-propanone	119		164d	125
2,4-Dimethyl-3-pentanone	125		160	95
2-Hexanone	129		122	110
4-Methyl-3-penten-2-one	130		164	203
Cyclopentanone	131		203	146
2,4-Pentanedione	139		122(mono); 209(di)	209
4-Heptanone	145		133	75
5-Methyl-2-hexanone	145		147	95
3-Heptanone	148		103(152)	
2-Heptanone	151		127	89
Cyclohexanone	156		167	162
6-Methyl-3-heptanone	160		132	
5-Methyl-3-heptanone	160		102	
4-Hydroxy-2-pentanone	166			203
2,6-Dimethyl-4-heptanone	169		122	92
2-Octanone	173		123	58
Acetophenone	203	20°	199	240
l-Menthone	209		189	146
3,5,5-Trimethylcyclohexen-1-one	215		200	
Benzyl methyl ketone	216	27	198	156
Propiophenone	218	20	174	191
p-Methylacetophenone	226	28	205	258
2-Undecanone	228		122	63
Carvone	230		163(143)	191
Dibenzyl ketone	330	34	146	100
p-Methoxyacetophenone	258	38	198	220
Benzophenone		48	167	239
p-Bromoacetophenone		51	208	235
Desoxybenzoin	320	60	148	204
Benzalacetophenone	348	58(62)	168	245
Benzoin		133	206d	245
p-Hydroxybenzophenone		135	194	242
d,l-Camphor		178	248	164

TABLE 25.3 AROMATIC HYDROCARBONS

Name of Compound	Boiling Point	Melting Point	Derivatives: Nitration Product Position	Derivatives: Nitration Product Melting Point	Derivatives: Oxidation Product
Benzene	80°		1,3	89°	
Toluene	111		2,4	70	Benzoic acid
Ethylbenzene	136		2,4,6	37	Benzoic acid
p-Xylene	138		2,3,5	13	Terephthalic acid
m-Xylene	139		2,4	83	Isophthalic acid
o-Xylene	142		4,5	71	Phthalic acid
Isopropylbenzene	153		2,4,6	109	Benzoic acid
t-Butylbenzene	169		2,4	62	
m-Diethylbenzene	181		2,4,6	62	Isophthalic acid
Tetralin	206		5,7	95	Phthalic acid
Cyclohexylbenzene	237		4	58	Benzoic acid
Diphenylmethane		25°	2,4,2′,4′	172	
Dibenzyl		53	4,4′	180	
Biphenyl		70	4,4′	237	
Naphthalene		80	1	61	
Triphenylmethane		92	4,4′,4″	206	
Acenaphthene		96	5	101	
Fluorene		115	2,7	199	

TABLE 25.4 ARYL HALIDES

Name of Compound	Boiling Point	Melting Point	DERIVATIVES			
			NITRATION PRODUCT		SIDECHAIN OXIDATION	
			Position	Melting Point	Product	Melting Point
Chlorobenzene	132°		2,4	52°		
Bromobenzene	156		2,4	75		
o-Chlorotoluene	159		3,5	63	o-Chlorobenzoic acid	140°
m-Chlorotoluene	162		4,6	91	m-Chlorobenzoic acid	158
p-Chlorotoluene	162		2	38	p-Chlorobenzoic acid	242
o-Dichlorobenzene	179		4,5	110		
o-Bromotoluene	182		3,5	82	o-Bromobenzoic acid	147
2,4-Dichlorotoluene	200		3,5	104	2,4-Dichlorobenzoic acid	160
α-Chloronaphthalene	259		4,5	180		
p-Bromotoluene	184	28°	2	47	p-Bromobenzoic acid	251
p-Dichlorobenzene		53	2	54		
β-Chloronaphthalene		56	1,8	175		
p-Dibromobenzene		89	2,5	84		

TABLE 25.5 ALCOHOLS

| Name of Compound | Boiling Point | Melting Point | DERIVATIVES | |
			3,5-Dinitro-benzoate	Phenylurethane
Methanol	65°		108°	47°
Ethanol	78		93	52
2-Propanol	82		123	88
2-Methyl-2-propanol	83		142	136
1-Propanol	97		74	51
2-Butanol	99		76	65
2-Methyl-2-butanol	102		117	42
2-Methyl-1-propanol	108		87	86
3-Pentanol	116		97	48
1-Butanol	118		64	57
1-Methoxy-2-propanol	119		85	
2-Pentanol	119		61	
4-Methyl-2-pentanol	131		65	143
3-Methyl-1-butanol	132		61	57
1-Pentanol	138		46	46
Cyclopentanol	141		115	132
2-Methyl-1-pentanol	148		51	
1-Hexanol	156		58	42
Cyclohexanol	161		113	82
Furfuryl alcohol	172		81	45
2,6-Dimethyl-4-heptanol	173			62
1-Heptanol	177		47	68
2-Octanol	179		32	114
1-Octanol	192		61	74
Ethylene glycol	198		169	157
Methylphenylcarbinol	203		95	94
Benzyl alcohol	206		113	78
β-Phenethyl alcohol	220		108	80
Cinnamyl alcohol		33°	121	90
1-Tetradecanol		39	67	74
(−)-Menthol		42	158	111
1-Hexadecanol		50	66	73
Neopentyl alcohol		53		144
1-Octadecanol		60	66	80
Benzoin		133		165
(−)-Cholesterol		148		168

TABLE 25.6   PHENOLS

Name of Compound	Boiling Point	Melting Point	DERIVATIVES		
			α-Naphthyl-urethane	Bromo Derivatives	
o-Chlorophenol	175°	7°	120°	Dibromo	76°
Phenol	180	42	133	Tribromo	95
o-Cresol	190	31	142	Dibromo	56
o-Bromophenol	195		129	Tribromo	95
p-Cresol	202	36	146	Dibromo	49
m-Cresol	202	3	128	Tribromo	84
2-Methoxyphenol	205	32	118	4,5,6-Tribromo	116
p-Chlorophenol	217	43	166	Dibromo	90
o-Nitrophenol		45	113	4,6-Dibromo	117
p-Ethylphenol		47	128		
p-Bromophenol		63	169	Tribromo	95
3,5-Dichlorophenol		68		Tribromo	189
3,5-Dimethylphenol		68	109	Tribromo	166
2,4,6-Trimethylphenol		69		Dibromo	158
2,5-Dimethylphenol		74	173	Tribromo	178
α-Naphthol		94	152	2,4-Dibromo	105
m-Nitrophenol		97	167	Dibromo	91
p-t-Butylphenol		100	110	Bromo	50
				Dibromo	67
o-Hydroxyphenol		105	175	Tetrabromo	193
m-Hydroxyphenol		110	275	Tribromo	112
p-Nitrophenol		114	151	2,6-Dibromo	142
p-Hydroxybenzaldehyde		115		3,5-Dibromo	181
β-Naphthol		122	157	Bromo	84
2,3-Dihydroxyphenol		133	230 (tri)	Dibromo	158

TABLE 25.7   CARBOXYLIC ACIDS

Name of Compound	Boiling Point	Melting Point	DERIVATIVES		
			Anilide	*p*-Toluidide	Amide
Ethanoic (acetic)	118°		114°	147°	82°
Propanoic	140		106	126	81
Butanoic	163		95	72	115
2,2-Dimethylpropanoic	164	35°	129	120	154
3-Methylbutanoic	176		109	109	135
Chloroacetic	185	63	134	120	118
2-Chloropropanoic	186		92	124	80
Dichloroacetic	189		118	153	98
Cyclohexylcarboxylic		30	146		186
3-Phenylpropanoic (hydrocinnamic)		48	92	135	82
Trichloroacetic		57	94	113	141
Phenylacetic		76	118	136	154
Glycolic		80	97	143	120
o-Methoxybenzoic		100	131		128
o-Toluic		104	125	144	142
m-Toluic		111	126	118	97
Benzoic		122	163	158	128
Cinnamic		133	153	168	147
meso-Tartaric		140			190
o-Chlorobenzoic		142	118	131	139
m-Nitrobenzoic		140	155	162	142
o-Bromobenzoic		150	141		155
Benzilic		150	175	190	154
Salicylic		158	136	156	139
m-Chlorobenzoic		158	122		134
p-Toluic		178	145	160	158
p-Anisic		185	169	186	162
Phthalic		208*d*	169	201(di)	149
p-Hydroxybenzoic		215	202	204	162
p-Nitrobenzoic		241	217	204	201
p-Chlorobenzoic		242	194		179

TABLE 25.8 PRIMARY AND SECONDARY AMINES

Name of Compound	Boiling Point	Melting Point	DERIVATIVES	
			Benzamide	Benzene-sulfonamide
Isopropylamine	33°		100°	26°
n-Propylamine	49		84	36
sec-Butylamine	63		76	70
n-Butylamine	77		42	
Piperidine	106		48	93
Di-n-propylamine	110			51
n-Hexylamine	129		40	96
Morpholine	130		75	118
Cyclohexylamine	134		149	89
N-Methylcyclohexylamine	147		86	
Aniline	183		163	112
Benzylamine	184		105	88
N-Methylaniline	196		63	79
o-Toluidine	199		146	124
m-Toluidine	203		125	95
N-Ethylaniline	205		60	
o-Ethylaniline	211		147	
p-Ethylaniline	216		151	
o-Anisidine	225		60	
m-Chloroaniline	230		120	
Tetrahydroisoquinoline	250	20°	75	
Dibenzylamine	300		112	68
p-Toluidine	200	45	158	120
α-Naphthylamine		50	160	167
2,5-Dichloroaniline		50	120	
Diphenylamine	302	54	180	124
p-Anisidine	240	58	154	95
2-Aminopyridine		60		165(di)
N-Phenyl-α-naphthylamine		62	152	
p-Bromoaniline		66	204	134
p-Chloroaniline	232	70	192	122
o-Nitroaniline		71	110	104
N-Phenyl-β-naphthylamine		108	148	
m-Nitroaniline	284	114	155	136
Anthranilic Acid		147	182	214
p-Nitroaniline		147	199	139
p-Nitro-N-methylaniline		152	112	121
Carbazole	351	246	98	

TABLE 25.9 TERTIARY AMINES

Name of Compound	Boiling Point	DERIVATIVES	
		Picric Acid	Methyl Iodide
Triethylamine	89°	173°	280°
Pyridine	116	167	117
2-Picoline	129	169	230
3-Picoline	143	150	92
4-Picoline	143	167	152
Tri-*n*-propylamine	156	116	208
2,4-Lutidine	159	183	113
2,4,6-Collidine	172	156	
2-Methyl-4-ethylpyridine	173	123	
N,N-Dimethylaniline	193	163	228*d*
Tri-*n*-butylamine	211	106	186
N,N-Diethylaniline	218	142	102
Quinoline	239	203	72(133)
Isoquinoline	243	222	159

TABLE 25.10 AROMATIC NITRO COMPOUNDS

Name of Compound	Boiling Point	Melting Point	ACYL DERIVATIVES OF REDUCED AMINE		Other Derivatives
			Benzene-sulfonamide	Benzamide	
Nitrobenzene	210°		112°	160°	m-Dinitro-benzene 90°
o-Nitrotoluene	224		124	147	2,4-Dinitro-toluene 70
m-Nitrotoluene	231	16°	95	125	m-Nitro-benzoic acid 140
o-Nitroanisole	265		89		2,4,6-Tri-nitroanisole 68
o-Chloronitrobenzene	246	32	129	99	2,4-Dinitro-chloroben-zene 52(50)
o-Bromonitrobenzene	261	43		116	2,4-Dinitro-bromobenzene 72
2,4-Dinitrochlorobenzene		52		178	2,4-Dinitro-phenol 114
p-Nitroanisole	258	54	95	154	2,4-Dinitro-anisole 89
β-Nitronaphthalene		78	136	162	
p-Chloronitrobenzene	242	83	121	192	p-Nitro-phenol 114
Picryl chloride		83	211		Picric Acid 122
m-Dinitrobenzene	302	90	194	240	m-Nitro-aniline 114
4-Nitrobiphenyl		114		230	Acetamide 171
p-Bromonitrobenzene	259	126	136	204	p-Nitro-phenol 114

TABLE 25.11   ESTERS

Name of Compound	Boiling Point	Melting Point	Name of Compound	Boiling Point	Melting Point
Methyl acetate	57°		n-Propyl benzoate	230°	
Ethyl acetate	77		Methyl m-chlorobenzoate	231	
Isopropyl acetate	91		Ethyl salicylate	234	
tert-Butyl acetate	98		n-Butyl benzoate	249	
n-Butyl acetate	126		Methyl p-toluate		33°
n-Pentyl acetate	149		Methyl cinnamate		36
Cyclohexyl acetate	175		Benzyl cinnamate		39
n-Heptyl acetate	192		Phenyl salicylate		42
Phenyl acetate	197		Methyl p-chlorobenzoate		44
o-Cresyl acetate	208		Ethyl m-nitrobenzoate		47
m-Cresyl acetate	212		α-Naphthyl acetate		49
p-Cresyl acetate	213		Ethyl p-nitrobenzoate		56
Ethyl benzoate	213		Phenyl benzoate		69
Methyl o-toluate	215		β-Naphthyl acetate		71
Benzyl acetate	217		p-Cresyl benzoate		71
Isopropyl benzoate	218		Methyl m-nitrobenzoate		78
Ethyl phenylacetate	229				

TABLE 25.12 NITRITES

Name of Compound	Boiling Point	Melting Point	Name of Compound	Boiling Point	Melting Point
Acetonitrile	81°		m-Chlorobenzonitrile		41°
Propionitrile	97		o-Chlorobenzonitrile	232°	47
Chloroacetonitrile	127		p-Methoxybenzonitrile		62
Benzonitrile	191		p-Chlorobenzonitrile		92
o-Tolunitrile	205		o-Nitrobenzonitrile		110
m-Tolunitrile	212		m-Nitrobenzonitrile		118
p-Tolunitrile	217	27°	p-Nitrobenzonitrile		147
Phenylacetonitrile	234				

TABLE 25.13 AMIDES

Name of Compound	Melting Point	Name of Compound	Melting Point
N-n-Propylacetanilide	50°	o-Bromobenzanilide	141°
m-Acetotoluide	66	Salicylamide	142
Propanamide	81	o-Toluamide	142
Acetamide	82	m-Nitrobenzamide	143
o-Nitroacetanilide	92	p-Toluanilide	145
n-Butyranilide	95	Cinnamanilide	151
Trichloroacetanilide	97	m-Nitrobenzanilide	154
N-Methylacetanilide	102		
Acetanilide	114	o-Bromobenzamide	155
Dichloroacetanilide	118	m-Nitroacetanilide	155
o-Toluanilide	125	p-Toluamide	160
m-Toluanilide	126	Benzanilide	163
p-Methoxyacetanilide	127	p-Bromoacetanilide	167
Benzamide	130	p-Chloroacetanilide	179
m-Chlorobenzamide	134	p-Chlorobenzanilide	194
m-Bromobenzanilide	136	p-Nitrobenzamide	201

26

THE LITERATURE OF ORGANIC CHEMISTRY

The purpose of this chapter is to assist the interested student and teacher in obtaining additional information on experimental organic chemistry—information that will be useful in amplifying, modifying, and extending the introductory organic laboratory course for which this book is designed. The chapter is not intended to be a comprehensive guide to the literature of organic chemistry; an excellent series of articles that serves this purpose has been written by Professor J. E. H. Hancock, and an updated review of the subject may be found in an appendix to an advanced textbook by Professor J. March. References to these sources are given at the end of this chapter.

Hancock divided all the literature of organic chemistry into 18 classes. Seven of these which should be of most value to the organic laboratory student (and probably also to the practicing organic chemist) are listed below. Selected examples of each are given with brief explanatory notes.

Class A: Primary Research Journals
Class B: Review Journals
Class C: Encyclopedias and Dictionaries
Class D: Abstract Journals
Class E: Advanced Textbooks
Class F: Reference Works on Synthetic Procedures and Techniques
Class G: Catalogs of Physical Data

CLASS A: PRIMARY RESEARCH JOURNALS

These journals publish original research, with theoretical discussion and experimental details.

1. *Journal of the American Chemical Society.* In recent years the articles on

organic chemistry have been limited to those which are especially timely (Communications to the Editor) or are of wide interest to all chemists.

2. *Journal of Organic Chemistry.* Articles, communications, and notes on organic chemistry.[1]

3. *Tetrahedron.* Articles and reviews on organic chemistry, some in German and in French, as well as in English.[2]

4. *Tetrahedron Letters.* Brief communications, some in German and in French, as well as in English.

5. *Journal of the Chemical Society, Perkin Transactions.* Articles on organic and bioorganic chemistry published by the British Chemical Society. Brief articles are published in a separate journal entitled *Chemical Communications;* prior to 1964 these were published as *Proceedings of the Chemical Society.*

6. *Angewandte Chemie, International Edition in English.* An English-language version of a German journal; the two versions are published almost simultaneously. Some articles approach the length and scope of a review.

7. *Journal of Organic Chemistry, U.S.S.R.* An English translation of a Russian-language journal. The English edition appears approximately 6 months after the publication of the original Russian issue.

CLASS B: REVIEW JOURNALS

Some review journals publish reviews in all areas of chemistry, but many cover only specific areas. All give references to primary journal articles.

1. *Chemical Reviews.* Since 1924 published bimonthly by the American Chemical Society. General in scope.

2. *Annual Reports on the Progress of Chemistry.* Published by the Chemical Society, London. Annual reviews of all areas of chemistry, well classified so that organic work can easily be identified. It is being supplemented by a series of *Specialist Periodical Reports,* covering many specific areas such as alkaloids, nuclear magnetic resonance, and organometallic compounds.

3. *Journal of Chemical Education.* Often contains reviews written by experts at a level that students and others unfamiliar with the subjects may understand them. New, tested experiments and modifications of old experiments suitable for organic laboratory courses are frequently published in the monthly issues.

4. *Accounts of Chemical Research.* Published monthly by the American Chemical Society. Concise reviews of active research areas.

5. *Angewandte Chemie, International Edition in English.* (See 6 under Class A.)

6. *Synthesis.* Reviews and communications of organic synthetic methods published monthly in English and in German.

Other reviews may be found in more specialized publications such as *Advances in Carbohydrate Chemistry, Annual Reviews of Biochemistry,* and *Progress in Stereochemistry.*

---

[1] Communications are published with less delay than the longer articles and notes.

[2] Unless otherwise specified, all publications mentioned are in English.

Beginning with the issue of July 21, 1978, the *Journal of Organic Chemistry* initiated a section entitled *Recent Reviews,* which will appear twice a year and will be an indexed list with titles of major reviews and monographs in organic chemistry.

CLASS C: ENCYCLOPEDIAS AND DICTIONARIES

1.   *Beilstein's Handbuch der Organischen Chemie,* first published in 1883, is perhaps the most complete reference work in any branch of science.[3] The subsequent fourth edition contains data on the 140,000 organic compounds known in 1909. This edition is referred to as the *"Hauptwerk,"* or main work. Instead of printing further editions, the German Chemical Society has issued supplements (*"Ergänzungwerke"*) covering certain periods. The first supplement (EI) covers the years 1910–1919, and the second supplement (EII) covers the years 1920–1929. The main work, as well as each of the supplements, has a volume covering each of 27 different classes of organic compounds. When the second supplement was issued, two additional volumes were included: Volume 28 of EII is the *"Sachregister,"* or name index, and Volume 29 is the *"Formelregister,"* or formula index. Thus the main work and EI have 27 volumes and EII has 29 volumes. These indexes are comprehensive for the main work and each of the supplements, so that in them can be found the name and formula for any organic compound known through 1929. The name index is not as simple or as reliable as the formula index, so the beginner is advised to use the formula index when it is feasible.

The fourth supplement is still in preparation. EIII and EIV were each originally intended to cover the periods 1930–1949 and 1950–1959, respectively. However, while these supplements were in preparation, a decision was made to combine them beginning with Volume 17 (recall that Volumes 1–27 each cover a specific class of compound). Thus, when complete, there will be Volumes 1–16 in EIII and Volumes 1–29 in EIV. The 16 volumes of EIII cover the years 1930–1949 for the respective type of compound included in each volume. The first 16 volumes of EIV will cover the years 1950–1959 for the same types of compounds, and Volumes 17–27 will cover the years 1930–1959 for the types of compounds in these volumes. The name and formula indexes of EIV and Volumes 28 and 29 will cover the period 1930–1959 for all classes of compounds. The 16 volumes of EIII and Volumes 1–4, part of Volume 5, Volumes 17 and 18, and part of Volume 19 of EIV were available in 1978. When the work is ultimately complete, Volume 29 of EII and/or EIV will provide the molecular formula of all organic compounds characterized through 1959, and consultation of the main work and supplements I–IV will provide all published data for a given compound through that date.

There is good reason for the delay in bringing this encyclopedic coverage of organic chemistry up to date. The total number of organic compounds known in 1968 was estimated to be 2.5 million, increasing by about 800 new ones every day. The organic chemical literature is so well organized that despite this staggering statistic it

---

[3] Although *Beilstein* is written in German, many organic chemical terms are the same in both German and English, so only a rudimentary knowledge of German will suffice for practical use of this work.

is possible for a practiced student to determine in no more than an hour or so in a good library whether a particular organic compound has ever been prepared. If it has, the student may also learn how it was synthesized and the physical properties of the pure compound.

For more information about the organization of *Beilstein* and the use of its indexes, refer to the third of the articles by Hancock.

Searches for organic compounds that have appeared in the literature since 1929 are best made by using the abstract journals discussed in the Class D section.

2. *Heilbron's Dictionary of Organic Compounds,* 4th ed., J. R. A. Pollock and R. Stevens, editors, Oxford University Press, New York, 1965. The fifth and tenth supplements published in 1969 and 1974 are cumulative for the preceding five-year periods, respectively. A formula index for the main work and the first five supplements was published in 1971, and the tenth supplement contains a formula index for the sixth through tenth supplements. Compounds are listed alphabetically, with physical properties, selected derivatives, reactions, and references for more than 40,000 compounds.

3. *Handbook of Chemistry and Physics,* annual or biennial editions, Chemical Rubber Publishing Company, Cleveland, Ohio. Gives physical properties of about 14,000 organic compounds.

4. *Lange's Handbook of Chemistry,* various editions. The most recent edition was edited by J. A. Dean, McGraw-Hill Book Company, New York, 1973.

5. *Handbook of Tables for Identification of Organic Compounds,* 3d ed., A. Rapoport, editor, Chemical Rubber Publishing Company, Cleveland, Ohio, 1967. Gives physical properties and derivatives for over 4,000 compounds; organized according to functional groups.

6. *CRC Atlas of Spectral Data and Physical Constants for Organic Compounds,* J. G. Grasseli, editor, Chemical Rubber Publishing Company, Cleveland, Ohio, 1973.

7. *Merck Index of Chemicals and Drugs,* 9th ed., Merck and Company, Rahway, N. J., 1976. Gives concise summaries of physical and biological properties of over 10,000 compounds, with some literature references. Organization is alphabetical by name; synonyms and trade names are provided.

CLASS D: ABSTRACT JOURNALS

Abstract journals provide concise summaries of articles of Class A, listings of reviews (Class B), and announcements of new books (Classes E and F), with references to the original articles or books.

1. *Chemical Abstracts.* This journal began publication in 1907. It now abstracts articles from about 14,000 scientific publications worldwide. Author, subject, and formula indexes appear semiannually. A collective formula index covers the years 1920–1946; the next one covers the 10-year period 1947–1956; subsequent collective, subject, and formula indexes have appeared at 5-year intervals. The most recent one covers the years 1972–1976. In the years between the issuances of the 5-year collective indexes, semiannual indexes must be consulted.

To use the subject indexes of *C.A.* effectively, one must acquire a working knowledge of the system used for naming and indexing of organic compounds. This is described fully in the subject index of Volume 56 (1962). In particular, it should be noted that the subject indexes are compiled under chief headings, followed by subheadings; for example, 1-benzyl-3,4-dihydroisoquinoline is listed under *Iso-quinoline* (which appears at the top of the page): ————, 1-benzyl-3,4-dihydro, **43**; 3742e, 5026a. In this reference the number 43 refers to the volume of *C.A.* which contains two individual abstracts concerning the subject compound; 3742e and 5026a. In these latter designations, the numbers 3742 and 5026 refer to the columns in Volume 43 where the abstracts will be found. (Since 1934 there have been two columns per page, each individually and sequentially numbered.) The letters e and a which are affixed to the column numbers indicate the approximate distance down each column where the abstract may be found on a scale from a to i. Thus the first abstract is about halfway down column 3742, and the second is at the top of column 5026. This use of *letters* began in 1947; previously a terminal superscript *number* was used to indicate position within the column.

Beginning in 1967 a new system was introduced. Abstracts are now coded by separate numbers assigned consecutively through each volume. Letters are still appended to the abstract number, however, not to denote position in the column but rather as a computer check character. For example, two consecutive abstracts are numbered 6865w and 6866d; the letters w and d were calculated by the computer on the basis of the sequence of digits in the abstract number. The letter serves as a check character to prevent errors such as digit transposition or miscopying in the computer handling of abstract numbers. The letter has no bearing on the information in an abstract or on the arrangement of abstracts within sections.

Chemical Abstracts Subject Indexes were substantially reorganized in 1972 and now include the following sections: (1) *Index Guide,* whose introductions should be consulted first; (2) *General Subject Index,* whose headings do not refer to specific chemical substances, but are more general such as "amines," "chromatography," or "ferrite substances"; (3) *Chemical Substances Index,* whose headings are for specifically defined chemical substances (by name) such as elements, chemical compounds, specific minerals, and alloys; (4) *Formula Index,* whose entries are for specific compounds by molecular formula rather than by name (as in the *Chemical Substances Index*); and (5) *Index of Ring Systems,* whose entries are organized according to the types of ring(s) in their structure, whether carbocyclic or heterocyclic. A companion to the *Chemical Abstracts Subject Indexes* is the *Chemical Abstracts Service Registry Index,* which involves a computer-generated numbering system for unique unambiguous chemical substances, providing a kind of Social Security number for each compound.

In the 1970s Chemical Abstracts Service (C.A.S.) and other abstracting agencies have increasingly relied on sophisticated computer technology to manage the huge volume of information generated by scientific disciplines in general and by chemists in particular. Consequently a new approach to locating information in scientific literature has been developed: the use of remote data bases for information retrieval.

C.A.S. regularly provides readable files derived from current issues of *C.A.* to licensed Information Centers or "Vendors" located throughout the United States and the rest of the world. Such files (or data bases) at these centers may be reached ("accessed") and read, usually for a modest fee, by remote terminals at any location serviced by telephone. This remote access is practiced at many colleges and universities as well as at government and commercial organizations.

Fundamentally the approach to using remote data bases derived from *C.A.* is similar to use of the conventional indexes. Key words involving subject topics, compound names, and/or authors are identified, and the index files are scanned by the Vendor's computer to determine if relevant information is present. Moreover, the data bases are constructed in such a way that complex search terms consisting of several "linked" key words (such as "naphthalene/oxidation/phthalate") may be employed. Thus the scope of information obtained may be *inclusive* or *exclusive*, depending on the way in which the search terms are linked. Assistance of a trained search analyst in devising a proper linking strategy facilitates both the quality and the cost effectiveness of the search.

Remote data bases derived from *C.A.* during the period from 1970 to the present may be searched in minutes. Indeed, up-to-date information may be obtained in this manner well before it is available in the hardcover editions of the *C.A.* indexes. In addition to *C. A. S.*, a number of other indexing services offer rapid search capabilities on such specific but diverse topics relating to organic chemistry as patents, pharmacology, toxicology, agricultural research, and environmental science.

2. *Chemisches Zentralblatt* is the second major abstract journal; it is also published in German. It predates *Chemical Abstracts* by over 50 years, and prior to 1940 was more complete and more reliable than the latter. For this reason, in an exhaustive search for an organic compound in the years between 1929 (the last year covered by *Beilstein's* formula index) and 1940, it would be well to consult the collective formula indexes of *Chemisches Zentralblatt,* one for the period 1929–1934 and one for the period 1935–1939. *Chemisches Zentralblatt* discontinued publication in 1970.

3. *Chemical Titles,* published biweekly by Chemical Abstracts Service since 1961, lists *titles* from over 700 chemical journals. The unique value of this publication derives from the fact that not only the title but also every significant word in the title is listed in alphabetical order. Although it is not actually an abstract journal, it serves a similar function, and a title appears with much less delay than an abstract.

CLASS E: ADVANCED TEXTBOOKS

The advanced textbooks are subdivided according to subject and function.

1. General

a. *Chemistry of Carbon Compounds,* E. H. Rodd, editor, Elsevier Publishing Company, New York, Vols. 1–5, 1951–1962. Although the second edition, edited by S. Coffey, is not yet completed, Volumes I–IIIE appeared in 1976. A comprehensive survey of all classes of organic compounds, giving properties and syntheses for many individual compounds.

b. *Comprehensive Organic Chemistry,* Sir Derek Barton, chairman of the editorial board, Pergamon Press, Oxford, U.K., 1978. A six-volume treatise on the synthesis and reactions of organic compounds, written by more than 100 authors with over 20,000 references to the original literature. Intended to fill the gap between existing multivolume series published in parts over many years, such as Rodd, and smaller books, such as the following.

    c. *Advanced Organic Chemistry: Reactions, Mechanisms, and Structures,* 2d ed., by J. March, McGraw-Hill Book Company, New York, 1977. Many references to the original work are given.

    d. *Advanced Organic Chemistry; Part A: Structure and Mechanisms; Part B: Reactions and Synthesis,* by F. A. Carey and R. J. Sundberg, Plenum Publishing Corporation, New York, 1977.

    e. *Modern Synthetic Reactions,* 2d ed., by H. O. House, W. A. Benjamin, Menlo Park, Calif., 1972. Three general classes of reactions—those used for reduction, for oxidation, and for the formation of new carbon-carbon bonds—are surveyed in terms of scope, limitations, stereochemistry, and mechanisms.

    f. *Basic Principles of Organic Chemistry,* 2d ed., by J. D. Roberts and M. C. Caserio, W. A. Benjamin, Menlo Park, Calif., 1977. Intended for a first course in organic chemistry but extensive in coverage. Innovative in consistent use of systematic nomenclature.

    2. Identification and analysis of organic compounds (see references at end of Chapter 25.3).

    3. Instrumental techniques of analysis (see references at end of Chapter 4).

CLASS F: REFERENCE WORKS ON SYNTHETIC PROCEDURES AND TECHNIQUES

    1. *Survey of Organic Syntheses,* by C. A. Buehler and D. E. Pearson, Wiley-Interscience, New York, Vol. 1, 1970; Vol. 2, 1977. This extensive two-volume work covers the principal methods of synthesizing the main types of organic compounds. The limitations of the reactions, the preferred reagents, the newer solvents, and experimental conditions are considered.

    2. *Organic Syntheses,* by Henry Gilman, editor-in-chief, John Wiley & Sons, New York, 1932–present, 56 volumes through 1977. Every 10 volumes have been collected, indexed, and published as *Collective Volumes.* Detailed directions for the synthesis of over 1000 compounds. Procedures have all been thoroughly checked by independent investigators before publication. Many of the general methods may be applied to synthesis of related compounds other than those described. The collective volumes contain indexes of formulas, names, types of reaction, types of compounds, purification of solvents and reagents, and illustrations of special apparatus. A cumulative index to the five collective volumes was published in 1976.

    3. *Organic Reactions,* by various contributors, John Wiley & Sons, New York, 1942–present, 25 volumes through 1977. Each volume contains from 5 to 12 chapters, each of which deals with an organic reaction of wide applicability. Typical experimental procedures are given in detail, and extensive tables of examples with references are included. Each volume contains a cumulative author and chapter-title index.

4. *Reagents for Organic Synthesis,* by Mary and Louis Fieser, Wiley-Interscience, New York, six volumes published between 1967 and 1977. In these six volumes 5300 reagents and solvents are described in terms of methods of preparation or source, purification, and utilization in typical reactions. Ample references to primary literature are given.

5. *Organicum, Practical Handbook of Organic Chemistry,* by many authors, English translation by B. J. Hazzard, Addison-Wesley Publishing Company, Reading, Mass., 1973. A standby of European chemists, which has been translated from the German.

6. *Technique of Organic Chemistry,* 3d ed., A. Weissberger, editor, Interscience Publishers, New York, 1959. Revised volumes have appeared at regular intervals; in 1970 the general title was widened to *Techniques of Chemistry.* Examples of titles of individual volumes are Vol. III, *Separation and Purification;* Vol. IV, *Distillation;* Vol. VIII, *Investigation of Rates and Mechanisms of Reaction;* Vol. XII, *Thin-layer Chromatography;* Vol. XIII, *Gas Chromatography;* Vol. IV, 2d ed., Part 1, *Elucidation of Organic Structures by Physical and Chemical Methods.*

CLASS G: CATALOGS OF PHYSICAL DATA

1. PMR spectra

a. *High Resolution NMR Spectra Catalog,* compiled by the staff of Varian Associates, Palo Alto, Calif., Vol. I, 1962; Vol. II, 1963. Hydrogen nmr spectra of 587 representative organic molecules are depicted, and the peaks are assigned to the hydrogen nuclei responsible for the absorptions.

b. *Nuclear Magnetic Resonance Spectra,* published by Sadtler Research Laboratories, Philadelphia. Hydrogen nmr spectra of over 26,000 compounds had been published by 1977, and about 1000 are being added annually. Assignment of peaks are made as in the Varian spectra, and integration of the signals is shown on many of the spectra.

2. IR spectra

a. *Sadtler Standard Spectra, Midget Edition,* published by Sadtler Research Laboratories, Philadelphia. In 1977 about 53,000 prism spectra had been published and about 38,000 grating spectra. New grating spectra are added each year. Volume 53 was published in 1977.

b. *Aldrich Library of Infrared Spectra,* 2d ed., by C. J. Pouchert, Aldrich Chemical Company, Milwaukee, Wis., 1975. Contains over 10,000 spectra.

3. UV spectra

Sadtler Research Laboratories, Philadelphia, has published over 25,000 spectra. Three volumes containing 1000 spectra were published in 1977.

Some of the texts of Class E, such as those listed at the end of Chapter 4, contain numerous pmr and ir spectra with molecular assignments. Several "problem books" of spectroscopic analysis have been published more recently; these give various combinations of ir, pmr, uv, and mass spectra of "unknown" organic compounds, with answers provided. Among these are the following: *Exercises in Organic Spectroscopy,* 2d ed., by R. H. Shapiro and C. H. DePuy, Holt, Rinehart and Winston, New York, 1977; and *Organic Spectral Problems,* by J. R. Dyer, Prentice-Hall, Englewood Cliffs, N. J., 1972.

NOTES ON USE OF THE LITERATURE OF ORGANIC CHEMISTRY IN AN INTRODUCTORY LABORATORY COURSE

The literature outline given in this chapter may be used in a variety of ways, according to the aims and needs of different courses and the library facilities available, ranging from no use at all to extensive application. Even in those cases where the pressure of time and/or lack of facilities preclude the use of literature beyond the pages of this textbook itself, we feel that this chapter may be valuable to the serious students who may decide to go farther in the study of organic chemistry.

In many organic laboratory courses instructors are interested in making part of the experimentation open-ended—encouraging the students to plan and carry out experiments with some independence. Although this is highly desirable, it has an element of danger unless the plans are checked and the work is monitored carefully. In several chapters of this text additional or alternative experiments are provided or suggested. The inclusion of the literature outline of this chapter now provides a wide source of information for additional experiments.

The most likely class of literature to yield appropriate synthetic experiments is Class F. The experiments from *Organic Syntheses* are particularly suitable; although they are usually on a large scale, they can easily be scaled down. Also deserving special mention are the experiments that appear from time to time in the *Journal of Chemical Education* (Class B). Useful improvements or modifications of experiments are often found first, however, in the primary literature (Class A); for example, the modification of the experiment on the preparation of triptycene (Chapter 11), which avoids the dangerous handling of the dry diazonium salt intermediate, is based on a report in the *Journal of the American Chemical Society* that appeared after the first edition of this text was published. For more experience in identification of unknown organic compounds, the books of Class C.2 will be most useful. The catalogs of spectra listed in Class G represent a vast reservoir from which to draw for paper unknowns and problems. They should be used with discretion, however, because many molecules give ir and pmr spectra that are not easily interpreted by beginners. If you wish to learn how to make a comprehensive search for a specific compound in the literature, to learn its properties or a preferred method of synthesis, refer to the second and third articles by Hancock for a fuller introduction to the use of the Class C and Class D literature.

The following example is given as an illustration of how one might proceed to solve problems such as those proposed in Exercise 11. "Mustard gas" is one of the names that has been applied to the compound $ClCH_2CH_2$—S—CH_2CH_2Cl. Find the answers to the following questions: (1) Has this compound been synthesized or isolated? (2) By whom? (3) When? (4) Where can the most recent information on this compound be found?

First, write the molecular formula as $C_4H_8Cl_2S$, and look in *Beilstein's General Formelregister, Zweites Ergänzungswerk.* On page 65 will be found the entry "β,β'-Dichlor-diäthylsulfid, Senfgas **1**, 349, I 175, II 348, 940," and just below it the entry "α,α'-Dichlor-diäthylsulfid **1** II 685." The first entry will be recognized as that pertaining to the subject compound. The references are to page 349 in Volume 1 of the main work, page 175 in Volume 1 of the first supplement, and to pages 348 and 940 in Volume 1 of the second supplement. Although the third supplement was published after the general index, by noting the "System No." of the subject compound, 23, one may locate it on page 1382 of the third supplement.

Referring to page 349 in Volume 1 of the main work, one finds "β,β"-Dichlor-diäthylsulfid $C_4H_8Cl_2S$ = $(CH_2Cl \cdot CH_2)_2S$. B. Aus Thiodiglykol S $(CH_2 \cdot CH_2 \cdot OH)_2$ und PCl_3 (V. Meyer, *B.* **19**, 3260).-." This translates: "B. = Bildung, Formation. From thiodiglycol and PCl_3 (V. Meyer, *Berichte,* **19**, 3260)." Looking up the reference in the *Berichte der Deutschen Chemischen Gesellschaft* (Vol. 19, p. 3260, published in 1886) one finds that Victor Meyer first prepared this compound in two steps as follows:

$$2\ Cl-CH_2CH_2-OH \xrightarrow{\ K_2S\ } HO-CH_2CH_2-S-CH_2CH_2-OH$$

$$HO-CH_2CH_2-S-CH_2CH_2-OH \xrightarrow{\ PCl_3\ } Cl-CH_2CH_2-S-CH_2CH_2-Cl$$

The entry in the main work of *Beilstein* (Vol. 1, p. 349) gives in a total of six lines the boiling point and solubility properties of the subject compound and one chemical reaction. The final two words are *Sehr giftig,* "very poisonous"—a terse commentary on a material that was much feared as a lethal military weapon in World War II but was never used. (It is interesting to note the statement by Meyer in his *Berichte* article that although his laboratory assistant developed skin eruptions and eye inflammation after preparing this compound, he himself suffered no ill effects even though he took no precautions in handling it!)

In the first supplement (p. 175) 15 lines are devoted to "β,β'-diäthylsulfid" and in the second supplement (p. 348), five and one-half pages, indicating the increased interest in this compound during 1920–1929. Turning next to the *Chemical Abstracts Collective Formula Index* for 1920–1946, under $C_4H_8Cl_2S$ is found the entry "(See also Sulfide, bis(chloroethyl).) Sulfide, 1-chloroethyl 2-chloroethyl, **25**: 2114[8]." Since "Sulfide, 1-chloroethyl 2-chloroethyl" is not the compound of interest, we look in the *Chemical Abstracts Decennial Subject Index,* 1917–1926, for the entry "Sulfide, bis (β-chloroethyl)," which is followed by the names *mustard gas; yperite* and by six general references and two columns of more specific references beginning with "absorption by skin, mechanism of, **14**: 300[4]" and ending with "toxicity and skin-irritant effects of, **15**: 1943[6]."

The *Chemical Abstracts Decennial Subject Indexes* could presumably be used for more recent decades, but one must be on guard for changes in nomenclature. For this reason it is usually advantageous to use formula indexes first. For example, when we go to the January–June 1972 *Formula Index,* under $C_4H_8Cl_2S$ we find "Ethane,

1,1'-thiobis[2-chloro-" as the name for our compound, with seven references to abstracts. Under this name in the *Chemical Substance Index* for the same period, we find the same seven references, but with specific subject headings: for example, "DNA, cross-linking induced by, 95344w."

EXERCISES

1. Find the melting points of the following crystalline derivatives (none of these are listed in the tables of Chapter 25): (a) 2,4-dinitrophenylhydrazone of isovaleraldehyde, (b) semicarbazone of methyl vinyl ketone, (c) 3,5-dinitrobenzoate of 2-methyl-2-pentanol, (d) *p*-toluidide of isobutyric acid, (e) benzamide of ethylamine.

2. Locate an article or a chapter on each of the following types of organic reactions: (a) the aldol condensation, (b) the Wittig reaction, (c) reactions of diazoacetic esters with unsaturated compounds, (d) hydration of alkenes and alkynes through hydroboration, (e) metalation with organolithium compounds.

3. Give a reference for a practical synthetic procedure for each of the following compounds and state the yield that may be expected: (a) 1,2-dibromocyclohexane, (b) α-tetralone, (c) 3-chlorocyclopentene, (d) 2-carbethoxycyclopentanone, (e) norcarane, (f) cycloheptatriene, (g) 1-methyl-2-tetralone, (h) adamantane.

4. Locate descriptions of procedures for the preparation or purification of the following reagents and solvents used in organic syntheses: (a) Raney nickel catalysts, (b) sodium borohydride, (c) dimethyl sulfoxide, (d) sodium amide, (e) diazomethane.

5. Find ir spectra for the following compounds: (a) N-cyclohexylbenzamide, (b) 4,5-dihydroxy-2-nitrobenzaldehyde, (c) benzyl acetate, (d) diisopropyl ether, (e) 3,6-diphenyl-2-cyclohexen-1-one, (f) 4-amino-1-butanol.

6. Find pmr spectra of the following compounds: (a) benzyl acetate, (b) diisopropyl ether, (c) 4-amino-1-butanol, (d) 1-propanol, (e) indane.

7. N-mesityl-N'-phenylformamidine

was first synthesized between 1950 and 1960. Find the primary research article in which this compound is described, and write an equation for the reaction used to prepare it.

8. N-phenyl-N'-*p*-tolylformamidine[4]

---

[4]The German name for this compound is the same as in English except that the final "e" is omitted.

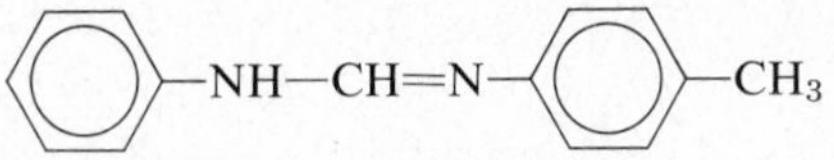

is reported in *Beilstein* to have a melting point of 86°. If you check the first reference given in *Beilstein,* however, you will find the surprising fact that the same chemist who reported this pure compound to have a melting point of 86° had described it as melting at 103.5–104.5° two years previously. The discrepancy between these reports was not explained until the ambiguity was reexamined in the period 1947–1956.

Find the article which solved this mystery.

9. The benzoyl derivative of α-phenylethylamine (formula **3** in Chapter 19) was first described in the form of the optically active (−)-isomer in 1905. Using the formula index of *Beilstein* and given the German name "benzosäure-1-α-phenyläthylamid," find the first reference to this compound in a primary research journal. If you can read the German, give (a) the method of preparation by writing the equation, (b) the melting point of the pure compound and the recrystallization solvent, and (c) the $[\alpha]_D$. For a description in English, find an article published between 1910 and 1920.

10. The name used for the compound described in Exercise 9 in the formula indexes of *Chemical Abstracts* is "benzamide, N-methylbenzyl" or "benzamide, N-α-methylbenzyl." (a) Find a second reference to the (−)-isomer which was published between 1930 and 1940, and compare the physical constants given there with the earlier data. (b) Find a reference to a paper published in Czechoslovakia between 1950 and 1960 giving data on the racemic form of the compound. (c) Find a reference to data on the (+)-isomer of the compound in a paper published between 1960 and 1970.

11. Determine whether or not each of the following compounds has ever been synthesized and, if it has, give the reference to its first appearance in the literature.

(a) Vitamin A   (b) Strychnine   (c)

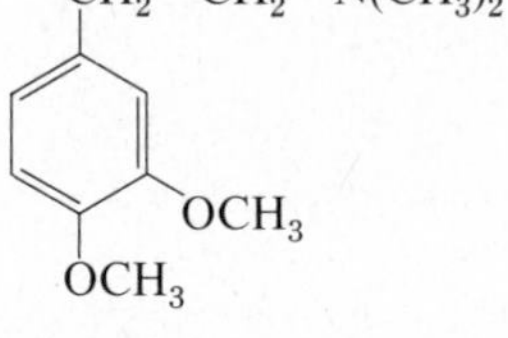

(d)

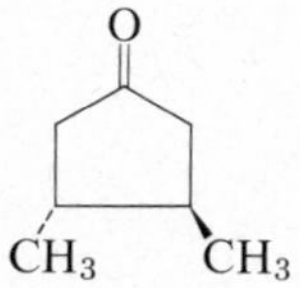

(e) Testosterone   (f)

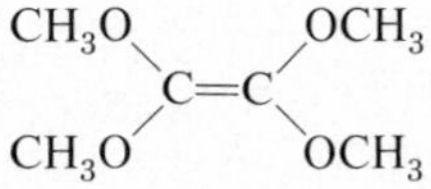

(g) Penicillin-G   (h)

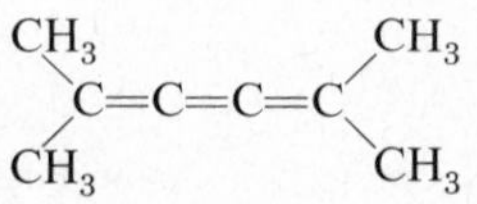

(i)

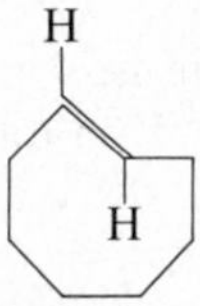

(j) 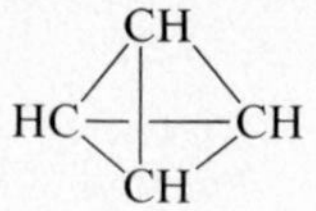 HC—CH / CH / HC——CH / CH

(k) 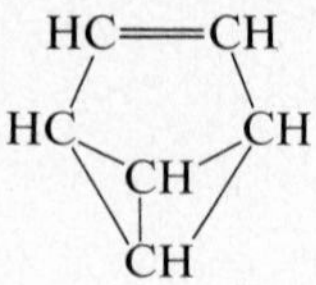 HC=CH / HC—CH / CH / CH

(l) 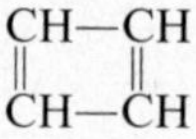 CH—CH ‖ ‖ CH—CH

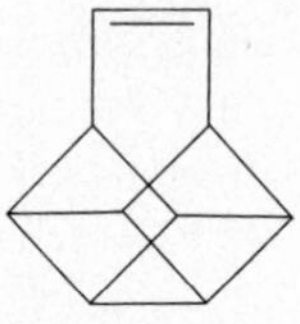

(m) Basketene

(n) Prostaglandin E_2

REFERENCES

1. J. E. H. Hancock, "An Introduction to the Literature of Organic Chemistry," *Journal of Chemical Education,* **45,** 193–199, 260–266, 336–339 (1968).
2. J. March, *Advanced Organic Chemistry: Reactions, Mechanisms, and Structure,* 2d ed., McGraw-Hill Book Company, New York, 1977, Appendix A.

1

DRYING AGENTS
Desiccants

The important procedure of drying either a reagent, solvent, or product will be encountered at some stage of nearly every reaction performed in the organic chemistry laboratory. The techniques of drying solids and liquids and some of the drying agents which are commonly used will be described in this appendix.

Solids. It is important to dry solid organic compounds, because water and organic solvents not only may affect melting points and quantitative elemental analyses, but also may cause difficulties if the wet solid is used in a reaction whose success depends on the absence of small amounts of water or other liquids. If the solid has been recrystallized from a volatile organic solvent, it can usually be dried to an extent satisfactory for most purposes by air-drying at room temperature. The process can be accelerated by spreading the solid on a piece of filter paper or on a clay plate, either of which will serve to absorb traces of water or any excess solvent present. Another useful method to enhance the rate of drying is to collect the solid on a Büchner funnel held in a filter flask, to press the solid as dry as possible with a clean cork, and then to pull air through the filter cake by means of the aspirator (water pump).

If the solid compound is hygroscopic or has been recrystallized from water or a high-boiling solvent, it is usually necessary to dry the substance at atmospheric pressure or under vacuum in an oven operating at a temperature below the melting or decomposition point of the solid. For air-sensitive solids, drying must be done either in an inert atmosphere, such as that provided by nitrogen, or under vacuum. For samples that are to be submitted for quantitative elemental analysis, the solid is normally dried to constant weight by heating under vacuum.

It is often convenient to store dried organic solids in desiccators containing desiccants such as silica gel, phosphorus pentoxide, and calcium chloride, although tightly stoppered bottles make good storage vessels. Of course, it is also possible to use a desiccator to dry an organic solid wet with water. A solid contaminated with a hydrocarbon solvent can be dried by placing it in a desiccator containing a block of paraffin, which serves to absorb the hydrocarbon.

Liquids.   The drying of organic liquids is particularly important because their most common method of purification is by distillation. Any water present in an organic compound being distilled may react with the compound at or below the temperature of the distillation, or it may co- or steam-distil with the liquid and thereby contaminate the distillate.[1]

Two of the more important general requirements for a drying agent are that neither it nor its hydrolysis product will react chemically with the organic liquid which is to be dried, and that it can be easily and *completely* separated from the dried liquid. Another important consideration for a drying agent is that it be efficient in its action so that most or all of the water will be removed by the desiccant.

The more commonly used drying agents and some of their properties are listed in the accompanying table. Of the agents listed, calcium chloride, sodium sulfate, and magnesium sulfate will generally suffice for the needs of the basic organic chemistry laboratory course.

It should be noted that the desiccants given in the table function in one of two ways: either the drying agent interacts reversibly with water, by the process of adsorption or absorption (equation 1), or it reacts irreversibly, with water serving as an acid or a base. In the case of reversible hydration a certain amount of water will remain in equilibrium with the hydrated drying agent; the lower the amount of water left at equilibrium, the greater the *efficiency* of the desiccant. Thus in a recently published[2] study of various desiccants it was found that sodium sulfate is more efficient in its drying power than is magnesium sulfate but is less efficient than calcium sulfate for drying *p*-dioxane. Desiccants that remove water by irreversible chemical reaction clearly have very high efficiencies but are generally more expensive than the other types of drying agents.

$$\text{Drying agent} + H_2O \rightleftarrows (\text{drying agent}) \cdot x\, H_2O \qquad (1)$$

With those drying agents which operate by formation of hydrates (equation 1), it is imperative that the drying agent be *completely* removed by filtration or decantation *before* distillation of the dried liquid, since most of the hydrates decompose with loss of water at temperatures above 30–40°. For those desiccants, such as calcium hydride, sodium, and phosphorus pentoxide, which react vigorously with water and for those such as calcium sulfate, which have a low capacity for removing water, a preliminary drying using a less reactive and efficient desiccant with a high capacity is normally required. When a drying agent which reacts with water to evolve hydrogen is used, appropriate precautions must be taken to vent the hydrogen and thereby prevent buildup of this highly flammable gas.

---

[1] The drying of some organic solvents makes profitable use of the codistillation of water with the solvent in a process termed azeotropic distillation (see Section 2.4). For example, benzene and toluene can be dried reasonably well in this manner (see footnote 2). Solids can also sometimes be dried by dissolving them in a suitable solvent, removing any water by azeotropic distillation and then recovering the solid by removal of the solvent.

[2] D. R. Burfield, K-H. Lee, and R. H. Smithers, *Journal of Organic Chemistry,* **42,** 3060 (1977). These authors also found that relative efficiencies of desiccants in their abilities to remove water from *gases* cannot be extrapolated with confidence to drying of liquids. Many discussions of desiccant efficiencies are based on just such extrapolations.

It is poor technique to use an unnecessarily large quantity of drying agent when drying a liquid, since the desiccant may adsorb or absorb the organic liquid along with water. Moreover, mechanical losses on filtration or decantation of the dried solution may become significant. Of course, the amount of drying agent required depends upon the quantity of water present and upon the capacity of the desiccant. In general, a portion of drying agent that covers the bottom of the vessel in which the liquid is contained should suffice. If additional desiccant is required, more can be added. Swirling the container of liquid and desiccant enhances the rate of drying when desiccants such as calcium chloride and magnesium sulfate are being used, because it hastens the establishment of the equilibrium for hydration. Agitation appears also to accelerate the rate of drying even in cases in which reaction with water is irreversible, for example, in using calcium hydride (see footnote 2). This is thought to be due to a breakdown in the size of the desiccant particles which results in an increase in the net surface area of the drying agent.

Drying Agent	Acid-Base Properties	Product(s) with Water	Comments[3]
$CaCl_2$	Neutral	$CaCl_2 \cdot H_2O$ $CaCl_2 \cdot 2 H_2O$ $CaCl_2 \cdot 6 H_2O$	High capacity and fast action; reasonable efficiency; good preliminary drying agent; readily separated from dried solution because $CaCl_2$ is available as large granules; cannot be used to dry either alcohols and amines (because of compound formation) or phenols, esters, and acids (because drying agent contains some $Ca(OH)_2$); hexahydrate decomposes (loses water) above 30°
Na_2SO_4	Neutral	$Na_2SO_4 \cdot 7 H_2O$ $Na_2SO_4 \cdot 10 H_2O$	Inexpensive, high capacity; relatively slow action and low efficiency; good general preliminary drying agent; physical form is that of a powder, so filtration required for removal of drying agent from dried solution; decahydrate decomposes above 33°
$CaSO_4$	Neutral	$CaSO_4 \cdot \frac{1}{2} H_2O$	Low capacity but somewhat higher efficiency than Na_2SO_4 and $MgSO_4$; preliminary drying of solution with drying agent of higher capacity strongly recommended; hemihydrate can be dehydrated by heating at 235° for 2–3 hr

[3]Capacity, as used in this table, refers to the amount of water which can be removed by a given weight of drying agent; efficiency refers to the amount of water, if any, in equilibrium with the hydrated desiccant.

Drying Agent	Acid-Base Properties	Product(s) with Water	Comments
K_2CO_3	Basic	$K_2CO_3 \cdot 1\frac{1}{2} H_2O$ $K_2CO_3 \cdot 2 H_2O$	Fair efficiency and capacity; good for esters, nitriles and ketones; cannot be used with acidic organic compounds
$MgSO_4$	Weakly acidic	$MgSO_4 \cdot H_2O$ $MgSO_4 \cdot 7 H_2O$	About equivalent to Na_2SO_4 as a general drying agent; requires filtration for removal of drying agent from dried solution; heptahydrate decomposes above 48°
H_2SO_4	Acidic	$H_3O^+HSO_4^-$	Good for alkyl halides and aliphatic hydrocarbons; cannot be used with even such weak bases as alkenes and ethers; high efficiency
P_2O_5	Acidic	H_2PO_3 $H_4P_2O_7$ H_3PO_4	See comments under H_2SO_4; also good for ethers, aryl halides, and aromatic hydrocarbons; generally high efficiency; preliminary drying of solution recommended; dried solution can be distilled from drying agent
CaH_2	Basic	$H_2 + Ca(OH)_2$	High efficiency with both polar and nonpolar solvents, although inexplicably it fails with acetonitrile (see footnote 2); somewhat slow action; good for basic, neutral, or *weakly* acidic compounds; cannot be used for base-sensitive substances; preliminary drying of solution is recommended; dried solution can be distilled from drying agent
Na or K	Basic	$H_2 + NaOH$ or KOH	Good efficiency but slow action; cannot be used on compounds sensitive to alkali metals or to base; care must be exercised in destroying excess drying agent; preliminary drying *required;* dried solution can be distilled from drying agent
BaO or CaO	Basic	$Ba(OH)_2$ or $Ca(OH)_2$	Slow action but high efficiency; good for alcohols and amines; cannot be used with compounds sensitive to base; dried solution can be distilled from drying agent
KOH or NaOH	Basic	Solution	Rapid and efficient but use limited almost exclusively to drying of amines; has potential for other nonacidic solvents such as *p*-dioxane (see footnote 2)

Drying Agent	Acid-Base Properties	Product(s) with Water	Comments
Molecular Sieve #3A or #4A[4]	Neutral	Water strongly adsorbed	Rapid and generally highly efficient; preliminary drying recommended; dried solution can be distilled from drying agent if desired; Molecular Sieve is the trade name for aluminosilicates whose crystal structure contains a network of pores of uniform diameter; the pore sizes of sieves #3A and 4A are such that only water and other small molecules such as ammonia can pass into the sieve; water is strongly adsorbed as water of hydration; hydrated sieves can be reactivated by heating at 300–320° under vacuum or at atmospheric pressure

[4]The numbers refer to the nominal pore size, in Ångstrom units, of the sieve.

APPENDIX 2

THE LABORATORY NOTEBOOK

A notebook used in the laboratory is to be a *complete* record of the experimental work. An 8×10-inch bound book is normally satisfactory for this purpose. The criterion used in judging what should be in the notebook is that the record should be so thorough and so well organized that anyone who reads the experiment can understand it and can see exactly what has been done and thus can repeat it, if necessary, in precisely the same way the original work was done.

All data are to be recorded in the notebook *at the time they are obtained.* There is no reason for recording anything on odd pieces of paper to be transcribed into the notebook later. Neatness is desirable, but it is less important than a complete notebook. Recopying experimental data costs time, cannot possibly improve them, and may sometimes even worsen the data because errors may be made in copying. The notes made at the time of performance necessarily constitute the primary record.

In setting up the notebook at the beginning of the laboratory work, the following general structure should be used:

1. Leave room at the beginning of the notebook for a table of contents and keep it up to date.
2. Number the pages, if they are not already numbered.
3. Start every new experiment on a fresh page. Make all entries *in ink.*
4. It is not necessary to copy the details of experimental procedure if you make no variation whatsoever from it. However, every experiment should have some *reference* (so that if the procedure is a standard one, it can be checked). In essence, your notebook should be a log of your laboratory operations; dates, times, and other pertinent conditions should be entered regularly.

It is particularly important that any *variations* from standard procedure and the reasons for them be noted; if this is not done, it is not possible to reconstruct later just what was done and why it was done. Nothing should ever be deleted from the notebook; merely draw a line through something *considered* wrong and amend it appropriately. The reasons that seem adequate at a particular time may later be found to be inadequate, and a result rejected at an early stage because it seemed impossible may eventually turn out to be the only worthwhile outcome of a particular experiment.

The notebook should not only contain a complete record of any observations but should also reflect precisely what was *concluded* from these observations and the *reasons* for these conclusions. In laboratory work it is often found that data have been interpreted incorrectly, and only a careful and complete record of the experimental observations will reveal discovery of the error in interpretation.

In organic chemistry two distinctly different "types" of experiments are normally carried out in the laboratory. The first of these is the "preparative type" of experiment, in which one compound is converted into another. The second of these is the "investigative type" of experiment, in which one studies physical properties such as boiling point or melting point, or chemical properties such as qualitative tests for various functional groups in a molecule. These two types of experiments require slightly different types of notebook write-up. As an aid in preparing an acceptable notebook, suggested formats for both preparative- and investigative-type experiments are presented here. However, the amount of advance preparation and the notebook form to be used will probably be discussed by the instructor. The formats given are illustrative only.

NOTEBOOK FORMAT FOR PREPARATIVE-TYPE EXPERIMENTS

The important information that should be included in describing preparative-type experiments is the following.

1. *Introduction.* Prepare a brief statement regarding the work to be done.
2. *Main reaction(s) and mechanisms.* Give the main reaction(s) leading to the preparation of the desired compound. Where possible, include the mechanisms of the reactions.
3. *Table of reagents and products.* List in tabular form the molecular weights of each reagent and product, and calculate and enter the number of moles of each reagent to be used. Use the chemical equation(s) for the main reactions to derive the theoretical molar ratio of reagents and products, and enter these values in the list.

Determine which reagent, if any, is used in less than the theoretical molar ratio required; if more than one such reagent is found, find which of these deviates most from the molar ratio needed. This will be the *limiting reagent* in the reaction, that is, the reagent which will determine the maximum amount of product that can be formed. After the limiting reagent has been consumed, the desired reaction will cease, no matter how much of the other reagents remain.

4. *Yield data.* Using the table of reagents and products, calculate the maximum theoretical yield of product that can be expected from the starting materials.

The theoretical yield of product (in moles) can be calculated by determining the moles of product formed per mole of limiting reagent used:

theoretical yield (in moles) = (theoretical molar ratio of product
to limiting reagent) × (moles of limiting reagent actually used)

Multiplying this theoretical yield by the molecular weight of the product gives the expected yield (in grams).

Determine in grams the actual yield of product that is obtained in the experiment. Use the actual yield and the theoretical yield to determine the percent yield, which is

$$\text{percent yield} = \frac{\text{actual yield (grams)}}{\text{theoretical yield (grams)}} \times 100$$

5.   *Observed properties of the product.*   Enter the physical properties of the product that is obtained from the preparation. Important data include melting point or boiling point, color, crystalline form, and related data. It is important to record these properties accurately.

6.   *Side reactions.*   List all possible side reactions that may occur in the reaction. To help determine what these might be, it is often necessary to consult additional sources, such as lecture notes or the lecture textbook.

7.   *Other methods of preparation.*   List alternate methods of preparation that could be used to prepare this same compound. Consult a textbook, and comment briefly on the advantages of the other methods as opposed to the one used. Suggest a reason why the present method of preparation is preferable, if it is.

8.   *Method of purification.*   A very important phase of synthetic organic chemistry is the purification of the desired product. Because of the many and varied reactions that organic compounds undergo, frequently the most difficult step in the preparation of a compound is not its actual formation but its isolation in pure form from side products and unchanged starting materials. In order to accomplish this, list all the compounds that could possibly be present on the basis of a consideration of the main and side reactions. Devise a purification scheme, using a flow sheet, to show how various experimental procedures eliminate the undesired substances and yield the pure product. By examining a purification scheme in this manner, the purpose of each step in the procedure becomes clear, and one may be able to predict what impurities, if any, may contaminate the product.

9.   *Answers to exercises.*

This type of format is shown in Figure A2.1 using an experiment involving the chlorination of cyclohexane (Section 5.1) as an illustration. The figure represents an open notebook; it should be noted that a new experiment is normally started on the left-hand page.

NOTEBOOK FORMAT FOR INVESTIGATIVE-TYPE EXPERIMENTS

A format that can be used for investigative-type experiments is similar to that shown at the end of this paragraph. The description of a new experiment should normally start on the left-hand page, and each new experiment should contain a title and a reference. The important information which should be included in the notebook is the following.

1. *Introduction.* Give a brief introduction to the experiment, and state the purpose(s) of it. This should take up no more than one-half page.

2. *Experiments and results.* Give a brief statement (one or two lines) of each experiment that is to be performed, and leave sufficient room to enter the results as they are obtained. Do *not* recopy the experimental details from the text—just identify each experiment that is to be done. Continue this section as far as needed.

3. *Conclusions.* After completing the assigned experiments, state briefly the conclusions reached on the basis of the results. If the experiment was the identification of an unknown, summarize the results of the findings here.

4. *Answers to exercises.*

FIGURE A2.1   Sample for preparative-type experiment.

Notebook page 3

CHLORINATION OF CYCLOHEXANE

Reference: Section 5.1 in Roberts, Gilbert, Rodewald and Wingrove, page 129
1. INTRODUCTION
    Chlorocyclohexane is to be prepared from cyclohexane using sulfuryl chloride as reagent.

2. MAIN REACTION(S) AND MECHANISM(S)
    $C_6H_{12} + SO_2Cl_2 \rightarrow C_6H_{11}Cl + HCl + SO_2$
        (Mechanism intentionally omitted)

Notebook page 4

3. TABLE OF REAGENTS AND PRODUCTS

Compound	M.W.	Wt. Used (g)	Moles Used	Ratio of Moles: Theory	Used	Other Data
Cyclohexane	84	33.6	0.4	1	2	Density = 0.779 bp 81°
Sulfuryl chloride	135	27.0	0.2	1	1	Density = 1.667 bp 69°
Chlorocyclohexane	118.5			1		Density = 1.016 colorless bp 142.5°
Azobisisobutyronitrile	—	0.1	—	(Initiator)		

Limited reagent: *Sulfuryl chloride*

4. YIELD DATA
    (To determine the *theoretical yield*, it is necessary first to determine the *limiting reagent*. The equation for the main reaction shows that 1 mol of cyclohexane reacts with 1 mol of sulfuryl chloride. Observe, however, that in this experiment 0.4 mol of cyclohexane is allowed to react with 0.2 mol of sulfuryl chloride, so that sulfuryl chloride is the *limiting reagent*. Since the theoretical molar ratio of chlorocyclo- hexane to sulfuryl chloride is 1:1, the number of moles of chlorocyclohexane produced will equal the number of moles of sulfuryl chloride used.)
    Theoretical yield of chlorocyclohexane = (mol $C_6H_{11}Cl$) (M. W. $C_6H_{11}Cl$) = (0.2 mol) (118.5 g/mol) = 23.7 g. If, in an acutal experiment, 15.0 g of pure chlorocyclohexane were obtained, the percent yield would be

$$\frac{15.0 \text{ g}}{23.7 \text{ g}} \times 100 = 63.3\%$$

FIGURE A2.1 (continued)   Sample for preparative-type experiment.

Notebook page 5

5. OBSERVED PROPERTIES OF THE PRODUCT
bp 138–140°; colorless liquid; insoluble in water.

6. SIDE REACTIONS
$C_6H_{11}Cl + SO_2Cl_2 \rightarrow C_6H_{10}Cl_2 + SO_2 + HCl$

   + other polychlorinated cyclohexyl compounds

7. OTHER METHODS OF PREPARATION
$C_6H_{10} + HCl \rightarrow C_6H_{11}-Cl$
$C_6H_{11}-OH + HCl \rightarrow C_6H_{11}-Cl + H_2O$
$C_6H_{11}-OH + SOCl_2 \rightarrow C_6H_{11}-Cl + HCl + SO_2$

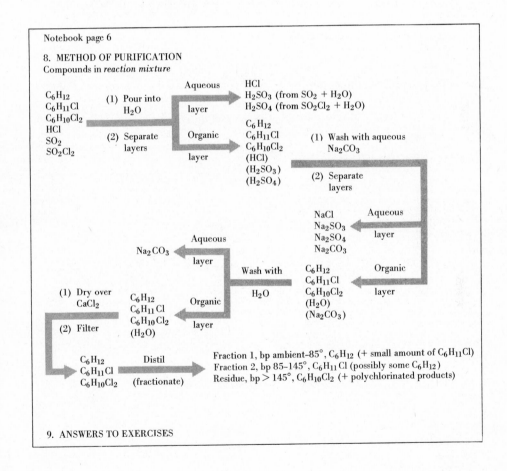

Notebook page 6

8. METHOD OF PURIFICATION
Compounds in *reaction mixture*

9. ANSWERS TO EXERCISES

3

COMPILATION OF PMR ABSORPTIONS

Listed below are the chemical shifts observed for the protons of a number of different types of organic compounds. The shifts are classified according to whether they are methyl, methylene, or methine types of hydrogen atoms. The italicized hydrogen is that responsible for the absorptions listed.

METHYL ABSORPTIONS

Compound	Chemical Shift (ppm)		Compound	Chemical Shift (ppm)	
	τ	δ		τ	δ
CH_3NO_2	5.7	4.3	CH_3CHO	7.8	2.2
CH_3F	5.7	4.3	CH_3I	7.8	2.2
$(CH_3)_2SO_4$	6.1	3.9	$(CH_3)_3N$	7.9	2.1
$C_6H_5COOCH_3$	6.1	3.9	$CH_3CON(CH_3)_2$	7.9	2.1
$C_6H_5-O-CH_3$	6.3	3.7	$(CH_3)_2S$	7.9	2.1
CH_3COOCH_3	6.4	3.6	$CH_2=C(CN)CH_3$	8.0	2.0
CH_3OH	6.6	3.4	CH_3COOCH_3	8.0	2.0
$(CH_3)_2O$	6.8	3.2	CH_3CN	8.0	2.0
CH_3Cl	7.0	3.0	CH_3CH_2I	8.1	1.9
$C_6H_5N(CH_3)_2$	7.1	2.9	$CH_2=CH-C(CH_3)=CH_2$	8.2	1.8
$(CH_3)_2NCHO$	7.2	2.8	$(CH_3)_2C=CH_2$	8.3	1.7
CH_3Br	7.3	2.7	CH_3CH_2Br	8.4	1.6
CH_3COCl	7.3	2.7	$C_6H_5C(CH_3)_3$	8.7	1.3
CH_3SCN	7.4	2.6	$C_6H_5CH(CH_3)_2$	8.8	1.2
$C_6H_5COCH_3$	7.4	2.6	$(CH_3)_3COH$	8.8	1.2
$(CH_3)_2SO$	7.5	2.5	$C_6H_5CH_2CH_3$	8.8	1.2
$C_6H_5CH-CHCOCH_3$	7.7	2.3	CH_3CH_2OH	8.8	1.2
$C_6H_5CH_3$	7.7	2.3	$(CH_3CH_2)_2O$	8.8	1.2
$(CH_3CO)_2O$	7.8	2.2	$CH_3(CH_2)_3Cl$, Br, I	9.0	1.0
$C_6H_5OCOCH_3$	7.8	2.2	$CH_3(CH_2)_4CH_3$	9.1	0.9
$C_6H_5CH_2N(CH_3)_2$	7.8	2.2	$(CH_3)_3CH$	9.1	0.9

METHYLENE ABSORPTIONS

Compound	Chemical Shift (ppm)		Compound	Chemical Shift (ppm)	
	τ	δ		τ	δ
EtOCOC(CH$_3$)=CH_2	4.5	5.5	EtCH_2Cl	6.6	3.4
CH_2Cl$_2$	4.7	5.3	(CH$_3$CH_2)$_4$N$^+$I$^-$	6.6	3.4
CH_2Br$_2$	5.1	4.9	CH$_3$CH_2Br	6.6	3.4
(CH$_3$)$_2$C=CH_2	5.4	4.6	C$_6$H$_5$CH_2N(CH$_3$)$_2$	6.7	3.3
CH$_3$COO(CH$_3$)C=CH_2	5.4	4.6	CH$_3$CH_2SO$_2$F	6.7	3.3
C$_6$H$_5$CH_2Cl	5.5	4.5	CH$_3$CH_2I	6.9	3.1
(CH$_3$O)$_2$CH_2	5.5	4.5	C$_6$H$_5$CH_2CH$_3$	7.4	2.6
C$_6$H$_5$CH_2OH	5.6	4.4	CH$_3$CH_2SH	7.6	2.4
CF$_3$COCH_2C$_3$H$_7$	5.7	4.3	(CH$_3$CH_2)$_3$N	7.6	2.4
Et$_2$C(COOCH_2CH$_3$)$_2$	5.9	4.1	(CH$_3$CH_2)$_2$CO	7.6	2.4
HC≡C−CH_2Cl	5.9	4.1	BrCH$_2$CH_2CH$_2$Br	7.6	2.4
CH$_3$COOCH_2CH$_3$	6.0	4.0	Cyclopentanone α=CH_2	8.0	2.0
CH_2=CHCH_2Br	6.2	3.8	Cyclohexene α=CH_2	8.0	2.0
HC≡CCH_2Br	6.2	3.8	Cycloheptane	8.5	1.5
BrCH_2COOCH$_3$	6.3	3.7	Cyclopentane	8.5	1.5
CH$_3$CH_2NCS	6.4	3.6	Cyclohexane	8.6	1.4
CH$_3$CH_2OH	6.4	3.6	CH$_3$(CH_2)$_4$CH$_3$	8.6	1.4
			Cyclopropane	9.8	0.2

METHINE ABSORPTIONS

Compound	Chemical Shift (ppm)		Compound	Chemical Shift (ppm)	
	τ	δ		τ	δ
C$_6$H$_5$CHO	0.0	10.0	C$_6$H$_5$Cl	2.8	7.2
p-ClC$_6$H$_4$CHO	0.1	9.9	CHCl$_3$	2.8	7.2
p-CH$_3$OC$_6$H$_4$CHO	0.2	9.8	CHBr$_3$	3.2	6.8
CH$_3$CHO	0.3	9.7	p-Benzoquinone	3.2	6.8
Pyridine (α)	1.5	8.5	C$_6$H$_5$NH$_2$	3.4	6.6
p-C$_6$H$_4$(NO$_2$)$_2$	1.6	8.4	Furan (β)	3.7	6.3
C$_6$H$_5$CH=CHCOCH$_3$	2.1	7.9	CH$_3$CH=CHCOCH$_3$	4.2	5.8
C$_6$H$_5$CHO	2.4	7.6	Cyclohexene	4.4	5.6
Furan (α)	2.6	7.4	(CH$_3$)$_2$C=CHCH$_3$	4.8	5.2
Naphthalene (β)	2.6	7.4	(CH$_3$)$_2$CHNO$_2$	5.6	4.4
p-C$_6$H$_4$I$_2$	2.6	7.4	Cyclopentyl bromide	5.6	4.4
p-C$_6$H$_4$Br$_2$	2.7	7.3	(CH$_3$)$_2$CHBr	5.8	4.2
p-C$_6$H$_4$Cl$_2$	2.8	7.2	(CH$_3$)$_2$CHCl	5.9	4.1
C$_6$H$_6$	2.7	7.3	C$_6$H$_5$C≡C−H	7.1	2.9
C$_6$H$_5$Br	2.7	7.3	(CH$_3$)$_3$C−H	8.4	1.6

4

TABLE OF CHARACTERISTIC IR FREQUENCIES

The hydrogen stretch region (3600–2500 cm^{-1}). Absorption in this region is associated with the stretching vibration of hydrogen atoms bonded to carbon, oxygen, and nitrogen. Care should be exercised in the interpretation of very weak bands because these may be overtones of strong bands occurring at frequencies one-half the value of the weak absorption, that is, 1800–1250 cm^{-1}. Overtones of bands near 1650 cm^{-1} are particularly common.

$\bar{\nu}$(cm^{-1})	Functional Group	Comments
(1) 3600–3400	O—H stretching Intensity: variable	3600 cm^{-1} (sharp) unassociated O—H, 3400 cm^{-1} (broad) associated O—H; both bands frequently present in alcohol spectra; with strongly associated O—H (CO$_2$H or enolized β-dicarbonyl compound) band is very broad (about 500 cm^{-1} with its center at 2900–3000 cm^{-1}).
(2) 3400–3200	N—H stretching Intensity: medium	3400 cm^{-1} (sharp) unassociated N—H, 3200 cm^{-1} (broad) associated N—H; an NH$_2$ group usually appears as a doublet (separation about 50 cm^{-1}); the N—H of a secondary amine is often very weak.
(3) 3300	C—H stretching of an alkyne Intensity: strong	The *complete* absence of absorption in this region 3300–3000 cm^{-1} indicates the absence of hydrogen atoms bonded to C=C or C≡C and

$\bar{\nu}(\text{cm}^{-1})$	Functional Group	Comments
(4) 3080–3010	C—H stretching of an alkene Intensity: strong to medium	*usually* indicates the lack of unsaturation in the molecule. Because this absorption may be very weak in large molecules, some care should be exercised in this interpretation. In addition to the absorption at about 3050 cm^{-1}, aromatic compounds will frequently show *sharp* bands of medium intensity at about 1500 *and* 1600 cm^{-1}.
(5) 3050	C—H stretching of an aromatic compound Intensity: variable; usually medium to weak	
(6) 3000–2600	OH strongly hydrogen-bonded Intensity: medium	A very broad band in this region superimposed on the C—H stretching frequencies is characteristic of carboxylic acids (see 1).
(7) 2980–2900	C—H stretching of an aliphatic compound Intensity: strong	Just as in the previous C—H entries (3–5), *complete* absence of absorption in this region indicates the absence of hydrogen atoms bonded to tetravalent carbon atoms. The tertiary C—H absorption is weak.
(8) 2850–2760	C—H stretching of an aldehyde Intensity: weak	Either one or two bands *may* be found in this region for a single aldehyde function in the molecule.

The triple-bond region (2300–2000 cm^{-1}). Absorption in this region is associated with the stretching vibration of triple bonds.

$\bar{\nu}(\text{cm}^{-1})$	Functional Group	Comments
(1) 2260–2215	C≡N Intensity: strong	Nitriles conjugated with double bonds absorb at lower end of frequency range; nonconjugated nitriles appear at upper end of range.
(2) 2150–2100	C≡C Intensity: strong in *terminal* alkynes, variable in others.	This band will be absent if the alkyne is symmetrical, and will be very weak or absent if the alkyne is nearly symmetrical.

The double-bond region (1900–1550 cm^{-1}). Absorption in this region is *usually* associated with the stretching vibration of carbon-carbon, carbon-oxygen, and carbon-nitrogen double bonds.

$\bar{\nu}(\text{cm}^{-1})$	Functional Group	Comments
(1) 1815–1770	C=O stretching of an acid chloride Intensity: strong	Conjugated and nonconjugated carbonyls absorb at the lower and upper ends, respectively, of the range.

$\bar{\nu}(\text{cm}^{-1})$	Functional Group	Comments
(2) 1870–1800 and 1790–1740	C=O stretching of an acid anhydride Intensity: strong	*Both bands* are present; *each band* is altered by ring size and conjugation to approximately the same extent noted for ketones (see 4).
(3) 1750–1735	C=O stretching of an ester or lactone Intensity: very strong	This band is subject to all the structural effects discussed in 4; thus a conjugated ester absorbs at about 1710 cm^{-1} and a γ-lactone absorbs at about 1780 cm^{-1}.
(4) 1725–1705	C=O stretching of an aldehyde or ketone Intensity: very strong	This value refers to the carbonyl absorption frequency of an acyclic, nonconjugated aldehyde or ketone in which no electronegative groups, for example, halogens, are near the carbonyl group; because this frequency is altered in a predictable way by structural alterations, the following generalizations may be drawn: (a) *Effect of conjugation:* Conjugation of the carbonyl group with an aryl ring or carbon-carbon double or triple bond lowers the frequency by about 30 cm^{-1}. If the carbonyl group is part of a cross-conjugated system (unsaturation on each side of the carbonyl group), the frequency is lowered by about 50 cm^{-1}. (b) *Effect of ring size:* Carbonyl groups in six-membered and larger rings exhibit approximately the same absorption as acyclic ketones; carbonyl groups contained in rings smaller than six absorb at higher frequencies, for example, a cyclopentanone absorbs at about 1745 cm^{-1} and a cyclobutanone absorbs at about 1780 cm^{-1}. The effects of conjugation and ring size are additive, for example, a 2-cyclopentenone absorbs at about 1710 cm^{-1}. (c) *Effect of electronegative atoms:* An electronegative atom (especially oxygen or halogen) bonded to the α-carbon atom of an aldehyde or ketone may raise the position of the carbonyl absorption frequency by about 20 cm^{-1}.
(5) 1700	C=O stretching of an acid Intensity: strong	This absorption frequency is lowered by conjugation as noted under entry 4.
(6) 1690–1650	C=O stretching of an amide or lactam Intensity: strong	This band is lowered in frequency by about 20 cm^{-1} by conjugation. The frequency of the band is raised about 35 cm^{-1} in γ-lactams and 70 cm^{-1} in β-lactams.

$\bar{v}(\text{cm}^{-1})$	Functional Group	Comments
1660–1600	C=C stretching of an alkene Intensity: variable	Nonconjugated alkenes appear at upper end of range, and absorptions are usually weak; conjugated alkenes appear at lower end of range, and absorptions are medium to strong. The absorption frequencies of these bands are raised by ring strain but to a lesser extent than noted with carbonyl functions (see 4).
1680–1640	C=N stretching Intensity: variable	This band is usually weak and difficult to assign.

The hydrogen bending region (1600–1250 cm⁻¹). Absorption in this region is commonly due to bending vibration of hydrogen atoms attached to carbon and to nitrogen. These bands generally do not provide much useful structural information. In the listing, the bands that are most useful for structural assignment have been marked with an asterisk.

$\bar{v}(\text{cm}^{-1})$	Functional Group	Comments
1600	—NH$_2$ bending Intensity: strong to medium	This band in conjunction with bands in the 3300 cm⁻¹ region is often used to characterize primary amines and unsubstituted amides.
1540	—NH— bending Intensity: generally weak	This band in conjunction with bands in the 3300 cm⁻¹ region is often used to characterize secondary amines and monosubstituted amines. In the case of secondary amines this band, like the N—H stretching band in the 3300 cm⁻¹ region, may be very weak.
* 1520 and 1350	NO$_2$ coupled stretching bands Intensity: strong	This pair of bands is usually very intense.
1465	—CH$_2$— bending Intensity: variable	The intensity of this band varies according to the number of methylene groups present; the more such groups, the more intense the absorption.
1410	—CH$_2$—bending of carbonyl-containing component Intensity: variable	This absorption is characteristic of methylene groups adjacent to carbonyl functions; its intensity depends on the number of such groups present in the molecule.
* 1450 and 1375	—CH$_3$ Intensity: strong	The band of lower frequency (1375 cm⁻¹) is usually used to characterize a methyl group. If two methyl groups are bonded to one carbon atom, a characteristic doublet (1385 and 1365 cm⁻¹) will be present.
1325	| —CH bending Intensity: weak	This band is weak and often unreliable.

The fingerprint region (1250–600 cm⁻¹). The fingerprint region of the spectrum is generally rich in detail, with many bands appearing. This region is particularly diagnostic for determining whether an unknown substance is identical with a known substance, the ir spectrum of which is available. It is not practical to make assignments to all these bands because many of them represent combination frequencies and therefore are very sensitive to the total molecular structure; moreover, many single-bond stretching vibrations and a variety of bending vibrations also appear in this region. Suggested structural assignments in this region must be regarded as tentative and are generally taken as corroborative evidence in conjunction with assignments of bands at higher frequencies.

$\bar{\nu}(\text{cm}^{-1})$	Functional Group	Comments
1200	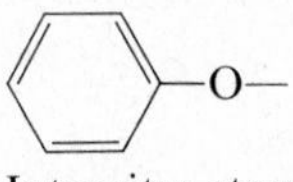 —O— Intensity: strong	It is not certain whether these strong bands arise from C—O bending or C—O stretching vibrations. One or more strong bands are found in this region in the spectra of alcohols, ethers, and esters. The relationship indicated between structure and band location is only approximate and any structural assignment based on this relationship must be regarded as tentative. Esters often exhibit one or two strong bands between 1170 and 1270 cm⁻¹.
1150	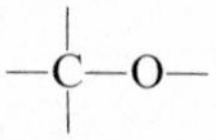 —C—O— Intensity: strong	
1100	—CH—O— Intensity: strong	
1050	—CH₂—O— Intensity: strong	
965	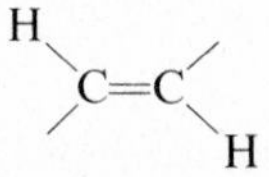 C=C H C—H bending Intensity: strong	This strong band is present in the spectra of *trans*-1,2-disubstituted ethylenes.
985 and 910	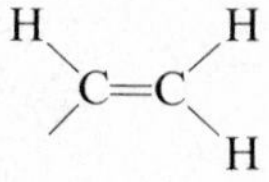 H H C=C H C—H bending Intensity: strong	The lower frequency band of these two strong bands is used to characterize a terminal vinyl group.
890	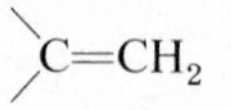 C=CH₂ C—H bending Intensity: strong	This strong band, used to characterize a methylene group, may be raised by 20–80 cm⁻¹ if the methylene group is bonded to an electronegative group or atom.
810–840	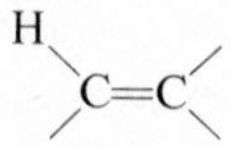 H C=C Intensity: strong	Very unreliable; this band is not always present and frequently seems to be outside this range, since substituents are varied.

$\bar{\nu}(\text{cm}^{-1})$	Functional Group	Comments
700	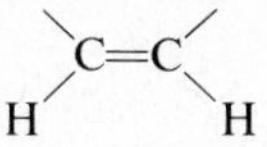 Intensity: variable	This band, attributable to a *cis*-1,2-disubstituted ethylene, is unreliable because it is frequently obscured by solvent absorption or other bands.
750 and 690	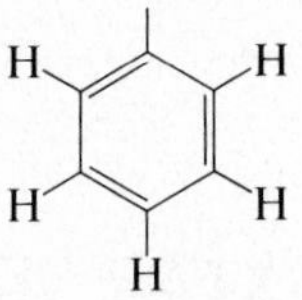 C—H bending Intensity: strong	These bands are of limited value because they are frequently obscured by solvent absorption or other bands. Their usefulness will be most important when independent evidence leads to a structural assignment complete except for position of aromatic substituents.
750	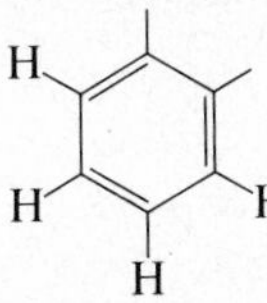 C—H bending Intensity: very strong	
780 and 700	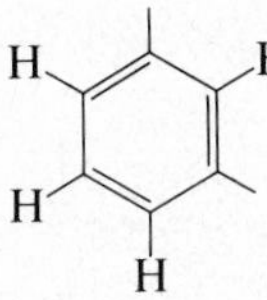 and 1, 2, 3 Intensity: very strong	
825	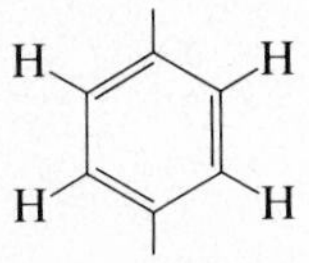 and 1, 2, 4 Intensity: very strong	
1400–1000	C—F Intensity: strong	The position of these bands is quite sensitive to structure. As a result, they are not particularly useful because the presence of halogen is more easily detected by chemical methods. The bands are usually strong.
800–600	C—Cl Intensity: strong	
700–500	C—Br Intensity: strong	
600–400	C—I Intensity: strong	

APPENDIX 5

HEATING AND STIRRING TECHNIQUES IN THE LABORATORY

Heating and stirring are such important laboratory techniques that they deserve special attention here. Heating serves a variety of functions, from increasing the rates of chemical reactions to effecting the distillation of liquids and the dissolution of solids during the course of recrystallization. Stirring also has many uses, two of which are the maintenance of homogeneity when a reagent is being added to a reaction mixture and the optimization of contact between reagents that are insoluble in one another in order to maximize the rate of reaction. Some of the techniques of heating and stirring are described in this appendix, along with their advantages and disadvantages.

HEATING METHODS

The primary sources of heat in the laboratory are (1) gas burners, (2) electrically heated mantles, oil baths, or hot plates, and (3) steam.

Burners. Almost all chemistry laboratories are supplied with natural gas for use with various types of burners. A burner provides the convenience of a rapid and reasonably inexpensive source of heat; since nearly all organic substances are flammable, however, care and good judgment should always be exercised when considering the use of a burner, especially when low-boiling, volatile solvents are to be employed (see Table 2.1). *Never* use a burner to heat open containers (beakers, Erlenmeyer flasks, and so forth) of flammable materials. *Doing so demonstrates a lack of common sense equivalent to using a match to look into one's automobile gas tank!* Burners may be used conveniently and safely to heat *aqueous* solutions containing no volatile and flammable solutes or to heat higher boiling flammable liquids contained in round-bottomed flasks fitted with a reflux condenser or equipped for distillation. In these instances it is safe practice to lubricate the joints of the apparatus with a hydrocarbon or silicone grease to avoid the leakage of vapors through the joints. In any event, if an alternative mode of heating sufficient to do the job is available, elect it in preference to using a burner.

Two common types of laboratory burners are pictured in Figure A5.1. Figure A5.1a shows the classical **Bunsen burner**, named after its inventor. The needle valve at the bottom of the burner serves as a fine adjustment of the gas flow. By turning the barrel of the burner, the air ports may be opened or closed to regulate air input into the flame. Proper adjustment of the gas and air flow will provide a flame in which there is a rather sharply pointed blue cone. The hottest portion of the flame is at the *top* of that cone.

Figure A5.1b shows what is commonly called a **microburner**. The smaller flame provided by the microburner makes it easier to control the rate of heat input to the medium being heated, particularly if the volume of that medium is small. Air flow is adjusted at the baffle on the bottom of the burner; gas flow is adjusted at the main valve on the laboratory bench.

The use of a burner to heat a flask directly will normally produce a "hot spot" on the flask; that is, most of the heat is supplied to a small area on the bottom of the flask, and it must then be dispersed throughout the liquid in the flask by convection or through the turbulence associated with boiling. Development of hot spots can lead to severe bumping in the flask being heated. It is good practice to avoid this situation by placing a wire gauze between the flame and the flask; in this way the dispersal of heat through the gauze will result in the flask being heated more uniformly over its surface. The gauze may be supported with an iron ring.

Another convenient way to use a burner is in connection with a hot-water bath. When temperatures from ambient to about 90° are desired, they may be attained by

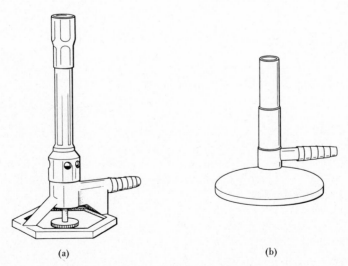

(a) (b)

FIGURE A5.1 Laboratory burners. (a) A Bunsen burner. (b) A microburner.

immersing the flask into a water bath held at the desired temperature by either intermittent or slow heating with a burner (or hot plate). The temperature may be monitored by means of a thermometer in the bath. Temperatures from 90–100° are best attained by means of steam heating, particularly if electrical heating is unavailable.

Electrical Heating.   Given the proper apparatus, resistance heating with electricity is very convenient for use in the laboratory. The rate of heat supply may be precisely controlled through adjustment of the applied voltage by means of a **vari**able **AC** transformer (a common brand name is Variac; see Figure A5.2). A widely used device for electrical heating of *round-bottomed flasks* is a *woven-glass heating mantle* such as the Glas-Col shown in Figure A5.2. These mantles have an electrical resistance coil imbedded in the woven-glass fabric. Although most mantles are constructed with hemispherical cavities *so that a different mantle is required for each differently sized flask,* some are conically shaped to accommodate flasks of different sizes. The initial financial outlay to supply each student with a transformer and mantles of two or three different sizes is fairly high; however, when available, these constitute probably the most convenient and safest form of heating in the undergraduate organic laboratory.

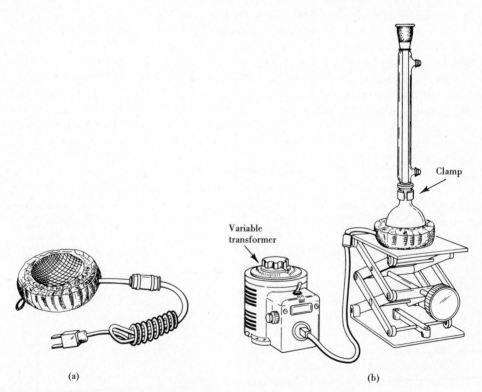

FIGURE A5.2   (a) Woven-glass heating mantle. (b) Use of a heating mantle and variable transformer in heating a reaction mixture.

Some inconveniences in the use of mantles are that they initially heat up rather slowly, they remain hotter than the contents of the flask, and they have a rather high heat capacity. Therefore it is not sufficient either to lower the voltage or to turn off the electricity if one suddenly needs to discontinue heating, for example, when a reaction begins to get out of control. The heating mantle should be *immediately removed from the flask* and the electricity turned off to allow the reaction mixture to cool, either on its own or by means of a cooling bath. Since it may be necessary to remove a heating mantle rapidly, apparatus should *not* be assembled with the mantle resting on the laboratory bench; rather, it should be supported above the bench by an iron ring or a laboratory jack (Figure A5.2) if one is available.

A further caution is in order. The temperature of the mantle is moderated by the boiling liquid within the flask being heated, since the refluxing vapors carry heat away from the mantle. If the flask becomes dry, or nearly so, the mantle may become sufficiently hot to "burn out" as a result of melting of the resistance wire. Most mantles are marked with a maximum voltage to be used with a dry flask, which should not be exceeded. A distillation that toward its end appears to require a voltage in excess of that maximum should be discontinued.

A convenience of the heating mantle is that it is constructed of nonferrous material, so that it may be used in conjuction with a magnetic stirrer for simultaneous heating and stirring of a reaction mixture. This is particularly convenient in allowing for stirring during the course of a distillation.

Electrically heated liquid baths are also commonly encountered in the laboratory. Since the liquid used in these baths is usually either mineral oil or silicone oil, they tend to be referred to as **oil baths**. However, other liquid media are also used. The liquid medium may be heated either by placing the bath container on a hot plate or, better, by immersing in the bath a coil of resistance wire, which is then attached to a variable transformer by means of an electrical cord and plug (Figure A5.3). In most instances a coil wound from about 3 m of 26 gauge Nichrome wire is satisfactory.

The use of a heating bath offers at least two important advantages. First, by monitoring the temperature of the bath with a thermometer and by careful adjustment of the transformer voltage, a desired reaction or distillation temperature may be accurately maintained. Second, heat is transferred smoothly and evenly to the full surface of the flask (to the depth of its immersion) so that there are no hot spots. This is particularly important in a carefully performed distillation where even ebullition is desirable.

The use of heating baths, however, also involves some inconveniences. The volume of the heating fluid is normally sufficiently large so as to require considerable time in reaching the desired equilibrium temperature. Further, if the desired temperature is "overshot," the high heat capacity of the bath is such as to require some time in cooling back to the desired temperature. This, however, may normally be counteracted by reducing the depth of immersion of the flask, since the rate of heat transferral is proportional to the surface area of contact between the flask and the heating fluid. A final minor nuisance associated with the use of oil heating baths is the cleaning of apparatus coated with mineral oil or silicone oil, both of which are water-insoluble; however, these can generally be removed by use of either hydrocarbon or chlorinated solvents. Alternatively, the polyethylene glycols can be employed

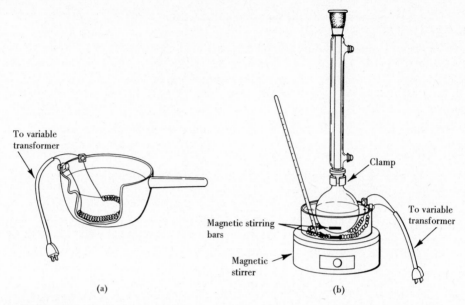

To variable
transformer

Clamp

Magnetic stirring
bars

To variable
transformer

Magnetic
stirrer

(a)

(b)

FIGURE A5.3   (a) An electrically heated oil bath. (b) Use of an oil bath with magnetic stirring of a reaction mixture.

as heating media. These have the advantage of being fully water-soluble, so that glassware is easily cleaned. One example is Carbowax 600, a polyethylene glycol which is liquid at ambient temperature and may be used up to about 180°.

Other factors are important with respect to the liquids used in heating baths. Silicone oils such as Dow Silicone 550 are more expensive, but they are generally preferable to mineral oils for use in heating baths because they can be used at temperatures well into the 200° range without danger of reaching the flash point or of thickening through decomposition. Mineral oil should not be used above about 200° because it will begin to smoke and there is then the potential of flash ignition of the vapors. This danger is most prevalent with darkened and used mineral oil. Mineral and silicone oils must also be protected against contamination by water. If droplets of water are present and these oils are heated about 100°, the water will boil and thus produce bumping in the bath and *spattering of hot oil*. Such baths should be examined regularly; if water droplets are clearly present, the heating fluid should be changed and the container cleaned and dried before refilling.

When materials are to be heated in a *flat-bottomed* container such as a beaker or an Erlenmeyer flask, **hot plates** are frequently convenient. The flat, upper surface of a hot plate is heated by electrical resistance coils to a temperature that is controlled by a rheostat on the front of the device. The use of a hot plate should generally be restricted to the heating of liquids, such as water or its solutions, and of nonflammable organic solvents such as chloroform and carbon tetrachloride. When organic solvents are being heated, the operation should be carried out in the hood to avoid filling the room with the vapors of these toxic materials. Under no circumstances should a hot plate be used to boil and/or to evaporate *highly flammable* organic solvents. The vapors of these solvents may ignite as they billow onto the hot surface of the hot plate or come into contact with the electrical resistance coils. In those circumstances when flammable solvents must be removed from a solution, either use a steam bath *in the hood* or set up distillation apparatus to accomplish this task. For added convenience, hot plates are also available in combination with built-in magnetic stirrers. This device is especially convenient when one wishes to heat and stir a solution or reaction mixture simultaneously. It should be noted that a *round-bottomed* flask cannot be heated effectively with a hot *plate,* since there is contact at one *small* point between the flask and the hot surface!

Steam Heating. Most chemistry laboratories are plumbed to supply steam from a central boiler; this can be a useful source of heat when temperatures up to 100° are desired. The steam outlet is connected to either a **steam bath** or a **steam cone** (Figure A5.4), both of which also have outlets at the bottom to serve as drains for condensed water. When the steam valve is first turned on, a minute or two is normally required for the condensed water to drain out of the lines; once the water has drained and steam is issuing smoothly, the valve should be adjusted to provide a *slow,* steady output. Steam baths and cones are typically fitted with a series of overlapping concentric rings which form their upper surface. These rings may be removed in succession until the central opening is the desired size. The proper size of this opening is normally dictated by the requirements of the operation being performed. For example, if a rapid rate of heat transfer is desired, rings are removed until perhaps up to one-half the surface of the flask can be immersed into the steam. If a slower rate of heating is desired, the opening should be smaller, so that less of the flask is in contact with the steam. Figure A5.4a shows the use of a steam bath for heating a round-bottomed flask. For beakers or Erlenmeyer flasks, the same principles hold except that the opening is always left small enough so that the container may sit directly on top of the steam bath with only its lower surface exposed to the steam.

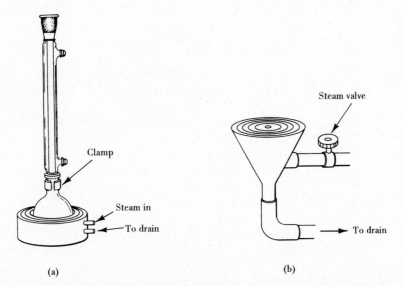

FIGURE A5.4 (a) Steam bath being used to heat a reaction mixture. (b) Steam cone.

If it is necessary to maintain *anhydrous* conditions within the flask being heated, the flask or apparatus to which it is attached should be protected with a drying tube. This will prevent moisture that has condensed on the flask or apparatus from running into the flask. Moreover, the humidity is very high in the vicinity of a steam bath which is in operation.

STIRRING METHODS

Laboratory stirring is most efficiently done by means of an electrical motor in one of the ways described in the following sections. Because of the cost factor, many undergraduate laboratories unfortunately cannot be equipped with enough electrical stirring devices for general use, so we will also describe an alternative manual procedure that may often provide satisfactory mixing.

When a reaction mixture is boiling, no additional stirring is usually necessary except in some cases of heterogeneous mixtures. The turbulence and motion of boiling may be sufficient to maintain reasonable mixing of the solution.

Swirling.   If electrical stirrers are not available, the best simple alternative for mixing the contents of a reaction flask is to loosen the clamp which supports the flask and attached apparatus and to swirl the contents by a manual rocking motion of the flask. This may be done periodically during the course of the reaction. Of course, if the entire apparatus is supported by clamps to a *single* ring stand, it is not necessary to loosen the clamp to the flask. Instead, *make sure all clamps are tight,* and pick up the ring stand to swirl the contents of the flask by gently moving the entire assembly.

Magnetic Stirring.   The equipment associated with magnetic stirring consists of a *stirrer,* which is basically a large bar magnet rotated by a variable-speed electric motor, and a *magnetic stirring bar,* which is placed in the flask whose contents are to be stirred. The stirring bar is usually covered with a chemically inert substance such as Teflon or glass. A flat-bottomed container may be placed directly on top of the stirrer (Figure A5.5), or a round-bottomed flask may be clamped in position above the center of the stirrer (Figure A5.3b). The magnetic stirring bar may be made to rotate in phase with the motor-driven magnet of the stirrer by interaction of their respective magnetic fields, and thus effect stirring of the contents of the flask. Figure A5.3b shows an example in which the contents of a flask are being simultaneously heated with an oil bath and stirred magnetically. Note that the contents of both the

Stirring
bar

FIGURE A5.5   Magnetic stirring of the contents of a beaker.

flask *and* the bath may be stirred by means of a single magnetic stirrer, provided a larger stirring bar is used in the bath than is used in the flask. Stirring of the heating bath is desirable to maintain thermal homogeneity in the heating fluid.

Mechanical Stirring. A variable-speed electric motor may also be used to drive a stirring shaft which extends directly into the flask whose contents are to be stirred. The paddle at the end of the shaft can provide vigorous agitation of the contents of the flask as the shaft is driven by the motor. This is the preferred and most efficient method for stirring *heterogeneous* reaction mixtures. The stirrer shaft is usually constructed of glass and is fitted with a paddle made of Teflon or glass. The paddle is normally removable from the shaft to facilitate cleaning and to make possible the use of different paddles of sizes convenient to the flasks being used. As a point of caution, a glass rather than a Teflon paddle must be used to stir a reaction mixture containing active metals such as sodium or potassium.

Figure A5.6 shows the most commonly employed type of stirrer, the *Trubore*

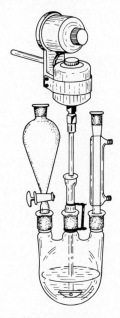

FIGURE A5.6 Flask equipped for mechanical stirring with a Trubore stirrer.

stirrer. The glass shaft and the inner bore of the standard-taper joint are ground to fit each other precisely to produce a bearing. A cup at the top of the bearing is used to hold a few drops of silicone oil to provide lubrication for the bearing. In this way an effective seal is also maintained. The stirrer shaft is connected to the motor by means of a short length of rubber tubing. Accurate alignment of the motor and the shaft must be maintained to avoid wear of the glass surfaces of the shaft and bearing and to minimize vibrations in the apparatus. It is also usually necessary to fasten the bearing to the flask by means of twisted copper wire and a rubber band, as shown in Figure A5.6. This is to keep the bearing from working loose while the motor is in operation.

INDEX

Abbé refractometer, 21
Absorbance, definition of, 99, 194
p-Acetamidobenzenesulfonyl chloride
  ammonolysis of, 372
    discussion of, 366
  ir spectrum of, 375
  preparation of, 365, 370
  sulfathiazole from, 367, 373
Acetanilide
  bromination of, 195–197, 377, 383
  chlorosulfonation of, 365, 370
  ir and pmr spectra of, 375
  preparation of, 364, 370
  recrystallization of, 58
Acetic anhydride
  in acetylation, 364, 395, 396
  as dehydrating agent, 401, 402
Acetogenins, definition and biosynthesis of, 482
Acetophenone, 310, 317, 434
  ir and pmr spectra of, 323
  reduction of, 434, 436
Acetylation, 364
Acetyl chloride
  in acetylation, 364, 399
  hydrolysis of, 338
Acetyl coenzyme A, role in biosynthesis, 483
Acetylferrocene
  acetylation of, discussion of, 395
  ir and pmr spectra of, 400, 401
  preparation of, 396
Acetylsalicylic acid, bromination of, 197
Achiral molecules, 407
Acid anhydrides (see Carboxylic acid
    anhydrides)
Acid halides (see Carboxylic acid halides)
Acidity
  of carboxylic acids, 329, 337
  of hydrocarbons, 392, 394
Acyl halides (see Carboxylic acid halides)
Addition (see Alkenes)
Adipic acid (see Hexanedioic acid)
Alcohols
  classification tests for, 513
  conversion to alkyl halides, discussion, 222

dehydration of, 151, 436
derivatives of, 514
in esterification, 330, 337
from esters, 330
oxidation of, 277, 279, 281
table of, and derivatives, 542
use in Dacron synthesis, 441
use in polyurethane synthesis, 446
Aldehydes
  aldol reactions of, 309
  with ammonia derivatives, 214, 290, 296
  Cannizzaro reactions of, 289
  classification tests for, 506
  conversion to alkenes by Wittig reaction, 293
  derivatives of, 507
    See also specific type
  haloform reaction with, discussion, 308
  hemiacetal formation in, 453
  oxidation of, 279, 281, 285, 287
  preparation of, 278, 281
  table of, and derivatives, 215, 538
Aldol condensation, 310, 318
Aldol reactions, theory of, 310
Alkaloids, definition of, 484
Alkenes
  addition reactions, concerted, 165, 200
    theory of, 164
  classification tests for, 507
  by dehydration, 151, 155, 435, 437
  by dehydrohalogenation, 141
  geometric isomerization of, 167, 168
  hydrogenation of, 167, 169
  preparation of, 140
  relative stabilities of, 153, 212
  by Wittig reaction, 293
Alkyl halides
  from alcohols, 222–225
  classification tests for, 509
  elimination of, 141
  with Lewis acids, 176
  with sodium iodide, 509
  with silver nitrate, 509
Alkylation reactions, 312
  use of dimethyl malonate in, 313

Aluminum trichloride, 176, 179, 399
Amides (*see* Carboxylic acid amides)
Amines
  acetylation of, 364
  from amides, 334
  classification tests for, 520
  derivatives of, 525
  diazotization of, 381
  from imines, 291, 296
  from nitro compounds, 362, 369
  preparation of, discussion, 291
  in preparation of Nylon, 442
  tables of, and derivatives, 545, 546
α-Amino acids
  as acids and bases, 467
  analysis of, 465, 468
  chromatography of, 467, 470
  2,4-dinitrophenyl derivatives, 474, 477
    thin-layer chromatographic analysis of, 478
  isatin reagent with, 466, 473
  isoelectric point, 467
  ninhydrin reagent with, 468, 472
  table of, 466
4-Aminoquinolines, structures of, 350
2-Aminothiazole
  ir, pmr, and uv spectra of, 348, 349
  preparation of, 345, 347
  sulfathiazole from, 367, 373
Analysis, qualitative
  of amino acids, automatic, 468
    paper chromatography, 472
  classification tests (*see* class of compound)
  derivatives in (*see* specific derivative)
  discussion of, 496
  elementary analysis in, 501
  separation of mixtures, 498
  solubility tests in, 503
  spectroscopy in, 533
Anhydrides (*see* Carboxylic acid anhydrides)
Aniline, 369
  acetylation of, 364, 370
  ir and pmr spectra of, 341, 342
  preparation of, 362, 369
  purification scheme for preparation, 496
  with benzoyl chloride, 338
Anisole, bromination of, 195
Anthracene
  in Diels-Alder reaction, 254
    discussion of, 252
  ir spectrum of, 256
Anthranilic acid, 252, 254
Apparatus, lubrication of, 5, 145
Arenes (*see* Hydrocarbons, aromatic)
Aromaticity, discussion of, 391
Aryl halides
  classification test for, 510
  derivatives of, 512
  table of, and derivatives, 541
Arynes, 251
Asymmetric carbon atom, 406
Atebrin (quinacrine), 350

Automatic amino acid analyzer, 468
Azeotropic distillation
  drying of liquids by, 33
  drying of solids by, 564
  drying of styrene by, 437
Azeotropic mixture
  maximum boiling, 34
  minimum boiling, 33
Azobenzene, separation of isomers, 87, 92
Azobisisobutyronitrile, as initiator, 127, 129, 130
Azulene, 392

Baeyer test, 508
Beer-Lambert law, 103, 120, 194
Benzalacetophenone
  with aniline, 311, 318
  with bromine, 311, 318
  ir, uv, and pmr spectra of, 323, 324
  preparation of, 318
  reactions of, discussion of, 311
Benzaldehyde, 297
  in Cannizzaro reaction, 287
  imines from, 297
  ir and pmr spectra of, 289
Benzamide
  benzonitrile from, 334, 339
  ir and pmr spectra of, 341
  preparation of, 338
Benzanilide, 338
  ir and pmr spectra of, 342
Benzene
  in Diels-Alder reaction, 200
  drying of, 564
Benzenediazonium-2-carboxylate
  benzyne from, 252
  preparation of, 254
Benzenesulfonyl chloride, with amines, 521, 523
Benzoic acid
  from acetophenone, 317
  acid chloride from, 337
  from benzaldehyde, 287
  from benzoyl chloride, 338
  from ethylbenzene, 336
  ir and pmr spectra of, 267, 268
  from phenylmagnesium bromide, 260
  purification scheme for Grignard reaction, 497
  recrystallization of, 57
Benzoic anhydride
  ir spectrum of, 343
  preparation of, 339
Benzonitrile
  ir spectrum of, 343
  preparation of, 339
Benzophenone, photochemical dimerization of, 418
p-Benzoquinone
  as dienophile, 203
  ir spectrum of, 208

Benzoyl chloride
  alcoholysis of, 338
  ammonolysis of, 338, 519
  benzoic anhydride from, 339
  hydrolysis of, 338
  preparation of, 332, 337
  reaction with toluene, 418, 425
Benzyl alcohol, 287
  ir spectrum of, 289
N-Benzylanilinium chloride, 297
Benzyl chloride, 299
  ir spectrum of, 305
N-Benzylideneaniline, 297
N-Benzylidene-m-nitroaniline, 297
  ir spectrum of, 304
N-Benzyl-m-nitroaniline, 297
  ir and pmr spectra of, 304, 305
Benzyne
  with anthracene, 252
  preparation of, discussion, 251, 252
Betaines, 293
Biosynthesis, definition and examples of, 482
Biphenyl, 259
Blocking group (see Protective group)
Boiling points
  definition of, 19
  determination of, 19
  effect of pressure on, 18
  micro, 19
  theory of, 18
Boiling point–composition diagram, 28
Bromination
  of acetanilide, 377
  free radical substitution, 135
  test for unsaturation, 507, 508
    discussion of, 165, 507, 508
Bromination, electrophilic, alkenes, theory of, 165
Bromination, electrophilic aromatic
  rate expressions for, 192
  substituent effects on rate of, 191
Bromine
  precautions for, 135
  test for, 502
Bromine water, with phenols, 517
4-Bromoacetanilide
  chlorination of, 379
  ir and pmr spectra of, 386
  preparation of, 377, 383
Bromobenzene
  Grignard reagent from, 259
  ir spectrum of, 269
  nitration of, 184, 187
1-Bromobutane
  from 1-butanol, 222
  Grignard reagent from, 261, 264
  preparation of, 222, 226
1-Bromo-3-chloro-5-iodobenzene
  ir and pmr spectra of, 389
  preparation of, 381, 385
  synthesis of, chart, 378

p-Bromophenol, bromination of, 195
1-Bromopropane, alkylation with, 178, 179
Buffers, in reactions of carbonyl compounds, 214
Bunsen burner, 582
Burner, gas, 582
1,3-Butadiene
  with maleic anhydride, 204
  preparation of, 204
1-Butanol
  conversion to 1-bromobutane, 222, 226
  ir spectrum of, 228
  pmr spectrum of, 116
n-Butyl alcohol (see 1-Butanol)
t-Butyl alcohol (see 2-Methyl-2-propanol)
t-Butylbenzene
  bromination of, 136
  ir and pmr spectra of, 139
n-Butyl bromide (see 1-Bromobutane)
1-Butyl butanoate, 279
t-Butyl chloride (see 2-Chloro-2-methylpropane)
t-Butyl peroxybenzoate
  as source of radicals, 432
  in synthesis of polystyrene, 437

d,l-Camphor, structure of, 483
Cannizzaro reaction, 287
Caprolactam, 446
Carbamic acids, reactions of, 239
Carbenes, 238
Carbocations
  in dehydration of alcohols, 151
  as intermediates in elimination reactions, 142, 151
  as intermediates in nucleophilic substitution, 222, 224
  rearrangement of, 152–154, 177
  stability of, 152
3-Carboethoxy-7-chloro-4-hydroxyquinoline,
  preparation and reaction of, 350, 354
Carbon, activated
  as catalyst support, 160
  as decolorizing agent, 52
Carbon dioxide, 260
Carbohydrates
  in living systems, 451–453
  photosynthesis of, 451
  specific rotation of, procedure for, 458
    See also Disaccharides, Monosaccharides,
    Polysaccharides
Carbonium ions (see Carbocations)
Carbonyl compounds (see specific class of
  compound)
Carboxylic acid amides
  from carboxylic acid halides, 338
  from carboxylic acids, 521
  classification test for, 532
  dehydration of, 335, 339
  derivatives of, 532
  discussion of, 334

Carboxylic acid amides (*cont.*)
  hydrolysis of, 490, 532
  Nylon, 442
  table of, 532
Carboxylic acid anhydrides
  in acylation, 364, 395
  in Diels-Alder reaction, 204
  discussion of, 334
  preparation of, 339
Carboxylic acid esters
  classification tests for, 528
  in Dacron, 441
  derivatives of, 530
  discussion of, 330
  with Grignard reagents, 260
  hydrolysis, 331
  by oxidation of alcohols, 279
  saponification equivalent, discussion, 332
    procedure, 529
  table of, 548
Carboxylic acid halides
  in acetylation, 399
  from carboxylic acids, 337, 520
  discussion of, 332
  hydrolysis of, 338
Carboxylic acids
  acidity of, 337
    discussion of, 330
  from aldehydes, 279
  from arenes, 329, 336
  from carboxylic acid halides, 338
  classification test for, 519
  as derivatives, 512
  discussion of, 329
  derivatives of, 519
  equivalent weight, procedure, 519
  esterification, 337
    discussion of, 330
  from esters, 331
  from Grignard reagent, 260, 265
  table of, and derivatives, 544
Carvone, ir and pmr spectra of, 536
Ceric nitrate
  with alcohols, 514
  with phenols, 516
Chemicals, disposal of, 5
Chemical shift, 110, 111
  effect of resonance on, 537
  effect of ring current on, 394
  table of, 111, 574
Chiral molecules, 407
Chlorination, of 4-bromoacetanilide, 379, 383
Chlorination, free radical
  procedures for, 129
  theory of, 126
Chlorine
  preparation of, 379, 383
  test for, 502
Chloroacetaldehyde
  preparation of, 347
  with thiourea, 346, 347

m-Chloroaniline
  with 4,7-dichloroquinoline, 350, 355
  with diethyl ethoxymethylenemalonate, 350
4-*m*-Chloroanilino-7-chloroquinoline
  preparation of, 357
  synthetic scheme for, 351
2-Chloro-4-bromoacetanilide
  ir and pmr spectra of, 387
  preparation of, 379, 383
2-Chloro-4-bromoaniline
  iodination of, 380, 384
  ir and pmr spectra of, 387, 388
  preparation of, 380, 384
2-Chloro-4-bromo-6-iodoaniline
  diazotization of, 381, 385
  ir and pmr spectra of, 388
1-Chlorobutane
  chlorination of, 130
  ir and pmr spectra of, 133, 134
Chlorocyclohexane
  ir spectrum of, 134
  preparation of, 129
7-Chloro-4-hydroxyquinoline
  preparation and reaction of, 355
  structure of, 353
2-Chloro-2-methylbutane
  dehydrohalogenation of, 145, 146
  ir and pmr spectra of, 148, 149
  preparation of, 227
  solvolysis of, 235
2-Chloro-2-methylpropane
  from 2-methyl-2-propanol, 224, 226
  solvolysis of, 234
2-Chloronaphthalene
  formation of, 241
  ir spectrum of, 250
Chloroquine, structure of, 350
Chlorosulfonation, 365, 370
Chlorosulfonic acid, 365, 370
Cholesterol, 483
Chondroitin A, 452
Chromatography, gas-liquid (*see* Gas-liquid
  chromatography)
Chromatography, liquid-liquid (*see* Extraction,
  Paper chromatography)
Chromatography, liquid-solid (*see* Column
  chromatography, Dry-column
  chromatography, High-pressure column
  chromatography, Thin-layer
  chromatography)
Chromic acid
  as classification reagent, 285
  with 2-methyl-1-propanol, 279
  as oxidant, discussion, 278
  as oxidant for side-chains, 513
  preparation of, 278, 281
Cinnamaldehyde
  imines from, 296, 297
  ir and pmr spectra of, 301, 302
  in Wittig reaction, 299
N-Cinnamylaniline, 297

ir and pmr spectra of, 303, 304
N-Cinnamylideneaniline, 297
  ir spectrum of, 303
N-Cinnamylidene-*m*-nitroaniline, 296
  ir spectrum of, 302
N-Cinnamyl-*m*-nitroaniline, ir and pmr spectra
  of, 302, 303
Citral
  components of, 485, 489
  ir and pmr spectra of, 487
  isolation of, from lemon-grass oil, 485
  uv spectrum of, 488
Classification tests (*see* Analysis, qualitative)
Column chromatography
  adsorbents for, 85
  application of, 87, 188, 397
  methods of detection in, 85
  preparation of column, 87
  solvents for, 86
  theory of, 83
Condensation polymers
  Dacron, 441
  Nylon, 442
Conjugate addition to α,β-unsaturated ketones,
  272
Copolymers, 432
Coupling constants, 112
Coupling reactions, 259
Crystallization, 54
Cyclobutanecarboxylic acid
  ir spectrum of, 326
  preparation of, 316, 320
Cyclobutane-1,1-dicarboxylic acid, 316
  ir spectrum of, 106
  pmr spectrum of, 326
  preparation of, 316, 320
1,3,5-Cycloheptatriene, 401
  ir and pmr spectra of, 403, 404
Cycloheptatrienyl cation (*see* Tropylium ion)
Cyclohexane
  bromination of, 136
  chlorination of, 129
  ir spectrum of, 134
cis-Cyclohexane-1,2-dicarboxylic acid
  ir spectrum of, 173
  preparation of, 169
Cyclohexanol
  dehydration of, 156
  ir and pmr spectra of, 163
  oxidation of, 279
Cyclohexanone, 214
  ir and pmr spectra of, 219
  oxidation of, 280
  preparation of, 279, 282
Cyclohexanone semicarbazone, formation of,
  214
Cyclohexene
  ir and pmr spectra of, 163, 164
  preparation of, 156
  reaction with dichlorocarbene, 245, 247
4-Cyclohexene-*cis*-1,2-dicarboxylic acid

  hydrogenation of, 169
  ir spectrum of, 172
  preparation of, 204, 205
4-Cyclohexene-*cis*-1,2-dicarboxylic anhydride
  hydrolysis of, 205
  ir and pmr spectra of, 209
  preparation of, 204, 205
Cyclopentadiene
  acidity of, 392
  with *p*-benzoquinone, 203
  ir and pmr spectra of, 207
  with maleic anhydride, 201, 202
  preparation of, 202
Cyclopropane, as an intermediate, 241

Dacron, 441
Dalton's law, 19, 27, 45
Decanedioyl chloride, 442, 444
Dehydration
  of cyclohexanol, 156
  of 4-methyl-2-pentanol, 155
  of 1-phenylethanol, 435, 437
Dehydrohalogenation (*see* Elimination
  reactions)
Delta (δ), in pmr, 110
Density, determination of, 22
Derivatives (*see* specific type and Table of)
Desiccants (*see* Drying, agents for)
Desiccator, vacuum, suggestion for, 56
D family (configuration), 451, 454, 465
1,1'-Diacetylferrocene
  detection of, 398
  structure of, 395
Diastereomers, 408
  definition of, 408
  of D-glucose, 454
  formation of, in resolution, 412
Diazotization
  of anthranilic acid, 252, 254
  discussion of, 381
  procedure for, 385
1,3-Dibromopropane, 313
  ir and pmr spectra of, 325
7,7-Dichlorobicyclo[4.1.0]heptane
  discussion of, 245
  ir and pmr spectra of, 251
  preparation of, 247
Dichlorocarbene
  from chloroform, 242, 247
  reactions of, 239
  from sodium trichloroacetate, 240
4,7-Dichloroquinoline
  preparation and reaction of, 356
  synthetic scheme for, 351
Dicyclopentadiene, cracking of, 202
Diels-Alder reaction
  benzyne as dienophile, 252, 254
    discussion of, 252
  dienes in, 200
  theory of, 199
Dienes, in Diels-Alder reaction, 200

Diethyl ethoxymethylenemalonate
  with *m*-chloroaniline, 352
  pmr spectrum of, 358
4,4'-Dimethylbenzopinacol
  ir and pmr spectra of, 430
  preparation from 4-methylbenzophenone,
    419, 426
  rearrangement of, 420, 426
2,3-Dimethylbutane, ir spectrum of, 102
Dimethyl cyclobutane-1,1-dicarboxylate, 313
  ir spectrum of, 326
  preparation of, 319
  side reactions in preparation of, 314–316
α,α'-Dimethyldibenzylamine (*see* Di-α-
    phenylethylamine)
Dimethyl fumarate
  from dimethyl maleate, 167, 168
  ir and pmr spectra of, 172
Dimethyl maleate
  ir spectrum of, 171
  isomerization of, 167
Dimethyl malonate, 313
  alkylation of, 313
  ir and pmr spectra of, 324, 325
  preparation of cyclobutanecarboxylic acid
    from, 313
Dimethyl sulfoxide, as solvent, 423
3,5-Dinitrobenzoic acid, 516
3,5-Dinitrobenzoyl chloride
  with alcohols, 515
  preparation of, 516
2,4-Dinitrofluorobenzene, with peptides, 474,
    475, 477
2,4-Dinitrophenylhydrazine, as classification
    reagent, 293
2,4-Dinitrophenylhydrazones
  as derivatives of aldehydes and ketones, 293
  preparation of, 298
1,4-Diphenyl-1,3-butadiene
  ir and pmr spectra of, 307
  preparation of, 295
Diphenyl ether, bromination of, 195, 196, 197
Di-α-phenylethylamine
  discussion of, 415
  specific rotation of as hydrochloric acid salt,
    416
Disaccharides
  hydrolysis of, 458, 461
  See also Carbohydrates
Dissymmetry, 407, 409
Distillation
  precautions for, 35, 36
  See also Azeotropic distillation
Distillation, fractional
  apparatus for, 29
  column packing, 30
  columns, 30
  procedure for, 37
  temperature gradients in, 30
  theory of, 27
Distillation, simple

  apparatus for, 26
  procedure for, 37
  theory of, 24
Distillation, steam
  apparatus for, 47, 48
  application of, 246, 486
  limitations of, 46
  theory of, 44
  utility of, 44, 48, 49
Distillation, vacuum
  apparatus for, 41–44
  application of, 320
  discussion of, 40
  procedure for, general, 43
Distribution coefficients, 65
Dry-column chromatography
  advantages of, 94
  application of, 397
  discussion of, 94
  procedure for, 397
Dry ice, 265
Drying
  agents for, table of, 565–567
  discussion of, 563
  of solids, 56, 371
  See also Azeotropic distillation

Electrical heating, 584
Electrophilic aromatic substitution (*see*
    Substitution, electrophilic aromatic)
Elementary analysis (*see* Analysis, qualitative)
Elimination reactions
  base-induced, 141
  competition with nucleophilic substitution,
    142, 151, 221
  in formation of carbenes, 238
  rate laws for, 141, 142
  theory of, 141, 151
  transition states in, 143, 151
Emulsions, remedy for, 71
Enantiomers
  definition of, 406
  resolution of, 411
Enolate ions, 312
  alkylation of, 312
Equilibrium control, theory of, 212
Equivalent weight, 330
  procedure for, 330
Esters (*see* Carboxylic acid esters)
Ethanol, 33
Ethylbenzene
  bromination of, 136
  ir and pmr spectra of, 138
  oxidation of, 336
Ethylene, 431
Eutectic point, 10
Extraction
  acid-base, 66, 72
  continuous, liquid-liquid, 68
    solid-liquid, 69, 491
  effect of pH on, 66

emulsions, 71
layer identification in, 71
selection of solvent for, 66
technique of, 70, 72
theory of, 64

Ferric chloride, with phenols, 518
Ferrocene
acetylation of, 396
ir and pmr spectra of, 400
structure and properties of, 395
Ferrous hydroxide, with nitro compounds, 527
Filter-aid, 54, 58, 283
Filtration, vacuum, apparatus for, 55
Fluoboric acid, 401
Fluted filter paper, 53
Fractional distillation (*see* Distillation, fractional)
Free radical reactions, 126, 433, 437
selectivity in, 135
Friedel-Crafts acylation, 395, 418
discussion of, 395, 418
preparation of 4-methylbenzophenone, 418, 425
Friedel-Crafts alkylation
apparatus for, 180
kinetic and equilibrium control in, 213
procedure for, 179
rearrangements in, 177
theory of, 176
D-Fructose, in sucrose, 458, 459
2-Furaldehyde, 214, 295
ir, pmr, and uv spectra of, 218
2-Furaldehyde semicarbazone, formation of, 214, 215
Furan, 345, 392

Gas-liquid chromatography
apparatus for, 75
application of, 38, 132, 146, 157, 181, 489
correction factors in, 81
table of, 82
preparative uses of, 82
quantitative analysis in, 79
retention time, factors affecting, 76, 78, 79
solid supports in, table of, 77
stationary phases for, table of, 77
theory of, 75
Geranial, structure of, 486
Glassware
assembly of, 36
care and cleaning of, 4
lubrication of, 4
α-D-Glucose
in sucrose, 451, 458
mutarotation of, 456
discussion, 456
preparation of, 456
D-Glucose, 454
Grignard reagent
with aldehydes, 261, 266

carbonation of, 265
discussion of, 260
discussion of, 258
with esters, 264
discussion of, 259
preparation of, 263
side reactions in preparation of, 259
solvents for, 258

Haloform reaction
procedure for, 317
theory of, 308
Halogenation (*see* Bromination, Chlorination)
Halogens, test for, 502
Hazards (*see* Safety)
Heating mantle, 584
Heating techniques, 582
Hemiacetal
glucose as, 454
in oxidation of aldehydes, 279
Hemiketal, fructose as, 459
Hempel column, 30
Heptane, chlorination of, 130
Heterocyclic compounds, 344
HETP, 30
1,6-Hexanediamine, 442
Hexanedioic acid
ir spectrum of, 285
in Nylon, 442
preparation of, 279, 282
High-pressure liquid chromatography
advantages in, 88
detectors in, 89
discussion of, 88
Hinsberg test
discussion of, 521
procedure for, 523
Hot filtration, 53
Hot plate, 586
Hückel's rule, 393, 401
Hydrobromic acid, preparation of, 223, 226
Hydrocarbons, aromatic
bromination of, 135
chlorination of, 126, 129
classification test for, 510
derivatives of, 512
side-chain oxidation of, 329, 336
table of, and derivatives, 540
See also Substitution, electrophilic aromatic
Hydrogen, generation and use of, 168, 169
Hydroxamic acids
from amides, 532
from esters, 528
from nitriles, 531
Hydroxylamine hydrochloride
with amides, 532
with esters, 528
with nitriles, 531

Ideal solution, definition of, 26
Ignition test, 501

Imidazole, structure of, 345
Imines
    formation of, 296–298, 415
        discussion of, 291
Indene
    with dichlorocarbene, 241, 242
    ir and pmr spectra of, 250
Index of refraction, 21
Infrared spectroscopy
    analysis of unknowns by, 533
    cells for, 105
    functional group frequencies, table of, 101,
        576
    identification using, 100, 102, 533, 576
    KBr pellets in, 104
    "neat" samples, 103
    Nujol mull, 104, 106
    perfluorokerosene mull, 104, 106
    with pmr spectroscopy, 113, 533
    practical considerations in, 103
    sample preparation, 103
    theory of, 99
Inverse addition, 262
Invert sugar, from sucrose, 459
Iodination, of 2-chloro-4-bromoaniline, 380,
    384
Iodine, test for, 502
Iodoform, 317
Iodoform test
    application of, 317, 506
    procedure for, 317
    theory of, 308
Isatin spray reagent, with amino acids, 466, 473
Isobutyl isobutyrate, 281
Isobutyraldehyde (see 2-Methylpropanal)
Isocyanates, in synthesis of polyurethanes, 447
Isomerization reaction, 167, 168
Isomers
    cis, trans, 407
    configurational, 405, 406
    conformational, 406
    constitutional, 405
    functional, 406
    geometric, 407
    positional, 405
    skeletal, 405
Isopentenyl pyrophosphate, role in
    biosynthesis, 483
Isopentyl nitrite, 253, 255
Isophorone, 273
    ir and pmr spectra of, 275, 276
Isopropylbenzene (cumene)
    bromination of, 136
    ir and pmr spectra of, 138, 139
Isopropyl-p-xylene, ir and pmr spectra of, 182,
    183

Ketones
    1,4-addition to, 311, 312
    aldol reactions of, 310, 318
    with ammonia derivatives, 214, 215, 290

classification tests for, 506
conversion to alkenes by Wittig reaction,
    293, 298
derivatives of, 507
    See also specific type
haloform reaction with, 308, 317
hemiketal formation in, 459
oxidation of, 279, 282, 308
preparation of, discussion, 277
reduction of, 434, 436
table of, and derivatives, 539
Ketoses, 454, 458
Kinetic control, theory of, 210
Kinetics
    for electrophilic aromatic substitution, 192,
        193, 194
    first-order rate expression for, 193, 229, 455
    least-square treatment in, 231
    for mutarotation, 455
    for nucleophilic substitution, discussion, 229
    for solvolysis, 230

Lambert-Beer law, 103, 120, 194
L family (configuration), 465
Library, use of, 550
Limiting reagent
    considerations in synthesis, 368
    definition of, 569, 572
Literature, chemical, 550
    abstract journals, 553
    encyclopedias and dictionaries, 552
    physical data, catalogs of, 557
    primary research journals, 550
    review journals, 551
    synthetic procedures and techniques,
        reference works, 556
    textbooks, advanced, 555
    use of, 558
Lithium aluminum hydride, potential use of,
    435
Lubrication, 5
Lucas test, 513

Magnetic stirring, 586, 588
Maleic anhydride
    as dienophile, 201, 203, 253
    ir spectrum of, 207
Markownikoff's rule, 165
Mechanical stirring, 587
Melting points
    apparatus, 13, 14, 15
    depression of, 10
    effect of heating rate on, 16
    effect of impurities on, 9
    effect of symmetry on, 186, 379
    mixture, 11, 16
    phase diagram, 10, 12
    theory of determination of, 8
Meso compounds, 408, 415
2-Methoxyethyl acetate, ir and pmr spectra of,
    535

Methyl benzoate
    from benzoic acid, 337
    from benzoyl chloride, 338
    ir and pmr spectra of, 268
    with phenylmagnesium bromide, 259, 264
4-Methylbenzophenone
    conversion to 4,4′-dimethylbenzopinacol, 419,
        426
    ir, pmr, and uv spectra of, 429
    photochemical dimerization of, 419, 426
    preparation of, 418, 425
2-Methylbutane, ir spectrum of, 102
2-Methyl-2-butanol, 227
    2-chloro-2-methylbutane from, 227
    ir and pmr spectra of, 228, 229
2-Methyl-1-butene
    by dehydrohalogenation, 143
    ir and pmr spectra of, 150
2-Methyl-2-butene
    by dehydrohalogenation, 143
    ir and pmr spectra of, 149
Methyl copper, preparation of, 274
Methylcyclohexane, bromination of, 136
3-Methyl-2-cyclohexenone, ir spectrum of, 100
2-Methyl-3-heptanol
    ir and pmr spectra of, 269, 270
    preparation of, 261, 266
Methyl iodide
    with amines, 526
    preparation of methyl copper from, 274
    preparation of methylmagnesium iodide
        from, 274
Methylmagnesium iodide, preparation of, 274
4-Methyl-2-pentanol
    dehydration of, 155
    ir and pmr spectra of, 159
    products from dehydration of, 154
2-Methyl-1-pentene, ir and pmr spectra of, 162
2-Methyl-2-pentene, ir and pmr spectra of, 160
4-Methyl-1-pentene, ir and pmr spectra of, 159,
        160
cis-4-Methyl-2-pentene, ir and pmr spectra of,
        161, 162
trans-4-Methyl-2-pentene, ir and pmr spectra
        of, 161
1-Methyl-3-penten-2-one, uv spectrum of, 119
2-Methylpropanal
    with Grignard reagent, 261, 266
    ir and pmr spectra of, 284
    preparation of, 281
2-Methyl-1-propanol
    ir and pmr spectra of, 114
    oxidation of, 281
2-Methyl-2-propanol, 224
    conversion to 2-chloro-2-methylpropane, 226
Microburner, 582
Migratory aptitudes in pinacol-pinacolone
    rearrangement, 422
Mixtures, separation of, in analysis, 498
Mole fraction, definition of, 26
Monosaccharides

elemental composition of, 451
    as hemiacetals, 454, 459
    mutarotation of, 454, 459
        rate of, procedure, 456
    from polysaccharides, 451
    structures of, 454
    thin-layer chromatography of, 461
    See also Carbohydrates
Multistep syntheses, 359
    of 1-bromo-3-chloro-5-iodobenzene, 377
    of cyclohexane-1,2-dicarboxylic acid, 390
    of 2-methyl-3-heptanol, 390
    of polystyrene, 432–434
    of sulfanilamide, 366
    of sulfathiazole, 362
Mutarotation
    discussion of, 454, 459
    kinetics of, 455

Naphthalene, recrystallization of, 59
α-Naphthol, bromination of, 195
α-Naphthyl isocyanate
    with alcohols, 514
    with phenols, 518
Natural products
    discussion of, 482
    isolation of, 461, 484
Neral, structure of, 486
Neutralization equivalent
    discussion of, 330
    procedure for, 519
Newman projection formula, 406
Nicotine, structure of, 484
Nitration
    apparatus for, 187
    in preparation of derivatives, 512
    procedure for, 187
    theory of, 184
Nitriles
    from carboxylic acid amides, 339
    classification test for, 531
    derivatives of, 531
    discussion of, 335
    hydrolysis of, 531
    table of, 549
m-Nitroaniline, 296, 297
Nitrobenzene
    ir spectrum of, 374
    reduction of, 362, 369
4-Nitrobromobenzene, ir and pmr spectra of,
        189, 190
Nitro compounds
    classification test for, 527
    as derivatives, 512
    derivatives of, 527
    table of, and derivatives, 547
Nitrogen, test for, 502
p-Nitrophenol, bromination of, 195
1-Nitropropane, pmr spectrum of, 109
Nonideal solution, definition of, 32

endo-Norbornene-*cis*-5,6-dicarboxylic acid,
    preparation of, 203
endo-Norbornene-*cis*-5,6-dicarboxylic anhydride
    hydrolysis of, 203
    preparation of, 202
Normal addition, 262
Notebook
    format for, 568
    importance of, 2
Nuclear magnetic resonance spectroscopy
    (*see* Proton magnetic resonance)
Nylon, 442
    discussion of, 442
    synthesis of, 444
Nylon Rope Trick, 444

Oil bath, heating with, 585
Optical activity, 409
Optical isomerism, 406
Optical rotation
    factors affecting, 409, 450
    measurement of, 409, 456
    See also Specific rotation
Orbital symmetry, 200, 352
Organocopper reagent, 272
Organocuprate, 272
Organolithium reagents, 270
Organometallic chemistry, 257
Overhead stirring, 587
Oxazole, structure of, 345
Oxidation
    of alcohols, 279, 282, 287
    of aldehydes, 285, 287
    of alkenes, 508
    aromatic side-chain, 336
    of ketones, 279, 282
    of methyl ketones, 317
    of side-chains, in preparation of derivatives,
        512, 513
Oximes
    as derivatives of aldehydes and ketones, 292
    preparation of, 298

Paper chromatography, 94
    of amino acids, 470, 471
    of inks, 95
t-Pentyl alcohol (*see* 2-Methyl-2-butanol)
t-Pentyl chloride (*see* 2-Chloro-2-methylbutane)
Peptides, 463
    hydrolysis of, 464, 465, 476, 477
    linkage in, 464
    structure, determination of, 474, 477
Percent yield, 570, 572
Petroleum ether, 50
Phase diagram
    liquid-vapor, azeotropic mixture, 33, 34
        typical, 28
    solid-liquid, examples of, 8, 9
Phase distribution, principle of, 64
Phase transfer catalysts, 243
    applications of, 244

types of, 244, 247
Phenol
    bromination of, 195
    esterification with, 339
Phenols
    bromine water with, 517, 518
    classification tests for, 516
    derivatives of, 518
    solubility properties, 516
    table of, and derivatives, 543
Phenyl benzoate, 339
1-Phenylethanol
    from acetophenone, 435, 436
    dehydration of, 435
    ir and pmr spectra of, 439, 440
α-Phenylethylamine
    derivatives of, 414
    resolution of, discussion, 411
        procedure, 413
    specific rotation, determination, 413
Phenylhydrazine, 293
Phenylhydrazones, as derivatives of aldehydes
    and ketones, 293
Phenyl isocyanate, 514
    with alcohols, 514
    with phenols, 518
Phenylmagnesium bromide
    carbonation of, 265
        discussion, 259
    with methyl benzoate, 264
    preparation of, 263
Photochemical reactions, 417
    dimerization in, 417
    dimerization-reduction of 4-
        methylbenzophenone, 419
    singlet states in, 419
    theory of, 418
    triplet states in, 419
Photochemistry (*see* Photochemical reactions)
Picric acid, with amines, 526
Pinacolone, 317
Pinacol-pinacolone rearrangements
    cleavage of pinacolones, 424, 427
    of 4,4'-dimethylbenzopinacol, 422
    migratory aptitudes in, 422, 427
    theory and discussion of, 420, 423
Piperic acid
    formation of, from piperine, 489
    ir spectrum of, 494
Piperidine
    formation of, from piperine, 490
    isolation of, as hydrochloride salt, 492
Piperine
    hydrolysis of, discussion, 489
        procedure for, 492
    ir and pmr spectra of, 493
    isolation of, from pepper, 489, 491
Planck's constant, 98, 109
pK_R^+, 394
Plexiglas, 432
Polarimeter

discussion of, 409, 410
  use of, 413, 415, 456, 458, 459, 461
Polarizability, 222
Polarized light, 409
Polyamides, 442
Polyesters, 441
Polyethylene, 431
Polymerization, 431
  addition, 431, 434
  condensation, 431, 441
Polymers, 431
Polysaccharides
  hydrolysis of, 458, 461
  See also Carbohydrates
Polystyrene
  ir spectrum of, 441
  from styrene, 432, 437
  synthesis of, chart, 433, 434
Polyurethanes
  discussion of, 446
  synthesis of, 449
Potassium t-butoxide, 147, 423
Potassium permanganate
  as oxidant, 280, 282, 330, 512
  test for presence of, 283
  test for unsaturation, 508
n-Propyl-p-xylene, ir and pmr spectra of, 182,
  183
Protective group, use in synthesis, 366, 378
Proteins, 463
  levels of structure in, 464
Proton magnetic resonance
  analysis of mixtures, 428
  analysis of unknowns by, 533
  application of, 150, 171, 428
  chemical shift in, 110
    table of, 106, 574
  coupling constants in, 112
  criterion for aromaticity, 394
  identification using, 113, 533
  integration in, 113
  with ir spectroscopy, 113, 533
  practical considerations in, 116
  sample preparation, 116
  solvents for, 116
  spin-spin splitting in, 112
  theory of, 107
Purine, structure of, 345
Pyridine, 345
  in anhydride formation, 334, 339
Pyridoxine, 452
Pyrilium ion, structure of, 345
Pyrimidine, structure of, 345
Pyrrole, 392

Quinacrine, structure of, 350
Quinine, structure of, 350
Quinoline, 345

Racemate, resolution of, 411
Racemic modification

definition of, 409, 412, 413
  resolution of, 412
Ramini test, 524
Raoult's law
  deviations from, 32
  liquid-vapor equilibria, 25, 27, 45
  solid-liquid equilibria, 10
Rate constant
  definition of, 229
  determination of by least-squares method,
    231
  graphical determination of, 193, 231
  quantitative determination of, 194, 197, 233
  semiquantitative determination of, 195
Recrystallization
  acetanilide, 58
  benzoic acid, 57
  discussion of, 49
  fluted filter paper, use in, 53
  hot filtration in, 52
  mixed solvents in, 50, 51
  naphthalene, 59
  oil formation in, 54, 58
  procedure for, 56
  selection of solvent for, 50, 57
  solvents for, 51
  vacuum filtration in, 55
Reduction
  of alkenes, 160
  of diazonium group, 381, 385
  of imines, 291, 292, 296
  of ketones, 434, 436
  of nitro group, 362, 369, 527
Reflux ratio, 32
Refraction, index of, 21
Resolution, of enantiomers (see Enantiomers,
  resolution of)
R_f values
  definition of, 91, 92
  use of, 91, 93, 398, 399, 470, 473, 477, 479,
    502
Rhamnetin, structure of, 483
Ring current, 394

Saccharides (see Carbohydrates)
Safety
  chemical precautions, 4
  explosion precautions, 36
  fire precautions, 3
Saponification equivalent
  discussion of, 332
  procedure for, 529
Saran, 432
Semicarbazide hydrochloride, 215
Semicarbazones
  dependence of formation on pH, 214
  as derivatives of aldehydes and ketones, 297
  preparation of, 215, 297
Separation of mixtures, 498
Separatory funnels, 70
Silver nitrate, with alkyl halides, 509

Simon test, 524
Simple distillation (*see* Distillation, simple)
Sodium bisulfite addition product, 281
Sodium borohydride
  in hydrogenation of alkenes, 168
  as reductant, 292, 296, 435, 436
Sodium ethoxide, preparation of, 147
Sodium fusion, 501
Sodium hypochlorite, 308, 317
Sodium hypoiodite, 308
  as classification reagent, 317
Sodium iodide, with alkyl halides, 509
Sodium methoxide, 294, 298
  preparation of, 146
Sodium nitroprusside test, 523
Sodium trichloroacetate
  preparation of, 245
  thermal decomposition of, 239
Solubility
  determination of, 57
  principles of, 50
Solubility tests (*see* Analysis, qualitative)
Solution
  ideal, definition of, 27
  nonideal, definition of, 32
Solvents, flammable, list of, 51
Solvolysis
  discussion of, 231
  factors affecting rate of, 232
  kinetics of, 233
Soxhlet extraction apparatus, 69
Specific rotation
  definition of, 410
  measurement of, 413–416, 456, 459, 461
Spectroscopy (*see* Infrared spectroscopy, Proton
  magnetic resonance, Ultraviolet-visible
  spectroscopy)
Spin-spin splitting, 112
Steam, generation of, 47
Steam bath, 587
Steam cone, 587
Steam distillation (*see* Distillation, steam)
Steam heating, 587
Stearic acid, structure of, 483
Stereoisomerism, discussion of, 406
Steroids, definition and biosynthesis of,
  483
cis-Stilbene, ir, pmr, and uv spectra of, 305,
  306, 307
trans-Stilbene
  ir, pmr, and uv spectra of, 306, 307
  preparation of, 299
Stirring motor, 589
Stirring techniques, 582
Strychnine, structure of, 484
Styrene
  from acetophenone, 436
    discussion of, 434
  ir and uv spectra of, 440
  polymerization of, 437
  polymerization of, discussion of, 432

2-Styrylfuran, preparation of, 299
Sublimation
  apparatus for, 61, 62
  limitations of, 62
  theory of, 60
Substitution, electrophilic aromatic
  acetylation, 395
  alkylation, 176
  bromination, 190
  discussion of, 174, 184
  examples of, table, 175
  halogenation, 380
  kinetics of, 174, 192
  nitration, 184
  relative rates of, discussion of, 190
    table of, 191
  substituent effects on, discussion of, 192
    table of, 191
  sulfonation, 365
Substitution, nucleophilic
  competition with elimination, 221
  factors affecting rate of, 229–233
  kinetics of, 229–233
  theory of, 229–233
Sucrose
  hydrolysis of, 459
    discussion of, 458
  in polyurethanes, 448
Sugars (*see* Carbohydrates)
Sulfanilamide
  ir and pmr spectra of, 376
  preparation of, 372
    discussion of, 366
Sulfathiazole
  ir and uv spectra of, 376, 377
  preparation of, 367, 373
  synthesis of, chart, 362
3-Sulfolene
  ir and pmr spectra of, 208
  as a source of 1,3-butadiene, 204
Sulfur, test for, 502
Sulfuryl chloride, 127, 129
Surfactant, 448

Table of derivatives
  alcohols, 542
  aldehydes, 538
  amides, 549
  amines
    primary and secondary, 545
    tertiary, 546
  aromatic hydrocarbons, 540
  aromatic nitro compounds, 547
  aryl halides, 541
  carboxylic acid esters, 548
  carboxylic acids, 544
  ketones, 539
  nitriles, 549
  phenols, 543
meso-Tartaric acid, 408
(+)-Tartaric acid, use of, in resolution, 412

Tau (τ), in pmr, 111
Teflon, 432
Terpenes, definition and biosynthesis of, 483
1,3,5,5-Tetramethyl-1,3-cyclohexadiene, 274
3,3,5,5-Tetramethylcyclohexanone, 273
 ir and pmr spectra of, 276
 preparation of, 275
Tetramethylsilane, as pmr reference, 110
Theoretical yield, definition of, 569, 572
Thermodynamic control, theory of, 211
Thermometer, calibration of, 17
Thiazole, structure of, 345
Thiele tube, 15, 16, 20
Thin-layer chromatography
 adsorbents for, 90
 application of, 92, 93, 188, 397, 479, 502
 R_f values in, 91, 479
 theory of, 90
Thionyl chloride, 337, 339, 516, 520
 discussion of use, 333, 335
Thiophene, 345, 392
Thiourea, with chloroacetaldehyde, 346, 347
Tollens' test
 discussion of, 506
 procedure for, 507
Toluene
 bromination of, 136
 relative rates of substitution on, table, 191
Transmittance, percent, definition of, 100
Trap
 aspirator, 55
 gas, 130
 safety, 41
 water from steam line, 48
α,α-Trehalose
 hydrolysis of, 461
 isolation of, 461
 natural sources of, 460
 structure of, 460
Trichloroacetic acid, 245
Triethylphosphite, 294, 299
3,3,5-Trimethylcyclohex-2-enone, 273
 with organocopper reagents, 273, 274
 See also Isophorone
Trioctylmethylammonium chloride, 247
 use as a phase transfer catalyst, 244, 247
Triphenylmethanol, 401
 pmr spectrum of, 269
 preparation of, 259, 264
 preparation of triphenylmethyl fluoborate
  from, 402
Triphenylmethyl fluoborate, preparation of, 402

Triptycene, 252
 ir spectrum of, 107
 pmr spectrum of, 256
 preparation of, 254
Tropylium fluoborate, preparation of, 402
Tropylium iodide, preparation of, 402
Tropylium ion, 393, 401
Trubore stirrer, 589

Ultraviolet-visible spectroscopy
 applications of, 123, 295
 Beer-Lambert law, 120
 chromophores in, 120
 electronic transitions in, 118
 molar absorptivity in, 122
 molecular requirements, 120
 sample preparation, 123
 solvents for, 124
 theory of, 117
 uv chromophores, table of, 121
Unsaturation, qualitative tests for, 507, 508

Vacuum distillation (see Distillation, vacuum)
Vapor pressure
 dependence on temperature, 18
 effect of nonvolatile impurities on, 25
 of nonideal solutions, 33
 relationship to boiling points, 18, 40
 relationship to melting points, 8
 of solids, 8
Variable transformer, 584
Variac, 584
Vigreux column, 30
Visible spectroscopy, 117, 194
 application of, 197
Vitamin A, structure of, 483
Vitamin B_1 (thiamine), structure of, 346
Vitamin B_6, 452, 453
Vitamin B_{12}, 360

Water, removal of by azeotropic distillation,
  292, 296
Wittig reaction
 phosphonate ester modification of, 293, 298
 theory of, 293

p-Xylene, alkylation with 1-bromopropane, 176

Yields, calculation of, 570
Ylides, 293

Zinc chloride, in Lucas test, 513